P9-CLH-039

LAP

DISCARD

LAP

Grzimek's
Animal Life Encyclopedia

Second Edition

••••

Grzimek's
Animal Life Encyclopedia

Second Edition

●●●●

Volume 4
Fishes I

Dennis A. Thoney, Advisory Editor
Paul V. Loiselle, Advisory Editor
Neil Schlager, Editor

Joseph E. Trumpey, Chief Scientific Illustrator

Michael Hutchins, Series Editor
In association with the American Zoo and Aquarium Association

GALE®

THOMSON
—★—
GALE

Detroit • New York • San Diego • San Francisco • Cleveland • New Haven, Conn. • Waterville, Maine • London • Munich

Grzimek's Animal Life Encyclopedia, Second Edition

Volume 4: Fishes I
Produced by Schlager Group Inc.
Neil Schlager, Editor
Vanessa Torrado-Caputo, Assistant Editor

Project Editor
Melissa C. McDade

Editorial
Stacey Blachford, Deirdre Blanchfield, Madeline Harris, Christine Jeryan, Kate Kretschmann, Mark Springer

Permissions
Margaret Chamberlain

Imaging and Multimedia
Randy Bassett, Mary K. Grimes, Lezlie Light, Christine O'Bryan, Barbara Yarrow, Robyn V. Young

Product Design
Tracey Rowens, Jennifer Wahi

Manufacturing
Wendy Blurton, Dorothy Maki, Evi Seoud, Mary Beth Trimper

For permission to use material from this product, submit your request via Web at http://www.gale-edit.com/permissions, or you may download our Permissions Request form and submit your request by fax or mail to: The Gale Group, Inc., Permissions Department, 27500 Drake Road, Farmington Hills, MI, 48331-3535, Permissions hotline: 248-699-8074 or 800-877-4253, ext. 8006, Fax: 248-699-8074 or 800-762-4058.

Cover photo of blue-spotted stringray by Jeffery L. Rotman/Corbis. Back cover photos of sea anemone by AP/Wide World Photos/University of Wisconsin-Superior; land snail, lionfish, golden frog, and green python by JLM Visuals; red-legged locust © 2001 Susan Sam; hornbill by Margaret F. Kinnaird; and tiger by Jeff Lepore/Photo Researchers. All reproduced by permission.

While every effort has been made to ensure the reliability of the information presented in this publication, The Gale Group, Inc. does not guarantee the accuracy of the data contained herein. The Gale Group, Inc. accepts no payment for listing; and inclusion in the publication of any organization, agency, institution, publication, service, or individual does not imply endorsement of the editors and publisher. Errors brought to the attention of the publisher and verified to the satisfaction of the publisher will be corrected in future editions.

ISBN   0-7876-5362-4 (vols. 1–17 set)
      0-7876-6572-X (vols. 4–5 set)
      0-7876-5780-8 (vol. 4)
      0-7876-5781-6 (vol. 5)

LIBRARY OF CONGRESS CATALOGING-IN-PUBLICATION DATA

Grzimek, Bernhard.
  [Tierleben. English]
  Grzimek's animal life encyclopedia.— 2nd ed.
    v. cm.
  Includes bibliographical references.
  Contents: v. 1. Lower metazoans and lesser deuterosomes / Neil Schlager, editor — v. 2. Protostomes / Neil Schlager, editor — v. 3. Insects / Neil Schlager, editor — v. 4-5. Fishes I-II / Neil Schlager, editor — v. 6. Amphibians / Neil Schlager, editor — v. 7. Reptiles / Neil Schlager, editor — v. 8-11. Birds I-IV / Donna Olendorf, editor — v. 12-16. Mammals I-V / Melissa C. McDade, editor — v. 17. Cumulative index / Melissa C. McDade, editor.
  ISBN 0-7876-5362-4 (set hardcover : alk. paper)
  1. Zoology—Encyclopedias. I. Title: Animal life encyclopedia. II. Schlager, Neil, 1966- III. Olendorf, Donna IV. McDade, Melissa C. V. American Zoo and Aquarium Association. VI. Title.
QL7 .G7813 2004
                      590'.3—dc21
                      2002003351

Printed in Canada
10 9 8 7 6 5 4 3 2 1

Recommended citation: *Grzimek's Animal Life Encyclopedia*, 2nd edition. Volumes 4–5, Fishes I–II, edited by Michael Hutchins, Dennis A. Thoney, Paul V. Loiselle, and Neil Schlager. Farmington Hills, MI: Gale Group, 2003.

Contents

Contents

Foreword

Earth is teeming with life. No one knows exactly how many distinct organisms inhabit our planet, but more than 5 million different species of animals and plants could exist, ranging from microscopic algae and bacteria to gigantic elephants, redwood trees and blue whales. Yet, throughout this wonderful tapestry of living creatures, there runs a single thread: Deoxyribonucleic acid or DNA. The existence of DNA, an elegant, twisted organic molecule that is the building block of all life, is perhaps the best evidence that all living organisms on this planet share a common ancestry. Our ancient connection to the living world may drive our curiosity, and perhaps also explain our seemingly insatiable desire for information about animals and nature. Noted zoologist, E.O. Wilson, recently coined the term "biophilia" to describe this phenomenon. The term is derived from the Greek *bios* meaning "life" and *philos* meaning "love." Wilson argues that we are human because of our innate affinity to and interest in the other organisms with which we share our planet. They are, as he says, "the matrix in which the human mind originated and is permanently rooted." To put it simply and metaphorically, our love for nature flows in our blood and is deeply engrained in both our psyche and cultural traditions.

Our own personal awakenings to the natural world are as diverse as humanity itself. I spent my early childhood in rural Iowa where nature was an integral part of my life. My father and I spent many hours collecting, identifying and studying local insects, amphibians and reptiles. These experiences had a significant impact on my early intellectual and even spiritual development. One event I can recall most vividly. I had collected a cocoon in a field near my home in early spring. The large, silky capsule was attached to a stick. I brought the cocoon back to my room and placed it in a jar on top of my dresser. I remember waking one morning and, there, perched on the tip of the stick was a large moth, slowly moving its delicate, light green wings in the early morning sunlight. It took my breath away. To my inexperienced eyes, it was one of the most beautiful things I had ever seen. I knew it was a moth, but did not know which species. Upon closer examination, I noticed two moon-like markings on the wings and also noted that the wings had long "tails", much like the ubiquitous tiger swallow-tail butterflies that visited the lilac bush in our backyard. Not wanting to suffer my ignorance any longer, I reached immediately for my *Golden Guide to North American Insects* and searched through the section on moths and butterflies. It was a luna moth! My heart was pounding with the excitement of new knowledge as I ran to share the discovery with my parents.

I consider myself very fortunate to have made a living as a professional biologist and conservationist for the past 20 years. I've traveled to over 30 countries and six continents to study and photograph wildlife or to attend related conferences and meetings. Yet, each time I encounter a new and unusual animal or habitat my heart still races with the same excitement of my youth. If this is biophilia, then I certainly possess it, and it is my hope that others will experience it too. I am therefore extremely proud to have served as the series editor for the Gale Group's rewrite of *Grzimek's Animal Life Encyclopedia*, one of the best known and widely used reference works on the animal world. *Grzimek's* is a celebration of animals, a snapshot of our current knowledge of the Earth's incredible range of biological diversity. Although many other animal encyclopedias exist, *Grzimek's Animal Life Encyclopedia* remains unparalleled in its size and in the breadth of topics and organisms it covers.

The revision of these volumes could not come at a more opportune time. In fact, there is a desperate need for a deeper understanding and appreciation of our natural world. Many species are classified as threatened or endangered, and the situation is expected to get much worse before it gets better. Species extinction has always been part of the evolutionary history of life; some organisms adapt to changing circumstances and some do not. However, the current rate of species loss is now estimated to be 1,000–10,000 times the normal "background" rate of extinction since life began on Earth some 4 billion years ago. The primary factor responsible for this decline in biological diversity is the exponential growth of human populations, combined with peoples' unsustainable appetite for natural resources, such as land, water, minerals, oil, and timber. The world's human population now exceeds 6 billion, and even though the average birth rate has begun to decline, most demographers believe that the global human population will reach 8–10 billion in the next 50 years. Much of this projected growth will occur in developing countries in Central and South America, Asia and Africa-regions that are rich in unique biological diversity.

Finding solutions to conservation challenges will not be easy in today's human-dominated world. A growing number of people live in urban settings and are becoming increasingly isolated from nature. They "hunt" in super markets and malls, live in apartments and houses, spend their time watching television and searching the World Wide Web. Children and adults must be taught to value biological diversity and the habitats that support it. Education is of prime importance now while we still have time to respond to the impending crisis. There still exist in many parts of the world large numbers of biological "hotspots"-places that are relatively unaffected by humans and which still contain a rich store of their original animal and plant life. These living repositories, along with selected populations of animals and plants held in professionally managed zoos, aquariums and botanical gardens, could provide the basis for restoring the planet's biological wealth and ecological health. This encyclopedia and the collective knowledge it represents can assist in educating people about animals and their ecological and cultural significance. Perhaps it will also assist others in making deeper connections to nature and spreading biophilia. Information on the conservation status, threats and efforts to preserve various species have been integrated into this revision. We have also included information on the cultural significance of animals, including their roles in art and religion.

It was over 30 years ago that Dr. Bernhard Grzimek, then director of the Frankfurt Zoo in Frankfurt, Germany, edited the first edition of *Grzimek's Animal Life Encyclopedia*. Dr. Grzimek was among the world's best known zoo directors and conservationists. He was a prolific author, publishing nine books. Among his contributions were: *Serengeti Shall Not Die*, *Rhinos Belong to Everybody* and *He and I and the Elephants*. Dr. Grzimek's career was remarkable. He was one of the first modern zoo or aquarium directors to understand the importance of zoo involvement in *in situ* conservation, that is, of their role in preserving wildlife in nature. During his tenure, Frankfurt Zoo became one of the leading western advocates and supporters of wildlife conservation in East Africa. Dr. Grzimek served as a Trustee of the National Parks Board of Uganda and Tanzania and assisted in the development of several protected areas. The film he made with his son Michael, *Serengeti Shall Not Die*, won the 1959 Oscar for best documentary.

Professor Grzimek has recently been criticized by some for his failure to consider the human element in wildlife conservation. He once wrote: "A national park must remain a primordial wilderness to be effective. No men, not even native ones, should live inside its borders." Such ideas, although considered politically incorrect by many, may in retrospect actually prove to be true. Human populations throughout Africa continue to grow exponentially, forcing wildlife into small islands of natural habitat surrounded by a sea of humanity. The illegal commercial bushmeat trade-the hunting of endangered wild animals for large scale human consumption-is pushing many species, including our closest relatives, the gorillas, bonobos and chimpanzees, to the brink of extinction. The trade is driven by widespread poverty and lack of economic alternatives. In order for some species to survive it will be necessary, as Grzimek suggested, to establish and enforce a system of protected areas where wildlife can roam free from exploitation of any kind.

While it is clear that modern conservation must take the needs of both wildlife and people into consideration, what will the quality of human life be if the collective impact of short-term economic decisions is allowed to drive wildlife populations into irreversible extinction? Many rural populations living in areas of high biodiversity are dependent on wild animals as their major source of protein. In addition, wildlife tourism is the primary source of foreign currency in many developing countries and is critical to their financial and social stability. When this source of protein and income is gone, what will become of the local people? The loss of species is not only a conservation disaster; it also has the potential to be a human tragedy of immense proportions. Protected areas, such as national parks, and regulated hunting in areas outside of parks are the only solutions. What critics do not realize is that the fate of wildlife and people in developing countries is closely intertwined. Forests and savannas emptied of wildlife will result in hungry, desperate people, and will, in the long-term lead to extreme poverty and social instability. Dr. Grzimek's early contributions to conservation should be recognized, not only as benefiting wildlife, but as benefiting local people as well.

Dr. Grzimek's hope in publishing his *Animal Life Encyclopedia* was that it would "...disseminate knowledge of the animals and love for them", so that future generations would "...have an opportunity to live together with the great diversity of these magnificent creatures." As stated above, our goals in producing this updated and revised edition are similar. However, our challenges in producing this encyclopedia were more formidable. The volume of knowledge to be summarized is certainly much greater in the twenty-first century than it was in the 1970's and 80's. Scientists, both professional and amateur, have learned and published a great deal about the animal kingdom in the past three decades, and our understanding of biological and ecological theory has also progressed. Perhaps our greatest hurdle in producing this revision was to include the new information, while at the same time retaining some of the characteristics that have made *Grzimek's Animal Life Encyclopedia* so popular. We have therefore strived to retain the series' narrative style, while giving the information more organizational structure. Unlike the original *Grzimek's*, this updated version organizes information under specific topic areas, such as reproduction, behavior, ecology and so forth. In addition, the basic organizational structure is generally consistent from one volume to the next, regardless of the animal groups covered. This should make it easier for users to locate information more quickly and efficiently. Like the original Grzimek's, we have done our best to avoid any overly technical language that would make the work difficult to understand by non-biologists. When certain technical expressions were necessary, we have included explanations or clarifications.

Considering the vast array of knowledge that such a work represents, it would be impossible for any one zoologist to have completed these volumes. We have therefore sought specialists from various disciplines to write the sections with which they are most familiar. As with the original *Grzimek's*, we have engaged the best scholars available to serve as topic editors, writers, and consultants. There were some complaints about inaccuracies in the original English version that may have been due to mistakes or misinterpretation during the complicated translation process. However, unlike the original *Grzimek's*, which was translated from German, this revision has been completely re-written by English-speaking scientists. This work was truly a cooperative endeavor, and I thank all of those dedicated individuals who have written, edited, consulted, drawn, photographed, or contributed to its production in any way. The names of the topic editors, authors, and illustrators are presented in the list of contributors in each individual volume.

The overall structure of this reference work is based on the classification of animals into naturally related groups, a discipline known as taxonomy or biosystematics. Taxonomy is the science through which various organisms are discovered, identified, described, named, classified and catalogued. It should be noted that in preparing this volume we adopted what might be termed a conservative approach, relying primarily on traditional animal classification schemes. Taxonomy has always been a volatile field, with frequent arguments over the naming of or evolutionary relationships between various organisms. The advent of DNA fingerprinting and other advanced biochemical techniques has revolutionized the field and, not unexpectedly, has produced both advances and confusion. In producing these volumes, we have consulted with specialists to obtain the most up-to-date information possible, but knowing that new findings may result in changes at any time. When scientific controversy over the classification of a particular animal or group of animals existed, we did our best to point this out in the text.

Readers should note that it was impossible to include as much detail on some animal groups as was provided on others. For example, the marine and freshwater fish, with vast numbers of orders, families, and species, did not receive as detailed a treatment as did the birds and mammals. Due to practical and financial considerations, the publishers could provide only so much space for each animal group. In such cases, it was impossible to provide more than a broad overview and to feature a few selected examples for the purposes of illustration. To help compensate, we have provided a few key bibliographic references in each section to aid those interested in learning more. This is a common limitation in all reference works, but *Grzimek's Encyclopedia of Animal Life* is still the most comprehensive work of its kind.

I am indebted to the Gale Group, Inc. and Senior Editor Donna Olendorf for selecting me as Series Editor for this project. It was an honor to follow in the footsteps of Dr. Grzimek and to play a key role in the revision that still bears his name. *Grzimek's Animal Life Encyclopedia* is being published by the Gale Group, Inc. in affiliation with my employer, the American Zoo and Aquarium Association (AZA), and I would like to thank AZA Executive Director, Sydney J. Butler; AZA Past-President Ted Beattie (John G. Shedd Aquarium, Chicago, IL); and current AZA President, John Lewis (John Ball Zoological Garden, Grand Rapids, MI), for approving my participation. I would also like to thank AZA Conservation and Science Department Program Assistant, Michael Souza, for his assistance during the project. The AZA is a professional membership association, representing 205 accredited zoological parks and aquariums in North America. As Director/William Conway Chair, AZA Department of Conservation and Science, I feel that I am a philosophical descendant of Dr. Grzimek, whose many works I have collected and read. The zoo and aquarium profession has come a long way since the 1970s, due, in part, to innovative thinkers such as Dr. Grzimek. I hope this latest revision of his work will continue his extraordinary legacy.

Silver Spring, Maryland, 2001
Michael Hutchins
Series Editor

• • • • •

How to use this book

Grzimek's Animal Life Encyclopedia is an internationally prominent scientific reference compilation, first published in German in the late 1960s, under the editorship of zoologist Bernhard Grzimek (1909–1987). In a cooperative effort between Gale and the American Zoo and Aquarium Association, the series has been completely revised and updated for the first time in over 30 years. Gale expanded the series from 13 to 17 volumes, commissioned new color paintings, and updated the information so as to make the set easier to use. The order of revisions is:

Volumes 8–11: Birds I–IV
Volume 6: Amphibians
Volume 7: Reptiles
Volumes 4–5: Fishes I–II
Volumes 12–16: Mammals I–V
Volume 3: Insects
Volume 2: Protostomes
Volume 1: Lower Metazoans and Lesser Deuterostomes
Volume 17: Cumulative Index

Organized by taxonomy

The overall structure of this reference work is based on the classification of animals into naturally related groups, a discipline known as taxonomy—the science in which various organisms are discovered, identified, described, named, classified, and catalogued. Starting with the simplest life forms, the lower metazoans and lesser deuterostomes, in volume 1, the series progresses through the more advanced classes, culminating with the mammals in volumes 12–16. Volume 17 is a stand-alone cumulative index.

Organization of chapters within each volume reinforces the taxonomic hierarchy. In the case of the volumes on Fishes, introductory chapters describe general characteristics of fishes, followed by taxonomic chapters dedicated to order and, in a few cases, suborder. Readers should note that in a few instances, taxonomic groups have been split among more than one chapter. For example, the order Cypriniformes is split among two chapters, each covering particular families. Species accounts appear at the end of the taxonomic chapters. To help the reader grasp the scientific arrangement, order and suborder chapters have distinctive symbols:

● = Order Chapter
○ = Suborder Chapter

The order Perciformes, which has the greatest number of species by far of any fishes order—and in fact is the largest order of vertebrates—has been split into separate chapters based on suborder. Some of these suborder chapters are again divided into multiple chapters in an attempt to showcase the diversity of species within the group. For instance, the suborder Percoidei has been split among four chapters. Readers should note that here, as elsewhere, the text does not necessarily discuss every single family within the group; in the case of Percoidei, there are more than 70 families. Instead, the text highlights the best-known and most significant families and species within the group. Readers can find the complete list of families for every order in the "Fishes family list" in the back of each Fishes volume.

As chapters narrow in focus, they become more tightly formatted. Introductory chapters have a loose structure, reminiscent of the first edition. Chapters on orders and suborders are more tightly structured, following a prescribed format of standard rubrics that make information easy to find. These taxonomic chapters typically include:

Scientific name of order or suborder
Common name of order or suborder
Class
Order
Number of families
Main chapter
    Evolution and systematics
    Physical characteristics
    Distribution
    Habitat
    Feeding ecology and diet
    Behavior
    Reproductive biology
    Conservation status
    Significance to humans
Species accounts
    Common name
    Scientific name

Family
Taxonomy
Other common names
Physical characteristics
Distribution
Habitat
Feeding ecology and diet
Behavior
Reproductive biology
Conservation status
Significance to humans
Resources
Books
Periodicals
Organizations
Other

Color graphics enhance understanding

Grzimek's features approximately 3,500 color photos, including nearly 250 in the Fishes volumes; 3,500 total color maps, including more than 200 in the Fishes volumes; and approximately 5,500 total color illustrations, including nearly 700 in the Fishes volumes. Each featured species of animal is accompanied by both a distribution map and an illustration.

All maps in *Grzimek's* were created specifically for the project by XNR Productions. Distribution information was provided by expert contributors and, if necessary, further researched at the University of Michigan Zoological Museum library. Maps are intended to show broad distribution, not definitive ranges.

All the color illustrations in *Grzimek's* were created specifically for the project by Michigan Science Art. Expert contributors recommended the species to be illustrated and provided feedback to the artists, who supplemented this information with authoritative references and animal specimens from the University of Michigan Zoological Museum library. In addition to illustrations of species, *Grzimek's* features drawings that illustrate characteristic traits and behaviors.

About the contributors

All of the chapters were written by ichthyologists who are specialists on specific subjects and/or families. The volumes' subject advisors, Dennis A. Thoney and Paul V. Loiselle, reviewed the completed chapters to insure consistency and accuracy.

Standards employed

In preparing the volumes on Fishes, the editors relied primarily on the taxonomic structure outlined in *Fishes of the World*, 3rd edition, by Joseph S. Nelson (1994), with some modifications suggested by expert contributors for certain taxonomic groups based on more recent data. Systematics is a dynamic discipline in that new species are being discovered continuously, and new techniques (e.g., DNA sequencing) fre-

quently result in changes in the hypothesized evolutionary relationships among various organisms. Consequently, controversy often exists regarding classification of a particular animal or group of animals; such differences are mentioned in the text.

Grzimek's has been designed with ready reference in mind, and the editors have standardized information wherever feasible. For **Conservation Status,** *Grzimek's* follows the IUCN Red List system, developed by its Species Survival Commission. The Red List provides the world's most comprehensive inventory of the global conservation status of plants and animals. Using a set of criteria to evaluate extinction risk, the IUCN recognizes the following categories: Extinct, Extinct in the Wild, Critically Endangered, Endangered, Vulnerable, Conservation Dependent, Near Threatened, Least Concern, and Data Deficient. For a complete explanation of each category, visit the IUCN web page at <http://www.iucn.org/themes/ssc/redlists/categor.htm>.

In addition to IUCN ratings, chapters may contain other conservation information, such as a species' inclusion on one of three Convention on International Trade in Endangered Species (CITES) appendices. Adopted in 1975, CITES is a global treaty whose focus is the protection of plant and animal species from unregulated international trade.

In the Species accounts throughout the volume, the editors have attempted to provide common names not only in English but also in French, German, Spanish, and local dialects. Readers can find additional information on fishes species on the Fishbase Web site: <http://www.fishbase.org>.

Grzimek's provides the following standard information on lineage in the **Taxonomy** rubric of each Species account: [First described as] *Acipenser brevirostrum* [by] LeSueur, [in] 1818, [based on a specimen from] Delaware River, United States. The person's name and date refer to earliest identification of a species, although the species name may have changed since first identification. However, the entity of fish is the same.

Readers should note that within chapters, species accounts are organized alphabetically by family name and then alphabetically by scientific name. In each chapter, the list of species to be highlighted was chosen by the contributor in consultation with the appropriate subject advisor: Dennis A. Thoney, who specializes in marine fishes; and Paul V. Loiselle, who specializes in freshwater fishes.

Anatomical illustrations

While the encyclopedia attempts to minimize scientific jargon, readers will encounter numerous technical terms related to anatomy and physiology throughout the volume. To assist readers in placing physiological terms in their proper context, we have created a number of detailed anatomical drawings. These can be found on pages 6 and 7, and 15–27 in the "Structure and function" chapter. Readers are urged to make heavy use of these drawings. In addition, selected terms are defined in the **Glossary** at the back of the book.

Appendices and index

In addition to the main text and the aforementioned *Glossary,* the volume contains numerous other elements. *For Further Reading* directs readers to additional sources of information about fishes. Valuable contact information for *Organizations* is also included in an appendix. An exhaustive *Fishes family list* records all recognized families of fishes according to *Fishes of the World*, 3rd edition, by Joseph S. Nelson (1994). And a full-color *Geologic time scale* helps readers understand prehistoric time periods. Additionally, the volume contains a *Subject index.*

Acknowledgements

Gale would like to thank several individuals for their important contributions to the volume. Dr. Dennis A. Thoney, subject advisor specializing in marine fishes, created the overall topic list for the volumes and suggested writers and reviewed chapters related to marine fishes. Dr. Paul V. Loiselle, subject advisor specializing in freshwater fishes, suggested writers and reviewed chapters related to freshwater fishes. Neil Schlager, project manager for the Fishes volumes, coordinated the writing and editing of the text. Finally, Dr. Michael Hutchins, chief consulting editor for the series, and Michael Souza, program assistant, Department of Conservation and Science at the American Zoo and Aquarium Association, provided valuable input and research support.

Library advisors

James Bobick
Head, Science & Technology Department
Carnegie Library of Pittsburgh
Pittsburgh, Pennsylvania

Linda L. Coates
Associate Director of Libraries
Zoological Society of San Diego Library
San Diego, California

Lloyd Davidson, PhD
Life Sciences bibliographer and head, Access Services
Seeley G. Mudd Library for Science and Engineering
Evanston, Illinois

Thane Johnson
Librarian
Oklahoma City Zoo
Oklahoma City, Oklahoma

Charles Jones
Library Media Specialist
Plymouth Salem High School
Plymouth, Michigan

Ken Kister
Reviewer/General Reference teacher
Tampa, Florida

Richard Nagler
Reference Librarian
Oakland Community College
Southfield Campus
Southfield, Michigan

Roland Person
Librarian, Science Division
Morris Library
Southern Illinois University
Carbondale, Illinois

Contributing writers

Fishes I–II

Arturo Acero, PhD
INVEMAR
Santa Marta, Colombia

M. Eric Anderson, PhD
J. L. B. Smith Institute of Ichthyology
Grahmstown, South Africa

Eugene K. Balon, PhD
University of Guelph
Guelph, Ontario, Canada

George Benz, PhD
Tennessee Aquarium Research
Institute and Tennessee Aquarium
Chattanooga, Tennessee

Tim Berra, PhD
The Ohio State University
Mansfield, Ohio

Ralf Britz, PhD
Smithsonian Institution
Washington, D.C.

John H. Caruso, PhD
University of New Orleans, Lakefront
New Orleans, Louisiana

Marcelo Carvalho, PhD
American Museum of Natural History
New York, New York

José I. Castro, PhD
Mote Marine Laboratory
Sarasota, Florida

Bruce B. Collette, PhD
National Marine Fisheries Systematics
Laboratory and National Museum of
Natural History
Washington, D.C.

Roy Crabtree, PhD
Florida Fish and Wildlife
Conservation Commission
Tallahasee, Florida

Dominique Didier Dagit, PhD
The Academy of Natural Sciences
Philadelphia, Pennsylvania

Terry Donaldson, PhD
University of Guam Marine
Laboratory, UOG Station
Mangliao, Guam

Michael P. Fahay, PhD
NOAA National Marine
Fisheries Service,
Sandy Hook Marine Laboratory
Highlands, New Jersey

John. V. Gartner, Jr., PhD
St. Petersburg College
St. Petersburg, Florida

Howard Gill, PhD
Murdoch University
Murdoch, Australia

Lance Grande, PhD
Field Museum of Natural History
Chicago, Illinois

Terry Grande, PhD
Loyola University Chicago
Chicago, Illinois

David W. Greenfield, PhD
University of Hawaii
Honolulu, Hawaii

Melina Hale, PhD
University of Chicago
Chicago, Illinois

Ian J. Harrison, PhD
American Museum of Natural History
New York, New York

Phil Heemstra, PhD
South African Institute for
Aquatic Biodiversity
Grahamstown, South Africa

Jeffrey C. Howe, MA
Freelance Writer
Mobile, Alabama

Liu Huanzhang, PhD
Chinese Academy of Sciences
Hubei Wuhan,
People's Republic of China

G. David Johnson, PhD
Smithsonian Institution
Washington, D.C.

Scott I. Kavanaugh, BS
University of New Hampshire
Durham, New Hampshire

Frank Kirschbaum, PhD
Institute of Freshwater Ecology
Berlin, Germany

Kenneth J. Lazara, PhD
American Museum of Natural History
New York, New York

Andrés López, PhD
Iowa State University
Ames, Iowa

John A. MacDonald, PhD
The University of Auckland
Auckland, New Zealand

Jeff Marliave, PhD
Institute of Freshwater Ecology
Vancouver, Canada

John McEachran, PhD
Texas A&M University
College Station, Texas

Leslie Mertz, PhD
Wayne State University
Detroit, Michigan

Elizabeth Mills, MS
Washington, D. C.

Katherine E. Mills, MS
Cornell University
Ithaca, New York

Randall D. Mooi, PhD
Milwaukee Public Museum
Milwaukee, Wisconsin

Thomas A. Munroe, PhD
National Systematics Laboratory
Smithsonian Institution
Washington, D.C.

Prachya Musikasinthorn, PhD
Kasetsart University
Bangkok, Thailand

John E. Olney, PhD
College of William and Mary
Gloucester Point, Virginia

Frank Pezold, PhD
University of Louisiana at Monroe
Monroe, Louisiana

Mickie L. Powell, PhD
University of New Hampshire
Durham, New Hampshire

Aldemaro Romero, PhD
Macalester College
St. Paul, Minnesota

Robert Schelly, MA
American Museum of Natural History
New York, New York

Matthew R. Silver, BS
University of New Hampshire
Durham, New Hampshire

William Leo Smith, PhD
American Museum of Natural History
and Columbia University
New York, New York

Stacia A. Sower, PhD
University of New Hampshire
Durham, New Hampshire

Melanie Stiassny, PhD
American Museum of Natural History
New York, New York

Tracey Sutton, PhD
Woods Hole
Oceanographic Institution
Woods Hole, Massachusetts

Gus Thiesfeld, PhD
Humboldt State University
Arcata, California

Jeffrey T. Williams, PhD
Smithsonian Institution
Washington, D.C.

Contributing illustrators

Drawings by Michigan Science Art

Joseph E. Trumpey, Director, AB, MFA
Science Illustration, School of Art and Design, University of Michigan

Wendy Baker, ADN, BFA

Brian Cressman, BFA, MFA

Emily S. Damstra, BFA, MFA

Maggie Dongvillo, BFA

Barbara Duperron, BFA, MFA

Dan Erickson, BA, MS

Patricia Ferrer, AB, BFA, MFA

Gillian Harris, BA

Jonathan Higgins, BFA, MFA

Amanda Humphrey, BFA

Jacqueline Mahannah, BFA, MFA

John Megahan, BA, BS, MS

Michelle L. Meneghini, BFA, MFA

Bruce D. Worden, BFA

Thanks are due to the University of Michigan, Museum of Zoology, which provided specimens that served as models for the images.

Maps by XNR Productions

Paul Exner, Chief cartographer
XNR Productions, Madison, WI

Tanya Buckingham

Jon Daugherity

Laura Exner

Andy Grosvold

Cory Johnson

Paula Robbins

· · · · ·

Topic overviews

What is a fish?

Evolution and systematics

Structure and function

Life history and reproduction

Freshwater ecology

Marine ecology

Distribution and biogeography

Behavior

Fishes and humans

What is a fish?

What is a fish?

The concept of "fish" certainly is more steeped in tradition than backed by scientists, despite the fact that countless ichthyologists (i.e., scientists who study fish) have written innumerable pages on the subject. The reality that fishes in the broadest sense have long played important roles in the promotion of industry and commerce, geographic exploration, politics, art, religion, and myth mandates that the definition of fish can vary according to human perspective and sometimes despite science. For example, from a chef's point of view, fishes come in two basic varieties—shellfish and finfish. Scientists eschew such groupings of distantly related creatures. However, lest they be hoisted with their own petards, ichthyologists might tread gently on the many concepts of fish, for they must acknowledge science's inability to form an absolute taxonomic definition of "fish" based on biological characteristics that are shared by all fishes and yet not shared with any "nonfish."

This whale shark (*Rhincodon typus*) measures 40 ft (12 m). Whale sharks are the largest fish on the earth today. (Photo by Amos Nachoum/Corbis. Reproduced by permission.)

Defining characteristics

Widespread views of the particular characteristics that define fishes, of course, are biased by general familiarity with extant (i.e., living) species and, in particular, with the widespread and well-known bony fishes. Thus, the notion of a fish as an aquatic ectothermic vertebrate possessing gills, paired and unpaired fins, and scales usually suffices as a casual definition of fish. Reasonable as this definition may seem, some of these characteristics are shared with other groups of animals that are not considered fishes, while others of them are not common to all fishes. For example, although most fish live in water, some fishes, such as the walking catfish (*Clarias batrachus*) or African lungfish (*Protopterus* species) can spend considerable periods out of water. Furthermore, other fishes may spend much briefer, yet highly significant periods out of water, which allow them to feed (e.g., mudskippers, *Periophthalmus*

A starry moray eel (*Gymnothorax nudivomer*) peering out from its home near the Philippines. (Photo by Robert Yin/Corbis. Reproduced by permission.)

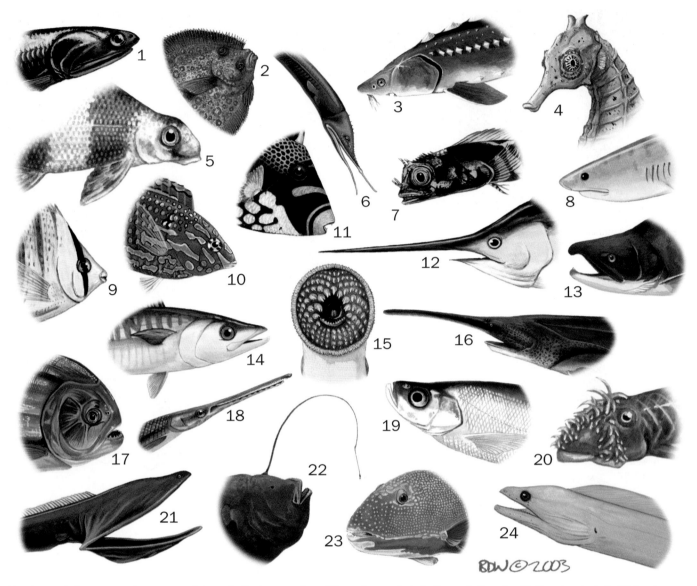

Mouth morphology comparison among fishes. 1. Northern anchovy (*Engraulis mordax*); 2. Peacock flounder (*Bothus lunatus*); 3. White sturgeon (*Acipenser transmontanus*); 4. Yellow seahorse (*Hippocampus kuda*); 5. Chinese sucker (*Myxocyprinus asiaticus*); 6. Bobtail snipe eel (*Cyema atrum*); 7. Secretary blenny (*Acanthemblemaria maria*); 8. Tiger shark (*Galeocerdo cuvier*); 9. Pebbled butterflyfish (*Chaetodon multicinctus*); 10. Blackspotted wrasse (*Macropharyngodon meleagris*); 11. Clown triggerfish (*Balistoides conspicillum*); 12. Swordfish (*Xiphias gladius*); 13. Sock-eye salmon (*Oncorhynchus nerka*); 14. King mackerel (*Scomberomorus cavalla*); 15. Sea lamprey (*Petromyzon marinus*); 16. Paddlefish (*Polyodon spathula*); 17. Red-bellied piranha (*Pygocentrus nattereri*); 18. Longnose gar (*Lepisosteus osseus*); 19. Minnow (*Culter alburnus*); 20. Catfish (*Ancistrus triradiatus*); 21. Pelican eel (*Eurypharynx pelecanoides*); 22. Krøyer's deep sea anglerfish (*Ceratias holboelli*); 23. Bicolor parrotfish (*Cetoscarus bicolor*); 24. Green moray (*Gymnothorax funebris*). (Illustration by Bruce Worden)

spp., and the arowanas, *Osteoglossum* spp.) or flee from predators (e.g., flyingfishes, Exocoetidae).

Similarly, whereas most fishes cannot control their body temperature other than through behavioral mechanisms involving migrations or local movements to and from waters of varying warmth, some lamnids (Lamnidae) and tunas (*Thunnus* spp.) and the swordfish (*Xiphias gladius*) can maintain body temperatures that are several degrees higher than the water that surrounds them for significant periods. Certainly, most fishes possess a well-developed vertebral column; however,

hagfishes (Myxinidae) lack well-defined vertebrae, and there is disagreement among scientists regarding whether this characteristic exists because the ancestors of these fishes were similar or, antithetically, because vertebrae were "lost" from this lineage through evolutionary modification. In fact, so different are hagfishes from other fishes that Aristotle considered them members of another, illegitimate taxonomic group—worms. Unlike worms, fishes are chordates (phylum Chordata), and they possess skeletal components that form a cranium (i.e., a brain case). This characteristic (as well as many others) distinguishes them from some fishlike chordates, such

The Hawaiian anthias (*Pseudanthias ventralis*) is one of many fishes that has vibrant, incredible colors. (Photo by Mark Smith/Photo Researchers, Inc. Reproduced by permission.)

A pike (*Esox lucius*) with a newly caught frog. (Photo by Animals Animals ©C. Milkins, OSF. Reproduced by permission.)

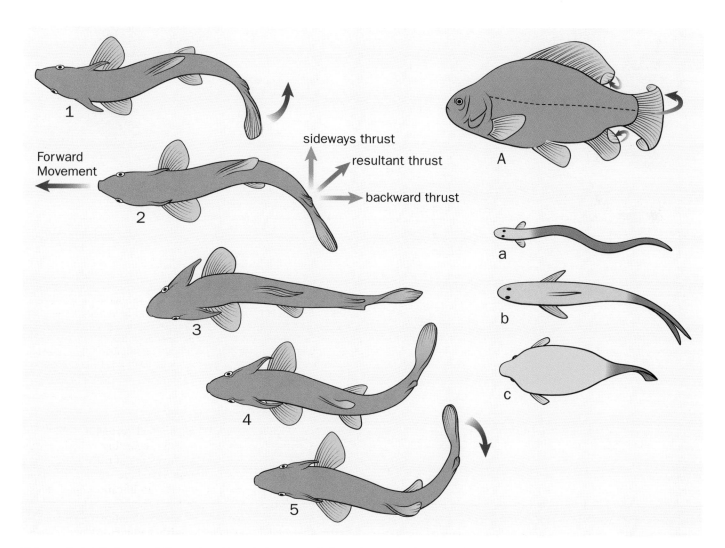

Fish use their tails to propel themselves through the water. A. A crucian carp's fin action for stabilizing and maneuvering. a. Anguilliform locomotion (eel); b. Carangiform locomotion (tuna); c. Ostraciform locomotion (boxfish). The blue area on these fish shows the portion of the body used in locomotion. (Illustration by Patricia Ferrer)

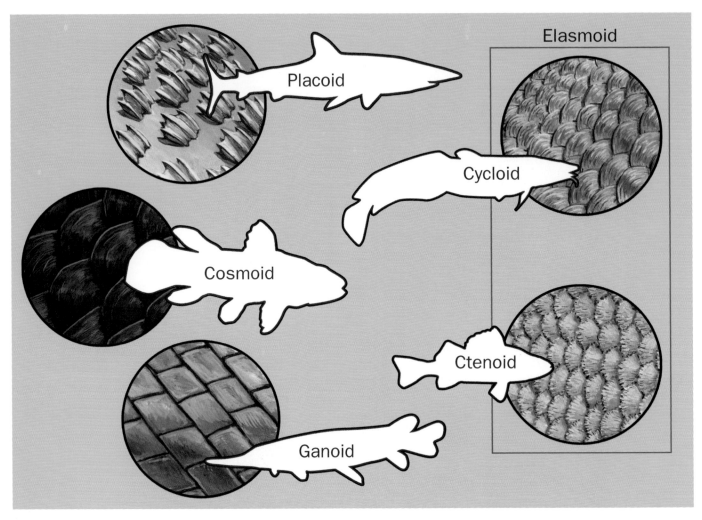

Scale types and patterns in fish. Clockwise from top: Placoid, cycloid, ctenoid, ganoid, cosmoid. (Illustration by Brian Cressman)

as the lancelets (Amphioxiformes), but, of course, amphibians, reptiles, birds, and mammals also have a cranium.

Gills cannot be used as an unequivocal characteristic defining fishes, because some amphibians have and use gills for at least a portion of their lives. Furthermore, whereas most fishes obtain oxygen from water through conventional gills, some fishes significantly supplement gill respiration by acquiring oxygen from the water or atmosphere via modified portions of the gills (e.g., the walking catfish) or skin (e.g., the European eel, *Anguilla anguilla*) or specialized tissues in the mouth (e.g., the North American mudsucker goby, *Gillichthys mirabilis*), gut (e.g., plecostomuses, *Plecostomus* species), swim bladder (e.g., the bowfin, *Amia calva*), or lungs (e.g., the Australian lungfish, *Neoceratodus forsteri*). Complicating matters still further, some fishes are obligate air breathers and must have access to the atmosphere or they will drown (e.g., the electric eel, *Electrophorus electricus* and the South American lungfish, *Lepidosiren paradoxa*).

At first glance, fins seem to define fishes. Several unrelated groups of nonfishes (e.g., lancelets, sea snakes, and some amphibians) possess finlike modifications associated with their tails that facilitate locomotion in water. Furthermore, although some fishes, such as hagfishes and lampreys (Petromyzontidae), lack paired fins, the paired appendages of amphibians, reptiles, birds, and mammals are considered homologous to the paired fins of fishes. Likewise, the scales that cover many common bony fishes are not a universally acceptable distinguishing feature, because numerous unrelated groups of fishes lack scales, for example, the hagfishes, the lampreys, and the North American freshwater catfishes (Ictaluridae). Moreover, those fishes that possess scales may be more or less covered by one of several basic scale types, for example, the placoid scales of sharks, the ganoid scales of gars, and the bony ridge scales of salmon and basses. These differences in the scales of fishes point to the fact that some other aquatic chordates, such as sea snakes, also have scales, even though the outer coverings of reptiles, birds, and mammals are heavily keratinized, whereas those of fishes are not.

Superclass Pisces as a polyphyletic group

Given that no one characteristic distinguishes all fishes from all other organisms, even the most committed ichthyologist must admit that the superclass Pisces (an assemblage that in-

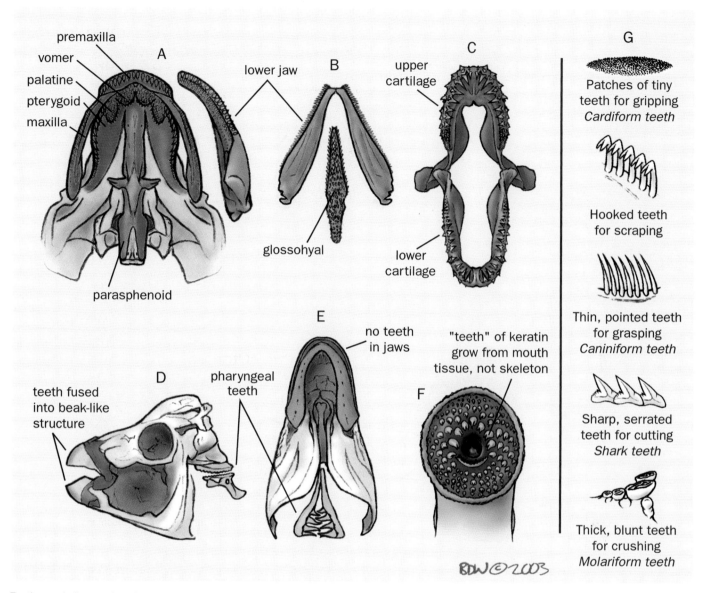

Tooth morphology and tooth-bearing structures typical of fishes: A. Bowfin (*Amia calva*); B. Mooneye (*Hiodon tergisus*); C. Sand shark (*Odontaspis taurus*); D. Parrotfish (*Scarus guacamaia*); E. Northern pikeminnow (*Ptychocheilus oregonensis*); F. Sea lamprey (*Petromyzon marinus*); G. Tooth forms and functions. (Illustration by Bruce Worden)

cludes all fishes) represents an unnatural or polyphyletic group. In fact, given our scientific understanding of fishes as of 2002, the only measure allowing them to stand together as a natural or monophyletic group requires the inclusion of all other craniates (i.e., amphibians, reptiles, birds, and mammals). Most biologists probably would agree that the consideration of all craniates as fishes would be of little scientific value and would betray the longstanding and widespread conception of a fish. In light of this situation, uncompromising cladists returning from a fishing trip for salmon are condemned to telling others of having been "salmoning" rather than "fishing."

General definition of fish

Despite the seemingly hopeless conundrum of defining "fish" scientifically, many scientists and non-scientists probably would agree that a general definition for this loose group of animals can be established. For these reasonable folks, a fish can be defined as an ectothermic chordate that lives primarily in water and possesses a cranium, gills that are useful virtually throughout life, and appendages (if present) in the form of fins. Those not willing to endorse this definition might rest easy by considering "fish" as the raison d'être for ichthyologists.

Resources

Books

Beard, J. A. *James Beard's New Fish Cookery*. New York: Galahad Books, 1976.

Bond, Carl E. *Biology of Fishes*. 2nd edition. Philadelphia, PA: Saunders College Publishing, 1996.

Bone, Q., N. B. Marshall, and J. H. S. Blaxter. *Biology of Fishes*. 2nd edition. Glasgow: Blackie Academic and Professional, 1995.

Helfman, Gene S., B. Bruce Collette, and Doug E. Facey. *The Diversity of Fishes*. Malden, MA: Blackwell Science, 1997.

Kurlansky, Mark. *Cod: A Biography of the Fish That Changed the World*. New York: Walker and Company, 1997.

Moyle, P. B., and J. J. Cech Jr. *Fishes: An Introduction to Ichthyology*. Upper Saddle River, NJ: Prentice Hall, 1996.

Nelson, J. S. *Fishes of the World*. 3rd edition. New York: John Wiley and Sons, 1994.

George W. Benz, PhD

· · · · ·

Evolution and systematics

Origin of fishes

Fishes are the most primitive members of the subphylum Craniata. The subphylum was previously called Vertebrata, but Janvier (1981) demonstrated that the most primitive members of the taxon possess a cranium but lack arcualia, or rudimentary vertebral elements. Thus the taxon is better termed Craniata than Vertebrata. Vertebrata is reserved for a subset of Craniata that possesses vertebral elements, in addition to a cranium. Two recently discovered fossils from China (Shu et al, 1999) extend the fossil record of fishes back to the early Cambrian, 530 million years ago (mya). These early forms are either the direct ancestors or the indirect ancestors to nearly all of the vertebrates, and their discovery suggests that vertebrates were part of the great explosion of metazoan life in the Cambrian. The fossils are small, about 0.98 to 1.1 in (25 to 28 mm) long, and possess a cartilaginous cranium, five to nine gill pouches, a large heart located behind the last pair of gill pouches and possibly enclosed in a pericardium, a notochord, zigzag-shaped muscle blocks or myomeres, and a dorsal fin (one of the two forms) supported by fin rays or radials. The more generalized fossil, *Myllokunmingia*, is thought to be the sister group of craniates except for the hagfishes. The other fossil, *Haikouichthys*, is considered to be a close relative of the lampreys. Unlike most other jawless fishes, these early forms lacked scales or bony armor. Until near the end of the twentieth century, evidence of vertebrates in the Cambrian was inconclusive. Carapace fragments thought to represent Ostracoderms, a group of armored, jawless fishes abundant in the Ordovician to the Devonian, were present in the Cambrian, but other experts consider the fragments to represent the carapaces of arthropods. Isolated, tooth-like elements, or conodonts, were common in the fossil record, but these elements could not be assigned to an organism. In the mid-1980s conodonts were discovered to be specialized feeding elements of soft-bodied, eel-like fossils possessing a notochord, dorsal nerve cord, V-shaped myomeres, and large eyes. Conodonts are considered to be the sister group of the remainder of vertebrates other than lampreys.

Other recent discoveries illustrate that craniates and vertebrates were rather diverse by the middle Ordovician (450 mya) (Young 1997, Sansom et al. 1996), with both jawless and jawed forms represented. Despite the occurrence of jawed fishes in the Ordovician, jawless forms dominated until the late Silurian. Ostracoderms are classified into about 10 to 12 major groups with poorly resolved relationships (Janvier 1999). However, it appears that this group is the sister group of the jawed fishes (gnathostomes). Most of the fossils representing these taxa possessed mineralized exoskeletons (except for the *Jamoytius*, and possibly *Euphanerops*), head shields (except for Anaspida, *Endeiolepis*, and Thelodonti) and multiple gill openings (except in Heterostraci), and lacked paired fins (except for the Anaspida, Osteostraci, Pituriaspida, and Thelodonti).

Modern jawless fishes

Myxiniformes (hagfishes) and Petromyzontiformes (lampreys) are modern jawless fishes that first appear in the Pennsylvanian (300 mya) and the late Mississippian (330 mya), respectively. Based on the structure of their pouch-like gills and several other characteristics, the two taxa were previously thought to form a monophyletic group, but as of 2002 they are considered to be paraphyletic, in that they do not share a common ancestor. Hagfishes are hypothesized to be the sister group to the remainder of the vertebrates. They are slender bodied and naked, lack fins with exception of the caudal fin, have degenerate eyes, four pairs of tentacles around the mouth and nasal openings, an esophago-cutaneous duct leading from the exterior to the esophagus, one semicircular canal in the inner ear, gill pouches posterior to the head, and ventrolateral slime glands. They are additionally distinguished from the vertebrates in lacking vertebral elements. Modern hagfishes are limited to soft bottom marine habitats and feed on soft-bodied, burrowing invertebrates and carrion. Lampreys are considered to be the next branch of the craniate tree and are slender bodied and naked, have two dorsal fins, a sucker surrounding the mouth, a rasping and sucker device (termed a tongue) that can be protruded from the mouth, two semicircular canals in the inner ear, and a dorsally located nasohypophysial opening on the head. Lampreys occur in fresh and marine waters, and some are anadromous, spending most of their adult lives in salt water and then migrating to freshwater streams and rivers to reproduce. Larvae are fossorial, or live in soft bottoms and filter feed on algae and detritus. Adults either are ectoparasites on ray-finned fishes or do not feed after metamorphosis from the larval to the adult stage. According to this phylogenetic scenario, the lack of a bony

A side-by-side view of a contemporary white shark tooth (left) and a megalodon tooth. The megalodon was a prehistoric great white shark. (Photo by Jeffrey L. Rotman/Corbis. Reproduced by permission.)

skeleton and scales in hagfishes and lampreys is primitive rather than a specialization related to their fossorial or parasitic life styles.

Origin of jawed fishes

Jawed fishes, Gnathostomata, possess true mandibular jaws, paired fins, inner ears with three semicircular canals, and gill arches internal to ectodermal gill filaments. Gnathostomes date back to the Ordovician (450 mya) but did not dominate aquatic regions of the world until the mid-to-late Devonian. Thus, jawless and jawed fishes coexisted for about 100 million years. The earliest jawed fish fossils are chondrichthyans, one of the five major groups of gnathostomes. Chondrichthyans today are represented by the chimaeras, sharks, skates, and rays, and are distinguished from the other four groups in lacking dermal bone, possessing cartilaginous rather than bony endoskeletons, and having distinctive gill filaments, multiple gill openings (except for the chimaeroids), horny unsegmented fin rays (ceratotrichia), and embryos encapsulated in leathery capsules. The Placodermi, likely the sister group of the chondrichthyans, appear in the early Silurian (420 mya). They assumed a wide variety of body forms and dominated fresh and salt waters in the Devonian before their extinction by the Mississippian. Some, such as Rhenaniformes and Ptyctodontiformes, were very similar in structure to modern chondrichthyans, such as rays and chimaeroids. Placoderms had bony head and shoulder plates, with the head shield movably articulating with the trunk shield, but lacked true teeth. Acanthodii first appeared in the early Silurian, reached peak diversity in the Devonian, and apparently became extinct in the Permian. They were small, slender, and elongate fishes that possessed dermal or endochondral bone, bony covering over gill slits, stout spines preceding fins, and scales covering most of the body. Some forms may have possessed endochondral bone, but most apparently had cartilaginous endoskeletons. The last two groups of fishes, the

Actinopterygii and Sarcopterygii, together comprise the Osteichthyes, and are thought to be the sister group of the acanthodians. The Osteichthyes have bony endoskeletons (endochondral bone) and lungs or swim bladders. The actinopterygians first appear in the late Silurian (410 mya) and today constitute the great majority of fishes. This group is distinguished by possessing ganoid or elasmoid scales, pectoral fin radials directly connected to scapulocoracoid or shoulder girdle, and nostrils located high on the head. Sarcopterygii are known from the Devonian (400 mya). They are distinguished in possessing true tooth enamel, reduction of branchial skeleton, and presence of a pulmonary vein. The reduced branchial skeleton and pulmonary vein suggest that they relied, at least in part on aerial respiration.

Chondrichthyans

Although fossil evidence in the form of scales has pushed the origin of chondrichthyans back to the Ordovician, cartilaginous fishes do not become abundant in the fossil record until the Carboniferous period. Throughout their history chondrichthyans have undergone several major radiations but today display only a modest radiation in body shape and are represented by a relatively small number of species. The Paleozoic cartilaginous fishes resemble recent forms but generally had terminal jaws, lacked vertebral centra, had fin radials that extended to the fin margins, and lacked skeletal connections between the halves of their pectoral and pelvic girdles. All but the earliest taxon, *Cladoselache* possessed male intromittent organs, suggesting that like their modern counterparts, they practiced internal fertilization and development. In the early Carboniferous, elasmobranchs underwent their second radiation. Male stethacanthid sharks, present from the late Devonian to the Permian, had bony, brush-like structures along the margin of the dorsal fin or modified dorsal fin spines bearing denticles, tooth-like scales (or placoid scales) that form distinctive patterns on the skin of various species of sharks. Some edestoid sharks had complex, coiled tooth whorls extending from their lower jaws that functioned in some unknown manner. Holocephalans, distinguished by possessing gill covers, upper jaw fused to the cranium, and crushing dentition, were described from the Upper Devonian and assumed a wide variety of forms in the Carboniferous, some resembling modern ray-finned fishes. Chondrichthyans suffered large number of extinctions at the end of the Permian, as did much of the world's biota, as the result of either extensive volcanic activity or a large asteroid's striking Earth. The final radiation of chondrichthyans began in the Jurassic, evolving from a lineage that survived the end of the Paleozoic extinction event(s). Nearly all modern families are represented in the fossil record by the end of the Mesozoic. Modern chondrichthyans generally possess subterminal jaws, vertebral centra, fin radials that fall short of the fin margins, and cartilaginous connections between the halves of their pectoral and pelvic girdles.

The recent chondrichthyans consist of two natural groups, the Holocephali, or chimaeroids, and the Neoselachii, or sharks and rays. In total, there are about 900 to 1000 living chondrichthyans. The chimaeroids number about 45 species and resemble some of their more conservative Carboniferous relatives. Anatomical research in the 1980s and 1990s concluded that the Neoselachii consisted of two basal groups, the Galeomorphii and Squalea, distinguished by a number of technical aspects of their skeletal structure and musculature. In this scenario the rays (Rajiformes) made up a terminal node of the Squalea. Molecular studies, however, suggest that sharks and rays are sister groups that in turn are the sister group of the chimaeroids. The division of sharks into Galeomorphii and Squalea is supported by the molecular studies.

The galeomorphs consist of four orders: Heterodontiformes (horn sharks), Orectolobiformes (wobbegons, nurse sharks, whale sharks), Lamniformes (sand tigers, basking sharks, thresher sharks, mackerel sharks), and Carcharhiniformes (catsharks, hound sharks, requiem sharks, and hammerheads). These sharks vary from benthic to pelagic and are best represented in tropical to warm, temperate seas. The Squalea consist of four major groups: Hexanchiformes (frill sharks, cowsharks), Squaliformes (sleeper sharks, dogfish sharks), Squatiniformes (angelsharks), and Pristiophoriformes (sawsharks). For the most part, these squalean sharks are associated with sea bottoms, often in deep water, and in temperate to boreal seas, although the Squatiniformes and Pristiophoriformes are exceptions in being distributed in tropical to warm temperate seas.

The rays differ from the sharks in being depressed to various degrees and having the pectoral fins attached to the cranium rather than free of the cranium, gills located on the ventral side of the body rather than laterally, and anterior trunk vertebrae fused into a tube (synarchial) rather than lacking a synarchial. Rays include the Torpedinoidei (electric rays), Pristoidei (sawfishes); Rhinidae, Rhinobatidae, Platyrhinidae (guitarfishes); Rajidae (skates); and Myliobatoidei (stingrays). Rays are, for the most part, associated with the sea bottoms in shallow to deep water.

Actinopterygians

Unlike the chondrichthyans, the ray-finned fishes have undergone significant morphological evolution since their first appearance in the Silurian. The earliest forms, in many respects, resembled the early chondrichthyans in possessing a heterocercal tail, having pectoral fins inserting low on the flank, and pelvic fins inserting behind the pectoral fins on the lower abdominal region. Unlike the early sharks, they possessed a single dorsal fin, endochondral bone, scales that grew throughout the life of the individual, and segmented and paired fin rays (lepidotrichia) rather than ceratotrichia. The scales had peg and socket articulations and consisted of a ganoine exterior, dentinous layer, and a basal spongy bone. The jaw teeth were set in sockets of the dermal jaw bones, the upper jaw was fused with the dermal bones covering the head, and the jaws were obliquely suspended to the cranium by the palatoquadrate bone. These primitive fishes are represented by several relic fishes today: Polypteridae (bichirs), Acipenseridae (sturgeons), and Polyodontidae (paddlefishes). The latter two groups, however, are highly modified from their Devonian ancestors.

By the end of the Paleozoic, numerous taxa of ray-finned fishes, classified as Neopterygii, appear in the fossil record.

Fossil of the primitive fish *Osteolepis macrolepidotys*, from the Middle Devonian period, around 30 million years ago. This specimen was found in Old Red Sandstone in the Sandwick fish beds at Quoyloo, Orkney, Scotland. (Photo by Sinclair Stammers/Science Photo Library/Photo Researchers, Inc. Reproduced by permission.)

Neopterygii differ in a number of respects from the earlier forms. The upper jaw is partially freed from the cheek bones, the jaws are perpendicularly suspended from the cranium, the number of fin rays are reduced to equal the number of supporting bones, and for the most part, ganoid scales are replaced by thin, elasmoid membranous scales. The elements of the upper jaw are fused medially. Branchial bones supporting the gill filaments develop pharyngeal teeth that assist in processing food. Today the basal neopterygians are represented by Lepisosteidae (gars) and Amiidae (bowfins). Gars have specialized jaws, but both they and the bowfin retain many primitive structures of the early neopterygians (upper jaw partially attached to dermal head bones, heterocercal or abbreviated heterocercal caudal fins, and lungs rather than swim bladders).

Teleostei, ray-finned fishes with an externally symmetrical or homocercal caudal fin and upper dermal jaw bones free of other dermal head bones, arose from a neopterygian ancestor in the mid-to-late Triassic (220–200 mya). There are a large number of fossil Triassic and Cretaceous teleosts, many with uncertain phylogenetic relationships, and four modern lineages of teleosts: Osteoglossomorpha, Elopomorpha, Clupeomorpha, and Euteleostei. The osteoglossomorphs are distinguished in that the primary jaw teeth are located on the parasphenoid bone along the middle of the roof of the mouth and on the tongue. In addition, the caudal fin skeleton is very specialized. The taxon includes Osteoglossidae (bony tongues), Hiodontids (mooneyes), Notopteridae (featherfin knifefishes), and Mormyridae (elephantfishes). All species are limited to freshwaters and most are tropical in distribution. The elopomorphs are distinguished by possessing ribbon-like leptocephalus larvae and numerous branchiostegal rays uniting the hyoid (second gill arch) with the opercular bones. The group includes Elopiformes (ladyfishes and tarpons), Albuliformes (bonefishes), Anguilliformes (eels), and Saccopharyngiformes (gulper eels). Nearly all of the elopomorphs are marine fishes and range from shallow to deepwater and from benthic to pelagic. The clupeomorphs are distinguished by their otophysic connection between extensions of the swimbladder and the inner ear within the cranium. The taxon includes the Ostariophysi (minnows, tetras, catfishes, and gymnotid eels) and the Clupeiformes (herring and anchovies). Ostariophysians dominate the freshwaters of the world. Clupeiforms range from marine to freshwater; the great majority are pelagic and consume plankton.

About half of living fishes are classified in Acanthopterygii that arose from an euteleostean ancestor in the Triassic or early Cretaceous. Because of the large number of taxa and vast morphological variation, the taxon is not well defined. Acanthopterygii are distinguished, in part, by having the pharyngeal teeth confined to the anterior gill arches, the swallowing muscle (retractor dorsalis) inserting on the upper segment of the third gill arch, and the ligament supporting the pectoral girdle attaching to the base of the cranium. Acanthopterygians comprise three major taxa: Mugilomorpha, Atherinomorpha, and Percomorpha. The mugilomorphs include the mullets, largely pelagic fishes found in shallow fresh and marine waters. Atherinomorphs include Atheriniformes (rainbowfishes and silversides), Beloniformes (needlefishes, sauries, halfbeaks, and flyingfishes), and Cyprinodontiformes (rivulines, killifishes, poeciliids, and pupfishes) and are surface swimming fishes found in both fresh and marine waters. The percomorphs comprise nine orders, 229 families, 2,144 genera, and over 12,000 species that dominate coastal marine waters, including coral reefs, but that are also well represented in most other aquatic habitats.

Sarcopterygii

Sarcopterygii, lobe-finned fishes, are the line of fishes that gave rise to the tetrapods (amphibians, reptiles, birds, and mammals), and the fishes of this lineage are better represented in the fossil record than in modern aquatic habitats. Sarcopterygians include Coelacanthimorpha (coelacanths), Porolepimorpha (including the Dipnoi or lungfishes), and Osteolepimorpha (rhipidistians). Coelacanths are well represented in the fossil record from the Devonian to the Upper Cretaceous, and two species are known in modern times. The taxon is distinguished by having a hinged cranium, possessing a three-lobed caudal fin, and lacking internal nares. Porolepimorpha first appear in the lower Devonian and are widespread in the fossil record until the end of the Carboniferous Period. The group, which includes the Dipnoi, is distinguished by either having slight mobility between the anterior and posterior sections of the cranium or by lacking mobility within the cranium, and in lacking true choanae or internal nares. Today Dipnoi are represented by six species in three families in freshwaters of Australia, South America, and Africa. Rhipidistians occur from the middle Devonian to the Lower Permian in the fossil record and are distinguished by possessing a hinged cranium, internal nares, and either a heterocercal or diphycercal caudal fin (fin equally developed above and below distal extension of body axis). Like some of their close sarcopterygian relatives, rhipidistians possessed pectoral fin bones that are homologues of the humerus, ulna, and radius of tetrapods.

Resources

Books

Bemis, William E., Warren W. Burggren, and Norman E. Kemp, eds. *The Biology and Evolution of Lungfishes.* New York: A. R. Liss, Inc. 1987.

Long, John A. *The Rise of Fishes: 500 Million Years of Evolution.* Baltimore, MD: Johns Hopkins University Press, 1995.

Maissey, J. G. *Santana Fossils: an Illustrated Atlas.* Neptune City, NJ: T. F. H. Publishers, 1991.

Nelson, Joseph S. *Fishes of the World.* New York: John Wiley and Sons, Inc., 1994.

Paxton, John R., and William N. Eschmeyer, eds. *Encyclopedia of Fishes.* San Diego, CA: Academic Press, 1995.

Schultze, Hans-Peter, and Linda Trueb, eds. *Origins of the Higher Groups of Tetrapods: Controversy and Consensus.* Ithaca, NY: Comstock Publishing Associates, 1991.

Stiassny, Melanie L. J., Lynne R. Parenti, and G. David Johnson. *Interrelationships of Fishes.* San Diego, CA: Academic Press, 1996.

Periodicals

Aldridge, R. J., et al. "The Anatomy of Conodonts." *Philosophical Transactions of the Royal Society London* 340 (1993): 405–421.

Chen, J.-Y, D.-Y Huang, and C.-W. Li. "An Early Cambrian Craniate-like Chordate." *Nature* 402 (1999): 518–522.

Cloutier, R. "Patterns, Trends, and Rates of Evolution Within the Actinistia." *Environmental Biology of Fishes* 32 (1991): 23–58.

Donoghue, P. C. J., P. L. Forey, and R. Aldridge. "Conodont Affinity and Chordate Phylogeny." *Biological Reviews Proceedings of the Cambridge Philosophical Society.* 75 (2000): 191–251.

Forey, P., and P. Janvier. "Agnathans and the Origin of Jawed Vertebrates." *Nature* 361 (1993): 129–134.

Janvier, P. "The Phylogeny of the Craniata, with Particular Reference to the Significance of Fossil 'Agnathans.'" *Journal of Vertebrate Paleontology* 1 (1981): 121–159.

———. "The Dawn of the Vertebrates: Characters Versus Common Ascent in the Rise of Current Vertebrate Phylogenies." *Palaeontology* 39, pt. 2 (1996): 259–287.

———. "Catching the Fish." *Nature* 402 (1999): 21–22.

Johnson, G. D., and W. D. Anderson Jr., eds. Proceedings of the Symposium on Phylogeny of Percomorpha, June 15–17, 1990, held in Charleston, South Carolina at the 70th Annual Meeting of the American Society of Ichthyologists and Herpetologists. *Bulletin of Marine Science* 52 (1993): 1–626.

Lauder, G. V., and K. F. Liem. "The Evolution and Interrelationships of the Actinopterygian Fishes." *Bulletin of the Museum of Comparative Zoology* 150 (1983): 95–197.

Resources

Maisey, J. G. "Heads and Tails: A Chordate Phylogeny." *Cladistics* 2 (1986): 201–256.

Sansom, I. J., M. M. Smith, and M. P. Smith. "Scales of Thelodont and Shark-like Fishes from the Ordovician of Colorado." *Nature* 379 (1996): 628–630.

Shu, D.-G, H.-L Luo, S.C. Morris, X.-L. Zhang, S.-X Hu, L. Chen, J. Han, M. Zhu, Y.Li, and L.-Z Chen. "Lower Cambrian Vertebrates from South China." *Nature* 402 (1999): 42–46.

Young, G. C. "Ordovician Microvertebrate Remains from the Amaseus Basin, Central Australia." *Journal of Vertebrate Paleontology* 17 (1997): 1–25.

Organizations

American Society of Ichthyology and Herpetology. Dept of Biological Sciences, College Of Arts & Science, Florida International University, North Miami, FL 33181 USA. Phone: (305) 919-5651. Fax: (305) 919-5964. E-mail: donnelly@fiu.edu Web site: <http://199.245.200.110>

John D McEachran, PhD

· · · · ·

Structure and function

Introduction

Fishes are phenomenally diverse in their anatomical and physiological characteristics. They have evolved spectacular and myriad anatomical and functional specializations to accomplish basic biological functions, such as feeding, moving, and reproducing. Through this diversity in structure and function, fishes have adapted to live successfully in a wide range of aquatic environments.

Body shape and external morphological features

Fishes vary in size by several orders of magnitude. The larval stages of many species are very small, many only a few millimeters in length. The largest fish species, the whale shark (*Rhincodon typus*), may reach a length of more than 59 ft (18 m). There is diversity in body shape as well, including elongate eels, fusiform tunas, dorsoventrally compressed fishes (such as skates and rays), and laterally compressed fishes (including flatfishes and many carangids). Body shapes are fascinating for their functional ingenuity. The flatfishes can lie against the bottom of the ocean, avoiding predators while being able to strike quickly at unsuspecting prey. The bullet-shaped bodies of tunas minimize water resistance for these high-performance cruising swimmers. Numerous unusual fish species have body shapes that make us wonder whether they are fish at all. Seahorses (genus *Hippocampus*), with their tapering prehensile tails and often spiky body armor, or related species, such as the leafy seadragon (*Phycodurus eques*), with leaflike projections from its fins and body surface, are camouflaged to blend in with the vegetation in which they live.

Fishes have many forms of locomotion that often correlate with external morphological features. The primarily axial movement patterns are classified into several swimming behaviors associated with backbone bending, including anguilliform, carangiform, and thunniform locomotion. Anguilliform movements, named after the eel genus *Anguilla*, typically are found in highly elongate species and involve axial bending along the entire axis of the fish. Carangiform movements, named after the family Carangidae (the jacks), entail more shallow body bends, with little bending near the head. Thunniform locomotion, named after the tuna genus *Thunnus*, involves movement of only the caudal backbone and caudal fin. Thunniform swimmers have many adaptations for

efficient swimming, such as fins that can be tucked into grooves on the body to minimize drag, a narrow and stiff caudal peduncle (the area just anterior to the caudal fin), and a crescent-shaped caudal fin. The latter morphological feature minimizes drag on the tail while generating strong propulsive forces.

Fin shape strongly affects the shape and function of a fish's external form as well as its locomotor ability. Fishes have two sets of paired fins, the pectoral fins and pelvic fins, and several fins on the body midline, including one or more dorsal fins on the dorsal midline, a posterior caudal fin, and an anal fin on the ventral midline. Many fishes swim primarily with fin movements rather than with waves of axial bending. Ostraciiform locomotion (named for the family Ostraciidae, or boxfishes) involves movement of the caudal fin without axial bending. Amiiform locomotion, named for the basal actinopterygian fish *Amia calva*, or bowfin, consists primarily of waves of oscillation of a long dorsal fin. In contrast, gymnotiform locomotion (named for the family Gymnotidae, or knifefishes) comprises similar oscillations along an elongate anal fin. Many fishes, including the triggerfishes (family Balistidae) and pufferfishes (family Tetraodontidae), coordinate movements of the dorsal and anal fins. The pectoral fins are diverse in their morphological features and their function in locomotion. In swimming, most fishes use pectoral fins to turn and maneuver. The coral reef fish family Labridae and its relatives mainly use pectoral fin locomotion, and thus this type of swimming has been named labriform swimming. Some labriform swimmers literally fly underwater with a graceful up-and-down flapping motion, leaving their bodies straight as they dart around the reef.

Fins perform many functions other than locomotion, among them, feeding, defense, camouflage, breeding, and social display. Many species, such as the lumpfish (*Cyclopterus lumpus*), have pelvic fins modified as suction disks to prevent detachment from the substrate. Fins also are used for feeding, or to deter predation, in a variety of ways. Sea robins (family Triglidae) use sensory cells on the pectoral fins to find marine invertebrates buried in sediment on the ocean floor. Anglerfishes (order Lophiiformes) are named for the structure of the first dorsal spine, which has been modified into a fishing pole and bait, called, respectively, the illicium and the esca, held over the anglerfish's mouth. With this complex

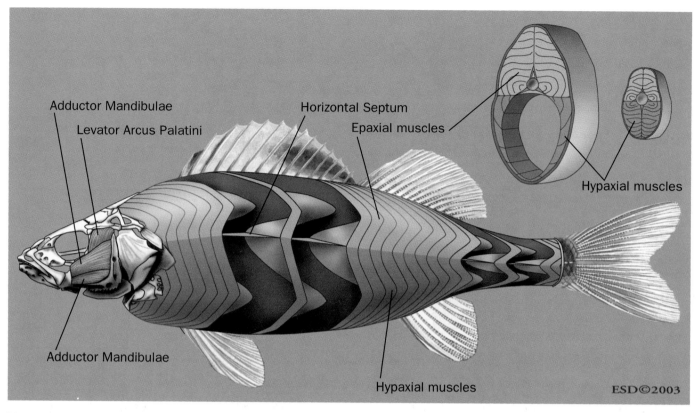

Musculature of the yellow perch (*Perca flavescens*). Axial muscles are organized into nested cones that extend across several vertebral segments. At the lateral midline they are divided by the horizontal septum into epaxial (dorsal) and hypaxial (ventral) regions. The jaw muscles include the large adductor mandibulae and the levator arcus palatini muscles in the cheek region. (Illustration by Emily Damstra)

modified fin structure, they lure potential prey fish within range of easy capture. Other fin adaptations are meant to deter predators. As many fishermen have experienced, fin spines of even small fishes, such as sunfish (genus *Lepomis*), can be painfully sharp.

Fin spines also may be associated with unpalatable or noxious toxins that deter predators. Many members of the scorpionfishes (family Scorpaenidae) and related groups have sharp and toxic spines that they use as defense. The stonefish (genus *Synanceia*) has strong neurotoxins in its dorsal spines that can be injected into predators. Fins are used in reproductive behavior as well. In sharks and other elasmobranches (sharks, skates, and rays), intromittent organs called claspers have evolved from the pelvic fins. Similarly, the live-bearing fishes (family Poeciliidae) and related groups have modified anal fins that function as an intromittent organ called the gonapodium.

The integument

The body's covering, the skin and scales, provides a protective barrier to the external environment. As in other vertebrates, the skin of fishes has a deep dermal layer and a superficial epidermis. In fishes, glands in the epidermis secrete mucus that coats and protects the surface of the animal. Scales are formed from the dermal and epidermal layers of the skin. Chondrichthyans have placoid scales that are homologous to vertebrate teeth. At the base of each scale there

is vasculature (blood vessels) covered with a dentine layer that is surrounded by enamel. Placoid scales are not replaced, but they increase in number with the growth of the fish. In some fish, such as the spiny dogfish (genus *Squalus*), placoid scales have been modified into large spines. The Osteichthyan fishes have several different scale types. Ganoid scales are found in basal groups of ray-finned fishes; they are formed from bony plates covered with a layer of ganoine and often create an interconnected armor over the surface of the body, as in bichirs (family Polypteridae) and gars (family Lepisosteidae). The scales of teleosts are derived from ganoid scales, losing the layer of ganoine to leave a thin plate of bone. Teleost scales are classified as ctenoid (toothed) and cycloid (circular) based on the shape of the outer edge. Unlike most placoid or ganoid scales, cycloid or ctenoid scales are arranged so as to overlap their more caudal neighbors. Scales protect the skin and deeper tissues from the environment. In many cases, as with ganoid scales, they form a tough armor against predators. Cycloid or ctenoid scales offer some protection from predators while not burdening the fish with the weight of heavy armor.

There is an amazing diversity in skin color patterns and their functions among fishes. Whereas some fishes use cryptic coloration to blend into their environments, others use bright colors or distinctive patterns to communicate. Cleaner fishes that pick parasites off other fish have bright colors and distinctive patterns that are recognizable by other species. Often color pattern is used to confuse or ward off predators. Toxic

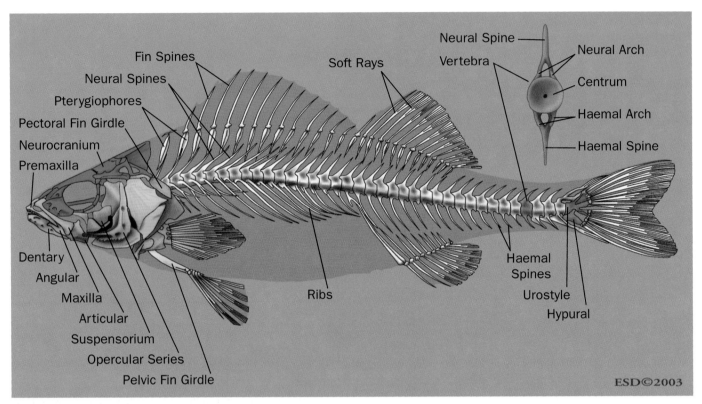

Lateral view of the skeleton of the yellow perch (*Perca flavescens*). The principal tooth-bearing jaw elements are the dentary and premaxilla. The dentary, angular, and articular bones together form the lower jaw while the premaxilla and maxilla form the upper jaw. The jaws are connected to the neurocranium, the region of the skull surrounding the brain, by a series of bones that together are called the suspensorium. The opercular series, caudal and ventral to the suspensorium, covers the gills. The axial skeleton consists of a series of vertebrae. Each vertebra is composed of a centrum, neural spine, and arch, through which the spinal cord runs, and hemal spine and arch, through which passes the dorsal aorta. At the caudal end of the vertebral column the urostyle and mondified hemal spines called hypural bones support the muscles of the caudal fin. Other median fins are supported by pterygiophores that extend toward the vertebrae from the fins. The paired pectoral and pelvic fins are supported by the fin girdles. The bones of the pectoral fin suspend the fins from the neurocranium and support the fin rays and muscles. The pelvic girdle is not attached to the skull or vertebrae and its position varies among species. Soft rays and fin spines project from the base of the fins to support the fin membranes. (Illustration by Emily Damstra)

lionfishes (genus *Pterois*) have distinctive red and white stripes that warn potential predators. Eyespots on the caudal fin of many species may confuse predators about a fish's orientation.

Internal morphological features

Cranial features and feeding

Unlike mammals, which have highly fused skulls with articulation only at attachments to the lower jaw and vertebral column, the cranium of fishes has more than 40 independently movable bony elements. These allow jaw protrusion, lateral expansion of the jaws, depression of the branchiohyoid apparatus and floor of the mouth, and movement of the gills and the operculum, which covers the gills. These movable elements are anchored to the neurocranium, which surrounds the brain and articulates with the vertebrae. The neurocranium corresponds to the chondrocranium, retained in chondrichthyan fishes, and additional dermal bones (the dermatocranium). As in other vertebrates, the neurocranium results from the fusion of many bones during development. The primary purpose of the neurocranium is to protect and support the brain. In addition, many species have a bone called the vomer, which forms part of the lower surface of the neurocranium, bears teeth, and aids in feeding.

The structure and function of fish jaws are astonishingly diverse, reflecting a wide array of feeding strategies and prey types that fishes exploit for food. This diversity results in large part from the increased mobility in the lower and upper jaws as well as the ability of fishes to incorporate other parts of their cranial anatomy into the feeding apparatus. In fishes both the upper and lower jaws articulate with the rest of the cranial skeleton, of which many elements are mobile. In bony fishes the primary jaws include the tooth-bearing premaxilla and, in more basal groups, maxilla in the upper jaw and the tooth-bearing dentary and the articular in the lower jaw. Dorsally, the premaxilla slides along the rostral end of the neurocranium. The upper and lower jaws connect caudally with each other and with the suspensorium, a group of bones suspended from the neurocranium. In addition, the lower jaw is connected to the series of opercular bones that covers the gills and to the hyoid apparatus in the floor of the mouth.

Several sets of bones form the floor of the mouth and wrap around the buccal cavity to connect to other cranial elements. The most rostral is the hyoid arch, involved in expansion of the buccal cavity. Following the hyoid arch are the branchial arches, which hold the gill structures, and, most caudally, the pharyngeal jaws, which bear teeth and help the fish eat prey. During feeding, jaw-depressor muscles rotate the lower jaw ventrally, causing the jaws to protrude forward. Suction forces are generated in the buccal cavity by dropping the floor of the mouth and flaring the suspensorium. The opercular bones seal the opercular opening to the gills. This combination of movements simultaneously leads to jaw protrusion, suction of water, and movement of the prey into the mouth.

The functional organization of jaw morphological features for feeding is a trade-off between the velocity of the movement and the force exerted. A striking example of a high-velocity feeding event involving extremely mobile jaws is illustrated by the slingjaw wrasse (*Epibulus insidiator*), aptly named for its ability to sling its jaws away from the rest of its head during the capture of evasive prey. This mechanism allows the rest of the body to remain still, minimizing the chances of being detected by the prey. An alternative strategy is seen in fishes that eat hard prey, such as mollusks, including the sheepshead (*Archosargus*). These fish do not feed on evasive prey, and so they do not need high-velocity jaw movements; instead, they maximize the force for crushing

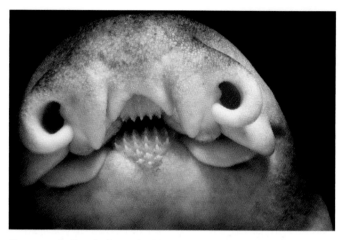

Close-up of a Port Jackson shark's (*Heterodontus portusjacksoni*) face. Port Jackson sharks feed primarily on invertebrates such as sea urchins, crabs, and starfish. (Photo by Jeffrey L. Rotman/Corbis. Reproduced by permission.)

shells. The difference in feeding strategy is reflected in the lengths of the lower jawbones, with long, slender bones being low force but high velocity and short, thick bones providing strong biting forces but less speed during jaw closing.

Teeth also vary markedly with a fish's prey. The teeth of predatory fish, including many carnivorous sharks, must cut through their prey and thus are triangular and serrated, providing effective blades for slicing through tissues. Other predators, including eels, which swallow prey whole, may have elongated backward-pointing teeth that are effective in grasping prey and preventing the prey from struggling out of the mouth. Many species have teeth adapted for biting or crushing hard material, such as shells or coral. Parrotfishes (family Scaridae) are named for their beaks, formed by fused teeth that function to bite hunks out of coral. Parrotfishes also have robust teeth on the pharyngeal jaws that contribute to the crushing of coral for digestion.

The action of the mouth and teeth (ingestion) is the first stage of digestion. From the buccal and pharyngeal spaces, food is moved through the esophagus to the stomach and intestine. The esophagus secretes mucus to help move food along its length, and it may stretch to accommodate large food items. The digestive enzyme pepsin and hydrochloric acid begin chemical digestion of the food and, in some groups, including mullets (family Mugilidae), the stomach may be modified into a grinding organ to continue physically processing food. The intestines vary in length among species, with the intestines of herbivores being substantially longer than those of carnivores. In addition, the surface area of the intestines may be increased for better internal absorption. Several organs are associated with the intestines. The pyloric caecae, liver, gallbladder, and pancreas produce enzymes and other substances that aid digestion in the intestines.

The axial system and locomotion

As suggested by their external morphological characteristics and functions, fins have diverse internal structures. Skeletal fin girdles support pectoral and pelvic fins, and the fins

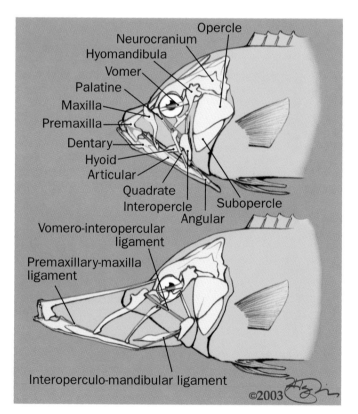

The slingjaw wrasse, *Epibulus insidiator*, has the greatest jaw protrusion known among fishes. The jaw's bone and ligament structure, depicted here, comprise the lever action responsible for it. (Illustration by Jonathan Higgins)

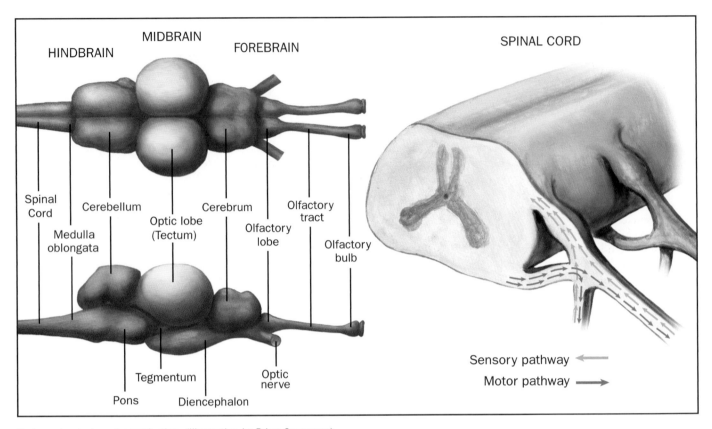

HINDBRAIN

MIDBRAIN

FOREBRAIN

SPINAL CORD

Spinal Cord

Medulla oblongata

Cerebellum

Optic lobe (Tectum)

Cerebrum

Olfactory lobe

Olfactory tract

Olfactory bulb

Tegmentum

Pons

Diencephalon

Optic nerve

Sensory pathway

Motor pathway

Brain and spinal cord organization. (Illustration by Brian Cressman)

themselves consist of spines or rays that ar e connected by the fin membrane. Muscles at its base actuate the fin and, particularly with the pectoral fins, allow complex movement. Median fins also have skeletal support, with muscles that raise or lower the fins and segmentally arranged muscles on individual fin segments that allow for a wave of muscle activity to propagate along the fin. The caudal fin, which generates much of the thrust in axial swimming, is supported by a series of laterally flattened bones and associated muscles in addition to

Southeastern African lungfish (*Protopterus amphibius*) using its pelvic fin as a "leg." (Photo by Tom McHugh/Steinhart Aquarium/Photo Researchers, Inc. Reproduced by permission.)

lateral body muscles. The caudal fin also is classified by the size of the dorsal and ventral caudal fin lobes. The most common designations are homocercal, in which the dorsal and ventral fin lobes are symmetrical (as in most bony fishes), and heterocercal, in which the dorsal and ventral lobes of the caudal fin are unequal in size. This type of tail is common in sharks but also is found in ray-finned fishes, such as the sturgeons and paddlefish (order Acipenseriformes).

Vertebrae have several components. The centrum is the central structural element of the vertebra. The neural arch above the centrum protects the spinal cord. In the trunk region, lateral processes extend from the ventrolateral centrum. In the tail, processes form the haemal arch, which encloses the large dorsal aorta. The arches extend to form spines in the dorsal and ventral midlines, which, with connective tissues, shape the vertical septum that divides the left and right sides of the fish. Similarly, extending left and right from the vertebrae is the sheet of connective tissue called the horizontal septum, which divides the epaxial (dorsal) and hypaxial (ventral) regions of the lateral muscles, called the myomeres.

Axial swimming movements are accomplished by contractions of the myomeres that connect through tendons to the vertebral column. The myomeres are organized into interdigitating cones and bands of muscle separated by the connective tissue myosepta. Myomere contraction transmits force to the network of tendons, which bends the vertebral column. Rostral-to-caudal (head-to-tail) propagating waves of muscle contraction generate the rostral-to-caudal waves of body bending during swimming.

The axial muscles often contain two common types of muscle fiber in fishes, each with a specific role in muscle function. Slow oxidative muscle functions in steady swimming. Slow muscles obtain energy through oxygen metabolism and are rich in myoglobin and vasculature that supplies oxygen, giving the muscle a red appearance. Because it is constantly supplied with oxygen, red muscle does not rapidly fatigue and thus can function in slow, sustained locomotion. Tunas and other large species that cruise steadily, searching for food, often have red muscle as a considerable proportion of their myomeres. In contrast to slow oxidative muscle, fast glycolytic muscle has a fast contraction time and uses glycogen stores as fuel. This type of muscle also is called white muscle, because with little vascularization and low levels of myoglobin, the muscle appears paler than oxidative fibers.

Because glycogen stores are used up quickly, fast muscle fatigues quickly and functions primarily in short swimming bursts as, for example, when a fish is startled. In most fishes the white muscle forms the major mass of the myomeres.

In addition to generating movement of axis, fins, or cranial structures, muscles perform other functions in fishes. In some species muscles have adapted to act as thermoregulatory, or "heater organs." In tunas (family Scombridae) muscle activity keeps the brain warm while they feed for squid in cold, deep waters. Electrical currents generated during muscle contraction have been harnessed by elephantnose fishes as a communication signal or by torpedo rays as a tool for disabling other species. This modification of muscle cells into electrogenerative organs has evolved independently many times in the evolution of fishes, and there is considerable diversity in the muscles that serve this function, including eye muscles (stargazers, family Uranoscopidae), jaw

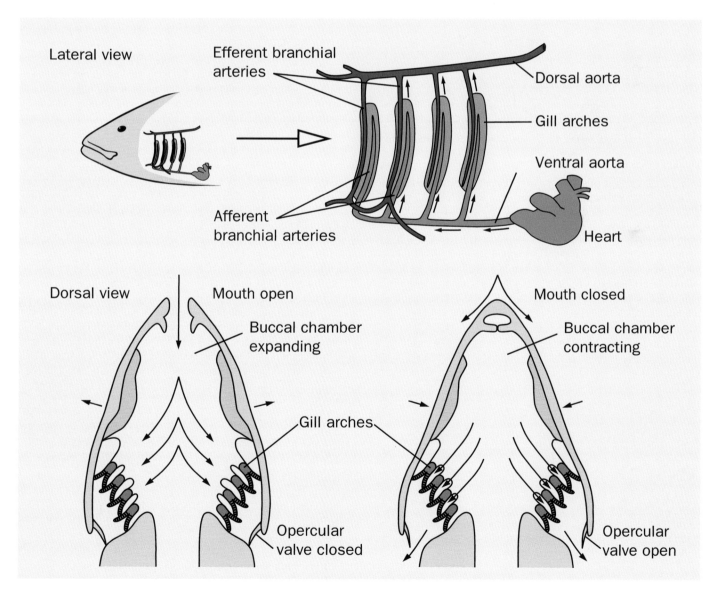

Flow of water and gas exchange through gills. (Illustration by Barbara Duperron)

The open mouth of a manta ray (*Manta birostris*) funnels food into its mouth while it swims, using two large, flap-like cephalic lobes that extend forward from the eyes. (Photo by Ivor Fulcher/Corbis. Reproduced by permission.)

muscles (torpedo rays), and axial muscles (electric eels, genus *Electrophorus*).

Neural control: The brain, spinal cord, and sense organs

The brain and spinal cord together form the central nervous system. The brain is subdivided into three regions—forebrain, midbrain, and hindbrain. The forebrain consists of the telencephalon and the diencephalon. At the rostral end of the telencephalon are the olfactory bulbs, which receive input from the olfactory receptors. The olfactory bulbs have neurons (the olfactory nerve, or cranial nerve I) that project into the olfactory regions of the telencephalon, also called the olfactory lobe because of its importance in this chemical sense. The olfactory bulb often is enlarged in fishes that rely heavily on olfaction, including many species of sharks. The diencephalon, which includes the epithalamus, thalmus, and hypothalamus, functions primarily in the regulation of the internal body environment. The pineal organ, which contains neurons and photoreceptors, is located at the distal end of the epiphyseal stalk and is part of the epithalamus, which projects from the dorsal surface of the diencephalon. In many species the pineal organ senses light through the cranium and may have numerous functions, including regulation of circadian rhythms. The optic nerve (cranial nerve II), which runs from the retina to the brain, enters the diencephalon and has inputs to the thalamus and hypothalamus as well as to the midbrain.

The midbrain consists of the optic lobe and tegmentum; both structures are involved in vision. The optic nerve has ex-

tensive connections to the optic lobe, and, as with the olfactory bulbs, a large optic lobe is associated with species that use vision extensively. The tegmentum functions in the control of intrinsic eye muscles to focus the visual image. The tegmentum also plays a part in motor control. For example, the midbrain locomotor region, which generates rhythmic swimming movements, is located in the tegementum.

The hindbrain includes the cerebellum, pons, and medulla oblongata. The cerebellum, unlike the more rostral brain regions, is a single structure rather than paired, bilateral lobes. The cerebellum's functions include maintaining equilibrium and balance. The pons and medulla form the brain stem. Many of the cranial nerves bring sensory information into the medulla and transfer motor signals to the muscles. Most of the cranial nerves enter the brain through the hindbrain. Cranial nerves III (oculomotor), IV (trochlear), and VI (abducens) control the six extraocular muscles that generate eye movements. Cranial nerve V (trigeminal) receives sensory input from and transfers motor signals to the mandible, and cranial nerve VII (facial) brings in sensory input from the hyoid arch and structures. Cranial nerve VIII (acoustic) contains sensory fibers that are involved in hearing and equilibrium. Cranial nerve IX (glossopharyngeal) serves the pharyngeal arch, providing both sensory information and motor output. Cranial nerve X (vagus) innervates the more caudal branchial arches as well as the lateral line and viscera.

The spinal cord runs the length of the vertebral column, protected by the neural arch. As with the myomeres and vertebrae, the spinal cord is organized segmentally. At each body segment, sensory neurons enter the cord through the dorsal roots, and motor neurons exit through the ventral roots. Interneurons, located entirely within the central nervous system, carry information between sensory and motor neurons and relay information to and from other interneurons in the brain.

The nervous system takes in sensory information and processes it to derive an appropriate response. Fishes use a wide array of senses to survey their environments. Vision is one of the best-understood sensory systems. The eyes of fishes are very similar in structure to the eyes of other vertebrates, with light coming in through the cornea and lens and projecting onto the retina, where rods, cones, and other nerve cells receive, process, and transmit the visual image. Instead of changing the shape of the lens, fishes focus the often near-spherical lens by moving it closer to or farther away from the retina. The organization and composition of the retina vary with a fish's visual environment. Deep-sea fishes have visual pigments that absorb light maximally in lower, blue wavelengths, while shallow-water species absorb a broader distribution of the light spectrum.

Eye position is also indicative of a fish's way of life. Bottom-dwelling predators that surprise prey from below, such as flatfishes or stonefishes, have eyes positioned upward and close together to provide binocular vision. Prey species generally have eyes positioned laterally to best survey for predators. *Anableps*, the four-eyed fish, lives at the surface of the water. It has four pupils to take in light dorsally through air and ventrally through water. The lens is shaped and positioned to focus the light from two sources on two regions of the retina, allowing for simultaneous input from both visual environments.

Both smell and taste permit fish to sense chemical signals. That fish may have an extremely well developed ability to sense chemical signals is illustrated in salmon and trout, which distinguish their natal streams based on chemical cues. Whereas olfactory receptors are localized in the olfactory epithelium within the bilateral nares, taste receptors, or taste buds, are more widespread, occurring not only in the mouth but also frequently on the gill structures and external surfaces, including the barbels, fins, and skin. Signals from the taste buds are transmitted to the brain through several cranial nerves. Cutaneous receptors have input through the facial nerve, whereas inputs

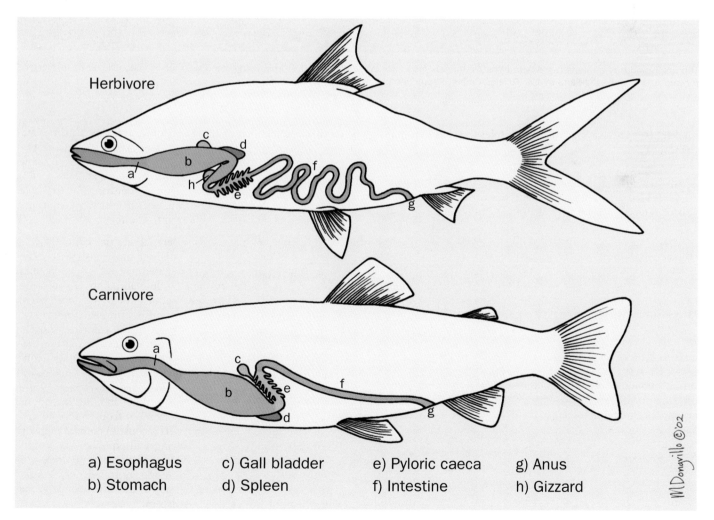

Herbivore

Carnivore

a) Esophagus   c) Gall bladder   e) Pyloric caeca   g) Anus
b) Stomach   d) Spleen   f) Intestine   h) Gizzard

Differences in digestive systems between herbivorous and carnivorous fishes. Although most structures are the same, the herbivore has a gizzard, as well as a longer intestine. (Illustration by Marguette Dongvillo)

Fin morphology of fishes: a. Sea robin (*Dactylopterus volitans*): b. Catfish (*Corydoras aeneus*); c. Piked dogfish (*Squalus acanthias*); d. Mosquitofish (*Gambusia affinis*); e. Anglerfish (*Lophius piscatorius*); f. Lumpfish (*Cyclopterus lumpus*). (Illustration by Marguette Dongvillo)

from receptors in the mouth travel to the brain through the glossopharyngeal and vagus nerves. These nerves lead to the brain's medulla, and, as with olfaction, fishes that use taste extensively to find prey have enlarged regions of the medulla corresponding to cranial nerves VII, IX, or X.

The primary mechanoreceptive systems are the ear, functioning in hearing and equilibrium, and the lateral line that senses contact at the surface of the body. The inner ears of elasmobranches and bony fishes are organized into three semicircular canals and three chambers, each containing an otolith, or ear stone, that rests on sensory hair cells associated with nerve cells. Two of the chambers, the saccule and the laguna, function in hearing. Vibrations from the environment lead to

movement of the chambers and the otoliths. The difference in movement is sensed by the hair cells and is processed as hearing. Ostariophysan fishes, including goldfish, catfishes, and others, have a series of bones called Webberian ossicles that connect the ear to the swim bladder. Vibrations are amplified through the swim bladder and improve hearing at high frequencies. Similarly, an otolith in the third chamber, the utricle, allows the fish to sense orientation in the water. This dense otolith lies upon sensory hair cells. When the body tips and the otolith moves, stimulation to those cells changes as well, signaling the change in orientation.

The sensory hair cells in the semicircular canals allow fishes to sense orientation and acceleration. Each canal is associated with an ampulla in which hair cells are located. Instead of an otolith, gel covers the cells. When fluid, called endolymph, in the semicircular canals moves as the result of a change in acceleration of the fish, the endolymph moves the gel and thus stimulates the sensory cells. The three semicircular canals are positioned approximately at right angles to one another to sense vertical, lateral, or forward movement.

Water movement on the surface of the fish is sensed through neuromasts. These structures can be found individually on the surface of the body, or they may sit below scales in canals called lateral lines. Neuromasts include a cupula of gel consistency and sensory hair cells, which project hairs into the cupula and synapse with nerve cells below the surface of the body. Movement of the cupula causes the hair cells to deflect, signaling a perturbation of the fluid around the fish.

Electroreception occurs in many groups of fishes and has numerous functions, including sensing murky environments; sensing prey, such as fishes that sleep buried in sand; or, in species that also generate electrical signals, receiving signals from other animals. Electrical input is received in pit organs on the surface of the body. Pit organs are filled with gel that conducts electrical current and, at their base, contain electroreceptor cells that synapse with sensory neurons.

Homeostasis

The autonomic nervous system and the endocrine system function together to regulate an animal's physiology. The autonomic nervous system, which includes a series of ganglia lateral to the spinal cord, receives input from the central nervous system to adjust the function of numerous tissues. Blood pressure is regulated through vasodilation or vasoconstriction, affecting numerous functions from digestion to oxygen uptake at the gills. The endocrine system similarly has broad effects on physiology, but through hormones rather than nerve activity. Controlled primarily through the hypothalamus and the pituitary, the endocrine system has many functions, including osmoregulation, growth, and metabolism.

Circulation and gas exchange

The circulatory system carries blood from the heart through the gills and to the body tissues before returning to the heart. The fish heart is unlike the mammalian heart, where left and right sides function to take deoxygenated blood from the body to the lungs and separately take deoxygenated blood from the lungs to the body. The heart of fishes is a single series of four chambers, with deoxygenated blood running through the heart to the gills and straight out to the body without returning to the heart. The four chambers of the heart are the sinus venosus, atrium, ventricle, and conus arteriosus. The chambers of the heart are separated by valves to prevent blood from flowing in the wrong direction during ventricular pumping.

The gills are the primary respiratory organs of fishes. Gills are located lateral to the mouth cavity. In bony fishes, they are covered by the opercula. Chondrichthyan fishes and lampreys do not have an operculum; instead, each gill vents to the surface of the body individually through gill slits. During ventilation, water flows into the mouth, across the gill, and through the gill slits or opercular opening. When negative

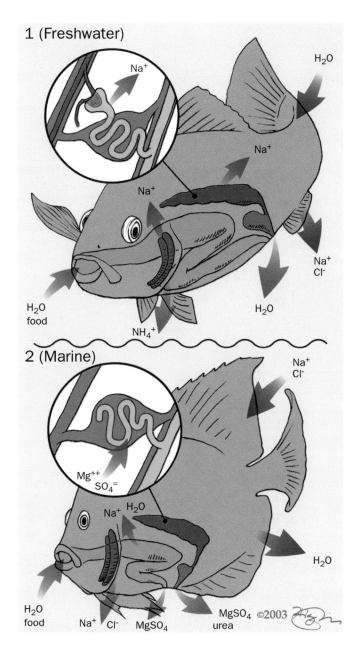

Osmoregulation/homeostasis in freshwater and marine fish. (Illustration by Jonathan Higgins)

pressure is generated in the mouth, the opercula or gill slits close over the gills to prevent water from flowing into the mouth through the opercular openings. The gills are formed from membranes and blood vessels lying over branchial arches. On each arch, lamellae project outside the buccal cavity. The lamellae have smaller processes called secondary lamellae, which are highly vascularized for oxygen exchange.

Fishes have a diverse array of other respiratory structures in addition to the gills. In larval fishes, gas exchange commonly occurs across the skin. Many fishes have accessory breathing organs. Numerous fishes have "lungs" in which air is stored. These fishes include many basal bony fishes, such as bowfins (order Amiiformes), gars (order Lepisosteiformes),

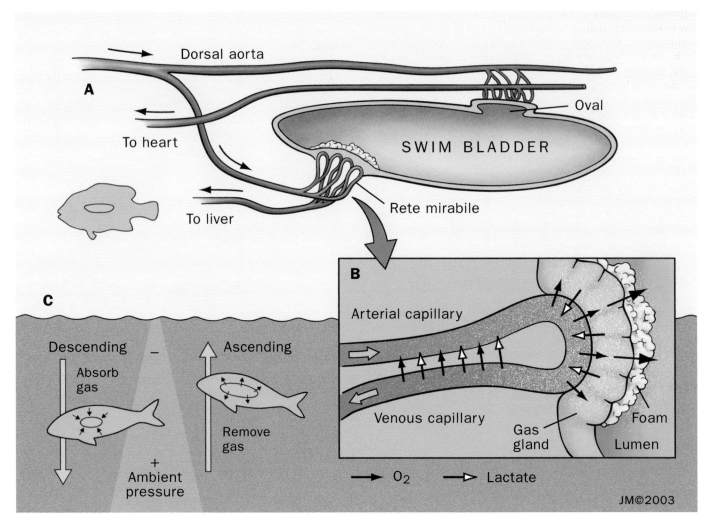

Structure and function of the swim bladder. The fish becomes less buoyant (descends) as gas is absorbed into the bloodstream and leaves the swim bladder. The fish ascends as gas is removed from the bloodstream and enters the swim bladder, enlarging it. (Illustration by Jacqueline Mahannah)

and reedfish (order Polypteriformes), which gulp air at the surface of the water. Several species that can breathe air include some catfishes (genus *Clarius*) and gouramis (family Osphronemidae), which have evolved structures associated with the gills for this function

Buoyancy control in the fluid environment

Gas exchange also occurs in the swim bladder, a sac full of gases that lies dorsally in the body cavity and functions primarily in buoyancy control. An increase in the amount of gases in this structure makes fish more buoyant, and a decrease makes them less buoyant. In many fishes, gas can enter the gas bladder only through the gas gland and rete mirabile ("wonderful net"), a highly vascularized tissue that, as in the gill filaments, provides a large surface area for gas exchange. The gas gland acts by acidifying the blood, decreasing the solubility of dissolved gases and thus increasing the available molecules for exchange into the swim bladder. A membrane called the oval controls the amount of gas in the bladder. Unlike the rest of the bladder, which is lined with

the amino acid guanine to prevent resorption of gases, the oval is highly permeable. The loss of gas from the bladder through the oval is controlled by muscles that can either obstruct the oval, preventing gas release, or free it for gas exchange. Some fishes have a pneumatic duct that runs from the alimentary canal to the swim bladder. This allows them to gulp air at the surface and store it in the gas bladder.

Many fishes have other adaptations to make them more buoyant, including morphological structures that are built for lightness. Some groups, including chondrichthyan fishes, do not have swim bladders and instead augment their buoyancy with fat stores. Another strategy in negatively buoyant fishes may be to alter body movements during locomotion to produce lift and upward thrust as well as forward thrust.

Osmoregulation and excretion

Living in water provides a set of challenges for osmoregulation, and fishes have developed a diverse array of strategies for manipulating their osmotic concentrations of various substances. Almost all fishes maintain osmotic levels that are lower (saltwater fishes) or higher (freshwater fishes) than the fluid in the environments around them. The lone exception is the hagfishes (family Myxinidae). Hagfishes, one of the most basal lineages of vertebrates, have internal salt concentrations at about the level of seawater, as do marine invertebrates. Marine elasmobranchs (sharks, rays, and skates) are isosmotic but with substantially lower salt concentrations in their bodies. They maintain this balance by retaining high concentrations of urea and trimethyl amine oxide (TMAO) in the blood. The urea increases the osmotic concentrations to the level of seawater. To keep salt concentrations low relative to the environment, elasmobranchs secrete salt through the kidneys and a special gland, the rectal gland, which connects to the alimentary canal. The rectal gland concentrates and eliminates both salt and chloride ions from the body tissues.

Teleost fishes are not isosmotic and have evolved mechanisms to regulate retention or elimination of ions. Marine teleosts with lower ionic concentrations than the fluid that surrounds them are constantly loosing water to the environment. They counter this loss by drinking and filtering saltwater. Salt and chloride ions are transported from the blood through the gill membranes, while magnesium and sulfates are filtered from the blood by the kidneys. Freshwater teleosts have the opposite problem of maintaining salts in an environment where the normal concentrations are low. In particular, water can move into the bloodstream through the alimentary canal and the gills, diluting internal concentrations. Again, the gills and the kidneys are critical to this balance. The gills actively take up some solutes from the water, and freshwater teleosts produce copious amounts of dilute urine.

Reproduction

Fishes demonstrate a wide range of reproductive strategies. Fishes may be males, females, or, in many species, hermaphrodites, with both male and female sex organs. There are several types of hermaphroditism, and simultaneous hermaphrodites may act as both male and female in a single breeding event. Hamlets, small species of sea bass (family Serranidae), breed in pairs, with individuals taking turns as male and female. Other simultaneous hermaphrodites, including numerous deep-sea species, are self-fertilizing. Serial hermaphrodites may be female at one time in their life history

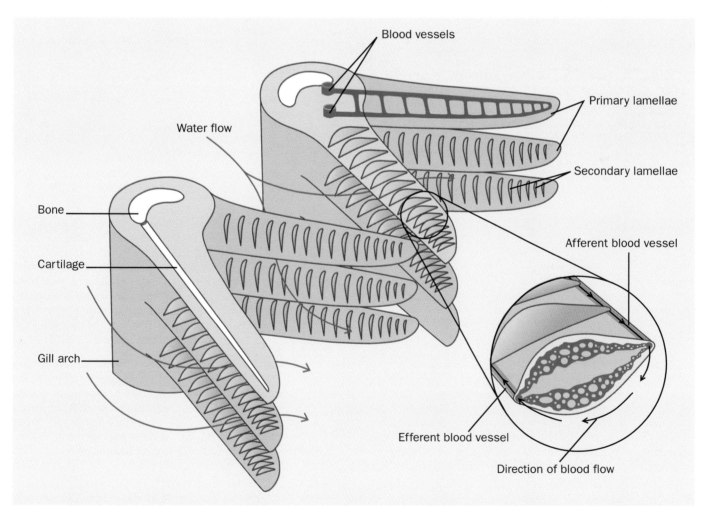

Microstructure of gills, showing water flow and blood flow. (Illustration by Barbara Duperron)

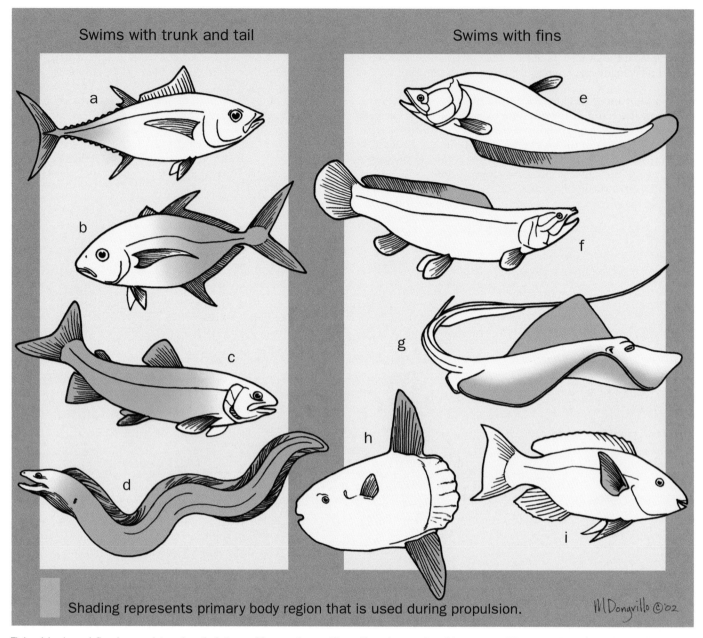

Swims with trunk and tail

Swims with fins

a

b

c

d

e

f

g

h

i

Shading represents primary body region that is used during propulsion.

Fishes' body and fin shapes determine their type of locomotion: a. Thunniform locomotion, bigeye tuna (*Thunnus obesus*); b. Carangiform, blue trevally (*Carangoides ferdau*); c. Subcarangiform, rainbow trout (*Oncorhynchus mykiss*); d. Anguilliform, green moray (*Gymnothorax funebris*); e. Gymnotiform, clown featherback (*Chitala ornata*); f. Amiiform, bowfin (*Amia calva*); g. Rajiform, southern stingray (*Dasyatis americana*); h. Tetraodontiform, mola (*Mola mola*); i. Labriform, bridled parrotfish (*Scarus frenatus*). (Illustration by Marguette Dongvillo)

and male at another. Protandrous species, including some damselfishes (family Pomacentridae), are first male and then become female as they age. Protogynous species, including many wrasses (family Labridae) and other perciformes, on the other hand, begin as females and become males. One of the more unusual strategies is that demonstrated by several families of anglerfishes. Smaller, parasitic males latch onto females with their mouths and fuse permanently with the female's body. The male obtains nutrition through the female' bloodstream and provides sperm for reproduction. Males of the seahorses and pipefishes

(family Syngnathidae) have pouches or specialized body surfaces that hold eggs while the embryos are developing.

Despite these variations in modes of reproduction, the basic reproductive structures are similar among taxa, with eggs being produced in the ovaries and sperm being produced in the testes. Ovaries and testes are held in places in the abdomen with mesenteries—the mesovaria for ovaries and mesorchia for the testes. The paths that the sperm and eggs take vary among species. Sperm or eggs may be released into the body cavity, as in Agnathans (lamprey and hagfish), and leave through pores in the abdomen. However, in most species, eggs are carried in the oviducts, which may be continuous with the ovary—as in many teleosts—or may be separated by a small space across which the eggs travel.

While in most fish species eggs and sperm are released by the parents and fertilization and development occur externally, a number of groups of bony fishes as well as sharks and other species in the class Chondrichthyes have internal fertilization. Sharks demonstrate a range of parental care prior to laying eggs or birthing pups. Many species, including skates (order Rajiformes), are oviparous, with the embryo relying completely on its yolk for sustenance. Oviparous species lay eggs that develop externally to the mother. In ovoviviparous species, such as the whale shark, the embryos depend on the yolk for nutrition but remain inside the mother through the embryonic period. Embryos hatch within the mother and are born free-swimming. Viviparous species similarly retain their embryos, but those embryos obtain nutrition both from the yolk sac and from the mother. Nutrition from the mother may be obtained via a placental structure connected to the mother's circulatory system, as in hammerheads (family Sphyrnidae), or from nutrient-rich fluids secreted by cells of the uterus, as in manta rays (*Manta birostris*).

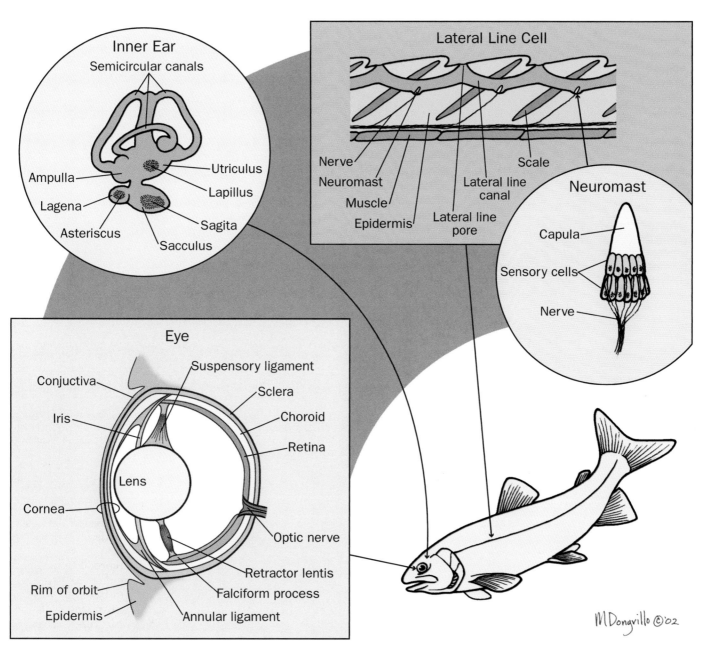

Sense organs of fishes visual, auditory, and lateral line systems. (Illustration by Marguette Dongvillo)

Resources

Books

Butler, Ann B., and William Hodos. *Comparative Vertebrate Neuroanatomy: Evolution and Adaptation.* New York: John Wiley and Sons, 1996.

Helfman, Gene S., Bruce B. Collette, and Douglas E. Facey. *The Diversity of Fishes.* Malden, MA: Blackwell Scientific, 1997.

Moyle, Peter B., and Joseph J. Cech. *Fishes: An Introduction to Ichthyology.* Englewood Cliffs, NJ: Prentice-Hall Inc., 2000.

Randall, John E., Gerald R. Allen, and Roger C. Steene. *Fishes of the Great Barrier Reef and Coral Sea.* Honolulu: Crawford House Publishing/University of Hawaii Press, 1997.

Melina Hale, PhD

● ● ● ● ●

Life history and reproduction

Types of reproduction

The vast array of adaptations that has evolved in fishes has given them the ability to inhabit a wide range of different habitats, including open seas and oceans, lakes, ponds, estuaries, rivers, tide pools, springs, deserts, forests, mudflats, and mountains. In addition, they have evolved the ability to exist in extreme areas with regard to temperature (e.g., the Antarctic), low oxygen, pH, and tremendous pressure. In fact, fishes exhibit the greatest vertical distribution of any group of vertebrates. As the result of selective pressures associated with these different environments, fishes have evolved three types of reproduction: bisexual, hermaphroditic, and parthenogenetic.

Bisexual

Bisexual reproduction is the most common form observed in fishes. In this type of reproduction, the sexes are separate ("dioecious") within the species. Species that are involved in bisexual reproduction may exhibit slight to very pronounced secondary sexual characteristics, or sexual dimorphism. Characteristic of these secondary sexual traits is that they usually are expressed in only one sex (typically the male), do not occur until maturation, may intensify during the breeding season, and generally do not enhance individual survival. Secondary sexual traits may consist of differences in body size, body parts (e.g., elongated fins), body ornamentation (e.g., nodules on the head), dentition, color pattern, and body shape and, possibly, differences in acoustic, chemical, and electrical attributes between the sexes. Bisexual mating systems include monogamy, polygamy, and promiscuity.

Hermaphroditic

The second type of reproduction in fishes involves sex reversal, where fishes function as male or female simultaneously or sequentially. Sequential hermaphrodites function as males during one part of their lives and females during another. There are two distinct forms of sequential hermaphrodites—protandric and protogynous. Protandric hermaphrodites are individuals that start out as male and later in life undergo internal morphological changes and become fully functional females. Protandric hermaphrodites are widespread among the sea basses (Serranidae). All wrasses (Labridae) appear to be protogynous hermaphrodites, in that all males are derived from females. Environmental factors or, more specifically, social cues influence sex change in wrasses. The social system

adopted by wrasses consists of a harem of females and one large male. The entire group is structured according to size, with the male at the top of this hierarchy. If a female is removed from the harem, the other females maneuver within the hierarchy. All the smaller females typically move up one position. If the male is removed or dies, the largest female in the harem attempts to fill the male's position by aggressively warding off neighboring males. If she is successful, within several hours she will display the male's behavior and will court and spawn (with no sperm released) with the subordinate females after two to four days. After approximately 14 days she becomes a fully functional male. In those taxa where sex reversal is mediated by social cues, the process varies widely, and a single individual can change from one sex to the other several times in response to these cues. On the other hand, there are many taxa (e.g., stripped bass, yellow perch, most groupers) of sequential hermaphrodites in which sex change occurs independently of social cues.

Simultaneous hermaphrodites possess a functional ovotestis and are capable of releasing viable sperm and eggs; hence, they have the potential to fertilize their own eggs. Only three species of cyprinodontiform fishes (*Cynolebias* species and *Rivulus marmoratus*) are known to be self-fertilizing hermaphrodites. Self-fertilization of *R. marmoratus* is internal and results in homozygous, genetically identical individuals. Because

California bullhead shark (*Heterodontus francisci*) eggcase. About 25% of all shark species lay eggs. (Photo by Tom McHugh/Steinhart Aquarium/Photo Researchers, Inc. Reproduced by permission.)

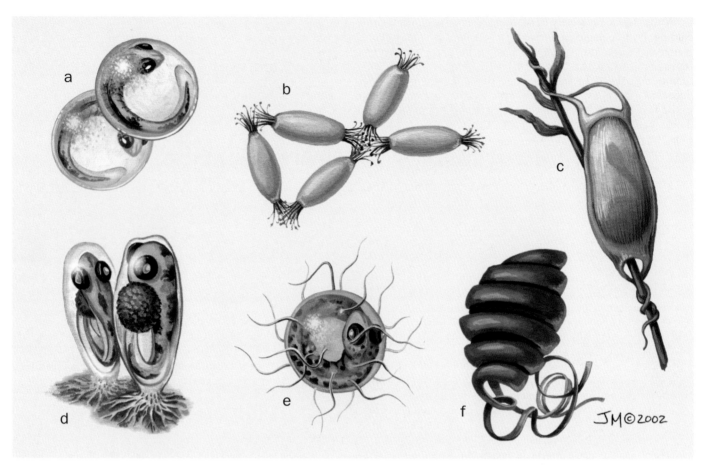

Diversity of external egg characteristics: a. Cod eggs; b. Hagfish eggs; c. Catshark eggcase; d. Gobie eggs; e. Flyingfish egg; f. Bull shark eggcase. (Illustration by Jacqueline Mahannah)

many cyprinodontiform fishes inhabit harsh habitats, self-fertilization may be a reproductive strategy to ensure mates in low-density and isolated populations. The more common pattern of simultaneous hermaphroditism is seen among the hamlets (*Hypoplectrus* and *Serranus*). Although these fishes are capable of producing both sperm and eggs at the same time, they function as only one sex at a time during a spawning event. Because some spawning events may last several hours (e.g., in hamlets), members of a pair may alternate sex roles, with one individual playing the part of the male and releasing sperm and later taking the role of the female and releasing eggs.

Parthenogenetic

Although it is rare in vertebrates, parthenogenetic reproduction does exist in a few species of fishes. By definition, parthenogenetic reproduction involves complete development of an egg without fertilization by a sperm of the same species. A variation on this type of reproduction exists in fishes. In fishes, mating with a heterospecific or conspecific male is required. The role of the male is in providing an active sperm, which comes in contact with the egg but which does not penetrate the egg membrane (chorion). The sperm acts as a stimulus for the egg to begin developing. The sperm does not contribute to the genetic makeup of the resulting fry. The fry is genetically identical to the female; hence, males are never produced by parthenogenetic reproduction. The

best-studied example of this phenomenon are live-bearing top minnows of the genus *Poeciliopsis.*

Modes of reproduction

In addition to the three different types of reproduction, there are three developmental modes of reproduction: oviparous, ovoviviparous, and viviparous.

Oviparous

Oviparous reproduction typically involves the release of both male and female gametes into the surrounding water, where fertilization takes place. Fertilization is internal in numerous fishes (e.g., rockfishes of the family Scorpaenidae and Neotropical catfishes of the family Auchenipteridae), however, and is followed by the often delayed release of embryonated eggs by the female into the surrounding environment. Upon fertilization, the developing embryo uses both yolk reserves and oil droplets as nutrients. In the marine environment, the eggs are generally buoyant and float in the upper water column for varying periods of time before undergoing metamorphosis and settling out of the plankton community. The resulting larva, in many cases, differs drastically from the adult and may spend considerable time floating in the water. Species that exhibit this mode of reproduction usually produce large quantities of small eggs owing to a high mortality rate.

While the vast majority of marine fishes are oviparous (including both pelagic and reef species), commonly with a planktonic stage, most freshwater fishes that have oviparous reproduction lack this stage, producing fry that closely resemble the adults. In addition, not all oviparous fishes produce buoyant eggs; instead, some produce demersal eggs (most freshwater fishes) that may have adhesive properties. The adhesive quality allows them to stick to rocks and plants, thus preventing them from washing away. Demersal eggs very often are associated with parental care, which is extremely widespread among fishes. Although oviparous species produce the greatest number of eggs, these eggs and the resulting larvae are very small.

Ovoviviparous

In ovoviviparous reproduction the eggs are retained by the female, and fertilization is internal. Although the eggs are retained, there is no placental or blood connection between the developing embryos and the female. Instead, the

A clown triggerfish (*Balistoides conspicillum*) guarding eggs (the yellow-green mass) near Menjangan Island, part of Bali Barat National Park in Bali, Indonesia. (Photo by Fred McConnaughey/Photo Researchers, Inc. Reproduced by permission.)

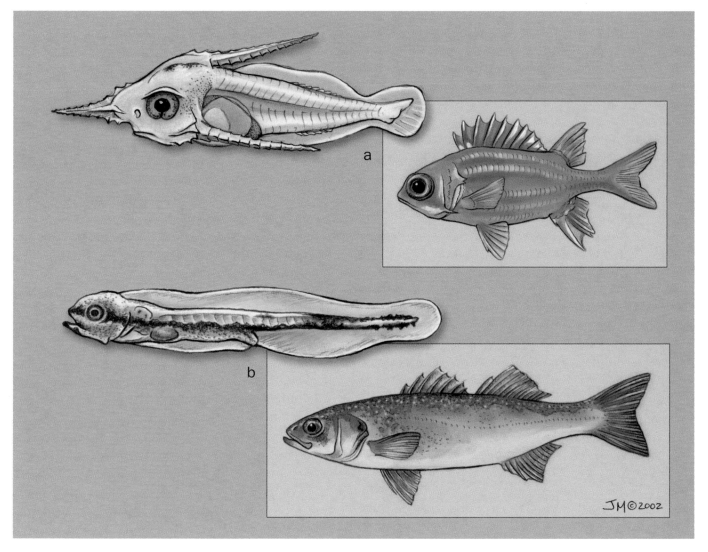

Larval diversity in fishes. a. Squirrelfish (*Sargocentron vexillarium*) larva, left, is 0.19 in (4.7 mm) in length. Adult is shown on the right. b. European sea bass (*Dicentrarchus labrax*) larva, left, is 0.24 in (6 mm) in length. Adult is shown to its bottom right. (Illustration by Jacqueline Mahannah)

Gravel flies as a female brook trout uses her tail fin to dig a redd (nest) in the Pleasant River in Windham, Maine. A male stands guard over the nest at rear. When the hole is sufficiently deep, the pair will come alongside each other. The female will then deposit her eggs while the male fertilizes them with his milt. (Photograph. AP/ Wide World Photos. Reproduced by permission.)

embryos develop completely within the egg, where all the necessary nutrients are present before hatching. Upon reaching full term, the embryos hatch inside the female, after which they are immediately born alive in the surrounding water. Those species that exhibit ovoviviparous reproduction do not have a pelagic stage, instead producing fry that closely resemble the adults. Consequently, ovoviviparous species have fewer eggs and larger fry than oviparous species. The most common ovoviviparous species are the poeciliids (e.g., guppies and swordtails), but the coelecanth also practices this reproductive method.

Viviparous

Viviparous reproduction is similar to ovoviviparous reproduction, but in the former method, there is a placental or blood connection between the mother and the eggs. Thus, the developing embryo acquires the necessary nutrients and oxygen from its mother. Once the developing embryo reaches full term, it is born alive. Viviparous species generally produce the smallest number of fry, but they typically are much larger than both oviparous and ovoviviparous fry. The most common viviparous species are sharks, but this form of reproduction also is found in the highland live-bearers of the family Goodeidae and the surfperches of the family Embiotocidae.

Reproductive strategies

Each reproductive mode, in combination with habitat, physiology, and behavior, plays an important role in the overall reproductive strategy. A reproductive strategy may dictate a large quantity of small eggs and high mortality or fewer large eggs with a greater chance of survival. These strategies must be designed such that a percentage of the eggs will survive through sheer numbers, camouflage, parental care, or re-

tention in the maternal or paternal body. In addition, time and location are important with respect to supplying ample food for the young and access to space occupied by the adults.

Gametogenesis

Gametogenesis refers to the origin and development of mature gametes through the processes of spermatogenesis and oogenesis.

Spermatogenesis

Spermatogenesis is the process in which spermatozoa are produced by follicles in the paired testes that undergo a series of meiotic and developmental transformations. Although spermatozoa consist of three parts (head, midpiece, and flagellum), they can differ between species. Despite the presence of the flagellum (tail), spermatozoa remain relatively inactive until they are released from the testes. At the time of spawning, the spermatozoa are combined with specialized secretions from the sperm duct (seminal fluid) to produce milt, which is released into the water during external fertilization. In the case of internal fertilization, the spermatozoa are transferred in packets called spermatophores. When combined with seminal fluid, the spermatozoa become very active. The life span of spermatozoa varies between species, and many other factors, such as temperature, have a profound effect. Spermatozoa remain viable for longer periods of time at lower temperatures. In addition, spermatozoa typically live longer when deposited inside the female compared with being exposed to the surrounding water. Long-term sperm storage and delayed fertilization of the eggs are characteristic of the reproductive biology of many ovoviviparous and viviparous fishes.

Oogenesis

The process in which ova (eggs) are produced in the paired ovaries is oogenesis. During oogenesis nutritional reserves are formed in the egg in the form of yolk material and oil droplets. The yolk is a source of protein, while the oil droplets provide fat and aid in buoyancy. Eggs can vary considerably in size, shape, outer shell characteristics, buoyancy, and adhesion. For example, most pelagic eggs are buoyant, whereas many demersal eggs are adhesive. The adhesiveness of an egg may facilitate fertilization, prevent unfertilized eggs from washing downstream, and allow eggs to attach to plants or rocks rather than falling onto the soft substrate, where they may suffocate. Eggs usually are deposited over a specific period of time rather than all at once. In addition, not all eggs present in the ovaries are deposited during spawning. Those eggs that are not deposited are resorbed by the ovaries, and the proteins, fats, and minerals are reused by the female for maintenance, growth, or egg production. In cases where there are too many eggs to be resorbed completely by the ovaries, the opening to the oviduct may become plugged with tissue, and the female becomes "egg bound." The obstructed oviduct prevents eggs from exiting the ovary during future spawning. In extreme cases, females that are severely egg bound may die.

Fecundity

The total number of eggs produced (fecundity) by a female may vary from one to two in some sharks to several hundred million in the ocean sunfish (*Mola mola*). In general,

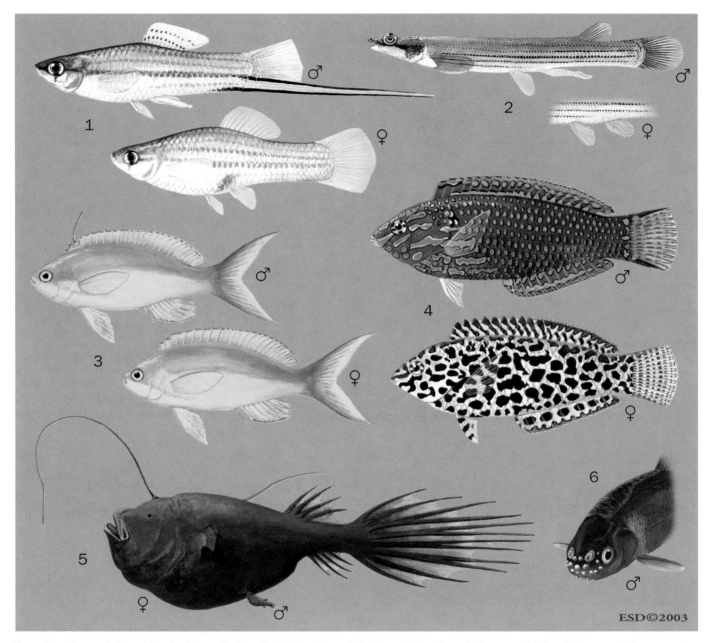

Examples of sexual dimorphism in fishes: 1. Swordtail, *Xiphophorus hellerii,* male has elongate lower caudal fin; 2. Largescale foureyes, *Anableps anableps,* male has modified anal fin (gonopodium); 3. Bicolor anthias, *Pseudanthias bicolor,* male has elongated dorsal spine; 4. Blackspotted wrasse, *Macropharyngodon meleagris,* different coloration and pattern; 5. Krøyer's deep sea anglerfish, *Ceratias holboelli,* parasitic male dwarfed by large female; 6. Fathead minnow, *Pimephales promelas,* male has breeding tubercles on its head. (Illustration by Emily Damstra)

fecundity declines with increasing egg size and parental care but increases with body size. In addition, fecundity is affected by numerous factors, including the species; age, size, and overall health of the female; food availability; time of year; and water temperature and quality.

Fertilization

In the vast majority of fishes, fertilization takes place outside the female's body. External fertilization may consist of varied types of spawning behavior, including paired or group broadcast spawning and oral fertilization. On the other hand, some fishes have internal fertilization. Males of those species that are involved in internal fertilization possess an intromittent organ, which is used for transferring the spermatophores. For example, in male sharks and poecilids, the pelvic and anal fins are modified into claspers and a gonopodium, respectively. Regardless of where fertilization takes place, fertilization is the point in time when a spermatozoa penetrates the chorion through a specialized opening called the micropyle. Once a spermatozoa enters the micropyle, the chorion hardens through a process known as water hardening, to prevent

A mouth brooding cichlid (*Haplotaxodon*) releasing fry in Lake Tanganyika, Africa. (Photo by Animals Animals ©Deeble & Stone OSF. Reproduced by permission.)

more than one spermatozoa from entering (polyspermy) and to aid in protecting the developing embryo. Some species, such as sturgeons, have multiple micropyle; thus, polyspermy does occur. Micropyle characteristics (e.g., size and shape) as well as other external egg characteristics (e.g., filaments, adhesive stalks, and tendrils) are quite varied among fishes. Once the spermatozoa enters the egg and the pronuclei and egg fuse, a zygote is created.

Embryology

Embryo development (embryogenesis) in fishes is similar to that of most vertebrates, with the embryo taking form on top of the yolk. Most teleost fishes, including elasmobranches and hagfishes, exhibit meroblastic cleavage, which involves cell division within a small disc-like region of the egg to produce the blastoderm (the disc of protoplasm where cleavage takes place) that eventually develops into a fish. On the other hand, lampreys have holoblastic cleavage, and bowfin, gar, and sturgeon have an intermediate form referred to as semi-holoblastic cleavage. Holoblastic cleavage is total, resulting in equal-size blastomeres (cells resulting from cleavage). From this point of embryogenesis, cell division and differentiation continue in a prescribed manner, although the time at which specific structures appear varies among species. Before hatching, many structures and organs develop at least partially, including body somites (metameres or body segments), kidney ducts, the neural tube, optic and auditory vesicles, eye lens placodes (thickening of the epithelium), head and body melanophores, a heart and functioning circulatory system, pectoral and median fin folds, opercular covers, lateral line sense organs, and the notochord. At the time of hatching, the mouth and jaw may be barely formed, little if any ossification (bone formation) exists, fin rays may be present, and a nonfunctional gut and gas bladder are present.

At this advanced stage, the embryo is curled around on itself within the tight confines of the egg. Owing to the harsh and extreme environments (e.g., deserts and other arid environments) occupied by some fishes, they have evolved a re-

productive strategy that involves a period of delayed development (diapause) for their embryos. For example, South American and African annual killifish fertilize and deposit their eggs in sand or peat moss in the rainy spring months. During the summer drought, when the pools of water dry up, the parent perishes, and the eggs enter a period of diapause. The rainy season of the following year triggers developmental completion and hatching.

The length of the developmental interval varies considerably, from a few days in surgeonfishes and several weeks in flatfishes up to several months in sharks. In general, the length of the developmental interval declines with increasing water temperature. When embryos are ready, hatching gland cells located on the head of some fishes secrete proteolytic enzymes, which breakdown proteins into simpler substances and aid in weakening the chorion. Erratic movements of the embryo's body and tail also aid in hatching.

Development

Larvae

Larval life generally begins when the young emerge from the egg and switch from internal yolk reserves to a diet consisting of plankton (e.g., diatoms, copepods, amphipods, ciliates, and larvaceans). The newly hatched, free-swimming individual, which may still have a large yolk sac attached, is referred to as a yolk-sac larva until the yolk is absorbed, after which it is referred to as a fry. At this point much larval development proceeds, including the axial skeleton, fins, organ systems, true and median fin rays and spines, scales, urostyle, hypural plate, and caudal rays. Characteristic pigmentation appears (including pigment in the eyes), and both the mouth and anus open and become functional. Before the development of the gill filaments, oxygen is absorbed across the membranous primordial fin folds through cutaneous respiration. The larval stage of fishes varies considerably in duration, ranging from one to two weeks in sardines (Clupeidae), about one month in many coral reef fishes, to several months or years in anguillid eels.

Egg cases containing lesser spotted dogfish (*Scyliorhinus canicula*) embryos. (Photo by Douglas P. Wilson: Frank Lane Picture Agency/Corbis. Reproduced by permission.)

Juveniles

The transition from larva to juvenile in many species (e.g., coral reef fishes) is associated with a change in habitat as juveniles settle out of the water column and take on a benthic, or reef, existence. In general, the juvenile phase is thought to begin when the larval characters are lost and the axial skeleton, organ systems, pigmentation, squamation (arrangement of scales), and fins become fully developed. It is at this time that the young take on the appearance of the adults. This transition in many species is very simple and can be completed in a short time (minutes to hours), as exhibited by damselfishes. This transition, however, is much more complex in some fishes and may involve significant alterations in the anatomy, physiology, or behavior of the species (e.g., metamorphosis in flatfishes and smoltification in salmon). In some fishes (e.g., Atheriniformes and Cyprinodontiformes) the immobile larval stage is bypassed, and the newly hatched fry are fully mobile and immediately capable of feeding actively.

Adults

By definition, an adult is a fully grown, sexually mature individual. Not surprisingly, the growth rate and age and size at which maturation occurs vary considerably in fishes. For example, the males of some surfperches are born with functional sperm, while it may take as long as 20 years for other fishes (e.g., sturgeon and some sharks) to mature. The spiny dogfish, which may have a life span of up to 70 years, may not become sexually mature until the age of 20 years. Among teleosts, American eels may not become sexually mature until they are 40 years old, at which time they participate in the spawning migration to the Sargasso Sea.

Senescence

In general, larger fishes live longer than smaller fishes, but the longevity of fishes varies considerably. Fishes that have a short life span (perhaps one year) include the South American and African annual killifishes as well as some North American minnows (*Pimephales* species), a silverside (Atherinidae), a stickleback (Gasterosteidae), and some gobies. Some of the oldest (90–140 years) fishes, whose ages have been determined through radioisotopic and otolith analyses, are scorpaenids from the northeastern Pacific. For the majority of all fishes, death is attributed to predation, accident, pathogens, accumulation of somatic mutations that cause a decline in health, alteration or loss of habitat, or commercial harvest. Among some fishes, death is attributed to senescence (old age), which refers to age-related changes in the body that have adverse effects on the organism. Over time these metabolic and anatomic processes make the organism more susceptible to death.

Resources

Books

Bond, Carl E. *Biology of Fishes.* 2nd edition. New York: Harcourt Brace College Publishers, 1996.

Deloach, Ned. *Reef Fish Behavior: Florida, Caribbean, Bahamas.* Jacksonville: New World Publications, Inc., 1999.

Helfman, Gene S., Bruce B. Collette, and Douglas E. Facey. *The Diversity of Fishes.* Malden, MA: Blackwell Science, Inc., 1997.

Hoar, W. S., and D. J. Randall, eds. *Fish Physiology.* Vols. 1–20. New York: Academic Press, 1969–1993.

Lagler, Karl F. *Ichthyology.* 2nd edition. New York: John Wiley & Sons, 1977.

Nelson, J. S. *Fishes of the World.* 3rd edition. New York: John Wiley & Sons, 1994.

Pitcher, Tony J., ed. *Behavior of Teleost Fishes.* 2nd ed. New York: Chapman & Hall, 1993.

Smith, C. L. *Patterns of Reproduction in Coral Reef Fishes.* NOAA Technical Memorandum NMFS-SEFC-80. Highlands, NJ: National Marine Fisheries Service, 1982.

Thresher, Ronald E. *Reef Fishes: Behavior and Ecology on the Reef and in the Aquarium.* Saint Petersburg, FL: Palmetto Publishing Company, 1980.

———. *Reproduction in Reef Fishes.* Neptune City, NJ: T.F.H. Publications, Inc., 1984.

Periodicals

Johannes, R. E. "Reproductive Strategies of Coastal Marine Fishes in the Tropics." *Environmental Biology of Fishes* 3 (1978): 65–84.

Jeffrey C. Howe, MS

Freshwater ecology

Diversity of freshwater fishes

Only a small portion (0.01%) of the surface water of the earth is freshwater, but these areas represent a variety of habitats, including swift-moving streams, deep glacial lakes, and ephemeral creeks. These freshwater habitats harbor diverse assemblages of fish, comprising more than 10,000 species in 23 orders. Much of this species richness is represented by Cypriniformes (2,662 species), Characiformes (1,343 species), Siluriformes (2,287 species), and Perciformes (2,185 species). The Amazon River alone is home to more than 1,300 species of freshwater fish, and more than 700 species of endemic haplochromine cichlids inhabit the East African rift lakes.

Distribution of freshwater fishes

On global and regional scales, the distribution of freshwater fishes is determined largely by historical circumstances. The world can be divided into distinct zoogeographic regions based on the distribution of organisms around the globe. Patterns noted at the global scale have been influenced over a long evolutionary time period by plate tectonics, including the movements and collisions of continental landmasses (continental drift). Major tectonic events played a large role in determining the families of fish that are present on a particular continent and that have the opportunity to become part of local fish assemblages. The isolation of fish caused by the separation of landmasses allowed for the diversification of species within major lineages.

Continental movements also affected climate, geologic, and drainage patterns across the landscape. The latitudinal location of landmasses influenced their susceptibility to glaciation during cooler periods of geologic history. During the Pleistocene epoch (11,000 to 1.8 million years ago), four major glacial periods resulted in the extirpation of fishes in areas covered by ice sheets, caused other species to move to nonglaciated refugia, and altered large-scale drainage patterns. These effects still influence the distribution of fish today, as many species no longer inhabit certain areas or are recolonizing portions of their previous range after seeking southern refugia during the Pleistocene.

Geologically, mountains that are pushed up from the collision of two landmasses can restrict the movement of fish.

For example, fish assemblages on one side of the Appalachian Mountains are substantially different from those on the other side of the divide. Mountains and other geologic features typically form the boundaries of drainage basins; the evolution of fish that are isolated in distinct drainages leads to further diversification and variance within species over time. Within drainage basins, a number of factors are associated with patterns of fish diversity. Fewer species tend to occur in headwater streams, while more species inhabit downstream portions of the watershed. The size and variety of local habitat types also affect fish diversity, with the diversity of fish increasing with habitat area and internal variability. In addition to natural controls on the range and composition of fish communities, it is important to recognize that human activities, including deforestation, construction of dams, introduction of non-native species, and pollution, have influenced the distribution of fish in many regions of the world throughout the course of recent history.

Acting within this broader context, physical and chemical characteristics of the environment regulate the composition and diversity of fish species that inhabit freshwater habitats. In all aquatic systems, light penetration and water temperature determine physical conditions that fish encounter. Fish also must be adapted to tolerate chemical attributes of freshwater systems, such as salinity, oxygen, and pH. Local species assemblages and species distribution within a habitat largely reflect the preferences of fish for different physical and chemical conditions.

Light

Light penetration directly and indirectly influences fish in freshwater habitats by warming the water, driving photosynthesis, and enabling visual activities. When light reaches the water surface, a small amount is reflected, and the remainder is absorbed as it enters the water column. Wavelengths of light are absorbed differentially with depth. Clear water in the upper few meters of the water column absorbs red wavelengths and converts the energy to heat. Only wavelengths between 400 and 700 nanometers can be used for photosynthesis, and these wavelengths penetrate deeper into the water column before they are absorbed. Light absorption is affected by particles and dissolved material in the water as well; for example, the presence of algae shifts absorption toward the green wavelengths. Light also enables fish to use vision to detect predators, prey, potential mates, and features of their habitat.

Temperature

Freshwater fish are ectotherms, and their internal temperature follows that of the surrounding water. Fish partition habitat space based on thermal gradients to avoid harmful temperatures as well as to take advantage of those that are optimal for a variety of physiological functions, including feeding, growth, and reproduction. Thus, seasonal movements and spawning are regulated strongly by temperature. Temperature varies on a geologic timescale, and shifts in the geographic distribution of fishes have been associated with major historical climatic changes. Temperature also varies locally and over short timescales, including diel and seasonal cycles, in freshwater aquatic systems. As light energy is converted to heat, the top portion of the water column warms first and to the greatest extent. Finally, latitude, altitude, and the velocity of water influence temperature. Due to large differences in water velocity, temperature patterns differ markedly in lakes versus running waters.

Salinity

Although rivers erode and transport some salts from geologic formations within the watershed, most bodies of freshwater lack the high concentration of dissolved salts that characterizes seawater. Exceptions occur in some desert areas of the southwestern United States and northern Mexico as well as in the Rift Valley of East Africa. In these areas, minerals accumulate when streams flow through underlying geologic salt formations or when evaporation leaves behind high concentrations of salts. While diverse fish communities inhabit these mineral-rich waters, most freshwater fishes cannot adapt physiologically to life in saline waters. Regulating internal salt concentrations poses a challenge to most fishes living in freshwater environments. Because salts are more concentrated inside their bodies than in the water, osmotic and diffusion processes work to bring in water and remove salts. To counter this situation, freshwater teleosts excrete large quantities of dilute urine and transport salts back into their blood using chloride cells. While this adaptation enables fish to osmoregulate in freshwater, the internal retention of salts also stresses most freshwater fish if they are exposed to saline conditions.

Oxygen

The concentration of oxygen in freshwater has serious implications for fish presence and distribution in a given area, and anoxia can result in the death of individuals. Oxygen enters water via diffusion from air at the water surface. Turbulence increases the surface area of water, such that moving waters contain more oxygen than stagnant waters. In addition, photosynthesis of plants, respiration of plants and animals, and the oxidation of organic materials drive diel changes in oxygen concentrations. Oxygen solubility in water is correlated negatively with water temperatures, and higher temperatures reduce dissolved oxygen levels. At the same time, fish metabolic rates and oxygen consumption levels increase with temperature, such that low oxygen conditions in warm water are particularly stressful for fish. To survive in low-oxygen waters, more than 40 genera of fish possess some capacity to breathe oxygen from the air; most of these species live in tropical freshwater habitats, where high temperatures

Tumultuous waters provide a very different environment for fishes compared to slow moving rivers. (Photo by Royalty-Free/Corbis. Reproduced by permission.)

and high rates of decomposition reduce the dissolved oxygen in the water.

pH

The relative acidity or alkalinity of a water body is measured as its pH. Hydrogen ions, which increase the acidity of water, are produced when carbonic acid dissociates from dissolved carbon dioxide in the water or from rainfall. The free hydrogen ions can be neutralized by carbonate minerals and buffered by calcareous compounds in geologic features surrounding bodies of freshwater. In poorly buffered systems, photosynthesis also can remove hydrogen ions and increase the pH of the water body. Metabolic functions of fish require pH within a certain range, and most fish cannot tolerate pH levels outside a range of approximately 4.0–10.0. High or low pH can be detrimental to reproductive success, gill function, and oxygen transport. Acidic pH appears to be most deleterious to fish. Acidic water dissolves metals, such as aluminum, that can be toxic to fish. In addition, the abundance and diversity of species, particularly of invertebrates that are eaten by freshwater fish, decline as water becomes acidic.

Major freshwater habitats

The variety of niches created by variation in physical and chemical factors within freshwater environments contributes to the great diversity of freshwater fishes. Many of these specific niches are organized within several major freshwater habitats. Most freshwater fishes inhabit streams, rivers, and lakes. Some fish prefer areas of swift-moving water in high mountain streams, others live at deep depths in lakes, and still others thrive in stagnant ponds.

heat. In temperate lakes the top portion of the water column heats during the summer, but temperature declines with depth. The warm surface water and cold bottom water are separated by the thermocline, a transition zone at the depth of greatest temperature change. Because water is most dense at 39.2°F (4°C) and becomes lighter by either cooling or heating, vertical temperature gradients and patterns of stratification vary with seasons and latitude. In cool portions of the temperate zone, differences in water temperature are minimal as the surface water of a lake warms to 39.2°F (4°C) in the spring; at this time, the lake mixes from surface to bottom. As warming continues, the lake stratifies in the summer, with a warm surface layer and cool deep waters. When the lake cools in the fall, stratification again breaks down. Reverse stratification occurs in the winter, however, as ice forms; colder, less dense water under the ice is suspended over warmer, more dense water around 39.2°F (4°C). The temperature gradient and stratification in lakes varies with latitude. In warmer temperate areas, the reverse stratification in winter does not occur, since ice rarely forms at these latitudes.

While it is common in temperate regions, this seasonal pattern of stratification driven by temperatures is not seen in all lakes. Some crater lakes may never stratify, because geothermal activity warms the deepest waters and minimizes temperature differences within the water column. Most tropical lakes stratify and mix on a daily basis. Annual variation in solar energy is minimal in the tropics, and daily changes in air temperature can establish and break down water column stratification. Wind is another factor that strongly affects stratification of many tropical lakes; wind adds kinetic energy to the lake and increases heat loss in surface waters. Some shallow tropical lakes, such as Lake Victoria, mix once a year when temperatures are lowest and winds are most persistent. Other deep tropical lakes may remain permanently stratified. For example, Lake Tanganyika reaches a depth of 4,823 ft (1,470 m), but the kinetic energy from wind cannot mix the waters below 820–984 ft (250–300 m). Polar or high-altitude lakes also may be stratified permanently if they remain frozen throughout the year.

Stratification of lakes caused by temperature gradients has two major effects on biological components of the lake ecosystem. First, stratification restricts mixing of nutrients within the lake to the area above the thermocline, unless wind or another turbulent force physically disturbs the lake waters. Thus, nutrients and other organic materials that enter the lake cannot be used to support production after they sink below the thermocline. Primary productivity of the lake is enhanced when vertical stratification breaks down. Spring blooms of plankton are common because sunlight for photosynthesis and nutrients from bottom depths are both available. In addition to affecting primary productivity, stratification can lead to oxygen depletion below the mixed zone, which affects the vertical distribution of many aquatic species. Oxygen is supplied in lakes by exchange with the atmosphere or from photosynthesis of green plants, both of which occur only in the upper portion of the water column. Organic matter eventually sinks to deeper waters below the photic and mixed zones, where it consumes oxygen through respiration and decomposition. The extent of oxygen deple-

The environment that fishes inhabit varies from the calm, slow waters of some rivers to the turbulant rapids of others. (Photo by Raymond Gehman/Corbis. Reproduced by permission.)

Lakes

Lakes are standing bodies of water surrounded by land, with small outflows relative to their internal volume. Lakes form in a variety of ways, including tectonic movements, volcanic activity, and glacial action. They also may originate as portions of rivers or bays that are cut off from the adjacent water body over time by deposition or sediment movement. Lakes receive inputs of water from the drainage basin, precipitation, and groundwater. These inputs are balanced by outflows of water to rivers, evaporation, and seepage into groundwater. The largest freshwater lake in the world is Lake Superior, which covers a surface area of 31,700 sq mi (82,103 km^2). Lake Baikal holds the largest volume of freshwater—14,292 cu mi (23,000 km^3). The 20 largest lakes contain over 67% of the total water in lakes worldwide, indicating that most lakes are small and shallow.

Ecological processes in many lakes are influenced greatly by a vertical temperature gradient that develops as sunlight warms the upper portion of the water column. Surface water warms when incoming radiation is absorbed and converted to

tion is greatest in highly productive lakes that are stratified for long periods of time.

Although abiotic factors establish the physical and chemical template of habitat conditions in lakes, biotic components and interactions also structure the ecology of lakes. Primary producers, such as phytoplankton, algae, and plants, form the basis of the food chain in lakes. As explained earlier, however, primary productivity follows seasonal patterns based on the availability of light and nutrients in the water column. Zooplankton graze on phytoplankton, but much biomass from primary producers eventually settles to the lake bottom, where it provides food for benthic detritivores, including insects, oligochaetes, and mollusks. Predation by fish has a strong effect on food webs in lakes. Although fish fill a wide variety of feeding niches, many species of fish, particularly as juveniles, consume large quantities of zooplankton. Pelagic fishes in lakes are typically strong swimmers that capture crustaceans and insects as the dominant component of their food supply, while others are predatory piscivores.

Rivers and streams

Rivers and streams flow downhill in defined channels from headwater streams to main river channels to estuaries. The Nile is the longest river in the world—4,180 mi (6,727 km). The Amazon drains the largest area (nearly 2.3 million sq mi, or 6.0 million km²) and carries the greatest flow, with a total discharge at its mouth of 6.4 million cu ft per second (180,000 m³ per second).

The nature of streams and rivers is determined largely by their setting in the watershed. The drainage basin is the total area drained by a river system, including all of its headwaters and tributaries, and the number of fish species typically increases directly with the area of the drainage basin. Streams within the drainage basin can be classified at tributary junctions to determine their stream order, a measure that serves as a useful indicator of stream size, discharge, and drainage area. As stream size increases, so does the order; thus, the smallest streams are termed "first order," and the confluence of two first-order streams is identified as a "second-order" stream. The process continues toward the main river channel until it has been assigned the appropriate order. Each increase in stream order represents three to four times fewer streams, each of which is roughly twice as long and drains approximately five times the area of a stream of the next smaller order.

Streams and rivers are considered physically open systems, meaning that physical factors, such as width, depth, velocity, and temperature, change continually along their course from source to mouth. The "river continuum concept" emphasizes the continuity of the structure and function of river communities from headwaters to lowland portions of river channels. This concept uses stream order as its basis and suggests that changes in physical conditions, functional feeding groups, and species diversity occur dynamically and continuously along the gradient from upstream to downstream portions of rivers.

The discharge of water increases from headwaters to the main stem of rivers and determines the size and habitat features of the channels. Low-order streams tend to flow alternately through riffles, pools, and runs. Riffles are shallow,

Rainbow trout (*Oncorhynchus mykiss*) are transferred from holding ponds through a tube into a fish stocking truck at the Leaburg Fish Hatchery before being placed in the McKenzie River, in Leaburg, Oregon. (Photograph. AP/Wide World Photos. Reproduced by permission.)

high-gradient stretches where fast-moving water flows over rocky substrates and creates turbulence. Pools are deep, low-gradient areas through which water moves slowly. In runs, water flows rapidly but smoothly. Slopes decrease in higher-order streams and rivers, such that they flow smoothly through their channels. River channels meander along a sinuous course. At each meander, the river deposits materials on the inward portion of the curve, where velocity is slowest, but high water velocities erode the outer portion of the curve. Velocity also affects the type of substrate and presence of vegetation in particular areas of flowing waters. For example, large particles, such as gravel, are transported only by fast-moving water. Finer particles, such as sand and silt, form the substrate in areas where current is slower.

Biotic patterns in streams and rivers are heavily influenced by the outcomes of physical processes. Many species of fish require specific substrates for spawning, while other substrate types and flow patterns support aquatic vegetation. Vegetation and woody debris are important to freshwater fish as well; both offer cover from predators, spawning areas, and food-rich foraging sites. In addition, as discharge varies with seasons and precipitation, rivers may overflow their channels and flood riparian wetlands or floodplains, temporarily expanding the vegetated habitats available to freshwater fish. In the tropics, where precipitation primarily occurs during one or two rainy seasons throughout the year, numerous fish communities are dependent on these seasonal floods to expand available habitats, increase feeding opportunities, and mobilize nutrients within the river.

In temperate regions, primary production is necessary as the basis of the food chain in streams and rivers, and algae attached to the sediment surface are the predominant primary producers. Some fish, such as loaches (Homalopteridae) and catfishes (Loricariidae), consume algae directly by scraping it off rocks in the stream. More commonly, aquatic insects and crustaceans are relied upon as intermediaries between the algae and fish—invertebrates graze on algae in small streams, and fish feed on the aquatic invertebrates. As the stream gradient decreases, mosses, rooted plants, and filamentous algae become important primary producers upon which invertebrates and fishes feed. Detritus also may constitute a major part of the food chain, particularly in streams or rivers with extensive cover of streamside vegetation. Scavenging invertebrates and fish feed on the organic detritus. In medium-size to large rivers, members of the fish fauna engage in diverse feeding strategies, including herbivory, invertebrate feeding, piscivory, omnivory, and detritivory.

While primary production from within the system forms the energy base in temperate streams, the large diversity of fish species and productivity exhibited in tropical streams is supported by organic matter from outside the system. Tropical streams and rivers generally flow through dense forested areas. Although large streams may be wide and open to the sun, small streams may be shaded completely by the forest canopy. Instead of relying on photosynthesis as an energy source, life in these streams depends on organic matter that enters in the form of leaves or detritus from the forest. The warm water of these tropical streams enhances colonization by bacteria and fungi, which break down the terrestrial organic matter. Some fish consume the detritus directly, but most species rely on decapod crustaceans and, to a lesser extent, insects as intermediate detritivores. In larger streams, fish are dependent on energy that enters during the rainy season. Rains produce an increase in suspended organic particles

and terrestrial insects that are washed from the land into streams. The most important energy sources become available to fish inhabiting tropical rivers when the river floods adjacent areas of land. Fish then can exploit food resources in the form of decaying vegetation, seeds, fruits, and insects that are available on the floodplain.

Other freshwater habitats

Although streams, rivers, and lakes provide the most abundant and important habitats, fish inhabit other bodies of freshwater as well. As mentioned briefly earlier, such wetlands as river floodplain marshes, shoreline marshes along lakes, and deepwater swamps provide an expanded foraging and refuge area for fish when they become inundated with water. Wetlands are particularly important as nursery habitats for a variety of fish species. Fish also utilize extreme habitats, such as underground caves and desert streams. Unique adaptations enable fish to survive in these environments. Many fish in caves have reduced eyes, and some are blind; instead of relying on vision, enhanced chemosensory and tactile abilities allow them to locate food, mates, and living space. In ephemeral streams, fish survive periods without water by resting in mud or another substrate during the dry season, depositing eggs that do not hatch until water inundates the streambeds in the following year, and utilizing respiratory adaptations to breathe atmospheric oxygen.

Interconnections of freshwater habitats

While physical, chemical, and biological features of different freshwater habitats have been distinguished here, it is important to recognize that all aquatic habitats truly are interconnected. Rivers flow into and out of lakes, rivers and lakes may spill over into wetlands, and rivers and streams may even flow through underground caves in the midst of a surface route. These linkages form a continuum of habitats that often extends to estuarine or marine systems. In addition to these physical connections, biotic connections are important among freshwater habitats. Organisms may disperse between habitat types directly or via a vector. Furthermore, sustaining a food web in one habitat may be dependent on nutrient inputs from another portion of the aquatic system or from terrestrial uplands. Because of the high level of interconnections between aquatic habitats, an action that is detrimental to one component may prove harmful to a much larger system. Recognizing these interconnections is essential for understanding the ramifications of human activities on aquatic environments.

Resources

Books

Allan, J. David. *Stream Ecology: Structure and Function of Running Waters.* New York: Chapman & Hall, 1995.

Cushing, Colbert E., and J. David Allen. *Streams: Their Ecology and Life.* San Diego: Academic Press, 2001.

Dobson, Mike, and Chris Frid. *Ecology of Aquatic Systems.* Essex, U.K.: Addison Wesley Longman, 1998.

Giller, Paul S., and Bjorn Malmqvist. *The Biology of Streams and Rivers.* Oxford: Oxford University Press, 1998.

Lampert, Winfried, and Ulrich Sommer. *Limnoecology: The Ecology of Lakes and Streams.* New York: Oxford University Press, 1997.

Matthews, William J. *Patterns in Freshwater Fish Ecology.* New York: Chapman & Hall, 1998.

Payne, A. I. *The Ecology of Tropical Lakes and Rivers.* New York: John Wiley & Sons, 1986.

Periodicals

Bootsma, H. A., and R. E. Hecky. "Conservation of the African Great Lakes: A Limnological Perspective." *Conservation Biology* 7, no. 3 (1993): 644–656.

Katherine E. Mills, MS

• • • • •

Marine ecology

Introduction

The ecology of marine fishes is a broad topic that may be addressed only in general terms here. The important factors for consideration are few, however. Quite simply, fishes interact with their physical environment; with other organisms, such as plants, invertebrates, reptiles, and mammals; and with other fishes. How and why these interactions occur is the focus of this discussion. In particular, we focus on aspects of the following: community ecology, population ecology, life history and reproductive ecology, habitat use, special habitats and adaptations, and feeding ecology.

Communities, assemblages, guilds, and niches

A community consists of all the organisms present and interacting within a given area. For example, a coral reef community consists of corals, benthic algae, phytoplankton, zooplankton (both demersal and pelagic), various micro- and macro-invertebrates, fishes, reptiles, marine birds, and marine mammals. There are numerous links, in terms of both habitat and trophic relationships, between members of each group.

Fishes and other organisms occur in assemblages within a community. A fish assemblage is composed of all the species populations within the community. Assemblages have order and structure, and both are maintained by interactions between species within the assemblage and with assemblages of other kinds of organisms within the community.

Within an assemblage are groups or species of fishes with similar patterns of resource use. These are called guilds. Although many guilds may consist of members that are taxonomically related to one another, membership is determined by ecological factors. For example, a guild of obligate *Pocillopora eydouxi* coral-dwelling fish species could include one or more hawkfishes (*Neocirrhites armatus*, *Paracirrhites arcatus*, and *Paracirrhites forsteri*—Cirrhitidae), a coral croucher (*Caracanthus maculatus*—Caracanthidae), a scorpionfish (*Sebastapistes cyanostigma*—Scorpaenidae), a goby (*Paragobiodon* species—Gobiidae), and a damselfish (*Dascyllus reticulatus*—Pomacentridae). In this example, only the hawkfishes are closely related to one another, although this relationship is not necessary for guild membership. Another example of a guild would be all those species that browse benthic algae,

pluck zooplankton from the water column, or hide beneath the sand. Furthermore, juveniles and adults of the same species might not be members of the same guild. For example, juveniles of numerous species may shelter in mangrove roots, but as adults some of those species are found living in association with corals.

The place a fish has within the community or assemblage is its niche. A niche simply defines habitat, microhabitat, and physical parameters within the two, as well as diet and feeding strategies, symbiotic relationships (if any), and other functional roles (i.e., its role in predator-prey interactions). Thus, within the guild of obligate coral-dwelling fishes described earlier, we would find coral crouchers living deep within the branches of the coral and feeding upon coral-dwelling microcrustaceans or passing zooplankton, while in another niche, the larger freckled hawkfish, *Paracirrhites forsteri*, would be perched on the outer branches of the same coral and ambushing smaller fishes or crustaceans that passed close by.

Considerable debate has taken place over the way in which fish assemblages, particularly those of reef fishes, are ordered and structured. This debate is centered on questions of how highly diverse assemblages are maintained, how so many species can coexist, what limits diversity and abundance, whether composition and structure is temporally and spatially predictable, and whether the processes involved are uniform across geographical scales. Essentially, assemblage structure was thought to be the outcome of deterministic or stochastic processes. Deterministic processes emphasize fine-scale ecological niches that encompass interactions, such as competition, cooperation, predator-prey, and so on between species. Larvae settling onto a given site of the reef would recruit successfully and become established only in the absence of conspecific adults in favored niches or in vacant niches otherwise. Stochastic processes are random, in that successful recruitment is dependent on chance. Actually, both kinds of processes operate on assemblage structure, and their relative importance is felt on different temporal and spatial scales.

Population ecology

Population structure is dependent upon rates of reproduction and survivorship among individuals within the population. Population structure is also influenced by rates of migration into and out of the population. Recruitment of larvae, both locally produced and from distant sources, is another major factor. One hypothesis about the effects of recruitment on population size and structure is the recruitment-limitation hypothesis. In this case, the number of adults per unit area is limited by the number of larvae available for recruitment. (This hypothesis also has been proposed to explain assemblage structure.) The rates of these various factors vary annually, and the overall structure of a given population may be denoted by year classes (ages) or cohorts. Some year classes are stronger or larger than others. This difference has implications for the management of populations under exploitation by fisheries, because if the largest or most successful year class of a given population reaches maximum age and subsequent year classes are not so successful, overfishing occurs, and the population could be in danger of collapsing.

With some species, age structure also may reflect size structure within a population. Thus, a population of a species with indeterminate growth may consist of different-sized fishes at different levels of abundance for each size class. In the case of many reef fishes, however, size becomes a poor indicator of age, because growth tapers off after a few years or less, depending on the species and certain environmental factors. Growth rates of individuals within populations determine how much biomass is produced for a given population during a given unit of time.

Natural mortality affects population structure too. Starvation; disease; predation on eggs, larvae, juveniles, and adults; cannibalism; and old age all contribute to natural mortality. Mortality caused by fishing is additive, and total mortality for any given population under exploitation is a matter of concern for fisheries and conservation managers.

The genetic structure of marine fish populations is determined by gene flow within and between populations of the same species. Interpopulation gene flow is dependent upon the level of connectivity between two populations. Populations that are relatively close together geographically and served by the same current regime are more likely to have higher levels of gene flow compared with distant populations. In contrast, isolated populations are more likely to diverge over time. Geographic or ecological variation in characters may result. If this variation is great and reproductive isolation occurs, speciation (the creation of new species) may ensue.

Competition exists if two or more fishes require the same resource and the abundance of that resource is limiting within a given area. Competition between fishes of the same species is termed "intraspecific," whereas competition between different species of fishes or between a fish and another organism, such as a sea urchin, using the same algal resource, is deemed "interspecific." Intraspecific competition is an important factor contributing to the success of one individual over another within a population of the same species. This success may be measured ultimately by the proportional reproductive contribution to the population that one individual makes. Interspecific competition is important for determining the structure, and hence diversity, of an assemblage of fishes.

This coral reef is near Townsville, Queensland, Australia. More than a quarter of the world's coral reefs have been destroyed by pollution and global warming and unless drastic measures are taken, scientists warn that most of the remaining reefs may be dead in 20 years. (Photograph. AP/Wide World Photos. Reproduced by permission.)

The relative success of a population of one species over another at securing a vital resource, such as microhabitat or food, determines which species persists and which does not. If so, then how can many fish assemblages be so diverse? Usually, competition is reduced or avoided altogether by resource partitioning among species that live together in as state known as "sympatry." In our coral-dwelling fish example, we see that coral crouchers and hawkfishes avoid competition for the same coral and food resources by living in different parts of the coral and eating different kinds of food. An example of a situation wherein competition probably functions is the case of two fish species that have identical food or habitat requirements but live apart in spatially or geographically distinct areas in a state known as "allopatry." If two or more allopatric species with the same ecological requirements come together, there probably will be two outcomes. The first is that only one species will "win out" and continue to use the contested resource while the other(s) will fail to become established. The second is that all of the species in question will become established, because there will be a shift in resource utilization, sometimes quite dramatic and including rapid morphological changes relevant to the resources available, with only one species using the original resource while the others adapt to using different resources.

The interaction between predators and prey in a given assemblage affects prey in many ways but also may have implications for the population of predators. With respect to prey, predators can cause mortality or injury, with obvious negative consequences. Or the steady influence exerted on prey species by predators results in changes in the way prey utilize habitat or food resources so as to avoid predation. These changes have a profound effect on how the prey population reproduces and sustains itself.

The size of a population of prey species affects the ability of the predator to influence that population. Thus, the number of

Icebergs float in the Arctic Sea near the coast of Barrow, Alaska, USA. The changing climate in Alaska is causing sea ice to freeze later in the winter and break up sooner in the spring, which is changing the habitat of sea animals that live there. (Photograph. AP/Wide World Photos. Reproduced by permission.)

predators in a population may increase in direct proportion to an increase in the size of a prey population. This is an example of a density-dependent response that is compensatory; that is, the predator compensates for the increased number of prey by increasing its own numbers. If the prey population becomes too large and the predator population does not keep pace, the ability to influence the size of the prey population is reduced. In other words, there is safety in numbers. This form of density-dependent response by the prey population is depensatory, which is defined as a decrease in the relative risk of predation or impact upon the population by predation because of an increase in prey numbers.

What happens to the predator population when the prey population disappears? This is an intriguing question, especially with respect to coral reef systems that have been affected negatively by natural or anthropogenic (caused by man) habitat destruction. For example, groupers (Serranidae: Epinephelinae) like to feed on their cousins, the fairy basslets (Serranidae: Anthiinae). Fairy basslets tend to recruit, as post-larvae, to corals. If a coral bleaching episode kills off the corals on a given reef, the fairy basslets have nowhere to recruit. In time, the population of fairy basslets will decline, and there will be few or none left for the groupers to eat. Will the grouper population decline, or will its members simply switch to another kind of prey and get by? Conversely, what happens to the fairy basslet population if the grouper population is reduced greatly by overfishing? Will the fairy basslet population increase significantly in size, or will some other factor come into play? These are important questions that require further attention in the study of reef fish assemblage interactions.

Life history and reproductive ecology

Marine fishes, like their freshwater counterparts, possess a number of life history and reproductive traits and strategies that enable them to live and reproduce successfully under a variety of environmental conditions. These traits and strategies may vary geographically, historically, and hence phylogenetically within or between species. Many of the most important traits and strategies are discussed here.

Body size varies among marine fishes, with both the largest, the whale shark (*Rhincodon typus*, Rhincodontidae), and the smallest, a dimunitive goby (Gobiidae), existing in the same environment. Different body sizes confer distinct advantages. Large body size is favorable for species that swim in the water column. Large size conveys greater protection against predation by all except the largest predators and also allows for greater storage of energy and longer and faster swimming abilities. The latter comes at a cost to reproductive effort, however, because energy that would be available for reproductive activities is required instead for somatic (body) growth. Small body size, on the other hand, allows for greater access to benthic shelter and the potential utilization of a wider spectrum of food items, but at a greater risk of predation. Naturally, there are exceptions to these examples. Small-sized baitfishes, such as anchovies (Engraulidae) or reef herrings (Clupeidae), swim openly in the water column, whereas large-sized morays (Muraenidae) or wolf eels (Anarhichadidae) are associated closely with benthic shelter. Within species, larger body size conveys distinct advantages in terms of territory size, the acquisition of mates, and reproductive success.

Age and size at maturation in marine fishes also vary. Generally speaking, a marine fish that matures at an early age and at a smaller body size has a greater opportunity to reproduce before dying but usually has relatively low fecundity and smaller eggs. However, for pelagic spawning fishes, the acts of courtship and spawning expose the participants to predation risk from lurking predators. Also, energy diverted toward reproductive effort typically means that growth is slower in these fishes. In contrast, older and larger fishes invest in growth, delay reproduction, and have a greater risk of death before reproduction first occurs. On the upside, the larger female fish are more fecund and produce either more eggs or larger eggs; the larger male fish produce more sperm and, within species, may have more opportunities to mate compared with smaller fishes. The relationship between age and size in marine fishes has long been thought to be linear for most species. Recent studies of numerous reef fishes, however, have shown that growth in many species may be rapid at first but tapers off after a few years, yet these fishes may live for several more years. Thus, body size cannot be used to predict age in these fishes.

Sex ratios may vary both within and between species of marine fishes. One reason may be the population size within a given area, and another may be the age of those individuals that make up the population. The sex ratio is important in relation to effective mating opportunities and the development of a mating system. Marine fishes may be gonochoristic, in that the sex is determined genetically and they begin life either as a male or as a female. A variation on this theme is known as "environmental sex determination." In this case, the sex is determined by some environmental factor, such as seasonal water temperatures. Thus, females of a given species are produced during one time of year at a given temperature, whereas males are produced later in the season at a different temperature.

Marine fishes also may be hermaphroditic, in that they are capable of changing from one sex to another and, in some species, back again. Alternatively, they may function as both a female and a male either sequentially or simultaneously. Protogynous sex change occurs when a female changes her sex to become male. Males generally are larger than females in this system. Protandrous sex change takes place when a male changes his sex and becomes a female. In this case, females are larger than males. The former strategy is more common than the latter. Control of sex change is largely social in relation to mating system dynamics, but age also may be a factor in many species. A third variation has been described for some highly site-attached species, such as coral-dwelling gobies of the genus *Paragobiodon* (Gobiidae). Here, a larger female changes sex, becomes a male, and realizes greater fitness by spawning with smaller resident females. If, however, a larger male joins this new male, the new male will be forced to compete with the larger male for access to the resident females and probably will lose. Thus, it will forfeit mating opportunities as well. The sex-changed male will change sex again, reverting back to being a female, but will still realize some measure of fitness by staying on and spawning with the larger male.

Sequential hermaphroditism occurs in some species, such as the hamlets, *Hypolectus* (Serranidae), in that a mating pair switches sexual roles during long bouts of courtship and spawning. First, one fish spawns eggs that are fertilized by the second fish. Afterward, the second fish spawns eggs that are fertilized by the first fish. Simultaneous hermaphroditism occurs when an individual is capable of producing both eggs and sperm at the same time. This less common strategy is practiced largely by fishes that dwell in deep waters (such as many members of the order Aulopiformes), where the probability of encountering a mate is relatively low. Some species, such as numerous wrasses (Labridae) and parrotfishes (Scaridae), have a dual strategy, in that males and females are determined genetically (primary phase) but females can undergo protogynous sex change and become males (terminal phase).

The number of mating partners a fish has during the course of a breeding season is known as the "mating system." Generally, marine fishes are monogamous, polygamous, or promiscuous. Monogamy consists of a single pair that may join together only for spawning but also may share a common territory or home range and remain together for one or more seasons. Polygamy occurs in two principal forms, polygyny and polyandry. Polygynous groups vary in form. For instance, a single male mates with two or more females, and mating may occur in a socially controlled group. Alternatively, males may form leks with other males in a specific area for the purpose of displaying to and attracting several females for spawning. Males also may defend nests or spawning sites within fixed territories and mate with two or more females in succession. In polyandry, females mate with more than one male over the course of a season. For some species, such as anenomefishes (Pomacentridae), a single female in an anenome exerts control over and spawns with two or more resident males and, through social interaction, delays the growth and maturation of additional males that also may reside there. Promiscuity occurs when males and females spawn together, with little or no mate choice.

There is some plasticity in the mating system in relation to local population size. For example, if a population of the humphead wrasse (*Cheilinus undulatus*, Labridae) is relatively large at a given locality, it will form a polygynous spawning aggregation. If the population level is quite low, however, it may reproduce in a single-male polygynous mating group. Similarly, the obligate coral-dwelling longnose hawkfish, *Oxycirrhites typus* (Cirrhitidae), is polygynous if the coral in which it dwells is large enough or near enough to neighboring corals to support a male plus two or more females. If the coral is capable of supporting only the male and one female, the pair is facultatively monogamous.

Sexual dimorphism in size or color pattern usually is found in polygamous and, to a lesser extent, some promiscuous species but seldom in monogamous species. Larger size or more distinct color patterns confer advantages for attracting mates and maintaining relationships with them. Regardless of the mating system used by a given species, if mates of one species are difficult to find, an individual may choose to spawn with a closely related species that is more common, and hybrids might result. Typically, these hybrids do not produce viable offspring should they have an opportunity to mate.

Marine fishes spawn eggs with external fertilization, lay eggs after internal fertilization, or have internal fertilization with the release of fully developed young. There are at least four different modes of spawning and external fertilization. Demersal spawning includes the deposition and fertilization of eggs in nests or directly on the substrate; in pouches, such as those of male pipefishes and seahorses (Syngnathidae); or by oral brooding, such as in cardinalfishes (Apogonidae), in which the eggs are deposited and cared for within the mouth cavity of a parent. Pelagic spawning is the release of eggs and their subsequent fertilization at the peak of an ascent into the water column by a pair or spawning group. Numerous species of marine fishes spawn in this manner. Fishes that spawn pelagically in the water column but have eggs that sink to the bottom are known as egg-scatters. On the other hand, fishes that spawn pelagic eggs close to the bottom are known as benthic egg broadcasts. In contrast, some species, such as skates (Rajidae), are oviparous and have internal fertilization but deposit egg cases that develop and hatch externally. Live bearers have internal fertilization of eggs, and then the eggs develop inside the mother before the young are released. There are two forms of this trait. When eggs develop with nutrients contained in the yolk sac but without nourishment from the mother, it is called "ovoviviparity." Stingrays (Dasyatidae), for example, have this reproductive trait. "Viviparity" is when the young receive nourishment from the mother during their development. An example would be the tiger shark (*Galeocerdo cuvier*, Carcharhinidae).

The timing of spawning or breeding also varies; it may occur at dawn, dusk, during daylight, or at night. Factors include light level, tidal state, mating system, and reproductive mode. Spawning or breeding frequency and seasonal duration vary within species, because of local environmental conditions, and between species, because of phylogenetic differences. Frequency of spawning or breeding is controlled by physiological and phylogenetic constraints that limit the production of eggs or the ability to brood young. Other factors include lunar periodicity and access to mates. Seasonality is highly pronounced and is dependent upon annual variation in water temperature, the number of hours of daylight, and a host of other factors. On tropical reefs, where temperatures are generally warm and stable throughout the year, some hawkfishes (Cirrhitidae) court and spawn daily all year long. The same species at higher latitudes are limited to spawning only during warmer months. Groupers (Serranidae) that form spawning aggregations in the tropics or warm temperate regions, on the other hand, may spawn only once or twice a year in relation to lunar phase. Fishes living in cold temperate regions may be limited to spawning only when there is a shift in season, such as from winter to spring or summer to autumn, whereas others spawn strictly during the warmer summer months. Whether spawning or breeding frequency and seasonality favor adults or their progeny is a subject of considerable interest.

Fecundity or clutch size varies with species, body size, egg size, age, spawning frequency within a season, and latitude in relation to both the length of the season and the water temperature. Generally, there is a positive relationship between body size and egg number. Larger fishes produce more eggs compared with smaller fishes. The relationship is not always so neat, however. Because fecundity can be partitioned into three kinds—batch, seasonal, or lifetime—it is possible for smaller fishes to have relatively greater fecundity than larger fishes. For example, a smaller fish that spawns one or more batches of eggs per night for the course of a spawning season that could last all year in the tropics might have greater fecundity seasonally or over the course of its lifetime than a larger fish that spawns just once during a relatively short season and then dies.

Egg and larval sizes vary between species and also may vary within species, depending on body size, latitude, or other geographical or environmental factors. Generally speaking, large eggs mean that a greater amount of resources has been devoted to their production, and the result is large larvae, better equipped for survival. The advantage of smaller eggs is that more can be produced per unit time compared with larger eggs, and thus there are more opportunities to produce viable young. The drawback to small egg size is that, with fewer resources made available for development, the larvae also will be small and less equipped for survival.

Parental care in fishes includes the investment a parent makes before spawning or breeding, or prezygotic parental care, and the investment made after spawning or breeding, or postzygotic care. An example of the former strategy is nest building, while the latter includes nest guarding, oral incubation, pouch brooding, or internal brooding. Postzygotic parental care of eggs and larvae is practiced extensively by freshwater fishes but far less so by marine fishes. Among marine fishes, postzygotic parental care of eggs means that small, cheaply produced eggs could be afforded a benefit that increases their chance of survival. Ironically, most species that practice parental care in the marine environment tend to have eggs that are much larger than those produced by species that lack parental care. Pelagic spawning, egg scattering, and benthic broadcast spawning are practiced by a majority of marine species, and they do not engage in parental care.

The duration of time between egg fertilization and hatching varies with egg size. Small eggs spawned pelagically usually hatch rapidly compared with larger eggs that require tending. Similarly, eggs fertilized internally require a longer gestation time before the young emerge from the mother. As most marine species have pelagic larvae, the amount of time spent drifting passively or swimming weakly in the water column varies with species and also with environmental circumstances. Larval life duration depends on the growth rate of the larvae, which in turn is dependent upon its energy stores, rate of metabolism, and ability to feed before settling. Rapid growth to a larger size dictates that energy requirements and metabolism will be high, and thus the need to feed more often will be greater, or else starvation will occur, and death will result. Exposure to predation also is greater, and most larvae fall victim to predators or the effects of starvation before settlement takes place. Small larvae with short larval life durations and poor dispersal capabilities are more likely to settle and recruit locally. Short larval life duration means less risk from predation, because rapid settlement into favorable habitats can be accomplished. If settlement does not occur and the larvae are carried out to sea, however, they, too, will die from predation or starvation. Size and the growth rate potential do not always influence dispersal capabilities of larvae, however. Larvae of many tropical reef species, for instance, may be adapted to long larval life and hence possess long-distance dispersal capabilities. These same larvae, however, may be caught or trapped by local oceanographic conditions and disperse only short distances before settling into suitable habitats.

Lifetime reproductive effort typically is defined in two ways. Fishes are either semelparous or iteroparous. Semelparous species usually reproduce only once in a lifetime, with a spawning event that may be quite large, depending upon the species. Examples include the freshwater eels (Anguillidae) and various salmons (Salmonidae), although some individuals of the latter group may survive and return to spawn again. Iteroparous species spawn or breed frequently during their lives, either in the course of a single season or over many seasons, depending upon the species. Examples include groupers (Serranidae), hawkfishes (Cirrhitidae), or parrotfishes (Scaridae).

Habitat use

Marine fishes live virtually everywhere in the world's oceans. In general, marine fishes may be found at the upper limit of the intertidal zone down to the bathypelagic realm several thousand meters deep, from freshwater in the upper portions of an estuary system to the hypersaline waters of now landlocked bodies of water or shallow flats in arid regions, and from balmy tropical reefs to intensely cold polar seas. Their distribution in these diverse habitats is made possible by behavioral, anatomical, and physiological adaptations that meet the specific or unique demands of those habitats. The classification of habitats is complex and is the subject of considerable interest if not outright debate. This review adopts a simple approach and considers habitat in relation to patterns of zonation based on depth and substrate.

Marine fishes that live on the bottom or in association with some form of structure on the bottom, such as a rock or submerged mangrove root, are considered to be benthic or demersal species. Those that swim up into the water column, whether in a shallow estuary or bay, in the open ocean, or in the deep open ocean, are considered to be pelagic species. Some species are both, in that they live in close association with the bottom but frequently are found in the water column, either foraging or moving over a wider area away from shelter. These species are considered to be benthopelagic species. With respect to depth, these kinds of fishes can be found across a wide range.

Among benthic fishes, certain species are specialized to live—as juveniles, adults, or both—in shallow tide pools or splash zones at depths of less than 1.5 ft, or 0.5 m. Examples of tide-pool fishes include certain morays (Muraenidae), marine sculpins (Cottidae), blennies (Blenniidae), and gobies (Gobiidae) that shelter within the confines of the pool and may endure the effects of a falling tide. The clingfishes (Gobiesocidae) are well adapted to life in both tide pools and the splash zone above them. Their ventral

fins, modified into effective sucking discs, allow them to cling to stone walls above tide pools that are immersed only by the splash of waves. Some clingfishes, and indeed other tide-pool species, are especially well adapted to avoid desiccation and thermal extremes. Benthopelagic fishes in tide pools, such as damselfishes (Pomacentridae) or kuhlias (Kuhliidae), swim about in the depths of the tide pool. Generally, these kinds of fishes are less tolerant of the effects of low water levels or temperature shifts and must migrate out of the tide pool with the falling tide, only to return again as the tide floods.

Tidal effects also are pronounced in estuaries, mangroves, sea-grass flats, algal and shallow kelp beds, coral reef flats, and other kinds of flats dominated by mud, sand, rubble, cobble, larger rocks, or living organisms, such as oysters. In estuaries, benthopelagic and pelagic fishes move upstream with a flooding tide, often well into freshwater, and then move back downstream with a falling tide. These fishes are termed "euryhaline" species, in that they are tolerant of a wide range of salinities. Fishes living among mangrove roots, inshore sea-grass flats, or algal beds, whether in or adjacent to an estuary, are adapted similarly. Benthopelagic and shallow pelagic species, such as halfbeaks (Hemirhamphidae) or mullets (Mugilidae), simply move off their various flats with the falling tide. Benthic species often seek shelter in holes, under rocks in depressions, and in deeper pools, or they may migrate to adjacent channels.

The effects of tide upon the fishes' habitats are less pronounced in the subtidal zone. Here, the various habitats are submerged constantly. Tidal effects typically are limited to patterns of current flow and what may be carried to and from this zone with the current. Thus, food, in the form of prey moving off a shallow flat in the intertidal zone with the falling tide, may be brought into the subtidal zone. Similarly, sediments from intertidal habitats move with the tide, creating turbid conditions in deeper water. Subtidal habitats are quite diverse as well and include coral and rocky reefs, sea grasses, algal and kelp beds, and deeper flats of sand, mud, rubble, rocks and boulders, and hard bottom or pavement. Some of these flats may be dominated by certain kinds of organisms, such as sponges, soft corals, or oysters.

Some habitats have pronounced levels of zonation as well. Coral reefs are a good example. Typically, there are three kinds of coral reefs: fringing reefs, where the reef is adjacent to a shoreline; barrier reefs, where the reef is well offshore and usually runs parallel to the adjacent landmass; and atolls, which are reefs that grow and emerge as a landmass, typically a sea mount, sinks beneath it over time. Barrier reefs and atolls usually have lagoons. Fringing reefs, being much narrower, have back troughs or some other form of channel on the reef flat within the intertidal zone. Regardless, seaward from the edge of the reef, one would find the reef front or spur-and-groove zone, one or more reef terraces or benches, and the reef slope. The reef front may be a shallow wall that drops directly from the reef margin to the first terrace. Fishes living here generally are pelagic or benthopelagic, although a number of benthic species may be found among emerging corals, in holes, or around rocks. Alternatively, the spur-and-groove zone extends outward from the reef margin in a pattern resembling a human hand. The "fingers" of the hand represent spurs of coralline rock that extend outward from the face of the reef. Live corals resistant to the effects of wave action may grow upon these spurs. The spaces between the fingers represent the grooves, which are nothing more than surge channels between the spurs. These grooves are shallow at the reef face and deeper as the first terrace is approached, and the bottom of the channels consists of coral rock pavement, boulders, dead coral rubble, sand, or live corals.

Often, a complex network of holes, caves, and tunnels exists within this zone, with direct connections to the reef flat above. Elongated spur-and-groove zones, especially those that extend well out onto the deeper first terrace, are indicative of the effects of the rise and fall of sea levels historically. Numerous species of benthic, benthopelagic, and pelagic species are found in the spur-and-groove zone. For example, blennies and damselfishes utilize holes or corals on the spurs or in the grooves. Morays, lionfishes (Scorpaenidae), squirrelfishes and soldierfishes (Holocentridae), and sweepers (Pempheridae) employ the network of caves and tunnels within this zone. Certain hawkfishes (Cirrhitidae) or groupers (Serranidae) may perch on corals or hide on ledges or next to rocks and ambush passing prey. Reef herrings (Clupeidae) may swim in the water column above and flee the approach of predatory trevallys (Carangidae) that patrol this zone right to the reef margin.

Below the spur-and-groove zone is the reef terrace or bench; there may be one or more of these, depending on local geological history and changes in sea level. Coral development on the terrace, independent of geographic variation, is dependent upon the degree of exposure to wave action. In somewhat protected areas, the diversity and abundance of corals may be relatively high, whereas in areas exposed to heavy wave action and scouring, the diversity and abundance of corals may be low, and coral pavement predominates. Regardless, fishes in the terrace zone utilize what is available to provide shelter, food, and mating sites. Benthic fishes, such as damselfishes, hawkfishes, and scorpionfishes, make use of corals. Benthic species, such as sandperches (Pinguipedidae), blennies, and gobies, employ holes, sand-filled depressions, boulders, and pavement. Benthopelagic species, such as snappers (Lutjanidae), goatfishes (Mullidae), butterflyfishes (Chaetodontidae), angelfishes (Pomacanthidae), wrasses (Labridae), parrotfishes (Scaridae), and filefishes (Monacanthidae), move about home ranges or defend territories. Pelagic species, such as gray sharks (Carcharhinidae), trevallys, and barracudas (Sphyraenidae), patrol the terrace in search of prey.

Deep slope and wall habitats generally occur below the reef terrace. In some places, such as atolls, the transition between the spur-and-groove zone and the wall or deep slope occurs without the presence of a reef terrace. In other places, one or more terraces are present before the deep slope begins, and the slope may separate two terraces from each other. Wall and deep slope habitats often are characterized by the presence of various corals, sea fans and black corals, sponges, hydrozoans, and numerous other benthic invertebrates. These offer shelter and food to innumerable species of small, benthic fishes. The face of the wall or slope may be eroded with numerous holes and caves that provide shelter for many diurnal species, such as groupers and dottybacks (Pseudochromidae), and nocturnal fishes, such as bigeyes (Priacanthidae), soldierfishes, and squirrelfishes. Off the face of the wall or slope are found hovering fishes, such as certain butterflyfishes, angelfishes, damselfishes, and fairy basslets (Serranidae), that feed upon plankton in the water column.

In reef systems with lagoons, some of the same general habitat types may be present. At the back side of a barrier reef flat or in a pass connecting the outer reef with the lagoon, there often is a slope or wall that drops down into the depths of the lagoon. These habitats generally are protected, although they may be subject to intense tidal currents, may have a rich community of benthic invertebrates, and, correspondingly, support a wide variety of species. Damselfishes, butterflyfishes, angelfishes, wrasses, and triggerfishes (Balistidae) may hover in the water column but seek shelter in or along the wall or slope as necessary. As the slope gives way to sand or rubble, garden eels (Congridae), sanddivers (Trichonotidae), gobies, and peacock soles (Soleidae) are visible, but if threatened, they will rush into holes or bury themselves in the sand. Shallow portions of the lagoon often have patch reefs or coral bommies (large, isolated coral heads) that function effectively as islands in a sea of sand and rubble. These islands provide structure and, no matter how small, attract a remarkable number of species. Lagoons also may have seagrass beds in shallower areas and a corresponding suite of species, such as juvenile emperor fishes (Lethrinidae), snappers, goatfishes, and parrotfishes.

A temperate region analog of the coral reef is the kelp forest. Kelp is a marine plant that may grow as long as 65.6 ft (20 m). It provides a dense jungle that is utilized by numerous temperate species and, depending upon depth and proximity to kelp, may offer different microhabitats to members of the same fish family. As such, different species will be adapted to the surface, mid-reaches, and base of the kelp and to the water column surrounding it.

Among fishes, location is everything. The exact place where a fish lives is known as its "microhabitat." Fishes, especially small fishes, have remarkable plasticity in what they adapt to as a home. For example, scorpionfishes, coral crouchers, hawkfishes, damselfishes, wrasses, gobies, and numerous other species live within or atop the branches of corals. Moray eels, jawfishes (Opistognathidae), blennies, and gobies, among others, inhabit holes. Other water column–dwelling species, such as some triggerfishes, seek shelter in holes as well. Some species, such as pipefishes (Syngnathidae) and clingfishes, are specialized for living in crinoids and sea urchins. Similarly, various seahorses, hawkfishes, and gobies are specialized for life on sea fans and black corals. Pipefishes, seahorses, some juvenile wrasses, and filefishes mimic the leaves of sea grasses or fleshy algae. Blennies and many other small benthic species have adapted to life in empty seashells and worm tubes. Clingfishes, gobies, blennies, and labrisomids (Labrisomidae) live in sponges. Even the sand serves as a distinctive microhabitat. Snake eels (Ophichthidae) burrow under the sand and seldom emerge, except at night. Stonefishes (Scorpaenidae) and stargazers (Uranoscopidae) lie buried beneath the sand and ambush passing prey. There are numerous other examples of benthic fishes that utilize natural and man-made structures as microhabitats.

The pelagic zone is divided into different zones relative to depth as well. The epipelagic zone ranges from the surface down to a depth of 656 ft (200 m). Between 656 and 3,281 ft (200–1,000 m) is the mesopelagic zone, followed by the bathypelagic zone at 3,281–13,123 ft (1,000–4,000 m), the abyssal zone at 13,123–19,685 ft (4,000–6,000 m), and the hadal zone below 19,685 ft (6,000 m). Most pelagic fishes occur in the epipelagic (more than 1,000 species) and mesopelagic and bathypelagic zones (about 1,000 species combined).

The epipelagic zone is the limit at which photosynthesis takes place. Phytoplankton occur there and form the basis for a food chain that consists of consumers ranging from zooplankton to blue whales. Fishes of the epipelagic zone have bodies that are streamlined, to allow for greater speed in the pursuit of prey or the evasion of predators. Many epipelagic species, such as dolphinfishes (Coryphaenidae), tunas (Scombridae), and marlins (Istiophoridae), make seasonal migrations to feed and mate. Speed often is essential, and many species have independently evolved the ability known as countercurrent exchange, which effectively turns them into warm-blooded organisms. This trait, found in mackeral sharks and tunas, among others, is especially useful in cooler waters. Species associated more with inshore waters, such as striped bass (Moronidae) and bluefishes (Pomatomidae), also make seasonal migrations to track prey movements and to reproduce. Reef-associated species, such as trevallys and barracudas may migrate for spawning, but the distances traveled are far less.

Many epipelagic species are denoted by their silvery, bluish, or greenish blue body coloration, which makes them difficult to see in open water and thus decreases the risk of predation. This coloration also benefits predators, allowing them to approach prey fishes more easily. Large predators, such as marlins and many tunas, are more darkly colored, however, and dolphinfishes are among the most brightly colored species in the epipelagic zone. Some species of epipelagic fishes are especially adapted for life at the surface. Specializations may include enlarged pectoral and caudal fins that may be used to facilitate escape, modified snouts that allow for greater feeding efficiency on the surface, or enlarged eyes that promote detection of potential predators and prey at the water-air interface. Predators include the needlefishes (Belonidae), which are capable of sudden, powerful bursts of speed that allow them to jump repeatedly out of the water in pursuit of prey. Prey species have adapted to escape this pursuit. Halfbeaks and ballyhoos also may make repeated jumps to evade predators. Perhaps the most famous example of aerial evasion is that of the flyingfishes (Exocoetidae), whose modified pectoral fins resemble wings and whose modified caudal fin rudders are capable, when touching the surface of the water while the fish is airborne, of supplying additional thrust. Flyingfishes can travel up to 1,312 ft (400 m) in a single flight at a speed of more than 43.5 mi (70 km) per hour and can make flights repeatedly in succession. Pelagic fishes not particularly adapted for flight often seek shelter beneath flotsam, under jellyfishes, or attached to other larger fishes. The remoras (Echeneidae), which attach themselves to sharks,

rays, billfishes, whales, and even ships, are able to hitchhike around the epipelagic zone more or less under the protection of their hosts.

Those zones below the epipelagic have confounding physical and biological factors, the effects of which escalate with depth. Increasing water pressure, decreasing water temperature, little or no light penetration, seemingly vast spatial distributions, and the patchy distribution of food resources all heavily influence which fishes live where and how. There is a remarkable diversity in species, however, and, because many of these factors have similar effects upon unrelated species, there is also a extraordinary similarity in characters that have evolved through convergent evolution. Fishes of these zones may be large (more than 6.6 ft, or 2 m) or small (less than 2 in, or 5 cm), yet they possess large or elongated mouths and dagger-like teeth for grabbing prey. Others have tubular eyes that augment the efficiency of light detection. Conversely, many species lack functional eyes entirely and rely upon other senses. Numerous species possess photophores (light-emitting organs) to attract both prey and mates. They also may have modified dorsal fin rays or chin barbels, often with photophores, that are used to attract prey. Many species have thin bones and specialized proteins that allow for gas regulation and neutral buoyancy in the absence of swim bladders. Fishes distributed in relatively shallow mesopelagic waters at higher latitudes often are found at greater depths in the tropics. This phenomenon, known as tropical submergence, allows these species to expand their geographical distribution while remaining in cooler and more comfortable water temperatures. Pelagic fishes of these zones also migrate, but the direction is more vertical than horizontal. At night many species rise hundreds, if not thousands of meters in depth, some to the surface, before returning downward during daylight hours. These vertical migrations usually track the movements of prey during a 24-hour cycle. Other fishes, especially such deep benthopelagic and benthic species as the tripodfishes (Ipnopidae), never make the migration and depend solely on what they encounter or what "rains" down upon them from the water column above.

Special habitats and adaptations

As indicated earlier, marine fishes are specially adapted to living in extreme environments. Deep-water pelagic and benthopelagic fishes, as illustrated earlier, are prime examples of adaptation to extremes. There are numerous shallow-water examples, too. Some marine fishes, such as the reef cuskfishes (Ophidiidae), have adapted to the dim world of tunnels and caves beneath the reef front or spur-and-groove zone but emerge at night to hunt small fishes and invertebrates. Polar fishes have adapted to extremely cold water temperatures that may fall below 32°F (0°C). (It is a curious trick of physics that saltwater does not freeze at this temperature.) For example, the icefishes (Nototheniidae and others in the suborder Notothenioidei) of Antarctica and the Southern Ocean have evolved a specialized protein that acts as an antifreeze that prevents these fishes from freezing. These species have evolved to fill various niches too. Some, such as *Trematomus nicolai*, are benthic, whereas others, such as *Trematomus loennbergii*, are benthopelagic in deeper water; still others, such as *Pagothenia borchgrevinki*, are pelagic and swim and feed just beneath the ice.

Fishes also have adapted to aerial exposure. Examples include clingfishes and blennies in the upper reaches of the intertidal zone and mudskippers (Gobiidae) of the genus *Periophthalmus*, which retain water in their gill cavities and are capable of hopping and skipping across mud flats, rubble flats, and among the branches of mangroves. Marine fishes have adapted to hypersaline conditions as well. In arid regions, back bays, estuary sloughs, tide pools, and now-landlocked seas, all tend to have salinity levels far higher than that of seawater. Some species of fishes, such as clingfishes and gobies, have evolved mechanisms that allow them to regulate their osmotic pressure under these conditions. There is a limit, however. The landlocked Dead Sea in the Middle East, with a salinity in excess of 200 parts per thousand (ppt), versus an average of 36 ppt in seawater, is simply too salty for fishes to survive. Conversely, some seas, such as the Baltic in northern Europe, have such low salinity levels in some areas that such freshwater species as the pike *Esox lucius* (Esocidae) may coexist with euryhaline marine species.

Species that migrate between marine and freshwater, or vice versa, during both juvenile and adult phases of their respective life cycles have managed to conquer the problems associated with different salinity levels and osmotic regulation. For example, anadromous fishes, such as the salmons and sea trout (Salmonidae), live most of their adult lives in the ocean but migrate up a river or stream (often the one in which they were born) to spawn. Their juveniles live for part of their life cycle in freshwater before moving downstream and out to sea. (Some populations may become landlocked, however.) Catadromous fishes, such as freshwater eels (Anguilidae), live their adult lives in freshwater but migrate well out to sea to spawn and die. Their juveniles often return to their natal streams to begin their adult lives. Amphidromous fishes, such as some gobies (Gobiidae) and sleepers (Eleotridae), also live their adult lives in freshwater and spawn there as well. Their eggs and larvae are carried out to sea, however, and the post-larvae migrate back up stream, sometimes against formidable barriers, to begin their lives as adults. Other amphidromous species are born in saltwater but have young that migrate into freshwater to grow and then return to saltwater to grow more and to reproduce as adults.

Feeding ecology

Marine fishes have a wide range of diets and methods of feeding. They may be divided generally into herbivores, carnivores, detritivores, and omnivores. Herbivores are those that feed upon plants and plant materials. They do so by grazing or browsing upon benthic algae, sea grasses, or other plant life. Other herbivores may use specialized gill rakers to strain phytoplankton from the water column. Some species, especially certain damselfishes (Pomacentridae), act as farmers of the benthic algae they consume. For instance, they may kill a patch of coral that subsequently is used as a substrate for benthic algae to recruit upon. The farmer fishes then tend the algae, removing unwanted species and feeding upon desired ones, at the same time that they defend the algal patch against other herbivores. Other herbivores, such as parrotfishes (Scaridae), may simply bite off and crush corals in order to strain the symbiotic zooxanthellae algae resident within the coral polyp.

Carnivores feed upon a great variety of animals. Zooplanktivores strain or pluck zooplankton from the water column. Corallivores excise or pluck polyps from their coral skeletons; alternatively, they may crush the corals with strong teeth and strain the polyps through their gill rakers. During coral spawning season, numerous fish species, especially butterflyfishes (Chaetodontidae) and damselfishes, feed upon coral eggs as they float upward into the water column. Others may be specialized to pluck or nip the tips of anenomes, hydrozoans, or other coral-like organisms. Some fishes may be generalists when feeding upon invertebrates, but others are highly specialized for taking only certain kinds. Thus, some fishes specialize in microinvertebrates, such as diminutive worms, crabs, shrimps, or mollusks, whereas others target macroinvertebrates, such as squids, octopuses, lobsters, or large crabs.

Invertebrate prey may be benthic, such as clams, oysters, and tunicates, or they may be pelagic, such as squids and swimming shrimp. Prey may be sifted from sand or rubble, crushed, grabbed, bitten, or swallowed whole. Some fishes are able to feed on prey items that may be difficult, if not dangerous, to consume. Some triggerfishes (Balistidae) can bite and crush sea urchins bearing venomous spines without apparent damage to themselves. Similarly, the humphead wrasse (*Cheilinus undulatus*, Labridae) can feed on adult crown-of-thorns starfish without being damaged by this organism's strong, venom-tipped spines. Pelagic macroinvertebrates may include jellyfishes that are consumed by molas or ocean sunfishes (Molidae) as they drift in the water column. Prey also may reside out of water. For example, archerfishes (Toxotidae) are specialized for feeding upon insects by shooting a stream of water at them so as to knock them down from mangrove branches or other forms of structure; the fallen insect then is consumed on the surface.

Piscivores feed upon fishes exclusively, although many species also vary their diet by consuming invertebrates. These predators actively hunt, chase, herd, grasp, stun, club, shock, ambush, bite, or engulf other fishes. Various small species are specialized for feeding on fish scales or skin, either as juveniles or as adults. Others, such as cleanerfishes (certain Labridae, Gobiidae, Chaetodontidae, and so on), have evolved to remove ectoparasites or damaged tissue from "client" fishes that visit their cleaning stations. Still other species are specialized as parasites that feed upon host fishes. Host fish species include potential predators (e.g., sharks, moray eels, needlefishes, groupers, snappers) as well as numerous other species that are active during daylight hours (e.g., butterflyfishes, angelfishes, damselfishes, wrasses, parrotfishes). Larger fishes, especially sharks, may feed on floating sea birds, reptiles, and mammals in addition to fishes. Detritivores sift detritus from the bottom and strain it through their gill rakers. Omnivores eat both plant and animal material as adults. Many species of fishes undergo ontogenetic shifts in diet and feeding methods, meaning that their diet and feeding methods change with age and growth. Species that feed upon phytoplankton or zooplankton as juveniles may switch to fishes or large invertebrates as adults.

Through their trophic interactions, fishes have important direct and indirect effects upon the structure of the commu-

nities in which they live. They can influence, among other factors, rates of productivity and biomass turnover, nutrient cycling, sediment production, shifts in water quality, shifts in food web composition, or shifts in species composition and relative abundance within a given assemblage.

Resources

Books

Briggs, J. C., and J. B. Hutchins. "Clingfishes and Their Allies." In *Encyclopedia of Fishes*, edited by J. R. Paxton and W. N. Eschmeyer. 2nd edition. San Diego: Academic Press, 1998.

Donaldson, T. J. "Assessing Phylogeny, Historical Ecology, and the Mating Systems of Hawkfishes (Cirrhitidae)." In *Proceedings of the 5th International Indo-Pacific Fish Conference, Nouméa 1997*, edited by B. Séret and J.-Y. Sire. Paris: Societé Française, Ichthyologie 1999.

Eschmeyer, W. N., E. S. Herald, and H. Hammann. *A Field Guide to Pacific Coast Fishes of North America.* Boston: Houghton Mifflin Co., 1983.

Helfman, G. S., B. B. Collette, and D. E. Facey. *The Diversity of Fishes.* Malden, MA: Blackwell Science, 1997.

Kock, K.-H. *Antarctic Fish and Fisheries.* New York: Cambridge University Press, 1992.

Marshall, N. B. *Aspects of Deep Sea Biology.* London: Hutchinson, 1954.

Myers, Robert F. *Micronesian Reef Fishes: A Comprehensive Guide to the Coral Reef Fishes of Micronesia.* 3rd edition. Barrigada, Guam: Coral Graphics, 1999.

Pitcher, Tony J., ed. *The Behaviour of Teleost Fishes.* London: Chapman and Hall, 1993.

Potts, G. W., and R. J. Wootton, eds. *Fish Reproduction: Strategies and Tactics.* London: Academic Press, 1984.

Robertson, D. R. "The Role of Adult Biology in the Timing of Spawning of Tropical Reef Fishes." In *The Ecology of Fishes on Coral Reefs*, edited by Peter F. Sale. San Diego: Academic Press, 1991.

Sale, Peter F., ed. *Coral Reefs Fishes: Dynamics and Diversity in a Complex Ecosystem.* San Diego: Academic Press, 2001.

Thomson, Donald A., Lloyd T. Findley, and Aalex N. Kerstich. *Reef Fishes of the Sea of Cortez.* 2nd edition. Tucson: University of Arizona Press, 1987.

Thresher, R. E. *Reproduction in Reef Fishes.* Neptune City, NJ: T.F.H. Publications, 1984.

Periodicals

Donaldson, T. J. "Facultative Monogamy in Obligate Coral-Dwelling Hawkfishes (Cirrhitidae)." *Environmental Biology of Fishes* 26 (1989): 295–302.

———. "Lek-Like Courtship by Males and Multiple Spawnings by Females of *Synodus dermatogenys* (Synodontidae)." *Japanese Journal of Ichthyology* 37 (1990): 292–301.

Kuwamura, T., and Y. Nakashima. "New Aspects of Sex Change Among Reef Fishes: Recent Studies in Japan." *Environmental Biology of Fishes* 52 (1998): 125–135.

Mapstone, B. D., and A. J. Fowler. "Recruitment and the Structure of Assemblages of Fishes on Coral Reefs." *Trends in Ecology and Evolution* 3, no. 3 (1988): 72–77.

Moyer, J. T., and A. Nakazono. "Prototandrous Hermaphroditism in Six Species of the Anenomefish Genus *Amphiprion* in Japan." *Japanese Journal of Ichthyology* 25 (1978): 101–106.

Terry J. Donaldson, PhD

Distribution and biogeography

Biogeography

Biogeography is the study of the geographical distribution of plants (phytogeography) and animals (zoogeography). This science attempts to explain how distributions of organisms have come about. These explanations depend upon a thorough knowledge of the phylogeny of the group and the geological history of the region. Freshwater fishes are especially important in understanding distribution patterns, because they are tied to their river systems, which are surrounded by a land barrier, which in turn is surrounded by a saltwater barrier.

Fish distribution and salt tolerance

Fishes can be grouped into freshwater and saltwater families. Freshwater fishes can be subdivided into primary, secondary, and peripheral division families, depending upon their salt tolerance. The members of primary division freshwater fish families have little salt tolerance and are more or less restricted to freshwaters. Saltwater is a major obstacle for them, and their geographic distribution has not utilized dispersal through the sea. Such fishes include members of the minnow (Cyprinidae) and perch (Percidae) families. Fish families with some salt tolerance, whose distribution may reflect movement through coastal waters or across short distances of saltwater, are known as secondary division freshwater fishes. Examples of this group include the killifishes (Cyprinodontidae) and cichlids (Cichlidae). Peripheral division families derive from marine ancestors that used the oceans as dispersal routes. They have a wide range of salt tolerance. Some peripheral families spend part of their life cycle in freshwater (diadromous), whereas others are mostly marine but invade river mouths and may ascend into completely freshwater. The salmons (Salmonidae) and mullets (Mugilidae) are examples of peripheral division fishes. Many strictly saltwater fish families live only in the sea, but some may have a few species that occasionally enter brackish or freshwaters.

Habitat richness

Freshwater in lakes and rivers makes up less than one hundredth of one percent (0.0093%) of the total amount of water on Earth, whereas the oceans account for 97%. The remainder of the water is bound in ice, groundwater, atmos-

pheric water, and so on. It is surprising that 42% of all fish species live in 0.01% of the world's water. This remarkably high percentage reflects the degree of isolation and the diversity of niches possible in the freshwater environment. The percentage of fish species in various habitats is as follows:

- Primary freshwater 33.1%
- Secondary freshwater 8.1%
- Diadromous 0.6%
- **Total freshwater** **41.8%**
- Marine shallow warm 39.9%
- Marine shallow cold 5.6%
- Marine deep benthic 6.4%
- Marine deep pelagic 5.0%
- Epipelagic high seas 1.3%
- **Total marine** **58.2%**

Stingrays swim in shallow waters near Grand Caymen Island. (Photo by Animals Animals ©James Watt. Reproduced by permission.)

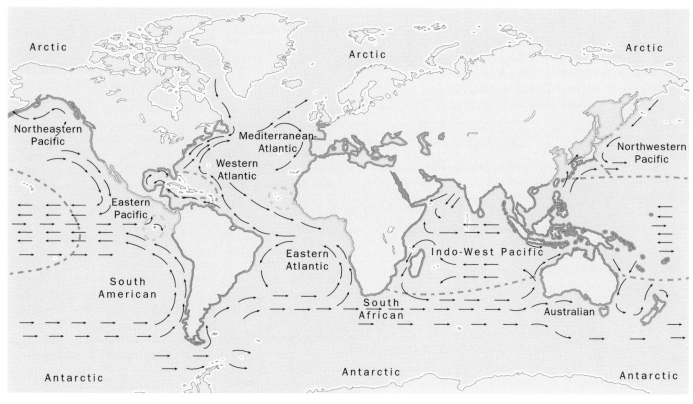

Marine regions and ocean currents. (Map by XNR Productions. Courtesy of Gale.)

Marine fishes and zoogeography

The salinity of the world's oceans is about 35 parts salt to 1,000 parts water. This ratio varies slightly around enclosed, hot regions with high evaporation rates, such as the Red Sea, but for the most part, salinity is not a major barrier for marine fishes, except at the local level. Temperature, on the other hand, is a substantial barrier. The four major temperature zones of the ocean surface are tropical, warm temperate, cold temperate, and cold. Structure, in the form of continental shelves and coral reefs, also must be considered; ocean currents are obviously important; and the vast distances across open oceans are barriers to the dispersal of coastal species. The interaction of these factors and the geological history of a region affect the distribution of marine fishes. Fishes that live along the relatively warm shoreline and on the continental shelf down to about 656 ft (200 m) make up 39.9% of the world's fishes. This includes tropical, warm temperate, and cold temperate waters. If we consider that there are approximately 27,300 fish species, this means that about 11,000 fish species inhabit the shallow, warm to cold temperate water zone.

Marine shallow, warm waters

Tropical waters harbor the greatest diversity of fishes, most likely owing to the presence of coral reefs, which provide habitat, structure, and food. Reef-building corals need clear, clean water of at least 68°F (20°C). The faunal regions of the oceans have been delineated in different ways, and boundaries often are imprecise and overlapping. Most authorities agree, however, on the following system (or a similar one). These re-

gions, which roughly correspond to Cohen's marine shallow, warm habitat, are the Indo-West Pacific, western Atlantic, eastern Pacific, eastern Atlantic, Mediterranean-Atlantic, northeastern Pacific, northwestern Pacific, South American, South African, and Australian. The remaining 18.3% of marine fishes are found in shallow, cold (Arctic and Antarctic), deep (cold), or epipelagic waters.

The Indo-West Pacific region extends from East Africa to northern Australia, southern Japan, and all of Polynesia. By virtue of its vast expanse, it has the richest fish fauna of any

Sheepshead minnows (*Cyprinodon variegatus*) are found in the salt marshes of New Jersey, USA. (Photo by Animals Animals ©Joe Mc-Donald. Reproduced by permission.)

Geographic distribution of *Galaxias maculatus* and relevant ocean surface currents. (Map by XNR Productions. Courtesy of Gale.)

marine region, estimated to be about 3,000 species. The Indo-West Pacific also has more coral and other reef-associated invertebrate species than any other region. All of the tropical fish families found elsewhere are represented in this area, and most show their maximum diversity in the Indo-West Pacific region. Many species in this region have pelagic larvae and therefore are wide-ranging species. Springer linked the distribution of many species with the Pacific Plate, and Briggs divided the Indo-West Pacific into provinces defined by their degree of endemism, especially around isolated oceanic island groups, such as Hawaii. The East Indies Triangle (from the Malay Peninsula and Philippines through the Indo-Australian Archipelago to the Bismark Archipelago) is an important evolutionary center and contains the greatest species diversity in the marine world.

The western Atlantic region extends from the temperate coast of North America through the Gulf of Mexico and the Caribbean Sea south through the tropical and then temperate coast of South America. Parts of this region, especially the West Indies, contain coral reefs but only about one-tenth as many coral species as the Indo-West Pacific. There are approximately 1,200 fish species in the western Atlantic region. The warm waters of the Gulf Stream provide a connection for northern and southern elements of this region and probably have prevented the differentiation of the tropical fish fauna of the otherwise northerly Bermuda Islands. There is a decrease in the number of species along the western Atlantic coast as one moves north or south from the tropics into colder waters. Some authorities consider the northwestern Atlantic region from Cape Hatteras north to be a separate province.

The eastern Pacific region extends from the Gulf of California to northern Peru. The cold Peruvian Current that runs

from the Antarctic up the west coast of South America abruptly ends the tropical conditions at the Gulf of Guayquil. The fish faunas of the western Atlantic and eastern Pacific are similar at the generic level, because they have been separated only recently by the elevation of the Isthmus of Panama about three million years ago. There are only a few coral species in the eastern Pacific, and the fish fauna likewise is much reduced from the numbers in the western Atlantic, with about 650 species. The Galápagos Islands fish fauna is relatively rich, with a large proportion of endemic species. A vast expanse of open ocean separates the extremely rich Indo-West Pacific fauna from the relatively depauperate eastern Pacific fauna. Only about 8% of the coastal species are shared between the central Pacific and the eastern Pacific.

The eastern Atlantic region is the smallest and least diverse of the four tropical inshore marine areas. This region runs along the West African continental shelf from Cape Verde to Angola. Coral reefs are nearly absent from this area. There are about 450 species of fishes, and roughly 25% of them are shared with the western Atlantic region. This may reflect a time when the South American and African coasts were much closer together. The region's remoteness probably accounts for the fact that approximately 40% of the species are endemic.

The warm temperate Mediterranean-Atlantic region is cooler than the inshore areas mentioned earlier. It includes the Atlantic and Mediterranean shoreline of Europe and Africa as well as the Black, Caspian, and Aral Seas. Northern and southern borders are ill defined, and warm waters of the Gulf Stream along the coast of Europe allow for a mixture of warm-water and cold-water species. There are about 680 species of fishes in the Mediterranean-Atlantic region, 540 of which are from the Mediterranean Sea, whose fauna is similar to the fauna of the Atlantic coasts of southern Europe and northern Africa.

The fish fauna of the northeastern Pacific region from Baja to the Aleutian Islands gradually changes from tropical to temperate to Arctic and is, therefore, very diverse. It is strongly influenced by the California Current, which brings cold water from the Gulf of Alaska down the California coast.

Coral reefs attract, feed, and house a large variety of fishes. (Photo by Animals Animals ©Mickey Gibson. Reproduced by permission.)

The northwestern Pacific or Asian-Pacific region extends from Hong Kong to north of the Kamchatka Peninsula. Its southern area overlaps with the Indo-West Pacific region. Temperate species reach their southern limit around Hong Kong, and the Bering Sea may be a barrier preventing exchange of coastal species between North America and Asia.

The South American region surrounds that continent from the coast of Peru around Tierra del Fuego and up the east coast to Rio de Janeiro. The borders are somewhat blurred, with tropical species in the north and Antarctic species in the south. The cold Peruvian Current and strong prevailing winds dominate the western coast of South America and produce upwellings of nutrient-rich waters that support large populations of anchoveta (*Engraulis ringens*) and other commercially important fishes as well as seabirds.

The South African region is warmer than the South American region, owing to its more northerly location and the warmer currents that surround Africa. Widespread tropical

Lungfish are able to survive when their pools dry up by burrowing into the mud and sealing themselves within a mucous-lined burrow. There are three genera surviving today and they are found in Africa, Australia, and South America. (Photo by Animals Animals © A. Root, OSF. Reproduced by permission.)

	Nearctic	Neotropical	Paleartic	Ethiopian	Oriental	Australian
# Primary Families	15	35	15	28	28	2
# Secondary Families	7	8	2	5	3	4
# Peripheral Families	9	8	13	3	11	14
TOTAL # FAMILIES	31	51	30	36	42	21
# Endemic Primary	8	32	0	17	14	1
# Endemic Secondary	1	2	1	1	0	2
# Endemic Peripheral	0	1	2	0	2	3
TOTAL ENDEMIC FAMILIES	9	35	3	18	16	7
% Endemic Primary	26	63	0	47	33	5
% Endemic Secondary	3	4	3	3	0	10
% Endemic Peripheral	0	2	7	0	5	14
% ENDEMIC TOTAL	29	69	10	50	38	33

Number of fish families and percent endemic in each biogeographical region, based upon 139 families. Widely distributed peripheral families are excluded. Data from Berra (2001). (Illustration by Argosy. Courtesy of Gale.)

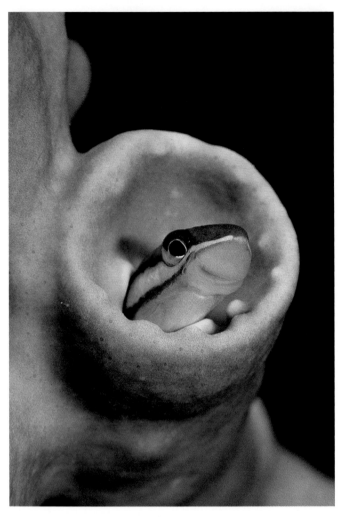

The saber-toothed blenny (*Meiacanthus migrolineatus*) makes its home in a tubeworm off the coast of Egypt. (Photo by Animals Animals ©Mark Webster, OSF. Reproduced by permission.)

families dominate the fauna, and most of the cold-tolerant families typical of the South American region are absent.

The Australian region includes western, southern, and eastern Australia as well as New Zealand. The southern coast is home to temperate fauna, with some cold-tolerant species, whereas the tropical component increases as one moves northward as the result of the effects of the warm Pacific and Indian Oceans.

Marine shallow, cold waters

Arctic and Antarctic shore fishes down to 656 ft (200 m) make up about 5.6% of all fish species. The Arctic region extends from about 60° north into the Arctic Ocean, Bering Sea, and the waters around Greenland. It shares some species with the surrounding regions. The Antarctic region and Southern Ocean contain less than 300 species. Many of these are endemic species of notothenioids adapted to the extremely cold waters. The Antarctic shares very few species with surrounding regions, probably because the more temperate species did not survive as Antarctica drifted into its present position.

Marine deep waters

Deep benthic (bottom-dwelling) fishes are found below 656 ft (200 m) along the continental slope or on the deep-sea floor. These fishes constitute 6.4% of all fishes. Deep pelagic (free swimming in the open ocean) fishes, below 656 ft (200 m), make up 5% of all fishes. These fishes are not tied to continental shelves, and many have a worldwide distribution. The habitat of the deep open sea is relatively poor in nutrients and niches. Fish living from about 656 to 3,281 ft (200–1,000 m) are mesopelagic, and those dwelling deeper than 3,281 ft (1,000 m) in the water column are bathypelagic. Pelagic faunas can be divided into various regions, such as north and south, temperate, subtropical, tropical, Arctic, and Antarctic.

Epipelagic

Fishes living from the surface to 656 ft (200 m) on the high seas are termed epipelagic. These are highly mobile fishes, such as the tunas, and many species are worldwide. Few niches are present in this habitat, and epipelagic fishes make up only 1.3% of the total fish fauna.

Freshwater fishes and zoogeographic realms

In his classic work *The Geographic Distribution of Animals*, published in 1876, Alfred R. Wallace recognized six major

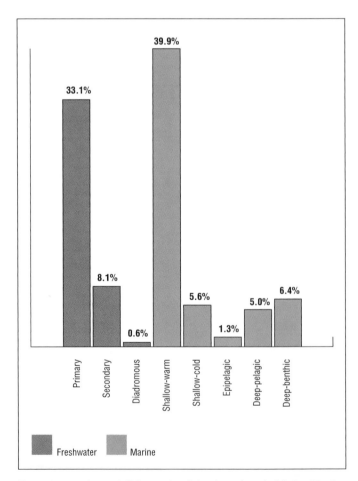

Percentages of recent fish species living in various habitats. (Illustration by Argosy. Courtesy of Gale.)

The palette tang (*Paracanthurus hepatus*), as well as other colorful fishes, are found in the Pacific and Indian Oceans. (Photo by Animals Animals ©M. Gibbs, OSF. Reproduced by permission.)

zoogeographical realms: Neartic (North America, except tropical Mexico), Neotropical (South and Central America, including tropical Mexico), Paleartic (nontropical Eurasia and the north tip of Africa), Ethiopian (Africa and southern Arabia), Oriental (tropical Asia and nearby islands), and Australian (Australia, New Guinea, New Zealand, and Celebes and nearby small islands to the east). Wallace proposed a hypothetical boundary between the Oriental and Australian faunas. This line, which became known as Wallace's Line, passes between Bali and Lombok, between Borneo and the Celebes (Sulawesi) and south around the Philippines. It is not an exact boundary between the Oriental and Australian realms for all animal groups, but freshwater fishes, for whom saltwater is a barrier, have not crossed Wallace's Line to any significant degree. There is an important faunal break separating the depauperate Australian island fauna from the rich Oriental continental fauna. Of the 23 families of primary division freshwater fishes on Borneo, only the bonytongues (Osteoglossidae) have crossed Wallace's Line without the help of humans.

Freshwater distribution patterns

In the book *Freshwater Fish Distribution*, Berra analyzed the distributions of 139 families of primary, secondary, and peripheral division freshwater fishes. Each zoogeographic realm has its endemic families, which occur only in that particular region, and some regions are richer than others. The Neotropical realm has the most diverse fish fauna, with 51 families. Of these families, 32 (63%) are endemic primary division families, two are secondary, and one is peripheral. The 35 endemic families (69%) comprise river stingrays (Potamotrygonidae), South American lungfish (Lepidosirenidae), 13 families of characid-like fishes, 12 families of catfishes, six families of electric fishes, the Middle American killifishes (Profundulidae), and the foureyed fishes (Anablepidae).

The Oriental realm has 42 families, 14 (33%) of which are endemic primary division, none are secondary, and two are peripheral. The 16 endemic families (38%) include the Sundasalangidae, Gyrinocheilidae, seven families of catfishes, Indostomidae, Chaudhuriidae, Asian leaffishes (Nandidae),

Pristolepidae, pikehead (Luciocephalidae), kissing gourami (Helostomatidae), and giant gouramies (Osphronemidae).

The Ethiopian realm includes 36 families, 17 (47%) of which are endemic primary division, and one is secondary. There are no peripheral endemic species. The 18 endemic families (50%) consist of African lungfishes (Protopteridae), bichirs (Polypteridae), butterflyfish (Pantodontidae), elephantfishes (Mormyridae), aba-aba (Gymnarchidae), denticle herring (Denticiptidae), Kneriidae, Phractolaemidae, four families of characid-like fishes, five families of catfishes, and Bedotiidae. For our purposes, Madagascar is considered part of this realm.

The Nearctic realm is home to 31 families, eight (26%) of which are endemic primary division, and one is secondary. No endemic peripheral division fishes are present. The nine endemic families (29%) include the bowfin (Amiidae), mooneyes (Hiodontidae), bullhead catfish (Ictaluridae), splitfins (Goodeidae), troutperch (Percopsidae), pirateperch (Aphredoderidae), cavefish (Amblyopsidae), sunfishes (Centrarchidae), and pygmy sunfishes (Elassomatidae).

The Palearctic realm has 30 families, none of which are endemic primary division fishes. There are one endemic secondary and two endemic peripheral families. The three endemic families (10%) include the Valenciidae, Comephoridae, and Abyssocotidae.

The Australian realm has the fewest families in freshwater, with 21. Only one family is an endemic primary division, two are secondary, and three are peripheral; the division of one family is not certain. The seven endemic families (33%) are the Australian lungfish (Ceratodontidae), Australian smelts (Retropinnidae), salamanderfish (Lepidogalaxiidae), rainbowfishes (Melanotaeniidae), blue eyes (Pseudomugilidae), Celebes rainbowfishes (Telmatherinidae), and torrentfish (Cheimarrichthyidae). For our purposes, Sulawesi is considered part of the Australian realm.

The distribution of fishes within each biogeographic realm can be subdivided by drainage basin. These subdivisions are dis-

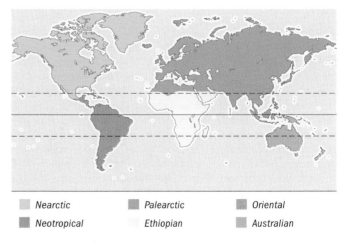

	Nearctic		Palearctic		Oriental
	Neotropical		Ethiopian		Australian

Freshwater fish families: the global pattern. (Map by XNR Productions. Courtesy of Gale.)

The shy damselfish peers out from swaying rods of soft coral of a Papua New Guinea reef. (Photo by Animals Animals ©Bob Cranston. Reproduced by permission.)

cussed by Banarescu, Bond, Moyle and Cech, Helfman et al., and Matthews. Berra summarized the biogeography of freshwater fish faunas and provided an entrance into the literature.

Families that occur in both the Nearctic and Palearctic regions are said to have a holarctic distribution. Examples include the sturgeons (Acipenseridae), pike (Esocidae), salmon and trout (Salmonidae), and perches (Percidae). The minnow family, Cyprinidae, is the largest fish family, with more than 2,000 species. It is the dominant fish family in the world's freshwaters, with a continuous distribution in Eurasia, Africa, and North America. It is absent from South America, where characiform fishes dominate, and it has not been able to penetrate the saltwater barrier surrounding Australia.

Continental drift

Between 1915 and 1936 Alfred Wegener published six editions of *The Origin of Continents and Oceans*, in which he proposed the idea that the continents originally were one large mass that subsequently fragmented, with the fragments drifting apart. He noted the congruence of the east coast of South America to the west coast of Africa and compared this pattern to a giant jigsaw puzzle. This idea met resistance until modern geophysical evidence of plate tectonics and paleomagnetics provided a mechanism for the way in which continents could move. About 200 million years ago all the continents were part of a huge landmass, Pangaea. This supercontinent began to fragment about 180 million years ago, resulting in a northern portion (Laurasia—North America and Eurasia) and a southern portion (Gondwana—South America, Africa, India, Antarctica, Australia, and New Zealand). Gondwana separated into the southern continents beginning about 90 million years ago and, propelled by convection currents in the earth's mantle generated by the heat of radioactivity, have been drifting into their present positions ever since. This, of course, has implications for the fish faunas carried on the continental plates, especially ancient groups such as the lungfish, which have families on three of the continents of Gondwana: South America, Africa, and Australia. Other more recent groups, such as the 18 families of characid-like fishes, show a South America–African connection. (Bond, Moyle and Cech, Helfman et al., Matthews, Lundberg, and Lundberg et al. comment on the relationship between continental drift and freshwater fish distribution.)

Dispersal

To understand the patterns of distribution, it is necessary to know the phylogeny and fossil record of the group under consideration as well as the geological history of the region. Lundberg provided an excellent analysis of South American–African fish distribution. In addition to vicariant events, such as drifting continents, freshwater fish also may disperse over a continent. Fishes may move from one drainage system into another during times of flooding. Stream capture due to lowland meanders or headwater piracy may provide access from one drainage system to another. Freshwater fish with some salt tolerance may swim through brackish water or saltwater from one river mouth to another and then ascend upstream. Streams may become connected by glacial meltwaters, thereby allowing passage of formerly isolated species. Some fish species, for example, walking catfish or eels, can wiggle through damp grass from one system to another. Mud attached to the feet of aquatic birds may contain fish eggs that hitch a ride to another drainage system. Waterspouts may pick up small fishes or their eggs and deposit them into another stream or lake. Lowered sea levels caused by glacial advance may isolate streams, or the elevation of a mountain range may divide river systems. This may lead to the divergence of initially similar faunas on both sides of a barrier. These are the sorts of factors that must be considered when trying to explain the distribution patterns of freshwater fishes.

Galaxias maculatus—a case study

The Southern Hemisphere peripheral division family Galaxiidae has a distribution that, at first glance, seems to reflect vicariance (continental drift). Most members of this family occur in Australia and New Zealand. Several species are in southern South America, however, and one species occurs at the southern tip of Africa. Whether vicariance or dispersal across the sea accounts for this pattern has been debated for decades. One species in this family, *Galaxias maculatus*, has received the most attention because it has one of the most extensive and disjunct distributions of any freshwater fish. It occurs in western and eastern Australia, New Zealand, Lord Howe Island, Chatham Island, southern Chile, Argentina, and the Falkland Islands. Berra has reviewed this "poster child" of vicariance versus dispersal.

The vicariance supporters suggested that because the galaxiids occur on all Gondwanan continents except India, their distribution reflects an ancient Pangaean pattern followed by continental drift. The dispersalists attributed the wide geographic distribution to transoceanic dispersal of the marine larval stage, called "whitebait." It was suggested that *G. maculatus* originated in Australia and dispersed eastward past Tasmania to New Zealand and on to South America via the East Australian Current and the West Wind Drift. The dispersalists reasoned that if *G. maculatus* predates the breakup of Gondwana 65 million years ago, as the vicariance advocates suggest, there would be much greater differences between South American and Australian populations. Since these populations exhibit very few physical differences, they could not have been isolated for 65 million years. Proteins from muscle extracts of *G. maculatus* from western and eastern Australia, New Zealand, and Chile were used to test the

hypothesis that populations from the western Pacific and the eastern Pacific do not differ genetically. Researchers found no fixation of alternative genes and only minor differentiation in gene frequency between western and eastern populations and concluded that the populations were part of the same gene pool, indicating that gene flow via dispersal through the sea is taking place. Using mitochondrial DNA variation, Waters and Burridge likewise supported the dispersal argument but reported greater population differentiation than was detected by Berra et al. with proteins. Waters et al. showed that intercontinental marine dispersal between New Zealand and Tasmania occurs but is insufficient to prevent mitochondrial DNA differentiation among continents.

Resources

Books

Banarescu, P. *Zoogeography of Fresh Waters.* Vol. 1. *General Distribution and Dispersal of Freshwater Animals.* Wiesbaden, Germany: AULA-Verlag, 1990.

———. *Zoogeography of Fresh Waters.* Vol. 2. *Distribution and Dispersal of Freshwater Animals in North America and Eurasia.* Wiesbaden, Germany: AULA-Verlag, 1992.

———. *Zoogeography of Fresh Waters.* Vol. 3. *Distribution and Dispersal of Freshwater Animals in Africa, Pacific Areas and South America.* Wiesbaden, Germany: AULA-Verlag, 1995.

Berra, Tim M. *An Atlas of Distribution of the Freshwater Fish Families of the World.* Lincoln: University of Nebraska Press, 1981.

———. *Freshwater Fish Distribution.* San Diego: Academic Press, 2001.

Bond, Carl E. *Biology of Fishes.* 2nd edition. Fort Worth, TX: Saunders College Publishing, 1996.

Briggs, John C. *Marine Zoogeography.* New York: McGraw-Hill, 1974.

———. *Global Biogeography.* Amsterdam: Elsevier, 1995.

Darlington, Philip J. Jr. *Zoogeography: The Geographical Distribution of Animals.* New York: John Wiley and Sons, 1957.

Eschmeyer, W. N., ed. *Catalog of Fishes.* 3 vols. San Francisco: California Academy of Sciences, 1998.

Groombridge, B., and M. Jenkins. *Freshwater Biodiversity: A Preliminary Global Assessment.* Cambridge, U.K.: World Conservation Monitoring Centre—World Conservation Press, 1998.

Helfman, G. S., B. B. Collette, and D. E. Facey. *The Diversity of Fishes.* Malden, MA: Blackwell Science, 1997.

Lundberg, John G. "African–South American Freshwater Fish Clades and Continental Drift: Problems with a Paradigm." In *The Biotic Relationships Between Africa and South America,* edited by P. Goldblatt. New Haven, CT: Yale University Press, 1993.

Matthews, W. J. *Patterns in Freshwater Fish Ecology.* New York: Chapman & Hall, 1998.

Moyle, P. B., and J. J. Cech Jr. *Fishes: An Introduction to Ichthyology.* 3rd edition. Upper Saddle River, NJ: Prentice Hall, 1996.

Periodicals

Berra, Tim M., L. E. L. M. Crowley, W. Ivantsoff, and P. A. Fuerst. "*Galaxias maculatus*: An Explanation of Its Biogeography." *Marine and Freshwater Research* 47 (1996): 845–849.

Cohen, Daniel M. "How Many Recent Fishes Are There?" *Proceedings of the California Academy of Sciences* 38, no. 17 (1970): 341–346.

Horn, M. H. "The Amount of Space Available for Marine and Freshwater Fishes." *Fisheries Bulletin* 70 (1972): 1295–1297.

Lundberg, J. G., M. Kottelat, G. R. Smith, M. L. J. Stiassny, and A. C. Gill. "So Many Fishes, So Little Time: An Overview of Recent Ichthyological Discovery in Continental Waters." *Annals of the Missouri Botanical Garden* 87 (2000): 26–62.

McDowall, R. M. "The Galaxiid Fishes of New Zealand." *Bulletin of the Museum of Comparative Zoology* 139 (1970): 341–431.

Myers, G. S. "Fresh-water Fishes and West Indian Zoogeography." *Annual Report of the Smithsonian Institution for 1937* (1938): 339–364.

Springer, Victor G. "Pacific Plate Biogeography with Special Reference to Shorefishes." *Smithsonian Contributions to Zoology* 367 (1982): 1–182.

Waters, J. M., and C. P. Burridge. "Extreme Intraspecific Mitochondrial DNA Sequence Divergence in *Galaxias maculatus* (Osteichthys: Galaxiidae), One of the World's Most Widespread Freshwater Fish." *Molecular Phylogenetics and Evolution* 11 (1999): 1–12.

Waters, J. M., L. H. Dijkstra, and G. P. Wallis. "Biogeography of a Southern Hemisphere Freshwater Fish: How Important Is Marine Dispersal?" *Molecular Ecology* 9 (2000): 1815–1821.

Tim M. Berra, PhD

Behavior

Introduction

Fish behavior is often varied and complex within and between species. Sensory stimuli, cyclic influences, population density and structure, habitat quality, the availability and use of space, the potential for competition and coexistence, the need to avoid predators, foraging and diet, reproduction, and other factors all contribute towards the evolution of patterns of behavior and their use. Despite the great diversity of fish species, their wide patterns of geographical and spatial distribution, and their highly variable ecological requirements, there are a number of patterns of behavior common to all fishes, as well as unique adaptations that occur only in a few.

Sensory systems and behavior

The behavior of marine fishes is shaped by sensory information provided by any one of their senses, both singly and in combination. They use vision to detect prey, avoid predators, identify species, choose mates, communicate, engage in social and territorial interactions, select and use habitat, and navigate. Fishes use their inner ear to detect sounds made by conspecifics during communication, by approaching predators, or by other fishes as they feed. If the fish has a swim bladder, these sounds can be amplified. Low frequency sounds made by movement, including struggling, are detected by the lateral line. Fishes may communicate by rasping mouthparts, gill arches, or other organs, and amplifying the sounds with their swim bladders. Touch is important in prey detection, predator avoidance, social interactions, and courtship and spawning behavior. Olfaction (smell and taste) is important in many predators for the detection of prey. The barbels beneath the mouth of a goatfish are highly sensitive and allow this fish to taste, as well as feel, potential prey. The sense of taste also allows a fish to determine quickly where a prey item is palatable or toxic. Chemical cues are also utilized for navigation. For example, salmon utilize chemical cues to detect the natal stream, where they will return to reproduce and die. Some fishes, especially sharks, skates, and rays, and the electric freshwater fishes of the families Gymnotidae and Mormyridae, are capable of detecting minute electrical currents discharged by their prey. The patterns they detect allow them to pinpoint the location of the prey, even if lies buried under benthic sediments or is obscured by turbid or muddy water. Electroreception also allows some migrating fishes to determine their geographical position relative to Earth's magnetic field. Electricity is also used for communication in gymnotids, mormyrids, and some catfishes (Synodontidae and Ictaluridae).

Activity cycles

Fish behavior is influenced by various cycles that govern such activities as habitat use, feeding, migration, and reproduction. Circadian rhythms derived from internal or endogenous 24-hour clocks control hormone releases and subsequent behaviors. Changes in light levels on a daily or seasonal basis are a principle factor influencing rhythms. Lunar periods control tidal cycles that influence patterns of local migration, feeding, and reproduction. Seasonal shifts in water temperature or other climatic variables trigger migratory and reproductive behaviors.

Fishes and other organisms possess an internal, or endogenous, clock that is set to a period of approximately 24 hours for a given day. This clock can be adjusted daily by some sort of trigger or stimulus. Two common stimuli are the onset of daylight and the constant progression of low and high tides. The clock governs a number of basic behaviors, such as the onset of movement, feeding, or courtship, along with the hormonal activity that influences or triggers these behaviors.

Most freshwater and marine fishes are diurnal, or active in daylight, during a 24-hour period. As dusk approaches, diurnal species seek shelter in which to rest or sleep and are replaced by nocturnal species that are active during the night. At dawn, these fishes retire to shelter or simply rest until dusk approaches again. Some fishes ignore the changeover between day and night and are more or less active for 24 hours. The dawn and dusk changeover periods, also known as crepuscular periods, also trigger pronounced reproductive or predatory activities in a number of species. Vertical migrations, in which deep-dwelling species rise hundreds or even thousands of feet in the water column at night, only to descend when daylight approaches, are also triggered during these times.

Tidal shifts, either the onset of low tide or high tide, and the corresponding movement of water off or onto a flat, tide pool, or other type of habitat, govern the movements of fishes within or between affected habitats. For example, as the tide falls in a tide pool, residents must move out of the pool and seek shelter elsewhere to avoid desiccation or thermal shock. As the tide returns, so do the fishes. Similarly, predators cue on the outgoing tide and move to locations where prey will gather or pass through as they move out of an affected habitat. Tidal shifts also trigger courtship and spawning behaviors that favor the movement of pelagic eggs and larvae off the reef or flat to avoid benthic predators or, alternately, to allow pelagic larvae to move back onto the reef or flat for settlement.

Temperate marine and freshwater fishes often spawn on a seasonal basis, usually during spring or summer, although others spawn in the autumn prior to the onset of winter. Spawning in spring or summer provides an opportunity for larvae to feed and grow before the falling temperatures of autumn and winter slow growth and activity rates. Fishes spawning in the autumn have eggs that may overwinter and hatch with the onset of spring. Spawning in river species is often timed to coincide with annual or seasonal flood cycles that trigger migrations, but that also provide feeding opportunities for juveniles and adults and the increased dispersal of young.

Tropical, and some temperate, species court and spawn in relation to phases of the moon. Some species are semilunar, in that they spawn every other week in relation to the new and full moon. Others are lunar, in that they spawn just once a month, either on the new or full moon. The actual day of spawning relative to moon phase may be variable, as a number of species spawn on the days on either side of the new or full moon, but reach a peak at a the height of the phase. Semilunar and lunar spawning may also be seasonal. For example, a number of groupers (Serranidae) form spawning aggregations once or twice a year, with the time of formation centered around a specific phase of the moon.

Many reef fishes, particularly those tropical species resident at low latitudes, court and spawn daily. Their reproductive cycle is regulated by daily shifts in light. Some species spawn at dawn, others at dusk or into the night, and still others during daylight, but the time of spawning shifts daily in relation to tidal phase.

Migratory behavior of marine and freshwater fishes may be controlled by annual, seasonal, lunar, or daily cycles that trigger movement from one location or depth to another. Pelagic fishes, such as marlins (Istiophoridae) or dolphinfishes (Coryphaenidae), migrate great distances annually. These species, and numerous others, track changes in water temperature and move from winter to summer grounds, or vice versa, for feeding and reproduction. Many river species, especially in larger rivers prone to flooding, migrate annually or seasonally to take advantage of the new spawning habitats and food sources made available when bottomlands are flooded. Fishes may migrate from one body of water to another for reproduction, and their progeny often migrate back from where their parents came. Diadromous, catadromous, and amphidromous migrations and subsequent recruitment of young may be triggered by annual or seasonal stimuli. Vertical migrations allow fishes to track the movements of potential prey as they migrate up and down in the water column.

Atlantic manta ray (*Manta birostris*) with cleaner fish in the Gulf of Mexico. (Photo by Animals Animals ©Joyce & Frank Burek. Reproduced by permission.)

Communication

Communication is an important component of fish behavior. The transmission and reception of information by a number of means facilitates social interaction, the partitioning of space, cooperative feeding, predation avoidance, and reproduction. Visual communication is important in all but the darkest or most turbid environments. Many freshwater and marine species possess color patterns that are helpful for species recognition, sex recognition, age determination, and for assessments of agonistic and reproductive states. Both black and white coloration and bright colors are utilized. Coral reef fishes, for example, are famous for their bright color patterns, or poster coloration. (Poster coloration, a term coined by Nobel laureate Konrad Lorenz, refers to the conspicuousness and potential advertising or function of bright color patterns in coral reef fishes. Such coloration is useful in intra- and interspecific communication during territorial interactions, aggregation formation and maintenance, or mating, and facilitates species recognition.) Color patterns may be permanent or temporary. The latter is under hormonal control in relation to the expression of certain behaviors. For example, some groupers assume temporary color patterns during social interactions. The detection of bioluminescent signals at night or in low-light habitats is another component of vision-based communication. Numerous deep-sea and deep-slope fishes utilize light flashes to communicate with conspecifics. Fishes also employ body and fin displays to communicate intentions during territorial encounters, courtship, and predation avoidance.

Several species of fishes communicate with sound. Sound production is used to warn predators, warn of predators, attract mates, attract conspecifics in school formation and maintenance, and to communicate intentions during agonistic, reproductive, and parental care interactions. Sound production also places fishes at risk from predation, as some predators have learned to locate sound producers and prey upon them.

Blue striped grunts (*Haemulon sciurus*) fighting. (Photo by Michael Patrick O'Neill/Photo Researchers, Inc. Reproduced by permission.)

Fishes produce chemical secretions known as pheromones, which may be detected by taste or smell. Chemoreception is significant for the recognition of conspecifics in catfishes (Ictaluridae), minnows (Cyprinidae), and other species. This recognition is important in establishing and maintaining social relationships, such as dominance hierarchies or territorial interactions. Parents and young in species that practice parental care of fry and juveniles, such as the cichlids (Cichlidae), employ chemical reception to identify each other.

Fishes make use of touch when communicating intentions during aggressive behavior, courtship behavior, and parental care. Electric communication in gymnotids, mormyrids, and some catfishes (Mochokidae and Ictaluridae) is also used for aggressive and courtship behaviors, and is especially helpful in waters where visual detection is greatly reduced or nonexistent. Electrical discharges made by these fishes are species specific. Variations in production properties, such as pulse length, interpulse length, frequency, and amplitude, allow these fishes to communicate or assess information about species identity, individual identity, sex, size, reproductive readiness, and level of agonistic behavior. The fishes also obtain information on the location of and distance between communicators in this way.

Behavior and habitat use

How fishes select and make use of habitat is determined by their behavior. In marine systems, especially coral and rocky reefs, pelagic larvae actively swim shoreward as they prepare to settle into a habitat. Prior to and during settlement they assess the suitability of that habitat. For example, damselfish (Pomacentridae) larvae settling onto a portion of a reef have been observed to reject this habitat, swim back up into the water column, and search for more suitable one. Post-larvae, juveniles, and adults all utilize a variety of patterns that allow them to compete or coexist with others already using a habitat. Agonistic behavioral displays are common. Sometimes, the behaviors involved are cryptic, in that fishes "sneak" into the habitat and become established without drawing at-

tention to themselves. Size-structured schooling species likely join the school as larvae and exploit a repertoire of innate patterns that allow them to function cohesively with conspecifics. In shoals or mixed-species schools, members use a similar repertoire to join, maintain, or leave the aggregation. Solitary pelagic species employ a repertoire of behavioral patterns that allows them to swim, feed, and avoid predation in a habitat that provides little or no cover. If cover is present in the form of drifting pelagic algae, logs, or other flotsam or jetsam, small pelagic species seek shelter there. Some species, such as the sargassumfishes (Antennariidae), are adapted to life in floating sargassum, where they shelter from predators, ambush prey, and mate. Other fishes, such as juvenile butterfishes (Stromateidae), shelter within the tentacles of pelagic jellyfishes. Many small pelagic fishes recruit to floating structures, and larger predators are attracted in turn.

Benthic freshwater and marine species often adapt to specific conditions and make use of seemingly novel structures that provide shelter, feeding sites, or places for reproduction. In moderate or fast-moving streams, trouts and charrs learn to make use of rocks, logs, holes, undercut banks, and other forms of structure as shelter. Their swimming behaviors allow them to move up into the current, feed or chase off intruders, and return to their shelter sites. Sand-dwelling darters (*Ammocrypta*; Percidae) in slower moving streams rest on the sand, but bury into it to avoid predators. On coral reefs, various species use specific behaviors to burrow into sand, rubble, cobble, or mud in order to avoid predation, ambush prey, or rest. Fishes that employ burrows use those dug by other organisms or dig and maintain their own structures, often building multiple entrances or exits. They swim, hover, or rest near these burrows as they feed or engage in social interactions, but will dart quickly into an entrance if threatened. If resting inside a burrow, they may quickly escape via an alternate route if a predator is blocking or has entered one of the entrance points. These burrows are sometimes shared by more than one fish species or by invertebrates, such as snapping shrimps, in a symbiotic relationship. Other benthic fish use abandoned tubes made by polychaete worms or other invertebrates. These fishes swim out of these tubes to feed or mate, but return and move backward into them if threatened. Some tube-dwelling species use these structures as cryptic ambush sites from which they attack passing fishes. Shrublike corals, sea fans, and black corals all provide structures for a number of small reef species. These structures are shared because intra- and interspecific behavioral interactions define the use of space and reinforce order and structure within the coral head.

Social behavior

Agonistic behavior is employed by fishes to establish social dominance, defend territories, and ward off potential predators, and involves the use of displays given at increasing levels of intensity. The displays are fixed or modal action patterns, and the sequence of their use is often highly ritualized. The information communicated by a pattern or series of patterns in sequence is therefore recognized in the context of its use, and aggressive behavior leading to the injury of one or more parties in the interaction is often averted.

Dominance of one or more individuals by the agonistic behavior towards another of the same species occurs among groups of individuals, in shoals, and in schools. Dominance is expressed in either of two ways. First, a single individual may dominate all others who hold equal rank under the dominant fish. More commonly, a dominance hierarchy forms linearly, with a single alpha individual dominating others. The alpha is followed by a beta individual that dominates the remaining individuals, and so on. Dominance hierarchies such a this are often ordered on the basis of body size, with larger fishes dominating smaller fishes. They may also be ordered by different levels of aggressive behavior between individuals, with the more aggressive fishes dominating less aggressive fishes. Mating groups of sex-changing fishes also have hierarchies. For example, the mating group of the hawkfish *Cirrhitichthys falco* (Cirrhitidae) consists of a single dominant male, a large dominant female, and two or more subdominant females of variable size, which dominate one another on the basis of greater body size.

Agonistic behavior is used to defend living space, food resources, or mating groups and sites. This is known as territorial behavior. Territorial displays include the use of body displays, erect fins, color changes, sound production, chasing, or a combination of patterns. Defense may be against both intra- and interspecific intruders. The latter include competitors for food and space, but may also include potential predators of a defender's nest of eggs or free-swimming offspring. Territories usually consist of an area of relatively fixed size. The size of a territory varies within species and between species, usually as a function of size or sex, but will also vary in relation to the give and take of interactions with neighbors. A territory will often be nested in a much larger home range that is utilized by the fish or a group of fishes. Only that space that is actively defended by an individual is considered a territory. Territories may be permanent or temporary. For example, territories needed for courtship and spawning of a number of fishes are formed only during the breeding season or at certain times during the breeding season. Agonistic behavior increases during these times, but will often be absent during nonmating periods. Fishes may leave their home ranges or territories to form temporary multimale territories at courtship sites, known as leks, at which to attract females for mating. Some of these leks are "floating," in that their position may change relative to the location of females in the area. Territories required for feeding may be quite large, especially for larger predatory fishes, and may often greatly exceed the area defended for shelter space. Permanent territories often involve the defense of a shelter site. Males in single-male, multifemale, mating groups will defend their territories and those of the females contained within. Females, in turn, defend their smaller territories from intrusions by neighboring females within the same group. Mating sites of these fishes are defended in the same way. At the extreme end of territorial behavior is the defense of personal space, as seen in shoaling or schooling fishes that display to, or ward off, neighboring fishes who swim too close.

Territorial defense is also practiced by monogamous pairs of fishes. The best-known examples are the butterflyfishes (Chaetodontidae) on coral reefs. Pairs of butterflyfishes patrol

Salmon leap up a waterfall to return to the place where they hatched to spawn. (Photo by Randy Wells/Corbis. Reproduced by permission.)

a territory and may encounter potential competitors for food or space. These competitors include conspecifics (also usually in pairs), as well as other species that may utilize the same resources. Defense involves an exchange of ritualistic displays culminating in the departure of the combatants. Highly territorial butterflyfishes, not usually in pairs, chase intruding competitors away; however, noncompetitive species are usually ignored. Clustering behavior, in which one or more individuals within a group, or mosaic, of territories rises up into the water column to assess others in the mosaic without engaging in agonistic interactions, is an interesting offshoot of territorial behavior, and has been observed in damselfishes.

Fishes form shoals or schools for protection from potential predators, foraging, overcoming the territorial defenses of individuals migration, and reproduction. Shoals are unorganized aggregations and may often be temporary in nature. They may consist of different species with a changing membership (heterospecific shoal), are relatively unstable, and may be dissolved and reformed quickly. Schools are organized or polarized aggregations that form permanently or temporarily. Generally, they are monospecific (contain only one species),

Tropical angler (*Antennarius* sp.) "fishing" for food using its elongated dorsal fin, wiggling its tip to simulate a small fish or worm. (Photo by Tom McHugh/Steinhart Aquarium/Photo Researchers, Inc. Reproduced by permission.)

and membership is often age specific. Schools may dissolve at night for diurnal species, during daylight for nocturnally active species, or remain constant over a 24-hour period. Some schools dissolve under heavy attacks from predators, but reform afterward if sufficient numbers of fishes survive. The movements of schools are governed by a unique set of behaviors that determines position, synchronized movements, swimming speed, evasion, and flight.

Fishes have been found to be capable of transmitting traditional information socially. For example, certain benthic reef fishes follow predictable long-term routes between feeding and sheltering grounds as either darkness or dawn approaches. Experimental manipulations of the composition of a shoal of these fishes have shown that resident fishes can transfer information about the location of resting shelters to fishes new to the shoal during a dusk or dawn migration.

Fishes form symbiotic relationships with other fishes, or with invertebrates that share some common form of microhabitat, food, or need. An excellent example of shared microhabitat is seen in the anenomefishes (Pomacentridae) and the burrowing shrimp gobies (Gobiidae). Anenomefishes live in close association with anenomes, flowerlike benthic invertebrates related to corals that have poisonous nematocycts in their tentacles and are capable of inflicting an injurious or deadly sting. The function of these nematocysts is to deter predators and immobilize prey. Anenomefishes have developed defenses against the effects of the sting and make use of the tentacles for shelter and nest sites. Part of the anenomefish's defense is behavioral, in that its undulating movements within the tentacles communicate to the anemone that it is neither a threat nor some form of prey to be stung and ingested. In exchange, the anemonefish defends the anenome against potential predators, such as butterflyfishes, that feed upon the anemone's tentacles. Anenomefishes may also "groom" the anemone by removing foreign matter and feces, as well as consuming potential parasites. Shrimp gobies share burrows with one or more species of snapping shrimps of the

family Alpheidae. The shrimp, which is usually blind, constructs a burrow that may be shared with one or two individuals of a given species. The goby, or pair of gobies, guards the burrow at its entrance while the shrimp maintains it. If danger approaches, the shrimp is alerted as the goby or gobies signal it with flicks of the caudal fin. If the threat continues, the goby or gobies will dive headfirst into the burrow and the shrimp will retreat. Often there is species specificity between shrimps and gobies, and there is some evidence that both the goby and the shrimp settle out of the plankton together as post-larvae, prior to the construction of their shared burrow. The behavior exhibited by the anenomefishes and anenomes, and the shrimp gobies and the burrowing shrimps, is known as a mutualistic symbiosis, in that parties benefit from the relationship. Other examples of mutualistic behavior relating to cleaning and the control of parasites, and commensalistic behavior, such as the relationship between remoras and large pelagic fishes, are discussed in the section on feeding behavior.

Reproductive behavior

Behavior is an essential component in the reproduction of fishes, and is essential for the identification of conspecifics, the attraction and selection of mates, the process of courtship and spawning, and, in a number of fishes, parental care. To perform these functions, fishes have developed a repertoire of behavioral patterns. These are often used in conjunction with some physical trait, such as a larger body size, elongated fin rays, larger hook-shaped lower jaws (kypes), well-developed humps on the forehead, and unique color patterns (temporary or permanent). These physical traits are coupled with one or more behavior patterns, whose use and intensity of use accentuate the traits. The products of sexual selection, these patterns and traits confer reproductive advantages to individuals in a given species population. Other patterns are essential for the physical act of spawning or breeding to occur, or for parental care to be successful.

Fishes identify conspecifics as potential mates by the recognition of species-specific morphological shape and form, color patterns, scents, or various behaviors. This recognition is not as easy as it seems because of variation in each as a function of sexual dimorphism or individual variation. Sometimes "mistakes" in species identification are made, and interspecific mating occurs. Most of these mating attempts likely fail, but occasionally functional offspring, or hybrids, result. Once species identity is recognized, the next step is to determine if the potential mate is of the opposite sex. Again, differences in morphology and color pattern or other factors accentuated by sexual dimorphism are recognized. If no such differences exist, then the use of behavioral patterns becomes an essential tool for the recognition of sex. Once sex is recognized, potential mates have to determine if they are attractive or suitable for one another and if they are ready to mate. Sexually selected traits and patterns are utilized to determine this. In many mating systems, females select mates. Females are attracted to males that appear stronger, fitter, and capable of providing the greatest investment for her offspring. Males utilize a series of behavioral patterns to convey this impression.

They may also emphasize the quality of their resources, such as feeding or sheltering territories, nest sites and qualities, or spawning site locations. Females assess these attributes and select the most attractive male or males, accordingly. Males, for their part, may wish to mate with the largest female or the most females possible. Larger females tend to be more fecund, thus increasing a male's opportunity to fertilize more eggs, pass down his genes, and achieve greater fitness. Mating with more than one female also increases fitness. Males compete with other males for access to females. The quality of their behavioral and physical traits and their use in intermale interactions, convey an advantage to one male over another.

Courtship patterns are used to attract mates, assess spawning readiness, and facilitate spawning or mating. A male employs a series of patterns to attract passing females or initiate courtship with females that form part of his mating group, and courtship bouts ensue with variable success. For example, in single male, multifemale mating groups (called "harems" by some authors), a male courts and spawns with each of his females in succession. Unfortunately for the male, this is not always easy. Individual variation in readiness to spawn among females means that a male may have to make repeated visits to females within his group before spawning takes place, and there is no guarantee that he will be successful in spawning with all females during a given spawning period. Sometimes the male is so busy attempting to spawn with one female that too much time passes and the "spawning opportunity window" closes before he can mate with others in his group. Alternatively, especially at dusk when light levels are falling, the male may become the target of a predator as he moves between females to court. Predator avoidance is costly because spawning opportunities may be lost. Worse, from the male's perspective, is that he might become lost, and control of the mating group will pass to a sex-changing female within his group or to another male from outside the group. Females are also at risk while waiting to court and spawn, and this threat may affect spawning readiness.

If courtship is successful and the male is able induce the female to spawn or mate, another series of patterns comes into play, which help the pair (or group, if spawning is in an aggregation) synchronize their activities so that the spawning or mating event is successful. Some of these patterns are as gentle as a male nuzzling the female's abdomen during a paired pelagic ascent, as in some of the marine angelfishes (Pomacanthidae). Other patterns, such as a sharp-toothed male shark grabbing a female's flank so that he can insert his clasper inside her and attempt internal fertilization of her eggs, are more forceful.

Modes of reproduction classify how fishes reproduce physically. The mode itself is not specific to a single taxonomic group, but may be shared by many taxa regardless of their phylogenetic affiliation. Criteria are defined based upon the degree of parental care, if any, invested. Thus, fishes may fall into three general guild categories: those that do not guard offspring, those that guard offspring, or those that bear offspring. Nonguarding species spawn openly upon substrates, either pelagically or benthically, or they hide their broods. Open pelagic spawning occurs in the water column. Fishes swimming in the water column engage in courtship and release eggs. Benthic fishes swim upward into the water column and release eggs and sperm at the apex of the spawning rise or rush. Spawning may be paired or in groups. Open benthic spawning includes the release of eggs onto the substrate (rocks, cobble, etc.), with the resultant larvae being either pelagic or benthic. Benthic spawning may also occur on plants as an obligatory or nonobligatory function, or on sand. Brood hiders deposit their eggs on the bottom, in caves, on or in invertebrates (such as corals, bivalve mollusks, and crinoids), or on beaches during a tidal cycle. They may also deposit eggs on a substrate that is prone to annual desiccation, in which case the eggs are adapted to resist this and hatch out when wet conditions return.

Spawning aggregations are a specialized behavior, and are formed by migrations of fishes to specific sites for courtship and spawning. There are two general types of spawning aggregations: resident and transient. Resident aggregations occur locally, in that members of the aggregation are drawn from the general area in which the aggregation forms. These aggregations usually form on a daily, semilunar, or lunar frequency. Transient aggregations consist of fishes that are drawn from a much larger area and population. Some species of groupers, for instance, migrate hundreds of miles along coastal waters in the Gulf of Mexico and Caribbean, and the number of individuals forming the aggregations can be in the hundreds or even thousands. These aggregations usually form annually or seasonally in relation to the lunar period. Both kinds of aggregations form at sites that appear to facilitate mating and the dispersal of eggs and larvae.

Guarders include those fishes that choose the substrate they spawn upon with subsequent guarding of the offspring. The substrates chosen include rocks, plants, terrestrial structures (such as overhanging leaves or flooded grasses), or the water column. Nest spawners construct simple or complex nests to attract females, deposit eggs, and provide parental care. Nests are made from a variety of materials, including gravel and rock, sand, holes, plant materials (with or without a "glue" secreted by the male), anenomes, or bubbles. In the latter case, males of some species make the nest by blowing bubbles of air and mucous onto an object or even the underside of the surface of the water. Miscellaneous materials are also utilized in nest construction. Generally speaking, nest construction is more common in freshwater than in marine species.

Fishes that bear young are either external or internal bearers. External bearers include mouth brooders, pouch brooders, gill chamber brooders, forehead brooders, skin brooders, or brooders that transfer offspring between one individual and another. Fertilization may be external or internal depending upon the taxon. Internal bearers have internal fertilization. Oviparous fishes deposit egg cases on the substrate; these eggs hatch externally. Ovoviviparous fishes retain fertilized eggs until they hatch and then release offspring "live." Viviparous fishes retain fertilized eggs that, as embryos, develop internally and are also released live. There are two forms of this strategy. The first is yolk-sac viviparity, in which the egg's yolk sac is attached to the digestive system of the developing embryo. The second is placental viviparity, in which a placental connection between the mother and the developing embryo occurs. Not all internally fertilizing fishes have internal bearing, however. For example, glandulocaudine characins (family Characidae) and many catfishes of the family Auchenipteridae have internal fertilization but are egg scatters. Similarly, members of the genus *Campellolebias* (family Rivulidae) have internal fertilization but are egg hiders.

One specialized form of live bearing is parthenogenesis, in which young are produced by a female without the fertilization of eggs by males. There are two forms, gynogenesis and hybridogenesis. In gynogenesis, an egg is activated following mating with a male of another species, but no fertilization of that egg occurs. The egg develops within its mother and is born as a female identical to the mother. In hybridogenesis, mating with a male of another species also occurs, but the egg is fertilized. The male's genetic component is discarded and the egg develops into a female identical to the mother.

Fishes select sites for reproduction in a variety of ways. For instance, salmons, trouts, and charrs make short or long-distance migrations to spawning habitat. Spawning occurs in gravel beds or other suitable substrates and includes the preparation of a redd, a depression made into the substrate where eggs are deposited, fertilized, and buried. In other fishes, nest sites may be placed within the territory of a male (or less commonly, a female) and are defended after spawning takes place between the nest owner and one or more mates. Nest sites may also be located within a home range of a temporary or permanent pair of fishes. There the eggs are deposited and fertilized, after which the site is abandoned. In some species, there is no nest at all, and eggs are merely broadcast over a suitable substrate. Spawning sites of pelagic species follow similar rules. Sites may be within a territory or cluster of territories, and are defended against intruders (usually same-sex rivals). Sites may also be used on a regular basis, due to some physical feature that may favor pelagic spawning, but are not defended.

Parental care of eggs and offspring is most often practiced by males than females, and is more commonly seen in freshwater rather than marine species. In some species of African cichlids, such as members of the genus *Lamprologus*, care is provided by helpers that are usually related to the parents and offspring. These helpers forego the opportunity to breed at this time, but manage to realize some sort of evolutionary fitness by learning parenting behaviors and by protecting offspring that carry a fraction of their own genes.

Care behaviors provide defense and maintenance of offspring. Parents protect the eggs and larvae in their mouths, brood pouches, or other structures, but also attack intruders that attempt to prey upon them. Defense also includes herding or shielding larvae or post-larvae from attacks. Maintenance behaviors include blowing on eggs to provide them with oxygen and to remove detritus or other undesirable objects (including dead eggs). Parents also provide alternative food sources for growing larvae. For example, some cichlids secrete a skin mucous that provides nutrition for their young, who ingest this by "glancing off" the flanks of their parents where the mucous is deposited. Intertidal species may wrap their bodies around an egg mass to shield it from desiccation dur-

ing low tide. A freshwater tetra, *Copella* sp. (Lebiasinidae) from South America, lays its eggs on the leaves of overhanging terrestrial plants to avoid predation, and splashes water on the egg mass to provide it with oxygen and to prevent desiccation. Electric eels and some bagrid catfishes produce infertile trophic eggs to feed their free-swimming young.

A number of fishes practice alternative mating tactics that exploit certain behaviors to a strategic advantage. For example, in mating systems with paired spawning, the pair consists of a dominant or parental male and a female. Other, smaller, males in the vicinity, known as satellite males, mimic females both in color and behavior. The satellite males approach a mating pair and, if successful at "fooling" the male, will not be chased off. Then the intruder inserts itself into the dominant male-female courtship bout and attempts to fertilize at least some of the female's eggs. In other cases, smaller "sneaker" males use stealth to approach a mating pair, then quickly streak or dart into the pair's spawning bout as eggs are released to fertilize a portion of them. Among nesting fishes, both tactics have been observed in the bluegill sunfish (*Lepomis macrochirus*; Centrarchidae), whereas the latter has been seen in salmons.

In pelagic spawning fishes, smaller satellite males will streak into the water column to join a pair during their spawning ascent and fertilize part of the female's eggs. This has been observed in a number of groups, and variations on the theme occur. Two independent teams of researchers studying the reproductive behavior of the "haremic" Japanese sandperch (*Parapercis snyderi*; Pinguipedidae), observed repeated sneaking behavior by dominant males from neighboring mating groups instead of by satellite males. One team found that these males sneak fertilizations in neighboring groups after spawning with all the females in their own groups had been completed for the night. The other research group also observed this pattern, but found that dominant males temporarily abandoned courtship with their own females and carried out "sneak" fertilizations in a neighboring group in close proximity to their own location, when the opportunity presented itself. Males also spent considerable amounts of time and effort defending their groups from sneaking neighbors. The downside of these behaviors was that the opportunity to mate with their own females was lost and the females mated with other males while their males were busy.

Lizardfishes (*Synodus dermatogenys*; Synodontidae) have two strategies that depend upon local population size and sex ratio. If the population is relatively low, and the numbers of female and males are approximately equal, paired courtship and spawning occurs. If, however, the population size is larger and males outnumber females, then a different strategy prevails. In this case, males form floating leks at sunset to display to larger females as the latter move about the spawning site. One male may be more successful and joins the female as she rises into the water column to release her eggs. As the female and male ascend, however, they are joined by other males, who all contribute sperm toward the fertilization of the eggs. Females do appear to exercise control over the timing of the spawning and the composition of the group. Unlike females of many other pelagic spawning reef species, female lizardfishes may spawn more than once during an evening. As courtship continues, the female rises to spawn again and is joined by more than one male. If, however, only one male joins her, she will abort the spawning ascent, return to the bottom, and wait for courtship from the group of males to resume. Females seem to pursue this strategy of group spawning in order to assure complete fertilization of their egg masses. The male closest to her in an ascent, the one she chooses from the others as the most attractive, will likely fertilize a significant proportion of her eggs, but the contributions of the other males may promote both genetic diversity and completion of the job!

Feeding behavior

Fishes use a variety of behaviors while feeding to detect and capture prey, extract plant materials, or to sift sediments to extract objects of nutritional value. Planktivory by fishes occurs during the day and at night. During daylight, hovering planktivores such as fairy basslets (Serranidae: Anthiinae), some freshwater sunfishes, damselfishes, angelfishes, certain triggerfishes (Balistidae), and many other species, feed upon zooplankton or phytoplankton by plucking individual plankton out of the water column. Plucking is accomplished with specialized mouthparts or by rapid movements of the mouth. The mola (Molidae), and other species that feed upon macroplankton or mesoplankton in the water column, forage in a similar fashion. Garden eels (Congridae) and other burrow-dwelling planktivores emerge partially from their burrows to feed upon plankton that drift past in the current. Schools of fusiliers (Caesionidae) dart erratically in the water column as they grab and feed upon plankton they detect there. Alternately, fishes such as whale sharks (Rhyncodontidae), basking sharks (Cetorhinidae), manta rays (Mobulidae), herrings, anchovies (Engraulidae), scads (*Decapterus* spp.; Carangidae), and similar species open their large mouths and strain the plankton from the water column as they swim. Fishes with smaller mouths, such as reef herrings and flyingfishes (Exocoetidae), also strain plankton in the water column. At night, a new set of planktivores emerges from shelter to feed upon pelagic and demersal plankton in the water column. These include squirrelfishes and soldierfishes (Holocentridae), cardinalfishes (Apogonidae), bigeyes (Priacanthidae), along with a host of deep-dwelling species. These fishes use their large eyes and other well-developed senses to detect and feed upon plankton.

Herbivory in marine fishes is more pronounced in tropical than in temperate species. Among the tropical fishes, halfbeaks (Hemirhamphidae), sea chubs (Kyphosidae), damselfishes, parrotfishes (Scaridae), blennies (Blenniidae), rabbitfishes (Siganidae), and surgeonfishes (Acanthuridae), are the prominent groups. In freshwater systems, numerous species feed upon benthic algae, emergent plants, and even seeds and fruits from terrestrial plants. Some marine groups, such as the butterflyfishes, angelfishes, filefishes (Monacanthidae), and triggerfishes, include species that are omnivorous and feed upon benthic algae. In temperate marine waters, some members of the Sparidae (porgies), Kyphosidae, Aplodactylidae (sea carps), Odacidae (rock or weed whitings), and

Pack of whitetip reef sharks (*Triaenodon obesus*) hunting at night in the Pacific Ocean, near Cocos Island off Costa Rica. (Photo by Jeff Rotman/Photo Researchers, Inc. Reproduced by permission.)

Stichaeidae (pricklebacks) are more dependent upon plant life. Freshwater omnivorous species, such as the carp and its relatives (Cyprinidae), and a number of characins (Characidae), often include plant materials as a significant part of their diet.

Herbivores feed by grazing or browsing, and are capable of learning what species of plant are edible and what are toxic. Access to edible benthic algae, sea grasses, or other plant materials may be as simple as swimming into a given area and stopping to graze or browse. Parrotfishes feed on zooxanthellic algae contained in coral skeletons by using their specialized beaks scrape algae off rocks or dead corals. They also bite off chunks of the coralline skeleton as they graze. The parrotfish's pharyngeal teeth crush the chunk within the mouth cavity, swallow and extract the algae, expel the resulting coralline sand. Territorial species, such as certain damselfishes known as "farmer" fishes, actively maintain patches of desirable species of algae that they tend, defend, and feed upon. Other herbivores attempting to feed upon this patch (or patches of algae not tended, but within the territory of any herbivorous species) are turned away. Fishes defending the patches must be overwhelmed before other herbivores can gain access to this resource. One way to accomplish this is for the grazing or browsing species to form schools that can move into a territory and easily outnumber the defender. One form of this type of school, the heterospecific or mixed-species shoal or school, consists of fishes of a number of species (including nonherbivorous fishes) that move about the bottom. Membership in the school is temporary, and its members not only gain protection from schooling but, more importantly, are able to breech the defenses of a territorial herbivore to feed upon the algae it attempts to protect.

Feeding by herbivorous monospecific schools takes place both at night and during the day. For example, unicornfishes (Acanthuridae) form schools that graze on sea grass flats at night. Their efficiency at feeding upon sea grass and benthic algae there is analogous to a giving the flat a good haircut.

Predators rely heavily upon one or more sensory systems in their search for prey. Among benthic species, ambush predators such as scorpionfishes (Scorpaenidae), flatheads (Platycephalidae), and sculpins (Cottidae) use vision to detect passing prey. Scorpionfishes and many sculpins remain motionless until they detect prey and estimate the distance to it. The predator then engulfs the prey by rapidly opening its mouth and sucking it in. Flatheads grab the prey with a mouthful of teeth, and manipulate and swallow it. Other ambush predators, such as lizardfishes or hawkfishes, launch themselves into the water column or down to the substrate to grab prey. Freshwater pikes and pickerels (Esocidae), groupers, basses (Centrarchidae), the Murray cod (Percicthyidae), and the barramundi (Centropomidae) utilize structures, such as submerged trees and stumps or weed beds, to mask their presence as they ambush passing prey. Their acute vision allows various trout species in streams to identify and assess aquatic insects carried by surface and subsurface currents. At night, the enlarged eyes of many nocturnal predators allow them to detect, track, and attack prey on the bottom or in the water column.

Other sensory systems, such as taste, touch, and chemoreception, allow fishes that prey on benthic invertebrates and smaller fishes, such as catfishes (various families), goatfishes, threadfins (Polynemidae), freshwater eels (Anguillidae), and moray eels (Muraenidae), to detect prey buried just beneath the surface of the substrate. Similarly, electrical receptors known as ampullae of Lorenzini allow elasmobranches, such as sharks and rays, to detect minute electrical currents generated by prey buried beneath sand, rubble, or mud; locate their position; and feed upon them. Knifefishes (Gymnotiformes) of South America and the Mormyridae of Africa use other organs to generate weak electric fields that allow them to detect prey in murky water.

In the pelagic realm, vision, olfaction, touch, and sound detection are important sensory components of predatory behavior. Swiftly moving predators, such as tunas (Scombridae) and billfishes (Istiophoridae), rely upon keen eyesight to track and hunt their swiftly moving prey. Predators, especially sharks, also rely upon the smell of prey, and in particular, the smell given off by injured prey, to detect them. Low frequency sounds generated by the movement of prey, whether swimming or struggling, are detected by the predator's lateral line system. These vibrations are felt, rather than heard, by the lateral line receptors. Higher frequency sounds generated by prey are detected by the inner ear and direct it to the location of the prey.

In deepwater environments, visual capabilities may be reduced in favor of other senses in some species, but accentuated in others. Deepwater predators with large eyes can detect bioluminescent flashes generated by potential prey. Chemoreception, hearing, and lateral line senses are also important in prey detection in waters where darkness prevails and little or no light penetrates from above.

The pursuit of prey varies among predatory species. Ambush predators frequently employ camouflage and position themselves among rocks, corals, or marine plants to conceal themselves until they can detect and ambush prey. Some ambush predators use lures fashioned from modified fin rays or other body parts. These lures, which may resemble a small fish or invertebrate in shape, are waved about to attract the prey. This behavior, sometimes referred to as aggressive mimicry, occurs in a few shallow and deepwater predatory families of fishes. In deepwater species, the lure may be bioluminescent. Other predators, such as juvenile snappers (Lutjanidae) or hamlets (Serranidae), mimic nonthreatening species such as damselfishes to get closer to prey. Predators, such as trevallys (Carangidae), often rapidly patrol the bottom throughout the day and night in search of prey that cannot retreat to shelter quickly enough to escape being eaten. These predators may also attack schooling baitfish species in the water column singly, in pairs, groups, or schools. Barracudas (Sphyraenidae) rest motionless in the water column, then strike rapidly, sometimes over a distance of several meters, at an unwary fish in the water column or on the bottom. Schools of salmon, striped basses (Moronidae), amberjacks (Carangidae), bluefishes (Pomatomidae), and tuna (Scombridae) rapidly chase and slash schools of fleeing baitfishes, and may also herd them while attacking. Other species, such as some barracudas and lionfishes (Scorpaenidae), hunt in packs and often utilize structures, including reef and cave walls and even suspended fishnets, to act as barriers to aid in escape. At sunset, a pack of lionfishes assembles and gathers near a school of sweepers (Pempheridae) that is emerging for the night. Then the lionfishes extend their large pectoral fins and use them to push the sweepers into an increasingly small aggregation that is ultimately trapped between the pack of lionfishes and the reef wall. The predators then rapidly inhale the sweepers with their large mouths as they try to escape. Other predators, such as groupers and large soapfishes (Serranidae), take advantage of this behavior to ambush escaping sweepers.

Many species detect prey hidden in bottom sediments and then dig or sift them out. A number of these predators have specialized teeth, mouths, or gills that allow them to do this; others fan the sediments to expose the prey. Some fishes, such as certain wrasses (Labridae) and trevallys, allow other fishes, such as stingrays and goatfishes, to do the digging for them. These predators follow the bottom-foraging species and feed opportunistically upon whatever may be exposed. Barracudas have been observed hanging motionless in the water column above nests of large triggerfishes (*Pseudobalistes* spp.). The triggerfishes constantly tend these nests by rearranging the substrate and turning over rocks, and in doing so they expose or startle prey fishes or other organisms. When this happens, the barracudas quickly rush down to the exposed prey and grab it before the triggerfish can respond.

Pelagic deepwater species ascend the water column as night falls and may rise hundreds of meters toward the surface as they track their prey. Some species are following the similar movements of pelagic invertebrates upon which they feed. Others track these fishes as prey, and some species, such as the snake mackerels or oilfishes (Gempylidae), have feeding behaviors that appear to be similar to shallow water species, such as barracudas (Sphyraenidae) or wahoos (Scombridae).

A number of predators employ mechanisms or behaviors that allow them to stun or otherwise immobilize their prey before they eat it. For example, torpedo rays (Torpedinidae), electric eels (Electrophoridae), electric catfishes (Malapteruridae), and some stargazers (Uranoscopidae) discharge a strong electrical current that stuns passing prey. Some wrasses (Labridae) smash their invertebrate prey with or against rocks. Archerfishes (Toxotidae) stun and knock down insect prey resting on mangrove branches or emergent grasses above the water surface by shooting a stream of water "bullets" at them. Billfishes and the swordfish (Xiphiidae) utilize their bills to herd, stun, spear, or slash their prey before eating them. Hammerhead sharks use their unique cephalic lobes to pin down stingrays, a favored prey. Sharks and barracudas use sharp teeth to capture and cut prey into smaller pieces before they are eaten. Sharks also take bites out of larger prey, such as whales, without capturing them. Relatively diminutive fishes, such as the poison-fang and fang blennies also practice this behavior. These small predators hover in the water column and launch themselves at passing fishes to bite off a scale or small piece of flesh. Fishes in the genus *Aspidontus* mimic bluestreak cleaner wrasses (*Labroides dimidiatus*), approaching fishes that may be fooled into thinking they will be cleaned, but end up being bitten. Juveniles of larger predators, such as the leatherback, *Scomberoides lysan* (Carangidae), also engage in this behavior. Scale-eating or biting off small pieces of flesh is also practiced in tropical freshwater fishes of Africa and South America. Specialized genera of scale-eating African cichlids (Cichlidae) include *Perridodus* and *Plecodus* from Lake Tanganyika, and *Corematodus* and *Genyochromis* from Lake Malawi. Fin-eating is practiced by characoid fishes of the families Citharinidae (genera *Ichthyoborus*, *Mesoborus*, and *Phago* from Africa) and Characidae (subfamily Serrasalminae, the piranhas and their relatives; genera *Catoprion* and *Serrasalmus* from South America).

Cleanerfishes, including the cleaner wrasses *Labroides* and *Labropsis* (Labridae), cleaner gobies (*Gobiosoma*), and some butterflyfishes establish cleaning stations along coral or rocky reefs, where they attract and clean the "client" fishes that approach. The cleaners swim in a regular pattern of movements. The client fishes, responding to the swimming behavior and distinctive color pattern of the cleaner, as well as to the location of the cleaning station, assume a posing posture to signal that cleaning is required and may begin. The cleaners then forage along the client's body, feeding on parasitic copepods and other parasitic organisms, along with any damaged tissues. Cleaner wrasses and gobies also enter the mouths and gill cavities of their clients, including large predators like groupers and moray eels, without being preyed upon. This type of behavior is a mutualistic symbiosis, because both the cleaner and the client benefit. Cleaning behavior has also been observed in freshwater fishes, including some members of the family Cichlidae. A pelagic variation of this behavior is practiced by remoras or sharksuckers (Echeneidae). These fishes attach themselves to large predators, such as sharks, billfishes, turtles, and whales, and hitchhike as their hosts swim. In turn, the remoras feed upon parasitic copepods on the host, but will also take advantage of scraps left over from the host's feeding bouts. This behavior is a commensalistic symbiosis, in that the cleaner gains while the client neither benefits or loses out.

Parasitism by marine fishes is not a common strategy, but it does occur in a number of different species. Internal parasites include the pearlfishes (Carapidae) that live in the gut cavities of seacumbers and large starfishes, where they feed upon gut tissue. The eel-like *Simonchelys parasitica* (Synaphobranchidae) has been found burrowed into the flesh of various bottom-dwelling fishes, but has also been recorded from the heart of a mako shark, a fast-swimming pelagic species. In freshwater fishes, however, catfishes of the tropical Amazon family Trichomycteridae are parasitic. Most species attack gill tissues of larger fishes, but members of the genus *Vandellia* are known to parasitize the urethra of mammals, including humans, causing considerable harm. Lampreys (Petromyzontidae) are external parasites, which attach themselves to the skin of their prey and feed upon tissue and body fluids with their specialized mouths.

Fishes that feed upon detritus use their mouths to scoop up sediments, from which they extract detrital materials with their pharyngeal teeth and expel sediments through their gills. Scavengers feed upon dead and dying fishes or other organisms and play an ecological role. Fishes bite or peck at the body of the organism to remove chunks of flesh. Some fishes, such as the deepwater hagfishes (Myxinidae), are specialized to enter the body cavity of dead fishes to feed upon them internally.

Predator avoidance behavior

Marine fishes have evolved a number of mechanisms and behavioral strategies to avoid predation. These include color patterns and modifications of body structure to provide camouflage, mimicry, or warnings of toxicity. Color patterns that disrupt the fish's outline, reduce its contrast against background coloration, or allow it to blend into the background all provide camouflage and protection from predation. These same attributes also favor predators who wish to hide from their prey. Countershading (dark color dorsally and pale or white ventrally) and reverse countershading (pale or white dorsally and dark ventrally) obscure the fish when it is viewed from above or beneath. A silvery or mirrorlike coloration, as seen in herrings, tarpons (Megalopidae), ladyfishes (Elopidae), smelts (Osmeridae), carps and minnows, and mullets (Mugilidae), reflects light and confuses potential predators. Fishes that are relatively transparent, such as in the glassfishes (Channidae) and some cardinalfishes, are difficult to see, especially in low light conditions. Similarly, modifications to the skin, fin rays, or other portions of the body can convey similar benefits. For example, the sargassumfish (Antennariidae) has modifications, which, in conjunction with its greenish brown coloration and hovering behavior, allow it to resemble sargassum algae. Pipefishes and seahorses (Syngnathidae), and some filefishes, have similar adaptations. Adult leaffishes (Nandidae) and juveniles of a number of species, including spadefishes and batfishes (Ephippidae), combine a color pattern with behavior to resemble inedible objects such as leaves.

Warning coloration informs potential predators that a fish is (or is giving the impression of) being toxic and should be ignored. The juveniles of some species of sweetlips (Haemulidae) have color patterns and behaviors that allow them to mimic toxic nudibranches and flatworms. Other fishes with toxins in their skin or organs, such as various tobys or sharpnose puffers (*Canthigaster* spp.; Tetraodontidae), have color patterns that advertise their toxicity. One species of filefish, *Paraluteres prionurus*, is nonpoisonous, yet is afforded protection from predation because it has a color pattern and behavior that exactly mimics the black saddled toby (*Canthigaster valentini*), and, to a lesser extent, the crowned toby (*C. coronata*).

Behavioral responses to perceived threats detected by senses such as vision, hearing, the lateral line, and smell may be rapid or subdued. These responses are used by solitary individuals but are accentuated in aggregations or schools. Many species react swiftly to the sight of an approaching predator or to the detection of noises or pressure waves generated by its approach. Flight avoidance, usually in the form of rapid or erratic swimming away from the predator, occurs in the water column. Some species, such as flyingfishes, halfbeaks, and needlefishes (Belonidae), leap above the surface of the water and may coast for a few to several meters in the air before entering it again. As a school or aggregation, reef herrings and other baitfishes leap repeatedly into the air as they flee approaching predators. Schooling behavior provides significant predation avoidance benefits because of the fact that there is safety in numbers. Most predators have to target a single individual to successfully prey upon it. The presence of many individuals within a school means that, with more prey to choose from, the chances of any one healthy individual of being preyed upon is reduced as the school takes flight.

Fishes with less energetic responses than the reef herrings, such as lionfishes and other scorpionfishes, merely erect pectoral and dorsal fins that have poison-tipped spines to detract predators. Other fishes with poisonous spines show a similar behavior. Adult pufferfishes (Tetraodontidae) and porcupinefishes (Diodontidae) inhale water and inflate their bodies to avoid predation by all but the largest predators. These fishes may also be poisonous or have spines that make ingesting them difficult. Other species, such as garden eels (Congridae), sanddivers (Trichonotidae), tube blennies, burrowing gobies, and triggerfishes duck into holes or burrows. Triggerfishes can "lock themselves in" their holes by extending their dorsal and anal fin spines. Similarly, soles (Soleidae), flounders (Bothidae), and their relatives bury themselves in the sand. At night, sleeping parrotfishes wedge themselves into the reef and secrete a somewhat gelatinous cocoon that allows it to detect predators by touch if the cocoon is violated.

Another behavior normally attributed to birds but also observed in some species of reef and freshwater fishes is mobbing. Mobbing serves to ward off an intruding predator by displaying to or attacking it in number, but also to warn conspecifics of the intruder. For example, the damselfish *Pomacentrus albifasciatus* maintains territories adjacent to conspecifics on reef flats. If a predator, such as a moray eel or scorpionfish, enters the area, those damselfishes closest to the intruder rise into the water column and display their erect fins as they dance in front of, behind, or along the flank of the intruder. These damselfishes are soon joined by others from the territorial matrix, and together they mob the intruder until it leaves the area. This damselfish also appears to be able to recognize which kind of danger the intruder may pose before mobbing it. If the intruder is a scorpionfish, the damselfishes avoid displaying near its head and focus their attacks along the posterior flank or tail, presumably because scorpionfishes have large mouths and are capable of extremely rapid feeding attacks. However, if the intruder is a moray eel, the damselfishes also mob in the region of its mouth because it is relatively smaller and the eel's feeding behavior is different.

Behavior, evolution, and conservation

The behavior patterns in fishes evolved over countless generations, largely in response to pressures from natural and sexual selection. These behavior patterns are distinct and measurable, and they are just as much characteristics or traits as those based upon morphology or biochemistry. Although general patterns among different, and often quite divergent, species exist, the applications of these patterns and their subtle differences are often unique. Similarity in patterns may be a function of convergent evolution, common descent among related groups, or an affinity among species. Differences in patterns may be the result of a lack of affinity or may be subtle changes in application within a group of closely related species. The study of phylogenetic relationships among species by the comparative method provides an understanding of the patterns of behavior observed, as well as the processes that underscore their evolution within species. This field of study, known as historical ecology, has great utility in ascertaining patterns of character development between and within species, and has predictive power in instances where information on a particular species within a group of related species is relatively lacking. This method also allows for testing hypotheses that may confirm the validity of the prediction.

The methods of historical ecology in the study of fish behavior have considerable utility as tools for predicting how fishes behave under exploitation, habitat destruction, or other problems addressed in the science of conservation biology. Conservation efforts, especially in highly diverse systems, are often stymied because of a lack of information on the biology and behavior of some fishes within those systems. As detailed studies of most groups are lacking, some considerable effort has been expended upon understanding the biology and behavior of a relatively few species within these systems. Generalizations are feasible for attempting to understand species that are less well studied. However, only the steady collection of data coupled with the predictive advantages of historical ecology allows scientists to understand the larger picture so they can convey to fishery managers the information necessary for the design and implementation of effective conservation and management plans.

Resources

Books

Briggs, J. C., and J. B. Hutchins. "Clingfishes and Their Allies." In *Encyclopedia of Fishes*, edited by J. R. Paxton and W. N. Eschmeyer. San Diego: Academic Press, 1998.

Brooks, D. R., and D. A. McLennan. *Phylogeny, Ecology, and Behavior*. Chicago: University of Chicago Press, 1991.

Donaldson, T. J. "Assessing Phylogeny, Historical Ecology, and the Mating Systems of Hawkfishes (Cirrhitidae)." In *Proceedings of the 5th Indo-Pacific Fish Conference, Noumea 1997*, edited by B. Seret and J. Y-Sire. Paris: Societé Française, Ictyologie, 1999.

Helfman, G. S., B. B. Collette, and D. E. Facey. *The Diversity of Fishes*. Oxford: Blackwell Science, 1997.

Moyle, P. B., and J. J. Cech Jr. *Fishes: An Introduction to Ichthyology*. 3rd edition. Upper Saddle River, NJ: Prentice-Hall, 1996.

Myers, R. F. *Micronesian Reef Fishes*. 3rd edition. Barrigada, Guam: Coral Graphics, 1999.

Pitcher, T. J., ed. *The Behaviour of Teleost Fishes*. London: Chapman and Hall, 1993.

Potts, G. W., and R. J. Wootton, eds. *Fish Reproduction: Strategies and Tactics*. London: Academic Press, 1984.

Thomson, D. A., L. T. Findley, and A. N. Kerstich. *Reef Fishes of the Sea of Cortez*. 2nd edition. Tucson: University of Arizona Press, 1987.

Thresher, R.E. *Reproduction in Reef Fishes*. Neptune City, NJ: T.F.H. Publications, 1984.

Wootton, R. J. *Ecology of Teleost Fishes*. London: Chapman and Hall, 1990.

Periodicals

Balon, E. K. "Reproductive Guilds of Fishes: A Proposal and Definition." *Journal of the Fisheries Research Board of Canada* 32 (1975): 821–864.

————. "Additions and Amendments for the Classification of Reproductive Styles in Fishes." *Environmental Biology of Fishes* 6 (1981): 377–389.

Donaldson, T. J. "Mobbing Behavior in *Stegastes albifasciatus* (Pomacentridae), a Territorial Mosaic Damselfish." *Japanese Journal of Ichthyology* 31 (1984): 345–348.

————. "Lek-Like Courtship by Males and Multiplespawnings by Females of *Synodus dermatogenys* (Synodontidae)." *Japanese Journal of Ichthyology* 37 (1990): 292–301.

————. "Mating Group Dynamics and Patterns of Sneaking by Dominant Male Sandperches, *Parapercis snyderi* (Pinguipedidae)." Unpublished manuscript in review.

Lassuy, D. R. "Effects of 'Farming' Behavior by *Eupomacentrus lividus* and *Hemiglyphidodon plagiometapon* on Algal Community Structure." *Bulletin of Marine Science* 30, Special Issue (1980): 304–312.

Thresher, R. E. "Clustering: Non-Agonistic Group Contact in Territorial Reef Fishes, with Special Reference to the Sea of Cortez Damselfish, *Eupomacentrus rectifraenum*." *Bulletin of Marine Science* 30, Special Issue (1980): 252–260.

Terry J. Donaldson, PhD

• • • • •

Fishes and humans

Overview

Fishes have figured prominently in human lives, cultures, and economies since ancient times. Fish themes appear in diverse aspects of human culture, including mythology, religion, literature, and art. In addition, fishes are important to humans as a source of food and income; thus, the quest for fishes played a large role in historical patterns of exploration, settlement, and even war. Fishes are the central focus of recreational activities enjoyed by many as well. Despite their value to humans, fishes are often negatively affected by the direct and indirect consequences of human actions. As a result, many fish species are threatened or endangered, and some have become extinct in recent years.

Fishes in human culture

Mythology and religion

Throughout history, fishes appeared in the legends, myths, and folklore of a variety of cultures. In many societies, fishes were associated with deities, perhaps indicative of the value and mystery of fishes in ancient cultures. In Iran and Babylon, archeological evidence revealed a deity with human legs covered by the full body of a fish. In Syrian culture, the mythical goddess of generation and fertility, Atargatis, was represented as the body of a woman with a tail of a fish—a depiction that gave rise to the image of mermaids. The ancient Egyptian deities Isis and Hat-Mehit were associated with fishes; due to the abundance of fishes available during the spawning season in the Nile Valley, these goddesses symbolized fertility. Many other indigenous cultures recognized deities that were believed to protect fish stocks and those persons that harvested fishes.

In some myths, fishes interact with deities in other ways. For example, two fishes derived from Greek mythology are visible in the sky each night—those of the constellation Pisces. According to this myth, Aphrodite, the goddess of love and beauty, was walking along the Euphrates River with her son, Eros, when they encountered the monster Typhon. One story suggests that Aphrodite and Eros jumped into the river, where they were transformed into fishes and fled. In another version of this myth, two fishes carried the mother and son to safety. Both versions imply that fishes can confer protection to deities.

Fishes continue to serve as important symbols in modern Christianity. The Greek word for fish, *ichthys*, is derived as the acronym for the biblical phrase *Iesous Christos Theou Hyious Soter*, which translates to "Jesus Christ, Son of God, Savior." The activity of fishing plays a central role in a variety of encounters between Jesus and his disciples, as he instructs them to be "fishers of men." It is believed that the fish symbol formed by two half-crescents arose as a way for persecuted Christians to identify themselves to one another in ancient Rome; this symbol remains widely used by Christians today and has been adopted in modified forms by those advocating alternative beliefs to certain Biblical teachings.

Literature and art

Many legends in classical and medieval literature convey tales of fishes as monstrous sea creatures that invoke fear into even the bravest humans. In his epic Roman poem *The Pharsalia*, Lucan suggests that large remoras, or shark suckers, could impede the progress of sailing and naval ships. Other stories tell of menacing sea serpents; some legends, such as that of the Loch Ness Monster, are perpetuated still today. In reality, many of the legendary "sea serpents" turned

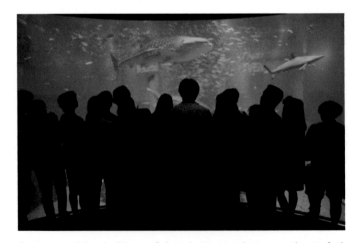

From a small bowl with one fish to huge aquariums, people are fascinated by the underwater world. This huge fish tank is at the Osaka Aquarium in Osaka, Japan. (Photo by Will & Deni McIntyre/Photo Researchers, Inc. Reproduced by permission.)

out to be nothing more than large eels, cuttlefish, squid, or sharks, and others were dismissed as figments of the imagination created by the interplay of light and water. Initially regarded as a folk legend, reports of "rains of fishes" date back to the third century A.D. Such incidents have been reliably documented in more recent years, and it is believed that violent storms may sweep fishes from water bodies and drop them back to land as air currents dissipate.

Fishes have been depicted in artwork throughout cultural history as well. Hieroglyphics of the ancient Egyptians show precise details of many fishes from the Nile River. Many of these carvings remain preserved on the walls of tombs, as fishes were believed to lead people to and sustain them in the afterlife. Native Americans of the Northwest carve fishes, specifically salmon, on totem poles, thereby conveying myths and spiritual connections of their societies to fishes. Fishes have long served as popular subjects in Asian art, particularly that of the Chinese, through which they are often displayed on pottery, screens, and paintings.

Fishes continue to appear in many forms in literature and media of modern culture. Novels such as Ernest Hemingway's *Old Man and the Sea* and nonfiction works such as Sebastian Junger's *Perfect Storm* recount the challenges and rewards of the pursuit of fishes, while childhood stories such as Dr. Suess's *One Fish, Two Fish, Red Fish, Blue Fish* are widely recognized and adored. In addition, movies periodically revitalize myths and fears of fishes. In the movie *The Little Mermaid*, the myths of mermaids, serpents, and sea gods again come to life. In contrast, *Jaws* preys upon general unfamiliarity with shark behavior to instill viewers with fear of these dominant ocean predators.

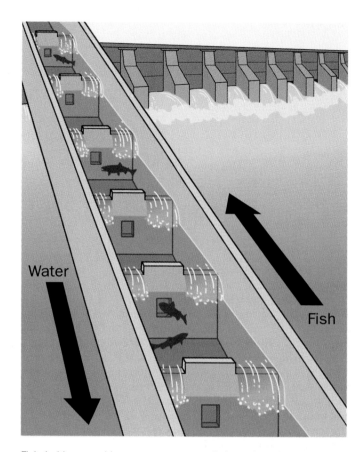

Fish ladders provide passageway around dams for migratory adults headed up-river for breeding, such as salmon. (Illustration by Amanda Humphrey)

Human uses of fish

Historical fisheries

Fishes have been utilized by humans throughout history for food, income, and other purposes. Archeological records indicate that Egyptians exploited fishes in the Nile River from prehistoric times. Carvings record the types of fishes caught, fishing techniques, preparation methods, and the trade of fishes. The Egyptians used spears, hooks, weirs, and nets to capture fishes in the wild, but residents of Mesopotamia constructed ponds in the fertile crescent of the Tigris and Euphrates rivers to maintain regular, accessible supplies of fishes. Pliny the Elder recommended fishes as medicinal remedies for a variety of ailments. Still today, cod-liver oil and castor oil are used to relieve internal and external ailments; and in some parts of the world, otoliths, the small ear bones of fishes, are believed to prevent colic.

The quest for fishes played a large role in early exploration and settlement patterns. Access to herring in the Baltic Sea conferred prosperity upon the Hanseatic League during the twelfth century. Disputes over fishing rights and profits led to conflicts, sometimes escalating to war, that continued among European nations long after the demise of the league. By the fourteenth century, Europeans traversed the sea to fish for cod off of Iceland. They preserved the fish in dried or salted forms to supply European markets, thereby rendering it a valuable commodity, and disputes over access to cod fishing grounds sparked further wars. Cod constituted a major portion of the diet and trade base of later British colonists that explored and settled on the northeast coast of the United States. By the late 1600s, the international cod trade involved other commodities as diverse as salt, sugar, molasses, cotton, tobacco, and even slaves.

Modern fisheries

Fisheries continue to provide a vital source of food, employment, and income in many countries today. World fisheries catch increased rapidly during the 1950s and 1960s. Following the collapse of the Peruvian anchoveta fishery in the 1970s, total catches appeared to level off, but worldwide catch began growing again in the 1980s. In 2000 marine fisheries captured nearly 95 million tons, and aquaculture production added another 35 million tons of fishes that were consumed by humans in some form—most commonly as food, oil, or fertilizer. Despite the value and varied uses of fishes, relatively few species dominate the catch. In 1997 marine fisheries exploited 186 species, but seven species accounted for 50% of the total biomass harvested.

A variety of gear types are used to catch fishes in modern commercial endeavors. Trawls (nets) may be towed along the ocean bottom to target demersal fish (bottom-dwellers) such

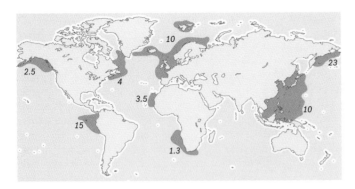

Fishery centers—catches are given in millions of tons (numbers from 2000). (Map by XNR Productions. Courtesy of Gale.)

as cod, or suspended in the water column to catch pelagic species such as herring. Purse seines surround migratory schools of fishes such as tuna. And hooks on long lines—baited lines stretched sometimes for miles through the water column—capture large predatory species, including sharks and swordfish. A variety of diverse techniques are utilized by many people throughout the world to harvest fishes for subsistence or local consumption. Much fishing occurs near shore using traps, nets, and spears. However, as fish populations are depleted, particularly in areas with few alternative sources of protein, humans resort to small mesh nets or destructive fishing techniques, including dynamite fishing, to capture as many remaining fishes as possible.

Recreation

Many people throughout the world enjoy recreational activities centered around fishes. Recreational anglers pursue fishes for sport in freshwater, coastal, and deep sea settings. In the United States alone, the number of anglers is estimated at over 50 million. Many of these anglers enjoy the sport for the thrill of hooking and landing a feisty fish; since the catch is of secondary importance, anglers often release their catch to reduce mortality effects on the fish population.

Another recreational pursuit involving fishes is that of maintaining aquariums. The idea of the aquarium was first developed in the mid-1800s, and the first public aquarium opened at the London Zoological Gardens in 1846. Today, the keeping and breeding of ornamental fishes is a popular hobby worldwide. Many individuals maintain smaller, personal aquariums, ranging from a single goldfish in a bowl to elaborate tropical marine tanks. This activity was the third most popular hobby in the United States in 2001. Worldwide, it generates millions of dollars of economic activity.

Conservation of fishes

Despite their value to our society, fishes suffer the direct and indirect consequences of human actions. Habitat alteration including inadequate water supplies and declining water quality, overharvesting, and introduced species all threaten the viability of fish populations. Over time, many humans have recognized the negative impacts of certain actions and initiated efforts to ameliorate the consequences of these activities.

Habitat alteration

Human actions alter fish habitats in a variety of ways, often resulting in the decline of fish populations. Dredging of channels and clearing of debris to maintain navigation in waterways removes substrates and rich food sources that fishes prefer. In addition, the loss of wetlands eliminates spawning and nursery habitats for numerous fish species. In some cases, wetlands are destroyed to create urban living space or agricultural land; in other instances, tapping into groundwater supplies lowers the water table so that wetlands are no longer inundated. While hydroelectric power provides energy to many areas, the construction of dams across rivers substantially alters the nature of flowing-water habitats and limits upstream migrations of fishes. Upstream of the dam, the river essentially becomes a lake, and dam operations control downstream water flows. Further, despite the fact that most large dams are now equipped with fish ladders to facilitate upstream migrations, significant declines in the populations of many fish species are commonly noted following dam construction. Indirect human effects on fish habitat are also important. In many tropical regions, deforestation can destroy a watershed's ability to store seasonal rainfall and release it slowly over an extended period of time. Such alterations to hydrological patterns can induce extreme cycles of flood and drought in streams that result in catastrophic consequences for the local fish fauna.

Humans consume large amounts of water for drinking and irrigation, and much of the water used is obtained from rivers and lakes inhabited by fishes. In many regions, reservoirs of dams offer convenient holding areas from which water can be withdrawn for human uses. In other settings, streams and rivers are diverted to cities for human consumption or to agricultural areas for irrigation. Either scenario results in less water to maintain natural flow patterns in rivers and streams; thus, the flooded habitat area available to fishes shrinks, many species move out of the area or become locally extinct, and

Not only do we like to watch fish swim, we sometimes join them underwater. Here, a diver swims with a lemon shark (*Negaprion brevirostris*). (Photo by Jeff Rotman/Photo Researchers, Inc. Reproduced by permission.)

An appetizer plate holds three kinds of caviar: golden whitefish, keta salmon, and American sturgeon. (Photo by Macduff Everton/Corbis. Reproduced by permission.)

others suffer mortality from crowding and disease. The dramatic collapse of the Aral Sea ecosystem after two major tributaries were tapped for agricultural irrigation in the 1980s demonstrates the potential consequences of water diversion on aquatic systems and their fish faunas. This problem is exacerbated in areas that draw their water from underground aquifers. Water held in aquifers is often extracted at rates that exceed natural replenishment. Mining these aquifers for human use can lower the water table sufficiently to completely eliminate aquatic habitats that are dependent on artesian spring flows, a situation of particular concern in arid regions of the southwestern United States and northern Mexico. As water shortages become more prevalent throughout the world and the human demand for water escalates, the severity of threats to fish populations will likely increase, and decisions concerning the allocation of water will become more complex and contentious.

In addition to concerns about water quantity, declining water quality also has serious implications for fishes. Humans have dumped industrial wastes, agricultural chemicals, and sewage directly into water bodies for much of history, and similar discharges continue in many areas today. Indirect runoff further reduces water quality. For example, deforestation increases siltation in adjacent streams, and nutrient runoff from the watershed may cause algal blooms in the receiving lake. These activities all impair water in ways that may be harmful to fishes. Waste effluents contribute to disease; chemicals may prove toxic or impair reproductive success; and algal blooms and the decay of materials may deplete oxygen to inadequate levels for sustaining fish life.

Overfishing

Overfishing poses another substantial threat to fish populations worldwide. The pattern noted in the history of many fisheries involves discovery, high levels of exploitation until the stock collapses, and then switching to a new stock. Even in classical times, stock collapses due to overfishing were common, and as early as the twelfth century, Edward II banned the use of a specific type of trawl net in the Thames Estuary.

In recent years, we have witnessed the collapse of major fisheries, including those for Peruvian anchoveta, California sardine, and Georges Bank cod. The failure of California sardine and Atlantic cod populations to recover demonstrates that intense fishing may deplete stocks to a point that the surviving population may not be large enough to assure recovery, even if fishing effort is eliminated entirely. In addition to the demise of exploited stocks, overfishing may have consequences at the ecosystem level as well. Fishing activities may destroy habitat if inappropriate fishing gear or techniques are utilized. Further, it has been suggested that as humans deplete stocks at high trophic levels, we move to species lower on the food chain; over time, this pattern of "fishing down the food web" threatens collapse of the whole ecosystem.

Introduced species

Finally, human actions affect fish populations and communities by introducing certain species to areas beyond their native range. In many cases, the introduction to a new environment frees a species from natural controls on its population growth. The introduced species may prey upon native fishes, infringe upon habitats and food supplies used by other species, hybridize with native species and reduce genetic diversity of the stocks, or act as vectors of exotic pathogens and parasites to which native species have no resistance. Some introductions are intentional and others accidental, but both may result in severe consequences to native fish stocks. Intentional introductions often follow collapses of native fish stocks due to overexploitation. As an example, the Nile perch (*Lates niloticus*) was introduced into Lake Victoria to expand protein production and enhance fisheries after the demise of an endemic tilapia, the ngege (*Oreochromis esculentus*), due to overfishing. Over three decades, predation by Nile perch resulted in the loss of up to 70% of species in the diverse flock of haplochromine cichlids that evolved in the lake.

Status and future of fishes

The human actions described above may result in fish population declines; in some cases, extinction of the species follows.

Shark fins drying on a fishing vessel. The fins are to be used for such delicacies as shark's fin soup. Federal regulators are trying to determine the impact on shark populations. (Photograph. AP/Wide World Photos. Reproduced by permission.)

As of 2002, 115 distinct species and subspecies of fishes were protected by the U. S. Endangered Species Act, and many other species are threatened or endangered in countries throughout the world. Despite the protection afforded to these species, over 40 species have already been lost to extinction in North America alone since the early 1900s. Estimates suggest that approximately 20% of all freshwater fish species are in serious decline or already extinct. Relatively few marine species are considered at risk of extinction, despite high levels of utilization of these stocks in fisheries. The much higher threat to freshwater fishes likely reflects their utilization of restricted habitats that are intertwined with land-based human populations. It is likely that even higher numbers of species are threatened in many tropical countries due to the rich species diversity and small geographical ranges of many species.

Despite the diversity and immensity of the threats identified above, many people now recognize the negative consequences of environmental decisions and actions. Thus, efforts have been initiated to mitigate impacts to fish species and populations. Local activities often include the rearing of fishes in hatcheries to supplement wild stocks and restoration of habitat areas that have been degraded by human activities. Other efforts involve a broader group of people and often require government mandates. Such endeavors include setting aside reserves to preserve highly diverse areas, protecting critical habitats of threatened species, and developing regulations to reduce overfishing and pollution. Internationally, countries have adopted several treaties that advance conservation of fishes and their habitats, including the Convention on Biological Diversity, the Ramsar Convention on Wetlands, and the Convention on International Trade in Endangered Species (CITES). Locally and globally, it seems that we are beginning to recognize that the futures of fishes and of humans may be closely linked.

Resources

Books

Committee on Ecosystem Management for Sustainable Marine Fisheries, Ocean Studies Board, Commission on Geosciences, Environment, and Resources, National Research Council. *Sustaining Marine Fisheries.* Washington, DC: National Academy Press, 1999.

Committee on Protection and Management of Pacific Northwest Anadromous Salmonids, Board on Environmental Studies and Toxicology, Commission on Life Sciences. *Upstream: Salmon and Society in the Pacific Northwest.* Washington, DC: National Academy Press, 1996.

Dobson, Mike, and Chris Frid. *Ecology of Aquatic Systems.* Essex, England: Addison Wesley Longman, 1998.

Helfman, Gene S., Bruce B. Collette, and Douglas E. Facey. *The Diversity of Fishes.* Malden, MA: Blackwell Science, 1997.

Kurlansky, Mark. *Cod: A Biography of the Fish that Changed the World.* New York: Walker and Company, 1997.

McGinn, Nature A., ed. *Fisheries in a Changing Climate: Proceedings of the Sea Grant Symposium Held at Phoenix,* *Arizona, USA, 20–21 August 2001.* Bethesda, MD: American Fisheries Society, 2002.

Periodicals

Pauly, Daniel, Villy Christensen, Johanne Dalsgaard, Rainer Froese, and Francisco Torres Jr. "Fishing Down Marine Food Webs" *Science* 279 (1998): 860–863.

Organizations

American Fisheries Society. 5410 Grosvenor Lane, Bethesda, MD 20814 USA. Phone: (301) 897-8616. Fax: (301) 897-8096. E-mail: main@fisheries.org Web site: <http://www.fisheries.org>

American Sportfishing Association. 225 Reinekers Lane, Suite 420, Alexandria, VA 22314 USA. Phone: (703) 519-9691. Fax: (703) 519-1872. E-mail: info@asafishing.org Web site: <http://www.asafishing.org>

Other

Food and Agricultural Organization of the United Nations. "FAO Fisheries" (cited 5 February 2003). <http://www.fao.org/fi/default.asp>

Katherine E. Mills, MS

Myxiniformes

(Hagfishes)

Class Myxini
Order Myxiniformes
Number of families 1

Photo: A Pacific hagfish (*Eptatretus stoutii*) with secreted slime. (Photo by Tom McHugh/Photo Researchers, Inc. Reproduced by permission.)

Evolution and systematics

Modern vertebrates are classified into two major groups, the Gnathostomes (jawed vertebrates) and the Agnathans (jawless vertebrates). The Agnathans are classified into two groups, myxinoids (hagfishes) and petromyzonids (lampreys). The Gnathostomes constitute all the other living vertebrates, including the bony and cartilaginous fishes and the tetrapods. The hagfishes are considered the most primitive vertebrates known, living or extinct.

Hagfishes are members of the family Myxinidae, which is the only surviving family of the class Pteraspidomorphi. The species are divided into two primary genera: *Eptatretus* and *Myxine*. The genus *Eptatretus* (found in the Pacific Ocean) has 37 species; the genus *Myxine* (found in the Atlantic Ocean) has approximately 18 species.

Hagfishes are the products of a long evolutionary history and can be considered as primitive, specialized, and degenerative. The hagfish lineage extends over 530 million years and is clearly monophyletic in its origin. Hagfishes are the oldest lineage of vertebrates and are thus very important to evolutionary studies. However, hagfishes are not well represented in the fossil record due to their lack of bony structures. The fossil record consists of a single fossil representing one species of one genus, *Myxinikela siroka*. The discovery of this fossilized hagfish, in sediments deposited roughly 330 million years ago, put the significance of the hagfishes into new light. The hagfishes are an important link between invertebrates and vertebrates, and thus are of interest to evolutionary biologists in regard to both their anatomy and physiology, because they may retain characteristics of ancestral extinct species that are common to their closest relatives, the primitive fossil fishes.

Physical characteristics

All hagfish species have a cartilaginous skeleton with no vertebrae, true fin rays, paired fins, or scales. Hagfishes lack jaws, but have two laterally biting dental plates with keratinous cusps. The mouth is an oval slit, with four fleshy barbels, and a strong tooth on the tongue. The single nostril is surrounded by another four sensory barbels that allow the hagfish to acutely scent food. The eyes of the Pacific hagfishes (*Eptatretus stoutii*) exist as degenerative eyespots covered with thick skin. Atlantic hagfishes (*Myxine glutinosa*) have more degenerative eyespots than Pacific hagfishes. Atlantic hagfishes range in size between 17.7–23.6 in (45–60 cm), but not exceeding 30.7 in (78 cm), in length. Pacific hagfishes are slightly smaller, not exceeding 25.6 in (65 cm) in length. Hagfishes have six to 10 pairs of internal gill pouches, which may open separately to the exterior or unite to form a single exterior opening on each side of the animal, depending on the species. In Pacific hagfishes, short efferent ducts lead to 10–14 external gill openings; in Atlantic hagfishes the efferent ducts discharge through a common external opening. Color ranges from reddish brown to grayish pink.

One interesting feature of the hagfishes that is unique among other fishes or vertebrates is the production of copious quantities of slime. Hagfishes have approximately 150 to 200 slime glands along the side of the body. When a hagfish is attacked or handled, it will secrete small amounts of slime. When the slime comes in contact with the surrounding seawater, the mucous component of the slime expands greatly as it is hydrated with the water, increasing its volume several fold. In order for the hagfish to rid itself of the slime, it literally ties itself in a knot and scrapes itself clean by moving the knot down the body. The slime is used as a defense mechanism and may be involved in reproduction.

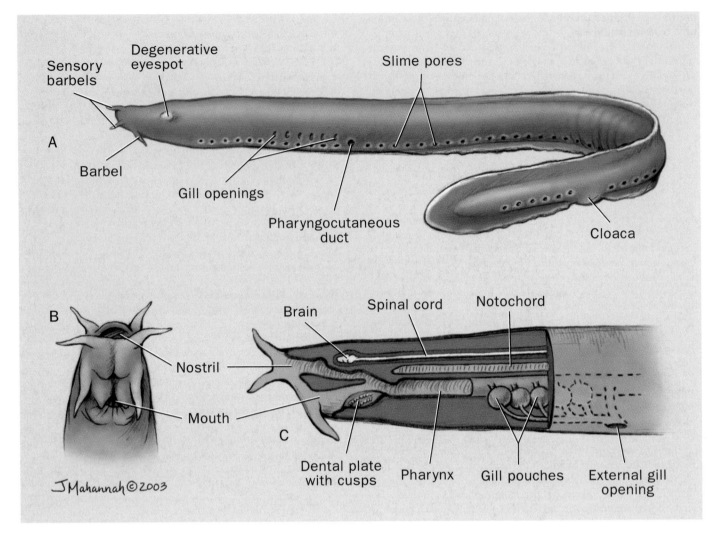

A. External anatomy of hagfish subfamily Eptatretinae; each gill pouch has a separate opening to the exterior. B. Ventral view of closed mouth and single nasal opening. C. Anterior section of subfamily Myxininae; gill pouches have a single common external opening on each side. (Illustration by Jacqueline Mahannah)

Distribution

Hagfishes have been reported from the Atlantic, Pacific, Indian, Arctic, and Antarctic Oceans, and from the Bering, Mediterranean, and Caribbean Seas. Hagfishes do not occur in the Red Sea or the Gulf of Thailand. Atlantic hagfishes are found on both sides of the North Atlantic and in Arctic Seas in deep water of 3,937–11,811 ft (100–300 m) on soft muddy bottoms. Their distribution is varied and patchy, being confined to areas with a suitable bottom. Pacific hagfishes are found along the Pacific coast of North America, from southern California to southeast Alaska. They are found on diverse substrates from muddy bottoms to sand/gravel and boulder/ sand substrates.

Habitat

Hagfishes mostly inhabit deep marine environments that are relatively free of circadian or seasonal changes. Temperature and salinity are thought to be two of the most important factors to influence hagfish distribution. The fishes are most often found in waters that are cooler than 71.6°F (22°C) and have salinities between 32–34 parts per thousand. Although they are most commonly found at the bottom of the ocean—the deepest reported hagfish sighting was 196,850 ft (5,000 m)—some species have been reported in depths as shallow as 394 ft (10 m). Pacific hagfishes occupy a wider range of substrate types than Atlantic hagfishes. Pacific hagfishes occur on substrates ranging from soft muddy bottoms to boulder/sand substrates, and are often found in a coiled position nestled among the rocks. Atlantic hagfishes are most often found on soft muddy substrates in which they form burrows. Atlantic hagfishes burrow by first orienting the body in a vertical position above the substrate and then swimming head first into the substrate. Once the anterior half is below the surface of the mud, the anterior portion pulls the posterior portion below the surface.

Hagfishes are an important part of the benthic marine environment. They are a substantial proportion of the benthic biomass; are critical for substrate turnover and the clean-up and processing of carrion falls; they prey on benthic invertebrates as well as provide prey for marine mammals and large predatory invertebrates.

Behavior

The Japanese hagfish (*Eptatretus burgeri*) is the only known species that undertakes an annual migration, which is thought to be associated with the reproductive cycle.

Feeding ecology and diet

Hagfishes are chiefly scavengers and feed on crustaceans, small marine worms, and vertebrate remains. Using its strong teeth, the hagfish pierces the fish's skin and bores into the body, eating the flesh and eventually only leaving the bone and skin. Gut analyses of Atlantic hagfishes have shown a diet consisting primarily of invertebrate organisms, including polychaetes, hermit crabs, and shrimps.

The Atlantic hagfish has few known predators. Small hagfish have been found in the stomachs of codfish, harbour porpoise, octopus, Peale's dolphin, and sea lions. Hagfish eggs have also been found in the stomachs of male hagfish.

Reproductive biology

The reproductive patterns of most hagfishes are unknown. Females produce a small number (20–30) of large yolky eggs 0.8–1 in (20–26 mm) long. The eggs are enclosed in a tough shell with threads at each end, which act as anchors in the mud. Males produce a small amount of sperm. As neither sex has a copulatory organ, the mode of fertilization is thought to be external.

Sex is often difficult to determine in hagfishes. Atlantic hagfishes have been considered functional hermaphrodites, with their single unpaired sex organ developing sperm in the posterior portion and eggs later in the anterior portion. Other investigations have shown that hagfishes are not hermaphrodites, but that the gonads undergo differentiation into male and female gonads. More recent studies suggest that Atlantic hagfishes could indeed function as hermaphrodites for part of their life cycle, reproducing as either male or female at other times.

The spawning behavior and frequency is unknown. The Japanese hagfish appears to have an annual reproductive cycle associated with its migration into deep waters. At least two species of hagfishes spawn throughout the year, for ripe Atlantic hagfish females and those nearing ripeness have been recorded during all seasons.

It has been noted that Pacific hagfish females in an aquaria ceased to feed when they approached sexual maturity, as do many other fishes. Hagfishes lay their eggs in clutches, strongly supporting evidence that hagfishes do not die after spawning. Although there are no documented answers as to how hagfishes reproduce, considerable data have led to the following conclusions: reproduction takes place at a depth in excess of 164 ft (50 m), there is no marked season of sexual activity (except in the Japanese hagfish), and the eggs are fertilized externally and anchor themselves by their hooks not far from where they were extruded. The last fertilized hagfish eggs reported were obtained by Julia Worthington in 1903.

Conservation status

No species are listed on the IUCN Red List. During the past 40 years, hagfishes have constituted a valuable fishery off the coasts of Japan and Korea, for both meat and skins. A commercial fishery for hagfishes begin in 1987 on the West Coast of the United States, and moved to the East Coast in early 1992 when catches on the West began to decline. There are currently few regulations on the commercial hagfish industry in the United States. Catches from the Gulf of Maine have increased steadily since the mid 1990s, as the Atlantic hagfishes were targeted by U.S. and Canadian fishermen to meet the South Korean demand for "eel" skin, used to manufacture leather goods. Since the fishery began along the New England coast there has been an apparent decline in the number of hagfishes caught in the nearshore fishery.

Significance to humans

Atlantic hagfishes are considered an important species in the Gulf of Maine because they play a significant role in the benthic ecosystem throughout the gulf; have both important direct and indirect effects on commercial fisheries in the gulf, consuming by-catch and providing food; and are targeted by U.S. and Canadian fishers to meet the South Korean demand for "eel" skin. It is likely that all hagfishes have a crucial and significant role in the benthic ocean ecosystem, and the loss of hagfishes could have a major impact on nutrient recycling in the world's oceans.

Species accounts

Atlantic hagfish
Myxine glutinosa

FAMILY
Myxinidae

TAXONOMY
Myxine glutinosa Linnaeus, 1785, Europe; Mediterranean to Murmansk.

OTHER COMMON NAMES
English: Slime eel.

PHYSICAL CHARACTERISTICS
Between 17.7–23.6 in (45–60 cm) in length, but not exceeding 30.7 in (78 cm). Jawless, single nasal opening, single pair of external gill openings, degenerative eyespots covered with thick skin. Grayish or reddish brown.

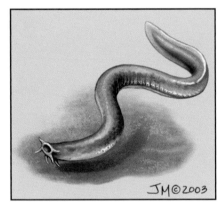

Myxine glutinosa

DISTRIBUTION
Widely distributed in European seas from Murmansk to Mediterranean Sea. Absent in eastern Mediterranean and Black Seas. Present in northwest Atlantic.

HABITAT
Deep waters of 328–984 ft (100–300 m) on soft muddy substrates in which they form burrows.

BEHAVIOR
Burrows by first orienting the body in a vertical position above the substrate and then swimming head first into the substrate. Once the anterior half is below surface of the mud, the posterior portion is pulled below the surface by the anterior, leaving only the nasal opening extruding above the mud.

FEEDING ECOLOGY AND DIET
Feeds on dead and dying fishes, crustaceans, and other small benthic organisms.

REPRODUCTIVE BIOLOGY
Females produce 20–30 large yolky eggs. Fertilization is thought to be external, but has never been observed. It is not known if there is a seasonal reproductive cycle.

CONSERVATION STATUS
Not listed as a threatened or endangered, but fishermen have reported reduced catches in recent years in the Gulf of Maine.

SIGNIFICANCE TO HUMANS
Hagfish skin is processed into various leather goods and marketed as "eel" skin. ◆

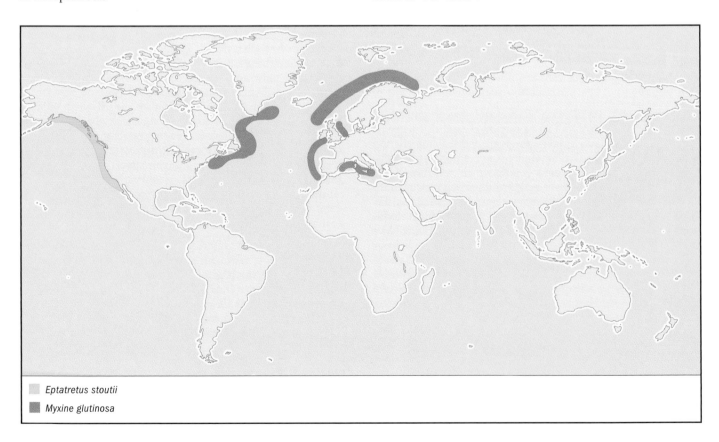

☐ *Eptatretus stoutii*
■ *Myxine glutinosa*

Pacific hagfish
Eptatretus stoutii

FAMILY
Myxinidae

TAXONOMY
Bdellostoma stoutii Lockington, 1878, West Coast of North America, South China Sea, Philippines.

OTHER COMMON NAMES
English: Slime eel.

PHYSICAL CHARACTERISTICS
Does not generally exceed 25.6 in (65 cm) in length. Jawless, single nasal opening; 10–14 gill pouches open directly to external gill openings. Dark brown, tan, gray, or brownish red, often tinted with blue or purple, never black, lighter ventrally,

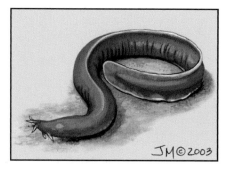
Eptatretus stoutii

rarely with large patches of white. Eyes exist as degenerative eye spots covered with thick skin.

DISTRIBUTION
Widely distributed in the Eastern Pacific: southeastern Alaska to central Baja California, Mexico.

HABITAT
Occupies a wider range of substrate types than Atlantic hagfishes. Occurs at over 330 ft (100 m) on substrates ranging from soft muddy bottoms to boulder/sand substrates.

BEHAVIOR
Burrows in soft sediments and is often found in a coiled position nestled among rocks in boulder/gravel substrates.

FEEDING ECOLOGY AND DIET
Feeds on dead and dying fishes, crustaceans, and other small benthic organisms.

REPRODUCTIVE BIOLOGY
Females produce 20–30 large yolky eggs. fertilization is thought to be external, but has never been observed. it is not known if there is a seasonal reproductive cycle.

CONSERVATION STATUS
Not threatened.

SIGNIFICANCE TO HUMANS
Hagfish skin is processed into various leather goods and marketed as "eel" skin. ◆

Resources

Books
Brodal, Alf, and Ragnar Fänge, eds. *The Biology of* Myxine. Oslo: Universitetsforlaget, 1963.

Hardisty, M. W., ed. *Biology of the Cyclostomes.* London: Chapman Hall, 1979.

Jørgensen, Jørgen Mørup, Jens Peter Lomholt, Roy E. Weber, and Hans Malte, eds. *The Biology of Hagfishes.* London: Chapman Hall, 1998.

Stacia A. Sower, PhD
Mickie L. Powell, PhD
Scott I. Kavanaugh, BS

Petromyzoniformes

(*Lampreys*)

Class Cephalaspidomorphi
Order Petromyzoniformes
Number of families 3

Photo: A sea lamprey (*Petromyzon marinus*) feeding on a fish. (Photo by Berthoula-Scott/Jacana/Photo Researchers, Inc. Reproduced by permission.)

Evolution and systematics

Modern vertebrates are classified into two major groups, the Agnathans (jawless vertebrates) and the Gnathostomes (jawed vertebrates). The Agnathans are classified into two groups, myxinoids (hagfishes) and petromyzonids (lampreys). The Gnathostomes include all other living vertebrates, including the bony and cartilaginous fishes and the tetrapods.

There are approximately 40 species of lampreys, which belong to the order of Petromyzoniformes. This order is divided into three families: the Petromyzonidae, the Northern Hemisphere lampreys (also referred to as the Holarctic species), and the two Southern Hemisphere families, Geotriidae and Mordaciidae. The Petromyzonidae consists of six genera: *Ichthyomyzon, Petromyzon, Caspiomyzon, Eudontomyzon, Tetrapleurodon,* and the *Lampetra*. The genus *Lampetra* is further divided into three subgenera: *Entosphenus, Lethenteron,* and *Lampetra*. The Geotriidae and Mordaciidae each consist of only one genus, *Geotria* and *Mordacia*, respectively.

The phylogeny of lampreys is based primarily on dentition and is justified by other shared anatomical traits, such as the proportional measurements of body parts, size of the adult, snout shape, eyes, and dorsal fins. The species of *Ichthyomyzon* are thought to be the most ancient of the lampreys because their simple teeth are arranged into rows throughout the entire oral disc. Of these species, the silver lamprey (*I. unicuspis*) is considered the most primitive.

Lampreys are representatives of the oldest lineage of vertebrates, the Agnathans. The agnathans probably arose as the first vertebrates about 550 million years ago, immediately after the evolutionary explosion of multicellular organisms in the Cambrian period. Paleontological analysis of extinct agnathans suggests that lampreys are more closely related to gnathostomes (the jawed vertebrates) than either group is to the hagfishes, although recent molecular analysis groups the hagfishes together with the lampreys in a single clade. Definite fossil records date back to the Upper Carboniferous, about 280 million years ago. Like hagfishes, lampreys are an important linkage between invertebrates and vertebrates and thus their anatomy and physiology are of interest to evolutionary biologists because they may retain characteristics of ancestral extinct species that are common to their closest relatives, the primitive fossil fishes.

Physical characteristics

Lampreys are scaleless, eel-like fishes that have skeletons of cartilage instead of bone. They have a notochord, but lack vertebrae. They also lack true fin rays and paired fins, but have one to two dorsal fins. Lampreys lack jaws but have teeth on the oral disc and tongue. Adult lampreys range in length from 7.9 in to 47.2 in (20 to 120 cm).

Lamprey species may be parasitic or nonparasitic. With a few exceptions, the nonparasitic species appear, based on characters and distribution, to have evolved from an extant parasitic lamprey. The four species of lampreys described in this chapter are parasitic lampreys: the sea lamprey (*Petromyzon marinus*), the silver lamprey (*I. unicuspis*), the pouched lamprey (*Geotria australis*), and the short-headed lamprey (*Mordacia mordax*). The sea lamprey and silver lamprey belong to the Northern Hemisphere family; the pouched lamprey and short-headed lamprey belong to the Southern Hemisphere family. Members of Petromyzonidae have the highest number of chromosomes (164–174) among vertebrates. Adult lampreys are distinct in their sex, either male or female. Lampreys spawn only once in their lifetime, after which they die. Parasitic lampreys are generally anadromous.

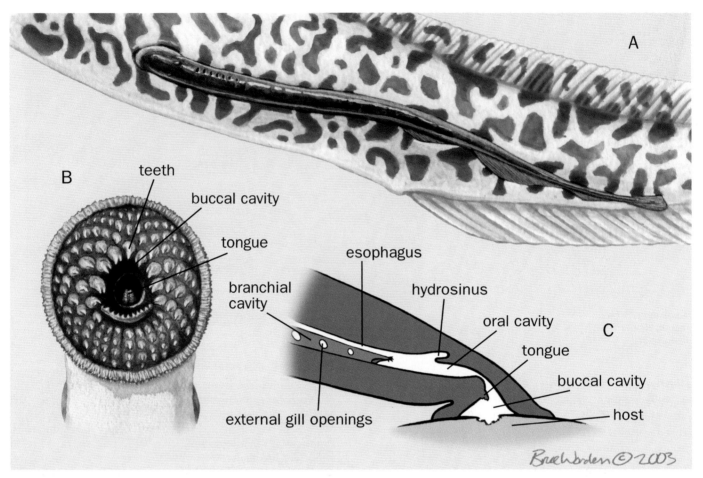

A. Sea lamprey (*Petromyzon marinus*) feeds on a host fish; B. Mouth of *P. marinus*; C. Sagittal section of *P. marinus* attached to host. (Illustration by Bruce Worden)

Distribution

Lampreys occur mainly in temperate zones. Parasitic species are generally of one of two types: those that are anadromous and feed at sea, and those that are restricted to river systems. There is little information on the marine distribution of anadromous species, although it is thought that the larger species move farther away from the coastline than the smaller species. The larger species may in turn give rise to forms that feed in lakes, such as the landlocked sea lamprey of the Great Lakes. Nonparasitic species are restricted to fresh water, most commonly to creeks and smaller rivers. The sea lamprey is found in coastal waters on both sides of the North Atlantic and in the western Mediterranean, and also in fresh waters of the Atlantic coasts of Europe and North America. The landlocked sea lamprey is found in the Great Lakes of North America. The silver lamprey is found in and around the states and provinces of the Great Lakes region. The distribution records of Southern Hemisphere lampreys are less well known due to lack of systematic investigations, but it is known that the pouched lamprey occurs in New Zealand, Western Australia, and Tasmania, and on both coasts of South America; the short-headed lamprey occurs only in southeast Australia and Tasmania.

Habitat

Larval lampreys are wormlike filter-feeding fishes that bury in the sand or mud of rivers. Toward the end of the larval phase, lampreys undergo an extensive metamorphosis in which they become free swimming and migrate to oceans or lakes, where they become parasitic. After one to three years in the parasitic phase, lampreys return to freshwater streams with sand, gravel or pebble substrates to spawn.

Behavior

Lampreys spawn only once in their lifetime, after which they die. The parasitic lampreys begin their lives as freshwater ammocoetes (larval lampreys), which are blind, filter-feeding larvae. After approximately three to seven years in freshwater streams, metamorphosis occurs, and the ammocoetes become free-swimming, sexually immature lampreys, which migrate to the sea or lakes. The actual time for the parasitic phase is not known for all species, but is generally thought to be one to two years. After this period, lampreys return to freshwater streams and undergo the final maturational processes resulting in mature eggs and sperm. The lampreys carry out specific spawning behaviors, including nest building and fanning behavior, after

A sea lamprey (*Petromyzon marinus*) showing underside of sucker-disc mouth. (Photo by Animals Animals ©Zig Leszczynski. Reproduced by permission.)

during, metamorphosis. In the parasitic sea lamprey, sexual maturation is a seasonal, synchronized process. During the parasitic sea phase, which lasts for approximately one to three years, the development of the gametes progresses. In males, spermatogonia proliferate and develop into primary and secondary spermatocytes; in females, vitellogenesis occurs. After this period, lampreys return to freshwater streams and undergo the final maturational processes that result in mature eggs and sperm, and finally spawn, after which the lampreys die. Both sexes of lampreys develop secondary sexual characters during the final weeks of reproduction and spawning activity.

Conservation status

No species of Petromyzoniformes are included on the IUCN Red List.

Significance to humans

Lampreys are important species in the ecosystems in which they reside, whether in streams as filter-feeding organisms

which they die. Prior to metamorphosis, the parasitic and nonparasitic lampreys are indistinguishable. After metamorphosis, the two are distinguished based on size, feeding behavior, and gonad structure, among other traits.

Feeding ecology and diet

During their larval phase, lampreys feed on microscopic plankton, algae and detritus filtered from the mud. During the parasitic phase, they attach to a host fish and extract the blood and/or muscle tissue. Lampreys do not feed in the final spawning phase. Natural predators of the nonparasitic and immature parasitic lampreys include a variety of species of fishes (e.g., eel sand trout) and birds (e.g., gulls).

Reproductive biology

The gonad in both sexes of lamprey sexes is unpaired and median, and is suspended from the dorsal wall of the body cavity by means of a mesentery containing connective tissue. Lampreys are among the few vertebrates, including teleost fishes, that have no intraperitoneal genital ducts. After hatching, for periods varying from six months to over two years in the larval phase, the undifferentiated gonad shows comparatively little further development. Throughout this stage, the germ cells divide only slowly, if at all, remaining solitary or arranged in small groups of slightly advanced cells. After this period, these cells continue to develop into primary oocytes, which occur in all ammocoete gonads regardless whether the lamprey is to become a male or female. Just before metamorphosis, the lampreys undergo sexual differentiation. In lampreys destined to become females, the gonad will continue with the process of oogenesis. In males, the oocytes undergo degeneration and atresia, and the remaining germ cells develop into nests of primary spermatogonia either shortly before, or

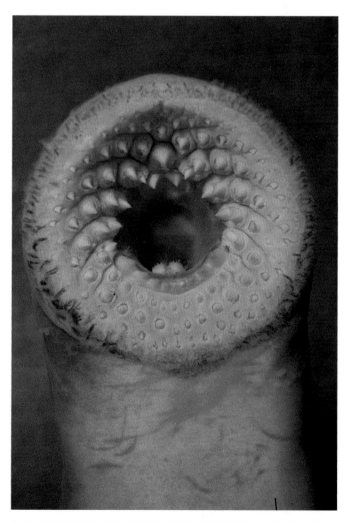

A sea lamprey (*Petromyzon marinus*) showing its circumeral teeth. (Photo by Gary Meszaros/Photo Researchers, Inc. Reproduced by permission.)

helping to recycle nutrients, or as food for predatory fishes and birds in streams and oceans. In certain parts of the world, such as New England in the United States, efforts are being made by state and federal agencies to maintain or increase the populations for this reason. However, in the Great Lakes Region and Lake Champlain, sea lampreys are considered a major deterrent to fish populations because of their parasitic phase, during which they feed on other fishes with their suctorial mouth, extracting body fluids, and often causing high mortalities. The extraordinary amount of damage to the fishery of the Great Lakes caused by the invasion of the sea lamprey has resulted in one of the largest and most intensive efforts to control a vertebrate predator ever attempted. The lampreys are believed to have invaded the Great Lakes beginning with the opening of the Erie Canal in 1819, and the Welland Canal in 1829, which allowed the movement of fishes from Lake Ontario into the Upper Great Lakes from the Atlantic Ocean. By the 1930s, the lampreys had established themselves in all the Great Lakes. The Great Lakes Fishery Commission (GLFC) was established in 1955 by a treaty between Canada and the United States. The two major responsibilities of this Commission were, and continue to be, to develop coordinated programs of research in the Great Lakes, and to formulate and implement programs to eradicate or minimize sea lamprey populations in the Great Lakes.

While lampreys are not presently regarded as food fishes, they were highly prized by both classical and medieval consumers of sea food.

1. Silver lamprey (*Ichthyomyzon unicuspis*); 2. Pouched lamprey (*Geotria australis*); 3. Sea lamprey (*Petromyzon marinus*); 4. Short-headed lamprey (*Mordacia mordax*). (Illustration by Emily Damstra)

Species accounts

Pouched lamprey
Geotria australis

FAMILY
Geotriidae

TAXONOMY
Geotria australis Gray, 1851.

OTHER COMMON NAMES
English: Wide-mouthed lamprey; Spanish: Anguila blanca, lamprea de bolsa.

PHYSICAL CHARACTERISTICS
Total length generally around 19.7 in (50 cm), but fishes up to 24.4 in (62 cm) have been reported. Eel-like, scaleless, lack jaws, have funnel-like mouths and cartilaginous skeletons. Grayish in color with bands of blue-green or brown, depending on stage of reproductive development. Gonad in both sexes is unpaired and median, and is suspended from the dorsal wall of the body cavity by means of a mesentery containing connective tissue.

DISTRIBUTION
Coastal waters of continents of the Southern Hemisphere; also upstream within freshwater tributaries.

HABITAT
Heads of freshwater streams in coastal areas. Lives in open waters for approximately two years, then returns to fresh waters to spawn.

BEHAVIOR
Anadromous; returns to fresh waters to reproduce, during which time it carries out spawning behaviors, including nest building and fanning behavior.

FEEDING ECOLOGY AND DIET
Larvae feed on microscopic plankton, algae, and detritus filtered from the mud. During the parasitic phase, this lamprey attaches to a host fish and extracts blood and/or muscle tissue. Does not feed during migration upstream to spawn in fresh water.

REPRODUCTIVE BIOLOGY
The spawning run lasts for approximately 16 months and takes place during the night, particularly during heavy rains and on nights with a dark moon. The female releases her eggs, which are fertilized by released sperm from the male. The adult lampreys die shortly after spawning.

CONSERVATION STATUS
Not listed by the IUCN.

SIGNIFICANCE TO HUMANS
Research on this species can provide insight into human biology and perhaps yield medicinal applications. ◆

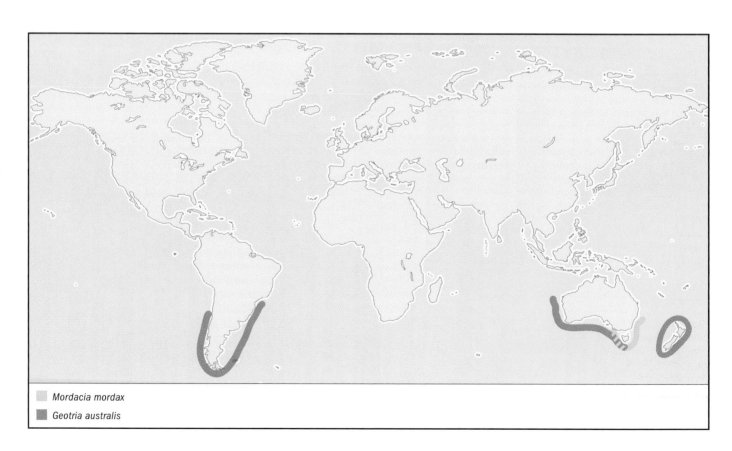

Mordacia mordax
Geotria australis

Short-headed lamprey

Mordacia mordax

FAMILY
Mordaciidae

TAXONOMY
Mordacia mordax Richardson, 1846.

OTHER COMMON NAMES
English: Australian lamprey, Murray lamprey.

PHYSICAL CHARACTERISTICS
Length generally around 20 in (50 cm). Eel-like, scaleless, lack jaws, have funnel-like mouths and cartilaginous skeletons. Body grayish brown in color, but turns blue during upstream spawning migration. The gonad in both sexes is unpaired and median, and is suspended from the dorsal wall of the body cavity by means of a mesentery containing connective tissue.

DISTRIBUTION
Coastal waters of southeastern Australia and Tasmania, as well as upstream within freshwater tributaries.

HABITAT
Heads of freshwater streams of coastal areas of southeastern Australia and Tasmania. Lives in the open waters around southeastern Australia, then returns to fresh waters to spawn.

BEHAVIOR
Anadromous; returns to fresh waters to reproduce, during which time it carries out spawning behaviors, including nest building and fanning behavior.

FEEDING ECOLOGY AND DIET
Larvae feed on microscopic plankton, algae, and detritus filtered from the mud. During the parasitic phase, adult attaches to a host fish and extracts blood and/or muscle tissue. Does not feed after migrating upstream to spawn in fresh water.

REPRODUCTIVE BIOLOGY
Female releases her eggs, which are fertilized by released sperm from the male. The adults die shortly after spawning.

CONSERVATION STATUS
Not listed by the IUCN.

SIGNIFICANCE TO HUMANS
Research on this species can provide insight into human biology and perhaps yield medicinal applications. ◆

Silver lamprey

Ichthyomyzon unicuspis

FAMILY
Petromyzonidae

TAXONOMY
Ichthiomyzon unicuspis Hubbs and Trautman, 1937.

OTHER COMMON NAMES
None known.

PHYSICAL CHARACTERISTICS
Total length 15.3 in (39 cm). Eel-like, scaleless, lack jaws, have funnel-like mouths and cartilaginous skeletons. Body grayish

brown in color. Gonad in both sexes is unpaired and median, and is suspended from the dorsal wall of the body cavity by a mesentery containing connective tissue. Considered the most primitive *Ichthyomyzon* species.

DISTRIBUTION
Hudson Bay and Great Lakes regions, as well as the St. Lawrence river system.

HABITAT
Heads of freshwater streams around the Great Lakes and Hudson Bay regions, as well as the St. Lawrence.

BEHAVIOR
Anadromous; returns to fresh waters to reproduce, during which time it carries out spawning behaviors, including nest building and fanning behavior.

FEEDING ECOLOGY AND DIET
Larvae feed on microscopic plankton, algae, and detritus filtered from mud. During the parasitic phase, adult attaches to a host fish and extracts blood and/or muscle tissue. Does not feed after migrating upstream to spawn in fresh water.

REPRODUCTIVE BIOLOGY
Female releases her eggs, which are fertilized by released sperm from the male. The adults die shortly after spawning.

CONSERVATION STATUS
Not listed by the IUCN.

SIGNIFICANCE TO HUMANS
Research on the species can provide insight into human biology and perhaps yield medicinal applications. ◆

Sea lamprey

Petromyzon marinus

FAMILY
Petromyzonidae

TAXONOMY
Petromyzon marinus Linnaeus, 1758.

OTHER COMMON NAMES
English: Eel sucker, Green sea lamprey, lamprey eel; French: Lamproie marine; German: Große lamprete; Spanish: Lamprea de mar.

PHYSICAL CHARACTERISTICS
Total length 47.2 in (120 cm). Eel-like, scaleless, lack jaws, have funnel-like mouths and cartilaginous skeletons. Body grayish brown in color. Gonad in both sexes is unpaired and median, and is suspended from the dorsal wall of the body cavity by a mesentery containing connective tissue.

DISTRIBUTION
Coastal waters on both sides of the North Atlantic, the western Mediterranean, also fresh waters of the Atlantic coasts of Europe and North America: landlocked in the Great Lakes of North America.

HABITAT
Immature fishes can be found in the mouths of freshwater streams of eastern North America, Northern Europe, and

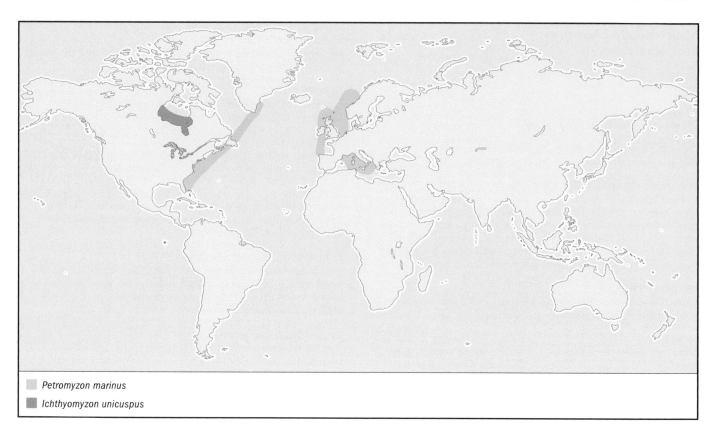

Petromyzon marinus
Ichthyomyzon unicuspus

western regions of the Mediterranean. Mature fishes live in the open waters of the Atlantic Ocean and Mediterranean Sea.

BEHAVIOR
Anadromous; returns to fresh waters to reproduce, during which time it carries out spawning behaviors, including nest building and fanning behavior.

FEEDING ECOLOGY AND DIET
Larvae feed on microscopic plankton, algae, and detritus filtered from mud. During the parasitic phase, adult attaches to a host fish and extracts blood and/or muscle tissue. Does not feed after migrating upstream to spawn in fresh water.

REPRODUCTIVE BIOLOGY
Female releases approximately 200,000 eggs, which are fertil-

ized by released sperm from the male. The adults die shortly after spawning.

CONSERVATION STATUS
Not listed by the IUCN. Considered a critical species in their natural ecosystems and efforts are being made by state and federal agencies to maintain or increase populations there. In the Great Lakes, where the species has been introduced, authorities are working to control their populations because of their detrimental impact on native fishes.

SIGNIFICANCE TO HUMANS
Very destructive to fish populations in the Great Lakes Region and Lake Champlain. During the parasitic phase, feeds on other fishes with its suctorial mouth, extracting body fluids and often causing high mortalities. ◆

Resources

Books

Fulton, W. "Tasmanian Freshwater Fishes." In *Fauna of Tasmania Handbook.* No. 7, edited by A. M. M. Richardson. Tasmania: University of Tasmania, 1990.

Hardisty, M. W., and I. C. Potter. *The Biology of Lampreys.* New York: Academic Press, 1971.

Sower, S. A., and A. Gorbman. "Agnatha." In *Encyclopedia of Reproduction.* Vol. 1, edited by E. Knobil and J. D. Neill. New York: Academic Press, 1999.

Periodicals

Hardisty, M. W., I. C. Potter, and R. W. Hillard. "Gonadogenesis and Sex Differentiation in the Southern Hemisphere Lamprey, *Geotria australis* Gray." *Journal of Zoology* 209 (1986): 477–499.

Smith, B. R., ed. "Proceedings of the Sea Lamprey International Symposium." *Canadian Journal of Fisheries and Aquatic Sciences* 37 (1980): 1,585–2,215.

Stacia A. Sower, PhD
Matthew R. Silver, BS

Chimaeriformes
(Chimaeras)

Class Chondrichthyes
Order Chimaerformes
Number of families 3

Photo: Spotted ratfish (*Hydrolagus colliei*) swim just above the sea floor and eat clams, worms, starfish, fish, and shrimp. (Photo by Brandon D. Cole/Corbis. Reproduced by permission.)

Evolution and systematics

The chimaeras are an ancient lineage of cartilaginous fishes related to sharks, skates, and rays. All chimaeras, sharks, skates, and rays are united in the class Chondrichthyes, which is further subdivided into two subclasses: Holocephali, consisting of the chimaeras characterized by a unique jaw and tooth morphology, and the Elasmobranchii, which includes the sharks, skates, and rays. Fossil evidence indicates that the chimaeras probably evolved nearly 300 million years ago. What is especially remarkable is that many of the modern forms look very much like their fossil ancestors.

Within the Order Chimaeriformes there are three families, each of which is distinguished by a unique snout morphology. The plow-nosed chimaeras of the family Callorhinchidae have a delicate flap of skin in the shape of a hoe projecting from the snout; the long-nosed chimaeras of the family Rhinochimaeridae are characterized by elongate, pointed snouts; and the ratfishes, family Chimaeridae, have blunt fleshy snouts. As of 2002 there are 33 described species of chimaeras with at least 10 additional species that are known, but not yet formally described.

Physical characteristics

Chimaeras are characterized by large heads and elongate bodies that taper to a whip-like tail. They range in size from small, slender-bodied fishes of 1–2 ft in total length (about 30.5–61 cm), to massive fishes, nearly 4 ft in length (122 cm), with gigantic heads and large girth. The skin is smooth and rubbery, completely lacking in scales or denticles. The four gill openings on each side of the head are covered by a fleshy operculum. The mouth is small with teeth that are formed into three pairs of non-replaceable tooth plates, two pairs in the upper jaw and one pair in the lower jaw. These tooth plates tend to protrude from the mouth like a rodent's incisors, suggesting the common names ratfish or rabbitfish for some of the species. The pectoral fins of chimaeras are broad and wing-like and serve to propel the fish through the water by a flapping motion much like underwater flying. All chimaeras have two dorsal fins; the first is preceded by a stout and often toxic spine, and the second is spineless. The lateral line canals are visible externally, and in many species are formed as open grooves. Chimaeras are sexually dimorphic. Adult males possess three unique secondary sexual characteristics: a bulbous, denticulate frontal tenaculum that rests in a pouch atop the head; blade-like prepelvic tenaculae that are hidden in pouches anterior to the pelvic fins; and pelvic claspers that extend from the posterior edge of the pelvic fins.

Distribution

These fishes are entirely marine and are distributed in all of the world's oceans with the exception of Arctic and Antarctic waters. Most species live in deep waters of the shelf and slope, generally at depths greater than 1,500 ft (457 m), and the deepest recorded capture was near 9,000 ft (2,743 m). Although most chimaeras are deepwater dwellers, several species occur in shallower waters, and some migrate inshore. Many species are known from a very widespread geographic range, sometimes throughout an ocean basin spanning the northern and southern hemisphere, while other species appear to be more restricted in their range both vertically and horizontally.

Habitat

Chimaeras usually live on or near muddy bottoms. They tend to remain close to the bottom and are not known to move into the pelagic zone. Most species occur near continental landmasses or off oceanic islands and on the slopes of seamounts and underwater ridges.

Behavior

Some species are locally migratory and congregate near the shore for breeding and spawning. It also has been observed that some chimaeras tend to aggregate into single-sex groups and to separate into groups based on age.

Feeding ecology and diet

The diet consists primarily of benthic invertebrates. The tooth plates are used to crush hard-bodied prey such as crabs, clams, and echinoderms. Chimaeras also are known to prey upon other fishes. Very little information exists with regard to predation of chimaeriformes; their main predators are sharks and humans.

Reproductive biology

Most species reach sexual maturity at about 18 in (45.7 cm) body length measured from the distal edge of the gill opening to the origin of the dorsal lobe of the caudal fin. Females are generally larger than males. Like their shark relatives, chimaeras have internal fertilization in which males, equipped with pelvic claspers, transfer sperm directly into the female reproductive tract. Males also possess two additional organs used in copulation. Unique to chimaeroids is the club-like frontal tenaculum that emerges from the top of the head in sexually mature males and has been observed to be used to grasp the posterior edge of the pectoral fin of the female dur-

ing courtship. Additionally, a pair of blade-like prepelvic tenaculae aid in anchoring the male during copulation. Sperm storage in females has been observed in one species and is likely to occur in all species.

All species of chimaeras are oviparous and reproduce by laying eggs. Fertilized eggs are encased in egg capsules and deposited onto the ocean floor. Egg capsules are laid in pairs with each egg capsule containing only a single egg. The egg capsules are generally spindle shaped, sometimes with broad lateral web-like flanges that vary in size and shape depending on the species. Embryological development may take six to twelve months, and fully developed hatchlings measure about 5 in (12.7 cm) in length and look like miniature adults. Very little is known about details of reproduction and development for most species of chimaeras.

Conservation status

There is insufficient data to determine if any species of chimaeras are threatened. However, chimaeras may be inadvertently subject to overexploitation from fisheries due to lack of understanding of their age, growth, and population structure, and seemingly low fecundity.

Significance to humans

A few species of chimaeras are fished commercially for human consumption, particularly in the southern hemisphere; however, most species of chimaeras are little used and are unlikely to become an important fishery resource. Chimaeras are sometimes taken as minor bycatch in trawls and can be processed for oil and fishmeal.

1. Female spotted ratfish (*Hydrolagus colliei*); 2. Female Pacific spookfish (*Rhinochimaera pacifica*); 3. Female ghost shark (*Callorhinchus milii*). (Illustration by Dan Erickson)

Species accounts

Ghost shark
Callorhinchus milii

FAMILY
Callorhinchidae

TAXONOMY
Callorhinchus milii Bory de St. Vincent, 1823, Australia.

OTHER COMMON NAMES
English: Elephant shark, whitefish; Maori: Reperepe.

PHYSICAL CHARACTERISTICS
Distinguished by a plow-shaped snout. Unlike other chimaeras, all of which have whip-like tails, callorhinchids have externally heterocercal tails with a large dorsal lobe and smaller ventral lobe. Body color is silvery and is black along the dorsal midline and top of the head with black saddle-like bands along the dorsal surface of the trunk.

DISTRIBUTION
Southern coasts of New Zealand and Australia.

HABITAT
Prefers coastal waters, living on or near sandy, muddy, or rocky bottoms.

BEHAVIOR
Known for seasonal migration inshore to spawn in shallow coastal waters.

FEEDING ECOLOGY AND DIET
Feeds on benthic invertebrates, particularly small bivalves. It may also eat other fishes.

REPRODUCTIVE BIOLOGY
This is an oviparous species, with eggs fertilized within the female reproductive tract. Females lay two eggs at a time, each

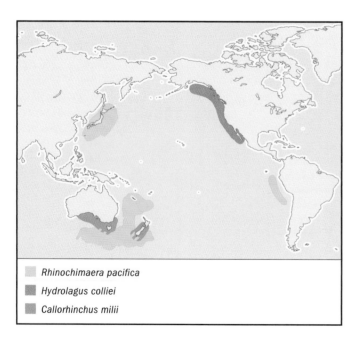

Rhinochimaera pacifica
Hydrolagus colliei
Callorhinchus milii

contained within its own egg capsule, over a period of months. Development appears to take 6–12 months.

CONSERVATION STATUS
Large fluctuations in population size have been recorded in New Zealand with a general trend toward decreasing numbers over the years. This species may be impacted by overfishing.

SIGNIFICANCE TO HUMANS
Commercially fished and used for human consumption in New Zealand and Australia. ◆

Spotted ratfish
Hydrolagus colliei

FAMILY
Chimaeridae

TAXONOMY
Hydrolagus colliei Lay and Bennett, 1839, Monterey, California.

OTHER COMMON NAMES
None known.

PHYSICAL CHARACTERISTICS
Head contains a bluntly pointed snout. Body color is a reddish to dark brown with silvery-blue and gold highlights, as well as numerous small white spots on the head and along sides and back of the trunk. Ventrally the color is an even pale cream or gray.

DISTRIBUTION
Southeastern Alaska to Baja, California, and the northern Gulf of California. It has been recorded at depths ranging from the surface to 2,995 ft (912.9 m).

HABITAT
Usually occurs near muddy, sandy, or rocky bottoms.

BEHAVIOR
Known to migrate from deeper to shallower waters. It tends to aggregate into groups based on age and sex.

FEEDING ECOLOGY AND DIET
Feeds on benthic invertebrates and other fishes.

REPRODUCTIVE BIOLOGY
Oviparous, with eggs fertilized within the female reproductive tract. Two egg capsules, each containing a single embryo, are laid every 7–10 days for a period of months. Development appears to take 6–12 months.

CONSERVATION STATUS
Not threatened.

SIGNIFICANCE TO HUMANS
At one time this species was fished locally for the oil extracted from the liver. There is no known commercial value, and it is considered a nuisance fish by local fishermen. ◆

Pacific spookfish
Rhinochimaera pacifica

FAMILY
Rhinochimaeridae

TAXONOMY
Rhinochimaera pacifica Mitsukuri, 1898, Japan.

OTHER COMMON NAMES
English: Knifenose chimaera; Spanish: Tucán; Japanese:
Tengu-ginzame.

PHYSICAL CHARACTERISTICS
A long, narrow conical snout extends forward from the head.
The body is elongate and tapering to a whip-like tail. Body
color is usually a uniform brown or grayish brown, with fins a
darker shade. The skin along the ventral side of the snout and
around the mouth is generally white in color.

DISTRIBUTION
Widely distributed throughout the western Pacific Ocean from
Japan to subantarctic waters. Also reported from the southeast-
ern Pacific of Peru.

HABITAT
Inhabits deep water slopes and seamounts usually associated
with muddy bottoms.

BEHAVIOR
Nothing is known.

FEEDING ECOLOGY AND DIET
Diet consists of a wide variety of benthic invertebrates and
possibly other fishes.

REPRODUCTIVE BIOLOGY
Oviparous, with eggs fertilized within the female reproductive
tract. Females lay two egg capsules at a time, each containing a
single embryo. Very few egg capsules and embryos have ever
been observed, and almost nothing is known of spawning and
embryological development in rhinochimaerids.

CONSERVATION STATUS
Insufficient information is available.

SIGNIFICANCE TO HUMANS
Not known to be commercially fished, although it may be
caught as bycatch and processed for oil or fishmeal. ◆

Resources

Books
Eschmeyer, W. N., E. S. Herald, and H. Hammann. *A Field
Guide to Pacific Coast Fishes of North America.* Boston:
Houghton Mifflin Company, 1983.

Hart, J. L. *Pacific Fishes of Canada.* Fisheries Research Board of
Canada, Bulletin No. 180, 1980.

Last, P. R., and J. D. Stevens. *Sharks and Rays of Australia.*
CSIRO Australia, 1994.

Paulin, C., A. Stewart, C. Roberts, and P. McMillan. *New
Zealand Fish: A Complete Guide.* National Museum of New
Zealand Miscellaneous Series No. 19. Wellington: New
Zealand, 1989.

Periodicals
Didier, D. A. "Phylogenetic Systematics of Extant Chimaeroid
Fishes (Holocephali, Chimaeroidei)." *American Museum
Novitates* 3119 (1995): 1–86.

———, E. E. LeClair, and D. R. Vanbuskirk. "Embryonic
Staging and External Features of Development of the
Chimaeroid Fish, *Callorhinchus milii* (Holocephali,
Callorhinchidae)." *Journal of Morphology* 236 (1998): 25–47.

Lund, R., and E. D. Grogan. "Relationships of the
Chimaeriformes and the Basal Radiation of the
Chondrichthyes." *Reviews in Fish Biology and Fisheries* 7
(1997): 65–123.

Mathews, C. P. "Note on the Ecology of the Ratfish,
Hydrolagus colliei, in the Gulf of California." *California Fish
and Game* 61 (1975): 47–53.

Quinn, T. P., B. S. Miller, and R. C. Wingert. "Depth
Distribution and Seasonal and Diel Movements of Ratfish,
Hydrolagus colliei, in Puget Sound, Washington." *Fishery
Bulletin* 78 (1980): 816–821.

Dominique A. Didier, PhD

Heterodontiformes
(Bullhead or horn sharks)

Class Chondrichthyes

Order Heterodontiformes

Number of families 1

Photo: A California bullhead shark (*Heterodontus francisci*) near San Clemente Island, California. (Photo by Animals Animals ©Randy Morse. Reproduced by permission.)

Evolution and systematics

The earliest fossil bullhead (or horn) shark known from articulated specimens is from the marine late Jurassic strata of Solnhofen, southern Germany (about 150 million years old). However, fragmentary remains are known from the early Jurassic of Germany, and most fossil bullhead species have been described from isolated teeth or finspines from Cretaceous and Tertiary deposits of Europe, North and South America, Australia, and Africa. These fossils indicate that bullhead sharks have occupied shallow marine environments throughout their long history.

Bullhead sharks are part of the superorder Galeomorphii, a group that also contains the carpet sharks (Orectolobiformes), mackerel sharks (Lamniformes), and ground sharks (Carcharhiniformes). These orders have the hyomandibular fossa closely adjacent to the orbit on the neurocranium (this fossa, or depression, anchors the hyomandibula, a cartilage that connects distally to the jaw joint, to the skull). Bullhead sharks were believed to be closely related to more primitive Mesozoic hybodont sharks (which also had dorsal finspines), and therefore considered to be living relics, but it is now well established that bullheads share a common ancestry with modern (galeomorph) sharks. However, the phylogenetic relationships among bullhead species have not been satisfactorily studied. All bullhead species are classified in the single family Heterodontidae.

Eight living species of bullheads are currently recognized, all in the single genus *Heterodontus*. Most species have been described in the mid- to late nineteenth century, but *H. portusjacksoni* was described in 1793, and two species have been

discovered and named in the twentieth century (in 1949 and 1972). Additionally, there is one undescribed species of bullhead shark off southern Oman, in the northwestern Arabian Sea. Most living species of bullheads have been relatively well characterized, but some species (such as *H. portusjacksoni*) are far better known than others (such as *H. ramalheira*).

Physical characteristics

Bullhead sharks have a tapered profile due to their large, bulky heads. Their snouts are blunt, short, and rounded. Bullheads also have prominent supraorbital ridges (elevated crests supporting the eyes), which provide a greater range of vision, possibly an advantage for bottom-dwelling sharks. Bullheads have two relatively large dorsal fins (the first is clearly larger than the second), each preceded by a short finspine. The finspine in embryos is blunt so as to not harm the mother, but relatively sharp in adults. The caudal fin is robust, with a prominent notch separating the upper and lower lobes. There are five pairs of gill slits. Bullheads are covered with large, abrasive dermal denticles, which are visible without magnification.

Bullheads are the only living sharks with a finspine preceding each dorsal fin in combination with presenting an anal fin. They also have unique dentitions, with small anterior teeth endowed with small cusps for clutching prey, contrasting to the more posterior tooth rows where the teeth are flattened and enlarged (up to 0.4 in/1 cm wide), adapted for grinding hard-shelled invertebrates (hence the generic name *Heterodontus*, meaning "having different teeth"). Their nostrils are also unique, being very large and circular, providing

A Port Jackson shark (*Heterodontus portusjacksoni*) swimming with fish. (Photo by Jeffrey L. Rotman/Corbis. Reproduced by permission.)

them with a well-developed sense of smell that is also important for bottom-dwelling species.

Coloration is helpful to distinguish among bullhead species. Three species present light-brown to grayish-brown background coloration with darker-brown spots on the head, body, fins, and tail (*H. francisci*, *H. quoyi*, and *H. mexicanus*) but the arrangement, number, and diameter of the spots is usually distinct for these species. *Heterodontus ramalheira* is unique in having a reddish-brown background with creamy-white, minute spots. *Heterodontus japonicus* and *H. galeatus* have dark saddlelike markings on their dorsal surfaces (and also over the eyes and underneath the first dorsal fin in the latter species), both with lighter background colors. *Heterodontus portusjacksoni* is unique in the genus in presenting a horizontal pattern of brownish-black stripes. However, the most spectacular of all bullhead species, and one of the most ornate sharks known, is *H. zebra*, with an intense, dark brownish-black vertical-stripe pattern from head to tail and over the pectoral fins, with some of the stripes coupled together along the sides of the trunk.

Bullheads are only average-sized sharks, reaching from 28–51 in (70–130 cm) long, but a few species may reach slightly larger sizes. Most species are sexually mature when between 15.7–28 in (40–70 cm) long for males, and slightly larger for females.

Distribution

Three species are present in the tropical eastern Pacific: *H. francisci*; *H. mexicanus*, distributed in the Gulf of California, along the Central American coast down to Colombia and possibly Peru; and *H. quoyi*, found in the Galápagos Islands and the coasts of Ecuador and Peru. Two species are Australian: *H. galeatus* (eastern Australia and perhaps in Tasmania and off Cape York) and *H. portusjacksoni* (southern [including Tasmania], western, and eastern Australia, and possibly in New Zealand). *Heterodontus ramalheira* occurs along the eastern African coast extending northward to the Arabian Peninsula; *H. japonicus* is a western Pacific species, occurring

around Japan, Korea, and off the Chinese coast; *H. zebra* is somewhat widespread in the western Pacific, distributed from Japan and Korea down to Vietnam, with records also in Indonesia and northwestern Australia.

Habitat

Bullheads inhabit the continental shelf, usually in shallow waters from the intertidal zone down to about 328 ft (100 m), and less frequently at greater depths (to 820 ft/250 m for *H. ramalheira* and *H. portusjacksoni*). They occur on hard and soft bottoms, including reefs, rocky, and sandy substrates, and commonly frequent caves, crevices, and kelp and sea grass beds.

Behavior

Bullheads are more active nocturnally, as are many benthic sharks. They are usually solitary, although recently born individuals may group together for a small period before going their separate ways, and aggregations of adults have been observed in some species. Their strong pectoral fins are used to "walk" over the substrate. In the most-studied species, (*H. portusjacksoni*), adults tend to occupy a restricted range, returning to the same resting location daily, and there is a certain degree of territoriality and competition for favored resting caves. Courtship patterns have been observed in *H. francisci*. At least one species, *H. portusjacksoni*, appears to be migratory, returning to breeding sites after periods spent in deeper waters.

Feeding ecology and diet

Bullheads consume abundant amounts of hard-shelled benthic invertebrates, including crabs, lobsters, shrimp, barnacles, starfish, urchins, gastropods, and polychaetes. Most species also eat fishes. Smaller individuals eat softer prey items while their molariform posterior teeth are still in development. Bullhead sharks commonly employ strong suction feeding. One bullhead shark has been found in the stomach of a tiger shark, but they are generally avoided as prey because of their finspines.

Reproductive biology

All bullhead sharks have internal fertilization and are oviparous (egg-laying), laying large, spiral-rimmed egg cases. The fully formed egg cases are expelled rather early by females, so that most fetal development occurs in the egg cases while in the environment, not inside the mother. Young hatch from between five and 12 months after being laid, one per egg case, and measure about 3.9–5.5 in (10–14 cm). The young often move into nursery areas or bays after hatching. The egg cases are laid in shallow water, sometimes in unguarded "nests" (*H. japonicus*), and usually in protected kelp beds or enclosed in protected areas (the egg may be carried and lodged by the mother, using her mouth, in a crevice). Adults have been observed to eat their own egg cases.

Conservation status

Not threatened.

Significance to humans

Their significance is mostly recreational (e.g., when observed by divers), as bullheads are not consumed on a regular basis. They are caught as bycatch in bottom trawls and usu- ally discarded, but they may be occasionally consumed or used as fishmeal (off eastern Mexico, for example). Various species of bullheads are commonly kept in public and private aquaria, where they can be maintained successfully for over a decade.

1. California bullhead shark (*Heterodontus francisci*); 2. Port Jackson shark (*Heterodontus portusjacksoni*). (Illustration by Dan Erickson)

Species accounts

California bullhead shark
Heterodontus francisci

FAMILY
Heterodontidae

TAXONOMY
Cestracion francisci Girard, 1854, California (Monterey Bay).

OTHER COMMON NAMES
English: Bullhead shark, horn shark; Spanish: Dormilón cornudo.

PHYSICAL CHARACTERISTICS
Background gray or light brown with smaller darker brown spots (smaller than eyes) scattered over body, head, fins, and tail. The young have more intense coloration, sometimes with darker bands in between eyes and on fins. The supraorbital ridges are moderately high; finspines relatively tall, but dorsal fins are not as tall as in *H. zebra* and *H. japonicus*.

DISTRIBUTION
California bullhead sharks occur off central and southern California and Mexico (Baja California and Gulf of California), extending south possibly to Ecuador and Peru. During warm water influxes, they may reach as far north as San Francisco Bay.

HABITAT
These sharks commonly inhabit from 6.6 to 33 ft (2–10 m), even though they can be found from the intertidal zone down to about 490 ft (150 m). Juveniles are usually in shallower waters, over sandy surfaces. These fishes occur on rocky and sandy bottoms, kelp forests, and in caves and crevices.

BEHAVIOR
These sharks are nocturnal, sluggish, and mostly solitary, preferring the protection of caves and shelters during the day, and hunting mostly at night. They can remain in a relatively small home range for much of the summer, moving into deeper waters in the winter, at least in some regions. Experiments demonstrate that their diel (24 hours, including one day and night) activity patterns appear to be regulated by light.

FEEDING ECOLOGY AND DIET
California bullhead sharks eat many different invertebrates, including sea urchins, crabs, shrimp, isopods, octopuses, anemones, bivalves, gastropods, polychaetes, and occasionally fishes (at least pipefishes [Syngnathidae] and blacksmiths [Pomacentridae]). Specimens have been filmed devouring purple-colored urchins, turning their teeth and spines into a strong shade of purple. Recently born pups take about one month to begin feeding. Adults in aquaria have been observed consuming their own egg cases.

REPRODUCTIVE BIOLOGY
The mating ritual of California bullhead sharks has been observed in captivity, particularly at the Steinhart Aquarium in San Francisco. Males pursue larger females until obtaining consent, and mating occurs on the bottom of the tank. The male grasps the female's pectoral fin with his teeth, and subsequently one clasper is inserted into the female after coiling around her. Copulation may last between 30 to 40 minutes, and in captivity the eggs are expelled one or two weeks later. In the wild, eggs can be expelled even after one to three months of copulation, as females can produce eggs for extended periods, and sperm is stored and utilized in stages. The young develop for between seven and nine months before hatching.

CONSERVATION STATUS
A decrease in numbers of individuals has been noticed in regions of southern California where there is substantial diving activity, but the species is not listed by the IUCN as Threatened.

SIGNIFICANCE TO HUMANS
These sharks are very common in public aquaria, where mating, egg-laying, and hatching have been observed. In the wild, they are not considered a threat to humans. However, despite their apparent calm demeanor, *H. francisci* have been known to infrequently swim after and bite divers after being harassed by them. ◆

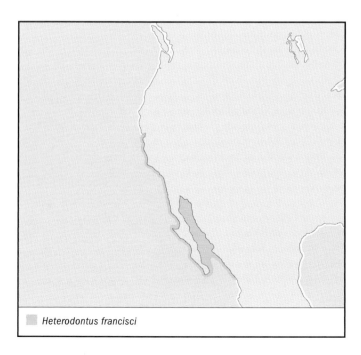

Heterodontus francisci

Port Jackson shark
Heterodontus portusjacksoni

FAMILY
Heterodontidae

TAXONOMY
Squalus portus jacksoni Meyer, 1793, Australia (Botany Bay, New South Wales).

OTHER COMMON NAMES
English: Bullhead shark, oyster crusher, tabbigaw.

PHYSICAL CHARACTERISTICS
Port Jackson sharks are distinguished by a gray to brownish background with a harnesslike pattern of darker brown stripes over the pectoral fins and below the first dorsal fin, with a dark stripe across the head and eyes and a few dark oblique stripes

Heterodontus portusjacksoni

along the trunk. The supraorbital ridges are only moderately high. The finspines do not reach the dorsal fin tips, being rather blunt and short; and the dorsal fins are not nearly as high as in *H. zebra* and *H. japonicus*.

DISTRIBUTION
These sharks occur around the southern, western, and eastern Australian coast, including Tasmania. A single record exists for New Zealand, but this is possibly a stray.

HABITAT
Port Jackson sharks are common on the temperate Australian continental shelf and upper slope, from close inshore down to 902 ft (275 m).

BEHAVIOR
Port Jackson sharks are nocturnal, occurring in caves and shelters during the day, and hunting at night. Seasonal migrations are frequent, and these may extend 528.2 mi (850 km) in one direction, after which they may return to the same localities each year. Migrations from Sydney to Tasmania have been reported. Captive juvenile Port Jackson sharks may grow 2–2.4 in (5–6 cm) per year; adults grow slightly less at 0.8–1.6 in (2–4 cm) per year. They take in water for respiration through the first gill slit while feeding, freeing their mouths for eating, as water exits from the remaining gill slits.

FEEDING ECOLOGY AND DIET
These sharks primarily eat benthic invertebrates, especially echinoderms, but prey items include crabs, shrimp, starfish, bivalves, gastropods, polychaetes, and small fishes.

REPRODUCTIVE BIOLOGY
These sharks segregate by size after hatching, but adults have been reported to segregate by sex. Adult males move into deeper water toward the end of the breeding season in winter, and adult females migrate shortly thereafter. Females lay from 10 to 16 egg cases in rocky substrates, in sheltered, shallow areas (ranging from 4 to 98 ft [1–30 m], but more commonly from 4 to 16 ft [1–5 m]). Females may utilize the same nesting sites repeatedly, and some have been observed to purposely wedge their egg cases into crevices. Adults may eat their egg cases. The young hatch after nine to twelve months, and quickly move into nursery areas. Maturity ages vary from eight to ten years for males and from eleven to fourteen years for females.

CONSERVATION STATUS
Not threatened.

SIGNIFICANCE TO HUMANS
Port Jackson sharks are taken as bycatch in benthic or demersal fisheries, but are not generally consumed. They are very common in public aquaria. The Port Jackson shark is perhaps the best known of all bullhead species, and is not dangerous to humans, although caution is necessary when approaching them. ◆

Resources

Books
Cappetta, H. *Chondrichthyes II, Mesozoic and Cenozoic Elasmobranchii.* Stuttgart: Gustav Fischer Verlag, 1987.

Compagno, L. J. V. *Sharks of the World. An Annotated and Illustrated Catalogue of Shark Species Known to Date.* Vol. 2, *Bullhead, Mackerel and Carpet Sharks (Heterodontiformes, Lamniformes and Orectolobiformes).* Rome: Food and Agriculture Organization of the United Nations, 2001.

Hennemann, R. M. *Sharks and Rays, Elasmobranch Guide of the World.* Frankfurt: Ikan, 2001.

Last, P. R., and J. D. Stevens. *Sharks and Rays of Australia.* Melbourne, Australia: CSIRO, 1994.

Nelson, J. *Fishes of the World,* 3rd ed. New York: John Wiley & Sons, 1994.

Springer, V. G., and J. P. Gold. *Sharks in Question. The Smithsonian Answer Book.* Washington, DC: Smithsonian Institution Press, 1989.

Whitley, G. P. *The Fishes of Australia.* Part 1, *The Sharks, Rays, Devil-Fish, and Other Primitive Fishes of Australia and New Zealand.* Sydney, Australia: Royal Zoological Society of New South Wales, 1940.

Periodicals
Compagno, L. J. V. "Phyletic Relationships of Living Sharks and Rays." *American Zoologist* 17 (1977): 303–322.

Edmonds, M.A., P.J. Motta, and R.E. Hueter. "Food Capture Kinematics of the Suction Feeding Horn Shark, *Heterodontus fancisci.*" *Environmental Biology of Fishes* 62 (2001): 415–427.

Resources

Luer, C. A., and P. W. Gilbert. "Elasmobranch Fish: Oviparous, Viviparous, and Ovoviviparous." *Oceanus Magazine* 34, no. 3 (1991): 47–53.

Maisey, J. G. "Fossil Hornshark Finspines (Elasmobranchii; Heterodontidae) with Notes on a New Species (*Heterodontus tuberculatus*)." *Neues Jahrbuch für Geologie und Paläontologie* 164, no. 3 (1982): 393–413.

Smith, B. G. "The Heterodontid Sharks: Their Natural History, and the External Development of *Heterodontus japonicus* Based on Notes and Drawings by Bashford Dean." In *Bashford Dean Memorial Volume: Archaic Fishes*, vol. VIII. New York: American Museum of Natural History, 1942: 649–770, plates 1–7.

Organizations

American Elasmobranch Society, Florida Museum of Natural History. Gainesville, FL 32611 USA. Web site: <http://www.flmnh.ufl.edu/fish/Organizations/aes/aes.htm>

Other

FishBase. August 8, 2002 (cited October 10, 2002). <http://www.fishbase.org/search.cfm>

The Catalog of Fishes On-Line. February 15, 2002 (cited October 17, 2002). <http://www.calacademy.org/research/ichthyology/catalog/fishcatsearch.html>

Marcelo Carvalho, PhD

Orectolobiformes
(Carpet sharks)

Class Chondrichthyes
Order Orectolobiformes
Number of families 7

Photo: Banded cat sharks (*Chiloscylium puncta-tum*) live around coral reefs and tidepools. This is a juvenile cat shark. (Photo by Mark Smith/Photo Researchers, Inc. Reproduced by permission.)

Evolution and systematics

The exact origin of this group is unclear. The Orectobi-formes have a Jurassic record and show relationships to the Hy-bodontiformes, Squatiniformes, and Squaliformes, with such common characters as a flange on the teeth and nasal barbels.

The order Orectolobiformes comprises seven families: the Rhincodontidae (whale shark), the Stegostomatidae (zebra shark), the Orectolobidae (the wobbegongs), the Gingly-mostomatidae (the nurse sharks), the Parascylliidae (the collared carpet sharks), the Brachaeluridae (the blind sharks), and the Hemiscylliidae (the longtail carpet sharks).

Physical characteristics

The Orectolobiformes are small to very large sharks with prominent nasoral grooves (grooves connecting the nostrils to the mouth), nasal barbels, two dorsal fins, an anal fin, and small terminal or subterminal mouths.

Distribution

The Orectolobiformes are mainly an Indo-Pacific species. Only two species are found in other oceans. The nurse shark (*Ginglymostoma cirratum*) is found in both the Pacific and At-lantic Oceans. The whale shark has worldwide distribution.

Habitat

With a few exceptions, most of the species of the order are found in the shallow waters of the continental shelves. They often are bottom dwellers in rocky areas or coral reefs. The whale shark is the only pelagic species in the order.

Behavior

Most orectolobiformes are sluggish bottom-dwelling sharks that hide among the bottom rocks or coral heads during the day. Almost nothing is known about the behavior of most species.

Feeding ecology and diet

Most of the orectolobiformes are small, sluggish sharks that feed on small invertebrates and fishes. The whale shark, the largest fish in the world, with a maximum length of 39 ft (12 m), feeds on plankton. Little information exists on creatures that are predatory towards fishes in this order, but carcharhinid sharks are known to prey on small orectolobi-formes.

Reproductive biology

Both oviparous and viviparous species have been reported. Little is known about the reproductive processes of most orectolobiformes. The reproductive processes of the nurse shark are probably the best known (see species account fol-lowing).

Conservation status

Most species in the order are not threatened by fisheries. Recently, there have been localized fisheries for whale sharks in the Philippines, India, and Taiwan. There are no fisheries for them in the Atlantic. The whale shark has been protected in several countries, primarily on aesthetic grounds. Two species are listed on the IUCN Red List: the blue-gray car-pet shark (*Heteroscyllium colcloughi*) and the whale shark (*Rhin-codon typus*). Both are categorized as Vulnerable. In addition, the whale shark was added to CITES Appendix II in De-cember 2002.

Significance to humans

Most Orectolobiformes have little, if any, commercial im-portance. Some species are found in the aquarium trade. The nurse shark has been used for its liver oil, hides, and meat. Nurse shark liver oil was used for various purposes in the past. In Jamaica, the nurse shark was fished solely for its liver, which was used in burning; a fish yielded about a gallon of

A carpet shark (*Orectolobus maculatus*) yawning. (Photo by Tom McHugh/ Photo Researchers, Inc. Reproduced by permission.)

oil. In the Florida sponge fisheries of the 1880s, fishermen used nurse shark liver oil to calm the water surface so that they could scan the bottom continuously. A teaspoon of oil was said to produce a smooth surface for as long as a small boat of fishermen worked in one spot. People in Key West killed nurse sharks in summer and extracted the oil, which, at the time, sold for one dollar per gallon.

The hides of the nurse shark were the most valuables hides in the Florida shark fishery of the 1940s. Nurse shark hides were bought by the shark leather industry at prices about 25% higher than those of other species. The price of a 90-in (230 cm) hide was about $3.10 in 1943. At present, nurse sharks are little utilized in Florida. Some nurse sharks are fished for crab bait. The fins are worthless in today's industries.

Although the nurse shark is edible, its meat is very seldom found in Florida markets at this time. This limited utilization affords little protection to the species—many shark fishermen kill nurse sharks caught in their longlines, because they consider them a nuisance species that takes baits intended for other species. The species is included in the Fishery Management Plan for Sharks of the Atlantic Ocean, which has regulated the shark fisheries of the East Coast of the United States since 1993. The ability of the nurse shark to survive in confinement and its hardiness have made it the most popular aquarium and laboratory shark. It is certainly the most commonly displayed shark in aquaria throughout the Americas.

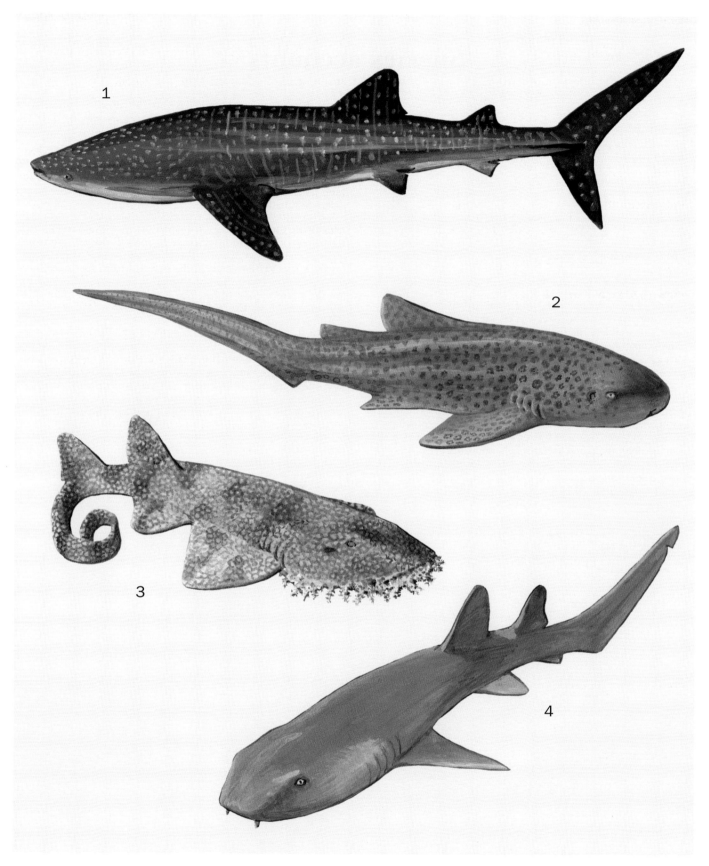

1. Whale shark (*Rhincodon typus*); 2. Zebra shark (*Stegostoma fasciatum*); 3. Tasseled wobbegong (*Eucrossorhinus dasypogon*); 4. Nurse shark (*Ginglymostoma cirratum*). (Illustration by Brian Cressman)

Species accounts

Nurse shark
Ginglymostoma cirratum

FAMILY
Ginglymostomatidae

TAXONOMY
Ginglymostoma cirratum Bonaterre, 1788, type locality not specified.

OTHER COMMON NAMES
None known.

PHYSICAL CHARACTERISTICS
Recognized by its conspicuous, long nasal barbels on the anterior margins of the nostrils and first dorsal fin originating over or posterior to the pelvic fin origin. Very wide head that gives it a tadpole appearance from above. Mouth is full of minute teeth, which are similar in both jaws. Teeth have one large cusp flanked on each side by two or three cusplets. Coloration ranges from rich yellowish to grayish-brown, with most specimens being reddish-brown. Yellow and even white specimens have been reported. Newborn nurse sharks have small black spots over the entire body, with an area of lighter pigmentation surrounding each spot and with bands of lighter and darker pigmentation alternating along the dorsal surfaces. The spots disappear by the time the specimens reach 20 in (50 cm) in length. Capable of limited color changes according to light intensity. Although the size and weight attained by the nurse shark often have been ex-aggerated in both the scientific and popular literatures, no specimen measured by researchers has exceeded 110 in (280 cm) in length. The largest specimen measured and weighed by Castro in Florida was 106 in (268 cm) long and weighed 243 lb (110 kg). The sizes of 132–144 in (335–365 cm) found in the literature must be considered exaggerations caused by Fowler's inaccurate estimates.

DISTRIBUTION
Distributed widely in littoral waters on both sides of the tropical and subtropical Atlantic. Ranges from tropical West Africa to the Cape Verde islands in the east and from southern Brazil to North Carolina and Rhode Island in the west. Also found on the western coast of America from the Gulf of California to Panama and Ecuador. Abundant in the shallow waters of tropical Florida and the Caribbean.

HABITAT
Small juveniles are found in shallow coral reefs and grass flats. Juveniles ranging in size from 47 to 67 in (1,200–1,700 mm) are found around shallow reefs and mangrove islands. Larger juveniles and adults are found near reefs and rocky areas at depths of 66–75 ft (20–75 m) during the daytime and in much shallower areas at night.

BEHAVIOR
Can be found resting on the bottom in small groups during the daytime, concealed under ledges or among boulders and rocks. These sharks often are in very close proximity to and

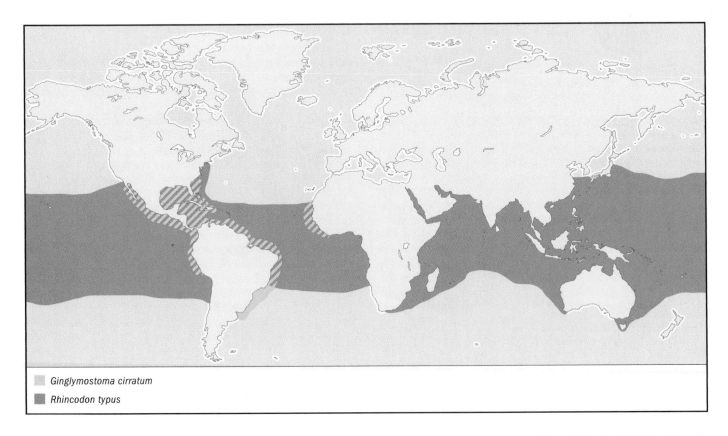

■ *Ginglymostoma cirratum*
■ *Rhincodon typus*

sometimes almost on top of one another. At dusk they disperse to search for food.

FEEDING ECOLOGY AND DIET
Said to feed chiefly on invertebrates (squids, shrimps, crabs, spiny lobsters, and sea urchins) and small fishes. A recent study of Florida nurse sharks showed that it is an opportunistic predator that consumes a wide range of species of small bony fishes, 3–9 in (8–23 cm) in length, primarily grunts. Prey typically is captured through a powerful sucking action. This suction accounts for the coral debris and solitary corals occasionally encountered in the stomach contents. Castro's study did not support the common belief that the nurse shark preys heavily on spiny lobsters and other crustaceans.

REPRODUCTIVE BIOLOGY
The nurse shark is a viviparous species without a placenta. In Florida and the Bahamas the mating period apparently extends from the last week in May to July, but most observations of mating have been made in mid-June. The embryos are enclosed in horny egg capsules for the first weeks of gestation. Embryos hatch out of the egg cases when they reach 8.9–9.2 in (21.8–23.3 cm). The embryos are lecithotrophic, that is they are fed from yolk stored in a yolk sac, and there is no evidence of any other mode of embryonic nourishment.

The embryos are in different developmental stages through the first four months of gestation. In females examined in October, some embryos measuring 8.5–9.2 in (21.5–23.3 cm) were still inside the egg cases, while others measuring 10.6–10.9 in (27–27.8 cm) had fully absorbed yolk sacs, open yolk sac scars and appeared ready for birth. During the last month of gestation all the embryos were in the same stage of development, that is, they all had absorbed the yolk sac and apparently were ready for birth. These embryos at different stages may have been a result of ovulation of the very large oocytes (2.3–2.4 in, or 5.9–6 cm, in diameter) and encapsulation of the eggs, lasting for two or three weeks, and of the very rapid development of the embryo thereafter. Females expel the empty egg cases after the embryos have hatched out of them.

The embryos measure 11.4–12 in (29–30.5 cm) at birth, after a gestation period estimated at about five to six months. Brood sizes are large, ranging from 21 to 50 young, with a median of 32. Aquarium observations on parturition suggest that the young are released over a period of a few days, usually at night. The reproductive cycle of the nurse shark encompasses a five- to six-month gestation period and a biennial reproductive cycle. After the gestation period of five to six months and birth in late November or early December, a female does not mate again until 18 months later, in June; thus reproduction is biennial.

CONSERVATION STATUS
Not listed by the IUCN.

SIGNIFICANCE TO HUMANS
The greatest value of the nurse shark probably lies in ecotourism. The nurse shark is the species of shark most often seen by recreational divers in Florida and the Caribbean. Because it is a large but harmless species, the nurse shark thrills divers that see one unexpectedly or at close range. In the last decade, numerous shark-watching operations have emerged in Florida and the Bahamas. In some locations nurse sharks have become habituated to being fed by divers. Although the long-term consequences and risks of these operations are still unclear, one can hope that public awareness and concern may result in some form of protection for these interesting animals. This species is one of the most common sharks in aquaria, be-

cause of its hardiness and its ability to survive in confinement for many years. According to Clark, one specimen survived 25 years at the Shedd Aquarium in Chicago, and another lived for 24 years at the Government Aquarium in Bermuda. The nurse shark is one of the most important species in shark research. ◆

Tasseled wobbegong
Eucrossorhinus dasypogon

FAMILY
Orectolobidae

TAXONOMY
Eucrossorhinus dasypogon Bleeker, 1867, Indonesia.

OTHER COMMON NAMES
English: Ogilby's wobbegong.

PHYSICAL CHARACTERISTICS
Bottom-dwelling, large shark with flattened body, wide mouth, and two dorsal fins set far back in the body. Head very flat, with a frill, or beard, of fleshy protuberances, known as dermal lobes, surrounding the outline of the head. These protuberances serve to break up the outline of the head against the bottom vegetation or rocks. Color is yellowish-brown or grayish-brown, with numerous reticulations and blotches. Coloration and frill of dermal lobes surrounding the outline of the head camouflage the shark very effectively against the bottom.

DISTRIBUTION
Western South Pacific in the shallow waters of Indonesia, Papua New Guinea, and northern Australia.

HABITAT
Found in shallow coral reefs. Poorly understood but the most commonly observed wobbegong in the Great Barrier Reef.

BEHAVIOR
Appears to be an ambush predator that waits for its prey while camouflaged against the bottom.

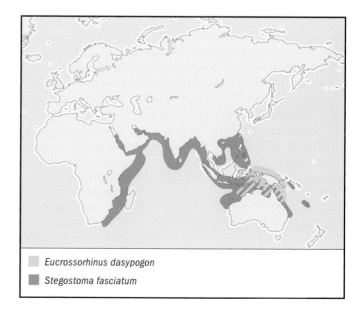

▨ *Eucrossorhinus dasypogon*

■ *Stegostoma fasciatum*

FEEDING ECOLOGY AND DIET
Details of its feeding ecology are not known.

REPRODUCTIVE BIOLOGY
Its reproductive processes are not known. It is believed to be ovoviviparous.

CONSERVATION STATUS
Not listed by the IUCN.

SIGNIFICANCE TO HUMANS
The species has little economic importance, other than as an attraction for fish watchers. Some attacks on divers have been attributed to this species. Its effective camouflage and large size make it potentially dangerous to divers, who may inadvertently approach it too closely, causing it to bite in self-defense. ◆

Whale shark
Rhincodon typus

FAMILY
Rhincodontidae

TAXONOMY
Rhincodon typus Smith, 1828, South Africa.

OTHER COMMON NAMES
None known.

PHYSICAL CHARACTERISTICS
Largest fish in the world. The average size for this species is 18–32.8 ft (5.5–10 m). The largest specimen measured was slightly over 39.4 ft (12 m) in length. It is said to grow larger, but no one has actually measured a whale shark over 40.03 ft (12.2 m). Very short snout, with a huge terminal mouth; its nostrils have short, blunt nasal barbels. Three pronounced longitudinal ridges along each side of the trunk, the lowermost ridges becoming strong caudal keels near the tail. Covered with white or yellowish dots and irregular bars. The spots, vertical bars, and longitudinal ridges along the flanks create a checkerboard appearance.

DISTRIBUTION
Cosmopolitan in tropical and subtropical waters.

HABITAT
Pelagic species that often approaches coastal areas.

BEHAVIOR
A sluggish shark, it often is seen swimming slowly on the surface, scooping up plankton and small fishes with its huge mouth.

FEEDING ECOLOGY AND DIET
Feeds on plankton. Plankton feeding is the most common and efficient feeding strategy of the largest sharks and most whales. Despite its being a ubiquitous species, we know very little about its feeding ecology.

REPRODUCTIVE BIOLOGY
This species is viviparous. Very little is known about reproduction of the whale shark, because only one gravid female has been examined. On 15 July 1995, a pregnant female was harpooned off the east coast of Taiwan. This female, estimated at about 35 ft (10.6 m) and weighing 17.6 tons (16 metric tons),

contained about 300 embryos in the uterui. As in the nurse shark, the embryos were in different stages of development. Some 237 of the embryos measured from 16.5 to 25.2 in (42–64 cm) in length; the largest embryos, at 22.8–25.2 in (58–64 cm), were probably ready to be born. This mode of reproduction is very similar to that of the nurse shark, where lecithotrophic embryos are found at different stages of development and hatch out of their egg cases at different times. This brood of 300 young is by far the largest reported for any elasmobranch.

CONSERVATION STATUS
The whale shark has been the subject of much attention; it is protected in some countries, although there are few actual threats to this ubiquitous species. The only active fisheries at the time of this writing are in India and Taiwan. The effects of these fisheries on the Pacific Ocean population are unknown. There are no whale shark fisheries in the Atlantic Ocean. This species is listed as Vulnerable by the IUCN.

SIGNIFICANCE TO HUMANS
Seeing this gigantic, harmless creature in the water is often an unforgettable experience. One ecotourism operation has developed around Ningaloo Reef, Australia, where whale sharks can be spotted at certain times of the year. Similar operations may develop elsewhere, given the general interest in whale sharks. ◆

Zebra shark
Stegostoma fasciatum

FAMILY
Stegostomatidae

TAXONOMY
Stegostoma fasciatum Smith, 1828, type locality not specified.

OTHER COMMON NAMES
English: Leopard shark.

PHYSICAL CHARACTERISTICS
Large species, reaching more than 9.8 ft (3 m) in length. Stout body with very long tail, almost as long as the rest of the body; short nasal barbels; and spectacular yellow coloration. The young are blackish-brown with vertical yellowish stripes and spots, hence the name zebra shark (although the coloration is the reverse of that of a zebra). Adults are yellowish with dark spots; hence the name leopard shark. Both juveniles and adults are easy to identify.

DISTRIBUTION
Found throughout the Indo-Pacific region.

HABITAT
Shallow coastal areas and coral reefs.

BEHAVIOR
Its behavior is poorly known, except that it often is seen resting on the bottom in coral reef areas.

FEEDING ECOLOGY AND DIET
Feeds primarily on gastropods, bivalves, and small fishes. Consumes large numbers of snails. Their spiral valve intestines often are full of the opercula (the horny or shell covering on a

snail's shell that closes the shell opening) of snails that they have digested, with the opercula stacking up like coins in the intestine.

REPRODUCTIVE BIOLOGY
Oviparous (egg layer). The egg cases are large, about 5.9–3.9 in (15–10 cm), and are provided with hairlike fibers that anchor them to the bottom.

CONSERVATION STATUS
Not listed by the IUCN.

SIGNIFICANCE TO HUMANS
Caught for food throughout its range. Small numbers are captured for the aquarium trade. The large adults are beautiful display fishes in large oceanaria. The species is of little economic importance.

Resources

Books

Anonymous. *Guide to Commercial Shark Fishing in the Caribbean Area.* Fishery leaflet no. 135. Washington, DC: U.S. Fish and Wildlife Service, 1945.

Bonaterre, J. P. *Tableau Encyclopédique et Méthodique des Trois Règnes de la Nature.* Paris: Panckoucke, 1789.

Cadenat, J., and J. Blache. "Requins de Méditerranée et de d Atlantique." In *Faune Tropicale.* Paris: ORSTOM, 1981.

Clark, E. "The Maintenance of Sharks in Captivity, with a Report on Their Instrumental Conditioning." In *Sharks and Survival,* edited by P. W. Gilbert. Boston: Heath and Co., 1963.

Dodrill, J.W. "A Hook and Line Survey of the Sharks Found Within Five Hundred Meters of Shore Along Melbourne Beach, Brevard County, Florida." Master's thesis, Florida Institute of Technology, 1977.

Gosse, P. H. *A Naturalist's Sojourn in Jamaica.* London: Longman, Brown, Green and Longmans, 1851.

Gudger, E. W. "The Breeding Habits, Reproductive Organs and External Embryonic Development of *Chlamydoselachus anguineus,* Based on Notes and Drawings by Bashford Dean." In *The Bashford Dean Memorial Volume: Archaic Fishes,* edited by E. W. Gudger. New York: American Museum of Natural History, 1940.

Masefield, J., ed. *Dampier's Voyages.* 2 vols. New York: E. P. Dutton, 1906.

Murdy, E. O., R. Birdsong, and J. A. Musick. *Fishes of the Chesapeake Bay.* Washington, DC: Smithsonian Institution, 1997.

Parra, A. *Descripción de Diferentes Piezas de Historia Natural.* Havana: Imprenta de la Capitanía General, 1787.

Poey, F. *Repertorio Físico-Natural de la Isla de Cuba.* Havana: La Viuda de Barcina y Comp., 1868.

Rathbun, R. "The Sponge Fishery and Trade." In *The Fisheries and Fishery Industries of the United States,* edited by G. B. Goode. Washington, DC: U.S. Government Printing Office, 1887.

Rivera-López, J. "Studies on the Biology of the Nurse Shark, *Ginglymostoma cirratum* Bonnaterre, and the Tiger Shark, *Galeocerdo cuvieri* Perón and Le Sueur." Master's thesis, University of Puerto Rico, 1970.

Wheeler, A. *The Fishes of the British Isles and Northwest Europe.* London: Macmillan, 1969.

Periodicals

Anonymous. "Reproducción del Tiburón Gato (*Ginglymostoma cirratum*)." *Boletín del Centro de Investigaciones, Educación, y Recreación "CEINER"* (Cartagena) 1, no. 10 (1992): 4.

Applegate, S. P. "A Revision of the Higher Taxa of Orectolobids." *Journal of the Marine Biology Association of India* 14, no. 2 (1972): 743–751.

Baughman, J. L., and S. Springer. "Biological and Economic Notes on Sharks of the Gulf of Mexico, with Special Reference to Those of Texas and with a Key for Their Identification." *American Midland Naturalist* 44, no. 1 (1950): 96–152.

Beebe, W. "External Characteristics of Six Embryo Nurse Sharks, *Ginglymostoma cirratum*." *Zoologica* 26, no. 1 (1941): 9–12.

Beebe, W., and J. Tee-van. "Fishes from the Tropical Eastern Pacific." Part 2: "Sharks." *Zoologica* 26, no. 2 (1941): 93–122.

Bigelow, H. B., and W. C. Schroeder. "Sharks." In *Fishes of the Western North Atlantic.* New Haven: Memoirs of the Sears Foundation of Marine Research, 1948.

Carrier, J. C., H. L. Pratt, Jr., and L. K. Martin. "Group Reproductive Behaviors in Free-Living Nurse Sharks, *Ginglymostoma cirratum*." *Copeia* 1994, no. 3 (1994): 646–656.

Castro, J. I. "Biology of the Blacktip Shark, *Carcharhinus limbatus,* off the Southeastern United States." *Bulletin of Marine Science* 59, no. 3 (1996): 508–522.

———. "The Biology of the Nurse Shark, *Ginglymostoma cirratum,* off the Florida East Coast and the Bahama Islands." *Environmental Biology of Fishes* 58 (2000): 1–22.

Clark, E., and K. von Schmidt. "Sharks of the Central Gulf Coast of Florida." *Bulletin of Marine Science* 15 (1): 13–83.

Coles, Russell J. "Notes on the Sharks and Rays of Cape Lookout, N.C." *Proceedings of the Biological Society of Washington* 28 (1915): 89–94.

Dahlberg, M. D., and R. W. Heard III. "Observations on Elasmobranchs from Georgia." *Quarterly Journal of the Florida Academy of Sciences* 32 (1969): 21–25.

Fowler, H. W. "Some Cold-Blooded Vertebrates from the Florida Keys." *Proceedings of the Academy of Natural Sciences Philadelphia* 58 (1906): 77–113.

Gudger, E. W. 1912. "Summary of Work Done on the Fishes of the Dry Tortugas." *Carnegie Institute of Washington* 11 (1912): 148–150.

Joung, S. J., C-T Chen, E. Clark, S. Uchida, and W. Y. P. Huang. "The Whale Shark, *Rhincodon typus,* Is a Livebearer: 300 Embryos Found in One 'Megamamma' Supreme." *Environmental Biology of Fishes* 46 (1996): 219–223.

Resources

Klimley, A. P. "Observations of Courtship and Copulation in the Nurse Shark, *Ginglymostoma cirratum*." *Copeia* 1980, no. 4 (1980): 878–882.

Pratt, H. L. Jr., and J. C. Carrier. "Wild Mating of the Nurse Sharks." *National Geographic Magazine* 187, no. 5 (1995): 44–53.

Regan, C. T. "A Classification of the Selachian Fishes." *Proceedings of the Zoological Society of London* (1906): 722–758.

———. "A Revision of the Family Orectolobidae." *Proceedings of the Zoological Society of London* (1908): 347–364.

Wourms, J. P. "Viviparity: The Maternal-Fetal Relationship in Fishes." *American Zoologist* 21 (1981): 473–515.

José I. Castro, PhD

Carcharhiniformes

(Ground sharks)

Class Chondrichthyes
Order Carcharhiniformes
Number of families 8

Photo: A whitetip reef shark (*Triaenodon obesus*) resting on coral rubble near Fiji. (Photo by Fred McConnaughey/Photo Researchers, Inc. Reproduced by permission.)

Evolution and systematics

The fossil history of ground sharks (Carcharhiniformes) is known from a handful of preserved skeletons, but long intervening periods exist in which fossil skeletons of ground sharks have not been recovered. Ground sharks first appear in the late Jurassic (some 150 million years ago) Solnhofen limestones of Germany. These early fossils (e.g., *Macrourogaleus*) are not well preserved, but they bear some resemblance to modern catsharks (family Scyliorhinidae). After a long absence from the fossil record, fossil ground shark skeletons reappear in the late Cretaceous chalk deposits of Lebanon (ranging in age from 84 to 95 million years ago, or mya). These sharks (e.g., *Pteroscyllium* and *Paratriakis*) are thought to be related to catsharks and hound sharks (Triakidae), respectively, but on scant evidence. A few species of fossil catsharks from Lebanon are even placed in the living genus *Scyliorhinus*, which would give it a remarkable longevity of some 90 million years. These fossils have been studied only superficially, however, and they probably represent extinct genera of uncertain affinity. Partial skeletons are present in the Monte Bolca beds of northeastern Italy (*Eogaleus* and *Galeorhinus*) of Eocene age (some 52 mya), again after a hiatus of more than 30 million years.

Many extinct species of ground sharks are known from isolated teeth, which are widespread and provide a fairly robust stratigraphic record. Tertiary ground shark fossils are relatively modern in their level of diversity. Ground shark fossils are present on every continent, indicating that they have been distributed widely for the past 65 million years at least. Remarkably, the fossil record of ground sharks parallels their phylogenetic history, where the most "primitive" family (Scyliorhinidae) also is the oldest.

Ground sharks are related closely to bullhead (Heterodontiformes), carpet (Orectolobiformes), and mackerel (Lamniformes) sharks among living elasmobranchs (sharks and rays), forming the larger group known as the Galeomor-phii. Galeomorph sharks are characterized by several evolutionary innovations, such as the unique placement of the hyomandibula (a cartilage supporting the jaws posteriorly) on the skull. Within the Galeomorphii, carcharhiniforms are related most closely to the mackerel sharks, as they share a tripodal rostrum (the anterior extension of the skull) supporting the snout internally. All carcharhiniforms have specialized secondary lower eyelids ("nictitating" eyelids, which are absent from all other sharks) as well as unique clasper skeletons. Similarly to lamnoids (a subgroup within mackerel sharks), there is also a group of "higher carcharhiniforms" characterized by plesodic pectoral fins (with internal supports reaching the fin margin).

There are approximately 216 species, 48 genera, and eight families in the Carcharhiniformes. This amounts to slightly more than half of all shark species and about half of all shark genera. The eight families are the Scyliorhinidae (catsharks, 15 genera and some 105 species—the largest shark family of any order), Proscylliidae (finback catsharks, three genera and five species), Pseudotriakidae (false catshark, monotypic), Leptochariidae (barbeled hound shark, monotypic), Triakidae (hound sharks, 10 genera and 39 species), Hemigaleidae (weasel sharks, four genera and seven species), Carcharhinidae (requiem sharks, 12 genera and 50 species), and Sphyrnidae (hammerhead sharks, two genera and eight species). New carcharhiniform species have been described in recent years, particularly of catsharks, and additional new species await formal description. Phylogenetic relationships among ground shark genera require further study, which may result in the merging of several currently monotypic genera and even of some of the families.

Physical characteristics

There are many different morphological and ecological trends within the Carcharhiniformes, which is to be expected from a large group that inhabits waters from the intertidal

A tiger shark (*Galeocerdo cuvier*) demonstrating aggressive behavior. (Photo by Jeff Rotman/Photo Researchers, Inc. Reproduced by permission.)

zone to the lower reaches of the continental slope. However, the morphological differences among ground shark families are not as great compared with the other orders of sharks, even though some families are quite distinctive. One of these families is the hammerhead shark (Sphyrnidae), which is unique among all sharks in having a laterally expanded head (the hammerhead, or cephalofoil, with eyes on the lateral extremes). Hammerhead sharks are otherwise very similar to requiem (carcharhinid) sharks. Catsharks (Scyliorhinidae) also are recognized easily, as their first dorsal fins are situated either on the same level as or behind the pelvic fins. The false catshark (Pseudotriakidae) is unique among sharks in having a first dorsal fin that is much longer than the caudal fin. The differences among the remaining families are subtle, and one must look at their teeth, labial furrows, and even intestines to identify them.

Carcharhiniforms are small, medium, or large sharks; adults usually range from 18 in (45 cm) to 20 ft (6 m) in length. The proscylliid *Eridacnis radcliffei* reaches only about 9.4 in (24 cm) in length and is one of the smallest known species of sharks. Ground sharks have two spineless dorsal fins (one species, *Pentanchus profundicolus,* with only one), the first larger than the second. There is a large caudal fin with a greater upper lobe, a prominent anal fin about as large as the pelvic fins (or even larger in some catsharks, especially *Apristurus* species), and moderately developed pectoral fins. There are five gill openings. The snout can be conical or broadly rounded (elongated in *Isogomphodon*). The eyes are elliptical, but they are rounded in some genera (e.g., *Rhizoprionodon*). The spiracles vary from a small pore to an opening just smaller than the eyes. When present, labial furrows (grooves along-side the mouth corners) vary from long to short. There is either a spiral or a scroll intestinal valve. The teeth may vary considerably between adults and juveniles of the same species, upper and lower jaws, and males and females and between species, genera, and families. A single broad, slanted, or erect cusp may be present, or there may be as many as five cusps per tooth (Proscylliidae). Many species are identified by their dental morphological features and formulas. The body is covered with small dermal denticles that do not form larger spines.

Carcharhiniforms vary widely in coloration. Shallow-water catsharks can be spectacularly colored, with many spots, saddle-like markings, and blotches, whereas deeper-water catsharks usually are drab or dark brown or black. Many species (requiem and hammerhead sharks) are gray or brown dorsally and laterally, with creamy or white ventral surfaces. Many triakids also are spotted or have other conspicuous markings similar to those of scyliorhinids. Many ground shark species have unique coloration.

Distribution

They are found worldwide in tropical to temperate waters, including cool boreal seas, but they are most abundant in tropical and warm temperate regions. Carcharhiniform sharks inhabit all major oceans except the Antarctic seas. (Deepwater catsharks of the genus *Apristurus* may inhabit Arctic waters.) They also are present in tropical freshwaters (rivers and lakes) in South, Central, and North America; Africa; Asia; and Australia.

Habitat

Carcharhiniforms are most abundant in tropical continental shelf regions. Most inshore, littoral habitats, including coasts, estuaries, river mouths, open bays and lagoons, atolls, and coral reefs (both coastal and barrier reefs), are occupied by ground sharks. They also are abundant offshore, off oceanic and continental islands, and are present in deeper waters along the upper continental slopes (especially species of *Apristurus* and *Pseudotriakis*). Some species are epipelagic in deeper ocean basins.

Behavior

Carcharhiniform sharks are present in many habitats, from the littoral to the oceanic; as a consequence they vary from sluggish, primarily bottom dwellers (such as many catsharks) to swift swimming, more active pelagic forms (e.g., blue shark and oceanic whitetip shark). The behavior of certain ground shark species has been studied in both captive and natural conditions (in particular, the lemon, gray reef, and bonnethead sharks). In general, pelagic species appear to cruise at low speeds, occasionally bursting into sudden activity, while bottom-dwelling species are more territorial and mostly nocturnal.

Some carcharhiniform species form aggregations, often by size or sex (except during mating season). Schooling is a very common social behavior in hammerhead sharks (e.g., the scalloped hammerhead, *Sphyrna lewini*, off the eastern Pacific coast of Mexico, where some 225 individuals may school together), but most carcharhiniforms spend much of their lives alone. Gray reef sharks (*Carcharhinus amblyrhynchos*) form schools in the Marshall Islands, as do lemon sharks (*Negaprion brevirostris*) in the Bimini Islands, Bahamas. Schools typically form during the day and break up for individuals to feed at night. Additionally, some species are now known to rest in caves for long periods; the whitetip reef shark, *Triaenodon obesus*, rests mostly during the day and sometimes in groups.

Sharks spend much of their lives as solitary predators and do not form family groups or cooperate with each other. Their types of behavior are less complex than those of marine mammals, but only a handful are known. In particular, aggressive display has been documented (e.g., for the gray reef shark), which may involve jerky head movements, arching of the back, and downward pointing of the pectoral fins. Social displays and social organization are particularly well known for the bonnethead, *Sphyrna tiburo*. Some of these displays are an outcome of size-dependent dominance hierarchies, such as swimming in a straight line. Bonnetheads do not show much aggressive behavior, a possible indication that more social species may be less aggressive.

Feeding ecology and diet

Ground sharks are voracious predators; none are filter feeders. Food items consist of numerous families of bony fishes, sharks and rays, marine mammals and marine mammal carrion, seabirds, marine reptiles (mostly turtles), and a wide range of invertebrates, including crustaceans, squid, octopi, cuttlefish, and shelled mollusks. Benthic ground sharks feed on items more readily available on or close to the bottom,

A copper shark (*Carcharhinus brachyurus*) near North Neptune Island, Australia. The shark is frequently found near Australia, but has been seen in the waters surrounding all continents except Antarctica. (Photo by Animals Animals ©James Watt. Reproduced by permission.)

such as hard-shelled invertebrates and certain fishes, whereas open-ocean forms feed intensely on pelagic fishes, such as tunas and their allies (Scombridae). Larger sharks prey on ground sharks, and larger species of ground sharks may feed on smaller ones. The tiger shark has the least selective diet of all sharks. Most species are not highly specialized in their feeding habits, but hammerheads are known to have a particular predilection for stingrays.

Reproductive biology

Carcharhiniform sharks are either oviparous (egg layers) or viviparous (giving birth to live young). Oviparous species deposit egg cases that contain the developing embryo along with its yolk reserves (in the yolk sac). These species include the majority of the catsharks (Scyliorhinidae, except *Cephalurus* and possibly some *Halaelurus* species) and *Proscyllium habereri* (Proscylliidae). The egg cases are secreted by the nidamental gland in the upper oviduct and usually are amber to greenish in color, with tendrils at the extremities that serve to anchor them to the substrate. In species with retained oviparity, the egg cases remain for a longer period in the uterus, with most embryonic development taking place inside the mother. In other oviparous species, the eggs are laid shortly after they are formed (less than one month in some cases), and most development of the fetus, which may take up to one year, occurs inside the egg cases in the environment.

Slightly more than half of carcharhiniform species are viviparous. Viviparous species can be yolk sac viviparous (ovoviviparous or aplacentally viviparous—the young deriving nourishment solely from the yolk sac, such as in the tiger shark, *Galeocerdo*), but many viviparous species form maternal-fetal connections in the form of yolk sac placentae. In these cases, the yolk sacs are modified into highly vascularized, nutrient-supplying structures fused to the internal uterine walls. Placentae are formed in the Hemigaleidae, Carcharhinidae (except *Galeocerdo*), Sphyrnidae, and some triakid species.

Gestation periods vary considerably; oviparous species lay eggs after a short gestation of just a few weeks, but some viviparous species retain the embryos for more than a year.

A scalloped hammerhead shark (*Sphyrna lewini*) swimming at night in Kanohe Bay, Hawaii. (Photo by Jeff Rotman/Photo Researchers, Inc. Reproduced by permission.)

Gray reef sharks (*Carcharhinus amblyrhynchos*) tend to be gregarious rather than solitary. (Photo by Animals Animals ©James Watt. Reproduced by permission.)

Litters vary from one to 135 per gestation. In many species, females give birth in shallow nursery areas. Males bite females during courtship, and mating has been observed in the wild for a few species (such as the whitetip reef shark, *Triaenodon obesus*). There is no parental care after birth.

Conservation status

The following species are listed by the IUCN: *Glyphis gangeticus* (as Critically Endangered); *Carcharhinus melanopterus, C. borneensis,* and *Glyphis glyphis* (as Endangered); *Galeorhinus galeus* (as Vulnerable); *Mustelus antarcticus* (as Lower Risk/ Conservation Dependent); *C. amboinensis* (southwestern Indian Ocean subpopulation only), *C. amblyrhynchoides, C. amblyrhynchos, C. brevipinna, C. leucas, C. limbatus, C. longimanus, C. melanopterus, C. obscurus, C. plumbeus* (northwestern Atlantic subpopulation only), *Furgaleus macki, Galeocerdo cuvier, Negaprion brevirostris, Poroderma africanum, Prionace glauca, Scoliodon laticaudus, Triaenodon obesus, Triakis megalopterus,* and *T. semifasciata* (as Lower Risk/Near Threatened); *C. brevipinna* (northwestern Atlantic subpopulation only), *C. hemiodon, C. limbatus* (northwestern Atlantic subpopulation only), *C. obscurus* (northwestern Atlantic and Gulf of Mexico subpopulation only), and *Galeorhinus galeus* (as Vulnerable); and *C. amboinensis* and *Smyrna mokarran* (as Data Deficient).

Significance to humans

Ground sharks are fished intensely, both for food and recreationally. Because they are abundant in shallow and oceanic waters, ground sharks frequently are fished by trawlers and longlines and are either targeted directly or captured as bycatch. Their flesh is marketed frozen, fresh, dried-salted, smoked, and even canned for human consumption. Their skin is used for leather products, their fins for the Chinese shark fin soup industry, their carcasses for fishmeal, and their liver oil for the extraction of vitamin A (in decline as vitamins are synthesized). Tourists often procure trophies, in the form of jaws and teeth. Recreational fisheries and angling tournaments capture large quantities of ground sharks, especially tiger sharks in shallow waters and blue sharks in oceanic settings. Internal fertilization, long gestation periods, production of few offspring, and relatively advanced ages at sexual maturity are all factors that constrain the exploitation of shark populations.

Ground sharks have been implicated in numerous shark attacks, especially the tiger and bull sharks, which account for more than 50% of shark attacks worldwide. This proportion is to be expected because of the high number of carcharhiniform species and their shallow-water predominance.

Carcharhiniforms also are very important in the growing ecotourism market. Many species can be encountered in the wild through commercial operations that specialize in taking tourists to areas where specific carcharhiniform species are common. These operations are worldwide, and surveys indicate that shark watching is a highly profitable enterprise. Certain ground shark species are common in public aquaria as well, especially *Triaenodon obesus, Carcharhinus plumbeus,* and *Triakis semifasciata.*

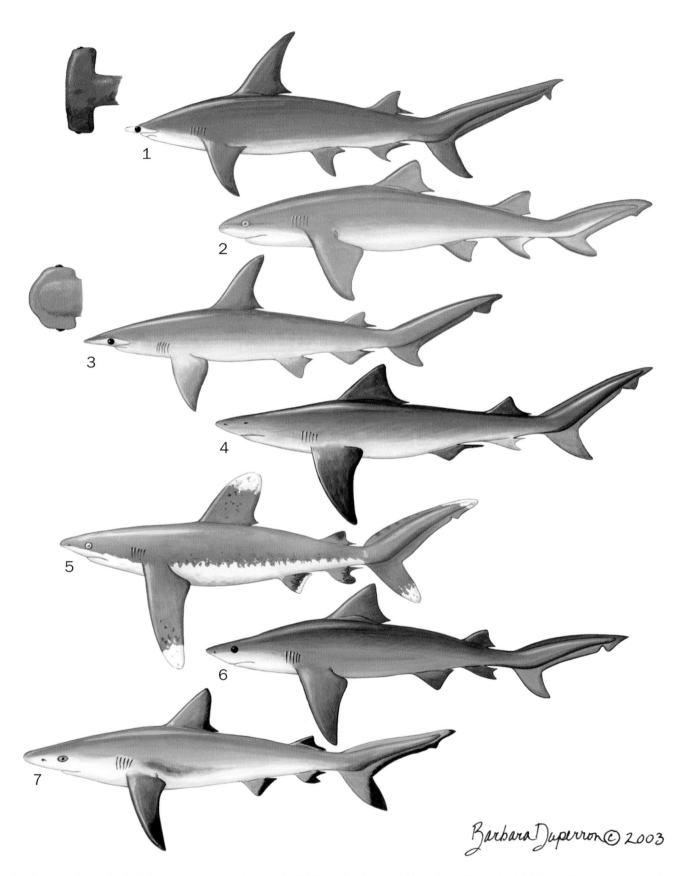

1. Great hammerhead shark (*Sphyrna mokarran*); 2. Lemon shark (*Negaprion brevirostris*); 3. Bonnethead shark (*Sphyrna tiburo*); 4. Ganges shark (*Glyphis gangeticus*); 5. Oceanic whitetip shark (*Carcharhinus longimanus*); 6. Bull shark (*Carcharhinus leucas*); 7. Gray reef shark (*Carcharhinus amblyrhynchos*). (Illustration by Barbara Duperron)

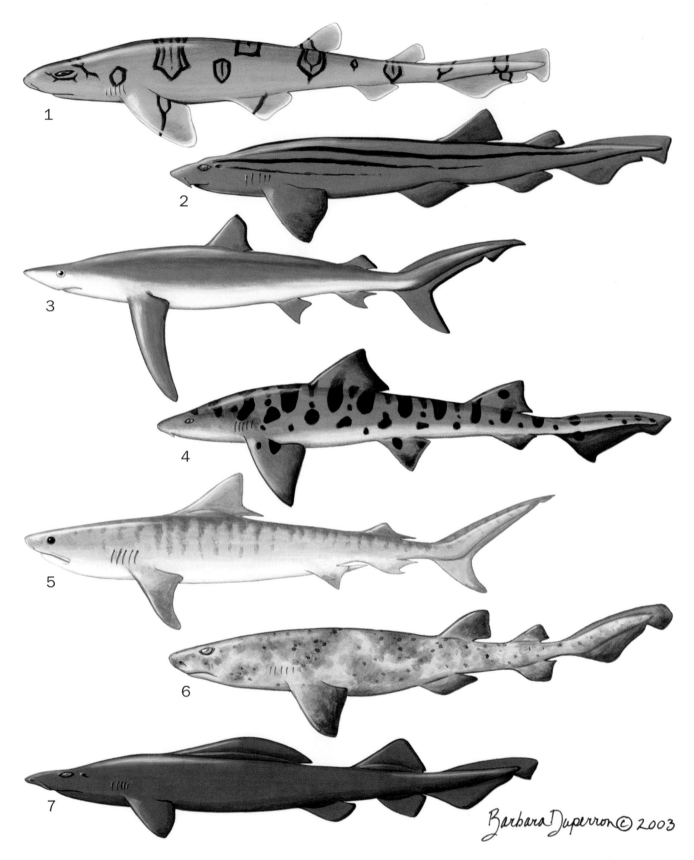

1. Chain catshark (*Scyliorhinus retifer*); 2. Pajama (*Poroderma africanum*); 3. Blue shark (*Prionace glauca*); 4. Leopard shark (*Triakis semifasciata*); 5. Tiger shark (*Galeocerdo cuvier*); 6. Swellshark (*Cephaloscyllium ventriosum*); 7. False catshark (*Pseudotriakis microdon*). (Illustration by Barbara Duperron)

Species accounts

Gray reef shark
Carcharhinus amblyrhynchos

FAMILY
Carcharhinidae

TAXONOMY
Carcharias (Prionodon) amblyrhynchos Bleeker, 1856, Java Sea, near Salambo Islands, Indonesia.

OTHER COMMON NAMES
French: Requin dagsit; Spanish: Tiburón de arrecifes.

PHYSICAL CHARACTERISTICS
Somewhat elongated snout but with a rounded anterior profile, first dorsal fin much larger than the second, and rounded eyes. Upper teeth have posteriorly slanted cusps with faint serrations; lower teeth are slender and more erect. Coloration is grayish dorsally and laterally, but with a distinctive black caudal fin margin. Reaches some 8 ft (2.5 m) in length.

DISTRIBUTION
Occurs in mostly tropical waters of the Indian and Pacific Oceans.

HABITAT
An inshore shark but also occurs pelagically and sometimes frequents oceanic waters from the intertidal zone down to about 330 ft (100 m). A common species in coral reefs, atolls, and shallow lagoons. Usually found close to the bottom but also may be seen near the surface.

BEHAVIOR
A curious shark that investigates "novel" circumstances while swimming. Approaches divers frequently but typically disappears shortly thereafter. Can be aggressive when in pursuit of prey. A particular threat display has been observed and documented, in which the gray reef shark arches its back, points its pectorals downward, lifts its head, moves its snout from side to side repeatedly, and even swims in a horizontal spiral. Considered to be a very social species.

FEEDING ECOLOGY AND DIET
Feeds on bony fishes (especially those that inhabit reefs and are shorter than 12 in, or 30 cm), octopi, squid, and a wide variety of crustaceans. Notable for being able to capture prey in tight crevices in reefs.

REPRODUCTIVE BIOLOGY
Viviparous, with a yolk sac placenta and litters raging from one to six young. Gestation period is about one year. Individuals are sexually mature by seven to seven and a half years old. Males are mature at a length of 51–57 in (130–145 cm) and females when they reach 48–54 in (122–137 cm).

CONSERVATION STATUS
Listed as Lower Risk/Near Threatened by the IUCN because of its relatively late age at maturity and increasing pressure from unmanaged fishing.

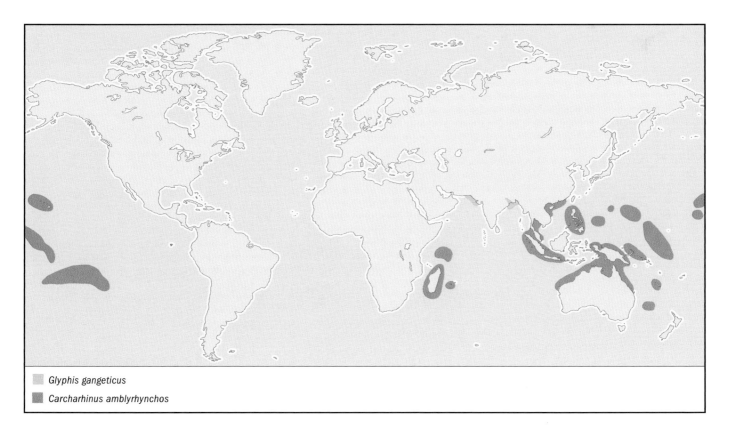

■ *Glyphis gangeticus*
■ *Carcharhinus amblyrhynchos*

SIGNIFICANCE TO HUMANS
A very common reef shark, with great potential for dive tourism, as is readily seen by ecotourists in many locations (e.g., Australia, Mauritius, and the Philippines). Fishing and utilization mostly unrecorded. ◆

Bull shark
Carcharhinus leucas

FAMILY
Carcharhinidae

TAXONOMY
Carcharias (Prionodon) leucas Valenciennes in Müller and Henle, 1839, Antilles.

OTHER COMMON NAMES
French: Requin bouledogue; Spanish: Tiburón sarda; Portuguese: Cabeça-chata.

PHYSICAL CHARACTERISTICS
Characteristically short and blunt snout, somewhat arched back, and relatively small eyes. Large first dorsal fin (much larger than the second dorsal fin). Triangular upper teeth with small cusplets. Upper teeth more broad than lower teeth, which are smooth laterally. Thirteen upper tooth rows and 12 lower rows. Gray to brownish dorsal and lateral coloration. Reaches 11.5 ft (3.5 m) in length.

DISTRIBUTION
Worldwide in tropical shallow waters but also ascending tropical rivers and freshwater lakes. Freshwater occurrences include the Amazon and Ucayali Rivers in South America, reaching

upriver as far as 2610 mi (4,200 km) from shore); the Mississippi and Atchafalaya Rivers of the United States; Lake Nicaragua and San Juan River (Nicaragua); Lake Izabal and Dulce River (Guatemala); and the Patuca River (Honduras). Also in freshwaters in Belize and probably elsewhere in other neotropical systems. Other freshwater occurrences include many African rivers (Gambia, Ogooué, and Zambezi Rivers), Middle Eastern rivers (the Tigris), Indian waters (Hooghly Channel of the Ganges River), New Guinea waters (Lake Jamoer), and systems in Australia (Brisbane River). Present in some oceanic islands (Fiji).

HABITAT
In the sea the bull shark is widespread in inshore, shallow waters, frequenting bays, estuaries, river mouths, and waters off piers and docks, usually down to a depth of 98.4 ft (30 m) but reaching 492 ft (150 m). Its capacity to penetrate freshwaters extensively and remain in them, tolerating great ranges in salinity, has been the subject of much scientific research. Freshwater populations are not believed to be landlocked, however, and migrate frequently to the sea, such as in the Lake Nicaragua system.

BEHAVIOR
Active both during the day and at night. May aggregate to migrate to cooler waters in the summer from equatorial latitudes, returning when water temperatures become too cool. Smaller, younger individuals may be more common close to shore, whereas larger individuals may inhabit slightly deeper waters. Appears somewhat sluggish but is capable of swift movements and sudden bursts of activity.

FEEDING ECOLOGY AND DIET
Feeds extensively on many different bony fishes as well as sharks and rays but is capable of consuming a wide range of

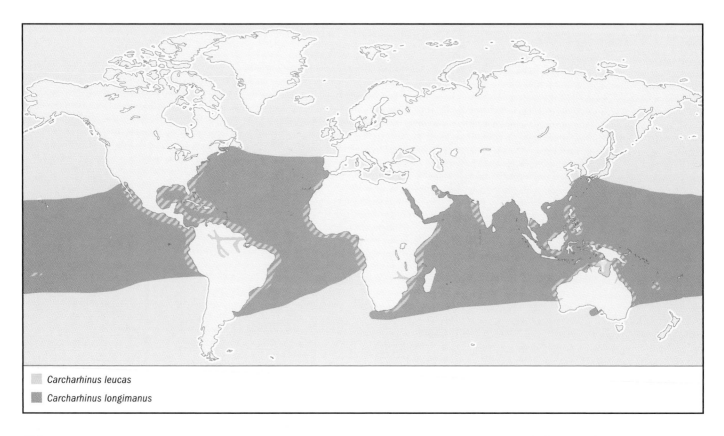

☐ *Carcharhinus leucas*
■ *Carcharhinus longimanus*

prey, including invertebrates and marine mammals, reptiles, and birds.

REPRODUCTIVE BIOLOGY
Viviparous with a yolk sac placenta; litters range from one to 13 young. Gestation periods range from 10 to 11 months. Breeding in freshwaters may occur (e.g., in Lake Nicaragua), but most breeding takes place in the sea. Sexual maturity is attained at about 98.4 in (250 cm) in length, after some six years. Pups frequently are born in sheltered nursing areas. Lengths at birth range from roughly 19.7 to 31.5 in (50–80 cm).

CONSERVATION STATUS
Listed as Lower Risk/Near Threatened by the IUCN, mainly because of its occurrence close to heavily populated areas and frequency of capture by local fisheries.

SIGNIFICANCE TO HUMANS
A hardy aquarium species; individuals have lived for 15 years. Captured as bycatch by fisheries in many places, leading to a concern that the bull shark may be threatened in some areas. Its meat is consumed fresh, dried/salted, and smoked, and its liver is particularly rich in oil. Also captured recreationally on hook and line in many regions. Considered a dangerous shark, with many attacks reported. It is believed that some attacks are caused by other species, such as the great white in temperate waters and the Ganges shark in the Ganges-Hooghly river system. Some attacks have occurred in freshwater (e.g., in Lake Nicaragua). The bull shark can be encountered in the wild in many places worldwide (e.g., the Bahamas, Cuba, and Belize). ◆

Oceanic whitetip shark
Carcharhinus longimanus

FAMILY
Carcharhinidae

TAXONOMY
Squalus longimanus Poey, 1861, Cuba.

OTHER COMMON NAMES
French: Requin océanique; Spanish: Tiburón oceánico.

PHYSICAL CHARACTERISTICS
Short and blunt snout. First dorsal fin is characteristically large, with a unique, broadly rounded apex; pectoral fins are long and also unique, with broadly rounded apex. Small second dorsal fin and large caudal fin. Broad upper teeth that are triangular with lateral serrations; lower teeth have straight, slender cusp. Gray dorsal and lateral color, with uniquely white extremities of pectoral and first dorsal fins, sometimes with darker blotches. Second dorsal, pelvic, anal, and caudal fins have dark tips; caudal extremities sometimes also white. Reaches 13 ft (4 m) in length.

DISTRIBUTION
Worldwide in tropical and temperate inshore and oceanic waters.

HABITAT
Typically occurs close to the surface, in offshore oceanic waters but may venture close to shore occasionally in waters as shallow as 121 ft (37 m). More abundant in the tropics.

BEHAVIOR
May segregate by sex and size, but little is known of its population structure. Slow moving but capable of quick bursts of

energy. Oceanic whitetips appear to cruise with their pectoral fins widely spread out. Inquisitive, the oceanic whitetip will investigate potential prey items by circling them repeatedly.

FEEDING ECOLOGY AND DIET
Feeds on oceanic bony fishes of numerous families as well as on sharks and pelagic stingrays and invertebrates, such as oceanic cephalopods. Has been noted to feed voraciously on schools of fish. Also may feed on marine mammal carrion, seabirds, and turtles.

REPRODUCTIVE BIOLOGY
Viviparous, with a yolk sac placenta and litters ranging from one to 15 young. Gestation periods of about one year have been reported, but little is known about the reproductive biology. Reproductive seasons may not strictly exist, at least in the central Pacific, where gravid females have been found year-round. Lengths at sexual maturity are between 71 and 78.7 in (180–200 cm) for females and 69 and 78 in (175 to 198 cm) for males; lengths at birth vary from 23.6 to 25.6 in (60–65 cm).

CONSERVATION STATUS
Listed as Lower Risk/Near Threatened by the IUCN, because it is captured frequently as by-catch in pelagic tuna fisheries and owing to its presumably low reproductive capacity.

SIGNIFICANCE TO HUMANS
A few verified attacks on people have occurred, and the oceanic whitetip has been regarded as somewhat aggressive when approaching divers or boats. Regularly captured by pelagic longlines. The flesh is consumed fresh, dried/salted, and smoked, and the fins are coveted by the shark fin soup industry. Can be seen in the waters off Hawaii, the Red Sea, and Australia. ◆

Tiger shark
Galeocerdo cuvier

FAMILY
Carcharhinidae

TAXONOMY
Squalus cuvier Peron and LeSuer, 1822, Australia.

OTHER COMMON NAMES
French: Requin tigre commun; Spanish: Tintorera.

PHYSICAL CHARACTERISTICS
Characteristically short and rounded snout. A large first dorsal fin (well anterior to the pelvic fin origin), well-developed caudal fin, and long upper labial furrows. Unique teeth, with posteriorly curved and serrated cusps. Coloration of dark vertical stripes (more apparent in juveniles) over a gray background. Large females reach 19.7 ft (6 m) in length, with unconfirmed records of up to 29.5 ft (9 m).

DISTRIBUTION
Worldwide in tropical to warm temperate, mostly continental waters but may occur pelagically in the western Pacific Ocean.

HABITAT
The tiger shark is mainly an inshore, warm-water species, occurring in continental waters as well as in remote oceanic islands from the intertidal zone down to about 459 ft (140 m). Very common in turbid waters, off river estuaries, near piers, and in coral reefs. May be found pelagically offshore but is not a truly oceanic shark like the blue shark.

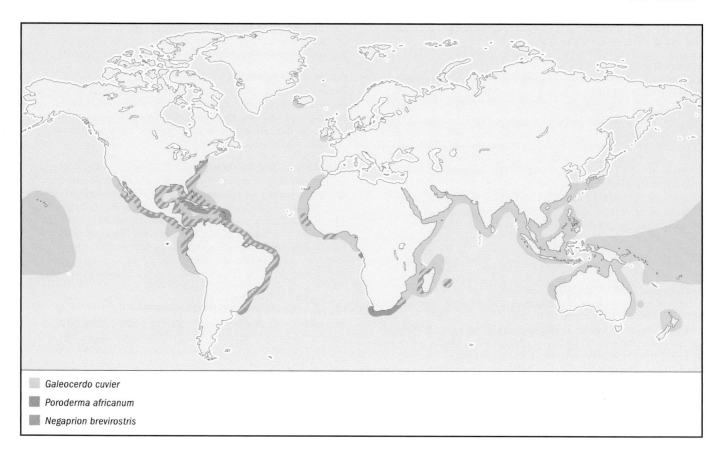

Galeocerdo cuvier

Poroderma africanum

Negaprion brevirostris

BEHAVIOR
A mostly nocturnal, active, and strong-swimming species, capable of frequenting very shallow lagoons. Usually solitary. Appears sluggish because it cruises at slow speeds near the surface. Can approach and display aggressive behavior toward divers, but in many instances tiger sharks have been turned away by strong retaliation.

FEEDING ECOLOGY AND DIET
Feeds on a wide variety of vertebrates and invertebrates, and is considered among the least specialized of sharks in relation to diet, scavenging opportunistically as well as being a top marine predator. Tiger sharks also have been known to ingest inedible objects (license plates, plastics, cans, and an amazing variety of trash). Prey items vary from large fishes (many sharks and rays as well as larger bony fishes, such as tarpon), marine reptiles, mammals, and birds to mollusks (octopi, squid, and cuttlefish) and crustaceans. Actively attacks birds resting on the surface, lifting its massive head out of the water to bite down on them as they attempt to escape.

REPRODUCTIVE BIOLOGY
Aplacentally viviparous. (The only member of its family in which maternal-fetal connections do not form.) Gestation is slightly longer than one year. Gives birth to 10–82 rather large young (20.1–29.9 in [51–76 cm]). Inshore nursing grounds are common. Matures sexually between four and six years old.

CONSERVATION STATUS
Considered to be Lower Risk/Near Threatened by the IUCN, as there is some evidence that several populations have declined where they are heavily fished.

SIGNIFICANCE TO HUMANS
Considered in many places to be a dangerous shark—many attacks on people and boats have been attributed to this species in tropical seas. Commonly hooked by fishermen and captured by longlines. The meat is utilized fresh, dried/salted, or frozen; the skin for leather; the fins for soup; and the oily liver processed for vitamin oil. Fished recreationally as well. Remains alive in aquaria for only short periods, not surpassing a few months. The tiger shark may be seen in the wild in Hawaii, Australia, and the Rangiroa Atoll (French Polynesia). ◆

Ganges shark
Glyphis gangeticus

FAMILY
Carcharhinidae

TAXONOMY
Carcharias (Prionodon) gangeticus Müller and Henle, 1839, Ganges River.

OTHER COMMON NAMES
French: Requin du Ganges; Spanish: Tiburón del Ganges.

PHYSICAL CHARACTERISTICS
Snout rounded anteriorly, with very small eyes, large first dorsal fin well anterior to pelvic fin origin, large pectoral and caudal fins, small upper labial furrow, upper jaw teeth broadly triangular with minute serrations, lower jaw teeth very acute

and slender with smooth edges. Gray coloration dorsally and laterally and white ventrally. Reaches more than 6.6 ft (2 m) in length.

DISTRIBUTION
Hooghly Channel of the Ganges River (India).

HABITAT
A freshwater species, the notorious Ganges shark is known from only two surviving specimens collected in the Ganges River. Its small eyes may indicate that this species lives exclusively in the murky Ganges. Needs confirmation by means of additional specimens from the Ganges and surrounding areas; unknown whether it is capable of tolerating marine waters.

BEHAVIOR
Nothing is known.

FEEDING ECOLOGY AND DIET
Presumably fishes, but presently unknown.

REPRODUCTIVE BIOLOGY
Unknown, but presumably reproduces in freshwater, as indicated by a newly born male specimen, 25.6 in (65 cm) long, collected from the Hooghly Channel of the Ganges River.

CONSERVATION STATUS
Listed as Critically Endangered by the IUCN because of the paucity of specimens and information and considered at extreme risk of extinction.

SIGNIFICANCE TO HUMANS
The Ganges shark has an unproven, folkloric reputation as a "man-eater" in the Ganges River. As the bull shark (*Carcharhinus leucas*) also occurs there, many of the attacks attributed to the Ganges shark may be a result of misidentification. The Ganges shark is one of the least known and most mysterious shark species. It is known originally from three museum specimens, collected from freshwaters of the lower reaches of the Ganges-Hooghly river system in the nineteenth century, but of which only one is extant. An additional specimen was found subsequently in India (the newly born male mentioned previously), but no further confirmed specimens are known. ◆

Lemon shark
Negaprion brevirostris

FAMILY
Carcharhinidae

TAXONOMY
Hypoprion brevirostris Poey, 1868, Cuba.

OTHER COMMON NAMES
French: Requin citron; Spanish: Tiburón galano.

PHYSICAL CHARACTERISTICS
Somewhat rounded and blunt snout, no labial furrows, first dorsal fin well anterior to pelvic fin origin, relatively large second dorsal and anal fins, and moderately large caudal fin. Teeth have slender, smooth, and triangular cusps. Coloration is gray to yellowish-brown dorsally and laterally, creamy white ventrally. Reaches about 11.5 ft (3.5 m) in length.

DISTRIBUTION
Tropical and warm temperate waters. Present in the eastern Pacific off the coasts of Mexico south to Peru; in the western

Atlantic from New England to southern Brazil, Gulf of Mexico, and Caribbean; and in the eastern Atlantic off Senegal and probably elsewhere in Africa.

HABITAT
An inshore, coastal species common in coral reefs, mangroves, and bays; around piers and docks; and at river mouths, occurring from the surface down to about 328 ft (100 m). May penetrate freshwater but not only for short distances.

BEHAVIOR
Lemon sharks are active both during the day and at night, with peaks of activity at dawn or dusk. They prefer shallow areas, and individuals may display site preferences, especially younger sharks. Their home range expands with growth, as young sharks may remain within a region encompassing 3.7–5 mi^2 (6–8 km^2), which may expand to 186 mi^2 (300 km^2) when they reach adulthood. Lemon sharks may remain active in low-oxygen environments because of their high respiratory efficiency. They typically are found resting on the bottom. Adults may undertake long seasonal migrations.

FEEDING ECOLOGY AND DIET
Feeds mostly on fishes (including many different rays), crustaceans, and mollusks. Also may consume seabirds occasionally. Young lemon sharks, 27.6 in (70 cm) in length, have been able to eat 3% of their body weight in captivity, with unlimited food available.

REPRODUCTIVE BIOLOGY
A viviparous species with a yolk sac placenta. Litters commonly have between four and 17 young, and gestation periods last from 10 to 12 months. Young are born in shallow nursery areas and remain there for a short period. Sexual maturity is reached after about six and a half years. Both courtship and mating have occurred in captivity.

CONSERVATION STATUS
Listed as Lower Risk by the IUCN, because young specimens inhabit coastal nursery regions that may be subject to development and habitat degradation.

SIGNIFICANCE TO HUMANS
A hardy species in captive conditions. Caught frequently on longlines and fished by anglers. The flesh may be consumed fresh or in other ways. There is evidence of population declines in the eastern Pacific and western Atlantic. Attacks on people have been recorded, largely owing to the lemon shark's preference for shallow water in areas that are heavily populated. The lemon shark, however, is not considered an aggressive species. May be seen in the wild in the Bahamas, Turks and Caicos, Florida, Belize, and many other Caribbean locations. ◆

Blue shark
Prionace glauca

FAMILY
Carcharhinidae

TAXONOMY
Squalus glauca Linnaeus, 1758, "Oceano Europaeo."

OTHER COMMON NAMES
French: Peau bleu; Spanish: Tiburón azul.

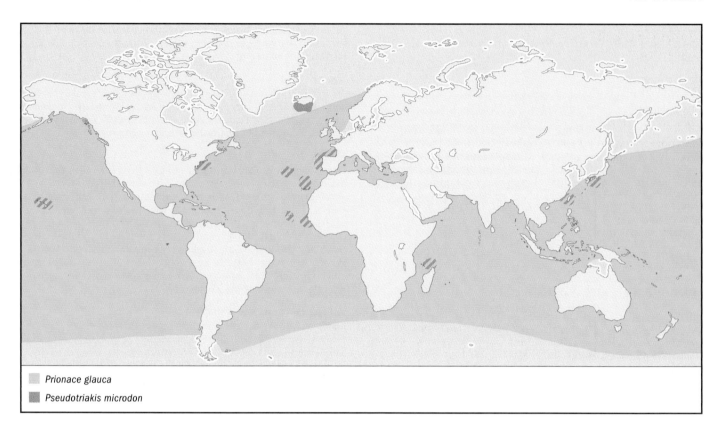

Prionace glauca
Pseudotriakis microdon

PHYSICAL CHARACTERISTICS
The blue shark has characteristically long pectoral fins, a slender body with a large caudal fin, an elongated snout, widely separated dorsal fins, and first dorsal fin much larger than the second. Teeth have small lateral serrations and are slightly different in the upper and lower jaws. Deep blue dorsal coloration, with lighter blue sides and white ventrally. Reaches some 13 ft (4 m) in length.

DISTRIBUTION
Worldwide in tropical and temperate waters. It is perhaps the widest-ranging chondrichthyan.

HABITAT
The blue shark is an oceanic, epipelagic shark that enters littoral regions with frequency. Usually found close to the surface but has been captured at depths slightly more than 722 ft (220 m) in warmer latitudes.

BEHAVIOR
May occur in loosely organized aggregations, cruising at slow speeds close to the surface, but capable of swift bursts of speed; may even jump out of the water. Known to circle potential prey items before attacking. May migrate seasonally, and tagged individuals have been recaptured at very distant locations (e.g., across the Atlantic). May bite objects or potential prey out of curiosity or in an attempt to "taste" them before committing to a feeding. Blue sharks are capable of making sharp, quick turns, an indication of their extreme versatility.

FEEDING ECOLOGY AND DIET
Consumes mostly pelagic bony fishes, and especially squid, that occur close to the surface but also may feed on other sharks (such as the piked dogfish, *Squalus acanthias*, and, in one case, the goblin shark, *Mitsukurina owstoni*), invertebrates, mammalian carrion, and even seabirds. Feeds massively on squid

during breeding aggregations and on flying-fish eggs during spawning in the Adriatic.

REPRODUCTIVE BIOLOGY
Viviparous with a yolk sac placenta; litters range from four to 135 young (the greatest range of any live-bearing shark). Gestation is from nine to 12 months. Sexual maturity is reached after some five years (slightly younger for males). Females are sexually mature at about 86.6 in (220 cm) in length, males at slightly smaller sizes. Females store sperm in the shell glands of their oviducts after copulation (which usually takes place from late spring to early winter in temperate regions and all year in tropical seas), utilizing it only when their ovaries ripen to produce and release eggs into the oviduct. Sperm may be stored in this manner for a period of one year. Courtship rituals involve biting, and the sexes can be distinguished readily according to the presence of scars on the body. Females may have skin three times as thick as males. Individuals may segregate by sex.

CONSERVATION STATUS
Listed as Lower Risk/Near Threatened by the IUCN, mostly owing to the lack of data concerning the effects of presently being the most fished shark species in the world and an important keystone predator in the oceanic realm.

SIGNIFICANCE TO HUMANS
Heavily fished in much of its range, usually by means of pelagic longlines. Consumed fresh, frozen, and dried/salted. The skin may be used for leather and the fins for shark fin soup. Also fished recreationally. Attacks on people have been attributed to this species, but there also have been many harmless encounters. Nevertheless, the blue shark should be approached with caution. There are many places where tourists may take day trips to see blue sharks in the wild (e.g., California, Portugal, and New Zealand). ◆

False catshark

Pseudotriakis microdon

FAMILY
Pseudotriakidae

TAXONOMY
Pseudotriakis microdon Capello, 1868, Setubal, Portugal.

OTHER COMMON NAMES
French: Requin à longue dorsale; Spanish: Musolón de aleta larga.

PHYSICAL CHARACTERISTICS
A unique shark, with an extremely elongated, low first dorsal fin; a tall, triangular second dorsal fin; a large anal fin, elongated, slitlike eyes; and a wide mouth with 200–300 rows of numerous minute teeth in each jaw. Covered in prickly denticles and with brownish-black coloration. Reaches 9.8 ft (3 m) in length.

DISTRIBUTION
Worldwide in tropical and temperate latitudes.

HABITAT
A deepwater, demersal species, occurring predominantly from 656 to 4,920 ft (200–1,500 m) along the continental slopes. May venture rarely into more shallow waters of the continental shelf.

BEHAVIOR
Nothing is known.

FEEDING ECOLOGY AND DIET
Largely unknown but presumably consumes demersal fishes and invertebrates. The huge mouth of the false catshark probably allows it to ingest prey of considerable size.

REPRODUCTIVE BIOLOGY
Aplacentally viviparous (ovoviviparous), with small litters (reportedly two to four young) but producing copious numbers of eggs (estimated at 20,000 in one ovary of a female 110 in, or 280 cm, in length).

CONSERVATION STATUS
Not threatened.

SIGNIFICANCE TO HUMANS
Not consumed in significant quantities but occasionally captured with bottom longlines. Not considered dangerous. ◆

Swellshark

Cephaloscyllium ventriosum

FAMILY
Scyliorhinidae

TAXONOMY
Scyllium ventriosum Garman, 1880, Valparaíso, Chile.

OTHER COMMON NAMES
Spanish: Pejegato hinchado.

PHYSICAL CHARACTERISTICS
Posterior margins of nasal flaps reaching the mouth, broadly rounded snout. The first dorsal fin behind the origin of the pelvic fins and larger than the second dorsal fin; relatively large anal fin and large eyes. Teeth have small lateral cusplets. Large brown blotches and saddles dorsally and laterally and small darker and lighter spots ventrally and laterally on a yellowish-brown background. Reaches about 3.3 ft (1 m) in length.

DISTRIBUTION
Occurs in the eastern Pacific Ocean from Monterey Bay, California, south to central Chile; not yet recorded from the south of Mexico to southern Peru.

HABITAT
Usually present in shallow waters, from 16.4 to 121.4 ft (5–37 m) but occasionally caught deeper on the upper continental slope, to 1,500 ft (457 m). Found on rocky bottoms but also on substrates covered by algae.

BEHAVIOR
A sluggish shark that remains mostly motionless, sheltered in caves or crevices during the day and becoming more active at night. Individuals may aggregate while resting. As their common name implies, swellsharks are capable of inflating their stomachs with water or air to escape predation (similarly to pufferfishes). They even may wedge themselves in crevices in this manner.

FEEDING ECOLOGY AND DIET
Feeds mostly on bony fishes but also may eat hard-shelled invertebrates.

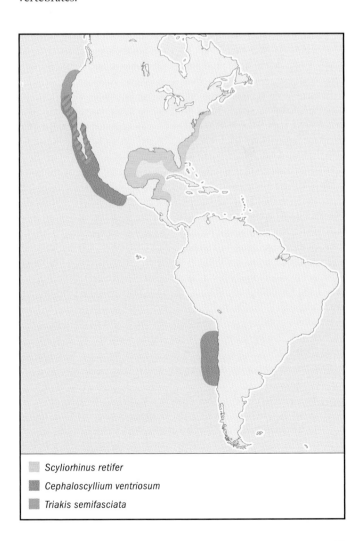

Scyliorhinus retifer
Cephaloscyllium ventriosum
Triakis semifasciata

REPRODUCTIVE BIOLOGY
Oviparous (egg laying), depositing amber to greenish egg cases with smooth surfaces and elongated tendrils. Young hatch after a period of seven and a half to 10 months, at about 5.1–5.9 in (13–15 cm) in length. Young have a row of enlarged denticles along the back that may aid them when leaving the egg cases. Males sexually mature between 32.3 and 33.5 in (82–85 cm) in length.

CONSERVATION STATUS
Not threatened.

SIGNIFICANCE TO HUMANS
Swellsharks do well in captivity and are featured in many aquaria. Females even lay egg cases in aquaria. These sharks are not consumed and probably are discarded if captured in trawls. Not considered dangerous to people but may become aggressive if harassed. ◆

Pajama catshark
Poroderma africanum

FAMILY
Scyliorhinidae

TAXONOMY
Squalus africanus Gmelin, 1789, "Mari Africanum," probably off South Africa.

OTHER COMMON NAMES
English: Striped catshark; French: Roussette rubanée; Afrikaans: Streep-kathaai.

PHYSICAL CHARACTERISTICS
An unmistakable shark, with longitudinal, broad stripes from head to tail on the dorsal and lateral sides. Elongated eyes, first dorsal fin posterior to pelvic fin origin and larger than the second dorsal, and relatively short narial barbels. Reaches about 3.3 ft (1 m) in length.

DISTRIBUTION
Found primarily off South Africa but also in the eastern Atlantic near the mouth of the Congo River. Records needing confirmation exist from Madagascar and Mauritius.

HABITAT
A shallow-water, inshore species, occurring in waters down to 328 ft (100 m) deep. Common in caves and over rocky substrates.

BEHAVIOR
A common, mostly nocturnal and somewhat sluggish shark, but behavior is poorly known.

FEEDING ECOLOGY AND DIET
Feeds on crustaceans, cephalopods, polychaetes, and many different bony fishes.

REPRODUCTIVE BIOLOGY
An oviparous species, laying a single egg case per oviduct. One egg laid in an aquarium hatched after five and a half months. Males are sexually mature between 22.8 and 29.9 in (58–76 cm) in length and females between 25.6 and 28.3 in (65–72 cm).

CONSERVATION STATUS
Listed as Lower Risk/Near Threatened by the IUCN, because of its restricted occurrence mostly in regions with high levels of human activity and fishing.

SIGNIFICANCE TO HUMANS
Readily kept in aquaria but not consumed or captured significantly. Occasionally captured as by-catch by bottom trawlers but usually discarded; fished recreationally. Not dangerous to people. ◆

Chain catshark
Scyliorhinus retifer

FAMILY
Scyliorhinidae

TAXONOMY
Scyllium retiferum Garman, 1881, off Virginia, United States.

OTHER COMMON NAMES
Spanish: Alitán mallero.

PHYSICAL CHARACTERISTICS
Coloration unique, composed of numerous brown saddles with a conspicuous internal network pattern that also is present over the pectoral and caudal fins. First dorsal fin behind origin of pelvic fins and larger than the second dorsal fin, back somewhat arched. Elongated, slitlike eyes. Reaches about 19.7 in (50 cm) in length.

DISTRIBUTION
Present in the western North Atlantic from southern New England to Florida; found in the Gulf of Mexico from Florida south to Nicaragua.

HABITAT
A mostly deepwater species, demersal on the outer continental shelf to the upper slope region, from 29 to 1,800 ft (73–550 m) in depth.

BEHAVIOR
Unknown.

FEEDING ECOLOGY AND DIET
Presumably feeds on fishes and invertebrates, as it lives mostly in close association with the bottom, but full stomach contents have yet to be examined. Cephalopod beaks were found in one specimen.

REPRODUCTIVE BIOLOGY
Oviparous (egg laying), but most details concerning its reproduction are unknown. Males mature sexually at about 14.6–16.1 in (37–41 cm) in length, females from 13.8 to 18.5 in (35–47 cm). Length at birth is about 3.9 in (10 cm).

CONSERVATION STATUS
Not threatened.

SIGNIFICANCE TO HUMANS
Captured occasionally as by-catch in bottom trawls but is not consumed and is discarded. Not dangerous, owing to its size and habitat. ◆

Great hammerhead shark
Sphyrna mokarran

FAMILY
Sphyrnidae

TAXONOMY
Zygaena mokarran Rüppel, 1837, Red Sea.

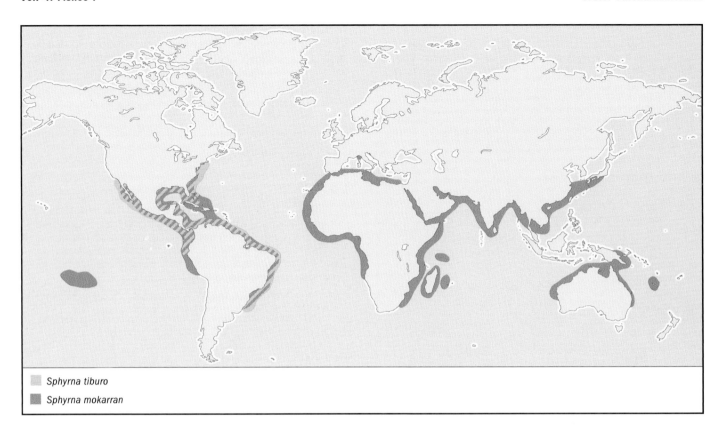

Sphyrna tiburo
Sphyrna mokarran

OTHER COMMON NAMES
French: Grand requin-marteau; Spanish: Cornuda gigante.

PHYSICAL CHARACTERISTICS
Cephalofoil (hammerhead) with somewhat straight anterior margin with a slight median indentation. Hammerhead not very broad proportionally. Very tall first dorsal fin entirely anterior to pelvic fin origins, pelvic and anal fins with strongly concave posterior margins, and well-developed caudal fin. Serrated teeth. Uniform gray coloration. The largest of the hammerhead sharks, reaching some 20 ft (6 m) in length.

DISTRIBUTION
Worldwide in coastal tropical to warm temperate waters.

HABITAT
Common in inshore tropical waters but also present offshore, occurring close to the water'surface to depths of 263 ft (80 m).

BEHAVIOR
A nomadic, migratory species, the great hammerhead shark can occur in great numbers, with individuals moving their heads from side to side as they cruise at midwater depth.

FEEDING ECOLOGY AND DIET
Eats a variety of bony fishes but seems particularly fond of stingrays, rays in general, groupers, and marine catfishes. Many fishes are commonly taken as prey (even smoothhound sharks). The venomous stings of stingrays and catfishes do not appear to harm the great hammerhead, as they frequently are found stuck in the mouth and pharynx. (One specimen apparently had more than 50 stings embedded in its mouth, throat, and tongue.) Also may feed on crabs and squid.

REPRODUCTIVE BIOLOGY
Viviparous with a yolk sac placenta and litters ranging from 13 to 42 young. Equal numbers of males and females usually are

born. Gestation may last seven months. Males are sexually mature at about 93–106 in (235–269 cm) in length and females from between 98 and 118 in (250–300 cm).

CONSERVATION STATUS
Currently listed as Data Deficient by the IUCN, even though it is not targeted specifically by the fishing industry. Listed mainly because of its capture as by-catch (which may be significant) and high-value fins.

SIGNIFICANCE TO HUMANS
The flesh is consumed commonly in the tropics and sold frozen, dried/salted, and smoked. Because of its large size, the great hammerhead was once believed to be dangerous. Attacks by hammerheads have occurred, but identifying the particular species is difficult; this species should be treated with caution. Important in the growing shark tourism industry, as it is observed by divers in many locations (e.g., in the Bahamas, Turks and Caicos, and Australia). ◆

Bonnethead shark
Sphyrna tiburo

FAMILY
Sphyrnidae

TAXONOMY
Squalus tiburo Linnaeus, 1758, "Habitat in America."

OTHER COMMON NAMES
French: Requin-marteau tiburo; Spanish: Cornuda tiburo.

PHYSICAL CHARACTERISTICS

Snout unique in being broadly rounded anteriorly (resembling a shovel) and without a median notch. Relatively tall first dorsal fin that is completely anterior to the pelvic fin origins; somewhat small second dorsal fin. Grayish-brown in color dorsally and on the sides and paler ventrally. Molariform posterior teeth. This is a small hammerhead species, reaching only about 5 ft (1.5 m) in length.

DISTRIBUTION

In the eastern Pacific from southern California south to Ecuador and in the western Atlantic, Caribbean, and Gulf of Mexico from North Carolina (exceptionally from New England) south to southern Brazil.

HABITAT

An inshore, coastal species, usually present in shallow waters down to depths of 82 ft (25 m) but occasionally down to 263 ft (80 m). Common in estuaries, sandy bottoms, coral reefs, shallow bays, and channels.

BEHAVIOR

Forms large schools in many regions of its range and is a social species, occurring typically in numbers from three to 15 individuals. Migrates seasonally and is found as far north as New England during the summer months. Sexual segregation has been noted in this species, with females remaining near shallow nursery areas to give birth. The behavior of this shark has been studied more intensely than that of others, and some 18 separate postures and behavioral patterns have been reported. Behaviors include patrolling (relatively straight-line swimming), maneuvering (systematic rapid turns), explosive gliding (rapid swimming initiated by tail beats), rapid head shaking, head snapping (strong vertical movements of the head and trunk), jaw snapping (opening and closing of the mouth), chafing (sudden rolling of the body), gill puffing (rapid expansion of the gills), clasper flexing (strong single clasper movements), circling head to tail (two sharks swimming in a tight circle, head to tail), and following (individuals closely following each other), among others.

FEEDING ECOLOGY AND DIET

Feeds abundantly on many different crustaceans (such as shrimp, isopods, crabs, lobsters, and even barnacles) as well as octopi, bivalves, and small fishes.

REPRODUCTIVE BIOLOGY

Viviparous with a yolk sac placenta and litters ranging from four to 16 young. Males are sexually mature at about 20.5–29.5 in (52–75 cm) in length and females by at least at 33 in (84 cm). Lengths at birth range from 13.4 to 15.7 in (34–40 cm).

CONSERVATION STATUS

Not listed by the IUCN.

SIGNIFICANCE TO HUMANS

Taken frequently by small fishing operations, usually as bycatch in bottom trawls, and utilized fresh, frozen, dried/salted, and in other forms. Not considered dangerous, owing to its small size, but it occurs close to heavily populated areas and should be approached cautiously. ◆

Leopard shark
Triakis semifasciata

FAMILY
Triakidae

TAXONOMY
Triakis semifasciata Girard, 1854, San Francisco Bay, California.

OTHER COMMON NAMES
Spanish: Tollo leopardo.

PHYSICAL CHARACTERISTICS

The coloration of the leopard shark is distinctive, with numerous dark gray saddle marks from nape to tail, dark blotches laterally, and a light gray background color. First dorsal fin anterior to the pelvic fin origins and large second dorsal fin ahead of the anal fin origin. Slitlike eyes and upper and lower labial furrows. Reaches about 71 in (180 cm) in length.

DISTRIBUTION

Restricted to the eastern Pacific from Oregon to central Mexico, including the Gulf of California.

HABITAT

Occurs in shallow waters, usually less than 33 ft (10 m) deep, but may be captured as deep as 328 ft (100 m). A bottom dweller, the leopard shark commonly is found over rocky, sandy, or muddy substrates. Also may enter bays and estuaries.

BEHAVIOR

Leopard sharks are highly active, and the wide variety of prey items and their habitats indicate that they employ a wide range of feeding behaviors, from removing burrowing mollusks from their hard shells (clams) to feeding on schooling fishes. An abundant shark, with limited movements (as concluded by tagging studies), that is sometimes observed resting on the bottom.

FEEDING ECOLOGY AND DIET

Eats fishes and invertebrates; reported to prefer invertebrates, such as crabs, shrimp, polychaete worms, and octopi. Also reported to have fed on sharks (the brown smoothhound shark, *Mustelus henlei*), guitarfish (*Rhinobatos productus*), bat rays (*Myliobatis californica*), and fish eggs. Diet may vary according to size and season, at least locally off California. Feeds on many mud-burrowing prey items (clams, certain shrimps, and polychaetes), which suggests that feeding takes place close to the bottom. Feeding also has been observed closer to the surface, on schools of anchovies.

REPRODUCTIVE BIOLOGY

Aplacentally viviparous, with four to 29 young per litter. Gestation lasts about 12 months. Reproductive maturity may take more than 10 years, as growth is very slow.

CONSERVATION STATUS

Listed as Lower Risk/Near Threatened by the IUCN, because of the lack of regulation with respect to fishing.

SIGNIFICANCE TO HUMANS

Leopard sharks are kept easily in aquaria, where they can live for more than 20 years. Seldom captured by anglers but caught somewhat more frequently by small commercial operations, especially in Mexico. The flesh is sold fresh or frozen and is reported to be of good quality. Not considered dangerous. ◆

Resources

Books

Bigelow, H. B., and W. C. Schroeder. "Sharks." In *Fishes of the Western North Atlantic*, edited by J. Tee-Van, C. M. Breder, S. F. Hildebrand, A. E. Parr, and W. C. Schroeder. New Haven, CT: Sears Foundation for Marine Research, Yale University, 1948.

Branstetter, S., ed. *Conservation Biology of Elasmobranchs*. NOAA Technical Report, NMFS 115. Seattle: U.S. Department of Commerce, 1993.

Cappetta, H. *Chondrichthyes II: Mesozoic and Caenozoic Elasmobranchii*. Handbook of Palaeoichthyology, vol. 3B. Stuttgart and New York: Gustav Fischer Verlag, 1987.

Carwardine, Mark, and Ken Watterson. *The Shark Watcher's Handbook: A Guide to Sharks and Where to See Them*. Princeton: Princeton University Press: 2002.

Compagno, L. J. V. *Sharks of the World: An Annotated and Illustrated Catalogue of Shark Species Known to Date*. FAO Species Catalogue, vol. 4, part 1. Rome: Food and Agriculture Organization of the United Nations, 1984.

———. *Sharks of the Order Carcharhiniformes*. Princeton: Princeton University Press, 1988.

Compagno, L. J. V., and V. H. Niem. "Families Scyliorhinidae, Proscylliidae, Pseudotriakidae, Triakidae, Hemigaleidae, Carcharhinidae, Sphyrnidae." In *Western Central Pacific Identification Sheets to Species*, edited by K. E. Carpenter and V. H. Niem. Rome: Food and Agriculture Organization of the United Nations, 1999.

Compagno, L. J. V., C. Simpfendorfer, J. E. McCosker, K. Holland, C. Lowe, B. Wetherbee, A. Bush, and C. Meyer. *Sharks*. Pleasantville, New York: Reader's Digest Association, Inc., 1998.

Hamlett, W. C., ed. *Sharks, Skates, and Rays: The Biology of Elasmobranch Fishes*. Baltimore: Johns Hopkins University Press, 1999.

Hennemann, Raof M. *Sharks and Rays: Elasmobranch Guide of the World*. Frankfurt: Ikan, 2001.

Last, P. R., and J. D. Stevens. *Sharks and Rays of Australia*. Melbourne, Australia: CSIRO, 1994.

Myrberg, Arthur A., Jr., and Donald R. Nelson. "The Behavior of Sharks: What Have We Learned?" In *Discovering Sharks*, edited by S. H. Gruber. Highlands, NJ: American Littoral Society, 1990.

Perrine, D. *Sharks and Rays of the World*. Stillwater, MN: Voyager Press, 1999.

Pratt, H. L. Jr., S. H. Gruber, and T. Taniuchi, eds. *Elasmobranchs as Living Resources: Advances in the Biology, Ecology, Systematics, and the Status of the Fisheries*. Proceedings of the Second United States–Japan Workshop East-West Center, Honolulu, Hawaii, 9–14 December 1987. NOAA Technical Report NMFS 90. Seattle: U.S. Department of Commerce, 1990.

Randall, J. E. "Review of the Biology of the Tiger Shark (*Galeocerdo cuvier*)." In *Sharks: Biology and Fisheries*, edited by J. G. Pepperell. Melbourne, Australia: CSIRO, 1992.

Springer, S. "Social Organization in Shark Populations." In *Sharks, Skates, and Rays*, edited by P. W. Gilbert, R. F. Mathewson, and D. P. Rall. Baltimore: Johns Hopkins University Press, 1967.

Springer, Victor G., and Joy P. Gold. *Sharks in Question: The Smithsonian Answer Book*. Washington, DC: Smithsonian Institution Press, 1989.

Tricas, T. C., and S. H. Gruber. *The Behavior and Sensory Biology of Elasmobranch Fishes: An Anthology in Memory of Donald Richard Nelson*. Developments in Environmental Biology of Fishes, vol. 20. Dordrecht, Netherlands: Kluwer Academic Publishers, 2001.

Whitley, G. P. *The Fishes of Australia*. Part 1. *The Sharks, Rays, Devil-fish, and Other Primitive Fishes of Australia and New Zealand*. Sydney: Royal Zoological Society of New South Wales, 1940.

Wetherbee, B. M., S. H. Gruber, and E. Cortes. "Diet, Feeding Habits, Digestion, and Consumption in Sharks, with Special Reference to the Lemon Shark, *Negaprion brevirostris*." In *Elasmobranchs as Living Resources: Advances in the Biology, Ecology, Systematics, and the Status of the Fisheries*. Proceedings of the Second United States–Japan Workshop East-West Center, Honolulu, Hawaii, 9–14 December 1987, edited by H. L. Pratt, S. H. Gruber, and T. Taniuchi. NOAA Technical Report, NMFS 90. Seattle: U.S. Department of Commerce, 1990.

Wourms, J., and L. Demski. *Reproduction and Development of Sharks, Skates, Rays and Ratfishes*. Developments in Environmental Biology of Fishes, vol. 14. Dordrecht, Netherlands: Kluwer Academic Publishers, 1993.

Periodicals

Johnson, R. H., and D. R. Nelson. "Agonistic Display in the Gray Reef Shark, *Carcharhinus menisorrah*, and Its Relationship to Attacks on Man." *Copeia* 1973, no. 1 (1973): 45–55.

———. "Copulations and Possible Olfaction-Mediated Pair Formation in Two Species of Carcharhinid Sharks." *Copeia* 1978 (1978): 539–542.

Motta, Phillip J., Robert E. Hueter, and Timothy C. Tricas. "An Electromyographic Analysis of the Biting Mechanism of the Lemon Shark, *Negaprion brevirostris*: Functional and Evolutionary Implications." *Journal of Morphology* 210 (1991): 55–69.

Myrberg, A. A., and S. H. Gruber. "The Behavior of the Bonnethead Shark, *Sphyrna tiburo*." *Copeia* 1974, no. 2 (1974): 358–374.

Nelson, D. R. "Aggression in Sharks: Is the Grey Reef Shark Different?" *Oceanus* 24, no. 4 (1981): 45–55.

Nelson, D. R., and R. H. Johnson. "Behavior of Reef Sharks of Rangiroa, French Polynesia." *National Geographic Society Research Reports* 12 (1980): 479–499.

Strong, W. R. Jr., F. F. Snelson Jr., and S. H. Gruber. "Hammerhead Shark Predation on Stingrays: An Observation of Prey Handling by *Sphyrna mokarran*." *Copeia* 1990, no. 3 (1990): 836–840.

Tricas, T. C., Taylor, L., and G. Naftel. "Diel Behavior of the Tiger Shark, *Galeocerdo cuvier*, at French Frigate Shoals, Hawaiian Islands." *Copeia* 1981, no. 4 (1981): 904–908.

Resources

Wourms, J. P. "Reproduction and Development in Chondrichthyan Fishes." *American Zoologist* 17 (1977): 379–410.

Organizations

American Elasmobranch Society, Florida Museum of Natural History. Gainesville, FL 32611 USA. Web site: <http://www.flmnh.ufl.edu/fish/Organizations/aes/aes.htm>

Other

"2002 IUCN Red List of Threatened Species." (26 Dec. 2002). <http://www.redlist.org>

"Fish-Base." 16 Dec. 2002 (26 Dec. 2002). <http://www.fishbase.org>

"Catalog of Fishes On-line." 12 Nov. 2002 (26 Dec. 2002). <http://www.calacademy.org/research/ichthyology/catalog/fishcatsearch.html>

"ReefQuest Expeditions." (26 Dec. 2002). <http://www.reefquest.com>

Marcelo Carvalho, PhD

Lamniformes

(Mackerel sharks)

Class Chondrichthyes

Order Lamniformes

Number of families 7

Photo: A basking shark (*Cetorhinus maximus*) feeding on plankton in the Irish Sea. (Photo by Jeff Rotman/Photo Researchers, Inc. Reproduced by permission.)

Evolution and systematics

Living lamniform sharks are mere remnants of a much greater lamniform lineage that has, for the most part, become extinct. The 15 surviving species pale in comparison to the countless hundreds that have been described from fossil remains; the genus *Carcharodon* alone is known from some 10 fossil species, in contrast to the single extant *Carcharodon carcharias*. However, the overwhelming majority of these fossils consist of isolated teeth, first appearing in the fossil record during the early Cretaceous period some 120 million years ago (mya). Fossil lamniform teeth are known from many widespread marine localities from all continents, and they resemble those of living mackerel sharks in usually being slender, with very sharp cusps and arched roots. Many living lamniform species have closely related fossil relatives, again known only from teeth, going back at least to the Paleocene epoch (some 62 mya). Some of these fossil species are even placed in genera that are still extant (e.g., *Carcharodon*, *Odontaspis*), corroborating that lamniform sharks have a remarkably long evolutionary history, as do most living shark groups.

Fossil lamniforms known from more complete remains are extremely rare and include preserved partial skeletons of goblin sharks (Mitsukurinidae) from Lebanon (about 90 million years old), and vertebrae of various taxa, such as the megalodon shark from Europe (of Miocene to Pliocene age, some16 to 2.6 mya). The late Cretaceous goblin shark (*Scapanorhynchus lewisi*) is similar to the living goblin species (*Mitsukurina owstoni*) in having a very elongated snout, but it differs in having a much longer anal fin and more angular dorsal fins. Moreover, some features of its teeth and denticles differ as well. The megalodon shark (*Carcharodon megalodon*) is the most notorious fossil lamniform. It is known from huge, triangular teeth (as large as 7.9 in [20 cm] in height), that are very similar to teeth of the living white shark (*Carcharodon carcharias*). The megalodon shark, however, was much larger (estimated to reach up to 49 ft [15 m] in length), some three times the size of the living white, and

was one of the greatest marine predators of all time (and the greatest macropredatory shark). Reconstructions of its jaws, believed to have been able to fit several people when agape, feature in many museum exhibits. Megalodon fossils are known from North and South America, the Caribbean, Europe, Australasia, Japan, and Africa.

Among living elasmobranchs (sharks and rays), lamniform sharks are more closely related to the ground sharks (Carcharhiniformes), bullhead sharks (Heterodontiformes), and carpet sharks (Orectolobiformes). These four orders, united in the larger group Galeomorphii, share various evolutionary innovations, such as the unique placement of the hyomandibula (a cartilage supporting the jaws posteriorly) on the skull. Within this group, lamniforms are most closely related to the ground sharks, as both orders share a tripodal rostrum supporting the snout internally.

Living lamniforms are among the most intensely studied and best-known sharks. Four of the living species were described in the eighteenth century, five in the nineteenth, and six in the twentieth century. (The last species described was the megamouth shark in 1983.) They are currently divided into seven families, 10 genera, and 15 species, and they were first recognized as a unique group by American ichthyologist David Starr Jordan (1851–1931) in 1923.

Phylogenetic (evolutionary) relationships among lamniform genera also have received much recent attention. The goblin shark (*Mitsukurina*) is considered the most basal, or primitive, living lamniform, followed by the sand tiger sharks (family Odontaspididae) and the crocodile shark (*Pseudocarcharias*). The remaining mackerel sharks have plesodic pectoral fin skeletons, in which the internal supports extend to the distal fin margin. Recent phylogenetic theories also support a common ancestry for a lamniform subgroup—comprising the basking shark and lamnids—with lunate caudal fins. Phylogenetic studies based exclusively on characters from the teeth disagree to some extent with those based on

A white shark (*Carcharodon carcharias*) approaches its prey from below where its gray topside camouflages its approach. (Photo by Corbis. Reproduced by permission.)

the skeleton, but teeth can often be misleading as indicators of evolutionary relationships in sharks and rays. Molecular phylogenies are also partly at odds with morphological ones, indicating that the evolutionary history of many lamniform genera is still in dispute.

Physical characteristics

Mackerel sharks are moderate to very large, ranging from about 3.3 ft (1 m) to 49 ft (15 m) in length. Some mackerel sharks, such as the great white and shortfin mako, are among the most popularly known and easily recognizable of all sharks. Other mackerel sharks are among the most bizarre and anatomically unique sharks, such as the megamouth, goblin, and thresher sharks. Lamniform sharks have unique teeth and intestines with a ring valve (with numerous, closely stacked turns).

There is some variation among lamniform species in relation to body and fin profiles, but all mackerel sharks have two dorsal fins (usually the first dorsal fin is very tall, while the

second is reduced in height), large pectoral fins (except in the goblin and crocodile sharks, and to a lesser degree in the sand tiger sharks), and a small anal fin (except in the goblin and sand tiger sharks). The caudal fin is lunate or semilunate (i.e., with a well developed lower lobe) and upright in some species (basking shark and lamnids). Thresher sharks have caudal fins about equal to the length of the body, and goblin, sand tiger, megamouth, and crocodile sharks have caudal fins with relatively small lower lobes. The snout is conical in most species (except in the megamouth shark), and paddle shaped in the goblin shark; internally the snout is supported by a tripodal rostrum, usually composed of three cartilaginous segments. The spiracles are extremely reduced. The eyes are black and round in most species and lack nictitating (protective) membranes. Five pairs of gill openings are present. Denticles along the body are very small and do not form larger spines.

Some lamniform species, particularly those of the family Lamnidae (white, porbeagle, salmon, and mako sharks) are capable of maintaining slightly elevated body temperatures in relation to the surrounding water. This is accomplished in a manner similar to tunas and mackerels (bony fishes of the family Scombridae), through a counter-current, vascular heat-exchange system. The body musculature, viscera, brains, and eyes remain at temperatures from 5.4°F (3°C) to 25.2°F (14°C) warmer than ambient water. This physiological mechanism enables lamnid mackerel sharks to maintain higher metabolic rates; hence they are capable of great bursts of activity.

Lamniforms are usually blue or blue-gray on their dorsal and lateral sides, but white to off-white ventrally. Well-defined spots and blotches are mostly absent, but the white shark has black ventral pectoral fin extremities, and some species may have whitish blotches on the tail; the salmon shark, *Lamna ditropis*, has brown blotches on its lateral and ventral aspects.

Distribution

Mackerel sharks are found worldwide in tropical and temperate marine waters. Some species penetrate boreal and subantarctic seas (basking shark and species of the genus *Lamna*), and other species are extremely wide-ranging, such as the shortfin mako and white shark. All species are somewhat widespread.

Habitat

Mackerel sharks are present in shallow, coastal waters, as well as epipelagically and mesopelagically in deeper oceanic waters. Most species, such as the mako, white, and sand tiger sharks, occur predominantly in shallow areas, while others are demersal inhabitants of continental slope regions (e.g., the goblin shark).

Behavior

The behavior of sharks that inhabit oceanic realms is generally not well known. Lamniform sharks, however, display different behaviors in relation to feeding (from filter feeding to predation), as well as in relation to metabolism. The more active species are laeterothermic (slightly warm-blooded); as a result they can swim at astounding speeds and are capable of great bursts of energy. The filter-feeding species, however, are relatively sluggish. Many species of lamniforms are known to leap completely out of the water (breaching), and not only as a result of being hooked on a line. The reasons behind this behavior are largely unknown but may have to do with escaping predators, snatching prey (as in the white shark), or ridding themselves of parasites. A lunate caudal fin may facilitate the strong upward swimming necessary to breach the water surface. Segregation by sex and size has been recorded in lamniform sharks, but much is yet to be learned about their population dynamics. More specific behavioral patterns have been described for particular species in the species accounts below.

Feeding ecology and diet

Almost all mackerel sharks are predaceous, extremely active eaters, feeding mostly on fishes belonging to numerous families (both bony fishes as well as sharks and rays), but also consuming large amounts of invertebrates (e.g., squids, octopi, gastropods, crustaceans) as well as marine mammals (pinnipeds, dolphins, and whales, as well as whale carcasses), marine turtles, and even oceanic birds. In contrast, two species, the megamouth and basking sharks, feed almost exclusively on zooplankton, and current evolutionary theories indicate that filter feeding evolved independently in both species, which also differ in their mode of filter feeding. Lamniform sharks are preyed upon by other shark species, including their own species.

Reproductive biology

As far as is known, all species of mackerel sharks are yolk-sac viviparous (ovoviviparous, aplacentally viviparous); i.e., they give birth to live young that develop in utero and that feed on the yolk contents of their yolk sacs. But in many lamniform species, intrauterine cannibalism has been confirmed or is suspected. This occurs when embryos prey on each other (adelphophagy) or on other eggs (oophagy) inside the uterus after their yolk reserves are depleted. This group is the only elasmobranch taxon in which this occurs. Adelphophagy is

known for certain in only one species, *Carcharias taurus*, but is suspected in others. Gestation periods vary among species and are comparatively poorly known. In some species, females are gravid from eight months to one year, while other species have longer gestations (up to 18 months). A period of reproductive inactivity may follow a gestation. Courtship patterns are presumably similar to those in many other sharks, with males biting females to subdue them prior to copulation and also during copulation.

Conservation status

The following species are listed by the IUCN: *Alopias vulpinus*, *Lamna ditropis*, *Megachasma pelagios*, and *Odontaspis noronhai* (as Data Deficient); *Carcharodon carcharias*, *Carcharias taurus*, and *Cetorhinus maximus* (as Vulnerable); *Lamna nasus*, *Isurus oxyrinchus*, and *Pseudocarcharias kamoharai* (as Lower Risk/Near Threatened).

Significance to humans

Many lamniform species are captured on longlines or trawls, either as bycatch or as specific targets, by the commercial fishing industry. The flesh is consumed fresh, frozen, smoked, or dried-salted, and their fins are procured by the destructive shark fin soup industry. Sport fishing for makos and other lamniforms is also common. This order contains what has been considered to be the most dangerous shark species, the white shark. But the misguided, anthropocentric perception that the white shark and other lamniform species are potential "man-eaters" has faded in the past decade; this negative image was given to this species mostly by sensationalistic media. Ironically, the roles are presently reversed, as it is now well understood that it is the sharks that are the victims of humankind, mostly through overfishing and the ruthless, cruel, shark fin soup fad, and not the other way around. In fact, many species of lamniforms and other sharks are quite valuable alive. The sand tiger shark is important as an exhibition fish in public aquaria, where it is relatively easily kept for long periods. Many lamniforms, such as the sand tiger, white, thresher, basking, and mako sharks, are even common ecotourist attractions in many places around the world.

1. Shortfin mako (*Isurus oxyrinchus*); 2.White shark (*Carcharodon carcharias*); 3. Sand tiger shark (*Carcharias taurus*); 4. Crocodile shark (*Pseudocarcharias kamoharai*); 5. Porbeagle (*Lamna nasus*); 6. Megamouth shark (*Megachasma pelagios*); 7. Thresher shark (*Alopias vulpinus*); 8. Goblin shark (*Mitsukurina owstoni*); 9. Basking shark (*Cetorhinus maximus*). (Illustration by Brian Cressman)

Species accounts

Thresher shark
Alopias vulpinus

FAMILY
Alopiidae

TAXONOMY
Squalus vulpinus Bonnaterre, 1788, Mediterranean Sea.

OTHER COMMON NAMES
French: Renard; Spanish: Zorro.

PHYSICAL CHARACTERISTICS
A very characteristic, large shark that may reach over 19.7 ft (6 m) in length, with an extremely elongated caudal fin (as long as the body), prominent first dorsal fin, minute second dorsal and anal fins, long pectoral fins, and small, conical snout. Coloration blue-gray to dark gray dorsally and laterally, with a white abdominal region, and white blotches laterally anterior to tail.

DISTRIBUTION
Circumglobal in both coastal and oceanic, tropical to temperate, waters.

HABITAT
Usually occurs over the continental shelf region, close to the surface, but also occupying oceanic waters down to a depth of 1,200 ft (366 m). Younger specimens are more commonly found inshore.

BEHAVIOR
Thresher sharks are swift, vigorous swimmers, capable of breaching. They segregate by sex and migrate seasonally off the western coast of North America.

FEEDING ECOLOGY AND DIET
Preys mostly on a wide variety of epipelagic, midwater, and demersal fishes, but known to feed also on squid, octopi, pelagic crustaceans, and even seabirds. Uses its long tail to stun prey, entrapping them by swimming in increasingly smaller circles around schools of fishes, sometimes even in tandem with another thresher shark.

REPRODUCTIVE BIOLOGY
Yolk-sac viviparous, embryos apparently are uterine cannibals (oophagy). Litter numbers range from two to six, most commonly four; three to seven have been recorded in the eastern Atlantic. Young remain for a short period in shallow water nursing grounds. A gestation period of nine months has been reported for California populations, where mating occurs in the summer. Individuals are sexually mature between three and eight years old. Individuals may live for 50 years.

CONSERVATION STATUS
Listed as Data Deficient by the IUCN.

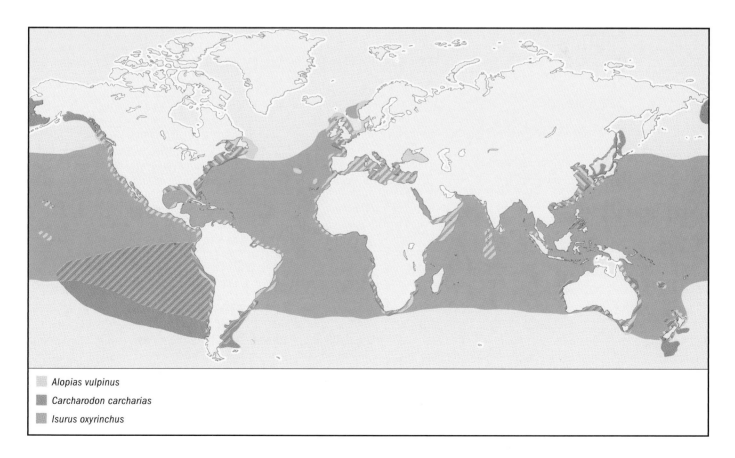

Alopias vulpinus
Carcharodon carcharias
Isurus oxyrinchus

SIGNIFICANCE TO HUMANS

Caught by both commercial (mostly with longlines) and recreational fisheries, and consumed (or have been) in somewhat regular quantities in many locations, from almost all major oceans. Heavily fished off the West Coast of the United States in the late 1970s, but because overfishing led to a significant decline in stocks, the targeted fishery was terminated in 1990, though they were still captured as bycatch. Not considered dangerous, but they command respect because of their large size and possible aggressiveness. ◆

Basking shark
Cetorhinus maximus

FAMILY
Cetorhinidae

TAXONOMY
Squalus maximus Gunnerus, 1765, Norway.

OTHER COMMON NAMES
French: Pélerin; Spanish and Portuguese: Peregrino.

PHYSICAL CHARACTERISTICS
An unmistakable, huge shark, with extremely elongated gill slits (reaching from the dorsal to the ventral side), a very wide gill region when gills are expanded during feeding, a large, capacious mouth, well-developed gillrakers on the inside of the gills to capture small food particles, very small teeth, elongated pectoral fins, and a large lunate caudal fin. Grayish in color all around. Reported to reach 40–50 ft (12.2–15.2 m) in length, but large specimens are more common at about 33 ft (10 m).

DISTRIBUTION
Worldwide in mainly coastal, cold, temperate waters, most abundant off both sides of the northern Atlantic, but also in warmer, subtemperate regions such as the Mediterranean Sea.

HABITAT
Usually found over relatively shallow, coastal, pelagic waters but can be caught in open seas over deeper waters. Basking sharks appear in regular periods in certain areas (probably to feed) but also disappear in what appears to be regular cycles. Where they "disappear" to is a mystery, and perhaps they "hibernate" or spend periods of relative inactivity on or close to the bottom of the ocean.

BEHAVIOR
Basking sharks have been seen to leap clear out of the water (as have other mackerel sharks). Usually they are observed cruising at about 2.3 mph (2 knots) near or at the surface, with their mouths open during feeding. They are highly migratory, and several individuals may swim in tandem.

FEEDING ECOLOGY AND DIET
A filter-feeding shark, capable of taking in massive amounts of zooplankton. It swims with its mouth open very wide, retaining food items on its gillrakers, which are covered by denticles, giving them a rough texture. The gillrakers are shed periodically, usually in the early winter. Basking sharks feed mostly in the summer months near the surface. They either feed by alternative means when the gillrakers are shed, or remain without feeding, inactive, until they are regenerated. Food is retained in the gillrakers, aided by secretions of mucus in the pharynx, and subsequently swallowed when the mouth is closed.

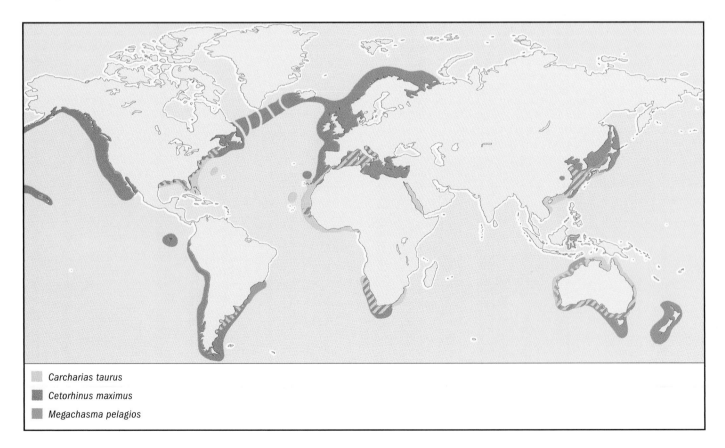

Carcharias taurus

Cetorhinus maximus

Megachasma pelagios

REPRODUCTIVE BIOLOGY
Basking sharks employ yolk-sac viviparity, giving birth to two to six pups per gestation. The pups are the largest of all shark species, ranging 59–67 in (150–170 cm) in total length. Gestation periods are mostly unknown but are estimated to be very long (more than 1 year). The ovaries produce huge quantities of eggs.

CONSERVATION STATUS
Listed as Vulnerable by the IUCN and protected in the United Kingdom, Malta, United States (East Coast), and New Zealand. Protection is pending in other areas (Mediterranean Sea, South Africa).

SIGNIFICANCE TO HUMANS
The basking shark has been captured in much of its range since the nineteenth century for its oily liver, which may contain up to 500 gal (1,893 l), meat (fresh and dried-salted), and skin, and also for its fins for the abhorrent shark fin–soup industry. Populations of the basking shark have declined significantly in many regions. The basking shark has some importance for the tourism trade, as it can be seen in many places, especially in the northern Atlantic (e.g., Bay of Fundy, Cape Cod, Isle of Man). Harmless, this shark poses no direct danger to people, but deserves respect because of its large size. ◆

White shark
Carcharodon carcharias

FAMILY
Lamnidae

TAXONOMY
Squalus carcharias Linnaeus, 1758, Europa.

OTHER COMMON NAMES
English: Great white shark; French: Grand requin blanc; Spanish: Jaquetón blanco.

PHYSICAL CHARACTERISTICS
Very large, reaching to 21.3 ft (6.5 m), more commonly to 18 ft (5.5 m), with a distinctive dentition comprised of large, triangular teeth with serrations on both edges, and with lateral cusps in embryos. They have a conspicuous white ventral coloration and a gray-to-bluish dorsal and lateral shade (the ventral and dorsal colorations are clearly separated on the sides), large gill slits, well-developed precaudal keels, a large first dorsal fin (much larger than the second), large and lunate caudal fin, pectoral fins with black tips ventrally, a conical snout, and a large, black eye.

DISTRIBUTION
Worldwide in coastal marine waters, and also around oceanic tropical islands, but more common in cold and warm temperate regions, and apparently rare or absent from most of the western Indian Ocean, Indonesia, and tropical Central America. Most common off California, Australia, and South Africa. Compared to other shark species, the white shark is relatively uncommon where it occurs.

HABITAT
The white shark is primarily a continental shelf inhabitant, cruising through relatively shallow waters either near the surface or close to the bottom. It also is found off oceanic islands and inshore bays and has even been captured on a bottom longline as far down as 4,199 ft (1,280 m). Capable of wide excursions in the pelagic realm.

BEHAVIOR
Whites are solitary and nomadic, and may occur in pairs, but feeding aggregations of some ten individuals also have been observed. It is known that they will leap completely out of the water (breaching) when capturing surface prey (or perhaps for other reasons). They are even capable of breaching vertically in a manner similar to dolphins. "Spy hopping" (when the shark will maintain its head out of the water as if to search the surroundings) and "repeated aerial gaping" (RAG; when the shark "bites" the air with its head clear out of the water) also have been observed. The white shark is known to satisfy its curiosity by circling intended prey items, or even boats and divers. It is capable of great bursts of speed. While feeding, their eyes roll back in their sockets. There may be segregation of individuals according to size.

FEEDING ECOLOGY AND DIET
The white shark is a formidable predator, feeding mainly on numerous families of bony fishes (as well as a large variety of sharks, even the basking shark), sea turtles, marine mammals (pinnipeds and whale carcasses), and even sea birds resting on the surface. Invertebrates also may be eaten (such as crabs), but most of its food comes from fishes and marine mammals taken from the surface or in the water column. White sharks are one of the top predators in the ocean; however, they sometimes fall prey to orcas (killer whales).

REPRODUCTIVE BIOLOGY
Embryos develop inside the uteri (yolk-sac viviparous), and intrauterine cannibalism (oophagy) is confirmed, as embryos have been found to have great amounts of yolk and egg membranes in their stomachs. Teeth also have been found in the stomachs of embryos, but embryos are believed to swallow their own teeth during development, as they undergo tooth replacement several times before birth. Gestation periods are mostly unknown. A litter of nine pups was reported for one pregnant female from the Mediterranean, and up to 10 embryos may reach term (data from gravid Japanese whites). The lack of knowledge concerning their reproduction is due to the scarcity of gravid females, perhaps an indication of pronounced segregation during gestation, or even of low fecundity. Size at maturity for females is between 13.1 ft (4 m) and 16.4 ft (5 m) long, and between 11.5 ft (3.5 m) and 13.1 ft (4 m) for males. Age at maturity ranges from 12 to 14 years for females and nine to 10 for males. Embryos measure 4 ft (1.2 m) to 5 ft (1.5 m) at birth, and can weigh up to 55 lb (25 kg). Courtship has been observed in one instance; the male bit the female into submission preceding a 40-minute-long copulatory embrace.

CONSERVATION STATUS
Presently threatened in many locations (e.g., Australia, South Africa) and heavily protected in Australia, South Africa, Namibia, Israel, Malta, and the United States. Australia is apparently the only country in which there is a detailed recovery plan for this species. Whites are listed as Vulnerable by the IUCN.

SIGNIFICANCE TO HUMANS
The white shark is perhaps the most notorious of all sharks, with an undeserved reputation as a "man-eater" and threat to humans. There are attacks on humans attributed to this species every year, but they average only about three per year from 1952 to 1992 (increasing slightly towards 1999). Attacks by the white are rare, however, when the whole phenomenon of

"shark attack" is taken into account. About 80% of all shark biting incidents have occurred in the tropics, where whites are far less common than in temperate zones. Attacks by whites are even more insignificant when one considers that more people have died from incidents with domestic livestock (e.g., pigs) than have died of attacks from this shark. Much of the maligned popular image is a result of the *Jaws* movies. However, it is the white shark that is in dire straits as a result of being slaughtered by recreational and commercial fishermen, either intentionally for trophies or as bycatch. Contrary to its folkloric, *Jaws* image, the white shark is worth more alive than dead and is an extremely valuable asset to ecotourism in many locations, attracting scores of interested onlookers who pay generously to see the creature from the protection of a submerged cage. Perhaps no other shark inspires as much fear and admiration as the white. A recent symposium volume (*Great White Sharks, The Biology of* Carcharodon carcharias) summarizes much valuable information concerning this species. ◆

Shortfin mako
Isurus oxyrinchus

FAMILY
Lamnidae

TAXONOMY
Isurus oxyrinchus Rafinesque, 1810, Sicily, Mediterranean Sea.

OTHER COMMON NAMES
French: Taupe bleu; Spanish: Marrajo dientuso.

PHYSICAL CHARACTERISTICS
A slender shark, with long, slightly curved teeth devoid of lateral cusps, slightly elongated pectoral fin, very lunate caudal fin with well developed lower lobe, conical snout, eyes not very large, very small second dorsal, pelvic, and anal fins. Reaches close to 13.1 ft (4 m) in length. Bluish dorsally and laterally, and white ventrally as well as on caudal fin.

DISTRIBUTION
Worldwide in tropical to temperate waters.

HABITAT
An oceanic and littoral shark, found from the surface down to 1,640 ft (500 m).

BEHAVIOR
The shortfin mako is probably the fastest and most agile of all sharks, jumping clear out of the water by several times its own length. Slightly endothermic, as its body musculature may reach up to 18° F (10° C) or more warmer than the temperature of the surrounding water. Highly migratory, capable of long-range migrations following warmer water masses.

FEEDING ECOLOGY AND DIET
Feeds mostly on fishes, both bony fishes as well as sharks and rays, squids, marine mammals, and turtles. Mako sharks are voracious feeders, consuming up to 3% of their body weight per day (compared to under 1% for many shark species), and digesting an average-sized meal in less than two days, whereas most sharks take some three to four days.

REPRODUCTIVE BIOLOGY
Yolk-sac viviparous, with uterine cannibalism (oophagy), and litters ranging from four to as many as 30 young (usually be-

tween 10 and 18). Gestation period long (possibly from 15 to 18 months). Size at birth ranging from 23 to 27 in (60 to 70 cm). Age of maturity may be from seven to eight years old.

CONSERVATION STATUS
Listed as Lower Risk/Near Threatened by the IUCN.

SIGNIFICANCE TO HUMANS
Shortfin makos are fished significantly in many areas worldwide, recreationally, artisanally, and industrially. Because of their highly active demeanor and size, shortfin makos may be dangerous, even though there are few incidents recorded, most of which are accidents while fishing and handling live individuals. ◆

Porbeagle
Lamna nasus

FAMILY
Lamnidae

TAXONOMY
Squalus nasus Bonaterre, 1788, probably Cornwall, England.

OTHER COMMON NAMES
French: Requin-taupe commun; Spanish: Marrajo sardinero.

PHYSICAL CHARACTERISTICS
A somewhat stout shark, with a conical snout, large dark eyes, tips of pectoral fins slightly rounded, lunate caudal fin, teeth with small accessory cusps, bluish dorsal and lateral coloration (posterior tip of first dorsal white), and white ventrally. Reaches slightly over 9.8 ft (3 m) in length.

DISTRIBUTION
Occurs in warm, temperate, to cold waters in both the northern Atlantic, Mediterranean Sea, and in the Southern Hemisphere in the Atlantic and Indian Oceans, and off southern Australia.

HABITAT
An epipelagic, littoral, and oceanic shark, most abundant on offshore fishing banks, usually in colder waters. Occurs from 3 to 2,296 ft (1 to 700 m) in depth.

BEHAVIOR
Porbeagles can be solitary or occur in schools. Usually migrates extensively at least in the northern Atlantic, and may aggregate by sex and size. An active, strong swimmer, capable of leaping out of water when captured.

FEEDING ECOLOGY AND DIET
Feeds mostly on fishes, both bony and cartilaginous, as well as on cephalopods. May be consumed by larger sharks.

REPRODUCTIVE BIOLOGY
Yolk-sac viviparous, with uterine cannibalism confirmed (oophagy). Litters vary from one to five young (usually four), and gestation periods are estimated to last between eight and nine months. Young inside uterus may have fang-like teeth specialized for tearing egg cases to release eggs for consumption. The fang-like teeth are then shed in utero. Young are born 23.6–29.5 in (60–75 cm) in length.

CONSERVATION STATUS
Listed as Lower Risk/Near Threatened by the IUCN.

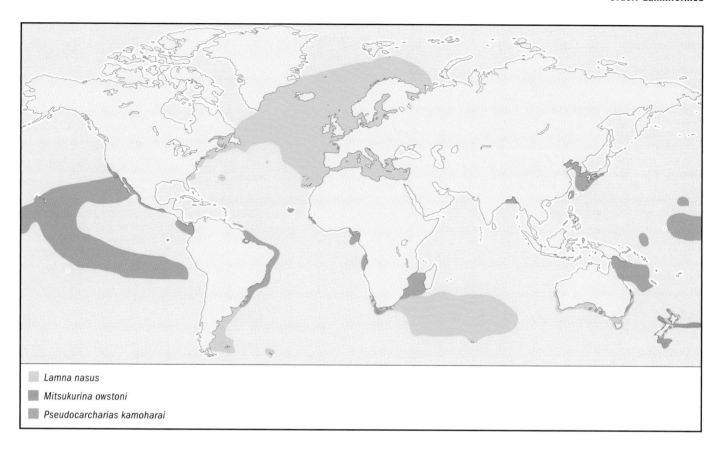

Lamna nasus

Mitsukurina owstoni

Pseudocarcharias kamoharai

SIGNIFICANCE TO HUMANS
Heavily fished for consumption, usually by longlining, but because of population declines, the porbeagle is now captured far
less frequently. Overfishing is a concern especially in the
North Atlantic, where the industry is regulated, but stocks may
not be able to rebound. Recreationally fished as well, but individuals must be released upon capture. Not considered particularly dangerous to people. ◆

Megamouth shark
Megachasma pelagios

FAMILY
Megachasmidae

TAXONOMY
Megachasma pelagios Taylor, Compagno, and Struhsaker, 1983,
Hawaii.

OTHER COMMON NAMES
French: Requin grand gueule; Spanish: Tiburón bocudo.

PHYSICAL CHARACTERISTICS
A remarkable, very large shark, up to 18 ft (5.5 m) long, with a
very wide mouth, extremely elongated jaws that reach to the
tip of the snout (the lower jaw is slightly longer than the upper
jaw), endowed with minute teeth, a large head region, relatively small eyes, dense, papillated gillrakers, very long pectoral
fins, low dorsal fins, and a large caudal fin. Coloration grayish
blue dorsally and whitish ventrally.

DISTRIBUTION
Known from some 17 specimens from Japan, the Philippines,
western Australia, California, southeastern Brazil, Senegal, and
South Africa. Probably circumglobal in warm temperate to
tropical waters.

HABITAT
A coastal and oceanic inhabitant, found in shallow waters some
16 ft (5 m) deep as well as in deep waters offshore, down to
545 ft (166 m) in waters up to 15,092 ft (4,600 m) deep. One
specimen was stranded on a beach after washing ashore.

BEHAVIOR
A sluggish, solitary shark, capable of vertical migrations (perhaps following the movements of euphasiid shrimps). The
megamouth shark is epipelagic but is capable of diving to great
depths.

FEEDING ECOLOGY AND DIET
Feeds on zooplankton, especially euphasiid shrimp, copepods,
and jellyfish, probably by taking in big gulps of water and retaining food on the very dense gillrakers. Luminous tissue lining the oral cavity may have a role in attracting prey items.
Megamouths are the only known sharks with bite marks of the
cookie-cutter shark (*Isistius brasiliensis*) and may be especially
vulnerable because of their rather soft skin and slow-swimming
behavior.

REPRODUCTIVE BIOLOGY
Mostly unknown, but probably yolk-sac viviparous, as are other
lamniforms. Pregnant females have not been captured. Small
oocytes, 0.2–0.4 in (5–10 mm) in diameter, were present in the
ovaries of an adult female. Bite marks matching the teeth of

males also were present, and these have been interpreted as courtship scars.

CONSERVATION STATUS
Listed as Data Deficient by the IUCN.

SIGNIFICANCE TO HUMANS
Caught occasionally as bycatch, but not consumed, though a Philippine specimen was apparently divided among fishermen. Of interest as a museum exhibit because of its extraordinary size and unusual features. The megamouth shark, only recently discovered, is one of the most spectacular ichthyological discoveries of the last 30 years and is the subject of a recent symposium volume, *Biology of the Megamouth Shark*, that provides much information on its anatomy and biology. ◆

Goblin shark
Mitsukurina owstoni

FAMILY
Mitsukurinidae

TAXONOMY
Mitsukurina owstoni Jordan, 1898, Japan.

OTHER COMMON NAMES
French: Requin lutin; Spanish: Tiburón duende.

PHYSICAL CHARACTERISTICS
A very unique shark, with soft, flabby flesh, an extremely elongated, paddle-shaped snout endowed with numerous pores of the ampullary sensory system, jaws that protrude greatly, very long, slender, sharp teeth without lateral cusps, a large anal fin, small dorsal fins, and long and low caudal fin. Captured specimens are off-white to pinkish white all over. Reaches some 12.5 ft (3.8 m) in length.

DISTRIBUTION
Scattered distribution on the continental slope, but probably worldwide.

HABITAT
A deep water, oceanic (found on seamounts), and continental slope shark, reaching depths of at least 4,265 ft (1,300 m).

BEHAVIOR
Mostly unknown, but a live specimen swam with its jaws tightly retracted, and not protruded as might have been expected.

FEEDING ECOLOGY AND DIET
Not well documented, but feeds on fishes and perhaps crustaceans.

REPRODUCTIVE BIOLOGY
Unknown, as no pregnant females have ever been recorded, but presumed to be yolk-sac viviparous, as are other lamniforms.

CONSERVATION STATUS
Not threatened.

SIGNIFICANCE TO HUMANS
Taken on longlines, mostly as bycatch, but not fished significantly. Not considered dangerous because of its deep-water habitat. ◆

Sand tiger shark
Carcharias taurus

FAMILY
Odontaspididae

TAXONOMY
Carcharias taurus Rafinesque, 1810, Mediterranean Sea.

OTHER COMMON NAMES
English: Gray nurse shark (Australia); French: Requin taureau; Spanish: Toro bacota.

PHYSICAL CHARACTERISTICS
A relatively large shark, up to 9.8 ft (3 m) in length, with characteristic slender, sharp teeth (with lateral cusps), appearing to project outside of mouth (functional rows may point slightly forward), dorsal fins almost equal in size (first slightly larger), large anal fin, slightly rounded pectoral fin tips, conical snout, somewhat depressed head, clearly demarcated lateral line, and a light brown coloration, usually with slightly darker blotches scattered on body.

DISTRIBUTION
Distributed worldwide in tropical and warm temperate waters, but absent at least from the eastern Pacific, and probably also from the Caribbean and eastern North Atlantic (north of Africa).

HABITAT
Mostly an inshore species, found in shallow waters, but recorded to occur down to 627 ft (191 m), and may occur either close to the bottom, at the surface, or midwater.

BEHAVIOR
Mostly nocturnal and solitary, but may form large schools, and capable of extensive migrations. Individuals aggregate for courtship, mating, feeding, and birth, and the sand tiger is capable of social interactions. Behavior in this species is known from aquarium observations, indicating that sand tiger sharks display specific patterns related to courtship and mating. These include specific movements of the claspers in males, submissive behavior by females, males poking the cloaca region of females with their snouts, and males biting females to establish dominance, among other behaviors. Sand tiger sharks will periodically gulp air into their stomachs from the surface, apparently as a buoyancy control.

FEEDING ECOLOGY AND DIET
Sand tiger sharks feed mostly on a wide range of fishes, including many families of pelagic and demersal bony fishes, as well as sharks and rays, cephalopods, crustaceans, and marine mammals.

REPRODUCTIVE BIOLOGY
Yolk-sac viviparous, with embryos consuming other embryos (adelphophagy) and eggs in uteri. Gestation varies from 9 to twelve months. Two young are born in a litter, one per uterus, but a cluster of 16 to 23 eggs are grouped together in egg cases within each uterus after fertilization. From this group, only one embryo will be dominant, feeding on other embryos and eggs and even moving vigorously within the uterus. Young are born at about 3.3 ft (1 m) in length, but already have sharp teeth at 6.7 in (17 cm). Breeds every other year.

CONSERVATION STATUS
Protected in Australian waters since 1984 because of steady declines in its populations as a result of overfishing, and listed as

Vulnerable by the IUCN. Local protection measures are in place in many regions, including the eastern coast of the United States since 1997.

SIGNIFICANCE TO HUMANS
Commonly displayed in aquaria, where specimens can live for extended periods, surpassing 30 years. Not considered particularly dangerous, but because it appears ferocious, sand tigers have been implicated in attacks off Australia that are probably the result of other shark species. Observations conducted in the wild and in aquaria indicate that this species is harmless and presents no real danger to divers. Important to the ecotourism industry, as many trips feature observations of wild sand tiger sharks. ◆

Crocodile shark
Pseudocarcharias kamoharai

FAMILY
Pseudocarchariidae

TAXONOMY
Carcharias kamoharai Matsubara, 1936, Japan.

OTHER COMMON NAMES
French: Requin crocodile; Spanish: Tiburón crocodilo.

PHYSICAL CHARACTERISTICS
A slender, relatively small shark up to 43 in (110 cm) in total length), with very large, blackish eyes, long, pointed teeth with minute lateral cusps only on lateral tooth rows (fewer than 30

total tooth rows in either jaw), pointed snout, low dorsal fins, and rounded pectoral fin tips.

DISTRIBUTION
Worldwide in tropical marine waters, but with a scattered distribution, needing confirmation from many areas.

HABITAT
Inhabits the tropical pelagic realm, but can occasionally be captured closer to shore. Reaches 984 ft (300 m) in depth.

BEHAVIOR
Mostly unknown, but is believed to be a fast-swimming shark, probably capable of leaping out of the water. Its large eyes might indicate either nocturnal activity or feeding at great depths.

FEEDING ECOLOGY AND DIET
Feeds mainly on pelagic or mesopelagic fishes (e.g., bristlemouths [Gonostomatidae], lanternfishes [Myctophidae]), and invertebrates such as squids.

REPRODUCTIVE BIOLOGY
Yolk-sac viviparous, giving birth to four pups at a time (two per uterus). Uterine cannibalism has been recorded, with more developed fetuses eating remaining eggs, and, uniquely, two individuals surviving per uterus. Gestation periods mostly unknown.

CONSERVATION STATUS
Considered at Lower Risk/Near Threatened by the IUCN.

SIGNIFICANCE TO HUMANS
Fished by longlines off Japan, but not significantly consumed. Because of its pelagic nature, there is little interaction with people, and hence the species is not considered dangerous.

Resources

Books
Applegate, S. P., and L. Espinosa-Arrubarrena. "The Fossil History of *Carcharodon* and Its Possible Ancestor, *Cretolamna*: A Study in Tooth Identification." In *Great White Sharks: The Biology of* Carcharodon carcharias, edited by A. P. Klimley and David G. Ainley, 19–36. San Diego, CA: Academic Press, 1996.

Bigelow, Henry B., and William C. Schroeder. "Sharks." In *Fishes of the Western North Atlantic*, Vol. 1, pt. 1 of *Memoir of the Sears Foundation for Marine Research*, 59–576. New Haven, CT: Yale University, 1948.

Branstetter, Steven, ed. *Conservation Biology of Elasmobranchs.* NOAA Technical Report, NMFS 115. Seattle, WA: U. S. Department of Commerce, 1993.

Burgess, G. H., and M. Callahan. "Worldwide Patterns of White Shark Attacks on Humans." In *Great White Sharks: The Biology of* Carcharodon carcharias, edited by A. Peter Klimley and David G. Ainley, 457–469. San Diego, CA: Academic Press, 1996.

Cappetta, Henri. *Chondrichthyes II, Mesozoic and Cenozoic Elasmobranchii.* Stuttgart, Germany: Gustav Fischer Verlag, 1987.

Carwardine, Mark, and Ken Watterson. *The Shark Watcher's Handbook: A Guide to Sharks and Where to See Them.* Princeton, NJ: Princeton University Press, 2002.

Compagno, Leonard J. V. "Relationships of the Megamouth Shark, *Megachasma pelagios* (Lamniformes: Megachasmidae), with Comments on Its Feeding Habits." In *Elasmobranchs as Living Resources: Advances in the Biology, Ecology, Systematics, and the Status of the Fisheries*, 357–379, edited by H. L. Pratt, Jr., S. H. Gruber, and T. Taniuchi. NOAA Technical Report, NMFS 90. Seattle, WA: U.S. Department of Commerce, 1990.

————. *Bullhead, Mackerel and Carpet Sharks (Heterodontiformes, Lamniformes and Orectolobiformes)*, Vol. 2 of *Sharks of the World: An Annotated and Illustrated Catalogue of Shark Species Known to Date*. Rome, Italy: Food and Agriculture Organization of the United Nations, 2001.

Compagno, Leonard J. V., and V. H. Niem. "Families Odontaspididae, Pseudocarchariidae, Alopiidae, and Lamnidae." In *Western Central Pacific Identification Sheets to Species*, edited by Kent E. Carpenter and Volker H. Niem, 1264–1278. Rome, Italy: Food and Agriculture Organization of the United Nations, 1999.

Demski, Leo S., and John P. Wourms, eds. *Reproduction and Development of Sharks, Skates, Rays and Ratfishes.* Boston, MA: Kluwer Academic Publishers, 1993.

Ellis, Richard, and John E. McCosker. *Great White Shark.* New York: Harper Collins, 1991.

Francis, M. P. "Observations on a Pregnant White Shark with a Review of Reproductive Biology." In *Great White Sharks:*

Resources

The Biology of Carcharodon carcharias, edited by A. Peter Klimley and David G. Ainley, 157–172. San Diego, CA: Academic Press, 1996.

Gottfried, M. D., Leonard J. V. Compagno, and S. C. Bowman. "Size and Skeletal Anatomy of the Giant 'Megatooth' Shark *Carcharodon megalodon*." In *Great White Sharks: The Biology of* Carcharodon carcharias, edited by A. Peter Klimley and David G. Ainley, 55–66. San Diego, CA: Academic Press, 1996.

Hamlett, William C., ed. *Sharks, Skates, and Rays: The Biology of Elasmobranch Fishes*. Baltimore, MD: Johns Hopkins University Press, 1999.

Hennemann, R. M. *Sharks and Rays, Elasmobranch Guide of the World*. Frankfurt, Germany: Ikan, 2001.

Klimley, A. Peter, and David G. Ainley, eds. *Great White Sharks: The Biology of* Carcharodon carcharias. San Diego, CA: Academic Press, 1996.

Last, P. R., and J. D. Stevens. *Sharks and Rays of Australia*. Melbourne, Australia: CSIRO Division of Fisheries, 1994.

Naylor, G. J. P., et al. "Interrelationships of Lamniform Sharks: Testing Phylogenetic Hypotheses with Sequence Data." In *Molecular Systematics of Fishes*, edited by Thomas D. Kocher and Carol A. Stepien, 199–218. San Diego, CA: Academic Press, 1997.

Nelson, J. *Fishes of the World*. 3rd ed. New York: John Wiley & Sons, 1994.

Perrine, Doug. *Sharks & Rays of the World*. Stillwater, MN: Voyageur Press, 1999.

Pratt, H. L. Jr., S. H. Gruber, and T. Taniuchi, eds. *Elasmobranchs as Living Resources: Advances in the Biology, Ecology, Systematics, and the Status of the Fisheries*. NOAA Technical Report, NMFS 90. Seattle: U. S. Department of Commerce, 1990.

Sibley, G., J. A. Seigel and C. C. Swift, eds. *Biology of the White Shark*, Vol. 9 of *Memoirs of the Southern California Academy of Sciences*. Los Angeles: 1985.

Springer, Victor G., and Joy P. Gold. *Sharks in Question. The Smithsonian Answer Book*. Washington, DC: Smithsonian Institution Press, 1989.

Stillwell, C. "The Ravenous Mako." In *Discovering Sharks*, edited by S. H. Gruber, 77–78. Highlands, NJ: American Littoral Society, 1990.

Whitley, G. P. *The Sharks, Rays, Devil-fish, and Other Primitive Fishes of Australia and New Zealand*. Pt. 1 of *The Fishes of Australia*. Sydney, Australia: Royal Zoological Society of New South Wales, 1940.

Yano, K., J. F. Morrissey, Y. Yabumoto, and K. Nakaya, eds. *Biology of the Megamouth Shark*. Tokyo, Japan: Tokai University Press, 1997.

Periodicals

Carey, F. G., et al. "Temperature, Heat Production, and Heat Exchange in Lamnid Sharks." *Memoirs of the Southern California Academy of Sciences* 9 (1985): 92–108.

Carey, F. G., et al. "The White Shark, *Carcharodon carcharias*, Is Warm-Bodied." *Copeia* 2 (1982): 254–260.

Eitner, B. J. "Systematics of the Genus *Alopias* (Lamniformes: Alopidae) with Evidence for the Existence of an Unrecognized Species." *Copeia* 3 (1995): 562–571.

Gilmore, R. G. "Reproductive Biology of Lamnoid Sharks." *Environmental Biology of Fishes* 38 (1993): 95–114.

Gilmore, R. G., J. W. Dodrill, and P. A. Linley. "Reproduction and Embryonic Development of the Sand Tiger Shark, *Odontaspis taurus* (Rafinesque)." *Fishery Bulletin* 81 (1983): 201–225.

Gruber, S. H., and Leonard J. V. Compagno. "Taxonomic Status and Biology of the Bigeye Thresher, *Alopias superciliosus*." *Fishery Bulletin* 79, no. 4 (1981): 617–640.

Hutchins, B. "Megamouth: Gentle Giant of the Deep." *Australian Natural History* 23, no. 12 (1992): 910–917.

Jordan, D. S. "A Classification of Fishes Including Families and Genera as far as Known." *Stanford University Publications: Biological Sciences* 3 (1923): 77–243.

Klimley, A. P. "The Areal Distribution and Autoecology of the White Shark, *Carcharodon carcharias*, off the West Coast of North America." *Memoirs of the Southern California Academy of Sciences* 9 (1985): 15–40.

———. "The Predatory Behavior of the White Shark." *American Scientist* 82, no. 2 (1994): 122–133.

Maisey, J. G. "Relationships of the Megamouth Shark, *Megachasma*." *Copeia* 1 (1985): 228–231.

Matthews, L. H. "Reproduction in the Basking Shark, *Cetorhinus maximus*." *Philosophical Transactions of the Royal Society of London*, Series B, Biological Sciences 234 (1950): 247–316.

McCosker, J. E. "White Shark Attack Behavior: Observations of and Speculations About Predator and Prey Strategies." *Memoirs of the Southern California Academy of Sciences* 9 (1985): 123–135.

———. "The White Shark, *Carcharodon carcharias*, Has a Warm Stomach." *Copeia* 1 (1987): 195–197.

Taylor, L. R., Leonard J. V. Compagno, and P. J. Strusaker. "Megamouth—A New Species, Genus, and Family of Lamnoid Shark (*Megachasma pelagios*, Megachasmidae) from the Hawaiian Islands." *Proceedings of the California Academy of Sciences* 43 (1983): 87–110.

Tricas, T. C., and J. E. McCosker. "Predatory Behavior of the White Shark (*Carcharodon carcharias*), with Notes on Its Biology." *Proceedings of the California Academy of Sciences* 43 (1984): 221–238.

Wourms, J. P. "Reproduction and Development in Chondrichthyan Fishes." *American Zoologist* 17 (1977): 379–410.

Organizations

American Elasmobranch Society, Florida Museum of Natural History. Gainesville, FL 32611 USA. Web site: <http://www.flmnh.ufl.edu/fish/Organizations/aes/aes.htm>

Marcelo Carvalho, PhD

Hexanchiformes

(Six- and sevengill sharks)

Class Chondrichthyes
Order Hexanchiformes
Number of families 2

Photo: The broadnose sevengill sharks (*Notorynchus cepedianus*) bear the most resemblence to prehistoric sharks and, therefore, are considered to be more primitive than the sharks with five or six gills. (Photo by Tom McHugh/Photo Researchers, Inc. Reproduced by permission.)

Evolution and systematics

Hexanchiforms are an ancient lineage, as well-documented fossil skeletons of *Notidanoides muensteri* date from the late Jurassic (some 150 million years ago, or mya) Solnhofen limestones of southern Germany. *Notidanoides* was a large shark, up to 9.8 ft (3 m) in length, with features suggestive of modern hexanchiforms, such as a single dorsal fin. Its teeth, with multiple cusps arranged in a series, indicate that it is related closely to the Hexanchidae. Most fossil hexanchiforms are known from isolated teeth found in all continents, ranging from the early Jurassic (180 mya) to the Tertiary, which makes the order one of the longest-surviving shark lineages. Many late Cretaceous to Tertiary species are even assigned to living genera, based on fossil teeth. The hexanchiform fossil record indicates that they were never very diverse, but more so than at present, as there are only five living species.

The bizarre frilled shark (*Chlamydoselachus anguineus*) was first described by the American zoologist and chondrichthyan taxonomist Samuel W. Garman in 1884. It has teeth that resemble those of some Paleozoic sharks ("cladodont teeth"), which led many early researchers to consider it a relic of Devonian seas. It is now well established, however, that *Chlamydoselachus* shares a more recent common ancestry with all living sharks and rays (the Neoselachii), only distantly related to Paleozoic forms.

Sixgill and sevengill sharks are the most basal ("primitive") members of the large group known as the Squalea, which includes the dogfishes and allies (Squaliformes), the angelsharks (Squatiniformes), the sawsharks (Pristiophoriformes), and the rays, or batoids (Batoidea). The Squalea group is characterized by numerous evolutionary specializations, such as complete hemal arches (ventral projections arising from the vertebral column) in the trunk region anterior to the tail. There has been debate as to whether the frilled shark is actually part of the Hexanchiformes or rather belongs in an order of its own, but derived features shared with other hexanchiforms support its placement within the order (e.g., the extra gill arch and more heart valve rows).

The five extant species of hexanchiforms are divided into two families: Chlamydoselachidae (*Chlamydoselachus anguineus*) and Hexanchidae (*Hexanchus griseus*, *H. nakamurai*, *Notorynchus cepedianus*, and *Heptranchias perlo*). The latter family also is known as "cowsharks."

Physical characteristics

The hexanchiform families are very distinct in their morphological characteristics. *Chlamydoselachus* is a highly modified and unique shark, with an eel-like body, an enlarged mouth, and well-delimited rows of teeth. (Its teeth are unlike those of any other living shark.) It shares with hexanchids a single, posteriorly located dorsal fin and a long caudal fin, an extra gill arch (hexanchiforms have either six or seven gill arches and gill slits on each side), small spiracles, a clearly demarcated lateral line along the trunk and precaudal tail regions, and a mouth extending posteriorly behind the level of the eyes. Hexanchids have unique teeth that are highly differentiated between the upper and lower jaws and also along either jaw. The upper jaw teeth are small and flattened, with either a single cusp or very small accessory cusps; the lower jaw teeth are very wide and flattened, with multiple prominent cusps in addition to a median (symphysial) tooth and smaller, blunt posterior teeth. Hexanchiforms are noteworthy for having mostly uncalcified vertebrae and notochords with little constriction.

Chlamydoselachus is a uniform dark brown to grayish brown in color, whereas hexanchids are mostly gray without strong color patterns. The exception is *Notorynchus*, which has darker spots. *Notorynchus* and *Hexanchus griseus* attain very large sizes, but the remaining species are more moderate in size, usually not surpassing 63 in (160 cm) in length.

Distribution

These fishes are found worldwide in tropical and temperate waters, but most species have a spotty distribution, that is, they are known from many isolated regions without records from intermediate areas. Because most hexanchiform sharks occur along the continental slopes, their scattered distribution may be only an artifact of sampling.

Habitat

Most species are deepwater inhabitants, occurring demersally along the continental slopes but sometimes venturing into more shallow pelagic or inshore waters. The primary exception is *Notorynchus cepedianus*, which also is a coastal species. There are shallow-water records for *Heptranchias perlo* and *Hexanchus griseus* as well, but these species are more common in waters deeper than 328 ft (100 m). Some species also have been recorded from oceanic islands.

Behavior

Little is known concerning their behavior, but individuals of *Notorynchus cepedianus* may hunt cooperatively. Sharks are almost exclusively solitary hunters, but broadnose sevengills have been observed hunting as a pack off the coast of Namibia. Individuals circled a large fur seal (which can weigh 770 lb, or 350 kg, more than the shark) and slowly closed in by tightening the circle. After one shark initiated an attack, the remaining sharks followed suit. Hexanchiforms may migrate vertically, entering more shallow waters at night. Some species migrate vertically in the water column, remaining closer to the bottom during the day and ascending to the surface to feed at night.

Feeding ecology and diet

Sixgill and sevengill sharks feed on a variety of bony fishes, sharks, and rays as well as invertebrates. Marine mammals (seals and dolphins) also are consumed. Bony fishes include numerous benthic and demersal families, but pelagic species also are eaten. Some hexanchiform species are known to have attacked individuals of the same species that have been hooked, biting off pieces while they were being reeled or towed in. Because of their mostly obscure habitats, very little is recorded concerning their feeding ecology. At least one species may hunt in packs. Hexanchiform sharks are presumably consumed by larger sharks, including of their own species, but little data exist in relation to their predators.

Reproductive biology

All species are ovoviviparous (aplacentally viviparous), and the young derive their nourishment exclusively from the yolk sac before birth. Litters can be very large in the two largest species. (More than 100 pups may be born at once in the case of the bluntnose sixgill shark.) Only up to 20 (usually about 12) young are present per litter, however, in the three remaining species (sharpnose sevengill shark, *Heptranchias perlo*; the bigeyed sixgill shark, *Hexanchus vitulus*; and the frilled shark, *Chlamydoselachus anguineus*). Females of the two largest species give birth to young in shallow-water nurseries (though not exclusively for the bluntnose sixgill shark, which also gives birth in other locations). Lengths at birth range from 10.2 to 25.6 in (26–65 cm). Almost nothing is known concerning gestation periods and other details of reproductive biology.

Conservation status

Two of the five currently recognized species of hexanchiforms are listed by the IUCN. *Hexanchus griseus* is considered Lower Risk/Near Threatened, and *Notorynchus cepedianus* is cited as Data Deficient.

Significance to humans

Certain hexanchiform species are fished commercially but not to a significant extent. They may be used fresh, frozen, or dried/salted, and the flesh of at least one species (*Notorynchus cepedianus*) is said to be of good quality. The skin of these sharks is used as leather (particularly in China). Hexanchiforms are not considered to be strictly dangerous, but because of the large sizes of at least some species, they should be approached with caution; very few attacks are attributed to sharks of this order. They are not hardy aquarium sharks. Two species (*Notorynchus cepedianus* and *Hexanchus griseus*) have been seen in the wild by tourists. The former species is seen in many areas of its shallow-water range (e.g., in Humboldt Bay and San Francisco Bay, California, United States). The latter was sighted through commercial operations that (until recently) took tourists out in small submersibles to see young sharks at depths of about 656 ft (200 m) off Hornby Island, British Columbia. Diving to observe *Hexanchus griseus* in the Strait of Georgia (between British Columbia, Canada, and the United States) also is possible during the summer months, at depths from 79 to 138 ft (24–42 m).

1. Frilled shark (*Chlamydoselachus anguineus*); 2. Bluntnose sixgill shark (*Hexanchus griseus*); 3. Broadnose sevengill shark (*Notorynchus cepedianus*). (Illustration by Brian Cressman)

Species accounts

Frilled shark
Chlamydoselachus anguineus

FAMILY
Chlamydoselachidae

TAXONOMY
Chlamydoselachus anguineus Garman, 1884, Japan.

OTHER COMMON NAMES
French: Requin lézard; Spanish: Tiburón anguila.

PHYSICAL CHARACTERISTICS
A slender, eel-like shark, with a single, low dorsal fin close to the caudal fin; long pelvic, anal, and caudal fins; six gill arches; very elongated gill slits reaching from the dorsal to the ventral side (the first gill slit is especially elongated); a huge mouth with clearly separated rows of numerous teeth (approximately 300), and ventral keels along trunk and tail. The teeth are similar in the upper and lower jaws; there are three long and very sharp, slender cusps with two minute cusplets between. Coloration is a uniform dark brown. Reaches perhaps 77.6 in (197 cm) in length, but more common at about 55 in (140 cm).

DISTRIBUTION
Known from many localities scattered in all major oceans but absent from the Mediterranean Sea.

HABITAT
A primarily deepwater species, demersal on the outer continental shelf and slope at depths from about 394 to 4,265 ft (120–1,300 m) but occasionally caught at the surface.

BEHAVIOR
Unknown.

FEEDING ECOLOGY AND DIET
Mostly unknown, but its teeth suggest that it feeds on deepwater demersal and benthic fishes and cephalopods or other soft invertebrates. The huge gape of its mouth indicates that this species is capable of swallowing large prey items. Predators of the frilled shark are unknown.

REPRODUCTIVE BIOLOGY
Ovoviviparous (yolk sac viviparous), with litters ranging from eight to 12 young. Reproduces year-round off Japan. Gestation periods are not known but are estimated at between 1 and 2 years. Size at birth is about 15.7 in (40 cm); sexual maturity for males is reached at about 39.4 in (100 cm) and for females at about 53 in (135 cm). Breeds from March to June (in Japan).

CONSERVATION STATUS
Not listed by the IUCN.

SIGNIFICANCE TO HUMANS
Taken incidentally as bycatch by trawls or bottom longlines; not considered a significant food item but may be utilized for

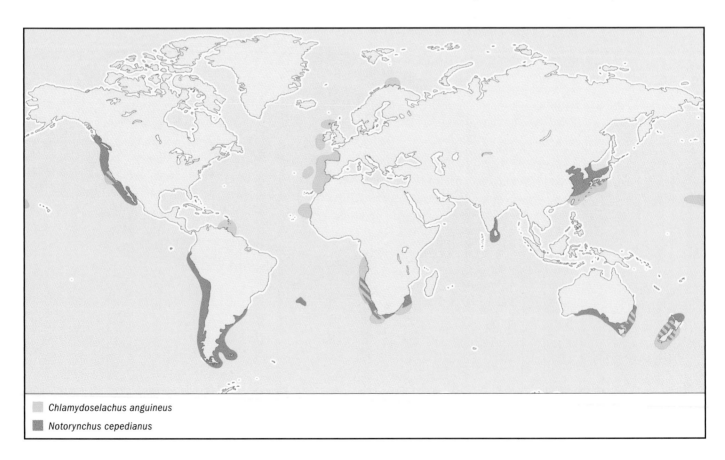

☐ *Chlamydoselachus anguineus*
■ *Notorynchus cepedianus*

fishmeal and human consumption. The remarkable appearance of this shark gives it a menacing and ferocious aspect, but because of its depth distribution, it is not considered dangerous. Unquestionably one of the most bizarre sharks known. ◆

Bluntnose sixgill shark

Hexanchus griseus

FAMILY
Hexanchidae

TAXONOMY
Squalus griseus Bonnaterre, 1788, Mediterranean Sea.

OTHER COMMON NAMES
French: Requin grisé; Spanish: Cañabota gris.

PHYSICAL CHARACTERISTICS
A stout-bodied, large shark that may reach 16.4 ft (5 m) in length, with a broad, blunt snout; relatively small eyes; six pairs of gill slits; a large caudal fin; a single dorsal fin situated close to the caudal fin, and broad and flattened teeth. There are eight to 10 posteriorly directed cusplets per tooth in the first six teeth of the lower jaw (except the symphysial tooth), and the upper teeth usually have a single cusp. Coloration is a uniform gray to dark brown.

DISTRIBUTION
Worldwide in temperate and tropical seas, including the Mediterranean Sea.

HABITAT
This species may be demersal along the continental slopes down to some 6,151 ft (1,875 m), but it also may occur pelagi-

cally at the surface. Young individuals are more common inshore, while larger adults are more common in deeper waters.

BEHAVIOR
A solitary, sluggish shark but also capable of strong swimming. Apparently sensitive to light. May migrate vertically to feed at night.

FEEDING ECOLOGY AND DIET
Feeds on a wide range of fishes, including swordfishes, marlins, dolfinfishes, herrings, grenadiers, cod, hake, ling, and flounders as well as sharks (including hooked individuals of its own species) and rays. Also eats invertebrates (squids, crabs, shrimps) and seals. Predators are unknown for this species, although presumably it may be eaten by larger sharks (including those of the same species).

REPRODUCTIVE BIOLOGY
Yolk-sac viviparous, with large litters that range from 22 to 108 young. Gestation periods are unknown. Gravid females may give birth in shallow bays. Size at birth is about 25.6 in (65 cm). Females are sexually mature at about 177 in (450 cm) and males at slightly smaller sizes.

CONSERVATION STATUS
Listed as Lower Risk/Near Threatened by the IUCN. As *Hexanchus griseus* is fished for both food and sport and also taken as bycatch, it may not be able to sustain target fisheries. Regional populations are already depleted (e.g., in the northeast Pacific), but fisheries data are generally lacking. Efforts to protect this species are not yet under way.

SIGNIFICANCE TO HUMANS
Not considered a dangerous species, but because of its size, it should be approached with caution. Young individuals are known to thrash about violently when captured. It is not

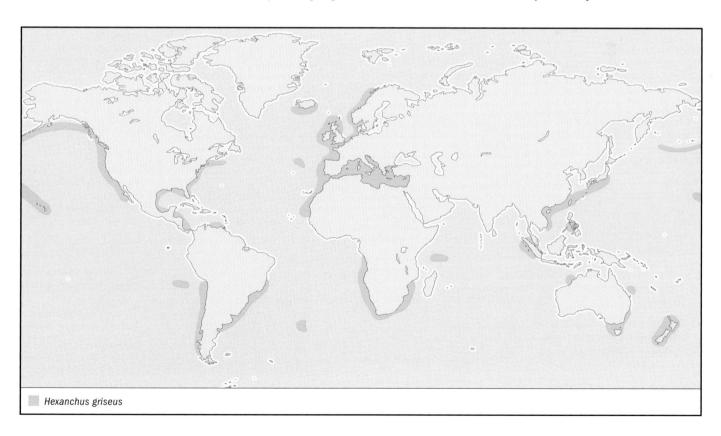

Hexanchus griseus

consumed much, but it is captured in pelagic and bottom trawls and by line and utilized locally for meat, oil, and fishmeal. Can be seen during the summer months off Hornby Island and also Vancouver Island (British Columbia), as young individuals penetrate into the shallow waters of the Strait of Georgia (located between British Columbia, Canada and the United States). They are observed more readily at night. ◆

Broadnose sevengill shark

Notorynchus cepedianus

FAMILY
Hexanchidae

TAXONOMY
Squalus cepedianus Peron, 1807, Tasmania, Australia.

OTHER COMMON NAMES
French: Platnez; Spanish: Cañabota gata.

PHYSICAL CHARACTERISTICS
A large shark, reaching at least 9.8 ft (3 m) in length, possibly 13 ft (4 m), with a broad head; seven pairs of gill slits; a moderately tall, single dorsal fin; and large pectoral and caudal fins. The first six pairs of teeth of the lower jaw are wide and flattened, with four to five posteriorly pointed cusps (except the symphysial tooth); the upper teeth usually have single cusp. Background coloration is gray to brown but mottled with numerous, small darker spots and blotches.

DISTRIBUTION
Widely distributed in temperate seas (unknown in the Indian Ocean and Mediterranean Sea) and more abundant in the Pacific Ocean. Common off the coasts of South Africa and Namibia.

HABITAT
A coastal species reaching depths of only some 443 ft (135 m) on the continental shelf and frequently found in very shallow water, bays, and close to shore.

BEHAVIOR
Considered to be an active, powerful shark that usually is observed cruising slowly near the surface but is capable of quick bursts of speed. May be aggressive if provoked and is known to snap vigorously when captured. Speculated to coordinate its entry into shallow bays with tidal fluxes. Cooperative hunting ("social facilitation") has been recorded for this species in addition to three other predatory strategies (stealth, ambush, and bursts of speed). Spy-hopping, in which an individual raises its head out of the water in a vertical position (possibly to see the surroundings), also has been recorded.

FEEDING ECOLOGY AND DIET
Feeds voraciously on many different families of fishes (including salmon, anchovies, and sturgeon) and also on sharks (dogfishes, houndsharks, and hooked individuals of its own species) and rays (such as eagle rays). Predators are unknown for this species, although presumably it may be eaten by larger sharks (including those of the same species).

REPRODUCTIVE BIOLOGY
Yolk-sac viviparous, with large litters of up to 82 young. Gravid females give birth in shallow bays during the warmer months. Gestation periods are unknown. Size at birth is between 17.7 and 20.9 in (45–53 cm). Males are sexually mature at 59–71 in (150–180 cm) and females at larger than 78.7 in (200 cm). One instance of copulation was observed at around noon in the shallow waters of Namibia, in which it was clear that the male held the female with his mouth. The eggs are relatively large, measuring some 3 in (7.5 cm) in diameter.

CONSERVATION STATUS
Listed as Data Deficient by the IUCN.

SIGNIFICANCE TO HUMANS
Fished for its high-quality flesh in many areas where it occurs. Also fished for sport (in the United States, Australia, and South Africa) and for its skin (in China) for the production of leather. Owing to its large size and putative aggressive behavior, this shark should be approached cautiously. It has been implicated in several attacks on people (in California and South Africa), but these attacks may have been by other species. It is known to have attacked divers while in captivity. ◆

Resources

Books
Bigelow, H. B., and W. C. Schroeder. "Sharks." In *Fishes of the Western North Atlantic*, edited by J. Tee-Van, C. M. Breder, S. F. Hildebrand, A. E. Parr, and W. C. Schroeder. New Haven, CT: Sears Foundation for Marine Research, Yale University, 1948.

Cappetta, H. *Chondrichthyes II: Mesozoic and Cenozoic Elasmobranchii*. Handbook of Palaeoichthyology, vol. 3B. Stuttgart and New York: Gustav Fischer Verlag, 1987.

Compagno, L. J. V. *Sharks of the World: An Annotated and Illustrated Catalogue of Shark Species Known to Date*. FAO Species Catalogue, vol. 4, part 1. Rome: Food and Agriculture Organization of the United Nations, 1984.

Compagno, L. J. V., C. Simpfendorfer, J. E. McCosker, K. Holland, C. Lowe, B. Wetherbee, A. Bush, and C. Meyer. *Sharks*. Pleasantville, NY: Reader's Digest, 1998.

Compagno, L. J. V., and V. H. Niem. "Family Hexanchidae." In *Western Central Pacific Identification Sheets to Species*, edited by K. E. Carpenter and V. H. Niem. Rome: Food and Agriculture Organization of the United Nations, 1999.

Daniel, J. F. *The Elasmobranch Fishes*. 3rd edition. Berkeley: University of California Press, 1934.

Ebert, D. A. "Aspects of the Biology of Hexanchid Sharks Along the California Coast." In *Indo-Pacific Fish Biology: Proceedings of the Second International Conference on Indo-Pacific Fishes*, edited by T. Uyeno, R. Arai, T. Taniuchi, and K. Matsuura. Tokyo: Ichthyological Society of Japan, 1986.

Gudger, E. W. "The Breeding Habits, Reproductive Organs and External Embryonic Development of *Chlamydoselachus* Based on Notes and Drawings by Bashford Dean." In *Archaic Fishes*, edited by E. W. Gudger. Bashford Dean Memorial Volumes, vol. 7. New York: American Museum of Natural History, 1940.

Hennemann, Ralf M. *Elasmobranch Guide of the World: Sharks and Rays*. Frankfurt: Ikan, 2001.

Last, P. R., and J. D. Stevens. *Sharks and Rays of Australia*. Melbourne, Australia: CSIRO, 1994.

Nelson, J. S. *Fishes of the World*. 3rd edition. New York: John Wiley & Sons, 1994.

Perrine, Doug. *Sharks and Rays of the World*. Stillwater, MN: Voyager Press, 1999.

Smith, B. G. "The Anatomy of the Frilled Shark (*Chlamydoselachus anguineus* Garman)." In *Archaic Fishes*, edited by E. W. Gudger. Bashford Dean Memorial Volumes, vol. 6. New York: American Museum of Natural History, 1937.

Springer, Victor G., and Joy P. Gold. *Sharks in Question: The Smithsonian Answer Book*. Washington, DC: Smithsonian Institution Press, 1989.

Whitley, G. P. *The Fishes of Australia*. Part 1. *The Sharks, Rays, Devil-fish, and Other Primitive Fishes of Australia and New Zealand*. Sydney: Royal Zoological Society of New South Wales, 1940.

Periodicals

Ebert, D. A. "Biological Aspects of the Sixgill Shark, *Hexanchus griseus*." *Copeia* 1986, no. 1 (1986): 131–135.

———. "Observation on the Predatory Behavior of the Sevengill Shark *Notorynchus cepedianus*." *South African Journal of Marine Science* 11 (1992): 455–465.

———. "Diet of the Sixgill Shark, *Hexanchus griseus*, off Southern Africa." *South African Journal of Marine Science* 14 (1994): 213–218.

———. "Biology of the Sevengill Shark, *Notorynchus cepedianus* (Peron, 1807), in the Temperate Coastal Waters of South Africa." *South African Journal of Marine Science* 17 (1996): 93–103.

Organizations

American Elasmobranch Society, Florida Museum of Natural History. Gainesville, FL 32611 USA. Web site: http://www.flmnh.ufl.edu/fish/Organizations/aes/aes.htm

Other

"ReefQuest Expeditions." (26 Dec. 2002). <http://www.reefquest.com>

Marcelo Carvalho, PhD

Squaliformes

(Dogfish sharks)

Class Chondrichthyes

Order Squaliformes

Number of families 7

Photo: The piked dogfish (*Squalus acanthias*) is the world's most abundant shark. (Photo by Monterey Bay Aquarium Foundation. Reproduced by permission.)

Evolution and systematics

The order Squaliformes contains 22 genera, 98 formally described species, and at least 17 known but undescribed species. However, higher-level systematics is not pure science, and hence it should not be surprising that experts may define the order Squaliformes somewhat differently. Some do not include the bramble sharks, Echinorhinidae, and others exclude additional groups, justifying their decisions on various fine morphological details. In this chapter, we will consider the order Squaliformes to comprise a more traditional and inclusive group of sharks, which includes the bramble sharks (Echinorhinidae), the dogfish sharks (Squalidae), the gulper sharks (Centrophoridae), the lantern sharks (Etmopteridae), the sleeper sharks (Somniosidae), the rough sharks (Oxynotidae), and the kitefin sharks (Dalatiidae). Fossils interpreted as representing Squaliformes have been laid down in deposits at least 150 million years old, and future discoveries will surely push this trail marker back deeper into the past. Amongst living elasmobranches, Squaliformes is usually accepted as a sister group to an evolutionary branch consisting of angel sharks (Squatiniformes), saw sharks (Pristiophoriformes), and rays. Defining membership within Squaliformes based on an unshared morphological character is not possible, and thus crafting a membership card for this order of fishes may have to await the results of molecular studies.

Physical characteristics

Squaliformes includes fishes that rank as the smallest and amongst the largest of all living sharks, from spined pygmy sharks (*Squaliolus*) growing only to about 10 in (25.4 cm) long, to Greenland sharks (*Somniosus microcephalus*) and Pacific sleeper sharks (*S. pacificus*) estimated at over 20 ft (6.1 m) long. These sharks are a mixed lot; however, and a general diagnosis of their physical characteristics includes: body obviously sharklike, with a cylindrical trunk; snout pointed to bluntly conical and possibly depressed; head not laterally expanded; small-to-large eyes on side of head without nictitating membranes; teeth in top and bottom jaws similar or different; teeth only moderately different along a jaw; bottom teeth always with sharp cutting edge; spiracles present and sometimes

large; five pairs of small gill openings just anterior to pectoral fins; pectoral fins small-to-moderate in size; two low-to-high dorsal fins with spines, as in *Squalus, Etmopterus,* and *Oxynotus* species or without spines as in *Isistius, Somniosus,* and *Eupotmicroides* species; pelvic fins small to moderate in size; anal fin absent; and caudal fin with or without dorsal notch and with ventral lobe shorter than dorsal lobe when present. At maximum size, males are generally smaller than their corresponding females. Species such as the piked dogfish (*Squalus acanthias*), which inhabit relatively shallow waters, may be counter shaded with lighter bellies, whereas those such as the pygmy shark (*Euprotomicrus bispinatus*) may be more uniform and dark. Luminous organs are possessed by a handful of species, representing genera such as *Oxynotus, Isistius, Etmopterus,* and *Centroscyllium.*

Distribution

Representatives of Squaliformes occur in all oceans. Most species are residents of temperate or tropical latitudes, where they inhabit nearshore waters associated with continental and insular shelves and slopes.

Habitat

Squaliformes are primarily found in marine environments; however, some species, such as the piked dogfish, can occasionally operate in estuarine waters. The Greenland shark has also been captured in rivers far from the sea. Such occurrences most likely represent sharks that have temporarily invaded upstream reaches by navigating the saltwater wedges associated with relatively deep tidal rivers. Squaliformes is a unique order of sharks in that at least one representative, the Greenland shark, is a common inhabitant of polar waters. Although some Squaliformes are found in the shallows, another general distinction of these sharks is their deep-water representation. In fact, most species of deep-water sharks are squaliforms, and some of them have been observed inhabiting abyssal depths. Because of this representation, it is likely that our understanding of the overall distribution of squaliforms will continue to expand as more resources are devoted to deep-sea exploration.

A newborn piked dogfish (*Squalus acanthias*) with yolk sac still attached. (Photo by Jeff Rotman/Photo Researchers, Inc. Reproduced by permission.)

Behavior

Given their deep-water habitats, observations of many dogfish sharks in their natural surroundings only exist as valuable glimpses obtained with remote cameras and manned or unmanned deep-sea vehicles. Some species, such as many *Squalus* dogfishes, commonly roam in sizeable packs; other deep-water species, such as the lantern sharks (*Etmopterus*), probably live more solitary lives. Shark species such as the piked dogfish appear quite purposeful in their hunting activities, schooling like a marine wolf pack and devastating a wide variety of prey. Others, such as some rough sharks (*Oxynotus*) observed in deep water, seem more calculating and less voracious. A handful of these sharks, including the Greenland shark, Pacific sleeper shark, and piked dogfish, are known from both deep water and the shallows, and the cookie-cutter shark (*Isistius brasiliensis*) is thought to undertake diurnal vertical migrations in its oceanic realm. Overall, little is known about the migrations or lack thereof of most of dogfish sharks; although individuals of species such as the piked dogfish have been known to travel great distances throughout their lives.

Feeding ecology and diet

The feeding ecology of squaliforms spans that of species that feed as generalists, such as many of the dogfishes that comprise the genera *Squalus*, *Centrophorus*, and *Etmopterus*, to specialists such as the cookie-cutter sharks (*Isistius*). Because of the small size of many squaliforms, it is not surprising that small fishes and invertebrates such as various squids often make up their diet. However, their diets do include some surprises, as species such as the cookie-cutter sharks take mouthfuls from large marine animals, and huge Greenland sharks have been found with hazelnut-sized snails in their stomachs. In addition, some species, such as the Greenland shark and Pacific sleeper shark, appear to specialize somewhat by occa-

sionally feeding on carrion, with the latter species having been filmed in deep water feeding on a decaying whale fall. The teeth of most squaliforms are not very impressive in regard to size; but they are typically well adapted for biting chunks from prey. The exceptions to this are the cookie-cutter sharks, whose teeth are proportionally some of the largest of any sharks. Little to nothing is known about the growth rate of most squaliforms, although some species grow less than 1 in (2.5 cm) per year. Similarly, the longevity of most squaliforms is unknown; however, some species are estimated to live for over 40 and possibly over 100 years. Dogfish sharks are preyed on by other sharks, teleosts, marine mammals, and humans.

Reproductive biology

The reproductive biology of these sharks is poorly known, with much knowledge stemming from several well-known species such as the piked dogfish. As in all elasmobranches, fertilization is internal, and it is suspected that all squaliforms are ovoviviparous, giving birth to live young that derive their embryonic nourishment from egg yolk, rather than from a more intimate and placenta-like maternal connection. Litters of these sharks typically yield six to 10 pups, with neonates of various species ranging from about 3–16 in (7.6–40.6 cm) long. Nursery areas are known or suspected for at least some species.

Conservation status

Three dogfish sharks are included on the IUCN Red List: the gulper shark (*Centrophorus granulosus*) is categorized as Vulnerable, the kitefin shark (*Dalatias licha*) as Data Deficient, and the piked dogfish (*Squalus acanthias*) as Lower Risk/Near Threatened. Because these sharks grow slowly and have a low reproductive output, fisheries should be able to overexploit them more easily than is possible with many other fishes. A report on the conservation status of 75 species of squaliforms published in 1999 noted 60% to be unexploited by fisheries. Of these, many are relatively small, deep-water forms such as the lantern sharks. Of the exploited species, 74% were categorized as species of unknown conservation status due to lack of information. Species in this group included many relatively deep-water forms such as gulper sharks (*Centrophorus*). The kitefin shark and the Greenland shark were considered exploited species that have limited reproductive potential and other life-history characteristics that make them especially vulnerable to overfishing. The bramble shark (*Echinorhinus brucus*) and the piked dogfish were considered the most imperiled of all squaliforms, falling into a conservation category comprised of species associated with historical declines in catches that have sometimes resulted in their being considered locally rare. Because many squaliforms live in relatively deep waters and catch statistics are lacking for many of them, the above conservation review must be deemed preliminary. But to be sure, there can be little doubt that heavily fished species such as the piked dogfish appear to be exploited to the maximum or overexploited throughout much of their range, and that fisheries management plans, some of which are being developed, will be needed to sustain fished populations. Little is known regarding the effects of pollution and habitat destruction on these sharks; however, several studies have indicated some squaliforms to be heavily polluted by such chemicals as PCBs and the pesticide DDT.

Significance to humans

In summing up the significance of squaliforms to humans, it is not an exaggeration to say that these sharks have probably been more fully utilized than any other group of sharks, and possibly more so than any other group of fishes. Historically squaliforms have played a role in myth, art, cultural rituals, and medicine. They have been utilized to feed people and domestic animals throughout the world. Oil extracted from their livers has served as lantern fuel, as a cosmetic additive, and as a machine-gun lubricant. The skin of squaliforms has been used as a natural type of sandpaper, and the hides of these fishes have been tanned into durable leather. Shark teeth have been used in various ceremonial art objects, as well as functioning as cutting tools. Studies of one species in particular, the piked dogfish, have facilitated the education of countless millions of biology students and have also been responsible for a great wealth of biological understanding regarding vertebrate physiology. And surprising to many, some of these species have and are being used as bait to catch lobsters and other more desirable species of sharks. There is no authenticated record of an attack on a human by any dogfish shark, although their vicious dorsal spines have certainly painfully lanced many fishermen. Furthermore, at some times and in some areas various squaliforms have posed a great nuisance to fishing operations and have destroyed valuable fishing gear.

1. Greenland shark (*Somniosus microcephalus*); 2. Piked dogfish (*Squalus acanthias*); 3. Cookie-cutter shark (*Isistius brasiliensis*). (Illustration by Dan Erickson)

Species accounts

Cookie-cutter shark
Isistius brasiliensis

FAMILY
Dalatiidae

TAXONOMY
Scymnus brasiliensis Quoy and Gaimard, 1824, Brazil.

OTHER COMMON NAMES
English: Luminous shark; French: Squalelet féroce; Spanish: Tollo cigarro.

PHYSICAL CHARACTERISTICS
A small shark; females about 22 in (56 cm), males seldom longer than 16.5 in (42 cm). Body shaped like torpedo cigar, widest at about midpoint and tapered most along posterior length, dorsal surface of head slightly depressed. Snout short, eyes large, mouth with fleshy lips, massive lower jaw bearing functional edge of 25–31 large triangular teeth in a row, large spiracles located atop head behind eyes, gill slits small, pectoral fins small, two small spineless dorsal fins of approximately same size set far back on body, anal fin lacking, caudal fin prominent with nearly symmetrical lobes. Body brownish, darker dorsally than ventrally, distinctive dark collar band encircling body at level of gills, trailing margins of fins translucent. Luminescent organs casting a bright greenish glow cover ventral surface of trunk with exception of fins and dark collar band.

DISTRIBUTION
Oceanic, reported from the Indian, Pacific, and Atlantic Oceans, scattered widely throughout tropical latitudes. Less frequently reported from warm temperate regions, sometimes in association with oceanic islands.

HABITAT
Based on fishing data, utilizes a fair vertical swatch of its potential environment. Has been captured at the surface, often at night, but also seems well represented at depths between 279–11,482 ft (85–3,500 m). Based on this information and a limited knowledge of its behavior, this species may undergo diurnal bathypelagic to epipelagic migrations associated with feeding.

BEHAVIOR
Few behavioral observations exist in nature. Thought to occasionally form loose aggregations of individuals and to be a slow swimmer. Based on characteristic craterlike wounds on organisms such as marine mammals and large fishes, as well as matching plugs of tissue from stomachs of captured cookie-cutter sharks, the species is thought to feed on large prey organisms by cleanly biting a mouthful of tissue from its victims before they swim off otherwise unharmed. Because of this, the species has been considered both micropredator and parasite; technically it should be labeled a facultative parasite because it also consumes smaller prey items in their entirety.

Explanations as to how these sharks attack large organisms have been advanced since at least the mid-1800s. As of 2002, a

Squalus acanthias
Somniosus microcephalus
Isistius brasiliensis

popular explanation points to a hunting method that employs deceit: the sharks use their luminescent abilities to counter-illuminate and camouflage their ventral surface so that the striking dark collar band near the head assumes the appearance of a small prey item. Large predators drawing near to investigate this potential meal are eventually met with a quick bite before they can speed off. As convincing as this explanation may seem, it must be considered speculation until supported by observation.

FEEDING ECOLOGY AND DIET
Feeds on a variety of invertebrates and vertebrates, presumably using two behaviors: eats relatively small prey items, such as squids, crustaceans, and bristlemouths, whole or entirely in pieces; eats large prey, such as megamouth sharks, marlins, tunas, wahoos, dolphins, and other cetaceans, dugongs, and pinnipeds, by removing a single mouthful. This flexibility works well, especially in light of its oceanic distribution, where similar predators profit from the vertical and horizontal movements of prey organisms of varying sizes. It is not known if this species is routinely preyed upon by others.

REPRODUCTIVE BIOLOGY
Males mature at 15 in (38 cm); females at 15.7–18.9 in (40–48 cm). Little first-hand information is known regarding the reproductive biology; but fertilization is presumably internal and development ovoviviparous. Although gestation period, number of embryos per litter, and size at birth are unknown, 6–12 eggs have been observed in the uteri of some specimens. Distributions of smaller sharks near oceanic islands have prompted some biologists to propose that young are born in coastal areas.

CONSERVATION STATUS
Not threatened.

SIGNIFICANCE TO HUMANS
Because of its small size and scattered distribution, this shark has had little fishery value. It has been reported that this vicious shark has attacked castaways and even a scuba diver; but there is little hard evidence for this. Nonetheless, given its habit of attacking prey much larger than itself, it is likely that some adventurous diver operating offshore at night will eventually be bitten. In addition, based upon characteristic bitelike holes in the neoprene boot of a hydrophone mast, this species has been implicated in otherwise uneventful attacks on a U.S. nuclear submarine. Such attacks aside, the primary significance of this species to humans lies in its ability to astound with its ferocious design and devious behavior. ◆

Greenland shark
Somniosus microcephalus

FAMILY
Somniosidae

TAXONOMY
Squalus microcephalus Bloch and Schneider, 1801, glacial seas ("Habitat in mari glaciali").

OTHER COMMON NAMES
English: Ground shark, gray shark, gurry shark, sleeper shark; French: Laimargue du Groenland, leiche; German: Eishai; Spanish: Tollo de Groenlandia; Dutch: Apekalle, havkel; Inuktitut: Ekalugssuak, eqaludjuaq, iqalugjuaq; Norwegian: Haakjaerring.

PHYSICAL CHARACTERISTICS
Large, stocky species; females at least 21 ft (6.4 m) and weighing 2,250 lb (1,020 kg), males typically smaller but at least 11 ft (3.3 m). Body "torpedo" shaped, head conical with blunt snout bearing large olfactory openings, body widest at about level of pectoral fins, eyes often infected by conspicuous parasitic copepods, spiracles obvious, mouth suctoral, lower jaw bearing a functional edge consisting of 48–53 teeth whose oblique cusps virtually overlap to form a continuous cutting edge, gill slits small, pectoral fins paddlelike, two spineless dorsal fins about equal size, anal fin lacking, caudal keels present, caudal fin semilunate and appearing powerful. Body brownish, purplish, or bluish gray, sometimes mottled by irregular spots, ventral surface minimally lighter than dorsum.

DISTRIBUTION
Polar and temperate seas, from high latitudes well above the Arctic Circle in waters adjoining the Atlantic Ocean to at least as far south in the Atlantic Ocean as 32°N on the Blake Ridge, approximately 230 mi (370 km) off the coast of Savannah, Georgia, United States. Most common north of Cape Cod, United States, in the western Atlantic and north of Great Britain in the eastern Atlantic. There are no incontrovertible records of Greenland sharks from the Southern Hemisphere or from the Pacific or Indian Oceans, although other sleeper sharks of the genus do inhabit those waters.

HABITAT
The only well-documented sharks to inhabit the Arctic, common in waters below 32°F (0°C) that are seasonally covered by sea ice. In northern portions of the range these sharks can be found lurking below landfast ice, or in open water at depths exceeding 500 ft (152 m), or in knee-deep shallows. At more southerly latitudes, they are usually only seen during the winter. However, this species probably goes unnoticed, residing year-round in cold, deep waters, where it has been seen at depths as great as 7,218 ft (2,200 m). These sharks may occasionally inhabit brackish waters, as indicated by reliable reports of individuals captured over 50 mi (80 km) upstream in the Saguenay River in Quebec, Canada.

BEHAVIOR
Somniosus means "sleeper" in Latin, and certainly this shark lives up to its name. Accounts of the lethargic nature of this fish are widespread, including accounts of them being hauled up on hooks with hardly a fight. In describing fresh-caught specimens, Hansen (1963) wrote, "owing to the sluggishness of this fish, it could sometimes be a little difficult to judge if it was dead or alive." How this most sluggish species is capable of catching swift prey such as chars and seals remains a mystery. Although it has never been observed, some Norwegian fisherman believed that a brightly colored parasitic copepod that infects the eyes of many of these sharks might serve as a lure to attract curious prey within striking distance. Nonetheless, those copepods do appear to cause enough damage to the eyes such that one might surmise that this shark does not rely on keen vision to capture its dinner. In some instances sound or olfaction may be the key, and fishing, sealing, and whaling operations have been noted to attract them.

FEEDING ECOLOGY AND DIET
The examination of stomach contents and studies of stable isotopes indicate that these sharks feed at a number of trophic levels, and while they can swallow smaller prey whole, they also feed in a piecemeal fashion by scooping large, melon ball–like chunks from larger items. Their diet has included jellyfishes, brittle stars, sea urchins, amphipods, crabs, snails,

squids, and seabirds; a variety of fishes such as small sharks, skates, herrings, chars, salmons, eels, capelins, redfishes, sculpins, lumpfishes, cods, haddock, wolfishes, and halibuts; as well as marine mammals such as seals and narwhals. As well as preying on live organisms, these sharks are well known to gluttonously consume animals compromised by fishing gear, along with almost any carrion. The latter habit prompted whalers to give them the unflattering name of gurry or "garbage" sharks. In some areas the flesh is toxic to both humans and dogs, and must be properly prepared before consumption. In addition, the flesh of some sharks is heavily contaminated with persistent organic pollutants that probably enter the sharks via their diet. Although this species will eat others of its kind that are compromised by fishing gear, it is not known whether it is normally preyed upon by species other than humans. Limited growth data on these sharks suggests they grow slowly and that some individuals may live for over 100 years.

REPRODUCTIVE BIOLOGY
Little is known about the reproductive biology of this species; but presumably fertilization is internal and development is ovoviviparous. Information regarding age at first maturity, gestation period, numbers of embryos per litter, and size at birth are nonexistent, except for some information from a single report of a 16 ft (5 m) long female that contained 10 embryos. The pups appeared full term and the single embryo measured was about 14 in (36.7 cm) long. Based on maximum size, males probably mature at a smaller size than females. Given the great size of this species and the tendency for it to eat anything in its path, it is possible that shallow nursery areas segregate neonates and juveniles from adults.

CONSERVATION STATUS
Not threatened.

SIGNIFICANCE TO HUMANS
Regardless of its large size, this species is inoffensive to humans, and no authenticated attacks have been reported. This fish has been employed in various ways by Inuit communities; the flesh consumed by humans and sledge dogs, the liver providing a fine oil, the teeth and rough skin used for cutting and scraping. Europeans, Greenlanders, and Icelanders have also fished the species for similar purposes. Directed commercial fisheries existed from about the mid-1800s until about the mid-1900s. Other than relatively small numbers taken by some traditionalists, most captured today are probably taken as bycatch in fisheries directed at other fishes such as turbots. In some areas these sharks have been so numerous and destructive to fishery equipment and operations that they have been looked upon as a nuisance and were sometimes targeted by operations aimed at reducing their abundance. ◆

Piked dogfish
Squalus acanthias

FAMILY
Squalidae

TAXONOMY
Squalus acanthias Linnaeus, 1758, European sea ("Oceano europaeo").

OTHER COMMON NAMES
English: Grayfish, spiny dogfish, spurdog; white-spotted spurdog; French: Aiguillat commun; German: Gemeiner dornhai;

Spanish: Mielga; Danish: Pighai; Dutch: Dornhaai; Norwegian: Pighaa.

PHYSICAL CHARACTERISTICS
A small shark; total length of females 4 ft (120 cm), weight up to 21.6 lb (9.8 kg); total length of males 3 ft (160 cm). Body quintessentially sharklike, snout narrow, eyes large, spiracles present, mouth with similarly shaped teeth in top and bottom jaws, 28 teeth in top jaw and 22–24 teeth in bottom jaw, first and second dorsal fins each bearing a dorsal spine, second dorsal spine more pronounced than first, anal fin lacking, caudal keels present, caudal fin without subterminal notch on upper lobe. Dorsal surface of body usually slate-colored, flanks each with a row of small white spots that are most conspicuous on younger fish, ventral surface light gray to white.

DISTRIBUTION
Antitropical marine and estuarine waters throughout the Atlantic and Pacific Oceans and adjoining seas, encompassing boreal to warm-temperate waters in the Northern Hemisphere and antiboreal to warm-temperate regions in the Southern Hemisphere.

HABITAT
Inhabits inshore and offshore waters associated with continental and insular shelves. Given its migratory nature and widespread abundance, this small dogfish utilizes a broad scope of habitats. Can be found as in waters as shallow as the intertidal zone, although may venture as deep and probably deeper than 2,952 ft (900 m). Seem to prefer waters about 42–46°F (6–8°C); however, tolerates up to 59°F (15°C). May make short upstream incursions utilizing the saltwater wedges of tidal rivers.

BEHAVIOR
Although they cannot be characterized as being swift, these sharks are strong swimmers and can maintain a steady cruising pace. Gregarious fishes, they typically gather and migrate in large schools consisting of equally sized individuals. Schools of large adult females tend to be more common inshore; schools of juveniles are more common offshore. In the northwestern Atlantic Ocean, their north-south seasonal migration usually places them in the northernmost part of their range near Labrador by early fall, and in the southernmost part of their range off the Carolinas to northeastern Florida by midwinter. In addition to these seasonal movements, they also exhibit general movements inshore during warmer months and movements offshore as it gets colder. Tagging studies have shown that some journey over long distances, with certain transoceanic wanderings in the North Atlantic and North Pacific Oceans totaling 994–4,039 mi (1,600–6,500 km).

FEEDING ECOLOGY AND DIET
Typically feeds in packs, often moves into an area and lays waste to or drives off most fishes. In any wide portion of their range, fishes usually constitute the largest percentage of the diet, followed by squids and other invertebrates. A list of stomach contents would mimic a relatively thorough inhabitant list for many waters. In the northwestern Atlantic Ocean, small schooling fishes such as herrings, menhadens, capelins, sand lances, and mackerels offer them ample opportunity to gorge. Although their relatively small size limits their lethal abilities, fishes as large as cods and haddocks can fall prey to these ravenous squaliforms. Because of their vast appetites, these fishes can grow about 0.6–1.4 in per year (1.5–3.5 cm), and estimates of their longevity based on the interpretation of growth rings on their spines are commonly about 40 years, but may be as long as 100 years. Because of their relatively small size, they

are preyed upon by large sharks and teleosts, seals, killer whales, and of course, humans.

REPRODUCTIVE BIOLOGY

More is known about the reproductive biology of this fish than about that of any other shark, skate, or ray. Males reach maturity at about 11 years old, females at 18–21 years. Fertilization is internal, development is ovoviviparous, and gestation may last from 18–24 months. Litter size ranges from 1–15 pups; however, six or seven is typical. At birth the neonates are usually about 10 in (26 cm) long.

CONSERVATION STATUS

Considered to be overexploited in many regions and listed by the IUCN as Lower Risk/Near Threatened. However, in many

parts of its range, this is probably the most abundant shark, and can be so common that within the Gulf of Maine, to "mention all the localities from which they have been reported would be simply to list every seaside village and fishing ground from Cape Cod to Cape Sable" (Bigelow and Schroeder, 1953).

SIGNIFICANCE TO HUMANS

Almost anyone who has taken an advanced course in biology has toiled to locate the cranial nerves of these fishes. Their flesh has been eaten fried (as in "fish and chips"), broiled, and baked by humans, and ground into meal for pet food. They have also been the bane of many commercial fisheries for driving valuable catches away, beating other fishes to the bait, and ruining nets full of keepers. Furthermore, they can administer a merciless sting with their sharp spines. ◆

Resources

Books

Bigelow, H. B., and W. C. Schroeder. "Sharks." In *Fishes of the Western North Atlantic: Lancelets, Cyclostomes, and Sharks. Sears Foundation for Marine Research Memoir* no. 1, pt. 1, edited by J. Tee-Van, C. M. Breeder, S. F. Hildebrand, A. E. Parr, and W. C. Schroeder. New Haven: Yale University Press, 1948.

———. *Fishes of the Gulf of Maine.* Fishery Bulletin 74. Washington, DC: U.S. Fish and Wildlife Service, 1953.

Burgess, G. H. "Spiny Dogfishes: Family Squalidae." In *Bigelow and Schroeder's Fishes of the Gulf of Maine,* edited by B. B. Collette and G. Klein-MacPhee. Washington, DC: Smithsonian Institution Press, 2002.

Castro, J., and R. L. Brudek. "A Preliminary Evaluation of the Status of Shark Species." In *Fisheries Technical Paper 380.* Rome: Food and Agriculture Organization of the United Nations, 1999.

Compagno, L. J. V. "Sharks of the World: Part 1: Hexanchiformes to Lamniformes." *FAO Fisheries Synopsis* 125, vol. 4, pt. 1. Rome: Food and Agriculture Organization of the United Nations, 1984.

———. "Systematics and Body Form." In *Sharks, Skates, and Rays: The Biology of Elasmobranch Fishes,* edited by W. C. Hamlett. Baltimore, MD: Johns Hopkins University Press, 1999.

———. "Checklist of Living Elasmobranchs." In *Sharks, Skates, and Rays: The Biology of Elasmobranch Fishes,* edited by W. C. Hamlett. Baltimore: Johns Hopkins University Press, 1999.

de Carvalho, M. R. "Higher-Level Elasmobranch Phylogeny, Basal Squaleans, and Paraphyly." In *Interrelationships of Fishes,* edited by M. L. J. Stiassny, L. R. Parenti, and G. D. Johnson. San Diego: Academic Press, 1996.

Last, P. R., and J. D. Stevens. *Sharks and Rays of Australia.* Melbourne, Australia: CSIRO Division of Fisheries, 1994.

Shirai, S. "Phylogenetic Interrelationships of Neoselachians (Chondrichthyes: Euselachii)." In *Interrelationships of Fishes,* edited by M. L. J. Stiassny, L. R. Parenti, and G. D. Johnson. San Diego: Academic Press, 1996.

Periodicals

Beck, B., and A. W. Mansfield. "Observations on the Greenland Shark, *Somniosus microcephalus,* in Northern Baffin Island." *Journal of the Fisheries Research Board of Canada* 26 (1969): 143–145.

Berland, B. "Copepod *Ommatokoita elongata* (Grant) in the Eyes of the Greenland Shark: A Possible Cause of Mutual Dependence." *Nature* 191 (1961): 829–830.

Borucinska, J. D., G. W. Benz, and H. E. Whiteley. "Ocular Lesions Associated with Attachment of the Parasitic Copepod *Ommatokoita elongata* (Grant) to Corneas of Greenland Sharks, *Somniosus microcephalus* (Bloch & Schneider)." *Journal of Fish Diseases* 21 (1998): 415–422.

Caloyianis, N. "Greenland Sharks." *National Geographic* 194, no. 3 (1998): 60–71.

Drainville, G., and L. Brassard. "Le requin *Somniosus microcephalus* dans la Rivière Saguenay." *Le Naturaliste Canadien* 87 (1960): 269–277.

Fisk, A. T., S. A. Tittlemier, J. L. Pranschke, and R. J. Norstrom. "Using Anthropogenic Contaminants and Stable Isotopes to Assess the Feeding Ecology of Greenland Sharks." *Ecology* 83 (2002): 2,162–2,172.

Francis, M. P., J. D. Stevens, and P. R. Last. "New Records of *Somniosus* (Elasmobranchii: Squalidae) from Australia, with Comments on the Taxonomy of the Genus." *New Zealand Journal of Marine and Freshwater Research* 22 (1988): 401–409.

Hansen, P. M. "Tagging Experiments with the Greenland Shark (*Somniosus microcephalus* Bloch and Schneider) in Subarea 1." *Special Publication, International Commission of Northwest Atlantic Fisheries* 4 (1963): 172–175.

Herdendorf, C. E., and T. M. Berra. "A Greenland Shark from the Wreck of the *SS Central America* at 2,200 Meters." *Transactions of the American Fisheries Society* 124 (1995): 950–953.

Jahn, A. E., and R. L. Haedrich. "Notes on the Pelagic Squaloid Shark *Isistius brasiliensis.*" *Biological Oceanography* 5 (1987): 297–309.

Resources

Johnson, C. S. "Sea Creatures and the Problem of Equipment Damage." *U.S. Naval Institute Proceedings* 1978 (August 1978): 106–107.

Koefoed, E. "A Uterine Foetus and the Uterus from a Greenland Shark." *Fiskeridirektoratets Skrifter, Serie Havundersøkelser* 11, no. 10 (1957): 8–12.

Nakano, H., and M. Tabuchi. "Occurrence of the Cookiecutter Shark, *Isistius brasiliensis*, in Surface Waters of the North Pacific Ocean." *Japanese Journal of Ichthyology* 37 (1990): 60–63.

Widder, E. A. "A Predatory Use of Counterillumination by the Squaloid Shark, *Isistius brasiliensis*." *Environmental Biology of Fishes* 53 (1998): 267–273.

George W. Benz, PhD

Squatiniformes

(Angelsharks)

Class Chondrichthyes
Order Squatiniformes
Number of families 1

Photo: The Pacific angelshark (*Squatina californica*) is well camouflaged when lying on the sandy ocean bottom. (Photo by David Hall/Photo Researchers, Inc. Reproduced by permission.)

Evolution and systematics

Angelsharks are an ancient lineage, first appearing in the fossil record about 150 million years ago during the late Jurassic period. The remains of articulated angelsharks are known from the marine deposits of Solnhofen, southern Germany (genus *Pseudorhina*); these are well-preserved specimens that resemble modern angelsharks in most morphological details. Most fossil angelshark species, however, are known from isolated teeth reported from the late Jurassic to the Pliocene epoch (some 5 million years ago) of many localities, including Europe, North America, Greenland, Japan, and Africa.

The evolutionary relationships of angelsharks have been intensely debated since their discovery. They are presently classified in a large group with the cow and frilled sharks (Hexanchiformes), the dogfishes and allies (Squaliformes), the sawsharks (Pristiophoriformes), and the rays (also known as batoids). Together, these groups comprise the Squalea, as all members have complete hemal arches (ventral projections arising from the vertebral column) in the trunk region anterior to the tail, among other anatomical innovations. Within the Squalea, the angelsharks are most closely related to a group that includes the sawsharks and the rays. Angelsharks have traditionally been thought to be closely related to rays, but rays are actually more closely related to sawsharks. Almost nothing is known about the phylogenetic relationships among angelshark species. All living angelsharks are classified together in one family, the Squatinidae.

Fifteen living species of angelsharks are currently recognized, all placed in the single genus *Squatina*. There may also be two undescribed species living off southern Australia. Most living species of angelsharks have not been very well characterized, and critical taxonomic studies are still needed. Much is yet to be learned about their population dynamics and reproductive patterns, which are essential because some species are of commercial importance locally (for example, in Australia).

Physical characteristics

Angelsharks are among the most distinctive living sharks, strongly resembling rays in being dorsoventrally flattened, with enlarged pectoral fins reminiscent of the batoid disc. The head of angelsharks, however, is not fused to the pectoral fins as it is in rays (where it forms the disc). Their pectoral fins have a unique anterior lobe that mostly conceals the five gill slits. The gill slits are laterally situated, but extend ventrally as well (contrasting to the exclusively ventral gill slits of rays). The head is exceptionally wide and depressed, entirely anterior to the pectoral fins, and with a very wide, terminal mouth. The eyes and spiracles are dorsally situated, close to the front margin of the head, and the spiracles are somewhat transverse. The anterior nasal flaps are wide and fringed, and the pattern of fringes is sometimes useful in identification. Angelsharks have two spineless dorsal fins posterior to the pectoral and pelvic fins, a rather slender, short tail that is abruptly demarcated from the trunk, a caudal fin that is unique in having a greatly elongated lower lobe (longer than upper lobe), and no anal fin.

Coloration is diagnostic for many angelshark species. Some have a relatively light background color, but others are darker, and most are marked with various spots, ocelli (eyelike spots), and blotches, sometimes of varying color and size. *Squatina tergocellata* from Australia has three pairs of characteristic dark ocelli on its back and pectoral fins, and *S. tergocellatoides* from Taiwan has similar spots dorsally, but not quite forming ocelli. *Squatina australis* is unique in presenting dark spots on the lower lobe of its caudal fin, with creamy white and darker spots dorsally. *Squatina japonica* is reddish dorsally (as is *S. squatina*) with darker red blotches over the trunk. The southwestern Atlantic species are similar to each other in coloration, composed of a dark or light tan background with yellowish and brownish spots and diffused blotches.

Angelsharks are mostly of medium size, reaching about 5 ft (1.5 m) long, but occasionally larger individuals, up to 6.6 ft (2 m), may occur. For most species, males mature when between 29.5–43.3 in (75–110 cm), and females at slightly larger sizes. Size at birth varies from 9.8–13.8 in (25–35 cm).

Distribution

Angelsharks occur almost worldwide, but most species are geographically restricted. *Squatina californica* occurs in the eastern Pacific; three species are Mediterranean (*S. squatina*, *S. oculata*, and *S. aculeata*), also occurring along the western African coast (the latter two down to Namibia); four species are found in the western Atlantic (*S. dumeril*, *S. argentina*, *S. guggenheim*, and *S. occulta*); *S. africana* occurs in the Indian Ocean; *S. tergocellata* and *S. australis* occur off southern Australia; and four species are western Pacific (*S. tergocellatoides*, *S. nebulosa*, *S. japonica*, and *S. formosa*).

Habitat

Angelsharks occupy the continental shelf and upper slope regions, in depths ranging from just several feet (1.5 m to 150 m) down to about 4,265 ft (1,300 m), but most species are found inshore. They usually occur on sand, mud, and gravel substrates.

Behavior

Angelsharks frequently bury themselves in sandy or muddy bottoms, as do many rays, in order to ambush their prey or to rest during the day (they are primarily nocturnal). Burial is accomplished by a vigorous "flapping" of the pectoral fins, which raises the sediment. Dorsal eyes and spiracles allow angelsharks to breathe regularly while waiting for their prey to pass by, as water may reach the gills through the enlarged spiracles. The top of the head and trunk are frequently left exposed. Angelsharks are ambush predators. As prey items have to be of the right size, angelsharks may remain concealed in an immobile state for long periods, waiting for the proper prey to pass close to their mouths. Capture of prey is done in a very swift, high-speed maneuver, in which the anterior part of the body is abruptly uplifted; the prey is ingested whole after being properly adjusted in the mouth.

Feeding ecology and diet

Angelsharks eat a wide variety of invertebrates and fishes. Their capacious mouths allow them to ingest prey items of substantial size. Crustaceans, cephalopods, bivalves, and bony fishes are commonly consumed, as well as other sharks and rays; one specimen was observed to spit out a newly born bullhead shark immediately after ingesting it, probably because of its finspines.

Reproductive biology

All angelsharks are ovoviviparous (aplacentally viviparous; giving birth to live young), and have internal fertilization, as do all sharks, rays, and chimaeroids (Chondrichthyes). Gestation periods are mostly unknown, but appear to be about 10 months for one species, *S. californica*. Litter sizes vary between one and 20 pups per gestation, and the young measure around 9.8 in (25 cm) when born. Southwestern Atlantic angelsharks employ cloacal gestation, in which the fully developed fetus remains, prior to birth, in a small chamber (posterior to the uterus) that opens into the cloaca; this occurs during the second half of the gestation period and lasts for several months.

Conservation status

Five species of angelsharks are listed by the IUCN. *S. occulta* is Endangered; the Brazilian subpopulation of *S. guggenheim* is listed as Endangered, while the rest of the species is Vulnerable; *S. squatina* is Vulnerable; *S. californica* is Lower Risk/Near Threatened; and *S. argentina* is Data Deficient.

Significance to humans

Angelsharks pose little direct threat, because of their size, habits, and vertical distribution. However, caution is necessary, as angelsharks have sharp teeth and strong jaws and have been known to defend themselves if provoked. They are consumed on a regular basis as food (fresh, frozen, or salt-dried), especially in the western Pacific (*S. japonica*), Australia (mainly *S. australis*), and in the eastern Pacific (*S. californica*). Their skin is also used for polishing wood surfaces and as sharkskin leather (consisting of the dried skin with denticles). They are caught as bycatch in bottom trawls.

1. Angelshark (*Squatina squatina*); 2. Pacific angelshark (*Squatina californica*). (Illustration by Dan Erickson)

Species accounts

Pacific angelshark
Squatina californica

FAMILY
Squatinidae

TAXONOMY
Squatina californica Ayres, 1859, California.

OTHER COMMON NAMES
English: California angelshark; Spanish: Pez ángel del Pacífico.

PHYSICAL CHARACTERISTICS
Pacific angelsharks present rather simple, spatulate, nasal flaps (not intensely fringed), and a light-brown background with small dark brown spots and blotches, mostly smaller than spiracle-diameter, scattered over the trunk and tail. There is a row of small denticles along the back and tail, and also in between the dorsal and caudal fins, and small spines are also present on the snout and around eyes.

DISTRIBUTION
These sharks occur from Alaska to the Gulf of California, and from Ecuador to Chile. South American occurrences may refer to a distinct species, presently not recognized. They are abundant close to the Channel Islands (California).

HABITAT
Pacific angelsharks inhabit shallow, nearshore waters, down to 164 ft (50 m) off the California coast, but to about 590.5 ft (180 m) in the Gulf of California, over rocky, muddy, and sandy bottoms, and near kelp forests.

BEHAVIOR
Pacific angelsharks are sluggish and mostly nocturnal. Adults are somewhat nomadic, spending brief periods in restricted areas of about 1 sq mi (2.6 sq km) before moving to a new region several miles distant. Although these sharks are mostly solitary, small aggregations have been observed.

FEEDING ECOLOGY AND DIET
These sharks eat fishes (croakers, blacksmith, and halibut), shrimp, squid, and squid egg cases. They also feed on the peppered catshark (*Galeus piperatus*) in the Gulf of California.

REPRODUCTIVE BIOLOGY
Male Pacific angelsharks mature at about 29.5–31.5 in (75–80 cm), females at slightly greater sizes. Pups are born at about 7.9–13.8 in (20–25 cm). Litter size ranges from a single offspring to 13, and the gestation period is reported to last 10 months. Birth occurs in March and June.

CONSERVATION STATUS
California angelsharks are listed as Lower Risk/Near Threatened by the IUCN.

SIGNIFICANCE TO HUMANS
These sharks are consumed fresh or frozen in California, and especially in areas near the Gulf of California. They are also consumed off western South America. Not usually kept in aquaria, Pacific angelsharks are one of the most commonly observed angelsharks in the wild. ◆

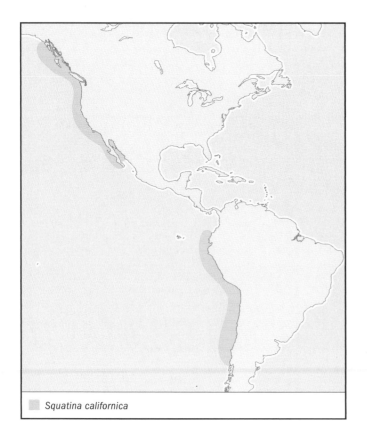
Squatina californica

Angelshark
Squatina squatina

FAMILY
Squatinidae

TAXONOMY
Squalus squatina Linnaeus, 1758, Mediterranean Sea ("Oceano Europaeo").

OTHER COMMON NAMES
English: Common angelshark; French: Ange de mer commun, Spanish: Angelote.

PHYSICAL CHARACTERISTICS
Squatina squatina is distinguished by having a long tentacle on its spatulate, unfringed nasal flap, and a dorsal coloration composed of a gray, yellow, or light-brown background, mottled in dark, minute spots. There are few denticles scattered along the midline, as well as on the snout and around the eyes.

DISTRIBUTION
Mediterranean, eastern Atlantic, bordering the United Kingdom to Scandinavia, also in the Shetland Islands, and south to Morocco and the Canary Islands.

Squatina squatina

HABITAT
These sharks inhabit shallow inshore waters, from 5 ft (1.5 m) down to 492 ft (150 m). Found on rocky, gravely, muddy, and sandy bottoms.

BEHAVIOR
Squatina squatina is mostly nocturnal. Adults are known to migrate northward in the summer in northern parts of the distribution.

FEEDING ECOLOGY AND DIET
Eats fishes (hake, argentines, flatfishes, and skates), crabs, shrimp, and mollusks.

REPRODUCTIVE BIOLOGY
Female angelsharks mature at about 49.2 in (125 cm), males at slightly smaller sizes. Pups are born at about 9.4–11.8 in (24–30 cm). Litter size ranges from seven to 25, depending on the size of the mother. Females give birth in the summer months in the colder part of the range, but during the winter in the Mediterranean.

CONSERVATION STATUS
Listed as Vulnerable by the IUCN.

SIGNIFICANCE TO HUMANS
Angelsharks are consumed fresh, frozen, or salt-dried; and are possibly utilized for fish oil and fishmeal. ◆

Resources

Books

Bigelow, H. B., and W. C. Schroeder. *Fishes of the Northwestern Atlantic*. Part I, *Lancelets, Cyclostomes, and Sharks*. New Haven: Sears Foundation for Marine Research, 1948.

Cappetta, H. *Chondrichthyes II, Mesozoic and Cenozoic Elasmobranchii*. Stuttgart: Gustav Fischer Verlag, 1987.

Compagno, L. J. V. *Sharks of the World. An Annotated and Illustrated Catalogue of Shark Species Down to Date*. Vol. 2, *Bullhead, Mackerel and Carpet Sharks (Heterodontiformes, Lamniformes and Orectolobiformes)*. Rome: Food and Agriculture Organization of the United Nations, 2001.

Hennemann, R. M. *Sharks and Rays, Elasmobranch Guide of the World*. Frankfurt: Ikan, 2001.

Last, P. R., and J. D. Stevens. *Sharks and Rays of Australia*. Melbourne, Australia: CSIRO, 1994.

Nelson, J. *Fishes of the World*. 3rd ed. New York: John Wiley & Sons, 1994.

Springer, V. G., and J. P. Gold. *Sharks in Question: The Smithsonian Answer Book*. Washington, DC: Smithsonian Institution Press, 1989.

Whitley, G. P. *The Fishes of Australia*. Part 1, *The Sharks, Rays, Devil-fish, and Other Primitive Fishes of Australia and New Zealand*. Sydney, Australia: Royal Zoological Society of New South Wales, 1940.

Periodicals

Compagno, L. J. V. "Phyletic Relationships of Living Sharks and Rays." *American Zoologist* 17 (1977): 303–322.

Luer, C. A., and P. W. Gilbert. "Elasmobranch Fish. Oviparous, Viviparous, and Ovoviviparous." *Oceanus Magazine* 34, no. 3 (1991): 47–53.

Natanson, L. J., and G. M. Caillet. "Reproduction and Development of the Pacific Angel Shark." *Copeia* 1986, no. 4 (1986): 987–994.

Shirai, S. "Phylogenetic Relationships of the Angel Sharks, with Comments on Elasmobranch Phylogeny." *Copeia* 1992, no. 2 (1992): 505–518.

Sunye, P. S., and C. M. Vooren. "On Cloacal Gestation in Angel Sharks from Southern Brazil." *Journal of Fish Biology* 50 (1997): 86–94.

Organizations

American Elasmobranch Society, Florida Museum of Natural History. Gainesville, FL 32611 USA. Web site: http://www.flmnh.ufl.edu/fish/Organizations/aes/aes.htm

Other

FishBase. August 8, 2002 (cited October 10, 2002). <http://www.fishbase.org/search.cfm>

The Catalog of Fishes On-Line. February 15, 2002 (cited October 17, 2002). <http://www.calacademy.org/research/ichthyology/catalog/fishcatsearch.html>

Marcelo Carvalho, PhD

Pristiophoriformes

(Sawsharks)

Class Chondrichthyes
Order Pristiophoriformes
Number of families 1

Illustration: Common sawshark (*Pristiophorus cirratus*). (Illustration by Dan Erickson)

Evolution and systematics

Sawsharks first appear in the fossil record during the late Cretaceous period (some 85 million years ago [mya]) in Lebanon in the form of more or less complete fossils, but almost all sawshark fossils consist of isolated remains of rostral spines from the Tertiary period (from between 65 and 5 mya). Sawshark fossils are not very common, but indicate that they were once more widespread, at least in the Pacific, southern and northeastern Atlantic, as remains have been found in Japan, Africa, Europe, New Zealand, and North and South America.

Sawsharks are closely related to the cow and frilled sharks (Hexanchiformes), the dogfishes and allies (Squaliformes), the angelsharks (Squatiniformes), and the rays (or batoids). Together these groups comprise the Squalea, as all members have complete hemal arches (ventral projections arising from the vertebral column) in the trunk region anterior to the tail, among other unique anatomical features. Within the Squalea, the sawsharks are most closely related to the rays. All sawsharks are classified in the family Pristiophoridae.

There are two genera of living sawsharks, *Pliotrema* and *Pristiophorus*. *Pristiophorus* has four included species, but *Pliotrema* is monotypic (that is, only a single species, *P. warreni*, is recognized). The phylogenic relationships among sawshark species have not been investigated. Their taxonomy is still poorly known, mainly as a result of the paucity of specimens available for study from many regions. Two putatively new species of sawsharks have been reported from Australia. Their general biology is also inadequately known.

Physical characteristics

Sawsharks are readily identified by their slender, slightly depressed body, which is preceded by a very elongated toothed "saw." The saw (or "rostral saw") is an anterior, hypertrophied extension of the snout region (rostrum). The saw has large, sharp, lateral rostral spines, which are replaced continuously through life, as well as a pair of long, ventrally extending barbels that are well anterior to the nostrils. The mouth and nostrils are entirely on the ventral surface. Sawsharks lack an anal fin, have relatively large pectoral and spineless dorsal fins, and a low, long caudal fin. The eyes and spiracles are large.

Sawsharks are sometimes confused with the sawfishes (Pristidae), a group of rays that also have elongated rostral saws. However, both groups are easily distinguished, as sawfishes (rays) do not present rostral barbels, have rostral teeth of equal size that grow continuously (as opposed to sawshark rostral spines which are replaced when broken off, and vary in size), and have ventral gill slits. There are numerous other anatomical differences, such as the arrangement of canals for the passage of vessels and nerves within the rostral saw, and the mode of attachment of rostral spines (which are not embedded in the saw in sawsharks, as they are in sawfishes).

The genera of sawsharks (*Pliotrema* and *Pristiophorus*) are easily distinguished. *Pliotrema* has six gill slits, whereas *Pristiophorus* has five (gills slits in all sawsharks are lateral, just anterior to the pectoral fins), and the rostral spines of *Pliotrema* bear small serrations on their posterior margins. Most species of sawsharks are drab and gray colored, with the exception of the more ornately colored *P. cirratus*.

Sawsharks are small to average in size, reaching about 55 in (140 cm) in length. Most species are born at about 9.8–13.8 in (25–35 cm). The size at sexual maturity is poorly known, but has been established to be around 32.7 in (83 cm) for males for at least one species, *P. warreni*, and slightly larger for females.

Distribution

As far as is known, sawsharks have a restricted distribution in tropical and warm-temperate waters, occurring in the western Pacific Ocean (Philippines, Japan, and China; *P. japonicus*), western Indian Ocean (southeastern Africa; *P. warreni*), southern, western, and eastern Australia (*P. nudipinnis* and *P. cirratus*), and also in the western North Atlantic, around the Bahamas, and between Cuba and Florida (*P. schroederi*).

Habitat

Most sawshark species inhabit the continental shelf region, but some species are also present along the continental slope in waters as deep as 3,002 ft (915 m), and one species, *P. schroederi*, is known only from deep waters. Some species present great depth variation, occurring in shallow bays but also along the upper slope. Sawsharks are generally found on soft gravely or sandy bottoms.

Behavior

Sawsharks, similar to the sawfishes (rays), utilize their saw to stun and kill fish by swinging it from side to side, and also to stir up bottom sediments when hunting. The rostral barbels may contain taste buds, and the saw is heavily endowed with sensory receptors (ampullary pores, which detect electrical fields, and lateral-line pores, which detect physical perturbations in the water). Some species are rather abundant, occurring in large groups, perhaps for feeding purposes, and may also present segregation of individuals according to age.

Feeding ecology and diet

Sawsharks feed mostly on fishes, but also on invertebrates such as squid and crustaceans.

Reproductive biology

All sawsharks have internal fertilization (as do all chondrichthyans), are ovoviviparous (or aplacentally viviparous), giving birth to from seven to 17 young, but little is known about the length of their gestation periods. Young feed on yolk from the yolk sac until birth. The rostral spines are concealed in embryonic sawsharks, oriented backward so as to not harm the mother (but still very sharp), and become perpendicularly oriented only after birth. Fetal individuals present elongated rostral barbels, proportionally much longer than adults.

Conservation status

The common sawshark (*P. cirratus*) is listed as Lower Risk/Near Threatened by the IUCN.

Significance to humans

Sawsharks are consumed locally as food at least in the western Pacific (*P. japonicus*), and in Australia (*P. cirratus*), being caught as bycatch during bottom trawls. They are not usually kept in aquaria. Sawsharks pose little threat to humans, due primarily to their more obscure habits and depth distribution, but the rostral spines are quite sharp even in late-term embryos or juveniles, and may easily cause injury if handled.

1. Common sawshark (*Pristiophorus cirratus*); 2. Sixgill sawshark (*Pliotrema warreni*). (Illustration by Dan Erickson)

Species accounts

Common sawshark
Pristiophorus cirratus

FAMILY
Pristiophoridae

TAXONOMY
Pristis cirratus Latham, 1794, Australia (Port Jackson, New South Wales).

OTHER COMMON NAMES
English: Longnose sawshark; French: Requin scie à long nez; Spanish: Tiburòn sierra trompudo.

PHYSICAL CHARACTERISTICS
Common sawsharks are uniquely pigmented, with a reddish brown background pattern; numerous irregular darker brown saddles along the trunk, fins, and head; dark brown stripes across rostral saw; as well as numerous small spots (smaller than the eye) scattered on body; the rostral spines have dark margins. Ventrally uniform creamy white. Rostral saw is relatively long and slender when compared to the other Australian species, *P. nudipinnis*, and the rostral barbels are located at about the center of the rostral saw, as opposed to slightly closer to the eyes as in the latter species.

DISTRIBUTION
Common sawsharks occur off southern and western Australia (reaching to about 30°S), including around Tasmania.

Pristiophorus cirratus

HABITAT
These sharks are found on the continental shelf and upper slope, from 131–1,017 ft (40–310 m), usually on sandy bottoms. They are also reported to occur in bays and estuaries, but are believed to be more abundant from 121–479 ft (37–146 m). Although present in the same general area as *P. nudipinnis*, *P. cirratus* apparently occupies deeper waters.

BEHAVIOR
Common sawsharks are believed to form schools (perhaps for feeding) and are abundant in trawls for benthic fishes. This species has been filmed swinging its saw from side to side in an attempt to injure small fishes. They commonly rest on the bottom, with the rostral saw slightly elevated and supported in a tripodlike stance by the rostral barbels. Common sawsharks are not frequently observed in the wild.

FEEDING ECOLOGY AND DIET
These sharks feed on fishes, including cornetfishes (Fistulariidae), and on invertebrates, particularly crustaceans.

REPRODUCTIVE BIOLOGY
Mostly unknown, but they are reported to breed during winter months. Their size at birth is about 15 in (38 cm), and males are sexually mature at about 3.3 ft (1 m) in length.

CONSERVATION STATUS
These fishes are listed as Lower Risk/Near Threatened by the IUCN.

SIGNIFICANCE TO HUMANS
Common sawsharks are captured as bycatch by trawlers and are available regularly in fish markets. The flesh is reported to be very good. The elongated rostral saw is sometimes coveted for its value as an ornament. ◆

Sixgill sawshark
Pliotrema warreni

FAMILY
Pristiophoridae

TAXONOMY
Pliotrema warreni Regan, 1906, South Africa (Natal and False Bay).

OTHER COMMON NAMES
English: Sawshark; French: Requin scie flutian; Spanish: Tiburòn sierra del cabo.

PHYSICAL CHARACTERISTICS
Sixgill sawsharks are unique in having six gill slits. The rostral saw is relatively elongated, and the rostral barbels are much closer to the mouth than to the rostral tip. The sharks are pale brown dorsally. At birth they measure 13.8 in (35 cm). They reach about 55 in (140 cm), the males maturing at about 32.7 in (83 cm), and the females at 43.3 in (110 cm).

DISTRIBUTION
Southeastern Africa, from Cape Agulhas (South Africa) to Mozambique and Madagascar. *P. warreni* is the only sawshark species in the western Indian Ocean.

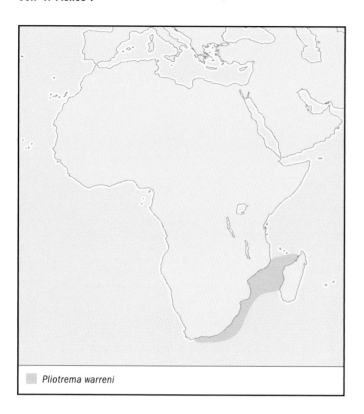

Pliotrema warreni

HABITAT
These sharks are found on the continental shelf and upper slope, from 197–1,481 ft (60–430 m). Reported to occur in deeper depths (below 361 ft/110 m) off Natal, South Africa.

BEHAVIOR
The behavior of these sharks is not well known, but they presumably use their rostral saw like other sawsharks—to stun, dig out, and kill prey. The adults are partially segregated from the young by occupying deeper waters in at least one area (off Natal).

FEEDING ECOLOGY AND DIET
Sixgill sawsharks feed on a variety of fishes (eels, hake, and gapers) as well as shrimp and squid.

REPRODUCTIVE BIOLOGY
The litter size is five to seven per gestation, but much is yet to be learned about its reproductive biology.

CONSERVATION STATUS
Not threatened.

SIGNIFICANCE TO HUMANS
Sixgill sawsharks are occasionally captured as bycatch and consumed locally.

Resources

Books
Cappetta, H. *Chondrichthyes II, Mesozoic and Cenozoic Elasmobranchii.* Stuttgart: Gustav Fischer Verlag, 1987.

Compagno, L. J. V. "Part 1: Hexanchiformes to Lamniformes." In *Sharks of the World: An Annotated and Illustrated Catalogue of Shark Species Known To Date.* FAO fisheries synopsis, no. 125; Vol. 4, part 1. Rome: United Nations Development Programme, 1984.

Hennemann, R. M. *Sharks and Rays, Elasmobranch Guide of the World.* Frankfurt: Ikan, 2001.

Last, P. R., and J. D. Stevens. *Sharks and Rays of Australia.* Melbourne, Australia: CSIRO, 1994.

Nelson, J. *Fishes of the World.* 3rd ed. New York: John Wiley & Sons, 1994.

Springer, V. G., and J. P. Gold. *Sharks in Question. The Smithsonian Answer Book.* Washington, DC: Smithsonian Institution Press, 1989.

Whitley, G. P. *The Fishes of Australia. Part 1. The Sharks, Rays, Devil-Fish, and Other Primitive Fishes of Australia and New Zealand.* Sydney, Australia: Royal Zoological Society of New South Wales, 1940.

Periodicals
Compagno, L. J. V. "Phyletic Relationships of Living Sharks and Rays." *American Zoologist* 17 (1977): 303–322.

Luer, C. A., and P. W. Gilbert. "Elasmobranch Fish. Oviparous, Viviparous, and Ovoviviparous." *Oceanus Magazine* 34, no. 3 (1991): 47–53.

Slaughter, B. H., and S. Springer. "Replacement of Rostral Teeth in Sawfishes and Sawsharks." *Copeia* 1968, no. 3 (1968): 499–506.

Springer, S., and H. R. Bullis, Jr. "A New Species of Sawshark, *Pristiophorus schroederi*, from the Bahamas." *Bulletin of Marine Science of the Gulf and Caribbean* 10, no. 2 (1960): 241–254.

Organizations
American Elasmobranch Society, Florida Museum of Natural History. Gainesville, FL 32611 USA. Web site: http://www.flmnh.ufl.edu/fish/Organizations/aes/aes.htm

Other
FishBase. August 8, 2002 (cited October 10, 2002). <http://www.fishbase.org>

The Catalog of Fishes On-Line. February 15, 2002 (cited October 17, 2002). <http://www.calacademy.org/research/ichthyology/catalog/fishcatsearch.html>

Marcelo Carvalho, PhD

Rajiformes

(Skates and rays)

Class Chondrichthyes
Order Rajiformes
Number of families 20

Photo: A largespot river stingray (*Potamotrygon faulkneri*) is well camouflaged on a river bottom in Brazil. (Photo by Dante Fenolio/Photo Researchers, Inc. Reproduced by permission.)

Evolution and systematics

Despite the fact that skates and rays greatly outnumber their shark relatives within Chondrichthyes, they have received far less recognition. As of 2002 there were about 513 recognized species of skates and rays, compared with about 390 species of sharks. Sharks are topics of books, television documentaries, and news coverage, whereas skates and rays get little press.

Skates and rays, sharks, and chimaeroids are members of Chondrichthyes, the cartilaginous fishes. The cartilaginous fishes are distinguished from other jawed fishes (Osteichthyes, or bony fishes) by several characters. (1) The endoskeleton consists of calcified cartilage. (2) The tooth-bearing jaws are the palatoquadrate and Meckel's cartilages, rather than being comprised of dermal bones. (3) Lungs and swim bladders are absent. (4) A single boxlike skull supports the brain and sense organs. (5) Males possess copulatory organs that are extensions of the pelvic girdle and internally fertilize the females. (6) Fins are supported by elastic connective tissue rays, or ceratotrichia. (7) The body is covered with dermal denticles or placoid scales, toothlike structures with enameloid crowns and dentine bases.

Rajiformes include the electric rays, sawfishes, guitarfishes, skates, and stingrays; their fossil record dates back to the Lower Jurassic (150 million years ago [mya]) (guitarfishes). All of the major taxa are known by the Upper Cretaceous (100 mya) to the Paleocene (50 mya). A majority of the early fossils come from northern Africa and southern Europe, areas that in the late mid-Mesozoic and Lower Cenozoic were part of the Tethys Sea, a shallow tropical sea that separated the northern and southern continents over much of this period. Although the fossil record spans more than 150 million years,

the record is very incomplete, owing to the paucity of hard body parts of skates and rays. Unlike the bony fishes, skates and rays (and all chondrichthyans for that matter) lack large bony external and internal structures that readily fossilize. In many cases, the fossil chondrichthyans are represented solely by teeth or enlarged scales. Teeth and scales serve to identify the fossils as skates and rays but provide little information on body structure or phylogenetic relationships.

Skates and rays have been classified variously within the cartilaginous fishes. Traditionally, they have been considered an equivalent group or sister group of the sharks. More recently, they have been grouped within a subsection of the sharks. Currently, based on morphological characters, they are considered to be a sister group of the angelsharks and sawsharks within the squalomorph sharks. The squalomorph sharks, in turn, are the sister group of the galeomorph sharks, a group that includes the horn sharks, carpet and nurse sharks, catsharks, mako sharks and white sharks, and requiem sharks. The squalomorph sharks include the sixgill and sevengill sharks and dogfish sharks in addition to the angelsharks, sawsharks, and skates and rays. These relationships make intuitive sense, because both squalomorph sharks, except for the most primitive members, and skates and rays lack an anal fin. Moreover, both groups are, for the most part, adapted for a benthic existence, and the sequential squalomorph sister taxa of skates and rays, angel- and sawsharks, share many anatomical characters with the skates and rays.

Recent molecular data, however, offer some support for the traditional relationship of skates and rays as a sister group of the sharks. If the molecular data are correct, suggesting that sharks and skates and rays shared a common ancestor, then squalomorph sharks and skates and rays independently acquired their adaptations for benthic habitats. As of 2002 the

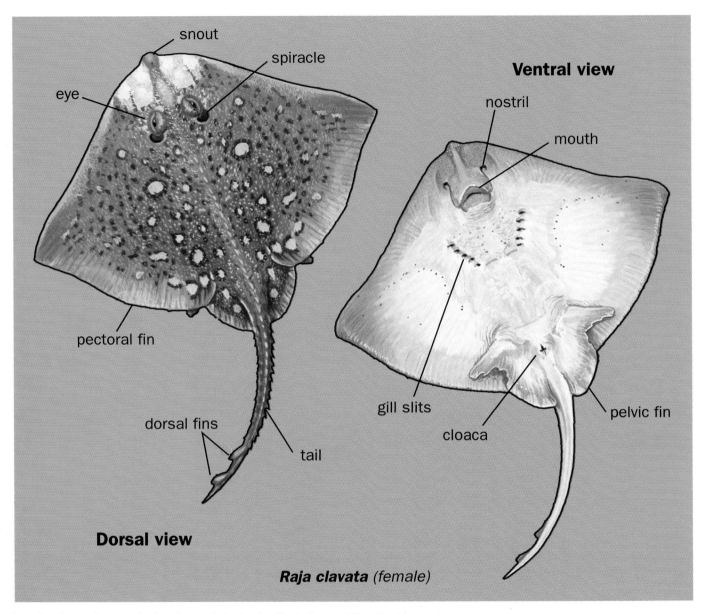

Dorsal and ventral views of a female ray, *Raja clavata*. (Illustration by Gillian Harris)

relationships of the skates and rays to the remainder of the cartilaginous fishes remain uncertain. Problems with classification of both the skates and rays and the sharks are due to their geological age and mediocre fossil record and, possibly, their parallel evolutionary trajectories.

As of 2002 the 513 species of skates and rays are classified within 63 genera and 20 families. The families are classified into eight suborders, although the higher-level classification is a work in progress. Most of the species are in the electric ray family Narcinidae (about 30 species), the guitarfish family Rhinobatidae (about 40 species), the skate family Rajidae (about 250 species), the round ray family Urolophidae (about 25 species), the freshwater stingray family Potamotrygonidae (about 25 species), and the stingray family Dasyatidae (about 63 species).

Physical characteristics

Skates and rays share a large number of characters, and as of 2002 there is little doubt that they form a natural group of fishes. They are defined largely by their adaptations for a benthic existence. In fact, they may have the ideal structure among vertebrates for such an existence. All taxa are flattened, at least anteriorly, with the pectoral fins joined to the head and trunk to form a disc. Eyes and spiracles are located on the upper side of the head; the nostrils, mouth, and gill slits are found on the ventral side of the head. Only a few sharks, orectolobids (carpet sharks), and squatinids, but no bony fishes, approach skates and rays in their degree of dorsoventral flattening. Flatfishes (bony fishes) are greatly laterally flattened or compressed, with both eyes on the same side of the head and the mouth contorted slightly to greatly. In other words, to achieve the degree of flatness of skates and rays, flatfishes had to become asymmetrical. Apparently, sharks and rays have the evolutionary potential to become depressed, whereas bony fishes are morphologically constrained from assuming such a posture.

Many of the distinguishing characteristics of skates and rays are concerned with the structural demands of their depressed shape. The vertebrae between the cranium and the shoulder girdle are fused into a tube (synarchial), the upper cartilage of the shoulder girdle (suprascapula) is either joined or articulated with the vertebral column or synarchial, the anterior pterygium cartilage of the shoulder girdle indirectly or directly joins the side of the cranium, and the upper jaw lacks an articulation with the cranium. All but the last of these adaptations provide the structure to support the expanded disc. The pectoral fin is supported anteriorly by the cranium, medially by the vertebral column, and posteriorly by the trunk, and this support system has allowed massive pectoral fins to develop in the majority of skates and rays. The massive pectoral fins made it possible for these ray fishes to swim by means of undulating or oscillating their pectoral fins. Freeing the upper jaw from the cranium provided greater versatility in feeding both on and in the benthic habitat.

Skates and rays vary in their degrees of flatness and disc development. The more primitive taxa, sawfishes (pristids) and guitarfishes (rhinids, rhinobatids, platyrhinids, and zanobatids), have rather small discs and stout, sharklike tails. These fishes swim by laterally undulating the trunk muscles like the sharks. Electric rays, thought to be the most primitive of the skates and rays, likewise have stout tails but rather expansive discs. They retain a large number of primitive characteristics, however. The large discs house the branchial electric organs that distinguish the group, and large discs may have been independently derived in this group to house the electric organs.

The skates (rajids) and stingrays (plesiobatids, hexatrygonids, urolophids, potamotrygonids, urotrygonids, dasyatids, gymnurids, myliobatids, rhinopterids, and mobulids) have very large, laterally expanded discs and slender to very slender tails. The tails of skates are slender, whereas those of stingrays are slender to very slender and mostly whiplike, and they bear one or more serrated spines. Both skates and stingrays swim by vertically undulating their discs or, in the case of the more derived stingrays, by vertically oscillating the discs like birds in flight. Stingrays have a ball-and-socket connection between the shoulder girdle and the vertebral column and an extra synarchial behind the shoulder girdle. Some of the derived forms of stingrays can generate enough speed to leap clear of the water. Skates lack the ball-and-socket connection, but they have bilobed pelvic fins with a finger-like anterior lobe. The anterior lobe can be used to "walk" or "punt" along the bottom. Punting is a unique locomotive gait of skates.

The arrangement of the eyes on the upper surface and the mouth on the lower surface means that skates and rays are unable to see their prey except at a distance. In fact, vision may play only a secondary role in feeding. Like sharks and some bony fishes, skates and rays have electric organs, ampullae of Lorenzini, symmetrically arranged around their mouths. The ampullae are at the end of pores and tubes filled with a

Two nostril-like features flank the mouth of a skate to guide it to food. The skate's mouth is lined with teeth capable of crushing small snails and crustaceans. (Photo by Jeffrey L. Rotman/Corbis. Reproduced by permission.)

conductive substance and are capable of responding to small electric currents, such as those produced by the muscle contractions of prey organisms. Skates and rays also have closed lateral line systems on the ventral surface that are sensitive to small jets of water pressure, such as those produced by bivalve mollusks that are often the prey of these fishes. When a jet of water strikes the surface in the vicinity of the canal system, it causes the fluid in the canal system to flow. Sensory cells lining the canal system perceive the moving current.

Members of two suborders, electric rays and skates, produce electric currents by means of modified muscle cells. Some of the gill arch or branchial muscles of electric rays are modified into electric cells that can produce up to 200 volts. These cells occupy most of the lateral area of the disc and are used to stun prey and defend against predation. Some electric rays have an auxiliary electric organ behind the main one that produces weak electric currents that may be used for communication among members of a population. Discharges of one individual can be perceived by the ampullae of Lorenzini of another individual. Skates have weak electric organs along the sides of the tail that apparently are used in communication among members of a population.

Skates and rays vary considerably in body size. Some electric rays (Narcinidae) mature at about 4 in (10 cm) in total length. Some skates (Rajidae) mature at about 6 in (15 cm) in total length and are probably the lightest of the chondrichthyans because of their very slender tails and thin discs. Sawfishes, on the other hand, can reach up to 23 ft (7 m) in length and have tooth-bearing rostra almost 6.6 ft (2 m) long. Manta rays (*Manta*) can reach 22 ft (6.7 m) in width and have been reported to be up to 29.8 ft (9.1 m) in width.

The coloration of skates and rays appears to be related largely to camouflage. Species that occur in shallow water tend to be dark yellow-brown, various shades of gray, or brown to black dorsally and cream to white ventrally. Those in mucky waters tend to be unpatterned dorsally, whereas those in clear water often are patterned with wavy lines,

A southern stingray (*Dasyatis americana*) with doctorfish (*Acanthurus chirurgus*) near Grand Cayman. (Photo by Charles V. Angelo/Photo Researchers, Inc. Reproduced by permission.)

stripes, bars, or ocelli. The patterning apparently functions in obscuring their outline and thus aids in making them unrecognizable to potential predators. There is little if any sexual dichromism. Species in deep water are typically dark colored dorsally and ventrally.

Distribution

Skates and rays are found worldwide in the marine environment from the shoreline to about 9,842 ft (3,000 m) and in many tropical freshwaters. The greatest diversity of species to subordinal taxa is in the tropical Indo-West Pacific, although not all of the higher taxa are represented in this region. The Indo-West Pacific region includes the tropical waters from the east coast of Africa to the east coast of Australia and Japan. With the exception of the stingrays, rajiforms are almost entirely absent from the coral islands of the central and western Pacific.

With few exceptions electric rays are limited to tropical and warm temperate seas over continental shelves. Torpeninid electric rays range into temperate latitudes, and some narcinid electric rays (*Benthobatis*) occur to depths of about 3,281 ft (1,000 m). Their eyes are minute and covered with skin, suggesting that they are either blind or respond only to light. Narkid and hypnid electric rays are limited to the tropical waters of the Indo-West Pacific.

Sawfishes are tropical and apparently limited to coastal, brackish, and freshwaters. Guitarfishes are tropical to warm temperate in coastal and brackish waters. One of the families is limited to the tropical Indo-West Pacific, and the other three families are most diverse in this region.

Skates (rajids) and stingrays (myliobatoids) largely have complementary distributions. In tropical waters skates are absent on inner continental shelves but are abundant in deeper water and in both the north and the south of tropical-subtropical regions. Stingrays, on the other hand, are limited primarily to shallow tropical seas. The stingray families Plesiobatidae (with one species) and Hexatrygonidae (with

one to five species) are exceptions and occur to depths of 1,640–3,281 ft (500–1,000 m) under tropical seas. Skates are the only rajiforms that occur at polar latitudes and that are common at great depths, to about 9,842 ft (3,000 m). Few species are found in estuaries, and only one species occurs in freshwater (*Dipturus* sp. from Bathurst Harbour, near Port Davey, Tasmania). Stingrays, on the other hand, are abundant in low-salinity regions, and a number are found strictly in tropical freshwaters. The stingray family Potamotrygonidae is limited to the freshwaters of South America.

Habitat

The majority skates and rays are benthic in marine habitats. The more sharklike forms, such as the sawfishes and guitarfishes, rest on the bottom and swim immediately over the bottom. The more depressed forms, such as the electric rays, skates, and most of the stingrays, rest and swim close to the bottom and often partially bury themselves in the bottom. They undulate their greatly expanded discs while lying over soft substrates to cover themselves partially. When partially covered with sediment, skates and rays ventilate by bringing water in through their spiracles and expelling the water through their gill slits. One species of Dasyatidae (the pelagic stingray, or *Pteroplatytrygon violacea*), the Myliobatidae (eagle rays), Rhinopteridae (cownose rays), and Mobulidae (manta rays) are largely pelagic or at least capable of sustained swimming. *Pteroplatytrygon violacea* and the mobulids spend most of their time and feed in the water column and are at least partially oceanic. The mobulids, however, appear to feed near continents, where upwelling of currents leads to high concentration of zooplankton. Myliobatidae are capable of leaping from the water like the mobulids, but they feed on the bottom.

As a group, chondrichthyans are uncommon in freshwater. Their absence in freshwater is related at least in part to

A smalltooth sawfish (*Pristis pectinata*) swims in the Atlantic Ocean. (Photo by Tom McHugh/Sea World/Photo Researchers, Inc. Reproduced by permission.)

The blue-spotted stingray (*Taeniura lymma*) is widespread in tropical coral reefs of the Indian and Pacific Oceans. (Photo by Jeffrey L. Rotman/ Corbis. Reproduced by permission.)

the retention of urea in their tissues. Urea acts as a salt and makes chondrichthyans about as salty as seawater. This reduces the costs of osmoregulation in marine waters but increases its costs in freshwaters. Chondrichthyans that enter freshwater apparently have the ability to reduce the urea content in their tissues. The ability to inhabit freshwater is more widespread among skates and rays than sharks. All species of sawfishes and many species of stingrays either move back and forth between saltwater and freshwater or reside permanently in freshwater. Some sawfishes become more freshwater tolerant with age. Numerous dasyatid stingrays move in and out of freshwater. Some dasyatid stingrays reside in freshwater, and the stingray family Potamotrygonidae is limited to the freshwaters of South America. In fact, the potamotrygonids have lost the ability to conserve urea.

Behavior

The majority of skates and rays are rather docile, both on a daily basis and over long periods of time. Some of the electric rays and stingrays that live in shallow water may limit daily excursions to moving in and out with the tide. During high tide they move shoreward and burrow into sandy bottoms; at low tide they abandon these depressions and construct similar abodes in deeper water. Limited data from tagging studies suggest that skates are rather provincial. Templeman found that most specimens of a particular skate (*Amblyraja radiata*) tagged off Newfoundland were recaptured within 60 mi (97 km) up to 20 years from the time that they were tagged. In temperate regions benthic species of skates and rays move northward and southward with vernal warming and cooling. More active species, such as eagle rays, cownose rays, and mantas, may be wide ranging, although there are reports that individuals of *Manta birostris* return to the same feeding areas on a yearly basis.

The social behavior and communication of rajiforms are poorly known, but some observations suggest that skates use electrical discharges of their tail electric organs for communication. The organs discharge posteriorly, and males follow directly behind females during mating. All skates have weak

electric organs, and the histologic characteristics of the electric cells vary among species. It is thus possible that different species have distinctive electric discharges and that these differences may be used as mate-recognition systems to aid in seeking mates of the same species. Such systems are well known for elephantnose fishes.

Little is known concerning the social relationships between skates and rays and other organisms. There are numerous observations, however, that *Manta birostris* enter shallow water reef areas to be cleaned of ectoparasites by cleaning bony fishes. One species of remora often hitches a ride on *Manta birostris* and even enters the ray's cloaca for extended periods of time. The remora probably feeds on ectoparasites and possibly on the ray's feces.

Feeding ecology and diet

The majority of skates and rays can be considered generalist benthic predators that feed on the more abundant benthic invertebrates and small to moderately sized bony fishes. Some groups, such as the electric rays and sawfishes, have specialized devices for capturing food. The branchial electric organs of electric rays are used to stun fishes and invertebrates, which then are quickly swallowed. Sawfishes use their tooth-bearing rostral blades to disable schooling fishes and to dislodge invertebrates from the substrate. Myliobatid and rhinopterid stingrays have jaw teeth that are fused into crushing plates that enable these fishes to crush bivalve clams, oysters, and mussels. Mobulids have specialized lateral extensions of their rostra (cephalic fins); large, oval-shaped mouths; and filter plates running between their gill arches that allow them to strain zooplankton from the water column. The cephalic fins direct the zooplankton into the mouth, and the filter plates separate the zooplankton from the water that flows over the gill slits. The ampullae of Lorenzini and the closed lateral line systems of skates and stingrays permit these fishes to sense infaunal organisms that can be sucked out of the substratum by means of their protrusible jaws and the sucking action of the mouth and gill cavities.

Winter skate (*Raja ocellata*) on the ocean bottom in the Gulf of Maine. (Photo by Andrew J. Martinez/The National Audubon Society Collection/ Photo Researchers, Inc. Reproduced by permission.)

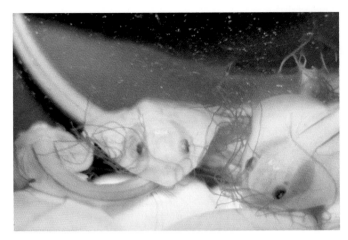

A mermaid's purse skate (*Raja binoculata*) egg sack with live embryos. (Photo by Animals Animals ©Joanne Huemoeller. Reproduced by permission.)

Skates and rays are preyed upon by sharks, and small skates and rays, including egg capsules of skates, are occasionally consumed by large skates and rays.

Reproductive biology

Mating has not been observed frequently in rajiforms, but all species practice internal fertilization. Males possess extensions of their pelvic fin cartilages (claspers) that are inserted into the cloaca of females and serve to transmit sperm into the oviduct. Only one of the pair of claspers is inserted at a time. Generally, copulation occurs between a single male and a single female. In many cases the males bite the anterior margin of the female's disc, to enable them to insert the clasper. Males of many taxa of skates and stingrays have sharp, pointed teeth, unlike the flat teeth of females; these teeth enable the males to remain in contact during copulation. Skates also have sharp, pointed and often barbed, clawlike thorns near the outer corners of their discs, which are used as additional points of contact during copulation.

Fertilization takes place in the anterior section of the oviduct of the female, and the fertilized egg then is encapsulated in the oviductal gland. The encapsulated egg descends into the uterus; in most rajiforms the egg is retained in the uterus, and development is termed "viviparity without a placenta," also termed ovoviviparity. The egg capsules are generally thin, and the embryos may be encapsulated only during the early stages of development. In addition to the yolk supplied with the encapsulated eggs, nutrients are available to the embryos from the uterine wall of the female. Internal development extends from several months to nearly a year among the various taxa of rajiforms. The young or neonates are immature copies of the adults at birth, and the mother provides no parental care.

In skates the encapsulated egg is shed to the environment, and development is termed "oviparity." As with the other rajiforms, the egg is fertilized and encapsulated in the oviductal gland, but the egg capsule is thick, collagenous, and rectangular shaped, with a horn at each corner. The young

remain in the capsule from several months to more than a year. Females do not offer any protection for the egg capsules, but it is possible that they seek special areas in which to release them.

Conservation status

Skates and rays have long been exploited by artisanal fisheries and small-scale fisheries in developing countries, but they have not been targeted by large-scale fisheries. Despite the lack of directed fisheries, humans have had a negative impact on populations of many species over the past half-century. Slow growth rates and low reproductive potentials make chondrichthyans, including skates and rays, vulnerable even to modest rates of exploitation. Chondrichthyans have much lower growth rates and fecundity than bony fishes. Thus a fishery directed at bony fishes may inadvertently negatively affect skates and rays before the fishery overexploits the targeted bony fishes. Inshore tropical habitats occupied by many skates and rays have been degraded by human activities, and commercial shrimp fishing has accidentally captured inshore species, such as guitarfishes and sawfishes. Bottom trawling (for shrimps, for example) unintentionally captures large quantities of skates and rays.

As of 2002 the IUCN listed 26 skates and rays as Vulnerable, Endangered, or Critically Endangered. These are species of sawfishes, guitarfishes, skates, or stingrays, and they occupy tropical freshwaters, inshore tropical waters, or continental shelf habitats in temperate regions that are under heavy fishing pressure. All seven species of sawfishes are listed as Endangered (5 species) or Critically Endangered (2 species). Two guitarfishes are listed as either Vulnerable or Critically Endangered. A total of six skates are listed as Near Threatened/Lower Risk (3 species), Vulnerable (1 species), or Endangered (2 species). Eleven stingrays are listed as Near Threatened/Lower Risk (1 species), Vulnerable (4 species), Endangered (5 species), or Critically Endangered (1 species). The Thailand population of *Himantura chaophraya* is Critically Endangered.

Significance to humans

For the most part, skates and rays are not considered high-quality food items. They do enter artisanal fisheries, however, and are landed by numerous commercial fisheries in the Far East and in Europe. Skates and rays are consumed fresh, dried, or salted.

The skin of skates and rays is very tough and can be used as leather. Handles of samurai swords may be covered with guitarfish skin. Various ethnic groups of the Indo-West Pacific once used the teeth of sawfishes and the serrated spines of stingrays as war clubs. Native Americans of the Amazon and Orinoco River drainages capture freshwater stingrays for food and use their serrated spines for arrowheads or as implements for self-mutilation. Today, dried, mutilated skates and rays are sold in seashore curiosity shops.

With the exception of stingrays and large electric rays, rajiforms are not harmful to humans. Bethnic stingrays, , often lie partially buried in the sand along beaches frequented by human bathers. Bathers who are unfortunate enough to step on a partially buried ray may receive a nasty wound and poison from a gland associated with the spine. This gland at the base of the spine releases neurotoxins and proteolytic toxins.

Some stingrays have contributed to the ecotourism industry. Tourists visit Stingray City in the Cayman Islands to feed large stingrays (*Dasyatis americana*). Scuba expeditions are conducted in Hawaii, the northern Gulf of Mexico, and various other areas, to observe manta rays at their feeding sites.

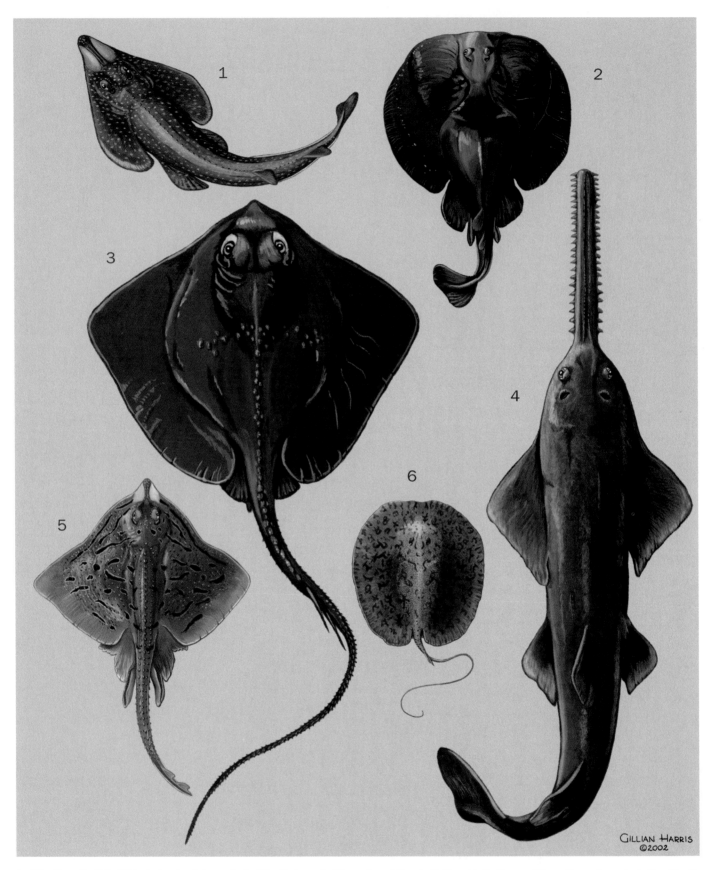

1. Atlantic guitarfish (*Rhinobatos lentiginosus*); 2. Atlantic torpedo (*Torpedo nobiliana*); 3. Roughtail stingray (*Dasyatis centroura*); 4. Smalltooth sawfish (*Pristis pectinata*); 5. Clearnose skate (*Raja eglanteria*); 6. Freshwater stingray (*Paratrygon aiereba*). (Illustration by Gillian Harris)

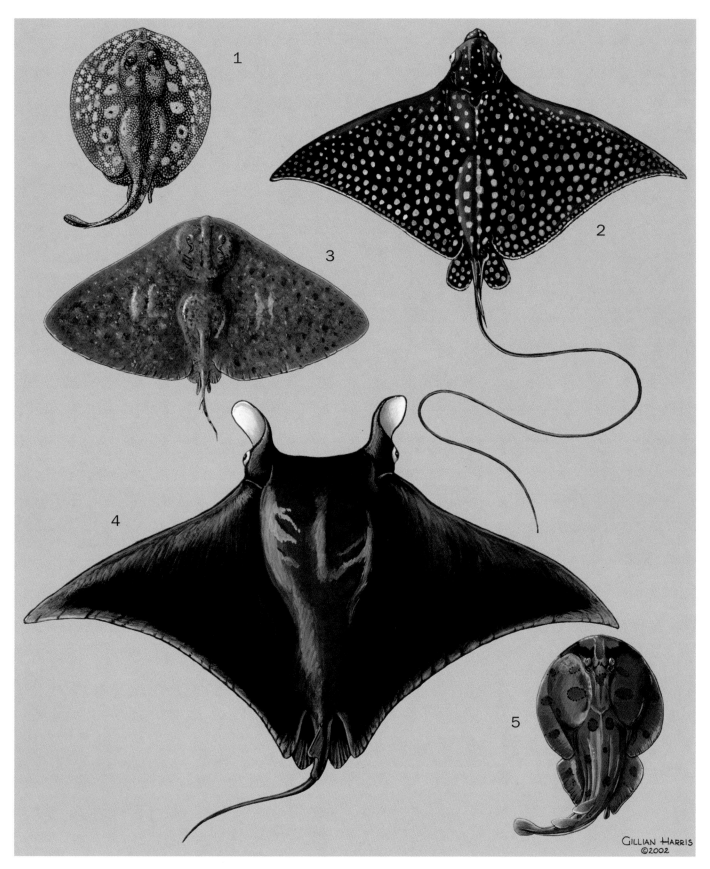

1. Yellow stingray (*Urobatis jamaicensis*); 2. Spotted eagle ray (*Aetobatus narinari*); 3. Spiny butterfly ray (*Gymnura altavela*); 4. Atlantic manta (*Manta birostris*); 5. Lesser electric ray (*Narcine bancrofti*). (Illustration by Gillian Harris)

Species accounts

Roughtail stingray
Dasyatis centroura

FAMILY
Dasyatidae

TAXONOMY
Raja centroura Mitchill, 1815, Long Island coast, New York.

OTHER COMMON NAMES
French: Pastenague épineuse; Italian: Trigone spinoso.

PHYSICAL CHARACTERISTICS
Size 83 in (210 cm) in disc width as an adult and 13–15 in
(34–37 cm) in disc width at birth. The head, pectoral fins, and
trunk are flattened and joined to form a broad, rectilinear disc.
The disc is about as broad as it is long, and the outer corners
are subangular. Tail 2.4 to 26 times the disc length, slender,
and whiplike. It lacks a fleshy dorsal keel and dorsal fins, and
the caudal fin but has one or more serrated spines and a low
ventral fold. Snout moderately long and very obtuse. Mouth
moderately wide and moderately arched. Specimens larger than
20 in (50 cm) in disc width have large tubercles or bucklers
along the midline and central area of the disc and along the
upper surface of the tail. Coloring brown to olive dorsally and
white to whitish ventrally.

DISTRIBUTION
Tropical to warm temperate regions in the Atlantic. Ranges
from Massachusetts to the northern Gulf of Mexico and
Uruguay in the western Atlantic and from the Bay of Biscay to
Angola, including the Mediterranean and Madeira.

HABITAT
Benthic habitats on soft bottoms from near shore to about 899
ft (274 m) in the western Atlantic and to about 197 ft (60 m) in
the eastern Atlantic.

BEHAVIOR
Moves northward and shoreward in the spring and southward
and offshore in the autumn.

FEEDING ECOLOGY AND DIET
Prey include polychaetes, cephalopods, crustaceans, and bony
fishes.

REPRODUCTIVE BIOLOGY
Viviparous, with litters ranging from two to six neonates.

CONSERVATION STATUS
Not listed by the IUCN.

SIGNIFICANCE TO HUMANS
Capable of inflicting painful wounds in waders and swimmers
that come into contact with them in inshore areas. ◆

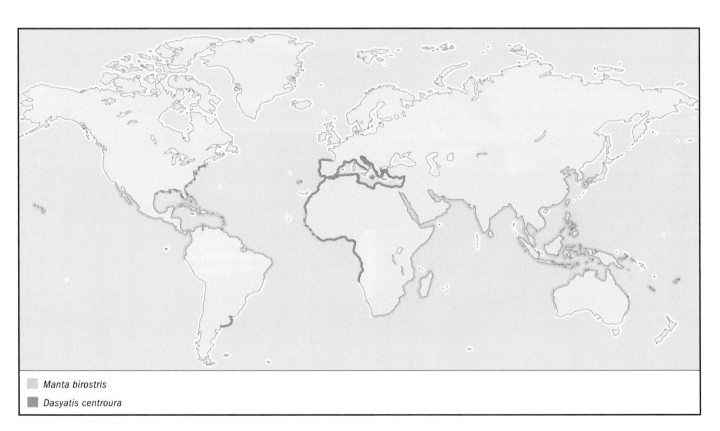

Manta birostris
Dasyatis centroura

Gymnura altavela

Torpedo nobiliana

Spiny butterfly ray
Gymnura altavela

FAMILY
Gymnuridae

TAXONOMY
Raja altavela Linnaeus, 1758, Mediterranean Sea.

OTHER COMMON NAMES
French: Raie-papillon épineuse; Spanish: Raya mariposa.

PHYSICAL CHARACTERISTICS
Size 80 in (202 cm) in disc width. The head, pectoral fin, and trunk are flattened and joined to form a very broad rectilinear disc. Disc 1.5 times broader than it is long, and the outer corners are abruptly rounded. Tail very short and slender, about one-fourth of the disc width. It lacks dorsal and caudal fins but has one or more serrated spines and dorsal and ventral ridges. Snout short and very obtuse. Mouth moderately arched. There is a tentacle-like structure on the inner posterior margin of the spiracle. The body is naked, except for small denticles over central areas of the disc. Coloring dark brown and white ventrally. Dorsal surface patterned with small dark and light spots.

DISTRIBUTION
Tropical to warm temperate waters of the Atlantic. Ranges from southern Massachusetts to Rio de la Plata, Argentina, in the western Atlantic and from Portugal to Angola, including Madeira, the Canary Islands, and the Mediterranean Sea, in the eastern Atlantic.

HABITAT
Benthic habitats on soft bottoms from near shore to about 197 ft (60 m).

BEHAVIOR
Nothing is known concerning the behavior of this species.

FEEDING ECOLOGY AND DIET
Prey include mollusks, crustaceans, and bony fishes.

REPRODUCTIVE BIOLOGY
Viviparous, with litters ranging from four to seven neonates. The gestation period is about six months.

CONSERVATION STATUS
Not listed by the IUCN.

SIGNIFICANCE TO HUMANS
Capable of inflicting painful wounds in waders and swimmers that come into contact with them in inshore areas. ◆

Atlantic manta
Manta birostris

FAMILY
Mobulidae

TAXONOMY
Raja birostris Walbaum, 1792, type locality not specified.

OTHER COMMON NAMES
English: Giant manta; French: Mante géante; Spanish: Manta voladora.

PHYSICAL CHARACTERISTICS
Size 22 ft (6.7 m) in disc width as an adult and 47 in (120 cm) in disc width at birth. The head, pectoral fin, and trunk is flattened and joined to form a broad rectilinear disc. The disc is

much broader than it is long, and the outer corners are slightly falcate. Tail very short and whiplike. It lacks folds, keels, and a caudal fin but has a dorsal fin at the base and either has or lacks a small serrated spine. The anterior section of the pectoral fin forms a narrow, vertically oriented cephalic fin that is attached to the head and is free of the remainder of the fin. Head slightly elevated and very broad. Mouth is terminal. Teeth are small and located only on the lower jaw. Brown to olive in color dorsally and white to whitish ventrally.

DISTRIBUTION
Tropical to warm temperate regions worldwide.

HABITAT
Pelagic species in near shore to oceanic waters but is most common in coastal waters.

BEHAVIOR
Performs somersaults during feeding. Occasionally leaps partially or completely out of the water. Enters shallow reef areas to be cleaned of ectoparasites by small bony fishes.

FEEDING ECOLOGY AND DIET
Prey include zooplankton and nektonic crustaceans.

REPRODUCTIVE BIOLOGY
Viviparous. Litters size unknown.

CONSERVATION STATUS
Listed by the IUCN as Data Deficient.

SIGNIFICANCE TO HUMANS
Harpooned or gill-netted for human consumption in some parts of the world. It is the focus of underwater scuba-based ecotourism. ◆

Spotted eagle ray
Aetobatus narinari

FAMILY
Myliobatidae

TAXONOMY
Raja narinari Euphrasen, 1790, Brazil.

OTHER COMMON NAMES
French: Aigle de mer léopard; Spanish: Chucho pintado.

PHYSICAL CHARACTERISTICS
Size 130 in (330 cm) in disc width as an adult and 7–14 in (18–36 cm) in disc width at birth. The head, pectoral fins, and trunk are flattened and joined to form a broad rectilinear disc. The head is elevated from the disc, and the anterior section of the pectoral fins forms a subrostral lobe above the mouth. Disc 2.1 times broader than long, and outer corners are slightly falcate. Tail very long, slender, and whiplike. It lacks folds, keels, and a caudal fin, but it has a dorsal fin at the base and one or more serrated spines. Snout moderately short; mouth is straight. Teeth flattened and pavement-like and aligned in series. The body is naked. Color ranges from olivaceous to dark brown dorsally and white ventrally, except for a dusky subrostral line and dusky pelvic fins. The dorsal surface is patterned with small white, greenish, or yellow spots.

DISTRIBUTION
Tropical to warm temperate regions worldwide. Species may represent a series of cryptic species, each limited to a particular location. It ranges from North Carolina to southern Brazil, including the Gulf of Mexico in the western Atlantic, Cape Verde to Angola in the eastern Atlantic, the entire tropical and subtropical region of the eastern Pacific, and throughout the tropical and subtropical Indo-West Pacific.

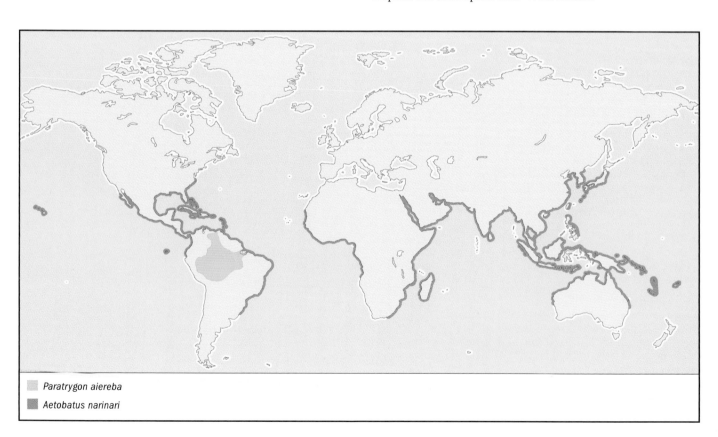

Paratrygon aiereba
Aetobatus narinari

HABITAT
Benthic. Occurs from near shore to about 197 ft (60 m).

BEHAVIOR
Spends much of its time actively swimming in the water column by the oscillatory action of its pectoral fins. It is capable of leaping clear of the water.

FEEDING ECOLOGY AND DIET
Prey include shellfish, such as clams, oyster, whelks, and other mollusks.

REPRODUCTIVE BIOLOGY
Viviparous, with litters ranging up to four neonates.

CONSERVATION STATUS
Listed by IUCN as Data Deficient. There is a possibility that future research may reveal that threatened classification is appropriate, but currently information is not sufficient to list the species.

SIGNIFICANCE TO HUMANS
Capable of inflicting painful wounds in waders and swimmers that come into contact with it in inshore areas. Occasionally, it is captured by gill net fisheries for human consumption. ◆

Lesser electric ray
Narcine bancrofti

FAMILY
Narcinidae

TAXONOMY
Torpedo bancrofti Griffith, 1834, Jamaica.

OTHER COMMON NAMES
Spanish: Raya eléctrica torpedo.

PHYSICAL CHARACTERISTICS
Adults up to 23 in (58 cm) in total length but only 3.5–3.9 in (9–10 cm) in total length at birth. The head, pectoral fins, and trunk are flattened and joined to form a fleshy disc. Tail is stout, and the caudal fin well developed. Snout is very blunt, with a narrow, greatly protrusible mouth that forms a short tube. Kidney-shaped electric organs are located on either side of head, giving the skin surface a honeycomb appearance. Coloring yellowish brown to grayish brown or dark brown dorsally and white to creamy white ventrally. The dorsal surface is patterned with dark blotches, spots, and crossbars.

DISTRIBUTION
Tropical to warm temperate waters of the western Atlantic. Ranges from North Carolina to Venezuela, including the northern and western Gulf of Mexico, the Greater and Lesser Antilles, Yucatan, Belize and northern South America.

HABITAT
Benthic habitats on soft bottoms in shallow water.

BEHAVIOR
Nothing is known concerning the behavior of this species.

FEEDING ECOLOGY AND DIET
Electric organs are used to stun prey, which consist of polychaetes, other invertebrates, and small fishes.

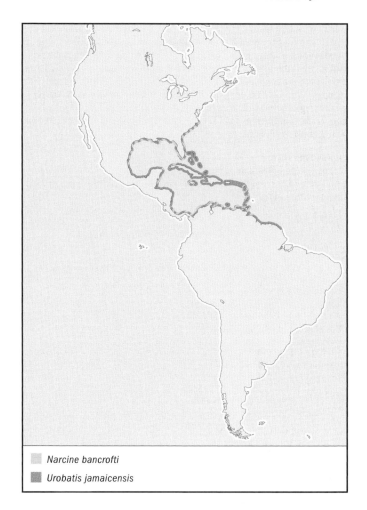

Narcine bancrofti

Urobatis jamaicensis

REPRODUCTIVE BIOLOGY
Viviparous without a placenta. Litter size can be as many as 18 embryos.

CONSERVATION STATUS
Not listed by the IUCN.

SIGNIFICANCE TO HUMANS
None known. ◆

Freshwater stingray
Paratrygon aiereba

FAMILY
Potamotrygonidae

TAXONOMY
Trygon aiereba Müller and Henle, 1841, Brazil.

OTHER COMMON NAMES
English: Discus ray.

PHYSICAL CHARACTERISTICS
Size 43 in (110 cm) in disc width. The head, pectoral fin, and trunk are flattened and joined to form an elliptical disc. The disc is longer than it is broad, indented anteriorly, and slightly broader across the anterior third than across the posterior

third. Pelvic fins are covered entirely by the disc. Tail relatively short, broad at the base, and filamentous distally. The preoral snout length is 30–38% of the disc width. The posterior outer margin of the spiracle bears a knoblike process. Eyes small and located just in front of the spiracles. Mouth small; teeth small and few in number. The dorsal surface is covered with small dermal denticles. Light brown dorsally and white ventrally. The dorsal surface is patterned with dark reticular or vermicular blotches.

DISTRIBUTION
Freshwaters of South America from northern Bolivia and eastern Peru to northern Brazil and Venezuela in the Amazon and Orinoco drainages.

HABITAT
Benthic habitats on soft bottoms in rivers.

BEHAVIOR
Nothing is known concerning the behavior of this species.

FEEDING ECOLOGY AND DIET
Prey include benthic organisms.

REPRODUCTIVE BIOLOGY
Viviparous.

CONSERVATION STATUS
Not listed by the IUCN.

SIGNIFICANCE TO HUMANS
Capable of inflicting very painful wounds in waders and swimmers that come into contact with it. ◆

Smalltooth sawfish
Pristis pectinata

FAMILY
Pristidae

TAXONOMY
Prisitis pectinata Latham, 1794, type locality not specified.

OTHER COMMON NAMES
French: Poisson scie tident; Spanish: Pejepeine.

PHYSICAL CHARACTERISTICS
Reaches 217 in (550 cm) in total length as an adult but only 24 in (60 cm) in total length at birth. Body elongated and moderately depressed, and the snout is prolonged into a long, narrow, flattened blade bearing 24 to 32 pairs of teeth. The blade is about 25% of the total length. The caudal fin is without a distinct ventral lobe. Color dark brownish gray dorsally and grayish white ventrally.

DISTRIBUTION
Tropical to warm temperate western Atlantic from New York to central Brazil, including the Gulf of Mexico, the Caribbean Sea, and Bermuda. It has been recorded from the eastern Atlantic, Mediterranean Sea, the eastern Pacific, Indian Ocean, and the Indo-Pacific region but these records need to be verified.

HABITAT
Lives near shore in bays, estuaries, and freshwater.

BEHAVIOR
Nothing is known concerning the behavior of this species.

FEEDING ECOLOGY AND DIET
The blade is used to dislodge invertebrates and to disable fishes.

REPRODUCTIVE BIOLOGY
Viviparous without a placenta. Litters range from 15 to 20 young.

CONSERVATION STATUS
Listed as Endangered by the IUCN. It is protected in Florida and Louisiana state waters.

SIGNIFICANCE TO HUMANS
None known. ◆

Atlantic torpedo
Torpedo nobiliana

FAMILY
Torpedinidae

TAXONOMY
Torpedo nobiliana Bonaparte, 1835, Italy, iconograph.

OTHER COMMON NAMES
French: Torpille noire; Spanish: Tremolina nigra.

☐ *Raja eglanteria*

▨ *Pristis pectinata*

▨ *Rhinobatos lentiginosus*

PHYSICAL CHARACTERISTICS

Reaches 71 in (180 cm) in total length in adulthood, and 7.9–9.8 in (20–25 cm) in total length at birth. The head, pectoral fin, and trunk are flattened and joined to form a fleshy disc. The tail is stout, and the caudal fin is well developed. The species has a very blunt snout, with a wide and slightly protrusible mouth. Kidney-shaped electric organs are located on either side of the head, giving the skin surface a honeycomb appearance. Coloring varies from brown to purplish brown dorsally and is white with a dark margin ventrally.

DISTRIBUTION

Tropical to temperate waters of the North Atlantic. Ranges from southern Nova Scotia to Venezuela, including the northern Gulf of Mexico, Cuba, and Trinidad in the western Atlantic and the British Isles and Mediterranean Sea to South Africa in the eastern Atlantic.

HABITAT

Benthic habitats on soft bottoms from the shoreline to the upper continental slope at 1,739 ft (530 m).

BEHAVIOR

Nothing is known concerning the behavior of this species.

FEEDING ECOLOGY AND DIET

Electric organs are used to stun prey, which consist of fishes.

REPRODUCTIVE BIOLOGY

Viviparous without a placenta. The litter size is unknown.

CONSERVATION STATUS

Not listed by the IUCN.

SIGNIFICANCE TO HUMANS

None known. ◆

Clearnose skate
Raja eglanteria

FAMILY

Rajidae

TAXONOMY

Raja eglanteria Bosc, 1800, Charleston Bay, South Carolina.

OTHER COMMON NAMES

French: Raie blanc nez; Spanish: Raya hialina.

PHYSICAL CHARACTERISTICS

Reaches 31 in (78.5 cm) in total length; at birth it measures 5–5.7 in (12.5–14.4 cm) in total length. The head, pectoral fins, and trunk are flattened and joined to form broad, spade-shaped disc. Tail moderately slender and makes up about half of the total length. The caudal fin is poorly developed. Snout moderately long and slightly obtuse, and mouth moderately wide and slightly arched. There are medium-size thorns on the head and in a row from the shoulder region to the first dorsal fin and irregular lateral rows located on either side of the tail. The dorsal surface is covered sparsely with dermal denticles. Coloring is brown to gray dorsally and whitish to yellowish ventrally. The dorsal surface is patterned with dark and light spots and transverse and diagonal dark bars.

DISTRIBUTION

Tropical to warm temperate regions of the western North Atlantic, ranging from Massachusetts to the northern Gulf of Mexico.

HABITAT

Benthic. Found on soft bottoms from near shore to about 390 ft (119 m).

BEHAVIOR

In the northern part of its range this species migrates north and inshore in the spring and south and offshore in the autumn.

FEEDING ECOLOGY AND DIET

Prey include polychaetes, amphipods, shrimps, crabs, and bony fishes.

REPRODUCTIVE BIOLOGY

Oviparous; neonates hatch from egg capsules after several months.

CONSERVATION STATUS

Not listed by the IUCN.

SIGNIFICANCE TO HUMANS

None known. ◆

Atlantic guitarfish
Rhinobatos lentiginosus

FAMILY

Rhinobatidae

TAXONOMY

Rhinobatus lentiginosus Garman, 1880, Florida.

OTHER COMMON NAMES

None known.

PHYSICAL CHARACTERISTICS

Reaches 30 in (76 cm) in total length. The head, pectoral fins, and trunk are moderately expanded and joined to form a wedge-shaped disc. Nostril length equals or is slightly longer than the distance between the nostrils. The rostral cartilage of the snout is moderately long and expanded and bears conical tubercles near the tip. Tail stout, and dorsal fins and caudal fin well developed. Caudal fin lacks a distinct ventral lobe. Color gray to olive or dark brown dorsally and white to pale yellow ventrally. The dorsal surface generally is freckled with many small white spots.

DISTRIBUTION

Tropical to warm temperate waters of the western Atlantic and ranges from North Carolina to the southern Gulf of Mexico.

HABITAT

Benthic habitats on soft bottoms from the shoreline to 59 ft (18 m).

BEHAVIOR

Nothing is known concerning the behavior of this species.

FEEDING ECOLOGY AND DIET

Feeds on benthic organisms.

REPRODUCTIVE BIOLOGY

Viviparous without a placenta. The litter size is unknown.

CONSERVATION STATUS

Not listed by the IUCN.

SIGNIFICANCE TO HUMANS

Found in local fish markets in the southern part of its range. ◆

Yellow stingray
Urobatis jamaicensis

FAMILY
Urotrygonidae

TAXONOMY
Raia jamaicensis Cuvier, 1816, Jamaica.

OTHER COMMON NAMES
None known.

PHYSICAL CHARACTERISTICS
Reaches 28 in (70 cm) in total length. The head, pectoral fin, and trunk are flattened and joined to form an oval-shaped disc. The disc is longer than it is broad, and the outer corners are broadly rounded. Tail stout and relatively short, less than half its total length. It lacks dorsal and ventral folds and keels and dorsal fins but has one or more serrated spines and a small caudal fin. Snout moderately short and rounded; mouth small and straight. The body is naked, except for small denticles along the midbelt of the disc and tail. Green to grayish brown dorsally and yellowish, greenish, and brownish ventrally. The dorsal surface is patterned with fine reticulations or light-colored spots.

DISTRIBUTION
Tropical to warm temperate waters of the western Atlantic, ranging from Cape Lookout, North Carolina, to southern Florida, including the Gulf of Mexico, Bahamas, and Greater and Lesser Antilles.

HABITAT
Benthic. Found on soft bottoms from near shore, including bays and estuaries.

BEHAVIOR
Docile. Spends much of its time partially buried in soft substrates.

FEEDING ECOLOGY AND DIET
Prey include benthic invertebrates and bony fishes.

REPRODUCTIVE BIOLOGY
Viviparous, with litters ranging from two to four neonates.

CONSERVATION STATUS
Not listed by the IUCN.

SIGNIFICANCE TO HUMANS
Capable of inflicting painful wounds in waders and swimmers that come into contact with it in inshore areas.

Resources

Books

Cappetta, H. *Chondrichthyes. II. Mesozoic and Cenozoic Elasmobranchii.* New York: Gustav Fischer Verlag, 1987.

Carroll, Robert L. *Vertebrate Paleontology and Evolution.* New York: W. H. Freeman and Company, 1988.

Hamlett, William C., ed. *Sharks, Skates, and Rays: The Biology of Elasmobranch Fishes.* Baltimore: Johns Hopkins University Press, 1999

Last, P. R., and J. D. Stevens. *Sharks and Rays of Australia.* Melbourne, Australia: CSIRO, 1994.

McEachran, John D., and Janice D. Fechhelm. *Fishes of the Gulf of Mexico.* Vol. 1, *Myxiniformes to Gasterosteiformes.* Austin, TX: University of Texas Press, 1998.

McEachran, John D., K. A. Dunn, and T. Miyake. "Interrelationships of the Batoid Fishes (Chondrichthyes: Batoidea)." In *Interrelationships of Fishes,* edited by M. L. J. Stiassny, L. R. Parenti, and G. D. Johnson. New York: Academic Press, 1996.

Paxton, J. R., and W. N. Eschmeyer. *Encyclopedia of Fishes.* New York: Academic Press, 1994.

Taylor, L. R., editor. *Sharks and Rays.* Alexandria, VA: Nature Company Guides, Time-Life Books, 1997.

Periodicals

Lovejoy, Nathan R., E. Bermingham, and A. P. Martin. "Marine Incursion into South America." *Nature* 396 (December 1998): 421–422.

McEachran, John D., and K. A. Dunn. "Phylogenetic Analysis of Skates, a Morphologically Conservative Clade of Elasmobranches (Chondrichthyes: Rajidae)." *Copeia* 1998, no. 2 (1998): 271–290.

Rosenberger, Lisa J. "Pectoral Fin Locomotion in Batoid Fishes: Undulation Versus Oscillation." *Journal of Experimental Biology* 204, no. 2 (2001): 379–394.

———. "Phylogenetic Relationships Within the Stingray Genus *Dasyatis* (Chondrichthyes: Dasyatidae)." *Copeia* 2001, no. 3 (2001): 615–627.

Rosenberger, Lisa J., and M. W. Westneat. "Functional Morphology of Undulatory Pectoral Fin Locomotion in the Stingray *Taeniura lymma* (Chondrichthyes: Dasyatidae)." *Journal of Experimental Biology* 202, no. 24 (1999): 3523–3539.

Organizations

American Elasmobranch Society, Florida Museum of Natural History. Gainesville, FL 32611 USA. Web site: <http://www.flmnh.ufl.edu/fish/Organizations/aes/aes.htm>

John D. McEachran, PhD

Coelacanthiformes

(Coelacanths)

Class Sarcopterygii

Order Coelacanthiformes

Number of families 1

Photo: Coelacanths (*Latimeria chalumnae*) are known from the fossil record dating back over 360 million years. (Photo by Planet Earth Pictures, Limited. Reproduced by permission.)

Evolution and systematics

The living coelacanths are often celebrated as the most unusual case and important example of animal evolution. The first fossil coelacanths were recognized in rocks between 380 and 75 million years old. More than 100 years ago Woodward published the first review on these fishes. Fossils younger than 75 million years were never found, as if all coelacanths had become extinct at that time, very much like the dinosaurs. The bony structures in these fossil crossopterygians, especially their paired (pectoral and pelvic) fins, placed them close to the ancestor of the first amphibians and all other land vertebrates.

It is of no surprise, therefore, that the December 1938 find of a living coelacanth, when announced to the world by J. L. B. Smith in March 1939, caused disbelief and created one of the greatest biological sensations of the last century. Finding a living coelacanth—morphologically so similar to the fossil specimens left in rocks more than 75 million years ago—was as inconceivable as meeting a living dinosaur on a weekend walk.

The living coelacanth, sometimes known by the common name "gombessa," is a single advanced life form that has survived with relatively little change for nearly 400 million years. While some of the coelacanth's relatives became implicated in the ancestry of all terrestrial vertebrates, the aquatic descendants developed structural solutions to life absent in other animals. For example, instead of the calcified vertebrae that normally reinforce the axial skeleton, the coelacanths evolved a strong-walled elastic tube that is as far transformed from the notochord as are the vertebrae. Instead of a solid braincase, they evolved a two-part neuro-

cranium with an intracranial joint that is operated by a special basicranial muscle. It is the only animal with that structure living today. This intracranial joint and other unique rotational joints in the head together with the rostral organs and the gular reticulate electrosensory system may explain the special suction feeding and head standing behavior observed by Hans Fricke.

In the synthesis of coelacanth evolution by P. Forey, a total of 83 species are recognized to have lived between 380 million years ago in the middle Devonian and 75 million years ago in the Upper Cretaceous. There is no fossil record of coelacanths from the past 75 million years. During that time coelacanths lived and died without leaving fossils. The highest diversity of coelacanths during their geological history was recorded in the Lower Triassic (16 species) and in the Upper Jurassic (8 species). While most fossil remnants were found in marine deposits, many were also found in freshwater deposits, especially in the Upper Carboniferous, Lower Permian, Upper Triassic, Jurassic, and Lower Cretaceous.

The first living coelacanth that came to the attention of scientists was named *Latimeria chalumnae* by J. L. B. Smith in 1939. Upon seeing a second specimen in 1952 and noting that it lacked the first dorsal fin, Smith considered it a new species and named it *Malania anjouanae*. When it was found later that except for the lack of the first dorsal fin all other structures were like the first specimen, it was concluded that this first dorsal fin was probably bitten off by a shark, and *Malania anjouanae* become a synonym of *Latimeria chalumnae*.

A second species of coelacanth, *Latimeria menadoensis*, was discovered near Sulawesi by M. V. Erdmann in 1998.

Scientists examine a coelacanth (*Latimeria chalumnae*) that was caught by a fisherman in Kenya. Until 1938 the coelacanth was thought to have vanished with the dinosaurs 75 million years ago. (Photo by George Mulala/Reuters NewMedia Inc./Corbis. Reproduced by permission.)

At the beginning of the twenty-first century, one family was recognized: Latimeriidae. It contains one genus (*Latimeria*) and two species: *Latimeria chalumnae* and *Latimeria menadoensis*.

Physical characteristics

The living coelacanth is often referred to as a "living fossil." A representative of an ancient group whose other members have all gone extinct, it has survived for millions of years with a virtually unchanged body form. Studies of the living coelacanth's soft anatomy and body fluids have shown various similarities with chondrichthyans. These characteristics are thus considered "primitive vertebrate features," but the coelacanth has also developed many specialized characteristics.

Coelacanths have several unique characteristics, the most obvious of which are the fleshy (lobate or pedunculate) fins. While these fins have some similarity with the lobate fins in fossil lungfishes, rhipidistians, and some polypterids, no other fish group has developed seven fleshy fins. The paired fins are supported by girdles that resemble the purported precursors of the pectoral and pelvic girdles of tetrapods. The axial skeleton of coelacanths evolved differently from that of other vertebrates, even those with a persistent notochord. Instead of developing vertebrae, the notochord of the living coelacanth developed into a tube, over 1.57 in (4 cm) in diameter in adults, which is stiffened by fluid under pressure.

The skull of coelacanths has an intracranial joint that divides the neurocranium into an anterior and a posterior half and that allows the mouth to open not only by lowering the lower jaw but also by raising the upper jaw. This increases the gape considerably, and by extending the buccal cavity creates a strong suction. No other living animal has this feature.

Adult coelacanths have a minute brain (occupying only 1.5% of the cranial cavity) in common with many deep-sea sharks and the sixgill stingray (*Hexatrygon bickelli*). The pineal complex, which is involved with photoreception in many vertebrates, is relatively primitive and undifferentiated in *Latimeria*, whereas the basilar papilla in the inner ear has some similarity to that of tetrapods. The electrosensory systems in the head and the gular plates, in addition to the rostral organs, might be useful for locating prey.

The coelacanth has a spiral valve with unique, extremely elongate, nearly parallel spiral cones in its intestine. The valvular intestine is a shared character with ancestral jawed fishes (Gnathostomata), progressively reduced in actinopterygians and replaced by an elongated intestine in teleosts and tetrapods. The heart is elongate but is not simple; it is as complex as in other fishes, and far removed from the superficial earlier interpretation as an S-shaped embryonic tube. Bogart, Balon, and Bruton reported in 1994 that a *Latimeria chalumnae* specimen caught in April 1991 at Hahaya, Grand Comoro, has a 48-chromosome karyotype. This karyotype is unlike those found in lungfishes but is very similar to the 46-chromosome karyotype of one of the ancient frogs, *Ascaphus truei.*

The complex dermal canals known only from fossil jawless and jawed fishes are combined in *L. chalumnae* with the common pit lines of superficial neuromasts (lateral line) of extant fishes. Therefore, retention and specialization of ancestral structures, no longer present in other living fishes, is one of the most significant attributes represented, along with their evolutionary persistence, in this true "living fossil."

The coloration of *Latimeria chalumnae* is bluish grey with large whitish marks scattered on the body, head, and the fleshy fin bases. The white markings are specific to each individual, so each single fish can be distinguished. The white marks simulate white sessile tunicates on the walls of caves where coelacanths aggregate and on the substrate over which they drift, so that the animals blend perfectly with their background. On a dying coelacanth the bluish hue turns brown, the color of all dead specimens. The Indonesian coelacanth was brown when still alive with much golden glitter in the whitish markings.

Females grow to 74.8 in (190 cm), males to 59.1 in (150 cm) and 110–198 lb (50–90 kg). Individuals are 13.8–15 in (35–38 cm) long at birth.

Distribution

Since 1998 the coelacanth distribution is known to be not only in the west Indian Ocean, but also 6,214 mi (10,000 km) east on the other side of the Indian Ocean. The specimen that was caught off the Chalumna River in 1938 was later thought to be a stray from the Comoran population around Grand Comoro and Anjouan. Captures near Malindi (Kenya) and at Sodwana Bay near St. Lucia estuary in South Africa extend the range of intermittent distribution along the East African coast. It has not yet been established whether these are discrete populations. Only the Mozambique specimen and the southwest Madagascar specimens were proven to be of Comoran origin. According to Victor Springer, *Latimeria menadoensis* in Indonesia is most likely isolated from the western population(s) by unsuitable habitats in the central Indian Ocean.

Habitat

The extant coelacanths are tropical marine fishes inhabiting inshore water below 328 ft (100 m) depth. They seem to prefer steep sloping areas with little coral sand deposits. The hemoglobin of *Latimeria chalumnae* has the best affinity to oxygen at 61–64.4°F (16–18°C). This temperature coincides with the isobaths of 328–984 ft (100–300 m) in most localities inhabited by coelacanths. As there seems to be very little prey at these depths, the coelacanth is forced to ascend at night to more shallow waters in order to feed, risking some respiratory discomfort. For the daytime, coelacanths descend back into more comfortable temperatures and hide in groups under overhangs and in caves. A sluggish locomotion and drifting instead of fast active swimming probably help to save energy. If this is the case, then a fish hauled to the surface often with a water temperature far above 68°F (20°C) is under

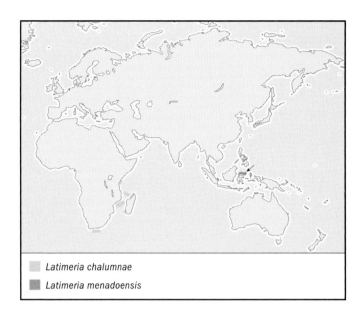

Latimeria chalumnae
Latimeria menadoensis

such respiratory stress that its survival is uncertain even if it is released back into cooler waters.

At Grand Comoro most coelacanth catches have occurred over the newest lava flows of the periodically erupting volcano Kartala. These lava fields under water consist of more cavities where prey can hide, and more caves for daytime aggregations of coelacanths than other less steeply sloping shores.

Behavior

Coelacanths aggregate in caves and overhangs about 328–656 ft (100–200 m) deep during the daytime. At Grand Comoro 19 adults were counted in one cave close together, gently moving their paired fins but never touching each other. Individuals distinguished by their specific white markings were found faithful to a particular cave for many months, although every day some strayed into other caves. At night the fish drifted individually close to the substrate.

After the first observations in 1987 from the submersible *GEO*, Hans Fricke noted that at night, all individuals took advantage of up- or down-wellings and drifted slowly with the current. The paired fins stabilized the drifting fish so that "all individuals seemed perfectly oriented in that they avoided obstacles in their environment, apparently detecting them well in advance." Fricke further commented that "all individuals irregularly performed a curious headstand, lasting up to 2 minutes."

During swimming, the coelacanth very slowly moves its paired fleshy pectoral and pelvic fins alternatingly in the manner of a trotting horse (left pectoral and right pelvic simultaneously and then right pectoral and left pelvic together). This pattern is also common to lungfish and a few other bottom-dwelling fishes and, of course, most tetrapods. The unpaired fleshy second dorsal and anal fins are sculled in synchrony from side to side and are the main organs for forward propulsion. This explains their similar shape and exact juxtaposition. The nonfleshy first dorsal fin is usually folded and flush with the dorsal surface in undisturbed fish, but when spread it appears to be used as a sail and/or for lateral display when the fish feels threatened. The large caudal fin (in fact the third dorsal, epicaudal, and second anal) is held rigid during drift-swimming as in weakly electric fishes (to enable interpretation of the electric field distortions), but provides powerful propulsion during rapid forward bursts. The small epicaudal lobe is bent to and fro when the coelacanth is swimming, drifting, or standing on its head, and may be implicated in electro-reception together with the rostral organ and the reticulate organs. The *GEO* team was able to induce headstands in the coelacanth by emitting weak electric currents from electrodes held in the submersible's remotely controlled arm.

Fin coordination probably developed to stabilize the bulky body of the coelacanth, but could in its extinct ancestors have facilitated the eventual transition to locomotion on land. When coelacanths have been observed coming in contact with the substrate, the paired, fleshy fins were not used for locomotion. The coelacanth probably never walks.

Feeding ecology and diet

Latimeria chalumnae is an opportunistic nocturnal bottom drift feeder. Prey items identified in several studies are benthic or epibenthic dwellers like some lanternfishes, deepwater cardinal fishes, cuttlefishes, snappers, cephalopods, and even a swell shark. Most of these are known to hide in bottom cavities.

The coelacanth prefers fresh lava rocks with cavities not filled by coral sand. Being a sluggish swimmer with a low metabolic rate, it regularly performs intermittent head stands during its nocturnal drifts along the bottom. *Latimeria* is able not only to move its lower jaw but, thanks to the intracranial joint, to lift its upper jaw. This feature, unique among extant vertebrates, allows for a considerable increase in the oral gape. In addition to the rostral organ the fish has a distinct reticular system in the gular bones under the head that probably also functions as an electro-sensory system.

The cranial morphology of *Latimeria chalumnae* suggests that it is a gape-and-suck predator whose anatomical specializations appear to permit it to extract prey from the crannies and cavities where it takes shelter.

Reproductive biology

Until 1975 *Latimeria chalumnae* had been considered to be an egg laying (oviparous) species because a 64.2 in (163 cm) long female caught at Anjouan in 1972 was found containing 19 eggs the size and color of oranges. But then another female, 63 in (160 cm) long, previously caught at Anjouan in 1962, preserved, and kept as an exhibit at the American Museum of Natural History (AMNH), was dissected in 1975. The curators of this museum were persuaded to cut open the specimen in order to sample needed tissue of some inner organs. The curators discovered in the female's oviduct five well-developed embryos, with a length of 11.8–13 in (30–33 cm), each with a large yolk sac. This finding meant that the living coelacanth is a livebearer (viviparous).

Later, John Wourms and Jim Atz studied these embryos and their mother's oviduct in detail and found that the heavily vascularized yolk sac surfaces were in very close contact with the equally densely vascularized oviduct walls, thus forming a simple placentalike structure. It seems, therefore, that in addition to the yolk, the embryos have a second, more direct, way to receive nutrients from the mother. A third way of obtaining nutrients suggested itself when more females were dissected. One 66-in (168-cm) long female contained 59 eggs the size of chicken eggs, another female had 65 eggs, and yet others had 62, 56, and 66 eggs. All these females produced more eggs than their oviduct would be able to accommodate as embryos. While the five embryos from the AMNH female still had large yolk sacs, the 26 fetuses from the female caught near Mozambique were close to term and had only a scar on the belly where the yolk sac once was. Both groups of embryos/fetuses had well-developed alimentary tracts and dentitions. It is thus possible that additional nutrient delivery occurred through the debris from the supernumerary eggs. After all, it is known that in some shark species the fetuses feed on eggs and siblings, so that at the end only one large predator is born. It is possible that such oophagy, as it is called, also occurs in *Latimeria*.

Further studies of these unborn fetuses revealed exceptionally wide gill-cover membranes full of cells capable of absorbing uterine milk (histotrophes) secreted by the oviduct walls. This type of nutrient transfer is known in some fishes. Finally, the carotenoid pigments in the yolk are also implicated in oxygen delivery. While more investigation is needed, it is clear that the coelacanth is a fish with a very advanced and complex style of reproduction. This is not surprising, given that the Jurassic coelacanth *Holophagus gulo* was probably a live-bearer, and the Carboniferous *Rhabdoderma exiguum*, although still oviparous, had eggs of relatively large size.

Circumstantial evidence suggests that the gestation time of the living coelacanth is very long (about 13 months), that the females become mature for the first time when older than 20 years (as in some sturgeons), and that a female does not deliver young every year but several years apart. Scientists do not know how the internal fertilization of a female is achieved and where the young live right after birth and in subsequent years. No young were noticed from the submersibles either drifting or in caves, and only one or two have been collected free swimming.

Conservation status

After the catch of the "second living coelacanth" became known to science in 1952, the Comoro Archipelago (then a French colony) was recognized as the "home" of the coelacanth. Soon, national ownership was declared for all subsequent specimens, and the second one declared stolen from its "rightful" owners. Only the French were allowed to collect them. J. Millot, a spider specialist on Madagascar, moved to Paris and with J. Anthony started detailed anatomical descriptions of the coelacanth. After close to 80 specimens were accumulated, some were used as diplomatic gifts. Eventually, other nationals joined the frenzy of working on the prestigious animal. Several special expeditions converged onto the

Comoros, but luckily the beast could not be caught at will by the methods employed.

The Japanese imported larger fiberglass boats called *japavas* to supplement the small single log dugout outriggers (*galavas*), and built a fishing school on Anjouan in the early 1980s. At the same time, rumor was started that the fluid from the "notochord" tube when ingested prolonged life. Soon, in addition to the demands for coelacanths for museum exhibitions, a black market for fresh or frozen specimens for "medicinal" purposes was started. The price soared to $3,000 or more per fish, especially during the rule of the white mercenary who called himself Colonel Baku.

The first deployment of Fricke's submersible *GEO* and the first sighting of coelacanths in their natural habitat coincided with the urgent need for conservation. The Coelacanth Conservation Council (CCC) was established by Eugene Balon, Mike Bruton, Christine Flegler-Balon, Hans Fricke, and Rafael Plante when they met in Moroni (Grand Comoro), the capital of the Federal Islamic Republic of the Comoros, in 1987. The CCC was inspired by the Desert Fishes Council that led to the most progressive conservation law in the world, the Endangered Species Act of the United States. Within the

next two years, members of the CCC managed to have the coelacanth included in Appendix I of CITES.

Subsequent dives by Fricke and his team with the new submersible *JAGO* at Grand Comoro revealed a serious decline in the number of coelacanths in each previously surveyed cave. Thus the initial estimates of the numbers of adults (200–500) became potentially invalid. In spite of the discovery of additional individuals off Sulawesi in 1998 and lately at Sodwana Bay (South Africa), the living coelacanth remains unique and highly vulnerable because of its narrow habitat range and very specialized physiology and life style. Although the 2002 IUCN Red List does not list *Latimeria menadoensis*, it lists *L. chalumnae* as Critically Endangered.

Significance to humans

Before the coelacanth's value for science was recognized in the mid-twentieth century, it was occasionally consumed for its presumed antimalarial properties. Because of its high oil content, the meat tastes foul and rancid and causes severe diarrhea when eaten. Since 1952 its interest to science has remained extremely high.

1. Coelacanth (*Latimeria chalumnae*); 2. Indonesian coelacanth (*Latimeria menadoensis*). (Illustration by Brian Cressman)

Resources

Books

Forey, P. *History of the Coelacanth Fishes.* London: Chapman & Hall, 1998.

Musick, J. A., M. N. Bruton, and E. K. Balon, eds. *The Biology of* Latimeria chalumnae *and Evolution of Coelacanths.* Dordrecht: Kluwer Academic Publishers, 1991.

Smith, J. L. B. *Old Fourlegs: The Story of the Coelacanth.* London: Readers Union, Longmans, Green, 1957.

Thomson, K. S. *Living Fossil: The Story of the Coelacanth.* New York: W.W. Norton & Company, 1991.

Walker, S. M. *Fossil Fish Found Alive: Discovering the Coelacanth.* Minneapolis: Carolrhoda Books, Inc., 2002.

Weinberg, S. A. *Last of the Pirates: The Search for Bob Denard.* New York: Pantheon Books, 1994.

———. *Fish Caught in Time: The Search for the Coelacanth.* London: Fourth Estate, 1999.

Periodicals

Anthony, J., and J. Millot. "Première capture d'une femelle de coelacanthe en estat de maturité sexuelle." *C.R. Acad. Sc. Paris* Sér. D224 (1972): 1925–1927.

Balon, E. K. "The Living Coelacanth Endangered: A Personalized Tale." *Tropical Fish Hobbyist* 38 (February 1990): 117–129.

———. "Prelude: The Mystery of a Persistent Life Form." *Environmental Biology of Fishes* 32 (1991): 9–13.

———. "Probable Evolution of the Coelacanth's Reproductive Style: Lecithotrophy and Orally Feeding Embryos in Cichlid Fishes and in *Latimeria chalumnae.*" *Environmental Biology of Fishes* 32 (1991): 249–265.

———. "Dynamics of Biodiversity and Mechanisms of Change: A Plea for Balanced Attention to Form Creation and Extinction." *Biological Conservation* 66 (1993): 5–16.

———. "See Also Other Recent Websites on the Coelacanth." *Environmental Biological of Fishes* 54 (1999): 466.

Balon, E. L., M. N. Bruton, and H. Fricke. "A Fiftieth Anniversary Reflection on the Living Coelacanth, *Latimeria chalumnae*: Some New Interpretations of Its Natural History and Conservation Status." *Environmental Biology of Fishes* 23 (1988): 241–280.

Bogart, J. P., E. K. Balon, and M. N. Bruton. "The Chromosomes of the Living Coelacanth and Their Remarkable Similarity to Those of One of the Most Ancient Frogs." *Journal of Heredity* 85 (1994): 322–325.

Bruton, M. N., A. J. P. Cabral, and H. W. Fricke. "First Capture of a Coelacanth, *Latimeria chalumnae* (Pisces, Latimeriidae), Off Mozambique." *South African Journal of Science* 88 (1992): 225–227.

Erdmann, M. V. "An Account of the First Living Coelacanth Known to Scientists from Indonesian Waters." *Environmental Biology of Fishes* 54 (1999): 439–443.

Erdmann, M. V., and R. L. Caldwell. "How New Technology Put a Coelacanth Among the Heirs of Piltdown Man." *Nature* 406 (2000): 343.

Erdmann, M. V., R. L. Caldwell, S. L. Jewett, and A. Tjakrawidjaja. "The Second Recorded Living Coelacanth from North Sulawesi." *Environmental Biology of Fishes* 54 (1999): 445–451.

Erdmann, M. V., R. L. Caldwell, and M. Kasim Moosa. "Indonesian 'King of the Sea' Discovered." *Nature* 395 (1998): 335.

Fricke, H. W., and J. Frahm. "Evidence for Lecithotrophic Viviparity in the Living Coelacanth." *Naturwissenschaften* 79 (1992): 476–479.

Fricke, H. W., and K. Hissman. "Natural Habitat of Coelacanths." *Nature* 346 (1990): 323–324.

———. "Locomotion, Fin Coordination and Body of the Living Coelacanth *Latimeria chalumnae.*" *Environmental Biology of Fishes* 34 (1992): 329–356.

———. "Home Range and Migrations of the Living Coelacanth *Latimeria chalumnae.*" *Marine Biology* 120 (1994): 171–180.

Fricke, H. W., K. Hissman, J. Schauer, O. Reinicke, and R. Plante. "Habitat and Population Size of the Coelacanth *Latimeria chalumnae* at Grande Comore." *Environmental Biology of Fishes* 32 (1991): 287–300.

Fricke, H. W., and R. Plante. "Habitat Requirements of the Living Coelacanth *Latimeria chalumnae* at Grande Comore, Indian Ocean." *Naturwissenschaften* 75 (1988): 149–151.

Fricke, H. W., O. Reinicke, H. Hofer, and W. Nachtigall. "Locomotion of the Coelacanth *Latimeria chalumnae* in Its Natural Environment." *Nature* 329 (1987): 331–333.

Fricke, H. W., J. Schauer, K. Hissmann, L. Kasang, and R. Plante. "Coelacanths Aggregate in Caves: First Observations on Their Resting Habitat and Social Behavior." *Environmental Biology of Fishes* 30 (1991): 282–285.

Gorr, T., T. Kleinschmidt, and H. W. Fricke. "Close Tetrapod Relationships of the Coelacanth *Latimeria* Initiated by Hemoglobin Sequences." *Nature* 351 (1991): 394–397.

Hensel, K., and E. K. Balon. "The Sensory Canal System of the Living Coelacanth, *Latimeria chalumnae*: A New Installment." *Environmental Biology of Fishes* 61 (2001): 117–124.

Hissmann, K., and H. W. Fricke. "Movements of the Epicaudal Fin in Coelacanths." *Copeia* 1996: 605–615.

Hissmann, K., H. W. Fricke, and J. Schauer. "Population Monitoring of a Living Fossil: The Coelacanth *Latimeria chalumnae* in Decline?" *Conservation Biology* 12 (1998): 759–765.

Holder, M. T., M. V. Erdmann, T. P. Wilcox, R. L. Caldwell, and D. M. Hillis. "Two Living Species of Coelacanths?" *Proceedings of the National Academy of Sciences U.S.A.* 96 (1999): 12616–12620.

McCabe, H. "Recriminations and Confusion over the 'Fake' Coelacanth Photo." *Nature* 406 (2000): 225.

McCabe, H., and J. Wright. "Tangled Tale of a Lost, Stolen and Disputed Coelacanth." *Nature* 406 (2000): 114.

Resources

Pouyaud, L., S. Wirjoatmodjo, I. Rachmatika, A. Tjakrawidjaja, R. Hadiaty, and W. Hadie. "Une nouvelle espèce de coelacanthe. Preuves génétiques et morphologiques." *C.R. Acad. Sci. Paris, Sciences de la vie* 322 (1999): 261–267.

Springer, V. G. "Are the Indonesian and Western Indian Ocean Coelacanths Conspecific: A Prediction." *Environmental Biology of Fishes* 54 (1999): 453–456.

Suzuki, N., Y. Suyehiro, and T. Hamada. "Initial Report of Expeditions for Coelacanth, Part I, Field Studies in 1981 and 1983." *Scientific Papers of the College of Arts and Sciences, Univ. Tokyo* 35 (1985): 37–79.

Wourms, J. P., J. W. Atz, and M. D. Stribling. "Viviparity and the Maternal-embryonic Relationship in the Coelacanth *Latimeria chalumnae.*" *Environmental Biology of Fishes* 32 (1991): 225–248.

Organizations

South African Coelacanth Conservation and Genome Resource Programme. South African Institute for Aquatic Biodiversity, Somerset Street, Private Bag 1015, Grahamstown, 6140 South Africa. Phone: +27 (0)46 636 1002. Fax: +27 (0)46 622 2403. Web site: <http://www.saiab.ru.ac.za/coelacanth>

Other

"Coelacanth Conservation Council/Conseil pour la Conservation du Coelacanthe Newsletter, No. 1." *Environmental Biology of Fishes* 23 (1988): 315–319

"Coelacanth Conservation Council/Conseil pour la Conservation du Coelacanthe Newsletter, No. 2." *Environmental Biology of Fishes* 30 (1991): 423–428.

"Coelacanth Conservation Council/Conseil pour la Conservation du Coelacanthe Newsletter, No. 3." *Environmental Biology of Fishes* 33 (1992): 413–417.

"Coelacanth Conservation Council/Conseil pour la Conservation du Coelacanthe Newsletter, No. 4." *Environmental Biology of Fishes* 36 (1993): 395–406.

"Coelacanth Conservation Council/Conseil pour la Conservation du Coelacanthe Newsletter, No. 5." *Environmental Biology of Fishes* 38 (1993): 399–410.

"Coelacanth Conservation Council/Conseil pour la Conservation du Coelacanthe Newsletter, No. 6." *Environmental Biology of Fishes* 54 (1999): 457–469.

Eugene K. Balon, PhD

Ceratodontiformes

(Australian lungfish)

Class Sarcopterygii

Order Ceratodontiformes

Number of families 1

Photo: An Australian lungfish (*Neoceratodus forsteri*) near Queensland. (Photo by Tom McHugh /Photo Researchers, Inc. Reproduced by permission.)

Evolution and systematics

The Australian lungfish is one of the most ancient living species of fishes (indeed, of vertebrates), as fossil remains belonging to *Neoceratodus forsteri* from the early Cretaceous of northern New South Wales are known, giving it a time span of close to 100 million years. The family Ceratodontidae, to which the Australian lungfish belongs, contains other fossil species (known mostly from toothplates), which date from the early Triassic. Many of these fossils, such as those of the genus *Ceratodus*, were more widespread than *Neoceratodus* and occurred circumglobally. *Neoceratodus* is therefore a survivor of a much more successful and ancient lineage. Current opinion diverges in relation to its ancestry, whether it is more closely related to extinct lungfishes of the family Ceratodontidae (the most likely scenario) or to the other living lungfishes of South America (*Lepidosiren*) and Africa (*Protopterus*), in which case *Neoceratodus* would be placed in its own family, Neoceratodontidae.

When first discovered and described by Johann L. G. Krefft in 1870 (as *Ceratodus forsteri*), the Australian lungfish was thought to be an ("gigantic") amphibian, similar to the circumstances surrounding the description of the South American lungfish 33 years before, even though many zoologists regarded *Lepidosiren* as a true "fish" by the mid-nineteenth century. *Neoceratodus* eventually found its place among the Dipnoi, the group containing all lungfishes, both fossil and living (established previously in 1844 by the German zoologist Johannes Müller). The Dipnoi presently contains some 280 fossil species and 60 fossil genera, originating in the early Devonian, in addition to the six species and three genera that are living today. Many of these fossils are known from well-preserved skeletons (some preserved three-dimensionally, such as *Griphognathus* and *Chirodipterus* from Gogo, Australia), but at least half of the species are known only from isolated toothplates. Lungfishes achieved their greatest diversity in the Devonian, when most (if not all) lungfish taxa were marine; living lungfishes are restricted to freshwater.

The systematic position of the Dipnoi among the vertebrates is still being debated. In a landmark study in 1981, Donn E. Rosen and collaborators placed the lungfishes as the closest relatives of the tetrapods (land vertebrates), challenging the widely held belief that certain "rhipidistians" (a heterogeneous group of fossil lobe-finned fishes) were their nearest ancestors. Current views on the ancestry of the tetrapods, based on morphological studies, indicate that lungfishes are their closest relatives if only living taxa are taken into account, but that other extinct groups of lobe-finned, fish-like vertebrates (rhizodonts, osteolepids, *Eusthenopteron*, and "elpistostegids" such as *Panderichthys*) are actually more closely related to tetrapods when all fossil evidence is considered; molecular data bearing on this issue are still controversial. The exclusively Devonian group Porolepiformes, which had pectoral fins anatomically similar to *Neoceratodus*; e.g. *Holoptychius*, is considered by many researchers to be the closest relative of the Dipnoi.

Physical characteristics

Neoceratodus forsteri is morphologically unique, presenting paddle-like or leaf-like pectoral fins that are fleshy and stout at their bases; the pelvic fins are similar, but smaller and not as fleshy as the pectorals. Their heads are wide and slightly depressed, with a terminal mouth. The nostrils are internal, composed of two small openings inside the labial cavity, which are followed by a pair of posterior openings in the roof of the mouth (choanae). The trunk is long and muscular and is laterally compressed, with a protocercal caudal fin that is posteriorly pointed and continuous with both the dorsal and anal fins. Their scales are remarkably large (but rather thin),

Neoceratodus forsteri

overlapping and posteriorly rounded. The teeth are fused into toothplates, two pairs of which are positioned in the roof of the mouth, and one pair on either side of the tongue on the mouth floor; the posterior toothplates are ridged, and in juveniles they are trilobed.

The single lung (a modified swim bladder) is large; highly vascularized and internally divided into two chambers; and connects with the esophagus ventrolaterally. The gills open into a large opercular chamber just ahead of the pectoral fins. Sensory canals of the lateral-line system are visible dorsally on the head and nape (but not as much as in the South American and African lungfishes), and the lateral line runs posteriorly at midheight to the tip of the tail. Large sensory pores are present on the snout and around the eyes. Their skeletons are mostly cartilaginous, in contrast to fossil lungfishes, which were more heavily ossified. Dorsally and laterally, the Australian lungfish is olive to dark greenish brown in color, but ventrally creamy-yellow or even pinkish. Many specimens also have darker blotches dorsolaterally on the tail, especially juveniles.

Distribution

Very restricted in distribution; present only in Queensland, Australia. When the continent was first colonized, this species was already restricted to the Mary and Burnett River systems, but it was subsequently introduced, successfully, into other rivers of southeastern Queensland, such as the Albert, Brisbane, Coomer, Fitzroy and Stanley Rivers, and also in the Enoggera Reservoir, where it is reported to be abundant.

Habitat

Often found in deep pools in still, slow moving rivers. Its ability to absorb oxygen periodically directly from the air enables it to live in rather stagnant waters. The rivers inhabited by *Neoceratodus* are typically calm and slow moving, with mud, sand, or gravel bottoms, and with plenty of marginal and aquatic vegetation, important for spawning.

Behavior

Mostly sluggish, but capable of quick bursts of speed in pursuit of prey or when threatened. Vision is reportedly poor, as captive specimens have been known to swim into obstacles, but they are known to hunt prey items mostly at night, using electroreception and a refined sense of smell. The single lung allows *Neoceratodus* to breathe air occasionally, but it breathes primarily through its gills and only ascends to the surface to gulp air when water conditions are poor or when the gills are clogged with mud or other debris. In their natural habitat, individuals have been observed to swallow air at intervals of 30 to 60 minutes, emitting a particular sound when air is exhaled. Juveniles and especially hatchlings are also capable of absorbing oxygen through the skin. *Neoceratodus* does not estivate (bury itself in a muddy burrow to wait for the rainy season), as do the South American and African lungfishes, and consequently it cannot remain alive out of water for periods greater than a few days, even if kept wet and in the shade.

Feeding ecology and diet

Essentially carnivorous, eating frogs, other fishes, and invertebrates such as insect larvae, earthworms, snails, and freshwater crustaceans. However, it has also been reported to eat both aquatic and terrestrial plants and even native fruits that have fallen into rivers. Prey items are captured through suction and crushed by the toothplates. The pectoral fins allow them to brace themselves when foraging for prey. Larvae and juveniles of *Neoceratodus* are preyed upon by insect larvae, fishes, and fish-eating birds.

An Australian lungfish (*Neoceratodus forsteri*). (Illustration by Brian Cressman)

The Australian lungfish (*Neoceratodus forsteri*) is also known as the Queensland lungfish, and has changed very little over the past 110 million years. (Photo by Animals Animals ©Fritz Prenzel. Reproduceed by permission.)

Reproductive biology

Reproduction occurs in shallow, warm waters before the summer rainy season. Complex courtship behaviors have been recorded; the pair is not easily distracted. A male and female remain in close association, as the male nudges the cloacal area of the female to stimulate her. Fertilization is external. Spawning may take up to one hour after the pair has chosen a site, usually a patch of aquatic plants. Fifty to 100 sticky eggs are laid on plants, to which they adhere. There is no guarding of the eggs or the young. The eggs are small, spherical, and enveloped by a gelatinous substance. Larvae emerge from the eggs after a period of some three weeks but remain close to them for shelter for some 10 days following. After 41 to 56 days, the yolk disappears and the larvae begin to feed, probably on insect larvae or other small invertebrates. The hatchlings do not have external gills but are capable of breathing air at a very small size, 0.98 in (2.5 cm). A size of 9.8 in (25 cm) is attained after six months, and 19.7 in (50 cm) after 20 months. They resemble adults after approximately six months.

Conservation status

This species is fully protected under CITES (Appendix 2) legislation and cannot be collected without special permit; it is not listed by the IUCN.

Significance to humans

Reported to adapt well to captivity, *Neoceratodus* is common in public aquaria. One specimen lived for more than 50 years in captivity at the Shedd Aquarium in Chicago. It is not consumed. In a more anthropocentric vein, *Neoceratodus*, along with the other living lungfishes and the living coelacanths, *Latimeria chalumnae* and *L. menadoensis*, are of special interest, and are given high profiles in museum exhibits and evolutionary biology textbooks because of their close ancestral ties to land vertebrates, including humans. They are more closely related to tetrapods than they are to the remaining fishes.

Resources

Books

Allen. G. R. *Freshwater Fishes of Australia.* Neptune City, NJ: T. H. F. Publications, 1989.

Bemis, William E., Warren W. Burggren, and Norman E. Kemp. *The Biology and Evolution of Lungfishes.* New York: A. R. Liss, 1987.

Berra, Tim M. *Freshwater Fish Distribution.* San Diego, CA: Academic Press, 2001.

Bruton, M. N. "Lungfishes and Coelacanth." In *Encyclopedia of Fishes,* edited by John R. Paxton and William N. Eschmeyer. San Diego, CA: Academic Press, 1994.

Cloutier, R., and P. E. Ahlberg. "Morphology, Characters, and the Interrelationships of Basal Sarcopterygians." In *Interrelationships of Fishes,* edited by Melanie L. J. Stiassny, Lynne Parenti, and G. David Johnson. San Diego, CA: Academic Press, 1996.

Conant, E. B. "Bibliography of Lungfishes, 1811–1985." In *The Biology and Evolution of Lungfishes,* edited by William E. Bemis, Warren W. Burggren, and Norman E. Kemp. New York: A. R. Liss, 1987.

Graham, Jeffrey B. *Air-breathing Fishes: Evolution, Diversity, and Adaptation.* San Diego, CA: Academic Press, 1997.

Janvier, Philippe. *Early Vertebrates.* New York: Oxford University Press, 1996.

Kemp, Norman E. "The Biology of the Australian Lungfish, *Neoceratodus forsteri* (Krefft, 1870)." In *The Biology and Evolution of Lungfishes,* edited by William E. Bemis, Warren W. Burggren, and Norman E. Kemp. New York: A. R. Liss, 1987.

Merrick, J. R., and G. E. Schmida. *Australian Freshwater Fishes: Biology and Management.* North Ryde, N.S.W., Australia: J. R. Merrick, 1984.

Nelson, Joseph S. *Fishes of the World.* 3rd ed. New York: John Wiley & Sons, 1994.

Periodicals

Bartsch, P. "Development of the Cranium of *Neoceratodus forsteri,* with a Discussion of the Suspensorium and the Opercular Apparatus in Dipnoi." *Zoomorphology* 114 (1994): 1–31.

Bemis, William E. "Paedomorphosis and the Evolution of the Dipnoi." *Paleobiology* 10, no. 3 (1984): 293–307.

Kemp, Norman E. "The Embryological Development of the Queensland Lungfish, *Neoceratodus forsteri* (Krefft)." *Memoirs of the Queensland Museum* 20, no.3 (1982): 553–597.

Kemp, Norman E., and R. E. Molnar. "*Neoceratodus forsteri* from the Lower Cretaceous of New South Wales, Australia." *Journal of Paleontology* 55, no. 1 (1981): 211–217.

Miles, R. S. "Dipnoan (Lungfish) Skulls and the Relationships of the Group: A Study Based on New Specimens from the Devonian of Australia." *Zoological Journal of the Linnaean Society* 61 (1977): 1–328.

Rosen, D. E., et al. "Lungfishes, Tetrapods, Paleontology, and Plesiomorphy." *Bulletin of the American Museum of Natural History* 167, no. 4 (1981): 163–275.

Other

"Sarcopterygii: Dipnomorpha." *Palaeos: The Trace of Life on Earth*. October 6, 2002 (cited January 19, 2003). <http://www.palaeos.com/Vertebrates/Units/Unit140/200.html>

Watt, Michael, Christopher S. Evans, and Jean M. P. Joss. *Use of Electroreception During Foraging by the Australian Lungfish*. October 6, 2002 (cited January 19, 2003). <http://galliform .bhs.mq.edu.au/Watt_et_al.html>

Marcelo Carvalho, PhD

Lepidosireniformes
(Lungfishes)

Class Sarcopterygii
Order Lepidosireniformes
Number of families 2

Photo: The South American lungfish (*Lepidosiren paradoxa*) resting in the waters of central South America. (Photo by Tom McHugh/Photo Researchers, Inc. Reproduced by permission.)

Evolution and systematics

The South American lungfish (*Lepidosiren paradoxa*) was the first lungfish species to be described (in 1837 by Leopold J. F. J. Fitzinger; 1802–1884). It was originally discovered and collected from the Amazon River by Johann Natterer (1787–1843), a dedicated and gifted Austrian naturalist who collected extensively in Brazil from 1817–1835. The first African lungfish was described shortly thereafter in 1839 by the British anatomist Richard Owen (1810–1890), who firmly believed that lungfishes were true "fishes" and not amphibians, mostly on the basis of his erroneous conviction that lungfishes did not have a second (internal) nostril (known as a choana) or a divided auricle. Zoologists of the mid-nineteenth century were divided about the ancestry of lungfishes, wondering whether they were more closely related to amphibians or to other bony fishes. It is now well established that lungfishes belong to a higher group, the Sarcopterygii, which also includes the tetrapods, coelacanths, and many other lobe-finned fossil fishes, and are therefore unequivocally more closely related to land vertebrates than to ray-finned bony fishes (Actinopterygii).

The genus *Lepidosiren* is monotypic, but detailed comparative studies of specimens from most of its extended range are needed (two other nominal species exist, but they are considered synonyms of *L. paradoxa*). Four species of African lungfishes (*Protopterus*) are recognized: *P. annectens, P. aethiopicus, P. dolloi,* and *P. amphibius.* Some of these species are difficult to distinguish, and are in need of critical taxonomic evaluation, including the validity of certain subspecies (a total of 10 nominal species have been described); their evolutionary relationships to each other have yet to be fully investigated. *Lepidosiren* and *Protopterus* are placed in the same order, but are classified in distinct families. They are closely related to the Australian lungfish (*Neoceratodus forsteri*) and its immediate fossil relatives, while the vast majority of remaining fossil lungfishes are more distantly

related (some 60 genera and 280 species and of lungfishes are known). Extant lungfishes are "living fossils," and belong to an ancient lineage, the Dipnoi, that was much more diverse in the Devonian (ca. 417–354 million years ago) and Triassic (ca. 248–205 million years ago) periods. Fossil relatives of *Lepidosiren* and *Protopterus* are known from the late Cretaceous of South America and Africa, respectively, and these genera, along with *Neoceratodus*, are among the oldest vertebrates living today.

Physical characteristics

South American and African lungfishes are morphologically similar, presenting elongated, eel-like bodies, with relatively small heads, and filamentous pectoral and pelvic fins. The pelvic fins are stouter than the pectorals in both genera; the pectorals are slightly more robust in *Protopterus*, and may resemble simple filaments in *Lepidosiren*. The caudal fin is confluent with the dorsal and anal fins (as in the Australian lungfish), tapering distally. The body is compressed laterally, especially at the anus, but not as much as in *Neoceratodus*. The eyes are minute and the mouth is terminal, with a lateral groove extending to the sides of the head. Sensory canals on the head and cheek appear as sinuous, deep lines that extend posteriorly at midbody height towards the tail; sensory pores are also present on the head. The anus is asymmetrical, situated laterally just posterior to the pelvic fins, and not directly in the middle as in *Neoceratodus* (the side may vary among individuals of both *Protopterus* and *Lepidosiren*). The scales are mostly embedded in the skin and are very thin, but clearly visible. The nostrils are on the internal lip margin, and the teeth are fused into sharp tooth plates.

Both *Lepidosiren* and *Protopterus* have two highly vascularized and separated lungs (modified swim bladders), positioned on each side of the gut and connected to the esophagus ven-

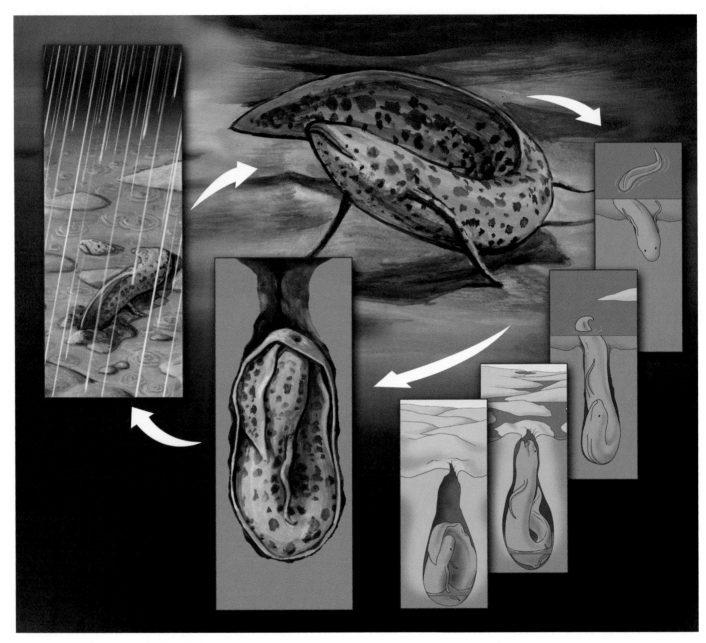

Lungfish estivation cycle. (Illustration by Brian Cressman)

trally, as in tetrapods and bichirs (Polypteriformes). The lungs have many alveoli, similar to the lungs of tetrapods. The gill openings are small, not nearly as large as in *Neoceratodus*. Newly hatched individuals have flared external gill filaments, absent in *Neoceratodus*, and these may persist vestigially above the pectoral fins in subadults and adults of *Protopterus*. The skeleton is mostly cartilaginous. Both *Protopterus* and *Lepidosiren* vary slightly in color, from dark brown to deep gray dorsally and laterally, with many varied blotches and spots; usually dark ventrally, although *Protopterus* may be lighter ventrally. *Lepidosiren* may reach 4.1 ft (1.25 m) in length, while *Protopterus* varies from between 17.7 in (45 cm) (*P. amphibius*) to 6.5 ft (2 m) in length (*P. aethiopicus*, which can weigh some 37.5 lb [17 kg]).

Distribution

Lepidosiren has the greatest distribution of any extant lungfish, occurring in many tributaries of the Amazon and Paraná-Paraguay River systems, as well as in French Guiana. Species of *Protopterus* are slightly more restricted. *P. annectens* is present in central and West Africa; *P. aethiopicus* occurs in central and East Africa; *P. dolloi* is restricted to the Congo basin; and *P. amphibius* occurs in coastal East Africa.

Habitat

The South American and African lungfishes are generally found in lentic (slow-moving) rivers, with plenty of associated vegetation and swampy, stagnant conditions (especially *L. paradoxa*). They can also be found in open lakes (e.g., *P. aethiopicus* in Lake Victoria); floodplains (e.g., *P. dolloi* in the Congo River basin, and *P. annectens* in the Senegal, Gambia, Niger, and Volta Rivers in West Africa); near river deltas (*P. aethiopicus* in Lake Tanganyika, *P. amphibius* in the Zambesi River delta); and in small pools.

Behavior

Lepidosiren and *Protopterus* species are sluggish, swimming through sinuous movements or by "crawling" on their pectoral and pelvic fins, especially to scavenge the bottom. Both genera are obligate air breathers, unlike *Neoceratodus*, which breathes primarily through the gills. *Lepidosiren* and *Protopterus* individuals will drown if forced to stay underwater, as the gill surfaces of these fishes are not large enough to satisfy their oxygen needs. Both genera also employ estivation, being capable of remaining inside a resting chamber for protracted periods during dry seasons and emerging when wet conditions return (estivation has been documented for Permian lungfishes, in the form of fossilized burrows). The degree of estivation varies among the species, but has been particularly well documented for *P. annectens*. The burrows are excavated by biting the soil and expelling mud through the gill openings. The fish will then turn around and remain with its head facing the burrow opening, from where it obtains oxygen. The individual suffers metabolic changes during this period to endure the lack of moisture (detailed below for *P. annectens*). One individual of *P. aethiopicus* remained in its cocoon for four years in captivity. Lungfishes do not feed during estivation. To sustain themselves, they initially metabolize fat reserves and then muscle mass.

Feeding ecology and diet

Lungfishes are mostly carnivorous, feeding mainly on invertebrates (insects, insect larvae, mollusks, crustaceans) but also on fishes and amphibians. Both genera may occasionally feed on aquatic plants. Lungfishes approach potential prey items through ambush or stalking, capturing them by quickly opening their mouths to create negative pressure that pulls them in. Little is known concerning their natural predators, but presumably larger carnivorous fishes and other vertebrates prey on lungfishes, especially when they are juveniles.

Reproductive biology

Spawning is usually seasonal, taking place during the wet season. Fertilization is external. In both genera the adult male guards and aerates the hatchlings and young temporarily. Female *Protopterus* usually lay eggs in burrows excavated by the males. The eggs are small (from 0.16–0.27 in/4–7 mm in diameter), and take one to two weeks to hatch, at which time they resemble tadpoles with slender, featherlike external gills. Only after a period from one month to 55 days do the larvae breathe air. At this stage, they range from 1 in (2.5 cm) to 1.6 in (4 cm) in length, and still have external gills. The larvae remain relatively inactive and are attached to the nest through their cement glands until their yolk reserves have been depleted, at which time they begin to forage for insect larvae and crustaceans and inhale air.

Conservation status

No species of Lepidosireniformes are listed by the IUCN.

Significance to humans

Lungfishes are common in both public and private aquaria. Although they are consumed as food in some parts of Africa, they are not important food fishes. They are harmless, but if provoked can inflict painful bites because of their strong jaws and sharp teeth.

1. African lungfish (*Protopterus annectens*); 2. South American lungfish (*Lepidosiren paradoxa*). (Illustration by Brian Cressman)

Species accounts

South American lungfish
Lepidosiren paradoxa

FAMILY
Lepidosirenidae

TAXONOMY
Lepidosiren paradoxa Fitzinger, 1837, Amazon River; Brazil.

OTHER COMMON NAMES
French: Anguille tété; German: Lurchfische; Portuguese: Pirambóia, peixe pulmonado.

PHYSICAL CHARACTERISTICS
To 4.1 ft (1.25 m) in length. Usually dark brown (sometimes gray) with darker and lighter spots and blotches dorsally and laterally.

DISTRIBUTION
Most of the Amazon basin, from Peru to the Amazon River delta, and in the Paraná-Paraguay Rivers basin as far south as the La Plata system. Recently reported in French Guiana, and probably occurs elsewhere in tropical South America.

HABITAT
Swamps, slow-moving waters, floodplains, and pools.

BEHAVIOR
An obligate air breather with reduced gills; can remain inactive for months during estivation, sometimes by closing the chamber opening so as to prevent further desiccation. This species is very intolerant of the close proximity of conspecifics under aquarium conditions. It also may be hyperdispersed in the wild.

FEEDING ECOLOGY AND DIET
Feeds on insects, insect larvae, other invertebrates and fishes, as well as algae; reported to masticate prey before swallowing.

REPRODUCTIVE BIOLOGY
Males present modified pelvic fins during reproduction, which develop featherlike protuberances that are highly vascularized and are believed to be accessory respiration organs, but it is not clear if they aid the adult or the larvae (or both). The male creates burrows in which the eggs are deposited and the larvae develop. Eggs are about 0.27 in (7 mm) in diameter. Hatchlings exhibit four pairs of external gills, and ventral adhesive glands anchor them in the burrow (both gills and adhesive glands are lost after six to eight weeks), after which they emerge to take their first breath of air, at about 1.6 in (4 cm) in length.

CONSERVATION STATUS
Not threatened.

SIGNIFICANCE TO HUMANS
Not consumed regularly as food. Often displayed in public aquaria, where it can live for many years. Not widely kept by amateur aquarists and does not figure prominently in the ornamental fish trade. ◆

Lepidosiren paradoxa

African lungfish
Protopterus annectens

FAMILY
Protopteridae

TAXONOMY
Lepidosiren annectens Owen, 1839, Congo River. Two subspecies sometimes recognized.

OTHER COMMON NAMES
German: Afrikanischer Lungerfische; Afrikaans: Longvis.

PHYSICAL CHARACTERISTICS
Reaches 3.3 ft (1 m) in length. Separated from other *Protopterus* species by its relatively more slender head; 40–50 scales between operculum and anus, 36–40 scales around body anterior to dorsal fin origin; and 34–37 pairs of ribs. Olive to dark brown dorsally, lighter underneath, usually with spots and blotches on dorsal and lateral aspects.

DISTRIBUTION
Numerous rivers and lakes throughout central, South, and West Africa, e.g., the Senegal, Niger, Gambia, Volta, and Chad basins; the Chari River in Western Sudan; Bandama and Camoé basins in Côte d'Ivoire; Congo basin; the Zambezi and Incomati Rivers in South Africa. Also in Sierra Leone and Guinea, and the upper Nile basin.

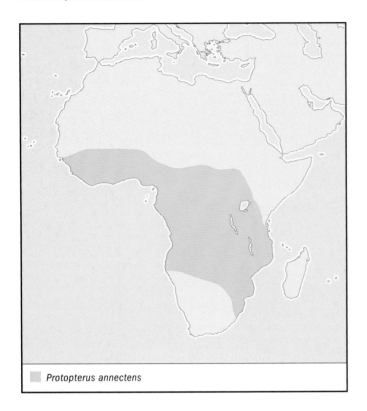

Protopterus annectens

HABITAT
Stagnant freshwater habitats, such as swamps and floodplains, also in more flowing rivers and streams.

BEHAVIOR
Estivation is well documented. Individuals will "chew out" a burrow 1.2–9.8 in (30–250 mm) deep, expelling mud through their gill openings. They eventually turn to rest facing the entrance, forming a bulb-shaped chamber that contains water in its lower portion. They periodically extend forward to breathe air from the opening, returning to rest in mucus secreted in the water-filled portion of the chamber. Their metabolic activity decreases progressively as the chamber becomes drier, and the chamber may eventually solidify into a hard cocoon. They may remain for up to seven or eight months in this resting state, until moist conditions return. This species is solitary and hyperdispersed in nature.

FEEDING ECOLOGY AND DIET
Feeds on insects, their larvae, other invertebrates, and fishes, also algae or aquatic plants. May masticate food items repeatedly, and even reported to spit out and intake prey repeatedly during feeding.

REPRODUCTIVE BIOLOGY
Males do not develop vascularized structures on their pelvic fins during breeding. Males usually guard larvae in nest sites that they dig out. Embryos hatch in one to two weeks, with conspicuous external gills, and will become obligate air breathers after about one month.

CONSERVATION STATUS
Not listed by the IUCN.

SIGNIFICANCE TO HUMANS
Kept in public aquaria, and consumed locally but not intensely. Not widely kept by amateur aquarists and does not figure prominently in the ornamental fish trade.

Resources

Books

Bemis, W. E., W. W. Burggren, and N. E. Kemp. *The Biology and Evolution of Lungfishes*. New York: A. R. Liss, 1987.

Berra, T. M. *Freshwater Fish Distribution*. San Diego: Academic Press, 2001.

Britski, H. A., K. Z. S. de Silimon, and B. S. Lopes. *Peixes do Pantanal, Manual de Identificação*. Brasília: Embrapa, 1999.

Cloutier, R, and P. E. Ahlberg. "Morphology, Characters, and the Interrelationships of Basal Sarcopterygians." In *Interrelationships of Fishes*, edited by M. L. J. Stiassny, L. Parenti, and G. D. Johnson. San Diego: Academic Press, 1996.

Conant, E. B. "Bibliography of Lungfishes, 1811–1985." In *The Biology and Evolution of Lungfishes*, edited by W. E. Bemis, W. W. Burggren, and N. E. Kemp. New York: A. R. Liss, 1987.

Gosse, J. P. "Protopteridae." In *Check-List of the Freshwater Fishes of Africa (CLOFFA)*, edited by J. Daget, J. P. Gosse, and D. F. E. Thys van den Audenaerde. Paris: ORSTOM; Tervuren: MRAC, 1984.

Graham, J. B. *Air-Breathing Fishes*. San Diego: Academic Press, 1997.

Greenwood, P. H. "The Natural History of African Lungfishes." In *The Biology and Evolution of Lungfishes*, edited by W. E. Bemis, W. W. Burggren, and N. E. Kemp. New York: A. R. Liss, 1987.

Janvier, P. *Early Vertebrates*. Oxford: Oxford University Press, 1996.

Lévêque, C. "Protopteridae." In *Faune des poissons d'eaux douces et saumâtres d'Afrique de l'Ouest*. Tome 1, edited by C. Lévêque, D. Paugy, and G. G. Teugels. Paris: ORSTOM, 1990.

Merrick, J. R., and G. E. Schmida. *Australian Freshwater Fishes, Biology and Management*. North Ryde, Australia: Macquarie University, 1984.

Nelson, J. S. *Fishes of the World*, 3rd edition. New York: John Wiley & Sons, 1994.

Planquette, P., P. Keith, and P. Y. LeBail. *Atlas des Poissons d'Eau Douce de Guyane*, Tome 1. Paris: Museum National d'Histoire Naturelle, 1996.

Skelton, P. *Freshwater Fishes of Southern Africa*, 2nd edition. Cape Town: Struik, 2001.

Periodicals

Atz, J. W. Narial. "Breathing in Fishes and the Evolution of Internal Nares." *Quarterly Review of Biology* 27, no. 4 (1952): 366–377.

Resources

Bemis, W. E. "Morphology and Growth of Lepidosirenid Lungfish Tooth Plates (Pisces: Dipnoi)." *Journal of Morphology* 179 (1984): 73–93.

———. "Paedomorphosis and the Evolution of the Dipnoi." *Paleobiology* 10, no. 3 (1984): 293–307.

Bertmar, G. "The Olfactory Organ and the Upper Lips in Dipnoi, a Comparative Embryological Study." *Acta Zoologica* 46 (1965): 1–40.

Carter, G. S., and L. C. Beadle. "Notes on the Habitat and Development of *Lepidosiren paradoxa*." *Journal of the Linnaean Society, Zoology* 37 (1930): 197–203.

Coates, C. W. "Slowly the Lungfish Gives Up Its Secrets." *Bulletin of the New York Zoological Society* 40 (1937): 25–34.

Cunningham, J. T., and D. M. Reid. "Pelvic Filaments of *Lepidosiren*." *Nature* 131 (1933): 913.

Dollo, L. "Sur la phylogénie des dipneustes." *Bull. Soc. Belge Geol., Paleont. Hydrologie.* 9, no. 2 (1896): 79–128.

Johnels, A. G., and G. S. O. Svensson. "On the Biology of *Protopterus annectens* (Owen)." *Ark. Zool. Stockholm* 7, no. 7 (1954): 131–164.

Littrell, L. "African Lungfishes." *Tropical Fish Hobbyist* 19, no. 8 (1971): 40–57.

McMahon, B. R. "A Functional Analysis of the Aquatic and Aerial Respiratory Movements of an African Lungfish, *Protopterus aethiopicus*, with Reference to the Evolution of the Lung-ventilation Mechanism in Vertebrates." *Journal of Experimental Biology* 51, no. 2 (1969): 407–430.

Miles, R. S. "Dipnoan (Lungfish) Skulls and the Relationships of the Group: A Study Based on New Specimens from the Devonian of Australia." *Zoological Journal of the Linnaean Society* 61 (1977): 1–328.

Poll, M. "Revision systématique et raciation géographique des Protopteridae de lÁfrique centrale." *Ann. Mus. R. Afr. Cent., Zool.*, 8, no. 103 (1961): 1–50, pls. 1–6.

Rosen, D. E., P. L. Forey, B. G. Gardiner, and C. Patterson. "Lungfishes, Tetrapods, Paleontology, and Plesiomorphy." *Bulletin of the American Museum of Natural History* 167, no. 4 (1981): 163–275.

Other

"FishBase" [cited January 15, 2003]. <http://www.fishbase.org/search.cfm>

"Catalog of Fishes On-Line" [cited January 15, 2003]. <http://www.calacademy.org/research/ichthyology/catalog/fishcatsearch.html>

"Palæos: Vertebrates" [cited January 15, 2003]. <http://www.palaeos.com/Vertebrates/Units/Unit140/200.html>

Marcelo Carvalho, PhD

Polypteriformes
(Bichirs)

Class Actinopterygii
Order Polypteriformes
Number of families 1

Photo: A Week's bichir (*Polypterus weeksii*) from Congo, West Africa. (Photo by Mark Smith/Photo Researchers, Inc. Reproduced by permission.)

Evolution and systematics

Polypteriforms are the most basal or "primitive" living actinopterygian group, according to many recent authors. There are two living genera: *Polypterus* (bichirs), and *Erpetoichthys* (reedfish or ropefish), with 11 to 16 species presently recognized (*Erpetoichthys* has a single species). Their sketchy fossil record suggests that the group has never been particularly diverse. Fossils have been found in both Africa and South America, indicating that they were in existence before the breakup of Gondwana in the early Cretaceous, some 118 million years ago (mya); living forms are restricted to Africa. However, if the most accepted evolutionary scheme is correct, then polypteriforms have been around since much earlier (from at least the late Devonian), as indicated by stratigraphic correlations with fishes more closely related to the remaining actinopterygians (e.g. *Mimia* and *Moythomasia*, from the late Devonian of Australia). The few fossil polypteriform occurrences, usually dermal remains, are from the late Cretaceous (Cenomanian, 100 mya) of Morocco (*Serenoichthys*), Niger (?Campanian, 84 mya) and Bolivia (Maastrichtian, 71 mya), and Paleocene (63 mya) of Bolivia (*Dagetella*). Hence, there is a tremendous gap in our knowledge of polypteriforms, as fossils are as yet unknown from the late Devonian to the Cenomanian, a period spanning some 270 million years. On the other hand, there is molecular evidence suggesting that polypteriforms are more closely related to neopterygians (gars, bowfins, and teleosts), which, if confirmed, slightly reduces the discrepancy with the fossil record.

Uncertainty regarding the evolutionary affinities of polypteriforms is not new, as they have been interpreted as being more closely related to either sarcopterygians or actinopterygians, or even lying somewhere in between (such as in Erik A. Stensiö's Brachiopterygii), at least until the influential 1928 study by Edwin S. Goodrich. Even though the current consensus is to place

them among the actinopterygians (following Goodrich), there is much room for refinement. The disagreement over their ancestry stems from their enigmatic amalgam of anatomical features. Some of these features are present in sarcopterygians (lobefins), such as fleshy pectoral fin bases (not the internal pectoral fin skeleton, however), feathery external larval gills, larval cement organs, and paired, vascularized swim-bladders (lungs) arising from a ventral esophageal pneumatic duct; the latter three features are present in lepidosireniform lungfishes and tetrapods. Other features are similar to those present in sharks and rays (e.g., intestinal spiral valve, pectoral fin skeleton). But all of these traits probably evolved independently in the Polypteriformes, which share various derived characters with actinopterygians (e.g., scales with ganoin, dermohyal, gill arch musculature), as summarized by British paleoichthyologist Colin Patterson in 1982. The structure of their eggs (with a single opening for the entry of sperm cells) also supports their evolutionary affinity with actinopterygians.

Evolutionary relationships among polypteriform species, as well as the taxonomic status of many of these (along with their respective subspecies) are in need of further evaluation. Species of *Polypterus* are usually identified by their color pattern and meristics (such as numbers of scales along the lateral line, number of dorsal finlets), but there is overlapping in many features among certain species. Both genera are easily separated, as *Erpetoichthys* lacks pelvic fins and is very elongate, eel-like, with posteriorly positioned, small, and widely separated dorsal finlets.

Physical characteristics

Polypteriforms are morphologically unusual, and as a result their anatomy has been intensely studied over the past 100 years. They are moderately large, ranging from 15.7 to 47.2 in (40 to 120 cm) in total length, and are readily identified,

The bichir (*Polypterus ornatipinnis*) is widespread across Africa. (Photo by Mark Smith/Photo Researchers, Inc. Reproduced by permission.)

presenting a slender body with a depressed head, wide terminal mouth with fleshy lips, and unique, subdivided spiny dorsal fins (finlets). There is a tubular pair of nostrils extending anteriorly beyond the mouth. The teeth are sharp, small, and numerous. The gill opening is large, with an extended skin covering ventrally; four functional gill arches are present. The arrangement of dermal bones of the head and cheek are visible externally. The pectoral fin is greatly rounded posteriorly, with a fleshy base. Each dorsal finlet is composed of a strong, sharp spine attached posteriorly to a dermal fold, which in turn is attached to the base of the succeeding spine. Spines vary from seven to 18 among species, and are bifid (double-edged) at their tips. The dorsal fin originates either shortly after the pectoral fins or farther posteriorly, and is confluent with the caudal fin origin. The caudal fin is posteriorly elongate and distally rounded, composed only of soft rays. Pelvic fins (*Polypterus*) are situated at the posterior third of the body, followed by the anal fin (in both genera) which is very close to the caudal fin (and is functionally correlated with it while swimming); the anal fin, unlike the dorsal fin, is separated from the caudal fin by a notch. The dorsal finlets are the only fins with spines.

Polypteriforms have a compact, dense covering of trapezoidal, shiny (ganoid) scales, arranged in numerous diagonal series, which give them a rigid texture (similar to gars). Scales along the lateral line vary from about 55 to 70. Internally, polypteriforms have paired, asymmetrical (right lobe larger than left), and highly vascularized swim bladders that function as air-breathing organs. Coloration is olive-brown to dark brown dorsally and laterally, and over the head, but creamy white ventrally. Numerous dark or clear spots and blotches and irregular stripes are present in many species, sometimes over pectoral fins (e.g., *P. ornatipinnis*), but others are more uniform in color (*P. senegalus*). The heads of most species have a mottled or reticulated appearance.

Distribution

Present in western and central tropical Africa, with three species also occurring in the Nile River. They are absent from rivers that drain into the Indian Ocean.

Habitat

Commonly found in both fast and slow moving rivers, floodplains, swamps, lakes, and pools. Because they are able to breathe air directly, bichirs and the reedfish are capable of living in stagnant waters. They enter rivers with associated marginal vegetation during the spawning season.

Behavior

Not many studies documenting polypteriform behavior have been conducted. They are reported to "walk" over land for small distances to feed on insects, as they are able to absorb oxygen directly from the air for at least a few hours. However, air breathing is not obligatory, as it is in lepidosirenid lungfishes. In aquaria their behavior varies from remaining motionless on the bottom for short periods to swimming about vigorously. Their pectoral fins function as paddles.

Feeding ecology and diet

Polypteriforms are carnivorous, feeding on invertebrates such as insect larvae, snails, earthworms, and freshwater crustaceans, as well as fishes and amphibians; they are primarily nocturnal predators. Polypteriforms are preyed upon by crocodiles and large, fish-eating birds.

Reproductive biology

Reproduction has been observed in aquaria for a few bichir species as well as for the reedfish. Males may compete for the attention of a female. The anal fins are sexually dimorphic, as males have a pronounced bulge at the anal fin origin (anal fin is broader and more muscular). This modification develops gradually with sexual maturity; otherwise the fin is identical in both sexes. The anal fin is important during spawning, as the male will use his anal and caudal fins to envelop the genital opening of the female, thereby forming a receptacle in which he will fertilize her eggs. Eggs are then released by the male, through vigorous shaking of the anal fin, and quickly adhere to vegetation. This behavior has been described for both polypteriform genera. The larvae have feathery external gills. Polypterids do not practice any form of parental care of their eggs or fry.

Conservation status

No species are presently threatened or protected under CITES legislation, and none are listed in the IUCN database.

Significance to humans

Imported with frequency in the ornamental fish trade. The larger bichirs are highly regarded food fishes in West Africa. Their firm, white flesh tastes very much like the freshwater prawns that constitute an important part of the human diet in this region. These are very long-lived fishes, with records of large bichirs living 50 years in captivity.

Species accounts

Bichir
Polypterus ornatipinnis

FAMILY
Polypteridae

TAXONOMY
Polypterus ornatipinnis Boulenger, 1902, Congo River.

OTHER COMMON NAMES
None known.

PHYSICAL CHARACTERISTICS
Maximum length 23.6 in (60 cm). Body protected by an armor of large, rhombic, bony scales. Moderately elongate, with nine to 10 independent dorsal finlets. Pelvic

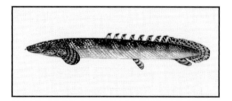

Polypterus ornatipinnis

fins located posteriorly. White belly with dark mottling on head, flanks and dorsum, with continuous parallel bands on fins.

DISTRIBUTION
Central and East Africa, found in the Congo Basin and in Lake Tanganyika.

HABITAT
Lakes, rivers, floodplains, and swamps, including waters with low oxygen content.

BEHAVIOR
Often sits motionless on the bottom, resting on its pectoral fins such that the head and anterior portion of the body are slightly elevated. Periodically gulps air from the surface in stagnant water.

FEEDING ECOLOGY AND DIET
Carnivorous; feed mostly at night on a variety of prey, including other fishes, frogs, insects, and crustaceans.

REPRODUCTIVE BIOLOGY
During courtship, their usual inactivity is abandoned, and both male and female engage in energetic twisting, turning, and darting movements. The male subsequently envelops the female's genital opening with his anal and caudal fin, fertilizing the eggs and then scattering them by thrashing his tail.

CONSERVATION STATUS
Not listed by IUCN.

SIGNIFICANCE TO HUMANS
Found in markets as a food fish; also captured for the aquarium trade. ◆

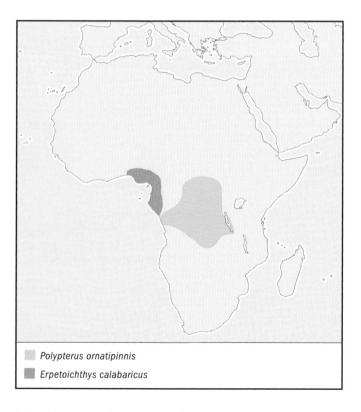

◻ *Polypterus ornatipinnis*
◼ *Erpetoichthys calabaricus*

Reedfish
Erpetoichthys calabaricus

FAMILY
Polypteridae

TAXONOMY
Erpetoichthys calabaricus Smith, 1866, Old Calabar, West Africa.

OTHER COMMON NAMES
English: Ropefish.

PHYSICAL CHARACTERISTICS
Maximum length 35.4 in (90 cm). Shares with *Polypterus* rhombic bony scales and distinct dorsal finlets. Unlike *Polypterus*, *Erpetoichthys* lacks pelvic fins and is very elongate and

Erpetoichthys calabaricus

eel-like in appearance. Uniform brown-olive color dorsally, with white underside and black spot on pectoral fins. Sexually active individuals develop an orange-red flush on the venter.

DISTRIBUTION
Coastal drainages of West Africa, from Nigeria to the Republic of the Congo.

HABITAT
Areas with aquatic vegetation in swamps and along rivers.

BEHAVIOR
Hunts along the bottom, moving in serpentine fashion.

FEEDING ECOLOGY AND DIET
Nocturnal predator, feeds on worms, crustaceans, and insects.

REPRODUCTIVE BIOLOGY
Very similar to that of *Polypterus*, in which the males wraps his anal fin around the female's genital pore, fertilizes the eggs,

and then scatters them into the surrounding vegetation with vigorous tail thrashing. In Benin and Nigeria, ripe individuals undertake mass movements overland into seasonally flooded swamp pools in order to spawn.

CONSERVATION STATUS
Not listed by IUCN.

SIGNIFICANCE TO HUMANS
A popular aquarium fish.

Resources

Books

Berra, Tim M. *Freshwater Fish Distribution.* San Diego, CA: Academic Press, 2001.

Boulenger, George A. *Les Poissons du Bassin du Congo.* Bruxelles, Belgium: Publication de l'État Indépendant du Congo, 1901.

———. *The Fishes of the Nile.* London, England: Hugh Rees, 1907.

Gosse, J.-P. "Polypteridae." In *Check-list of the Freshwater Fishes of Africa (CLOFFA),* edited by J. Daget, J.-P. Gosse, and D. F. E. Thys van den Audenaerde. Paris, France and Tervuren, Belgium: Orstom and MRAC, 1984.

———. "Polypteridae." In *Faune des Poissons d'eaux douces et saumâtres de l'Afrique de l'Ouest,* edited by Christian Lévêque, Didier Paugy, and Guy G. Teugels. Paris, France and Tervuren, Belgium: Orstom and MRAC, 1990.

Graham, Jeffrey B. *Air-breathing Fishes: Evolution, Diversity, and Adaptation.* San Diego, CA: Academic Press, 1997.

Janvier, P. *Early Vertebrates.* New York: Oxford University Press, 1996.

Kerr, J. Graham. "The Development of *Polypterus senegalus* Cuv." In *The Work of John Samuel Budgett,* edited by J. Graham Kerr. Cambridge, England: Cambridge University Press, 1907.

Nelson, J. S. *Fishes of the World.* 3rd ed. New York: John Wiley & Sons, 1994.

Patterson, C. "Bony Fishes." In *Major Features of Vertebrate Evolution,* edited by Donald R. Prothero and Robert M. Schoch. Knoxville, TN: Paleontological Society, 1994.

Stensiö, E. A. *Triassic Fishes from Spitzbergen.* Vol. I. Vienna, Austria: A. Holzhausen, 1921.

Wiley, E. O. "Bichirs and Their Allies." In *Encyclopedia of Fishes,* edited by John R. Paxton and William N. Eschmeyer. San Diego, CA: Academic Press, 1995.

Periodicals

Azuma, H. "Breeding *Polypterus endlicheri.*" *Tropical Fish Hobbyist* 44, no. 2 (1995): 116–128.

Bartsch, P., and S. Gemballa. "On the Anatomy and Development of the Vertebral Column and Pterygiophores in *Polypterus senegalus* Cuvier, 1829 ("Pisces," Polypteriformes)." *Zool. Jb. Anat.* 122 (1992): 497–529.

Bartsch, P., S. Gemballa, and T. Piotrowski. "The Embryonic and Larval Development of *Polypterus senegalus* Cuvier, 1829: Its Staging with Reference to External and Skeletal Features, Behavior and Locomotory Habits." *Acta Zoologica* 78 (1997): 309–328.

Bartsch, P., and R. Britz. "Zucht und Entwicklung von *Polypterus ornatipinnis.*" *Datz* 1 (1996): 15–20.

———. "A Single Micropyle in the Eggs of the Most Basal Living Actinopterygian Fish, *Polypterus* (Actinopterygii, Polypteriformes)." *Journal of Zoology* 241 (1997): 589–592.

Britz, R., and P. Bartsch. "On the Reproduction and Early Development of *Erpetoichthys calabaricus, Polypterus senegalus,* and *Polypterus ornatipinnis* (Actinopterygii, Polypteridae)." *Ichthyological Exploration of Freshwaters* 9, no. 4 (1998): 325–334.

Daget, J. "Révision des affinités phylogénétiques des polyptéridès." *Mem. L'Inst. Fran. D'Afr. Noire* 11 (1950): 1–178.

Dutheil, D. B. "First Articulated Fossil Cladistian: *Serenoichthys kemkemensis,* gen. et spec. nov., from the Cretaceous of Morocco." *Journal of Vertebrate Paleontology* 19 (1999): 243–246.

Gardiner, B. G., and B. Schaeffer. "Interrelationships of Lower Actinopterygian Fishes." *Zoological Journal of the Linnean Society* 97 (1989): 135–187.

Goodrich, E. S. "*Polypterus* a Paleoniscid?" *Palaeobiologica* 1 (1928): 87–92.

Nelson, G. J. "Subcephalic Muscles and Intracranial Joints of Sarcopterygians and Other Fishes." *Copeia* 1970, no. 3 (1970): 468–471.

Patterson, C. "Morphology and Interrelationships of Primitive Actinopterygian Fishes." *American Zoologist* 22 (1982): 241–259.

Poll, M. "Les tendances évolutives des polyptères d'après l'étude systématique des espèces." *Ann. Soc. R. Zool. Belg.* 72, no. 2 (1941): 157–173.

———. "Contribution à l'étude systématique des Polypteridae (Pisc.)." *Rev. zool. Bot. Afr.* 35 (1941): 143–179.

———. "Contribution à l'étude systématique des Polypteridae (Pisces)." *Rev. zool. Bot. Afr.* 35 (1942): 269–317.

Swinney, G. N., and D. Heppell. "*Erpetoichthys* or *Calamoichthys:* The Correct Name for the African Reedfish." *Journal of Natural History* 16 (1982): 95–100.

Marcelo Carvalho, PhD
Robert Schelly, MA

Acipenseriformes

(Sturgeons and paddlefishes)

Class Actinopterygii

Order Acipenseriformes

Number of families 2

Photo: Sturgeons (*Acipenser gueldenstaedtii* is shown here) produce cavier prized globally, and this has resulted in a decline of sturgeon populations. (Photo by Tom McHugh/Photo Researchers, Inc. Reproduced by permission.)

Evolution and systematics

The order Acipenseriformes includes 25 species of sturgeons in four genera (*Acipenser, Huso, Scaphirhynchus,* and *Pseudoscaphirhynchus*) in the family Acipenseridae and two living species of paddlefishes in the family Polyodontidae. The Acipenseriformes are primitive fish; recognizable fossils date to the early Cretaceous (144–65 million years ago). The Acipenseridae and Polyodontidae probably diverged from each other during the Jurassic (208–146 million years ago).

Physical characteristics

Acipenseriformes are some of the largest freshwater fishes, with species ranging in maximum size from 2.5 ft (0.76 m) to nearly 28.2 ft (8.6 m). Their bodies are elongate with large heads, small eyes, and fins positioned towards the posterior. A lateral line and scales are absent. Sturgeons and paddlefishes are dark on the tops of their bodies, but pigmentation fades to much lighter ventral colors, and many have white bellies. Species of sturgeon take on a variety of dull colors: gray, brown, dark blue, olive-green, and nearly black. Paddlefishes may appear bluish gray, brown, or black on their dorsal surface.

All Acipenseriformes share relict characteristics, including a largely cartilaginous endoskeleton and heterocercal caudal fin. The only ossified bones are found in the skull, jaws, and pectoral girdle. Other common anatomical features include an elongated snout with sensory barbels, a ventral mouth, an unconstricted notochord, and a lack of scales covering their skin.

Although they share many similar characteristics, anatomical and ecological distinctions exist between sturgeons and paddlefishes. Sturgeons have four barbels used for detecting prey, and the ventral mouth is protrusible. Paddlefishes have only two small sensory barbels and nonprotrusible mouths. Another major anatomical difference between sturgeons and

paddlefishes is in their body coverings. The skin of paddlefishes is largely naked, with patches of minute scales. In contrast, sturgeons are armored with five rows of bony shields along their bodies.

Distribution

Acipenseriformes are found throughout the Northern Hemisphere in North America, Europe, and Asia. Among the sturgeons, nine species inhabit North America, four are found in Europe, ten live in Asia, and four have Eurasian distributions. One species of paddlefish is found in North America; the other paddlefish species is endemic to China.

Habitat

Acipenseriformes inhabit seas, rivers, and lakes. Some species spend a large portion of their lives at sea but enter coastal rivers to spawn. Other species live strictly in freshwater rivers and lakes. Sturgeons are typically associated with sand, gravel, or rock substrates.

Behavior

Most sturgeons spend their lives in their native river or in nearshore areas of adjacent seas, but some individuals move long distances through coastal habitats. Sturgeons exhibit seasonal and spawning migrations. They may move from shallow to deep water in the summer and return to shallow areas in the winter. All sturgeons spawn in fresh water; thus, those that live in the sea migrate to fresh water for spawning. Paddlefishes swim constantly, both day and night, and migrate upstream to spawn.

Sturgeons are active primarily during the day, and many species congregate in discrete seasonal feeding areas.

Fish farm workers catch a 15-year-old Chinese sturgeon in order to inject it with hormones and make it produce eggs that will later be artificially fertilized, at Yichang, in China's Hubei province. The fish, which can grow up to 13 ft (4 m) long and weigh more than 1,000 lb (454 kg), are threatened by hydropower projects. The dams have affected the sturgeon's spawning grounds in the Yangtze, prompting authorities in 1982 to set up the "Chinese Sturgeon Park" in Yichang to breed the fish artificially. (Photograph. AP/Wide World Photos. Reproduced by permission.)

Observations of lab-reared juveniles suggest that certain species may establish a dominance hierarchy based on size, with large fish acting aggressively towards smaller fish in disputes over limited foraging space. Although sturgeons and paddlefishes are solitary for most of their life, some aggregation has been observed in larvae, which migrate in unorganized groups.

Feeding ecology and diet

Sturgeons locate food by swimming close to the bottom with their sensory barbels dragging the substrate. They selectively ingest slow-moving benthic invertebrates, including insects, worms, crustaceans, and mollusks, and feed on other fishes to a limited extent. Paddlefishes feed by swimming through the water with their mouths open and filtering large amounts of water through their gill rakers. Paddlefishes primarily consume microcrustaceans and insect larvae in the plankton, but they occasionally eat benthic invertebrates and other fishes.

Because of their large size and protective bony scutes, adult sturgeons and paddlefish have few predators except humans. However, sturgeons may be attacked, and possibly killed, by the parasitic sea lamprey, *Petromyzon marinus*.

Reproductive biology

Sturgeons typically spawn during spring and summer months. Prespawning activities involve rolling near the bottom and leaping out of the water. Spawning takes place in groups of two to three fish, with one or two males per female. Female sturgeons produce large quantities of eggs (up to several million), which are deposited over shallow shoals or rocky areas and fertilized by males. No nests are constructed, but the eggs are adhesive and stick to the substrate. Sturgeons do not devote any parental care to their offspring. Adults of some species spawn every year, but most species allow longer intervals between spawning events.

Paddlefishes spawn in the early spring as water levels are rising. They migrate from lakes and rivers into streams to locate spawning sites in shallow water. Males and females broadcast eggs and sperm over gravel substrates while swimming in groups. No parental care is provided to the offspring. Female paddlefishes produce very large numbers of eggs (up to 600,000) and do not spawn annually.

Conservation status

Overexploitation and habitat alteration, particularly the construction of dams, threaten and limit populations of Acipenseriformes throughout their range. The commercial landings of sturgeons exceeded 3,000 tons (2,721 tonnes) in 1890, but landings declined by 99% over the next century. Overfishing threatened many populations with local extinction, and stock enhancement programs have been introduced to maintain many sturgeon fisheries. Dams limit access to spawning sites and isolate populations. Other human activities on the shores of rivers increase siltation and contaminate rock or gravel spawning areas.

A lake sturgeon (*Acipenser fulvescens*) hovering over the sandy bottom. (Photo by Tom McHugh/Steinhart Aquarium/The National Audubon Society/Photo Researchers, Inc. Reproduced by permission.)

All Acipenseriformes are cited on the IUCN Red List. While some species are considered Lower Risk/Near Threatened (2 species) by the IUCN, most species are at greater risk and are classified as either Critically Endangered (6 species), Endangered (11 species), or Vulnerable (8 species). The international trade of Acipenseriformes is regulated through the Convention on International Trade in Endangered Species of Wild Flora and Fauna (CITES). The shortnose sturgeon (*Acipenser brevirostrum*) and the common sturgeon (*Acipenser sturio*) are considered threatened with extinction and are listed on Appendix I of CITES. All other species of sturgeon and paddlefish are listed on Appendix II of CITES. The shortnose sturgeon is listed as an endangered species in the United States.

Paddlefishes (*Polydon spathula*) do not have teeth, and eat by swimming through the water with their mouths open, scooping up plankton. (Photo by Daniel Heuclin/BIOS. Reproduced by permission.)

Significance to humans

Sturgeons have been valued for their caviar, the unfertilized eggs of the female, since the times of the ancient Persian, Greek, and Roman empires. The Chinese began trading caviar during the tenth century. It became popular as a luxury food in Europe during the seventeenth and eighteenth centuries and remained prized as a culinary delicacy at the end of the twentieth century. The smoked meat of sturgeons also is highly valued, particularly in European and Asian markets. In the late 1800s, paddlefish eggs and flesh also were sought commercially.

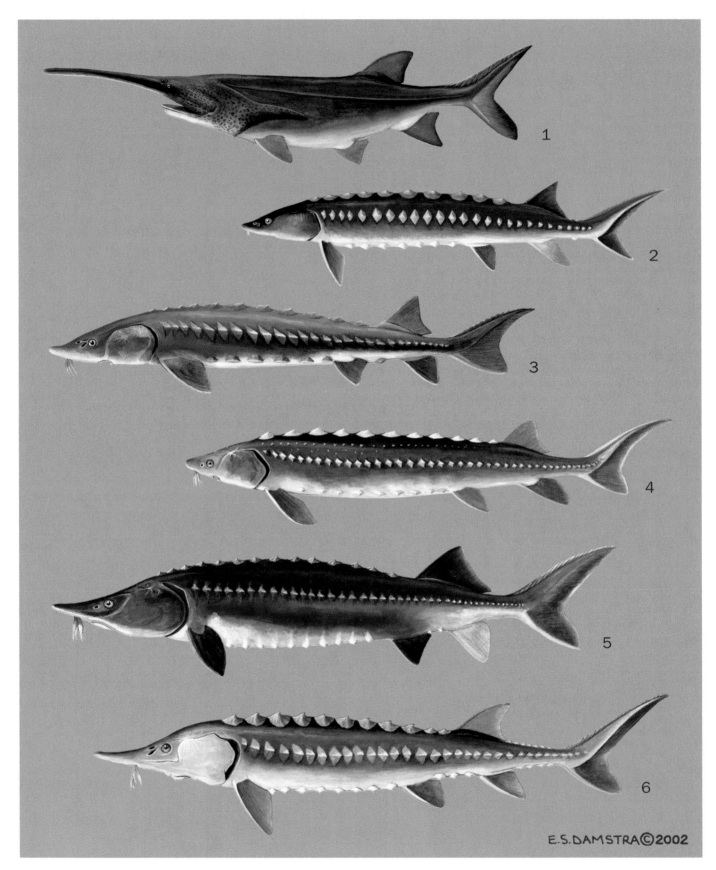

1. American paddlefish (*Polyodon spathula*); 2. Shortnose sturgeon (*Acipenser brevirostrum*); 3. Lake sturgeon (*Acipenser fulvescens*); 4. White sturgeon (*Acipenser transmontanus*); 5. Beluga sturgeon (*Huso huso*); 6. Atlantic sturgeon (*Acipenser oxyrhinchus*). (Illustration by Emily Damstra)

Species accounts

Shortnose sturgeon
Acipenser brevirostrum

FAMILY
Acipenseridae

TAXONOMY
Acipenser brevirostrum LeSueur, 1818, Delaware River, United States.

OTHER COMMON NAMES
English: Shortnosed sturgeon; French: Esturgeon à nez court; Spanish: Esturión hociquicorto.

PHYSICAL CHARACTERISTICS
At approximately 3 ft (0.9 m) in length, the shortnose sturgeon is the smallest species in the genus *Acipenser*. It has a shorter snout than other sturgeons and a wide mouth. Its upper body is dark brown or black, with lighter colors on the ventral portion. The bony plates are light in color.

DISTRIBUTION
Shortnose sturgeons occur along the East Coast of North America, from St. John River in New Brunswick, Canada, to Indian River, Florida.

HABITAT
Shortnose sturgeons live in the ocean, estuaries, and large coastal rivers.

BEHAVIOR
Shortnose sturgeons migrate upstream and downstream seasonally in coastal rivers. In southern portions of the range, these fishes spend longer periods of time at sea and migrate into rivers to spawn. Juvenile shortnose sturgeons may compete for limited foraging space, and larger individuals become aggressive to ward off encroaching individuals of smaller size.

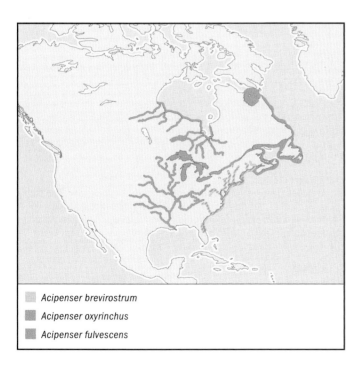

■ *Acipenser brevirostrum*
■ *Acipenser oxyrinchus*
■ *Acipenser fulvescens*

FEEDING ECOLOGY AND DIET
Shortnose sturgeons are opportunistic benthic feeders. Young individuals eat insects and crustaceans. Adults consume mollusks, benthic crustaceans, polychaete worms, and insect larvae.

REPRODUCTIVE BIOLOGY
Male shortnose sturgeons first spawn around three to four years of age, and females first spawn between six and fifteen years. Spawning takes place in the spring over gravel or rocky substrates. Females can produce 200,000 eggs per fish, and eggs hatch after approximately 13 days. Females spawn at intervals of three to five years, but males may spawn every year.

CONSERVATION STATUS
The shortnose sturgeon is listed as Vulnerable by the IUCN and protected under Appendix I of CITES. It also is recognized as an endangered species under the U.S. Endangered Species Act and as a vulnerable species by the Committee on the Status of Endangered Wildlife in Canada.

SIGNIFICANCE TO HUMANS
The caviar and flesh of shortnose sturgeons were commercially important during the 1800s and 1900s. Populations began declining in the 1800s due to industrial pollution of rivers and overfishing. As of 2002, all fisheries for this species are closed. ◆

Lake sturgeon
Acipenser fulvescens

FAMILY
Acipenseridae

TAXONOMY
Acipenser fulvescens Rafinesque, 1917, Lake Erie, North America.

OTHER COMMON NAMES
English: Freshwater sturgeon, Great Lakes sturgeon; French: Esturgeon jaune; Spanish: Esturión lacustre.

PHYSICAL CHARACTERISTICS
The back and sides of large lake sturgeon are olive-brown to dull gray in color; juveniles are light brown with dark blotches. Most lake sturgeons today are 3–5 ft (0.9–1.5 m) long and weigh 10–80 lb (4.5–36.3 kg), but a female of nearly 8 ft (2.4 m) and 310 lb (140.6 kg) has been documented.

DISTRIBUTION
Lake sturgeons occur in the following North American drainages: Great Lakes, St. Lawrence River, Hudson Bay, and Mississippi River.

HABITAT
Lake sturgeons inhabit large rivers and lakes.

BEHAVIOR
Lake sturgeons migrate seasonally between shallow and deeper waters, particularly in the northern extent of their range. They also undertake extensive migrations, typically of around 80 mi (128.7 km), to find suitable spawning grounds in rivers.

FEEDING ECOLOGY AND DIET
Lake sturgeons primarily consume insects, as well as other benthic invertebrates, such as snails, clams, and crayfishes. They occasionally feed on fish eggs, algae, and small fishes.

REPRODUCTIVE BIOLOGY
Lake sturgeons first spawn at 14–23 years for females and 12–20 years for males. Spawning intervals range from two to seven years in males and four to nine years in females. In the spring when ice clears, lake sturgeons migrate into smaller rivers and streams for spawning. Spawning typically takes place in swift-moving water 2–15 ft (0.6–4.6 m) deep. In the Great Lakes, lake sturgeons spawn along rocky shores in groups of two to three individuals. Females shed eggs in batches over a period of days. The eggs adhere to rocks for five to eight days before hatching.

CONSERVATION STATUS
Populations of lake sturgeons are threatened because of human exploitation, as well as habitat alteration and fragmentation that is caused by the construction of dams and roads. Lake sturgeons are listed as Vulnerable by the IUCN. They are protected by state and provincial fishing regulations and habitat restoration efforts in the United States and Canada.

SIGNIFICANCE TO HUMANS
Lake sturgeons were harvested for food by Native Americans before Europeans settled in North America, and commercial markets developed for the eggs and smoked flesh in the mid-1800s. Isinglass, a gelatin obtained from the swim bladder, was used to make jam and jellies, as a pottery cement, and as a waterproofing agent. Recreational fishing for lake sturgeons remains popular. ◆

Atlantic sturgeon
Acipenser oxyrinchus

FAMILY
Acipenseridae

TAXONOMY
Acipenser oxyrinchus oxyrinchus Mitchill, 1815, New York, United States. Two subspecies are recognized.

OTHER COMMON NAMES
English: Sea sturgeon, common sturgeon.

PHYSICAL CHARACTERISTICS
The Atlantic sturgeon is a large species that often grows to over 10 ft (3 m) long. Individuals are blue-black in color, with lighter shades on the sides. The head, ventral portion of the body, and fin edges are typically white.

DISTRIBUTION
Atlantic sturgeons are found along the Atlantic coast of North America from Ungava Bay, Quebec, to the St. John's River in Florida.

HABITAT
This species lives in the ocean and in bays, estuaries, and rivers.

BEHAVIOR
Atlantic sturgeons migrate between the sea and freshwater. Juveniles spend several years in freshwater before first entering the sea. Most individuals remain near their native river, but some travel long distances over the continental shelf. The migratory behavior of this species is typically associated with spawning ac-

tivities, but some individuals move into freshwater and do not spawn. Some evidence suggests that Atlantic sturgeons establish priority for foraging areas based on body size, with larger individuals dominant over smaller ones for feeding space.

FEEDING ECOLOGY AND DIET
Atlantic sturgeons consume bottom-dwelling plants and animals, such as insects, crustaceans, and mollusks. As adults, they also eat small fish.

REPRODUCTIVE BIOLOGY
Male Atlantic sturgeons typically reach sexual maturity around 12–24 years of age, and females are capable of spawning at 18–28 years. It is believed that females spawn in approximately four-year intervals, whereas males may spawn every year. The spawning season extends from late spring to early summer. Eggs are demersal and adhere to substrates near the spawning area.

CONSERVATION STATUS
Although populations have declined due to habitat alteration and fishing activities, Atlantic sturgeons are not considered threatened or endangered in the United States or Canada. They are listed as Lower Risk/Near Threatened by the IUCN.

SIGNIFICANCE TO HUMANS
Atlantic sturgeons are valuable for their flesh and roe, with colonial fisheries extending back to the 1600s. In the United States, commercial fisheries for Atlantic sturgeons were closed in 1998, although fishing had ceased in many states before that date. Commercial fishing continues in the St. Lawrence and St. John Rivers of Canada. ◆

White sturgeon
Acipenser transmontanus

FAMILY
Acipenseridae

TAXONOMY
Acipenser transmontanus Richardson, 1836, Vancouver, Washington, United States.

OTHER COMMON NAMES
English: Pacific sturgeon, Columbia sturgeon, Oregon sturgeon; French: Esturgeon blanc.

PHYSICAL CHARACTERISTICS
The white sturgeon is the largest North American sturgeon, attaining a maximum length of 20 ft (6.1 m). The upper body is gray, olive, or gray-brown, and its lower body is light gray to white.

DISTRIBUTION
Native distribution of the white sturgeon is along the Pacific coast of North America from the Aleutian Islands, Alaska, to Monterey, California. Landlocked populations occur in Montana and California. The species has also been introduced in the Colorado River in Arizona.

HABITAT
White sturgeons populate the ocean, estuaries, rivers, and lakes.

BEHAVIOR
White sturgeons spend most of their lives at sea but enter large rivers to spawn. Some individuals move long distances in coastal migrations.

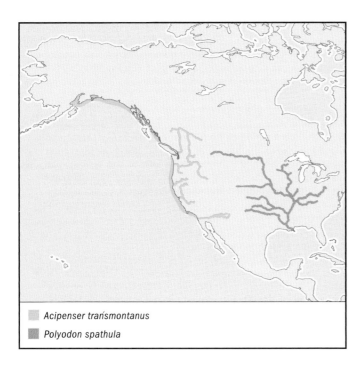

Acipenser transmontanus

Polyodon spathula

FEEDING ECOLOGY AND DIET
Juvenile white sturgeons feed on benthic invertebrates, such as chironomids, mollusks, and crustaceans. Adults primarily consume other fishes, shellfish, and aquatic invertebrates.

REPRODUCTIVE BIOLOGY
White sturgeons usually spawn in May or June in swift waters over rocky substrates. Males spawn initially between 11 and 22 years of age; females do not spawn until they are between 26 and 34 years. Younger females spawn every four years, while the interval increases to nine to eleven years for older females. The largest female spawners may produce three to four million eggs.

CONSERVATION STATUS
White sturgeons are classified as Lower Risk/Near Threatened by the IUCN. This species has been particularly affected by the damming of rivers. Populations were also severely overfished, but successful stocking programs and fishing regulations have enabled recovery.

SIGNIFICANCE TO HUMANS
White sturgeons have been used by Native Americans in the northwest region of the United States since long before the arrival and settlement of Europeans in the area. A commercial fishery for white sturgeons began on the Columbia River in the late 1800s, but the stock was depleted within a decade. Strict regulations put in place during the 1950s led to a population recovery by the late 1990s. By the early twenty-first century, commercial, recreational, and tribal fisheries actively targeted white sturgeons throughout their range. ◆

Beluga sturgeon
Huso huso

FAMILY
Acipenseridae

TAXONOMY
Huso huso Linnaeus, 1758, Danube and rivers of Russia.

OTHER COMMON NAMES
English: European sturgeon, great sturgeon.

PHYSICAL CHARACTERISTICS
The beluga sturgeon is the largest sturgeon species. It has been recorded to attain a length of 28.2 ft (8.6 m) and weight of 2,866 lb (1,300 kg), although such large specimens are rare. The body is gray or dark green in color with lighter sides and a white belly.

DISTRIBUTION
Beluga sturgeons occur in the Black, Caspian, and Adriatic Seas and in most of their tributaries.

HABITAT
This species inhabits nearshore areas of seas and large channels of rivers.

BEHAVIOR
Adult beluga sturgeons live at sea for most of the year but migrate up large river tributaries for spawning. The fry, or young fish, move downstream from rivers to the sea immediately after hatching.

FEEDING ECOLOGY AND DIET
Juvenile beluga sturgeons feed on benthic invertebrates, such as mollusks, worms, and crustaceans; adults eat other fishes.

REPRODUCTIVE BIOLOGY
Beluga sturgeons mature slowly and are extremely long lived (up to 150 years). Sexual maturity occurs around 14 years of age for males and 18 years for females. Females may produce over seven million eggs, but reproduction only occurs once every five to seven years. Beluga sturgeons spawn in late spring by scattering eggs and sperm in the water over rocky substrates.

CONSERVATION STATUS
The beluga sturgeon is listed as Endangered on the IUCN Red List. It may be extinct in the Adriatic Sea, and populations have declined throughout its range. The Caspian population is made up largely of fish from stocking programs.

SIGNIFICANCE TO HUMANS
Beluga sturgeons are valued throughout the world as the source of superior caviar. The caviar commands high prices, and the market demand has driven fisheries in eastern Europe to continue exploitation despite severe population declines. ◆

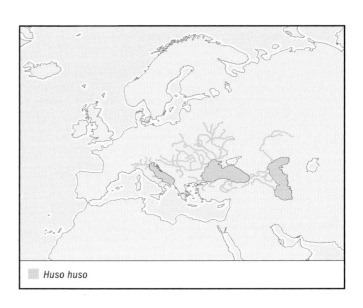

Huso huso

American paddlefish

Polyodon spathula

FAMILY
Polyodontidae

TAXONOMY
Polyodon spathula Walbaum, 1792, Louisiana, Mississippi River, United States.

OTHER COMMON NAMES
English: North American paddlefish, Mississippi paddlefish, spoonbill cat; French: Poisson spatule.

PHYSICAL CHARACTERISTICS
A defining characteristic of the American paddlefish is its large paddle-shaped rostrum, or snout. The paddle is covered with electroreceptors that enable paddlefish to sense objects and concentrations of planktonic prey. American paddlefish live up to 30 years and may attain lengths of 6.6 ft (2 m) and weights of 190 lb (86.2 kg).

DISTRIBUTION
American paddlefishes currently occur within the Mississippi River and Mobile Basin drainages in the United States, although the historical distribution included the Laurentian Great Lakes of Canada.

HABITAT
This species is found in large rivers and lakes.

BEHAVIOR
American paddlefishes swim continuously, often moving long distances. They typically are found near the water surface and frequently leap from the water.

FEEDING ECOLOGY AND DIET
American paddlefishes swim through the water with their mouths open and feed passively by filtering zooplankton and larvae of aquatic insects. Other fishes are occasionally found in stomach samples, indicating that paddlefishes are not strictly filter feeders.

REPRODUCTIVE BIOLOGY
Male paddlefishes mature between seven and nine years of age; females, between 10 and 12 years. Females may produce up to 600,000 eggs. Paddlefishes spawn in fast-flowing waters with clean gravel bottoms at intervals of two to five years. Spawning takes place in early spring in water depths of approximately 10 ft (3 m). Eggs and sperm are broadcast into the water column; eggs stick to the substrate and hatch within about seven days.

CONSERVATION STATUS
American paddlefishes are listed as Vulnerable by the IUCN. This species once occurred throughout the Mississippi River system, but habitats were fragmented by damming of the main stem of the Mississippi and its tributaries. Paddlefishes have been overfished, but state regulations and stocking programs are attempting to restore populations. Although fishing for paddlefishes is prohibited in most states, a few states allow commercial and recreational fisheries that target this species.

SIGNIFICANCE TO HUMANS
Like sturgeons, paddlefishes are valued for their flesh and roe. An important commercial fishery existed for paddlefishes in the Mississippi Valley following the decline of the sturgeon fishery in 1895, but this fishery reached its peak in 1900.

Resources

Books
Birstein, Vadim J., John R. Waldman, and William E. Bemis, eds. *Sturgeon Biodiversity and Conservation.* Dordrecht, The Netherlands: Kluwer Academic Publishers, 1997.

Periodicals
Billard, Roland, and Guillaume Lecointre. "Biology and Conservation of Sturgeon and Paddlefish." *Reviews in Fish Biology and Fisheries* 10 (2000): 355–392.

Jennings, Cecil A., and Steven J. Zigler. "Ecology and Biology of Paddlefish in North America: Historical Perspectives, Management Approaches, and Research Priorities." *Reviews in Fish Biology and Fisheries* 10 (2000): 167–181.

Kynard, B., and M. Horgan. "Ontogenetic Behavior and Migration of Atlantic Sturgeon, *Acipenser oxyrinchus oxyrinchus*, and Shortnose Sturgeon, *A. bervirostrum*, with Notes on Social Behavior." *Environmental Biology of Fishes* 63 (2002): 137–150.

Kynard, B., E. Henyey, and M. Horgan. "Ontogenetic Behavior, Migration, and Social Behavior of Pallid Sturgeon, *Scaphirhynchus albus*, and Shovelnose Sturgeon, *S. platorynchus*, with Notes on the Adaptive Significance of Body Color." *Environmental Biology of Fishes* 63 (2002): 389–403.

Peterson, Douglas L., Mark B. Bain, and Nancy Haley. "Evidence of Declining Recruitment of Atlantic Sturgeon in the Hudson River." *North American Journal of Fisheries Management* 20, no.1 (2000): 231–238.

Wilkens, L. A., D. F. Russell, X. Pei, and C. Gurgens. "The Paddlefish Rostrum Functions as an Electrosensory Antenna in Plankton Feeding." *Proceedings of the Royal Society of London B* 264 (1997): 1723–1729.

Other
"Lake Sturgeon Fact Sheet." *New York State Department of Environmental Conservation.* <http://www.dec.state.ny.us/website/dfwmr/wildlife/endspec/lakestur.html>. 30 Sept. 1999 (25 Oct. 2002).

"White Sturgeon." *Pacific States Marine Fisheries Commission.* 25 Oct. 2002 (16 Dec. 1996). <http://www.psmfc.org/habitat/edu_wsturg_fact.html>

"Fish: Paddlefish." *Tennessee Aquarium.* 25 Oct. 2002. <http://www.tnaqua.org/amazing/paddlefish.html>

Katherine E. Mills, MS

Lepisosteiformes

(Gars)

Class Actinopterygii

Order Lepisosteiformes

Number of families 1

Photo: The spotted gar (*Lepisosteus oculatus*) is frequently found in the pet trade. (Photo by Garold W. Sneegas. Reproduced by permission.)

Evolution and systematics

Lepisosteiformes contains the extant family Lepisosteidae (the gars). Extinct families that may belong in the order include the Semionotidae. The order Lepisosteiformes is sometimes classified within the division Ginglymodi.

The family Lepisosteidae includes seven living species contained in two genera, *Lepisosteus* and *Atractosteus*. Fossil gars are known as far back as the early Cretaceous period. The earliest fossil gars are represented only by scale, teeth, or bone fragments, but there are complete skeletons known as far back as 110 million years.

The gars have sometimes been included within the order Semionotiformes because of a presumed close relationship with the Semionotidae. More recently, this relationship has come into question. Consequently, the ordinal name Lepisosteiformes as used in this chapter contains the gars, and the Semionotiformes includes Semionotidae, but excludes the gars.

The gars comprise one of only five living actinopterygian families not contained within Teleostei (a group containing over 25,000 living species). They have often been referred to as "living fossils," and understanding their morphology is important to deciphering the evolutionary relationships of ray-finned fishes. Some authors have placed the gars within Holostei (together with bowfins), but they are thought by most systematic ichthyologists to comprise the living sister group to Halecostomi (a group containing bowfins and teleosts, but excluding gars). Whether to recognize Holostei (grouping gars with bowfins) or Halecostomi (placing gars outside of a bowfin/teleost group) remains controversial. Morphological data supports Halecostomi, whereas molecular data supports Holostei. It has even been suggested that gars and teleosts form a monophyletic group that excludes bowfins, although this phylogenetic hypothesis has not been widely accepted.

Physical characteristics

Extant lepisosteids and many of the fossils have a similar appearance. They have a highly elongate snout or "bill," well-armored elongate bodies covered with interlocking rhomboid-shaped ganoid scales, posteriorly positioned median fins with a dorsal fin set above the anal fin, a "tongue" supported by a number of bony basihyal tooth plates, and a jaw articulation anterior to the orbit. They also have an "abbreviate heterocercal" caudal fin, in which the hypurals (caudal ray supports) attach proximally with the ventral surface of the upturned end of the vertebral column. Gars also have a number of extremely diagnostic small features that enable the identification of even fragmentary fossils as gars. These features include plicidentine teeth (a peculiar folded dentine structure surrounding the pulp cavity) and opisthocoelous vertebrae (vertebral centra that are convex anteriorly and concave posteriorly). The ganoid scales are also diagnostic among living fishes, although the scales of African polypterids are superficially similar.

Distribution

Extant gars are restricted to freshwaters of eastern North America, as far north as Montana, United States, and southern Quebec, Canada, and as far west as Montana; Central America; and Cuba. When fossil (extinct) species are included, gars comprise a much greater diversity and geographic range. Well-preserved fossil gar material extends the geographic range of the family into what are now parts of western North America, Europe, Africa, Madagascar, India, and South America. Fossil and living gars are notably absent from East Asia. The one report of a living gar from China (*Lepisosteus sinensis* Bleeker, 1873) was in error, and was a belonid (Teleostei) rather than a lepisosteid.

Habitat

Gars are primarily freshwater fishes, although some species are known to occasionally swim into brackish or nearshore marine environments. The alligator gar, *Atractosteus spatula*, in particular, is frequently caught by shrimp trawlers in the salt marshes of Louisiana, and has often been observed in waters of the Gulf Coast.

Gars can withstand aquatic environments of low oxygen content because their swim bladder is highly vascularized and

Longnose gars (*Lepisosteus osseus*) can breathe through their air bladders and thus can tolerate poorer water conditions. (Photo by Garold W. Sneegas. Reproduced by permission.)

connected to the pharynx by an enlarged pneumatic duct, allowing them to breathe air.

Behavior

Gars are generally sluggish, but are capable of extremely quick movements for short periods of time. They often lie motionless near the surface until prey swims within reach. Then with a quick sideways thrust of its sharply toothed bill, the fish impales the food item and eventually swallows it. Although alligator gar reach a very large size (up to 9.8 ft/3 m total length) and have numerous, large, sharply pointed teeth and a head that superficially resembles that of a crocodilian, there are no authenticated records of any serious attacks on humans.

Feeding ecology and diet

Gars are primarily piscivorous, although most species supplement their diet with frogs, invertebrates, or even refuse that is dumped into the water. Gars are occasionally cannibalistic. The elongate, well-toothed jaws of gars facilitate the grasping of swimming prey with quick movements of the head. Large alligator gars also occasionally feed on water birds. Adult gars are well armored with their thick scales and dermal bones; consequently, they have few predators.

Reproductive biology

Gars spawn in freshwater generally in the spring (e.g., mid-May to mid-June in New York, United States). Fertilization is external, and large numbers of individuals concentrate in shoal areas and disperse quickly afterward. No parental care is given to the eggs or young. The eggs are black in color, adhesive, and stick to the substrate, rocks, or plants. After hatching, the larvae have adhesive suckers that enable them to stick to objects, even in moving water. The eggs are highly toxic.

Conservation status

No species of gars are included on the IUCN Red List. Most species are quite abundant, although the alligator gar is becoming very rare in some areas of its former range. Sport fishing for alligator gars is popular in some areas of the southeastern United States. Because the species are widely perceived by sport fishermen as being detrimental to game fishes, these fishes have received little sympathy. There have been

efforts to manage alligator gars, and they have been declared an endangered species in some of the southeastern states.

Significance to humans

Gars are often thought of as a nuisance fish detrimental to game fishes, and they often break up the nets of commercial fishermen in the southeastern states; but alligator gars are important predators in most aquatic ecosystems where they occur. The flesh of gars is extremely bony and not generally used for food. Exceptions include New Orleans and some other regions of the southeastern United States where alligator gar meat is sold, and the Pacific side of southern Mexico and Guatemala where the tropical gar is an important food item. The eggs of gars are toxic.

The ganoid scales of gars have historically been used for jewelry, arrowheads, and ornaments.

Alligator gars are popular sport fishes in the southern United States, and have inspired "fishing rodeos" and other tournaments. The Florida gar, *Lepisosteus platyrhincus*, has an attractive color pattern which makes it a popular aquarium fish. The longnose, *Lepisosteus osseus*, spotted, *Lepisosteus oculatus*, and alligator gars occasionally turn up in the pet trade as well.

Alligator gars (*Atractosteus spatula*) in Everglades waters, Everglades National Park, Florida. (Photo by Jim Zipp/Photo Researchers, Inc. Reproduced by permission.)

1. Cuban gar (*Atractosteus tristoechus*); 2. Tropical gar (*Atractosteus tropicus*); 3. Longnose gar (*Lepisosteus osseus*); 4. Shortnose gar (*Lepisosteus platostomus*); 5. Spotted gar (*Lepisosteus oculatus*); 6. Florida gar (*Lepisosteus platyrhincus*); 7. Alligator gar (*Atractosteus spatula*). (Illustration by Emily Damstra)

Species accounts

Alligator gar
Atractosteus spatula

FAMILY
Lepisosteidae

TAXONOMY
Atractosteus spatula Lacépède, 1803 type locality not specified. "*Lepisosteus ferox*" Rafinesque, 1820. The type species for the genus *Atractosteus*, is a subjective junior synonym of *A. spatula*, making this species the effective species type of the genus *Atractosteus*.

OTHER COMMON NAMES
French: Garpique alligateur; Spanish: Catán, gaspar baba, peje-lagarto.

PHYSICAL CHARACTERISTICS
Attains the largest size of any living gar, up to 9.8 ft (3 m) total length. One specimen has been reported at "9 feet 8.5 inches long and [weighing] 302 pounds" (2.8 m/137 kg). Adults usually have heavy ornamentation on exposed surfaces of the scales.

DISTRIBUTION
North America from Vera Cruz, Mexico, north through the Mississippi River drainage into southern Illinois, Indiana and Ohio, United States, and along much of the Gulf Coast. There is also a disjunct population reported from Lake Nicaragua and Rio Sapoá, Nicaragua. There are records of exotic introductions by humans as far west as California. Reported as a fossil from Pliocene deposits of Kansas and Pleistocene deposits of Texas and Florida, but fossils assigned to this species are isolated fragments (mostly scales) and somewhat tenuous in their assignment.

HABITAT
Freshwater river and swamp habitats, but also enters brackish and even marine waters. Of gar species, the most tolerant of salinity.

BEHAVIOR
Little is known besides feeding behavior.

FEEDING ECOLOGY AND DIET
Often portrayed as a voracious predator, although many reports are largely poorly documented sensationalism. As the largest, most solidly toothed gar, it is anatomically equipped to take a large variety of large prey, but this species is also a scavenger, and has been reported to compete with sharks for garbage at the wharves in Pensacola, Florida. The diet includes other fishes, blue crabs and other invertebrates, small mammals, and water birds such as ducks and water turkeys. Will prey opportunistically on alligator and crocodile hatchlings.

REPRODUCTIVE BIOLOGY
Very little is known about the reproductive habits. As in other gar species, the eggs are toxic to other animals.

CONSERVATION STATUS
Not threatened, although large individuals are taken in fish rodeos, spear fishing, and numerous annual contests.

SIGNIFICANCE TO HUMANS
Used for food and in sport fishing in the southern United States. Sometimes turns up in the pet trade. ◆

Atractosteus spatula
Atractosteus tropicus
Atractosteus tristoechus

Cuban gar
Atractosteus tristoechus

FAMILY
Lepisosteidae

TAXONOMY
Atractosteus tristoechus Bloch and Schneider, 1801, Cuba. Sometimes confused with the alligator gar. It is distinguishable as a separate species, but in the late nineteenth and early twentieth centuries, several authors considered the Cuban gar (*A. tristoechus*) and the alligator (*A. spatula*) to be synonyms. Consequently, some authors referred to the alligator gar as "*A. tristoechus*" or "*L. tristoechus*," and museum specimens of alligator gars are sometimes labeled as "*tristoechus*." In the author's experience, museum specimens collected in North America labeled "*L. tristoechus*" or "*A. tristoechus*" are actually *A. spatula*.

OTHER COMMON NAMES
French: Garpique cubain; German: Alligatorhecht, Kaimanfisch; Spanish: Manjuari.

PHYSICAL CHARACTERISTICS
Not as large as the alligator gar; largest known specimen is 36.6 in (93 cm) total length. Caudal fin has a distinctive color pattern, with the fin outlined with a thin line of dark pigment.

DISTRIBUTION
Western Cuba and the nearby Isle of Pines.

HABITAT
Very little is known.

BEHAVIOR
Similar to that for entire order.

FEEDING ECOLOGY AND DIET
Based on stomach-content analysis, evidently feeds on other fishes, but further studies are still needed.

REPRODUCTIVE BIOLOGY
Very little is known.

CONSERVATION STATUS
Not listed by the IUCN.

SIGNIFICANCE TO HUMANS
None known. ◆

Tropical gar
Atractosteus tropicus

FAMILY
Lepisosteidae

TAXONOMY
Atractosteus tropicus Gill, 1863.

OTHER COMMON NAMES
Spanish: Catán, gaspar, pejelagarto.

PHYSICAL CHARACTERISTICS
Small species; largest specimen known to the author is 49.2 in (125 cm) total length. Trunk is more pigmented than in other *Atractosteus* species.

DISTRIBUTION
Southern Mexico and Central America, with disjunct populations on both Atlantic and Pacific drainages

HABITAT
Freshwater rivers, streams, and near-shore lacustrine environments, but will occasionally enter brackish water.

BEHAVIOR
Visible on the surface and resemble floating logs.

FEEDING ECOLOGY AND DIET
Feeds mainly on fishes, but also may take copepods, insects, and plant material.

REPRODUCTIVE BIOLOGY
Reaches sexual maturity at about 14 in (36 cm) total length. Enters shallow lakes at beginning of dry season, spawning as rains cause flooding. Large schools form to lay eggs in a large gelatinous mass.

CONSERVATION STATUS
Not listed by the IUCN.

SIGNIFICANCE TO HUMANS
None known. ◆

Spotted gar
Lepisosteus oculatus

FAMILY
Lepisosteidae

TAXONOMY
Lepisosteus oculatus Winchell, 1864, Duch Lake, Calhoun Co., Michigan, United States.

OTHER COMMON NAMES
French: Garpique tachetée; German: Gefleckter Knochenhecht; Spanish: Gaspar pintado.

PHYSICAL CHARACTERISTICS
Maximum total length (known to the author) is 32.9 in (83.5 cm), although reported up to 44 in (112 cm). Has a profusion of dark spots on the body, head, and fins (although these spots are not generally as large and strong as in the Florida gar). Adults have a series of small bony plates on the ventral surface of the isthmus. Females have been reported to have proportionately longer snouts than males.

DISTRIBUTION
Great Lakes south to the gulf coast of Texas, United States, and northern Mexico, east to northwestern Florida, United States. Reported as a fossil from Pleistocene deposits of Texas, but the material is too fragmentary to be reliably included in the species.

HABITAT
Quiet, clear waters with abundant vegetation, also brackish waters along the Gulf of Mexico.

BEHAVIOR
Little is known.

FEEDING ECOLOGY AND DIET
Feeds mainly on fishes, but may also take crabs and crayfishes.

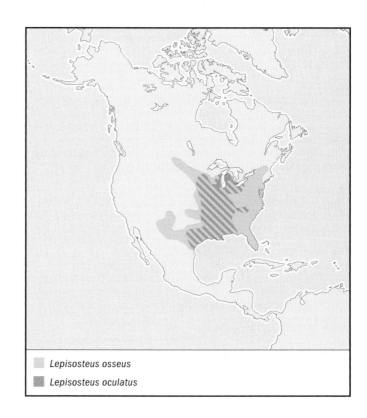

▨ *Lepisosteus osseus*
▨ *Lepisosteus oculatus*

REPRODUCTIVE BIOLOGY
Spawns in shallow freshwater. Like *L. osseus*, newly hatched larvae have adhesive pad on the head that allows them to adhere to the substrate or objects on the substrate. This organ is lost early in development. Sometimes hybridizes with *L. platyrhincus*.

CONSERVATION STATUS
Not threatened.

SIGNIFICANCE TO HUMANS
Turns up frequently in the pet trade. ◆

Longnose gar
Lepisosteus osseus

FAMILY
Lepisosteidae

TAXONOMY
Lepisosteus osseus Linnaeus, 1758, eastern United States.

OTHER COMMON NAMES
English: Billfish, common gar pike, needlenose gar; French: Garpique longnez; German: Gemeiner Knochenhecht, Gemeiner Langschnäuziger, Langnasen-Knochenhecht; Spanish: Catán, gaspar picudo, pejelagarto.

PHYSICAL CHARACTERISTICS
Maximum total length about 6 ft (183.4 cm), weight of about 50 lb (22.7 kg). Has the longest snout and most elongate body shape of any gar species. Color pattern variable. As a juvenile, it has a lateral stripe that disappears in the adult.

DISTRIBUTION
Southern Quebec, Canada, south to Florida, United States, westward from the Great Lakes region to Montana, United States, and from Florida to northern Mexico. Reported as a fossil from Pleistocene deposits of Kansas and North Carolina, United States, although species identification of this fragmentary material is tenuous.

HABITAT
Normally inhabits quiet, weedy, shallow-water lake environments or large rivers. Typically freshwater, but occasionally enters brackish water along its coastal distribution, particularly in the southern United States. Can also survive for weeks in oxygen-poor stagnant ponds and canals by breathing air at the surface with its functional lung (vascularized swim bladder).

BEHAVIOR
Often lies motionless near the surface until prey swim within reach. With a quick sideways thrust of its sharply toothed bill, it impales the prey.

FEEDING ECOLOGY AND DIET
Voracious predators; by the time they reach about 1 in (2.6 cm) total length, feed primarily on other fishes. Also feed on decapods, insects, and other invertebrates. Young are cannibalistic, sometimes feeding on siblings 70% of their own length.

REPRODUCTIVE BIOLOGY
Lake-dwelling; often migrates up streams and rivers to spawn. Some also spawn in nearshore lake shallows. Eggs in a female 40 in (102 cm) long can number more than 36,000. Males mature at three or four years, females at six. Longevity appears to be sexually dimorphic.

CONSERVATION STATUS
Not threatened.

SIGNIFICANCE TO HUMANS
Occasionally turns up in the pet trade. ◆

Shortnose gar
Lepisosteus platostomus

FAMILY
Lepisosteidae

TAXONOMY
Lepisosteus platostomus Rafinesque, 1820, type locality not specified but probably Mississippi River basin, United States.

OTHER COMMON NAMES
English: Duckbill garfish; Finnish: Pikkuluuhauki.

PHYSICAL CHARACTERISTICS
Maximum total length 34.6 in (88 cm). Has a reduced color pattern on the trunk region (i.e., spots are few and not as strong as in other *Lepisosteus* species) and two complete rows of premaxillary teeth.

DISTRIBUTION
Primarily within the low gradient regions of the Mississippi River basin in the United States, running from northeastern Texas north to Montana, east to Ohio, and south to Mississippi. It is absent from the Ozark plateau. Has been reported as a fossil from Kansas, but the material is too fragmentary to be reliably included in the species.

☐ *Lepisosteus platostomus*
▨ *Lepisosteus platyrhincus*

HABITAT
Quiet, sparsely planted backwater areas of rivers, lakes, and oxbows. Appears to be more tolerant of turbidity than most other gar species.

BEHAVIOR
Little is known.

FEEDING ECOLOGY AND DIET
Feeds mainly on crayfishes, fishes, and aquatic insects.

REPRODUCTIVE BIOLOGY
Reaches sexual maturity at about three years of age. Spawning takes place in spring. Eggs are scattered in quiet, shallow, freshwater and hatch in about eight days. Young become active (and feed) about seven days after hatching.

CONSERVATION STATUS
Not threatened.

SIGNIFICANCE TO HUMANS
None known. ◆

Florida gar
Lepisosteus platyrhincus

FAMILY
Lepisosteidae

TAXONOMY
Lepisosteus platyrhincus DeKay, 1842, Florida, United States.

OTHER COMMON NAMES
English: Florida spotted gar; Finnish: Floridanluuhauki.

PHYSICAL CHARACTERISTICS
Maximum total length 52.4 in (133 cm). Has numerous dark brown spots covering the anterior body and head; spots similar to those of *L. oculatus*, but more prominent and darker on the dorsal surface of the head and body. Also distinguished from *L. oculatus* by the lack of plates on the ventral surface of the isthmus.

DISTRIBUTION
Florida and the lowlands of southern Georgia, United States. Reported from Pleistocene deposits of Florida based on fragmentary material, but material is too fragmentary to be reliably included in the species.

HABITAT
Quiet lowland streams and lakes with heavy vegetation and a mud-sand bottom. The rarity of records of this species in brackish or salt water may reflect a very limited tolerance to salinity.

BEHAVIOR
Little is known.

FEEDING ECOLOGY AND DIET
Feeds primarily on fishes, but also on crustaceans and insects.

REPRODUCTIVE BIOLOGY
Little is known about the reproductive habits; may interbreed with *L. oculatus* in the Apalachicola River drainage.

CONSERVATION STATUS
Not threatened.

SIGNIFICANCE TO HUMANS
Turns up frequently in the pet trade.

Resources

Books

Breeder, C. M., and D. E. Rosen. *Modes of Reproduction in Fishes.* New York: Natural History Press, 1966.

Bussing, W. A. *Peces de las aguas continentales de Costa Rica.* San José, Costa Rica: Universidad de Costa Rica, Trejos Hermanos Sucesores, S.A., 1987.

Jordan, D. S. *A Guide to the Study of Fishes.* New York: Henry Holt and Co., 1905.

Lee, D. S., C. R. Gilbert, C. H. Hocutt, R. E. Jenkins, D. E. McAllister, and J. R. Stauffer, Jr. *Atlas of North American Freshwater Fishes.* Raleigh: North Carolina State Museum of Natural History, 1980.

Scott, W. B., and E. J. Crossman. *Freshwater Fishes of Canada.* Ottawa: Fisheries Resource Board of Canada, 1973.

Periodicals

Agassiz, A. "Embryology of the Gar-Pike (*Lepidosteus*)." *Science News* 1 (1879): 19–20.

Bleeker, P. "Mémoire sur la faune ichthyologique de Chine." *Nederlandsch Tijdschrift voor de Dierkunde* 4 (1873): 113–154.

Dugas, C. N., M. Konikoff, and M. F. Trahan. "Stomach Contents of Bowfin (*Amia calva*) and Spotted Gar (*Lepisosteus oculatus*) Taken in Henderson Lake, Louisiana." *Proceedings of the Louisiana Academy of Sciences* 39 (1976): 28–34.

Reséndez Medina, A., and M. L. Salvadores–Baledón. "Contribución al conocimiento de la biología del pejelagarto *Lepisosteus tropicus* (Gill) y la tenguayaca Petenia splendida Günther, del estado de Tabasco." *Biotica* 8, no. 4 (1983): 413–426.

Suttkus, R. D. "Order Lepisostei: Fishes of the Western North Atlantic." 3. *Memoirs of the Sears Foundation for Marine Research* 1 (1963): 61–88.

Uyeno, T. and R. R. Miller. "Summary of Late Cenozoic Freshwater Fish Records for North America." *Occasional Papers of the Museum of Zoology, University of Michigan, Ann Arbor* 631 (1963): 1–34.

Wiley, E. O. "The Phylogeny and Biogeography of Fossil and Recent Gars (Actinopterygii: Lepisosteidae)." *University of Kansas Museum of Natural History Miscellaneous Publication* 64 (1976): 1–111.

Lance Grande, PhD

Amiiformes
(Bowfins)

Class Actinopterygii

Order Amiiformes

Number of families 1

Photo: Bowfins (*Amia calva*) are found in lakes and slow moving rivers. (Photo by Tom McHugh/Shedd Aquarium/Photo Researchers, Inc. Reproduced by permission.)

Evolution and systematics

The order Amiiformes (*sensu* Grande and Bemis, 1998) contains one extant family, Amiidae, and three extinct families, Caturidae, Liodesmidae, and Sinamiidae.

The family Amiidae is another of the so-called living fossil families. In other words, it is a relatively basal neopterygian group with an old history (late Jurassic to present day), a formerly diverse distribution (with numerous genera and species in the Jurassic through the Eocene), and only a single extant species. The bowfin (*Amia calva*) is the only living species remaining from the order Amiiformes.

The Amiidae is one of only five extant families of actinopterygian fishes outside of Teleostei. Consequently, it is a family of great interest to fish systematists and evolutionary biologists. Even its most basic relationships remain controversial (i.e., is Amiidae the living sister group to Teleostei, or to Lepisosteidae?). Although there is only a single species in the family today, a rich fossil record going back over 150 million years to the late Jurassic indicates that in the past the family contained many more species, and was morphologically diverse and geographically widespread. In their revision of the Amiidae, Grande and Bemis (1998) divided the family into four subfamilies: Amiinae, Vidalamiinae, Solnhofenamiinae, and Amiopsinae. Amiopsinae was an extinct group with five valid species (some marine, some freshwater) ranging in age from late Jurassic to late Cretaceous (about 150 million to 100 million years ago). The group is known only from fossil deposits of western Europe. Solnhofenamiinae is an extinct group containing a single valid species and is known from late Jurassic marine deposits of western Europe. Vidalamiinae is an extinct group containing five genera and eight valid species (some marine, some freshwater) ranging in age from at least the early Cretaceous to the early Eocene (about 135 million to 55 million years ago). The group is known from deposits located in western Europe, North America, eastern South America, and western Africa. Amiinae is a freshwater subfamily containing two valid genera and about 11 valid species. This subfamily ranges in age from at least the late Cretaceous (about 95 million years ago) through to the pre-

sent. While the twenty-first century finds this family living only in North America, fossil members are widespread throughout the Northern Hemisphere.

Taxonomy is *Amia calva* Linnaeus, 1766, Charleston, South Carolina, United States.

Other common names: English: Blackfish, cottonfish, cypress trout, freshwater dogfish, grindle, grinnel, marshfish, mudfish, scaled ling, speckled cat; French: Choupique, poisson de marais.

Physical characteristics

The Amiidae are uniquely characterized by the condition of their caudal vertebral region. Most of the caudal centra are

Amia calva

Bowfin (*Amia calva*). (Illustration by Brian Cressman)

Behavior

Amia calva is a very hardy species. These fishes can withstand high temperatures and breathe air at the surface if necessary. They are even known to estivate. Specimens have been documented as being out of water for 24 hours without apparent harm.

Feeding ecology and diet

Bowfins are voracious predators. At the small postlarval stages (e.g., under 4 in/10 cm) total length) this species feeds on small animals such as insects, insect larvae, ostracods, and other zooplankton. Once the fishes start to get larger than about 4 in (10 cm), other fishes become its primary diet. Adults are also known to eat decapods. Observations of aquarium specimens indicate that the bowfin is a sluggish, clumsy, stalking predator that uses scent as much as sight in stalking food, which it captures by means of sudden intake of water. Other than humans, natural predators of adult *Amia calva* are unknown.

Reproductive biology

Bowfins spawn in spring. The males move into shallow waters of lakes and rivers, where they prepare a circular nest in areas of heavy vegetation or under logs. Once a female is attracted into the nest, spawning takes place, and four or five batches of eggs are laid. The eggs are adhesive and stick to the bottom of the nest. Females can lay up to 64,000 eggs. The young hatch in eight to 10 days and, like gars, have an adhesive organ on the tip of the snout by which they remain attached to vegetation (or other objects on the bottom) for seven to nine days. Then the young form a compact school which is guarded by the males for several weeks.

both solidly ossified and diplospondylous; in other words, the neural and haemal arches occur only on every other centrum rather than on every centrum. Like all halecomorphs, amiids have a peculiar double jaw articulation in which the jaw suspension bones articulate with the lower jaw at two separate places rather than one.

Amia calva is characterized, among living fishes, partly by its long, bow-shaped dorsal fin; hence the common name "bowfin." There are also several fossil species of bowfin (species in the genera *Amia* and *Cyclurus*) dating back as far as the late Cretaceous (about 65 million years ago). The living species reaches a total length of approximately 35.4 in (90 cm).

Distribution

The bowfin is restricted to eastern North America, inhabiting fresh waters over most of the eastern half of the continental United States, southern Ontario, and Quebec, Canada.

Habitat

The bowfin is known only from fresh waters. As an adult it generally inhabits swampy, sluggish water of vegetated bays of warm lakes and rivers. Young individuals are rarely seen after the postlarval schools break up, suggesting that they move into deeper water or dense vegetation.

Conservation status

Bowfins are not on the IUCN Red List and are currently quite common in areas of the southern United States.

Significance to humans

As with gars, bowfins are often considered to be pest fishes. They have little value as food fishes, and are of little commercial or recreational use. Yet they are important predators in some regions, controlling undesirable species.

Resources

Books

Eddy, S., and J. C. Underhill. *Northern Fishes*, 3rd edition. Minneapolis: University of Minnesota Press, 1974.

Grande, L., and W. E. Bemis. "A Comprehensive Phylogenetic Study of Amiid Fishes (Amiidae) Based on Comparative Skeletal Anatomy: An Empirical Search for Interconnected Patterns of Natural History." *Society of Vertebrate Paleontology Memoir 4*; Supplement to *Journal of Vertebrate Paleontology* 18, no. 1 (1998).

Lee, D. S., C. R. Gilbert, C. H. Hocutt, R. E. Jenkins, D. E. McAllister, and J. R. Stauffer Jr. *Atlas of North American Freshwater Fishes*. Raleigh: North Carolina State Museum of Natural History, 1980.

Scott, W. B., and E. J. Crossman. *Freshwater Fishes of Canada*. Ottawa: Fisheries Resource Board of Canada, 1973.

Periodicals

Neill, W. T. "An Estivating Bowfin." *Copeia* 1950, no. 3 (1950): 240.

Lance Grande, PhD

Osteoglossiformes

(Bony tongues and relatives)

Class Actinopterygii

Order Osteoglossiformes

Number of families 6

Photo: Clown knifefish (*Chitala chitala*) as juveniles have stripes, but adults lose the stripes, and have ringed spots instead. (Photo by M. H. Sharp/ Photo Researchers, Inc. Reproduced by permission.)

Evolution and systematics

The osteoglossiformes are an unusual group of teleost fishes comprising about 220 species of freshwater fishes, most of which are in one African family, the Mormyridae (19 genera; 182 species). The other species are scattered about the continents and are generally considered to be relicts of a once more abundant group. Fossil records of the family Osteoglossidae indicate these fishes to be between 38 and 55 million years old. However, the present distribution of members of the Osteoglossidae family suggests that the group was present on Gondwana prior to Gondwana's fragmentation. Biogeographic evidence thus suggests a considerably greater age than the 55 million years inferred from the fossil record. Most osteoglossiformes have most of their teeth located on the tongue and on the roof of the mouth. They also have a caudal fin with 16 or fewer branched rays (most bony fishes have more), lack intermuscular bones on the back, and have cycloid scales with ornate microsculpturing. The intestine curls around to the left side of the esophagus rather than to the right as in most other bony fishes.

Six living families are recognized. The monotypic family Gymnarchidae (*Gymnarchus niloticus*) together with the Mormyridae, comprises the superfamily Mormyroidea; this group is considered the sister group of the Notopteroidea (family Notopteridae; four genera and eight species). The position of the three remaining families is somewhat uncertain. The Osteoglossidae comprise seven species (four genera) and the Pantodontidae but one. The phylogenetic position of the Hiodontidae (two species) is not very clear. The two species of this family have a similar ear–swim bladder connection as do the clupeomorph fishes.

Recent data indicate the presence of a group of mormyrid fishes of the genus *Brienomyrus* in Gabon of uncertain taxo-

nomic status. Morphological, physiological (electric discharge), and molecular genetic data indicate these fishes represent a species flock. A comparable situation is known from the East African lakes Victoria, Malawi, and Tanganyika. A very limited number of riverine cichlid species adjusted to lacustrine conditions and have evolved into a species flock now comprising more than 200 species in each lake and dominating the fish fauna of these lakes.

Physical characteristics

The Mormyridae, the elephantfishes, are odd-looking fishes ranging from 1.6 in to 5 ft (4 cm to 1.5 m) in length. The head morphology varies considerably related to feeding specializations: some species possess prolonged heads, others trunklike snouts or appendages on the lower jaw, hence the common name. The tail is often deeply forked and the caudal peduncle very narrow. The skin is thick and of high electrical resistance; all species indeed are weakly electric. The electric organ is located in the caudal peduncle. Larvae possess a larval electric organ in the lateral muscle. The electric field set up around the fish is used for electrolocation and electrocommunication. Related to this sensory modality is the enlarged cerebellum; thus brain volume, relative to body size, is roughly the same size as that of humans. Male elephantfishes in most species can be distinguished from females by the lobed, enlarged front part of the anal fin. The sperm of mormyrids lacks flagellum. In the remaining osteoglossiform fishes, sexual dimorphism is not very pronounced or lacking.

Gymnarchus niloticus, the only species of the family Gymnarchidae, can reach 5 ft (1.5 m) in length. It possesses a long snout and a long dorsal fin used for locomotion; the anal, caudal, and pelvic fins are absent. The fish produces sinusoidal weakly electric discharges.

The Peter's elephantnose (*Gnathonemus petersii*) normally emits 800 electrical impulses per minute. (Photo by Wally and Burkard Kahl. Reproduced by permission.)

The eight species of knifefishes of the family Notopteridae have long, strongly compressed bodies tapering to a point. The long anal fin extends from just behind the head to the tiny caudal fin, which it joins. The dorsal fin, which is absent from *Xenomystus nigri*, is small and featherlike, so these fishes are commonly called featherbacks. The swim bladder is connected to the gut and is used for air breathing. The species of the African subfamily Xenomystinae, genera *Xenomystus* and *Papyrocranus*, possess cutaneous electroreceptors. Knifefishes range from 7.9 in (20 cm) in length (in *Xenomystus nigri*) up to 5 ft (1.5 m) (in *Chitala lopis*, the giant featherback).

Species of the family Osteoglossidae, the bony tongues, have heavy, elongate bodies covered with large scales. The dorsal and anal fins are long and placed on the rear part of the body. All these fishes can apparently breathe air with their lunglike swim bladders. *Arapaima gigas* can reach lengths of about 14.7 ft (4.5 m); other species attain lengths of about 3.3 ft (1 m).

The African freshwater butterflyfish, the only species of the family Pantodontidae, reach 3.9 in (10 cm) in length. The fishes possess a large gape and a straight dorsal profile. The pelvic fins with the prolonged fin rays are located under the greatly enlarged, winglike pectorals. The swim bladder can act as an air-breathing organ.

The two species of the family Hiodontidae, the mooneye (*Hiodon tergisus*) and the goldeye (*H. alosoides*), superficially resemble clupeid fishes. Their most distinctive external features are their large eyes, which have bright gold irises (goldeye) or gold/silver irises (mooneye). Goldeyes have only rods in their retinas and are known to feed mostly at night.

Distribution

Elephantfishes occur all over tropical Africa. The highest diversity is found in the Congo River basin, where mormyrids comprise about 20% of the total number of about 600 fish species. *Gymnarchus niloticus* is found in all large rivers of the Sahelo/Sudanean region in Africa.

African knifefishes inhabit coastal streams in West Africa (*Xenomystus nigri* and *Papyrocranus afer*), or the Congo basin (*P. congoensis*). The Asian knifefishes are found in the Indus, Ganges-Brahmaputra, and Mahanadi River basins in India (*Chitala chitala*), in Indonesia, Malaysia, and Thailand (*Chitala lopis*), and in Laos, Thailand, Cambodia, and Vietnam (*Chitala ornata*). *Notopterus notopterus* is very widely distributed, inhabiting rivers in India, Indochina, Thailand, Malaysia, and Indonesia.

The three South American bony tongues are either restricted to the Rio Negro (*Osteoglossum ferreira*), or occur in the Amazon River system and French Guiana (*Arapaima gigas* and *O. bicirrhosum*). *Heterotis niloticus* occurs in Africa in all river basins of the Sahelo/Sudanian region. The Asian bony tongue (*Scleropages formosus*) is native to Indonesia, Malaysia, Thailand, Cambodia, and Vietnam. *S. jardini* is found in New Guinea and Northern Australia; *S. leichardti* is restricted to northeastern Australia.

Pantodon buchholzi, the only representative of the family Pantodontidae, occurs in various rivers of central and West Africa.

The two species of *Hiodon* are found in the central part of North America, with *H. alosoides* being more widely distributed than *H. tergisus*.

Habitat

African elephantfishes are mainly riverine species and rarely occur in lakes. They are pelagic, midwater, or bottom-oriented fishes. The knifefishes inhabit stagnant backwaters of the large rivers, and are sometimes found in lakes; the smaller species prefer habitats with dense vegetation. Large bony tongues are found in open, slow-moving, or stagnant water and are surface-oriented hunters. *Pantodon buchholzi* prefers surface water of habitats with stagnant water. *Hiodon alosoides* lives in turbid waters in large lakes and muddy rivers, occasionally in swift current. *H. tergisus* is usually found in the clear waters of large lakes and streams.

Behavior

Elephantfishes are nocturnal, often hiding during the day in dense vegetation or under other kinds of cover. Aquarists have long admired mormyrids for their learning abilities and the fact that many species engage in apparent "play" behavior consisting of batting around small objects, including air bubbles, with the head. They usually swim slowly with their body rigid, presumably to avoid distorting the electrical field they are generating. The electric field is used for electrolocation and electrocommunication. The frequency of the electric signals can be modified to communicate with other fish, and thus can be used in courtship, aggressive behavior, and other intraspecific encounters. Because each species has its own set of electrical patterns, recognition and avoidance of other species is also possible. Mormyrids possess a well-developed sense of hearing; a part of the small swim bladder is in contact with the inner ear. Mormyrids use acoustic signals during courtship behavior. *Gymnarchus niloticus*, the close relative of the mormyrids, produces sinusoidal electrical discharges (the elephantfishes produce pulse-type discharges); this slow moving fish also uses its discharge for electrolocation. Knifefishes generally remain quietly in cover during daytime, but come out to prey in the evening. Bony tongued fishes are active during the day, spending most of the time patrolling very close to the surface. From aquaculture and aquarium observations, it has been deduced that the Australian spotted barramundi (*Scleropages leichardti*) can withstand water temperatures of between 44.6 and 104°F (7 and 40°C). During the summer, when surface temperatures exceed 87.8°F (31°C) in their natural habitat, surface cruising ceases and the fish remain in deeper, cooler areas. *Pantodon buchholzi* is a slow-moving fish of surface waters; it can jump out of the water and has been observed gliding over 13.1–16.4 ft (4–5 m). The pronounced tapetum lucidum of the two *Hiodon* species enables these fishes to hunt effectively at night.

Feeding ecology and diet

The mormyrid fishes eat various kinds of zooplankton or feed on a variety of benthic organisms such as insect larvae, crustaceans, oligochaets, and snails. The species with the long snouts find their prey in holes and crevices. Large *Mormyrops* species are piscivorous. Elephantfishes themselves are eaten by the large predator *Gymnarchus niloticus* (who also feeds on other fishes) and large piscivorous catfishes. Smaller knifefishes feed on insect larvae, crustaceans, worms, and snails; the larger species are mainly piscivorous. The bony tongues are midwater and surface feeders. Species of the genus *Osteoglossum* and *Scleropages* are carnivorous, feeding in roughly equal measure upon smaller fishes and terrestrial insects. While large specimens of both are known to take small terrestrial vertebrates opportunistically, these items do not constitute a significant portion of their diet in nature. The large *Arapaima gigas* prefers fishes. *Heterotis niloticus* has its fourth gill arch modified into a spiral-shaped filtering apparatus. This organ secretes mucus in which phytoplankton and bits of organic matter are trapped and then swallowed. The surface-oriented *Pantodon buchholzi* lives on crustaceans, insects, and small fishes. The two species of the family Hiodontidae feed on a variety of prey, including aquatic insects, crustaceans, mollusks, small fishes, frogs, shrews, and mice. They are preyed upon by birds, some mammals, and other fishes.

Reproductive biology

Most osteoglossiform fishes breed during the rainy season. Experimental studies with elephantfishes have shown that gonad maturation is triggered by decreasing water conductivity. About nine species have been bred in captivity. Parental care in the male is found in *Pollimyrus isidori*; the eggs are transported after oviposition by the male into the nest of about 2 in (5 cm) in diameter, generally made from plant material. After hatching (three days after fertilization) the male guards the embryos until the beginning of exogenous feeding (on days 13–14) and also during the larval period. Courtship behavior is characterized by acoustic and electrical behavior. Species of the genus *Stomatorhinus* probably also show parental care. The remaining species bred so far (*Petrocephalus*

African arawanas or "bony tongues" (*Heterotis niloticus*) live in nothern and western Africa, in the Nile, Senegal, Gambia and Niger Rivers. (Photo by Tom McHugh/Steinhart Aquarium/Photo Researchers, Inc. Reproduced by permission.)

soudanensis, Brienomyrus brachyistius, Marcusenius sp., *Mormyrus rume proboscirostris, Mormyrus* sp., *Hippopotamyrus pictus, Campylomormyrus cassaicus,* and *C. phantasticus*) do not show parental care.

Egg size ranges between 0.07 in (1.8 mm) in *P. soudanensis*) and 0.12 in (3 mm) in *H. pictus*. Eggs number between a few hundred in *P. isidori*, and more than a thousand in *C. cassaicus*. Spawning intervals range between a few days in *P. isidori* and several weeks in most other species.

Gymnarchus niloticus breeds in swamps during the high-water season. Prior to spawning, these fishes construct a floating nest of plant fibers in which the thousand or so eggs, each about 0.39 in (10 mm) in diameter, are laid. The newly hatched young have long gill filaments and an elongate yolk sac. They come to the surface for air. Young fishes feed on insects and other invertebrates.

Reproduction in knifefishes is not well known. *Xenomystus nigri* females lay 150–200 eggs of 0.08 in (2 mm) diameter; in *Notopterus notopterus*, eggs (1,000–3,000) are deposited in small clumps on submerged vegetation. *Chitala chitala* lays eggs on a stake or stump of wood, the male fans them with his tail and guards them against predators. *Arapaima gigas* males build a nest about 6 in (15 cm) deep and 20 in (50 cm) wide in sandy bottoms at the end of the dry season; the large eggs and young are guarded by the male and occasionally by the female. Parental care lasts up to 14 weeks. The two *Osteoglossum* species are male mouth brooders. The large eggs (about 0.6 in/16 mm diameter) are incubated for 50–60 days. At release, the juveniles measure 3.1–3.9 in (8–10 cm).

The *Scleropages* species are female mouth brooders. *S. leichardti* incubates 70–200 eggs about 0.4 in (10 mm) in diameter. Spawning occurs in small ponds during spring, when water temperatures rise to 68–73.4°F (20–23°C). Hatching takes place between one and two weeks after spawning; the embryos with their large yolk sac are about 0.6 in (15 mm) long. After the total incubation period of five to six weeks, the juveniles are released at a total length of about 1.4 in (35 mm). During a three-day period, the female shows a "release-and-recall" behavior. When the young become independent of the female, they take up territories around the edge of the pond. *Heterotis niloticus* is a nest builder, breeding in still waters close to the river, and excavating a nest some 3.9 ft (1.2 m) in diameter with thick walls of vegetation and mud. Within this nest, eggs about 0.1 in (2.5 mm) in diameter are laid; protected by the parents; they hatch in two days. The newly hatched embryos have external gills. *Pantodon buchholzi* has a prolonged spawning season, spawning 80–200 small buoyant eggs every day; the small embryos hatch after 36 hours. Goldeye (*Hiodon alosoides*) spawn in late spring on gravelly shallows of tributary streams. Their eggs are about 0.16 in (4 mm) in diameter, and are semibuoyant even after hatching, as the oil globule in the yolk buoys up the newly hatched 0.3 in (7 mm) embryo.

Conservation status

Four species are listed by the IUCN: *Arapaima gigas* is listed as Data Deficient; *Chitala blanci* is listed as Lower Risk/Near Threatened; *Scleropages formosus* is listed as Endangered; and *Scleropages leichardti* is listed as Lower Risk/Near Threatened. *Scleropages formosus* is also included on Appendix I of CITES, as a result of which international trade is banned (CITES Appendix I). For *Arapaima gigas*, international trade is restricted (CITES Appendix II).

Significance to humans

Most osteoglossiform fishes, particularly the larger species, are economically important food fishes. Even many of the medium-sized African elephantfishes, which measure approximately 7.9–23.6 in (20 to 60 cm) in length, are regularly fished for food. Some of the larger osteoglossiforms are used in aquaculture, including *Arapaima gigas*; *Scleropages leichardti* and *S. jardini*; *Heterotis niloticus*; *Chitala blanci* and *C. chitala*; and *Notopterus notopterus*. The larger species are important as food fishes, as well as for exhibition in public aquaria. The various color breeds of *Scleropages formosus* are favored as ornamental fishes in Asia. The elephantfish, *Gnathonemus petersii*, is a well-known species in the international aquarium trade. Several species of the weakly electric mormyrids are intensively studied by scientists.

1. Arapaima (*Arapaima gigas*); 2. Arawana (*Osteoglossum bicirrhosum*); 3. Clown knifefish (*Chitala chitala*); 4. Aba-aba (*Gymnarchus niloticus*). (Illustration by Bruce Worden)

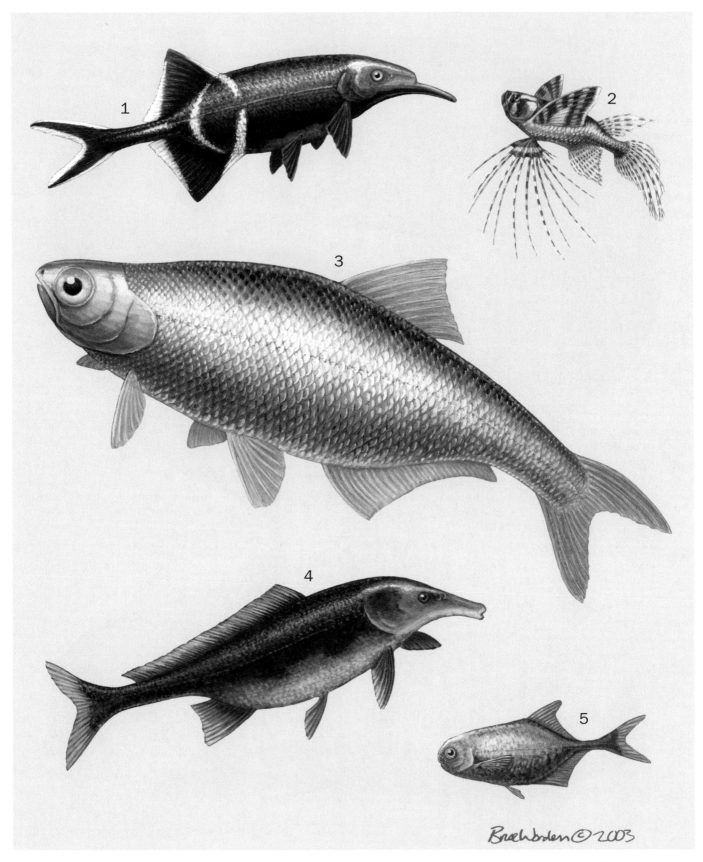

1. Elephantnose fish (*Gnathonemus petersii*); 2. Freshwater butterflyfish (*Pantodon buchholzi*); 3. Mooneye (*Hiodon tergisus*); 4. *Mormyrus rume proboscirostris*; 5. Elephantfish (*Pollimyrus isidori*). (Illustration by Bruce Worden)

Species accounts

Aba-aba
Gymnarchus niloticus

FAMILY
Gymnarchidae

TAXONOMY
Gymnarchus niloticus Cuvier, 1829, Nile River.

OTHER COMMON NAMES
German: Nilhecht.

PHYSICAL CHARACTERISTICS
Maximum size 5.48 ft (1.67 m) at 40.8 lb (18.5 kg). Heavy elongate fishes covered with small scales. Have long anal fin tapering in a short caudal appendage, lack dorsal, pelvic, and caudal fins. Caudal part can regenerate after injury. Anal fin used for propulsion. Weakly electric fish.

DISTRIBUTION
Large rivers and back waters of the Sahelo and Sudanian regions of Africa

HABITAT
Back waters with slow-moving or stagnant water, vegetation, and various kinds of cover.

BEHAVIOR
Nocturnal; hides during daytime. Very aggressive towards conspecifics. Weakly electric discharges of the sinusoidal type used for electrolocation.

FEEDING ECOLOGY AND DIET
Piscivorous, ecology not well known.

REPRODUCTIVE BIOLOGY
Breeds in swamps during the high-water season. Prior to spawning, a floating nest of plant fibers is created, most probably by the male. 1,000 or so eggs, each about 0.16 in (4 mm) in diameter, are laid in the nest. The newly hatched young have long gill filaments and an elongate yolk sac. They come to the surface for air. Young fishes feed on insects and other invertebrates. Parental males defend the nest very aggressively and do not hesitate to attack and bite human intruders. It is quite common to see fishermen in West Africa with the distinctive half moon–shaped scars left by an Aba-aba attack.

CONSERVATION STATUS
Not listed by the IUCN.

SIGNIFICANCE TO HUMANS
Very important food fish. Risk of overexploitation due to low reproductive capacity (low number of fry). ◆

Mooneye
Hiodon tergisus

FAMILY
Hiodontidae

TAXONOMY
Hiodon tergisus LeSueur, 1818, Lake Erie at Buffalo, New York, and Ohio River at Pittsburgh, Pennsylvania, United States.

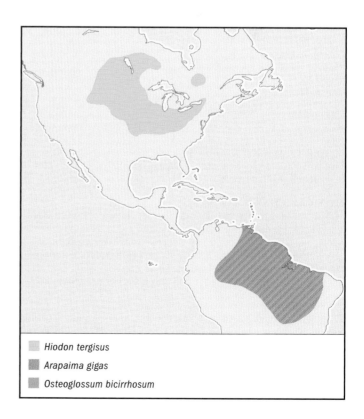

Gymnarchus niloticus
Gnathonemus petersii

Hiodon tergisus
Arapaima gigas
Osteoglossum bicirrhosum

OTHER COMMON NAMES
None known.

PHYSICAL CHARACTERISTICS
Maximum size 17 in (43 cm). Resembles clupeid or cyprinid fishes with large eyes and large oblique gape. The tapetum lucidum of the retina gives the silvery appearance of the eye.

DISTRIBUTION
St. Lawrence River and the Great Lakes (except Lake Superior) in Canada and United States; Mississippi River in the United States, Hudson Bay basins from Quebec to Alberta in Canada, and south to Gulf of Mexico. Gulf slope drainages from Mobile Bay in Alabama to Lake Pontchartrain in Louisiana.

HABITAT
Deep pools and backwaters of medium to large rivers, lakes, and impoundments; prefers clear water.

BEHAVIOR
The specialized eyes allow the fishes to forage at low light intensities.

FEEDING ECOLOGY AND DIET
Insects, insect larvae, and small fishes.

REPRODUCTIVE BIOLOGY
Reproduction biology probably similar to that of the related species *H. alosoides*. Spawning occurs in late spring on gravelly shallows of tributary streams. Eggs are about 0.16 in (4 mm) in diameter and are semibuoyant due to oil globules.

CONSERVATION STATUS
Not listed by the IUCN.

SIGNIFICANCE TO HUMANS
The species is locally exploited for food. ◆

Elephantnose fish
Gnathonemus petersii

FAMILY
Mormyridae

TAXONOMY
Gnathonemus petersii Günther, 1862, "Old Calabar, Westafrika."

OTHER COMMON NAMES
English: Peter's elephantnose; German: Tapirfisch, Elefanten-Rüsselfisch, Spitzbartfisch.

PHYSICAL CHARACTERISTICS
Maximum length 9.8 in (25 cm). Slender, laterally compressed fish with long dorsal and anal fin located at the rear part of the body. Narrow caudal peduncle houses the weakly electric organ. Caudal fin deeply forked. Body coloration dark brown to black. Two whitish transversal bands at the beginning and in the middle of dorsal and anal fins. Chin barbel on lower jaw.

DISTRIBUTION
West Africa from Niger to Congo River basins. Limited to the Lower Niger, in the Ogun, in the Cross River Basin and in the Upper Chari.

HABITAT
Habitat not very well known, but probably slow-moving waters of large rivers.

BEHAVIOR
Social and nocturnal. Often occurs in large schools. Weakly electric discharges of the pulse type used for electrocommunication. Captive animals appear to have a complex social structure, with a nonlinear "peck order."

FEEDING ECOLOGY AND DIET
Bottom-oriented, feeds on invertebrates of soft substrate.

REPRODUCTIVE BIOLOGY
Not known. Probably spawns during the rainy season.

CONSERVATION STATUS
Not listed by the IUCN.

SIGNIFICANCE TO HUMANS
Most important mormyrid species in the international fish trade. Known to aquarists for their "play" behavior. Often used in scientific studies concerning neuroanatomy, physiology, and behavior. ◆

No common name
Mormyrus rume proboscirostris

FAMILY
Mormyridae

TAXONOMY
Mormyrus rume proboscirostris Boulenger, 1898, "Upoto" Upper Congo.

OTHER COMMON NAMES
Chokwe (Angola): Sosha.

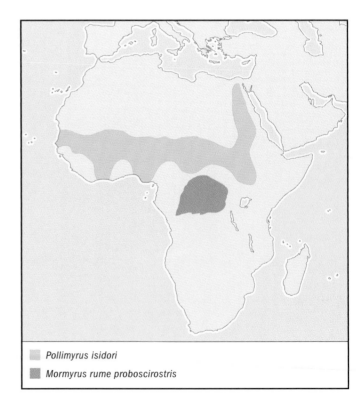

Pollimyrus isidori
Mormyrus rume proboscirostris

PHYSICAL CHARACTERISTICS
Maximum length about 13.7 in (35 cm). Elongate fishes with very long dorsal fin, short anal, and forked caudal fin. Snout prolonged and curved downward, dolphinlike. Electric organ in adults in the caudal peduncle; in the lateral muscle of young fish a larval electric organ is found. Mormyrus is the only mormyrid genus whose members produce weakly electric discharges of up to 30 V, which can be felt by touching the caudal peduncle of the fish. Body coloration dark gray.

DISTRIBUTION
Congo River basin in Africa.

HABITAT
Not known.

BEHAVIOR
Very active both day and night. Very social fish, however with many aggressive interactions that can cause death of subordinate specimens.

FEEDING ECOLOGY AND DIET
Feeds on insect larvae, crustaceans, and mollusks, as well as small fishes. Produces fine jets of water with the tubular mouth in search for prey in the substrate.

REPRODUCTIVE BIOLOGY
Reproduction occurs during the high-water season. Gonad maturation can be provoked experimentally by decreasing conductivity for several weeks. During each fractional spawning event several hundred eggs of 0.1 in (2.5 mm) in diameter are deposited. Hatching on day three. Exogenous feeding starts on day 10–11. Spawning intervals three to four weeks. No parental care.

CONSERVATION STATUS
Not listed by the IUCN.

SIGNIFICANCE TO HUMANS
All Mormyrus species are economically important as food fishes. ◆

Elephantfish
Pollimyrus isidori

FAMILY
Mormyridae

TAXONOMY
Pollimyrus isidori Valenciennes, 1847, "Westafrika."

OTHER COMMON NAMES
None known.

PHYSICAL CHARACTERISTICS
Maximum length 3.94 in (10 cm). Broad, laterally compressed fish with long dorsal and anal fins located at the rear part of the body. Slightly subterminal mouth. Weakly electric organ found in the narrow caudal peduncle. Larval electric organ in the lateral muscle of larvae up to 1 in (25 mm) long. Body coloration uniform gray and black.

DISTRIBUTION
Nile River, Upper and Middle Niger, Chari, and Lagone River systems, including Lake Chad. Disjunctly distributed in the coastal rivers of West Africa between the Niger and the Sénégal.

HABITAT
Slow-moving water and back waters of rivers; lakes.

BEHAVIOR
Nocturnal and territorial. Males occupy territories of 3.3–9.8 sq ft (1–3 sq m). Pronounced acoustic signaling during courtship behavior.

FEEDING ECOLOGY AND DIET
Insect larvae, crustaceans, and small mollusks.

REPRODUCTIVE BIOLOGY
Reproduction occurs during the rainy season. Parental care in the male. Thirty to 200 eggs 0.1 in/2.5 mm in diameter are deposited in a nest of plant material. Free embryos and larvae are also guarded; exogenous feeding starts on days 13–14. Spawning intervals five to 20 days.

CONSERVATION STATUS
Not listed by the IUCN.

SIGNIFICANCE TO HUMANS
Best-studied mormyrid fish in science concerning reproductive biology. ◆

Clown knifefish
Chitala chitala

FAMILY
Notopteridae

TAXONOMY
Chitala chitala Hamilton, 1822, type locality unknown (probably India or East Indies).

OTHER COMMON NAMES
English: Clown featherback; German: Indischer Messerfisch.

PHYSICAL CHARACTERISTICS
Up to 31.5–35.4 in (80–90 cm) long. Strongly compressed, tapering to a point. Very long anal fin continuous with the caudal, small dorsal fin. Dorsal profile is markedly convex. Small pelvic fins unite together at their base.

DISTRIBUTION
Indus, Ganges, Brahamaputra, and Mahandi River basins in India.

HABITAT
Rivers, canals, reservoirs, and ponds.

BEHAVIOR
Generally remains quietly in cover during daytime, but comes out to prey at night.

FEEDING ECOLOGY AND DIET
Aquatic insects, mollusks, shrimps, and small fishes.

REPRODUCTIVE BIOLOGY
Spawns once a year during May to August. Eggs usually laid on wooden substrate, male fans them with tail and keeps them aerated and silt free. Eggs are guarded against small catfishes and other predators. Embryos hatch after one week and are guarded by the male for some days.

CONSERVATION STATUS
Not listed by the IUCN.

SIGNIFICANCE TO HUMANS
Moderately important food and game fishes also used in aquaculture. Large specimens are often exhibited in public aquaria. ◆

Arapaima
Arapaima gigas

FAMILY
Osteoglossidae

TAXONOMY
Arapaima gigas Schinz, 1822, probably Amazon River.

OTHER COMMON NAMES
English: Pirarucu; German: Paiche; Spanish: Paíche; Portuguese: Piracuçu.

PHYSICAL CHARACTERISTICS
Heavy elongate fishes with large, ornate scales. One of the largest freshwater fishes, reaching 15 ft (4.5 m) in length and 441 lb (200 kg). Pelvic and unpaired fins located posteriorly.

DISTRIBUTION
Amazon River system and French Guiana.

HABITAT
Midwater fishes found in open, slow-moving, or stagnant water.

BEHAVIOR
Slow-moving, air-breathing fishes that surface every 10–20 minutes. This behavior makes it an accessible target for harpoon fishermen. Sometimes aggressive toward conspecifics.

FEEDING ECOLOGY AND DIET
Swallow fish and other large prey. The diet also includes heavily armored loricariid catfishes. Ecology in general not well studied.

REPRODUCTIVE BIOLOGY
Breed at the end of the dry season. Male builds nest about 6 in (15 cm) deep and 19.7 in (50 cm) wide in sandy bottoms at the end of the dry season. Large eggs and young are guarded by the male and occasionally by the female. Parental care lasts up to 14 weeks.

CONSERVATION STATUS
Listed as Data Deficient by the IUCN. Heavily overfished. The unsustainable and environmentally destructive practice of fishing for this species using dynamite during the breeding season has resulted in the loss of breeding pairs and their fry. This practice is one of the chief reasons for the dramatic decline of this species in western Amazonia. International trade restricted; listed on Appendix II of CITES.

SIGNIFICANCE TO HUMANS
One of the most important food and game fishes of Amazonia. Also used in aquaculture. Popular fish in public aquaria. ◆

Arawana
Osteoglossum bicirrhosum

FAMILY
Osteoglossidae

TAXONOMY
Osteoglossum bicirrhosum Cuvier, 1829, Amazon River.

OTHER COMMON NAMES
English: Silver aruana, aruana; German: Arabuana, Gabelbart.

PHYSICAL CHARACTERISTICS
Maximum size 3.3 ft (1 m). Large-scaled, elongate fishes, laterally compressed with straight dorsal provile and large gape. Very long anal and dorsal fins nearly joining the caudal fin. Prominent barbels at the tip of the chin.

DISTRIBUTION
Amazon River system and French Guiana.

HABITAT
Surface-orientated, live in open, slow-moving or stagnant water, preferably at the shore zone.

BEHAVIOR
Day active, spends most of the day patrolling very close to the surface.

FEEDING ECOLOGY AND DIET
Omnivorous, mainly eat invertebrates, or fishes to a lower percentage. Frequently jump out of the water to seize small vertebrates and large (particularly terrestrial) insects.

REPRODUCTIVE BIOLOGY
Reproduction takes place at the beginning of the floods, in general in December and January. About 200 large eggs 0.63 in (16 mm) in diameter are incubated in the males' mouth for 50–60 days. When released, juveniles are 3.15–3.93 in (8–10 cm) long.

CONSERVATION STATUS
Not listed by the IUCN.

SIGNIFICANCE TO HUMANS
An important food fish of Amazonia. Of special value in caboclo (person of mixed Brazilian, Indian, European, or African ancestry) folklore because it is one of the few species that women are allowed to eat postpartum; other species, especially catfishes, are thought to cause inflammation if eaten in times of illness and recovery. ◆

Freshwater butterflyfish
Pantodon buchholzi

FAMILY
Pantodontidae

TAXONOMY
Pantodon buchholzi Peters, 1877, Victoria River, Cameroon.

OTHER COMMON NAMES
French: Poísson papillon; German: Schmetterlingsfisch.

PHYSICAL CHARACTERISTICS
Small (up to 3.94 in/10 cm), surface-oriented fishes with straight dorsal profile and large, winglike pectorals; prolongated fin rays on the pelvic fins. Upper part of the body olive, ventral part silvery yellow amplified with red.

DISTRIBUTION
Lower Niger, Lake Chad, Cameroon, Ogooué, Congo River basin, and Upper Zambezi of Africa. A relict population might be present in Sierra Leone in Western Africa.

HABITAT
Surface water of habitats with stagnant water.

BEHAVIOR
Lives in schools underneath surface. Can jump out of the water for feeding or to escape predators. While this species has been observed gliding at distances between 13.1–16.4 ft (4–5 m), and even over 49.2 ft (15 m), this behavior needs to be documented and is highly controversial. The anatomy of the pectoral and ventral fins in this species display none of the anatomical modifications that would allow it to make long glides, nor does it possess the sort of hypertrophied pectoral musculature that permits powered flight.

FEEDING ECOLOGY AND DIET
Crustaceans, insects, and small fishes.

REPRODUCTIVE BIOLOGY
Prolonged spawning season. Spawning occurs preferably at night. Between 80–200 buoyant eggs, 0.12 in (3 mm) in diameter, are laid each night. The embryos hatch after 36 hours at 78.8°F (26°C) and are 0.43 in (11 mm) long. Raising is not easy as the fry need live food near the surface. Growth is rather quick; after one year individuals can reach 0.39 in (10 cm) in length and can begin to spawn.

CONSERVATION STATUS
Not listed by the IUCN.

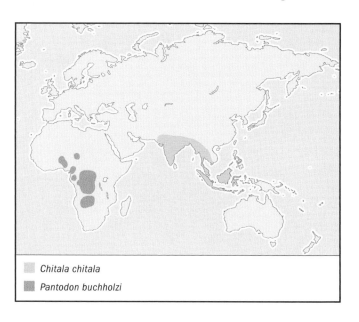

Chitala chitala

Pantodon buchholzi

SIGNIFICANCE TO HUMANS
None known.

Resources

Books

Bullock, T., and W. Heiligenberg. *Electroreception.* New York: John Wiley & Sons, 1986.

Daget, J., J. P. Gosse, and D. F. E. Thys van den Audenaerde, eds. *Check-List of the Freshwater Fishes of Africa* (CLOFFA), Vol. 1. Paris: ORSTOM, Tervuren: MRAC, 1984.

Merrick, J. R., and G. E. Schmida. *Australian Freshwater Fishes, Biology and Management.* North Ryde, Australia: School of Biological Sciences, 1984.

Moller, P., ed. *Electric Fishes: History and Behavior.* London: Chapman & Hall, 1995.

Periodicals

Kirschbaum, F. "Reproduction and Development of the Weakly Electric Fish *Pollimyrus isidori* (Mormyridae, Teleostei) in Captivity." *Env. Biol. Fishes.* 20 (1987): 11–31.

Kirschbaum, F., and C. Schugardt. "Reproductive Strategies and Developmental Aspects of Gymnotiform and Mormyrid Fishes." *J. Physiol.,* in press.

Roberts, T. R., "Systematic Revision of the Old World Freshwater Fish Family Notopteridae." *Ichthyological Explorations of Freshwaters* 2, no. 4 (1992): 361–383.

Frank Kirschbaum, PhD

Elopiformes

(Ladyfish and tarpon)

Class Actinopterygii

Order Elopiformes

Number of families 2

Photo: Ladyfish (*Elops saurus*) inhabit shallow marine waters from Cape Cod to Brazil. (Photo by Garold W. Sneegas. Reproduced by permission.)

Evolution and systematics

Tarpon, along with bonefish and ladyfish, are primitive fishes, and while tarpon and ladyfish are considered to be more closely related to each other than to any other elopomorph group, their distinct lineages extend more than 100 million years back in the fossil record. The structure of the skull, fin placement, and large thick scales are characteristic of ancient fishes.

Tarpon and ladyfish are united by the common possession of a leptocephalus larvae and a variety of primitive features. The leptocephalus larvae is shared with a diverse group of other elopomorph fishes including the eels; however, the leptocephalus larvae of tarpon and ladyfish are the smallest of all leptocephali and possess a forked tail. Leptocephali of some albuliformes also have a forked tail.

The order Elopiformes contains two families: the Elopidae and the Megalopidae. The family Megalopidae contains the single genus *Megalops*. Two species of tarpon exist worldwide. The Atlantic tarpon occurs in the eastern and western Atlantic, and the oxeye tarpon occurs in the Indian and Pacific oceans. Morphologically the two species are quite similar; however, the Atlantic tarpon reaches a much larger size and can exceed 220 lb (100 kg) and a length of over 6.6 ft (2 m). The oxeye tarpon is smaller and seldom exceeds 3.3 ft (1 m).

The family Elopidae contains the single genus *Elops*, which occurs worldwide. As many as six morphologically similar species of *Elops* are thought to exist. The genus is in need of revision, and the total number of species is unclear.

Physical characteristics

These are silver, elongate herring-like fishes with large upturned mouths, large eyes, and deeply forked tails. An important structural character is the presence of a long, bony gular plate between the branches of the lower jaw, a feature that the ladyfish shares with the tarpon but not with herring.

Distribution

The Megalopidae and Elopidae occur worldwide in tropical and subtropical seas.

Habitat

Tarpon and ladyfish are coastal in habitat and often occur in estuarine waters. Both tarpon and ladyfish are quite tolerant of low salinities. Tarpon commonly enter freshwater and often travel far up freshwater rivers and enter lakes far from sea.

Behavior

Tarpon and ladyfish are pelagic predators that feed principally on mid-water prey. Both have small sandpaper-like teeth, and their prey is swallowed whole. They often occur in large schools in shallow coastal and inshore waters.

Feeding ecology and diet

Small tarpon feed predominantly on cyclopoid copepods, fishes, caridean shrimp, and mosquito larvae. No detailed studies have examined the feeding habits of large tarpon, but anecdotal information suggests that a wide variety of fishes are consumed. Ladyfish feed principally in midwater on pelagic prey. Feeding is mainly on fish, but decapod crustaceans also are consumed.

Ladyfish are probably preyed upon by a wide variety of inshore predators including sharks, porpoises, snook, and tarpon. They are occasionally used as bait by recreational anglers for tarpon and other species. Juvenile tarpon are also likely preyed upon by a variety of species such as gar, snook, and larger tarpon. Because juvenile tarpon are most often found in poorly oxygenated waters, they are probably vulnerable to a more limited suite of predators than ladyfish. Large tarpon are preyed upon only by large coastal sharks including bull sharks and hammerheads.

An Atlantic tarpon (*Megalops atlanticus*) with a school of silversides in the Grand Caymans. (Photo by Animals Animals ©Clay Wiseman. Reproduced by permission.)

Reproductive biology

Both tarpon and ladyfish spawn offshore in high salinity oceanic waters. Precise spawning areas are unknown, and fertilized eggs are undescribed. Tarpon and ladyfish are broadcast spawners that produce large numbers of buoyant eggs that float in the surface waters of the ocean. The eggs hatch into the distinctive leptocephalus larvae characterized by an elongate, laterally compressed body consisting principally of an acellular mucinous material, large well-developed eyes, and large fang-like teeth. Larvae of tarpon and ladyfish reach a length of from 1.0 to 2.0 in (25–50 mm) before metamorphosis. Metamorphosis occurs as the larvae enter coastal waters and pass through inlets into the inshore waters where juveniles are found. Recruitment of tarpon through inlets appears to be pulsed and related to storm events.

Conservation status

Tarpon and ladyfish are abundant, and there is no evidence that stocks of these species have been depleted by overfishing. It is unknown to what extent habitat loss has affected stocks.

Significance to humans

Tarpon support important recreational fisheries in Florida and the Caribbean. Ladyfish are a food fish of minor importance in some areas.

1. Ladyfish (*Elops saurus*); 2. Atlantic tarpon (*Megalops atlanticus*). (Illustration by Jacqueline Mahannah)

Species accounts

Atlantic tarpon
Megalops atlanticus

TAXONOMY
Megalops atlanticus Valenciennes, 1847, Guadeloupe, Santo Domingo, Martinique, and Puerto Rico.

Anglers have long believed that the tarpon in some areas were different and larger than in other areas, but there is no genetic basis for this belief. While some areas may attract larger fish, these fish are not different genetically from those found elsewhere in the western Atlantic, and they all appear to interbreed freely. However, the tarpon of the eastern Atlantic do appear to be genetically distinct from their western Atlantic cousins. These populations have probably been isolated by the vast expanse of the Atlantic Ocean, and there is little or no interbreeding with western Atlantic tarpon. In is not known if the exceptionally large sizes attained by African tarpon have a genetic or environmental basis, but the isolation of the two stocks indicates that the differences could be genetically based.

OTHER COMMON NAMES
None known.

PHYSICAL CHARACTERISTICS
Elongate and highly compressed body. Eye large. Mouth oblique with a prominently projecting lower jaw. Large, thick, prominent scales. Teeth small and feel like sandpaper when touched. All fins are soft rayed. A single dorsal fin is located behind the pelvic fins but entirely before the anal fin; the dorsal fin has a distinctive and greatly prolonged final ray. The final ray of the anal fin is also somewhat elongate, but much less so than that of the dorsal fin. Deeply forked caudal fin. Tarpon are bright silvery all over, and the back is darker than the sides or belly.

▨ *Elops saurus*

▨ *Megalops atlanticus*

DISTRIBUTION
Both sides of the tropical and subtropical Atlantic Ocean. In the western Atlantic, tarpon regularly occur from the eastern shore of Virginia to central Brazil and throughout the Caribbean Sea and Gulf of Mexico, as well as off Central and South America. At least seven records exist from as far north as Nova Scotia, where a few large tarpon have been captured between August and October. Tarpon also are present in the eastern Atlantic off the coast of tropical Africa and are occasionally found as far north as Portugal and France. There is a single record of a tarpon from Ireland. African tarpon are known to reach exceptionally large sizes, and many recent world records have come from this area, including some unconfirmed reports of 330.7-lb (150-kg) fish. Tarpon are sexually dimorphic, and females reach much larger sizes than males.

HABITAT
Young-of-the-year tarpon occur in small stagnant pools and sloughs of varying salinity and have been reported from North Carolina, Georgia, Florida, Texas, Caribbean islands, Costa Rica, and Venezuela. In tropical areas, juvenile tarpon typically occur in mangrove habitats, often in water with low dissolved oxygen levels. Tarpon occur in salinities ranging from freshwater to more than 45 parts per thousand and are capable of surviving temperatures of at least 105°F (65.6°C), but they suffer mortalities at temperatures of 50–55°F (10–12.8°C). Large numbers of tarpon die during severe cold fronts off Florida.

BEHAVIOR
Anglers often detect the presence of schools of tarpon by observing individuals "rolling" at the surface. The tarpon's habit of rising to the surface and breathing air is unusual among marine species, although this practice is common among tropical freshwater swamp-dwelling fishes. Breathing air is accomplished by way of a highly vascularized swimbladder that functions as an air-breathing organ. The swimbladder is an elongate, balloon-like sac located above the viscera and just below the backbone. In most fish species, the swimbladder acts as a buoyancy control mechanism. The fish can adjust the volume of air in the bladder and remain neutrally buoyant. In tarpon, this swimbladder is connected to the gut by a duct enabling the tarpon to gulp air and ventilate the swimbladder. Young tarpon, when held in experimental chambers from which all of the dissolved oxygen has been removed, are able to meet their oxygen needs by breathing air. This adaptation allows tarpon to survive in water with low dissolved oxygen concentrations such as commonly encountered by juveniles in hot, stagnant mangrove marshes. Experimental work also suggests that tarpon are facultative air-breathers, and in well-oxygenated waters are able to meet their oxygen requirements without breathing air. Young tarpon can survive in well-oxygenated water when deprived of the opportunity to reach the surface and breathe air. However, after several unsuccessful attempts to reach the surface they have emptied their swimbladders and become negatively buoyant until allowed access to the surface again.

FEEDING ECOLOGY AND DIET
Small tarpon (0.6–3.0 in [16–75 mm]) feed predominantly on cyclopoid copepods, fishes, caridean shrimp, and mosquito larvae. No detailed studies have examined the feeding habits of large tarpon, but anecdotal information suggests that a wide

variety of fishes are consumed, including mullet (*Mugil* spp.), pinfish (*Lagodon rhomboides*), ariid catfishes, and clupeids, as well as crabs and shrimp.

REPRODUCTIVE BIOLOGY
Female tarpon are larger than males regardless of capture location, and average fish size varies geographically. Sexually mature Florida females average about 110 lb (50 kg) and can exceed 220 lb (100 kg). In contrast, sexually mature Florida males average only 66 lb (30 kg), and they rarely exceed 110 lb (50 kg).

Tarpon from Costa Rican waters are year-round spawners, unlike tarpon from other areas. Inactive or resting ovaries are rare in Costa Rica females, suggesting that females spawn repeatedly throughout the year and have no extended period of inactivity. In Florida, tarpon spawning is seasonal and peaks between May and July. By August, most females are finished spawning. In the Southern Hemisphere, off the northeast coast of Brazil, researchers have reported that tarpon spawn from October to January—during the Southern Hemisphere's spring and summer.

Ripe tarpon ovaries are large and can contain up to 20 million maturing oocytes and many more small resting oocytes. "Oocyte" is the proper name of a developing egg that has not ovulated and is not yet ready to be spawned. Although hundreds of mature female tarpon have been examined during the spawning season, none have been caught in the act of spawning. This is probably because spawning occurs in areas not typically fished. Even though the number of eggs released by a female in a single spawning event is unknown, the numbers of developing oocytes in the ovary suggests that their reproductive output is immense.

Tarpon are relatively long lived and can live more than 50 years. By age one, tarpon are about 1.5 ft (450 mm) long and are common in rivers and the upper reaches of estuaries, where they remain until reaching sexual maturity. In Florida, sexual maturity is reached at an age of about 10 years. After attaining sexual maturity, tarpon become more coastal in habitat and are most numerous around inlets and off beaches. Large tarpon targeted by anglers in Florida are typically from 15 to 35 years old.

CONSERVATION STATUS
Florida's fishery is intensely regulated, and anglers must purchase a $50 permit before harvesting a fish. Since the establishment of the permit system in 1989, the harvest of tarpon in Florida has declined to fewer than 100 fish per year, and the fishery is now mostly catch-and-release. Encouraging catch-and-release fishing for tarpon has been an effective management strategy, because the vast majority of released fish survive to be caught again. The sale of tarpon in Florida is prohibited, but in most of their range tarpon have never been considered a desirable food fish.

SIGNIFICANCE TO HUMANS
In Central America and South Florida, tarpon are the basis of economically important recreational fisheries. Tarpon occur in a variety of habitats ranging from freshwater lakes and rivers to offshore marine waters, but large tarpon targeted by Florida's fishery are most abundant in estuarine and coastal waters. In Florida, the fishery is seasonal; most tarpon are caught between May and July, although some fish are caught in all months. Tarpon are known for their spectacular leaps from the water when hooked and for their willingness to enter shallow water and eat artificial baits. Probably more than any other species, tarpon offer anglers in small boats an opportunity to pursue a large gamefish. Tarpon are pursued by a large for-hire charter boat fleet in Florida. ◆

Ladyfish
Elops saurus

TAXONOMY
Elops saurus Linnaeus, 1766, "Carolina." The taxonomic status of *Elops saurus* is unclear, and this name may be applied to more than one species.

OTHER COMMON NAMES
None known.

DISTRIBUTION
Abundant from North Carolina south through the Gulf of Mexico and into the Caribbean.

PHYSICAL CHARACTERISTICS
Ladyfish have a single, soft-rayed dorsal fin that originates about midway along the back. The pelvic fins are located midway between the tip of the snout and the fork of the deeply forked caudal fin. Scales are small and thin. Ladyfish are silvery all over; the back is bluish, and the lower parts of the sides and the belly are yellowish.

HABITAT
Common in estuaries and coastal waters of tropical and subtropical latitudes. Often occur in large schools. Tolerant of a wide range of salinities but seldom occur in freshwater.

BEHAVIOR
Little is known other than general descriptions of feeding habits and reproductive migrations. Ladyfish can be extremely abundant and most often occur in large schools. They are voracious predators.

FEEDING ECOLOGY AND DIET
Ladyfish feed principally in midwater on pelagic prey. Feeding is mainly on fish, but decapod crustaceans also are consumed.

REPRODUCTIVE BIOLOGY
Spawning appears to occur offshore. Larvae are common in the Gulf of Mexico and off the southern United States, where they have been reported as far north as Virginia. Fertilized eggs are undescribed. Spawning may occur throughout the year but probably peaks during fall in Florida and in the Gulf of Mexico.

CONSERVATION STATUS
Not threatened.

SIGNIFICANCE TO HUMANS
Ladyfish are often caught by recreational anglers but are seldom a targeted species. Ladyfish are voracious predators and will attack a variety of lures and baits. The species is fished commercially in Florida and sold both for human consumption and as bait to recreational anglers.

Resources

Books

Hildebrand, S. F. "Family Elopidae." In *Fishes of the Western North Atlantic,* edited by H. B. Bigelow. New Haven, CT: Sears Foundation for Marine Research, 1963.

Periodicals

Andrews, A., E. Burton, K. Coale, G. Cailliet, and R. E. Crabtree. "Radiometric Age Validation of Atlantic Tarpon, *Megalops atlanticus.*" *Fishery Bulletin* 99 (2001): 389–398.

Crabtree, R. E., E. C. Cyr, R. E. Bishop, L. M. Falkenstein, and J. M. Dean. "Age and Growth of Larval Tarpon, *Megalops atlanticus,* in the Eastern Gulf of Mexico With Notes on Relative Abundance and Probable Spawning Areas." *Environmental Biology of Fishes* 35 (1992): 361–370.

Crabtree, R. E., E. C. Cyr, D. Chacon, W. O. McLarney, and J. M. Dean. "Reproduction of Tarpon, *Megalops atlanticus,* from Florida and Costa Rican Waters and Notes on Their Age and Growth." *Bulletin of Marine Science* 61 (1997): 271–285.

Geiger, S. P., J. J. Torres, and R. E. Crabtree. "Air-breathing and Gill Ventilation Frequencies in Juvenile Tarpon, *Megalops atlanticus*: Responses to Changes in Dissolved Oxygen, Temperature, Hydrogen Sulfide, and pH." *Environmental Biology of Fishes* 59 (2000): 181–190.

Zale, A. V. and S. G. Merrifield "Species Profiles: Life Histories and Environmental Requirements of Coastal Fishes and Invertebrates (South Florida)—Ladyfish and Tarpon." *U.S. Fish and Wildlife Service Biological Report* 82 (1989).

Roy Eugene Crabtree, PhD

Albuliformes
(Bonefishes and relatives)

Class Actinopterygii

Order Albuliformes

Number of families 3

Photo: Bonefishes (*Albula vulpes*) are found in most of the warm seas of the world. (Photo by Tom McHugh/Shedd Aquarium/Photo Researchers, Inc. Reproduced by permission.)

Evolution and systematics

The order Albuliformes includes three extant families: Albulidae, Notacanthidae, and Halosauridae. The bonefish family (Albulidae) contains two genera: *Albula*, with possibly eight species, and *Istieus*, with two species. The halosaur family (Halosauridae) contains three genera: *Aldrovandia*, with six species; *Halosaurus*, with nine species; and *Halosauropsis*, with one species. The marine spiny eel family (Notacanthidae) contains three genera: *Lipogenys*, with one species; *Notacanthus*, with six species; and *Polyacanthonotus*, with four species. The fossil record for this order extends back almost 100 million years.

Physical characteristics

In Albulidae (bonefishes) the body is moderately slender. The snout is distinctively pointed and conical. The mouth is inferior, and the snout projects well beyond the tip of the lower jaw. All fins lack spines. The dorsal fin originates at about the midpoint of the body in *Albula*. In *Istieus* the dorsal fin origin is more forward, and the fin is elongate, extending nearly to the caudal fin. The anal fin is short and originates well behind the base of the dorsal fin. The pelvic fins are positioned below the last dorsal fin rays in *Albula* and under the middle of the dorsal fin in *Istieus*. Scales are small. Most fishes are less than 3.3 ft (1 m) in length. The back of *Albula* is blue-green in color, with narrow, dark horizontal lines that fade rapidly after death. The sides are silvery.

Halosaurs have an elongate, eel-like body, which tapers to a point. There is no caudal fin. The anal fin is elongate and extends along the posterior half of the body. There is a single short dorsal fin located just before the midpoint of the body. All fins have soft rays with no spines. The mouth is inferior, and the snout projects well beyond the tip of the lower jaw. Most fishes are less than 3.3 ft (1 m) in length. Colors range from tan to black.

The Notacanthidae, or marine spiny eels, have an elongate, eel-like body, which tapers to a point with little or no caudal fin. The anal fin is elongate and extends along the posterior half of the body. The anal fin consists of spines anteriorly grading to soft rays posteriorly. The dorsal fin in most species has from 26 to 41 isolated spines, from which the family's common name "spiny eels" derives. The mouth is inferior, and the snout projects beyond the tip of the lower jaw. Most fishes are less than 3.3 ft (1 m) in length. The coloring is typically tan.

Distribution

This order is worldwide in distribution. The family Albulidae occurs in shallow tropical waters worldwide. The Notacanthidae and Halosauridae are little-known families of deep-sea fishes that occur along the continental slope and rise of the world's oceans at depths from 3,281 to 9,843 ft (1,000–3,000 m). Bonefish (*Albula*) frequent coastal and inshore waters of tropical seas worldwide. In the western Atlantic, bonefish regularly occur in the Florida Keys and the Bahamas and throughout the Caribbean. Halosaurs and spiny eels are deep-sea fishes of worldwide distribution.

Habitat

Bonefish are common in tropical shallow-water areas. They are most abundant at depths of less than 115 ft (35 m) and often feed in water less than 3.3 ft (1 m) deep. Bonefish can be found over shallow grass flats and in sandy areas. Juvenile bonefish and metamorphic larvae occur along sandy beaches with scattered patches of sea grass in water from 1

Bonefish (*Albula vulpes*). (Illustration by Emily Damstra)

to 4.3 ft (0.3–1.3 m) deep. In southern Florida bonefish larvae recruit to sandy beaches during winter and early spring and are found in water temperatures ranging from 60.8 to 82.8°F (16.0–28.2°C) and salinity levels ranging from 10.4 to 37.0 ppt.

Halosaurs typically are found at depths of 1,640–13,123 ft (500–4,000 m) on the continental slope and rise. Spiny eels usually are found at depths of 656–11,483 ft (200–3,500 m) on the continental slope and rise. They generally hover just above the bottom.

Behavior

Bonefish are remarkable because of their common presence in extremely shallow water (less than 1.6 ft, or 0.5 m). The fisheries for bonefish in most areas require specialized boats capable of entering shallow water with little or no noise. Fish typically swim in small schools of five to 20 individuals, although larger schools of more than 100 individuals are not uncommon. Anglers searching for bonefish often detect their presence by spotting their tails protruding from the water as the fish dig in the bottom to feed.

Feeding ecology and diet

Bonefish feed on a variety of small benthic and epibenthic invertebrates and fishes. Feeding often takes place in shallow water, where foraging bonefish are seen with their fins protruding from the water. As they forage, bonefish schools frequently dig in the bottom and disturb considerable quantities of mud and sand. Xanthid crabs, toadfish (*Opsanus beta*), portunid crabs, alpheid shrimp, and penaeid shrimp make up most of the diet of populations in southern Florida. In some areas mollusks and small worms are important in the diet. Juvenile bonefish feed on a variety of polychate worms and small crustaceans, principally copepods, amphipods, and caridean shrimp. Bonefish are subject to occasional predation by sharks.

Halosaurs feed primarily on benthic prey, including worms and small benthic and epibenthic mollusks and such crustaceans as decapods and amphipods. Larger species also consume various fish. Spiny eels feed mainly on small benthic macrofauna, including worms and small crustaceans, such as amphipods and mysids. Species of the genus *Notacanthus* have specialized teeth that form a continuous serrated cutting edge probably used to crop sessile invertebrates. Little is known about which animals prey on halosaurs or spiny eels.

Reproductive biology

In Florida male bonefish reach sexual maturity at a fork length (measured from the tip of the snout to the fork in the tail) of about 15.7 in (400 mm) and an age of about 3.5 years. Florida females reach sexual maturity at a somewhat larger size, about 19.7 in (500 mm), and an age of about four. Gonadal activity is seasonal and peaks from November to May. Yolked oocytes are present in the ovaries in every month except August and September and are most abundant November to May. In Florida juvenile bonefish and post larvae recruit to sandy beach areas during winter and spring. Total fecundity ranges from 0.4 million to 1.7 million oocytes and increases with fish weight. Spawning areas of bonefish are unknown. Larvae reach a maximum size of about 3 in (76 mm). Bonefish live for at least 19 years. Growth of the bonefish is rapid until the age of about six years and then slows considerably.

Little is known about halosaur or spiny eel reproduction. In halosaurs spawning appears to be seasonal in some species, but others spawn year-round. It is unknown where the eggs and larvae develop. In spiny eels spawning occurs year-round. It is unknown where the eggs and larvae develop. Spiny eels have remarkable leptocephalus larvae that can reach lengths of 3.3–6.6 ft (1–2 m). Aside from their extremely large size, the larvae resemble those of eels in appearance.

Conservation status

No species of Albuliformes are included on the IUCN Red List.

Significance to humans

In many areas of the species' range, including the waters off the Florida Keys, bonefish are the basis of economically important recreational fisheries, among them a for-hire charter boat fishery in the Florida Keys. Bonefish are renowned by anglers for their wariness and fighting abilities and often are caught in water as shallow as 1 ft (0.3 m). In the Florida Keys fishing for bonefish is a year-round activity and provides a significant source of income to professional fishing guides. The commercial sale of bonefish in Florida is prohibited; the limits placed upon the recreational fishery for bonefish are a bag limit of one fish per angler per day and a minimum fish size of 18 in (457 mm) in total length. Bonefish are not considered a food fish in Florida, and most bonefish are released when caught. Halosaurs and spiny eels are of no commercial value.

1. Marine spiny eel (*Polyacanthonotus merretti*); 2. Halosaur (*Halosauropsis macrochir*). (Illustration by Emily Damstra)

Species accounts

Halosaur
Halosauropsis macrochir

FAMILY
Halosauridae

TAXONOMY
Halosaurus macrochir Gunther, 1878, off Gibraltar.

OTHER COMMON NAMES
None known.

PHYSICAL CHARACTERISTICS
Elongate, eel-like body, which tapers to a point with no caudal fin. The anal fin is elongate and extends along the posterior half of the body. There is a single short dorsal fin located just before the midpoint of the body. All fins have soft rays with no spines. The mouth is inferior, and the snout projects well beyond the tip of the lower jaw. Among the largest of halosaurs, reaching a length of almost 3.3 ft (1 m). Can be distinguished from other halosaurs by the deeply pigmented sheath of the conspicuous lateral line. Black in color. Occurs at depths of 3,281–9,843 ft (1,000–3,000 m) in the Atlantic and Indian Oceans. Also reported from waters off New Zealand.

DISTRIBUTION
Eastern Atlantic from Ireland to Mauritania and South Africa; western Atlantic, including Canada to 25°N, and off southern Brazil; western Pacific, including Australia, New Zealand, and Japan; and western Indian Ocean.

HABITAT
Found over the continental slope and rise. Little is known regarding specific habitat requirements. Appears to be widespread.

BEHAVIOR
Little is known. Usually seen moving slowly just over the bottom.

FEEDING ECOLOGY AND DIET
Feeds principally on benthic prey, including worms and small benthic and epibenthic mollusks and crustaceans, such as decapods and amphipods. Larger specimens also consume some fish.

REPRODUCTIVE BIOLOGY
Little is known regarding spawning. It is unknown where the eggs and larvae develop. Eggs develop into leptocephalus larvae.

CONSERVATION STATUS
Not listed by IUCN. Stocks probably have not been affected by human activities.

SIGNIFICANCE TO HUMANS
Because of its occurrence at great depths, the species is of no economic importance. ◆

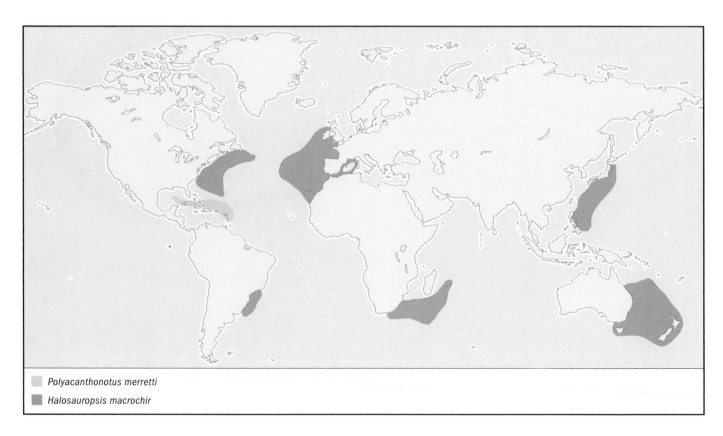

Polyacanthonotus merretti
Halosauropsis macrochir

Marine spiny eel
Polyacanthonotus merretti

FAMILY
Notacanthidae

TAXONOMY
Polyacanthonotus merretti Sulak et al., 1984, the Bahamas.

OTHER COMMON NAMES
None known.

PHYSICAL CHARACTERISTICS
Elongate, eel-like body, which tapers to a point with little or no caudal fin. The anal fin is elongate and extends along the posterior half of the body. It consists of spines anteriorly grading to soft rays posteriorly. The dorsal fin in most species has 28–36 isolated spines. The mouth is inferior, and the snout projects beyond the tip of the lower jaw. Attains a length of about 11.8 in (300 mm). Typically tan in color. Found on both sides of the North Atlantic, predominantly tropical in range. Occurs at depths from 1,969 to 6,562 ft (600–2,000 m); most at 3,281–4,921 ft (1,000–1,500 m).

DISTRIBUTION
Western Atlantic off the Bahamas.

HABITAT
Found over the continental slope and rise. Little is known regarding specific habitat requirements. Appears to be widespread.

BEHAVIOR
Little is known.

FEEDING ECOLOGY AND DIET
Feeds principally on small benthic macrofauna, including worms and small crustaceans, such as amphipods and mysids.

REPRODUCTIVE BIOLOGY
Appears to spawn year-round. It is not known where the eggs and larvae develop. Eggs develop into leptocephalus larvae.

CONSERVATION STATUS
Not listed by IUCN. Stocks probably have not been affected by human activities.

SIGNIFICANCE TO HUMANS
Because of its occurrence at great depths, the species is of no economic importance.

Resources

Books
Hildebrand, S. F. "Family Albulidae." In *Fishes of the Western North Atlantic*, edited by H. B. Bigelow. Vol. 3. New Haven: Sears Foundation for Marine Research, Yale University, 1963.

Periodicals
Colborn, J., R. E. Crabtree, J. B. Shaklee, E. Pfeiler, and B. W. Bowen. "The Evolutionary Enigma of Bonefishes (*Albula* spp.): Cryptic Species and Ancient Separations in a Globally Distributed Shorefish." *Evolution* 55, no. 4 (2001): 807–820.

Crabtree, Roy E., Christopher W. Harnden, Derke Snodgrass, and Connie Stevens. "Age, Growth, and Mortality of Bonefish, *Albula vulpes*, from the Waters of the Florida Keys." *Fishery Bulletin* 94 (1996): 442–451.

Crabtree, Roy E., Derke Snodgrass, and Christopher W. Harnden. "Maturation and Reproductive Seasonality in Bonefish, *Albula vulpes*, from the Waters of the Florida Keys." *Fishery Bulletin* 95 (1997): 456–465.

Crabtree, Roy E., K. J. Sulak, and J. A. Musick. "Biology and Distribution of Species of *Polyacanthonotus* (Pisces: Notacanthiformes) in the Western North Atlantic." *Bulletin of Marine Science* 36, no. 2 (1985): 235–248.

Roy E. Crabtree, PhD

Anguilliformes
(Eels and morays)

Class Actinopterygii
Order Anguilliformes
Number of families 15

Photo: The pale blue, mottled head of an American conger (*Conger oceanicus*) on a rock off the Island of Mahe, in the Seychelles. (Photo by Lawson Wood/Corbis. Reproduced by permission.)

Evolution and systematics

Fossil Anguilliformes are known from the Upper Cretaceous (about 93 million years ago) until the Pliocene (about two million years ago) and have been found in Africa, Europe, North America, the East Indies, Australia, and New Zealand. The Anguilliformes also are called Apodes ("limbless"), because of their lack of protruding fins, and true eels, because there are many other fishes (about 45 families) that do not belong to this group but have similar burrowing habits, and an eel-like shape as a result of convergent evolution. Anguilliformes are related to the Elopiformes (tarpons), the Albuliformes (spiny eels and halosaurs), and the Saccopharyngiformes (snipe and gulper eels) because they all have a leptocephalus, or ribbonlike, larval stage during development. The larval stage groups them into the subdivision or superorder called Elopomorpha. Some researchers, such as Filleul and Lavoué, have questioned this phylogenetic relationship based on molecular studies. Nelson divided this order into three suborders and 15 families (Anguillidae: 15 spp.; Heterenchelyidae: 8 spp.; Moringuidae: 6 spp.; Chlopsidae: 16 spp.; Myrocongridae: 2 spp.; Muraenidae: 200 spp.; Synaphobranchidae: 26 spp.; Ophichthidae: 250 spp.; Colocongridae: 5 spp.; Derichthyidae: 3 spp.; Muraenesocidae: 8 spp.; Nemichthydae: 15 spp.; Congridae: 150 spp.; Nettastomatidae: 30 spp.; Serrivomeridae: 10 spp.). Much more work is needed in this area to determine the exact phylogenetic relationships within this group.

Physical characteristics

In addition to their eel-like bodies, anguilliform species have widely varying coloration that ranges from black or dark gray in deep-sea species to rich colors and complex patterns in tropical reef species. Adult sizes range from about 4 in (10 cm) to 11.5 ft (3.5 m), as in the moray species *Thyrsoidea*

macrura. Systematists have emphasized numerous other morphological characteristics that have been found useful for phylogenetic purposes, including the lack of pelvic fins and the continuous dorsal, anal, and caudal fins that can have up to 650 soft rays, giving some individuals the appearance of having a pointed tail. Most species do not have pectoral fins, but when they are present, they lack bony connections to the skull. Most species also lack scales; in those species that have them, they are cycloid in type and embedded under the skin. The gill openings usually are narrow, with the gill region elongated and the gills displaced posteriorly. These species also have lost gill rakers. The skeleton is reduced, but the vertebrae may number as many as 700. They lack both pyloric caeca and oviducts but have retained the swim bladder. In summary, this order has many morphological simplifications or losses as a result of their evolutionary trend toward a wormlike configuration; the increased number of vertebrae is the result of the same phenomenon.

Distribution

Both the current distribution and the fossil record indicate that the members of this order always have occupied the same geographical areas, that is, tropical and temperate ocean. Anguilliformes are found in rivers draining into the North Atlantic, Baltic, and Mediterranean. They also have been introduced into Asia, South America, and Central America, but for the most part they have not reproduced in those areas. However, Anguillidae have a more restricted distribution, and do not inhabit the eastern Pacific and South Atlantic.

Habitat

The order Anguilliformes can be found in a wide variety of marine, brackish, and freshwater habitats, including streams,

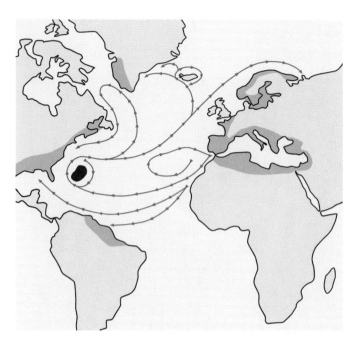

Migration pattern of the American eel and the European eel. (Illustration by Barbara Duperron)

lakes, deep-sea waters, and coral reefs. Some representatives of this order are catadromous, meaning that adults spend most of their lives in estuaries and freshwater and then move to the sea to spawn. The same species can be found in marine, estuarine, and freshwater environments. While some are pelagic, most are found living in small openings in coral reefs and rocks or burrowing in soft substrates. In general, morays and congers inhabit coral reefs and rock crevices, whereas certain congrids of the subfamily Heterocongrinae form vast colonies of up to several hundred individuals in tropical reef areas. Despite the fact that they favor these specific habitats as adults, all of the leptocephalus larvae form part of the marine plankton at one time or another in their life cycle.

Behavior

One of the most extraordinary aspects of their biology is their ability to migrate, yet they are slow swimmers. They swim by means of sinuous lateral movements of the body and median fins. Another interesting aspect of their swimming behavior is the ability of burrowing species to swim backward, which allows them to retreat rapidly into their burrows while still being able to look at any potential enemy. Although they can congregate in large numbers under specific circumstances, both larvae and adults do not form schools and thus can be considered to be solitary.

Feeding ecology and diet

The species of this order can be labeled generalists and opportunistic feeders, to the point that virtually any animal species they encounter can become a source of food for them, from aquatic insects in freshwater to crustaceans and many

other species of fishes. This flexibility toward food items and even feeding habits is evident during development: depending upon the stage of development, they will shift toward the most appropriate food source and capturing tactics. Extreme cases include the parastic snubnosed eel, *Simenchelys parasitica*, which burrows into the tissues of other species of fish. They can attach themselves to the heart of their host, from which they consume the blood. Other species feed on dead animals that lie on the bottom, including whales. This has led to a renewed interest in the ecological role played by some anguilliforms in benthic habitats, including the recycling of nutrients. Anguilliformes are preyed upon mostly by other types of fish. When they are in larval form, small fish and even some invertebrates will prey on them. As they grow larger, the size of their predators also increases.

Reproductive biology

The migratory and reproductive biological characteristics of anguilliforms are intertwined closely; thus, one cannot be explained without explaining the other. Although the life cycle of every anguilliform species has yet to be studied, it is believed that all of them undergo the same complex path of development, regardless of the final habitat they occupy. Fertilization among these fishes is external, and the eggs are relatively large (about 0.98 in, or 2.5 mm), which allows them to undergo extended development even before being able to

A yellow moray (*Gymnothorax prasinus*) in Wreck Bay, New South Wales, Australia. (Photo by Animals Animals ©A. Kuiter-OSF. Reproduced by permission.)

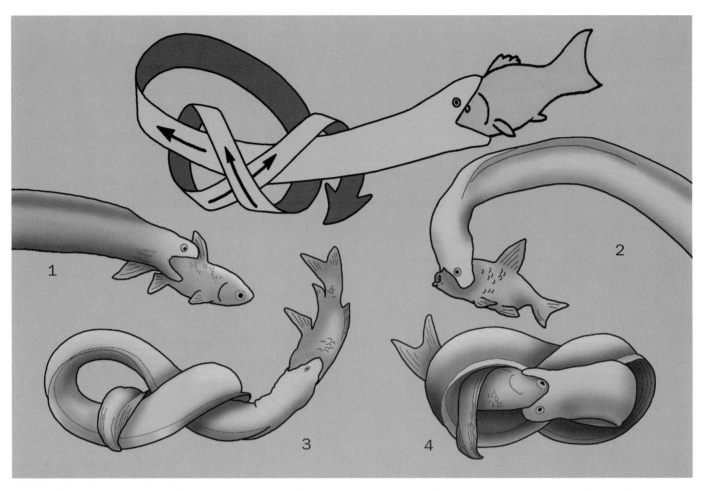

Certain eels possess the ability to tear apart prey items by tying themselves into a knot in order to obtain leverage against the prey item. The general procedure is this: (1) the eel grabs a fish that is too large to swallow whole, often by the head (2). The tail of the eel then turns back toward the eel's body and forms a series of interlacing loops which forms a knot similar to a square knot or a figure-eight knot (3). This knotting process continues until the heads of the eel and prey contact the knotted eel's body. The eel then pulls its own head through the knot and with it a mouthful of food (4). The prey fish is generally decapitated by this action. The eel then bites onto another section of the prey animal and the process continues. (Illustration by Dan Erickson)

feed. The eggs hatch, producing a prolarva, which in turn transforms into the leptocephalus larva.

The leptocephalus larva is so singular that biologists have studied it closely since the nineteenth century, when many researchers thought these larvae were actually adult fishes, given their complex morphologic features and behavior. They are elongated and laterally compressed while being transparent and gelatinous, which could make them difficult to detect. Although there is a great deal of morphological diversity among leptocephalus larvae, they all have a small and round caudal fin that is continuous with the dorsal and anal fins. This gives them varied shapes that are leaflike in appearance. In fact, the diversity of larval morphological features even within the same species is such that it is difficult to tell which larva belongs to which adult form. A couple of important characteristics of these larvae is their W-shaped myomeres (muscle packages) and prominent sharp teeth. These two features, together with their size, usually 2–4 in (5–10 cm) in length, make them sustained swimmers and powerful predators of other planktonic organisms. Some,

like the slender snipe eel, *Nemichthys scolopaceus*, can reach 18 in (45 cm) in length, undoubtedly very large for fish larva.

Leptocephalus larvae can be found at varying depths, from the surface of the ocean to 1,600 ft (500 m). As opportunistic feeders, they eat anything that is available, from diatoms to small crustaceans and other fish larvae. By the same token, they are preyed upon by different species of fish. It has been calculated that, on average, of six million eggs released by the European eel, *Anguilla anguilla*, only one survives to reach adulthood.

Leptocephalus larvae undergo metamorphosis in the open ocean after a period that ranges from six months to three years. In general, it can be said that the colder the waters, the longer the larval stage. The juveniles usually look like smaller versions of the adults. These juveniles are the product of many changes that can be summarized as follows: (a) reduction in the total body mass (up to 90% of weight) and body length, making the initial juvenile smaller than the larva itself; (b) transformation of the leaflike shape into a cylindrical shape;

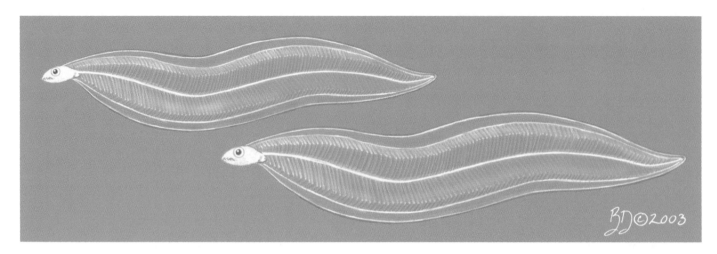

Eels go through a larval (leptocephalus) stage during which they are paper-thin, as shown by these American eel larvae (*Anguilla rostrata*). (Illustration by Barbara Duperron)

(c) loss of larval teeth; (d) loss of larval melanophores (pigment cells); (e) loss of pectoral fins; and (f) change in the position of the dorsal fin to much farther back.

The juveniles use oceanic currents to disperse; once they have occupied what is going to be their habitat as adults, they continue to grow and mature. This process can be quite lengthy, up to 10 years for some species. The complexity of this process also involves their sexual maturation, which includes phases of neutrality, precocious feminization, and juvenile hermaphroditism before they become adult males or females. As in some reptiles and other species of vertebrates, the sex ratio (proportion of males to females) can be the result of environmental factors (the more stressful the environmental conditions, the higher the proportion of females). Once true eels become fully adult, they undergo either a short-distance or a long-distance migration to a spawning area.

Conservation status

No anguilliform species have been listed by the IUCN under any category. With freshwater habitat modification and the threat posed to coral reefs all over the world, however, several species could be considered threatened in one way or another.

Significance to humans

Eels, whether "true eels" or otherwise, have been mentioned in mythology from ancient Greece to Polynesia. Today, only the freshwater eels (family Anguillidae) are of major economic importance in areas in which they are abundant, because of their value as food at both juvenile and adult stages. Some morays and congers are popular in public aquaria and among marine aquarists.

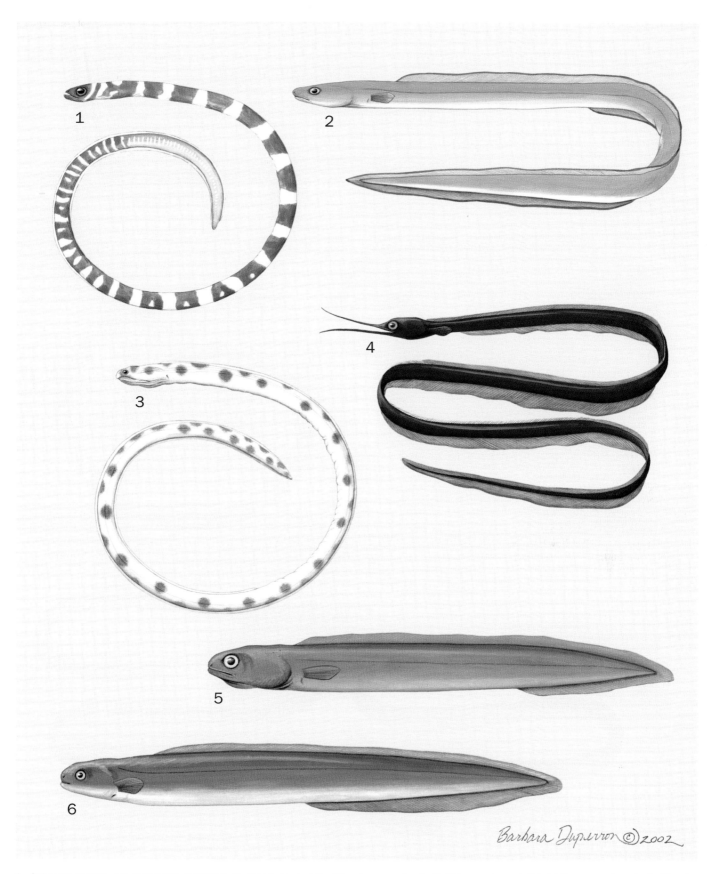

1. Splendid garden eel (*Gorgasia preclara*); 2. American conger (*Conger oceanicus*); 3. Tiger snake eel (*Myrichthys maculosus*); 4. Slender snipe eel (*Nemichthys scolopaceus*); 5. Froghead eel (*Coloconger raniceps*); 6. Snubnosed eel (*Simenchelys parasitica*). (Illustration by Barbara Duperron)

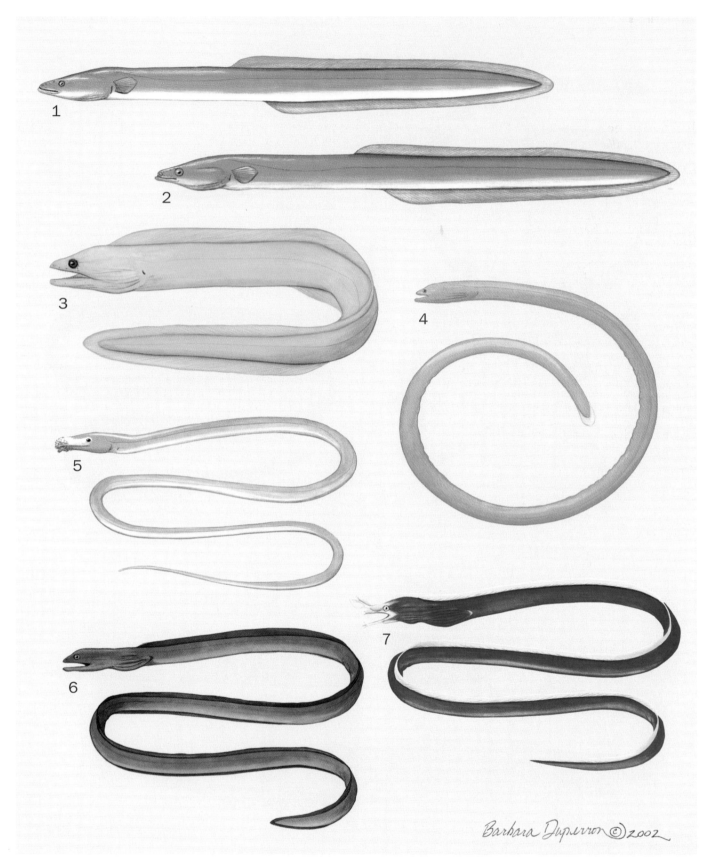

1. European eel (*Anguilla anguilla*); 2. American eel (*Anguilla rostrata*); 3. Green moray (*Gymnothorax funebris*); 4. Rusty spaghetti eel (*Moringua ferruginea*); 5. Pignosed arrowtooth eel (*Dysomma brevirostre*); 6. Slender giant moray (*Strophidon sathete*); 7. Ribbon moray (*Rhinomuraena quaesita*). (Illustration by Barbara Duperron)

Species accounts

European eel
Anguilla anguilla

FAMILY
Anguillidae

TAXONOMY
Muraena anguilla Linnaeus, 1758, "Europe." Tucker (1959) suggested that the European eel and the American eel, *A. rostrata*, are the same species.

OTHER COMMON NAMES
English: Common eel; French: Anguille; German: Aal; Spanish: Anguila.

PHYSICAL CHARACTERISTICS
Specimens have been reported to reach 52.36 in (133 cm) in length, with a weight of 14.548 lb (6.599 kg). Distinguished from other types of freshwater eels mostly by the number of vertebrae, which range from 110 to 119. Color greenish-brown to yellowish-brown. It has small vertical gill openings that are restricted to the sides. The lower jaw is slightly longer and projects. The dorsal fin originates far behind the pectoral fins, whereas the anal fin originates slightly behind the anus and well back from the origin of the dorsal fin.

DISTRIBUTION
Rivers of the North Atlantic, Baltic, and Mediterranean, along the coasts of Europe from the Black Sea to the White Sea. Its spawning area is the western Atlantic, specifically the Sar-

gasso Sea. It has been introduced successfully into Finland and Romania. Introductions in Norway, Israel, Japan, Taiwan, Brazil, Indonesia, California, Eritrea, and Jordan have not been successful.

HABITAT
Waters where the temperatures range from 32–86°F (0–30°C). Young eels grow in freshwater where they stay for 6–12 years (males) or 9–20 years (females). After becoming sexually mature, they migrate to the sea, where they can be found in deep waters living on the bottom, under stones, in the mud, or in crevices. Spawning takes place in the Sargasso Sea. The larvae are brought by the Gulf Stream to the coasts of Europe. They evolve into small eels before moving into freshwater basins.

BEHAVIOR
The European eel spawns in the Sargasso Sea in the subtropical northwestern Atlantic Ocean. Their larvae, leptocephali, are transported by the Gulf Stream and North Atlantic current system to Europe. Despite being an individualistic species, large groups of elvers and young eels can be observed from time to time in estuaries and rivers. An elver is a small cylindrical young eel, more advanced in development than a leptocephalus larva but less developed than an adult eel. Those congregations of elvers and juveniles are not fish schools in the real sense of the word (active assembling for selective advantages such as protection against predators or reproduction) but rather a mass response to environmental conditions. When elvers and young eels are observed in mass from time to

Anguilla anguilla
Anguilla rostrata

time in estuaries and rivers, it is because they are responding individually to particular ecological conditions and not because they are actually forming schools.

FEEDING ECOLOGY AND DIET
Opportunistic feeders. They include among their food items almost any species of aquatic fauna, freshwater as well as marine, that they encounter. Adults do not feed during migration to sea. Other eels, herons, cormorants, pikes, and gulls prey upon them.

REPRODUCTIVE BIOLOGY
The American eel (*A. rostrata*) and the European eel (*A. anguilla*) spawn in Sargasso Sea, located in the subtropical northwestern Atlantic Ocean, between January and May. Their larvae, leptocephali, are transported by the Gulf Stream and North Atlantic current system to North America and Europe, respectively.

Before entering the continental coastal zones and estuaries, the leptocephali transform into elvers. Once there, and before entering the freshwaters, they develop into small (juvenile) eels. The young eels spend their growing period in freshwater, where males stay for 6 to 12 years; females spend from 9 to 20 years there. While in freshwater, they live on the bottom under stones or in the mud or rock crevices. At the end of their growth period, the eels become sexually mature and migrate to the sea, where they inhabit deep waters.

There is a significant differential in time in the life cycle span between both species. The overall mean age of European elvers is 350 days at metamorphosis (from leptocephalus to glass eel) and 448 days at estuarine arrival, with 98 days between metamorphosis and estuarine arrival. These ages were all significantly greater than those of American elvers 200, 55, and 255 days, respectively. Also, growth rate of the American eel (0.008 in [0.21 mm] per day) is greater than that of European eel (0.006 in [0.15 mm] per day). This is a result of delayed metamorphosis in the European species, which allows the European eel larvae to be transported from North America to Europe by the oceanic current. Thus, the European eel evolves the strategy to delay metamorphosis by reducing growth rate, enabling it to segregatively migrate with the American eel. The differences in leptocephalus stage duration and growth rate are the principal factors determining the segregation of migrating American and European eels.

CONSERVATION STATUS
Not listed by the IUCN.

SIGNIFICANCE TO HUMANS
The European eel is consumed fresh, dried or salted, or smoked, and it can be fried, boiled, and baked. It is particularly popular among Mediterranean Europeans. This species has been raised by the aquaculture industry, particularly in Japan and Taiwan, with some success. ◆

American eel
Anguilla rostrata

FAMILY
Anguillidae

TAXONOMY
Muraena rostrata Lesueur, 1817, Cayuga Lake, New York.

OTHER COMMON NAMES
English: Common eel; French: Anguille américaine; German: Amerikanischer Aal; Spanish: Anguila Americana.

PHYSICAL CHARACTERISTICS
Males grow to 59.84 in (152 cm) and females to 47.24 in (120.0 cm); these eels weigh as much as 16.16 lb (7.330 kg). The major difference between the European eel and the American eel is the number of vertebrae, which is 110 to 119 and 103 to 111, respectively. Otherwise, the species are almost identical.

DISTRIBUTION
Western Atlantic from Greenland and the Atlantic coast of Canada and the United States to Panama and throughout much of the West Indies south to Trinidad. The range includes the Great Lakes, the Mississippi River, and the Gulf basin. It has been introduced to Guam and Japan.

HABITAT
At sea they are found over rather deep waters; in freshwater they are inhabit permanent streams with continuous flow.

BEHAVIOR
Individuals of this species are solitary and nocturnal. While in freshwater, they hide during the day in undercut banks and in deep pools near logs and boulders and sometimes bury themselves in the substrate, whether mud, sand, or gravel. At night they typically swim near the bottom in search of food. They can breathe through the skin along with their gills and are able to live for several hours outside water.

FEEDING ECOLOGY AND DIET
Like the European eel, food items vary with the stage of development and location. The leptocephalus larvae, for example, is planktivorous; the elver feeds on aquatic insects, small crustaceans, and dead fish; and the adult eats insects, crustaceans, clams, worms, fish, frogs, toads, and dead animal matter. Sharks are their main predator.

REPRODUCTIVE BIOLOGY
Despite many attempts to conduct direct observations, knowledge of reproductive behavior can only be inferred, based on circumstantial evidence. We know that during the autumn adults migrate to the Sargasso Sea to spawn, with spawning taking place in January. At that time, females lay up to four million buoyant eggs, dying shortly after. After fertilizing the eggs, the males also die. With the help of ocean currents, the leptocephalus larvae drift toward coastal waters for as long as 18 months. After becoming an elver, American eels undergo a slow transformation that includes increases in their size, eye diameter relative to body size, and in the amount of eye pigments. They also become darker along the body. They spend most of their lives (up to 20 years) in freshwater before returning to the sea for spawning.

CONSERVATION STATUS
Not listed by the IUCN. It has been listed as "rare" by a number of U.S. counties and states, but lacks specific legislation to protect it. Nonetheless, fishery authorities in the United States are taking measures to decrease the impact of fisheries, particularly at the larval and elver level. The Atlantic States Marine Fisheries Commission is preparing a Fishery Management Plan (FMP), requesting that the U.S. federal government include this species under some protection status under the supervision of the U.S. Fish and Wildlife Service.

SIGNIFICANCE TO HUMANS
They are consumed as food and prepared in many ways. Larvae and elvers (considered a delicacy) are captured using fine

mesh fyke nets and dip nets; adults are caught with eel pots and trot lines. Although they can be caught in considerable numbers, their handling can be difficult, because the adults exude a noticeable layer of slime over the entire body. Moreover, large eels actively bite when caught on a hook and line. ◆

Froghead eel

Coloconger raniceps

FAMILY
Colocongridae

TAXONOMY
Coloconger raniceps Alcock, 1889, Bay of Bengal.

OTHER COMMON NAMES
Japanese: Fusa-anago.

PHYSICAL CHARACTERISTICS
May grow to 19.7 in (50 cm). The body is stubbier (particularly in the anterior region) than the bodies of most true eels. They have numerous pores in short tubes.

DISTRIBUTION
Indo-West Pacific area, from East Africa and Madagascar in the west to the western Pacific in the east to southern Japan in the north.

HABITAT
Deep-sea species usually found at depths between 980 and 3,720 ft (300–1,113 m).

BEHAVIOR
Because of its deep-water habits, it is rarely observed except very occasionally by deep sea submersibles.

FEEDING ECOLOGY AND DIET
Apparently feeds on other fishes.

REPRODUCTIVE BIOLOGY
Nothing is known.

CONSERVATION STATUS
Not listed by the IUCN.

SIGNIFICANCE TO HUMANS
None known. ◆

American conger

Conger oceanicus

FAMILY
Congridae

TAXONOMY
Anguilla oceanica Mitchill, 1818, Atlantic.

OTHER COMMON NAMES
English: Conger eel, sea eel; French: Congre d'Amérique; Spanish: Congrio americano.

PHYSICAL CHARACTERISTICS
Specimens may reach 90.6 in (230 cm) and 88.2 lb (40 kg). The species has a long snout and a very large dorsal fin that originates much closer to the pectoral fins. It is gray on the dorsum and white on the venter.

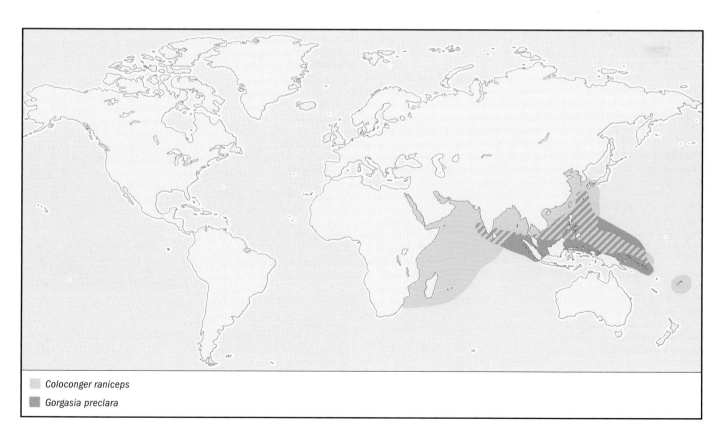

Coloconger raniceps

Gorgasia preclara

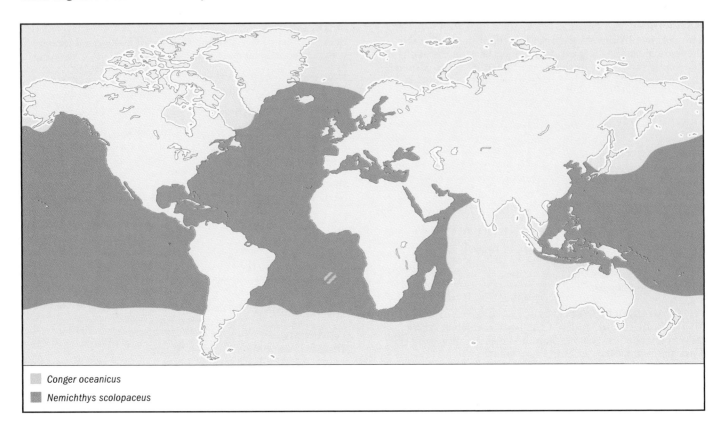

Conger oceanicus
Nemichthys scolopaceus

DISTRIBUTION
Western Atlantic from Cape Cod, Massachusetts, to northeastern Florida in the United States as well as in the northern Gulf of Mexico westward to the Mississippi delta. In the eastern Atlantic it has been reported around Saint Helena Island, South Carolina.

HABITAT
Usually inhabits shallow inshore waters to depths of 1,565 ft (477 m).

BEHAVIOR
Chiefly nocturnal feeder in shallow waters (60 ft [18 m] or less).

FEEDING ECOLOGY AND DIET
Feeds mainly on fishes but also on shrimps, worms, and other small benthic organisms.

REPRODUCTIVE BIOLOGY
Spawning occurs from June through August. The leptocephalus larva reaches a maximum length between 5.9 and 6.3 in (15–16 cm). Metamorphosis consists of thickening of the head and body and development of the swim bladder, permanent teeth, and pigment in the skin.

CONSERVATION STATUS
Not listed by the IUCN.

SIGNIFICANCE TO HUMANS
Anglers along piers, docks, and jetties in the mid-Atlantic states commonly catch this species. It is caught in baited fish and crab traps as well as on hook and line but seldom in nets, because the fish can squirm through them. It is difficult to remove them from hooks. They are marketed fresh and salted in the Chesapeake Bay region, but today the species is not subject

to commercial fishing in the United States and is rarely eaten. American congers are much more appreciated in parts of Europe, Africa, and Asia, where they are smoked. ◆

Splendid garden eel
Gorgasia preclara

FAMILY
Congridae

TAXONOMY
Gorgasia preclara Böhlke and Randall, 1981, Sumilon Island, Philippines.

OTHER COMMON NAMES
English: Orange-barred garden eel.

PHYSICAL CHARACTERISTICS
Individuals may reach 15.75 in (40 cm) in length. They have slender and elongated bodies with short mouths, anterior nostrils on the snout tip between restricted labial flanges, and small pectoral fins. The number of vertebrae ranges from 144 to 156.

DISTRIBUTION
Indo-West Pacific region from the Maldives in the west to Papua New Guinea in the east and from the Philippines and Ryukyu Islands in the north to the Coral Sea in the south.

HABITAT
Found in colonies on sand slopes exposed to current at depths usually below 90 ft (30 m).

BEHAVIOR
Live individually in burrows, forming large colonies. They hover above their sand burrows and retreat tail first when disturbed.

FEEDING ECOLOGY AND DIET
They feed on plankton that they capture while standing in their burrows.

REPRODUCTIVE BIOLOGY
Nothing is known.

CONSERVATION STATUS
Not listed by the IUCN.

SIGNIFICANCE TO HUMANS
None known. ◆

Rusty spaghetti eel
Moringua ferruginea

FAMILY
Moringuidae

TAXONOMY
Moringua ferruginea Bliss, 1833, Island of Mauritius.

OTHER COMMON NAMES
English: Slender worm-eel; Gela (Solomon Islands): Poli ni tahi.

PHYSICAL CHARACTERISTICS
May reach 55.1 in (140 cm) in length. It has a wormlike, very elongated body with yellow to reddish coloration. The dorsal and anal fins are reduced to low folds. It lacks scales and has greatly reduced eyes. The gill openings are low on the body. The rusty spaghetti eel has about 73 lateral-line pores before the anus.

DISTRIBUTION
Indo-Pacific, from East Africa in the west to Easter Island in the east and from the Ryukyu Islands in the north to Australia in the south. It is distributed throughout Micronesia as well.

HABITAT
Found in sandy bottoms.

BEHAVIOR
Fossorial in that it can burrow headfirst. The physical attributes, such as an elongated body and reduced eyes, allow for this behavior.

FEEDING ECOLOGY AND DIET
Feeds on small prey found either on the bottom or burrowed in the sand.

REPRODUCTIVE BIOLOGY
Little is known, except that rusty spaghetti eels seem to show sexual dimorphism in size and coloration.

CONSERVATION STATUS
Not listed by the IUCN.

SIGNIFICANCE TO HUMANS
None known. ◆

Green moray
Gymnothorax funebris

FAMILY
Muraenidae

TAXONOMY
Gymnothorax funebris and *Lycodontis funebris* Ranzani, 1840, Atlantic Ocean.

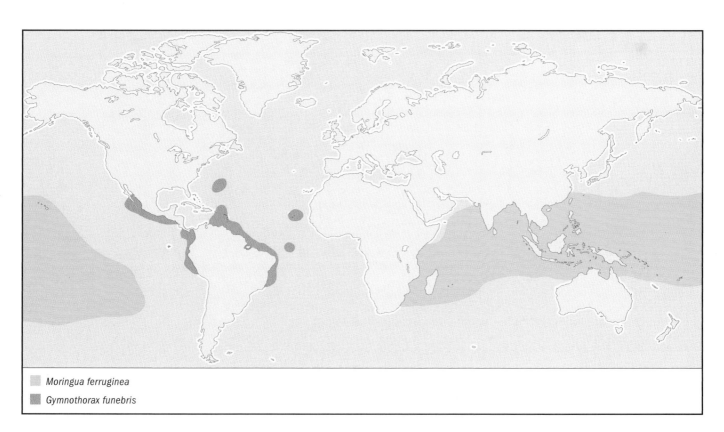

Moringua ferruginea
Gymnothorax funebris

OTHER COMMON NAMES
English: Black moray; French: Murène verte; Spanish: Culebra morena.

PHYSICAL CHARACTERISTICS
Grows to 98.5 in (250 cm) in length and weighs up to 64 lb (29 kg). It is considered the largest Atlantic moray. Individuals of this species are uniformly greenish to dark gray-greenish. The green moray's color is a result of a yellowish mucous over the animal's dark blue skin.

DISTRIBUTION
Distributed throughout the western and eastern Atlantic (from Nova Scotia, Canada, to Brazil, including the Gulf of Mexico and Bermuda) and the eastern Pacific.

HABITAT
Benthic and solitary species commonly seen along rocky shorelines, reefs, and mangroves, including dirty harbors, in waters shallower than about 90 ft (about 30 m).

BEHAVIOR
Cleaned by some species of gobies and other fish species, as observed on the coral reefs in Bonaire and the Netherlands Antilles and at the Fernando de Noronha archipelago in the western South Atlantic.

FEEDING ECOLOGY AND DIET
Feeds on fishes and benthic crustaceans.

REPRODUCTIVE BIOLOGY
Little is known about reproduction, except that green morays have external fertilization and, like any other anguilliform, they have a leptocephalus larval stage.

CONSERVATION STATUS
Not listed by the IUCN.

SIGNIFICANCE TO HUMANS
Green morays are consumed as food and are marketed both fresh and salted. Large individuals are ciguatoxic, however. Ciguatera is a type of food poisoning caused by the consumption of tropical marine species that harbor a heat-resistant, acid-stable toxin known as ciguatoxin. The green moray consumes certain species of microorganisms that form this toxin. This natural toxin can concentrate as it moves up the food chain, but its adverse effects appear to be limited to humans. Because of its large size and aggressiveness, the bites of this moray are particularly dangerous. ◆

Ribbon moray
Rhinomuraena quaesita

FAMILY
Muraenidae

TAXONOMY
Rhinomuraena quaesita Garman, 1888, Marshall Islands.

OTHER COMMON NAMES
English: Ribbon eel, black ribbon eel; French: Rhinomurène noire; Samoan: Pusi.

PHYSICAL CHARACTERISTICS
May reach 51.2 in (130 cm) in length. It has a very elongated body. Mature males are mostly blue, whereas mature females

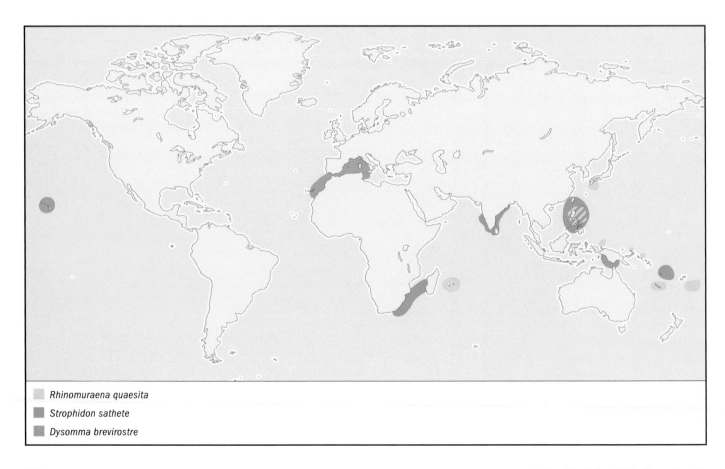

Rhinomuraena quaesita
Strophidon sathete
Dysomma brevirostre

are almost completely yellow. They have three fleshy tentacles on the tip of the lower jaw; a single fleshy, pointed projection at the tip of the snout; and tubular anterior nostrils ending in gaudy, fanlike expansions.

DISTRIBUTION
Indo-Pacific from East Africa in the west to the Tuamotu Archipelago in the east and from southern Japan in the north to New Caledonia and French Polynesia in the south, including the Marianas and Marshall Islands in Micronesia.

HABITAT
Lagoons and associated seaward reefs as deep as 180 ft (60 m).

BEHAVIOR
Secretive, nonmigratory species that normally hides in sand or rubble, sometimes with only its head protruding.

FEEDING ECOLOGY AND DIET
Feeds on small fishes.

REPRODUCTIVE BIOLOGY
Fertilization in this species is external. This may be the only moray that undergoes abrupt changes in coloration and sex. It is classified as a protandrous hermaphrodite, that is, functioning males reverse sex to become females.

CONSERVATION STATUS
Not listed by the IUCN.

SIGNIFICANCE TO HUMANS
It is acquired for display in aquaria because of its striking coloration and unusual morphological features. ◆

Slender giant moray
Strophidon sathete

FAMILY
Muraenidae

TAXONOMY
Muraenophis sathete Hamilton, 1822, Ganges River.

OTHER COMMON NAMES
English: Gangetic moray, giant estuarine moray; French: Murène fil géante; German: Süsswassermuräne; Spanish: Morenilla gigante; Tagalog (Philippines): Payangitan.

PHYSICAL CHARACTERISTICS
Specimens as large as 157.5 in (400 cm) have been recorded. Individuals of this species have a very elongated body. This species is brownish-gray dorsally and paler on the venter.

DISTRIBUTION
Indo-West Pacific Ocean, from the Red Sea and East Africa to the western Pacific.

HABITAT
Benthic muddy environments of marine and estuarine areas, including inner bays and rivers.

BEHAVIOR
The most interesting behavioral feature of this species is their ability to stand vertically from a burrow with the head kept horizontally beneath the surface, rising and falling with the tide.

FEEDING ECOLOGY AND DIET
Feeds on a wide variety of crustaceans and fishes.

REPRODUCTIVE BIOLOGY
Almost nothing is known about reproduction.

CONSERVATION STATUS
Not listed by the IUCN.

SIGNIFICANCE TO HUMANS
This species is consumed in India, the Philippines, Sri Lanka, South Africa, and other southeastern African countries as well as in Oceania. ◆

Slender snipe eel
Nemichthys scolopaceus

FAMILY
Nemichthyidae

TAXONOMY
Nemichthys scolopacea Richardson, 1848, type locality not available.

OTHER COMMON NAMES
English: Atlantic snipe eel, glass snake, threadfish; French: Avocette ruban; Spanish: Pez agazadicha.

PHYSICAL CHARACTERISTICS
Grows to 51.2 in (130 cm). They are extremely long eels whose posterior end is exceptionally narrow, to the point that it ends as a long filament. They have exceptionally long jaws that curve outward and do not close together except among fully mature males. They are also unusual because of their proportionally very large eye. In color they vary between dark brown and gray, often darker ventrally, with the anal fin and tips of the pectoral fins almost black. Males are quite different from females in that once they fully develop, their jaws shorten, and they lose their teeth. This feature led some researchers to believe that each sex was a separate species.

DISTRIBUTION
Worldwide in tropical and temperate seas. In the western Atlantic they range from Nova Scotia in Canada to the northern Gulf of Mexico and all the way south to Brazil. In the eastern Atlantic they are found from Spain to South Africa, including the western Mediterranean, although there are some reports from Iceland. In the northwestern Pacific they inhabit Japanese waters and the Arafura Sea. In the eastern Pacific they occur from Alaska to Chile, including the Gulf of California.

HABITAT
Pelagic, found mostly in middle to deep waters between 295 and 6,560 ft (91–2,000 m). The depth varies with the latitude—they occur in shallower waters at higher latitudes.

BEHAVIOR
As with many planktonic organisms, it is possible that their extremely elongated bodies are used to increase drag and therefore buoyancy in midwaters.

FEEDING ECOLOGY AND DIET
Feed on crustaceans while swimming with their mouths open.

REPRODUCTIVE BIOLOGY
They are oviparous, with external fertilization, buoyant eggs, and planktonic leptocephalus larva. The leptocephalus larva is

very elongated, with a filiform tail. The strong sexual dimorphism in the direction of degenerative changes in males and females suggests that they may display semelparity, that is, that they breed only once and then die immediately.

CONSERVATION STATUS
Not listed by the IUCN.

SIGNIFICANCE TO HUMANS
None known. ◆

Tiger snake eel
Myrichthys maculosus

FAMILY
Ophichthidae

TAXONOMY
Muraena maculosa Cuvier, 1816, Mediterranean Sea.

OTHER COMMON NAMES
English: Ocellated snake eel, spotted snake eel; Afrikaans: Swartogies-slangpaling; Tahitian: Puhi popooru.

PHYSICAL CHARACTERISTICS
Specimens may reach 39.4 in (100 cm). The young have black saddles. Adults are pale cream in color, with large and small black spots. All have a stiffened, pointed tail.

DISTRIBUTION
Indo-Pacific region from the Red Sea and East Africa in the east to the central Pacific in the west. The species does not

occur in the Hawaiian islands, where it has been replaced by *M. magnificus*.

HABITAT
Sandy areas of reef flats, lagoons, and seaward reefs. Lives buried in the sand.

BEHAVIOR
The most interesting behavioral characteristic of tiger snake eels is their ability to burrow tail first and then move equally forcefully forward and backward through the sediment. They may aggregate in large numbers under a light at night.

FEEDING ECOLOGY AND DIET
Feeds on small fishes and invertebrates.

REPRODUCTIVE BIOLOGY
Nothing is known.

CONSERVATION STATUS
Not listed by the IUCN.

SIGNIFICANCE TO HUMANS
None known. ◆

Pignosed arrowtooth eel
Dysomma brevirostre

FAMILY
Synaphobranchidae

TAXONOMY
Nettastoma brevirostre Facciolà, 1887, Sicily, Italy.

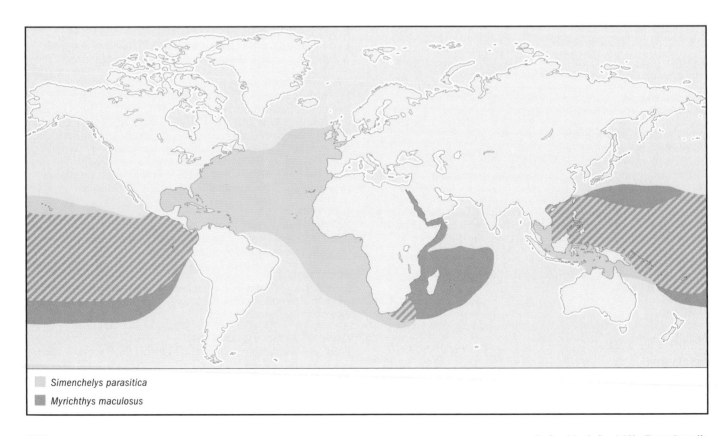

☐ *Simenchelys parasitica*
■ *Myrichthys maculosus*

OTHER COMMON NAMES
None known.

PHYSICAL CHARACTERISTICS
Individuals reach 11.8 in (30 cm) in length. Like most deep-sea fishes, they are pale in coloration. This species lacks pectoral fins as well as scales. The lower jaw is shorter than the upper jaw. It has between 193 and 204 vertebrae.

DISTRIBUTION
North Atlantic, from Madeira to the Gulf of Guinea, including the western Mediterranean. It is found all the way to Messina in Sicily. It also occurs off the coasts of Florida in the United States and has been recorded in Hawaii.

HABITAT
Muddy substrates of waters at a depth range between 1,150 and 2,130 ft (350–650 m).

BEHAVIOR
Burrowing, solitary species.

FEEDING ECOLOGY AND DIET
Probably feeds on small benthic fish and invertebrates.

REPRODUCTIVE BIOLOGY
Nothing is known.

CONSERVATION STATUS
Not listed by the IUCN.

SIGNIFICANCE TO HUMANS
None known. ◆

Snubnosed eel
Simenchelys parasitica

FAMILY
Synaphobranchidae

TAXONOMY
Simenchelys parasitica and *S. parasiticus* Gill, 1879, Massachusetts.

OTHER COMMON NAMES
English: Slime eel; French: Anguille à nez court.

PHYSICAL CHARACTERISTICS
Specimens may reach 24 in (61 cm). This species has a slimy body with a blunt, thick, round snout and a small mouth. The gill slits are broadly separated, and the scales are embedded in the skin. Coloration is gray to grayish-brown; it is darker on the fin edges and along the lateral line.

DISTRIBUTION
Worldwide species, particularly in tropical and subtropical waters.

HABITAT
Individuals of this species are found at depths of 1,200–8,700 ft (365–2,650 m), over muddy, deep-sea bottoms. They also parasite on other fishes.

BEHAVIOR
Little is known besides their feeding behavior. They are capable of homing in on dead animals.

FEEDING ECOLOGY AND DIET
Feeds on benthic invertebrates and fish, including dead tissue. It is parasitic on some fishes. Large, dead fishes may look as if they are alive as these eels feed inside their carcasses.

REPRODUCTIVE BIOLOGY
Nothing is known.

CONSERVATION STATUS
Not listed by the IUCN.

SIGNIFICANCE TO HUMANS
No significant economic importance but are of scientific value because their unusual ecological characteristics and feeding habits.

Resources

Books

Bertin, L. *Eels: A Biological Study.* New York: Philosophical Library, 1957.

Berra, Tim M. *Freshwater Fish Distribution.* San Diego: Academic Press, 2001.

Forey, P. L, D. T. J. Littlewood, P. Ritchie, and A. Meyer. "Interrelationships of Elopomorph Fishes." In *Interrelationships of Fishes,* edited by M. L. J. Stiassny, L. R. Parenti, and G. D. Johnson. New York: Academic Press, 1996.

Lee, D. S., C. R. Gilbert, C. H. Hocutt, R. E. Jenkins, D. E. McAllister, and J. R. Stauffer, Jr. *Atlas of North American Freshwater Fishes.* Raleigh: North Carolina State Museum of Natural History, 1980.

Moyle, Peter B., and Joseph J. Cech, Jr. *Fishes: An Introduction to Ichthyology.* Upper Saddle River, NJ: Prentice Hall, 2000.

Nelson, J. S. *Fishes of the World.* 3rd edition. New York: John Wiley and Sons, 1994.

Page, L. M., and B. M. Burr. *A Field Guide to Freshwater Fishes: North America North of Mexico.* Boston: Houghton Mifflin, 1991.

Randall, J. E., G. R. Allen, and R. C. Steene. *Fishes of the Great Barrier Reef and Coral Sea.* Honolulu: University of Hawaii Press, 1990.

Robins, C. Richard, and G. Carleton Ray. *A Field Guide to Atlantic Coast Fishes of North America.* Boston: Houghton Mifflin, 1986.

Tesch, F. W. *The Eel: Biology and Management of Anguillid Eels.* New York: John Wiley and Sons, 1977.

Periodicals

Bruun, A. F. "The Breeding of the North Atlantic Freshwater-Eels." *Advances in Marine Biology* 1 (1963): 137–170.

Resources

Costa, J. L., C. A. Assis, P. R. Almeida, F. M. Moreira, and M. J. Costa. "On the Food of the European Eel, *Anguilla anguilla* (L.), in the Upper Zone of the Tagus Estuary, Portugal." *Journal of Fish Biology* 41, no. 5 (1992): 841–850.

Deelder, C. L. "Synopsis of Biological Data on the Eel *Anguilla anguilla* (Linnaeus, 1758)." *FAO Fisheries Synopsis* 80, rev. 1 (1984): 1–73.

Filleul, A., and S. Lavoué. "Basal Teleosts and the Question of Elopomorph Monophyly: Morphological and Molecular Approaches." *Comptes Rendus de l'Académie des Sciences, Paris* 324 (2001): 393–399.

McCleave, J. D., P. J. Brickley, K. M. O'Brien, D. A. Kistner-Morris, M. W. Wong, M. Gallagher, and S. M. Watson. "Do Leptocephali of the European Eel Swim to Reach Continental Waters? Status of the Question." *Journal*

of the Marine Biological Association of the U.K.* 78 (1998): 285–306.

Romero, A., and J. Gimeno. "Las Anguilas, Eternas Pasajeras de las Aguas." *Algo* 286: 23-25.

Tucker, D. W. "A New Solution to the Atlantic Eel Problem." *Nature* 183 (1959): 495–501.

Wang, C.H. and W.N. Tzeng. "The Timing of Metamorphosis and Growth Rates of American and European Eel Leptocephali: A Mechanism of Larval Segregative Migration." *Fisheries Research* 46 (2000): 191-205.

Other

"Anguilliformes: Eels." (13 Nov. 2002). <http://www .floridasmart.com/subjects/ocean/animals_ocean_eels.htm>

Aldemaro Romero, PhD

Saccopharyngiformes

(Swallowers and gulpers)

Class Actinopterygii
Order Saccopharyngiformes
Number of families 4

Illustration: Gulper eel (*Saccopharynx ampulla-ceus*). (Illustration by Jacqueline Mahannah)

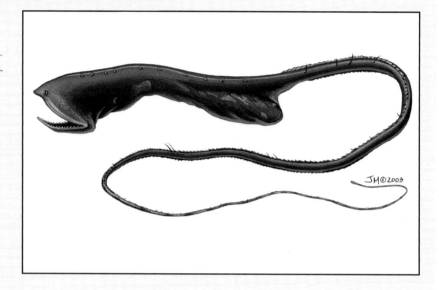

Evolution and systematics

The Saccopharyngiformes are divided into two suborders, the Cyematoidei, with the single family Cyematidae (with two monotypic genera), and the Saccopharyngoidei, which contains the other three families. Of these three families, the Monognathidae is the most diverse, with 14 identified species in the genus *Monognathus*. The Saccopharyngidae has 11 species in the single genus *Saccopharynx*; the closely related family Eurypharyngidae is monotypic. There is still controversy over the inclusion of the Cyematidae in this order, but they are placed here on the basis of reduction of skeletal features that are common among all four families. Systematists consider the Saccopharyngiformes to be quite different from anguilliform eels. The order is thought to consist of highly specialized fishes. All four families share numerous common features, most of which have to do with extreme loss of skeletal features, presumably the result of the extremely energy poor environment.

Within the Saccopharyngoidei, the eurypharyngids and saccopharyngids are superficially most similar in appearance and are considered the closest in taxonomic relationship. The Monognathidae represent a more advanced and highly specialized family, as evidenced by even greater reduction in skeletal components, that is, the loss of the upper jaw. The first fossil evidence for this order is reported to be from the Middle Cretaceous.

Physical characteristics

The loss of skeletal structures has resulted in fishes that are among the most unusual and striking in their appearance. Among other characteristics, all are scaleless, lack pelvic fins, and have very long dorsal and anal fins. All are rather "flabby" to the touch and presumably are very poor swimmers. In members of the Saccopharyngoidei, the mouths are very large to enormous, with distensible pharynges and stomachs, to allow for the capture of very large prey. Dentition varies among the families. Well-produced, posteriorly curved teeth are found in the Saccopharyngidae, with the other three families possessing small to minute teeth in the jaws.

Except for the enlarged head and mouth structure, the rest of the body of these fishes is elongated and very slender (filamentous in eurypharyngids and saccopharyngids). The body coloration varies from scattered pigment patches to a light uniform brown in monognathids, with dark brown to solid black in cyematids, eurypharyngids, and saccopharyngids. Thin white lines of unknown function extend from the head to the tail along the upper body in the saccopharyngids and eurypharyngids, and individuals in both families have luminous bulbs at the very tip of the filamentous tail. The filament may constitute 50% or more of the overall length of the fish. Overall lengths of the substantial part of the body in all saccopharyngiforms is small, not exceeding 19.6 in (50 cm).

Distribution

Saccopharyngids are most abundant and diverse in the Atlantic Ocean. *Eurypharynx pelecanoides* is well known from the Atlantic and central and eastern Pacific Oceans, and the monognathids are about equally diverse in the Atlantic (six species) and Pacific Oceans (seven species). Among the Cyematidae, *Cyema atrum* is widespread in the Atlantic, Pacific, and Indian Oceans, while *Neocyema erythrostoma* is only known from the eastern South Atlantic. Saccopharyngiformes have not been reported from the Mediterranean.

Habitat

These are primarily bathypelagic inhabitants, with the majority of adult specimens being collected at depths greater than 3,280 ft (1,000 m). Larvae and juveniles live in shallower waters, even into the upper mesopelagic zone below 656 ft (200 m).

Behavior

Owing to the extreme depths at which these fishes live, there are few reports of any behavior.

Feeding ecology and diet

All saccopharyngiform species are poor swimmers at best. There have been no reports on feeding in the cyematids, but it is thought that eurypharyngids and saccopharyngids draw prey close to them by means of luminescent lures on their tails and then quickly open their mouths to suck food in. Saccopharyngids are piscivorous (eat only other fish); eurypharyngids take a broader range of fish and invertebrate prey. An even more unusual form of feeding has been postulated for monognathids. It is thought that their prey (crustaceans) may be lured by scent released from glands around the mouth; when they come close enough, the fish bite them by means of a hollow fang in the mouth that injects venom, much like a rattlesnake. The fish then swallows the dead or dying shrimp whole. Little is known about the predators that feed in members of this order.

Reproductive biology

Nothing is known about the cyematoids, other than that they have separate sexes that do not appear to exhibit dimorphism. They also have leptocephalus larvae—a thin, largely transparent, ribbonlike stage that is common to several primitive orders of bony fishes (Elopiformes and Anguilliformes), including all saccopharyngiforms. In all three families of saccopharyngoids, sexually mature individuals are dimorphic. Males have greatly enlarged nasal structures and slightly enlarged eyes, and the jaws and stomachs in both males and females atrophy. It is widely believed that males locate their mates by following scent trails of pheromones released by the females and that spawning is a terminal event, with both individuals dying after mating. This reproductive pattern has been found in a number of shallow water eels and other fish species.

Conservation status

There are no known conservation measures specific to these families. No species from either family is listed on the IUCN Red List.

Significance to humans

Owing to their rarity and poorly studied biological characteristics, no significance can be attributed to saccopharyngiforms. They are objects of curiosity because of their extreme body specializations.

1. Bobtail snipe eel (*Cyema atrum*); 2. Pelican eel (*Eurypharynx pelecanoides*); 3. *Monognathus rosenblatti*; 4. Gulper eel (*Saccopharynx ampullaceus*). (Illustration by Jacqueline Mahannah)

Species accounts

Bobtail snipe eel
Cyema atrum

FAMILY
Cyematidae

TAXONOMY
Cyema atrum Günther, 1878, South Pacific, Challenger station 1,770 ft (539 m); Antarctic, Challenger station 948; 9,000; and 10,800 ft (289; 2,743; and 3,292 m).

OTHER COMMON NAMES
English: Bobtail eel, deepwater eel; Danish: Korthalet ål; Finnish: Nuoliankerias.

PHYSICAL CHARACTERISTICS
This species has a rather striking appearance that is quite different from that of other saccopharyngiforms. Adults are black in coloration. This species is scaleless, like all members of the order. The eyes are very small. The jaws are thin and long, with numerous very fine teeth, and the jaws curve slightly away from each other at their tips. The dorsal and anal fin rays become progressively more elongated toward the rear of the body and extend well past very short caudal rays; the effect is that in side view the fish looks like an arrow! It is a small species, with a maximum reported size of about 6.3 in (160 mm).

DISTRIBUTION
It has been reported from all oceans between about 70° north and 55° south. Most collections have been from the Atlantic and Pacific Oceans.

HABITAT
The species is oceanic, lower mesopelagic to bathypelagic. Although it has been reported from collections made as shallow as 1,148 ft (350 m), most records are from depths exceeding 4,921 ft (1,500 m).

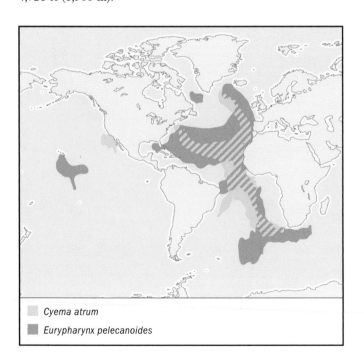

Cyema atrum
Eurypharynx pelecanoides

BEHAVIOR
Nothing is known.

FEEDING ECOLOGY AND DIET
There has been little research on the feeding habits of this eel. Because of its jaw structure, it is suggested that the species feeds on comparable prey types and in a fashion similar to that of the anguilliform eels of the family Nemichthyidae, commonly known as snipe eels. Nemichthyids use their thin, recurved jaws to feed on crustacean shrimps, especially those in the family Sergestidae. Predators of this species are unknown.

REPRODUCTIVE BIOLOGY
Unlike the other saccopharyngiforms, there is no apparent sexual dimorphism in adults. No other reproductive data have been reported for this species. The leptocephalus stage is rather distinctive; the deep oval body has a very small pointed head and a pointed caudal extension. These features grow a bit larger than in other saccopharyngiform leptocephali, with a maximum recorded total length of 2.8 in (70 mm).

CONSERVATION STATUS
Not threatened.

SIGNIFICANCE TO HUMANS
None known. ◆

Pelican eel
Eurypharynx pelecanoides

FAMILY
Eurypharyngidae

TAXONOMY
Eurypharynx pelecanoides Vaillant, 1882, off New England, United States, about 40°N, 68°W, 3 Albatross stations, 2,334–8,802 ft (711–2,683 m).

OTHER COMMON NAMES
English: Big mouth gulper eel, pelican gulper, pelican gulper fish, pelican fish, deep-sea gulper, umbrella mouth gulper; French: Grand-gousier pelican; German: Pelikanaal; Spanish: Pez pelicano; Danish: Pelikanål; Finnish: Pelikaaniankerias; Icelandic: Gapaldur; Japanese: Fukuro-unagi; Polish: Polykacz.

PHYSICAL CHARACTERISTICS
Superficially similar to species in the genus *Saccopharynx*, with which it shares the closest taxonomic relationship within the order, this species is coal black overall, except for a tiny white region on the caudal organ. It is scaleless. Probably the most striking differences between the pelican eel and *Saccopharynx* species are that the jaw length is extreme, almost 50% of the distance to the anus; the jaw teeth are very small; and there is a gradual narrowing of the body posterior to the abdomen. Other similarities to *Saccopharynx* species include small eyes that detect light rather than form visual images, the presence of a presumably luminous caudal organ at the end of a very long filamentous tail, an expansible stomach, and a weakly ossified and poorly muscled body. Because the delicate tail is usually broken, the maximum size is uncertain, but the largest

intact specimen ever collected measured 25.9 in (750 mm) in total length.

DISTRIBUTION
This a circumglobal species, found in temperate and tropical waters of all oceans. It is best known from the Atlantic and eastern and central Pacific Oceans.

HABITAT
The species is oceanic and bathypelagic. Although there are some shallow-water capture records at less than 1,640 ft (500 m), most individuals are collected between 3,281 and 9,842 ft (1,000–3,000 m).

BEHAVIOR
Nothing is known.

FEEDING ECOLOGY AND DIET
This species takes in a wider range of prey than do species in the genus *Saccopharynx*. Prey items include fishes, various crustaceans (especially caridean decapod shrimps), and cephalopod mollusks. In addition, there have been several reports of benthic prey items in the stomachs of pelican eels. Predators are unknown.

REPRODUCTIVE BIOLOGY
Reproduction is similar to that of species in the genus *Saccopharynx*, in that sexually mature males have greatly expanded nasal structures, accompanied by stomach atrophy, loss of dentition, and reduction in jaw structure. Reproduction is apparently a terminal event. Leptocephalus larvae are oval and deep-bodied, like *Saccopharynx* species, but they are smaller, with a maximum length of about 1.6 in (40 mm). They have several greatly elongated larval teeth in the upper jaw.

CONSERVATION STATUS
Not threatened.

SIGNIFICANCE TO HUMANS
None known. ◆

No common name
Monognathus rosenblatti

FAMILY
Monognathidae

TAXONOMY
Monognathus rosenblatti Bertelsen and Nielsen, 1987, Central North Pacific, 31°N, 159°W, 14,300–17,300 ft (4,853–5,266 m)—bottom is 19,000 ft (5,800 m).

OTHER COMMON NAMES
None known.

PHYSICAL CHARACTERISTICS
The monognathids are truly bizarre in appearance. Probably the two most striking features are the complete lack of an upper jaw (which gives the family its name) and the presence of a hollow rostral fang that extends downward from the roof of the mouth. The fang is associated with a glandular mass thought to secrete venom. The body is scaleless, very slender, and pale tan to light brown in color. As with other saccopharyngoids, the eyes are very small, the stomach is distensible, the body skeleton is poorly ossified, and the musculature is weak. The caudal region is flattened laterally but does not extend into the long filament seen in saccopharyngids and eu-

rypharyngids. Monognathid species have been divided into two groups based on the relative shape and length of their skulls; there is a "short-skulled" group and a "long-skulled" group, with *M. rosenblatti* belonging to the latter. The largest specimen collected measured 2.8 in (70 mm) in length, although the largest monognathid reported thus far was 11.4 in (290 mm).

DISTRIBUTION
This species is known only from the northeastern Pacific Ocean.

HABITAT
As with others in this genus, this species is found in oceanic, deep bathypelagic habitats. The shallowest record for *M. rosenblatti* is 6,889 ft (2,100 m). Due to their habitat, this family is exceptionally rare. All 14 species are known form a combined total of fewer than 80 individuals, about 50% of which belong to *M. rosenblatti*.

BEHAVIOR
Nothing is known.

FEEDING ECOLOGY AND DIET
No stomach contents have been reported from any *M. rosenblatti* specimens, but prey from other monognathid species have all been crustacean shrimps. All of the shrimps were quite large relative to the body size of the fish. It has been hypothesized that these weak fish inject their prey with venom using the rostral fang, in much the same fashion as certain venomous snakes overcome their prey. Predators are unknown.

REPRODUCTIVE BIOLOGY
As with the other saccopharyngoids, the sexually mature collected specimens of monognathids (none of which were *M. rosenblatti*) exhibit dimorphism and evidence that spawning is a one-time terminal event. Males possess greatly enlarged nasal structures, suggesting that locating of mates takes place by scent. Although it is believed that the larval form is a leptocephalus, as yet none has been positively identified as belonging to this family.

CONSERVATION STATUS
Not threatened.

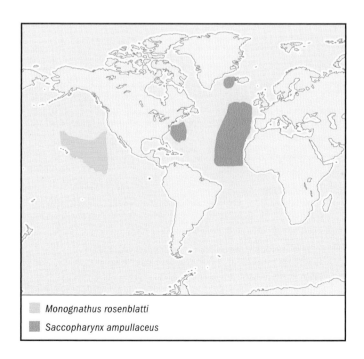

▨ *Monognathus rosenblatti*
▨ *Saccopharynx ampullaceus*

SIGNIFICANCE TO HUMANS
None known. ◆

Gulper eel
Saccopharynx ampullaceus

FAMILY
Saccopharyngidae

TAXONOMY
Ophiognathus ampullaceus Harwood, 1827, NW Atlantic Ocean, 32°20′N, 30°16′W, 0–6,234 ft (0–1,900 m). Neotype: ISH 3288/79. Original locality 62°N, 57°W. Neotype selected by Nielsen and Bertelsen (1985).

OTHER COMMON NAMES
English: Pelican fish; Danish: Slugål; Finnish: Ahmattiankerias; Icelandic: Pokakjaftur; Polish: Gardzielec.

PHYSICAL CHARACTERISTICS
The body is attenuated and very flabby, with poorly ossified bones and weakly developed muscles. The most striking attributes are tiny eyes that function as light detectors; a greatly enlarged mouth with numerous slightly recurved teeth; an elongated stomach region, with the posterior end of the abdomen clearly demarcated from the tail, and an extremely long tail (about 75% of body length), with an elongated caudal filament that terminates in a "caudal organ" believed to be luminescent. Because of the delicacy of the body, the filaments often are broken off in captured specimens. The body is scaleless. The largest intact specimen measured 5.2 ft (1.6 m), although much of the body length consists of the elongated whiplike tail and caudal filament.

DISTRIBUTION
This species is the best known of the genus. It has been collected only from the North Atlantic Ocean between 10° and 65° north latitude.

HABITAT
The gulper eel is oceanic and bathypelagic. Only juveniles have been captured at depths of less than 2,624 ft (800 m). It is believed that adults typically reside deeper than 6,561 ft (2,000 m).

BEHAVIOR
Because of the great depths of its habitat, aspects of the behavior of this species are largely the subject of conjecture.

FEEDING ECOLOGY AND DIET
The species is piscivorous. Relatively few saccopharyngids have been recovered with intact stomach contents, but in all cases various fish species were the prey. The gulper eel has an extremely distensible stomach, allowing it to ingest very large prey. Because of its weak skeleton and body muscles, it is believed to be a very poor swimmer. It is thought to lure prey within range by means of the luminescent caudal organ, which it may suspend in the water near its mouth. The jaw muscles are the only well-developed muscles and probably allow the gulper eel to suck its prey into the large mouth by quickly opening the jaws. Predators are unknown.

REPRODUCTIVE BIOLOGY
Males and females are sexually dimorphic. Sexually mature males show extreme degeneration of the jaws, along with a loss of teeth and reduction in abdominal size. In addition, the eyes become somewhat enlarged, and the nasal apparatus is significantly enlarged. It has been suggested that males locate females by tracking pheromone (scent) trails released by the females. As with numerous eel species as well as some other deep-sea fish species, reproduction is thought to be a terminal event. As with eels in general, larval gulper eels have a leptocephalus, a ribbon-like transparent stage. Relatively few leptocephali have been collected, but all are deep-bodied and small, with a total length of 1.47–1.9 in (40–50 mm).

CONSERVATION STATUS
Not threatened.

SIGNIFICANCE TO HUMANS
None known.

Resources

Books
Bertelsen, E., Jørgen Nielsen, and David G. Smith. "Families Saccopharyngidae, Eurypharyngidae, and Monognathidae." In *Fishes of the Western North Atlantic*, edited by Eugenia B. Böhlke. Part 9. New Haven: Sears Foundation for Marine Research, 1989.

Nelson, Joseph S. *Fishes of the World*. 3rd edition. New York: John Wiley and Sons, 1994.

Smith, David G. "Order Saccopharyngiformes, Family Cyematidae." In *Fishes of the Western North Atlantic*, edited by Eugenia B. Böhlke. Vol. 9, Part 1. New Haven: Sears Foundation for Marine Research, 1989.

———. "Families Cyematidae, Saccopharyngidae, Eurypharyngidae, and Monognathidae: Leptocephali." In *Fishes of the Western North Atlantic*, edited by Eugenia B. Böhlke. Vol. 9, Part 2. New Haven: Sears Foundation for Marine Research, 1989.

Periodicals
Bertelsen, E., and Jørgen G. Nielsen. "The Deep-Sea Eel Family Monognathidae (Pisces, Anguilliformes)." *Steenstrupia* 13, no. 4 (1987): 141–198.

Gartner, John V. Jr. "Sexual Dimorphism in the Bathypelagic Gulper Eel *Eurypharynx pelecanoides* (Lyomeri: Eurypharyngidae), with Comments on Reproductive Strategy." *Copeia* 2 (1983): 446–449.

Nielsen, Jørgen G., and E. Bertelsen. "The Gulper-Eel Family Saccopharyngidae (Pisces, Anguilliformes)." *Steenstrupia* 11 (1985): 157–206.

Other
"FishBase: A Global Information System on Fishes." 7 Nov. 2002 (12 Nov. 2002). <http://www.fishbase.org/home.htm>

John V. Gartner Jr., PhD

Clupeiformes
(Herrings)

Class Actinopterygii
Order Clupeiformes
Number of families 5

Photo: A school of threadfin shad (*Dorosoma pete-nense*). (Photo by Tom McHugh/Photo Researchers, Inc. Reproduced by permission.)

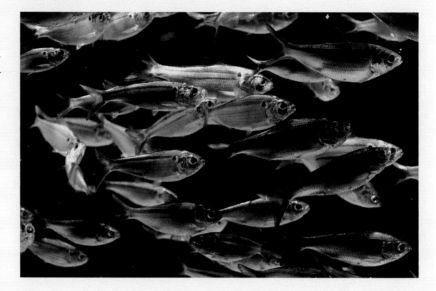

Evolution and systematics

Fishes in the order Clupeiformes are teleosts, a group of fishes characterized evolutionarily by the presence of true bone forming their skeletons and by specific bone structures in the tail and skull. Teleosts arose during the early Mesozoic era (approximately 200 million years ago). Four subsequent radiations gave rise to the current major groups of fishes, with one of these radiations producing the Clupeomorpha. A major evolutionary feature that distinguishes clupeomorphs is the extension of the gas bladder into the brain case so that it contacts the inner ear, and thereby increasing the hearing ability of the fish. Modern Clupeiformes also possess other evolutionary advances over their closest ancestors: a modified joint in the jaw caused by fusion of the angular to the articular and a reduction of the caudal skeleton.

Two suborders, Clupeoidei and Denticipitoidei, are recognized in the Clupeiformes. The Clupeiodei includes the families Chirocentridae (wolf herrings; 1 genus, 2 species), Clupeidae (herrings, menhadens, pilchards, sardines, shads, and sprats; 5 subfamilies, 56 genera, 214 species), Engraulidae (anchovies; 2 subfamilies, 16 genera, 145 species), and Pristigasteridae (sawbelly herrings; 2 subfamilies, 9 genera, 36 species). The Denticipitoidei includes the family Denticipitidae (denticle herring; 1 genus, 1 species).

Physical characteristics

Clupeoids are small fusiform (tapering toward each end) fishes with streamlined bodies that facilitate fast swimming in open water. They have dark shading on their backs and bright silvery sides. Except for the head, their bodies are completely covered in large scales. Most clupeoids lack a lateral line, and only in the deticipitoid herring does this line extend along the body. The fins of clupeoids lack spines. A single dorsal fin is located near the middle of the body, and the tail is forked. Many clupeoids have a row of scutes, modified scales that usually have sharp points towards the rear, along the medial line of the belly. The smallest cluepoid is the Sanaga pygmy herring (*Thrattidion noctivagus*), measuring only 0.83 in (2.1 cm) in standard length; male wolf herrings (*Chirocentrus* spp.) are the largest herring, attaining standard lengths of 39 in (100 cm).

Distribution

Clupeiformes are widely distributed worldwide between 70°N to 60°S latitude. They primarily live in oceans, but some species inhabit coastal margins and fresh water for at least a portion of their lives.

Habitat

Nearly all Clupeiformes are open-water, pelagic species. Four-fifths of all species are marine, with habitats ranging from nearshore littoral zones to nearly 100 mi (160 km) offshore. Many are found near the surface at times but often move to deeper waters during the day. Some Clupeiformes live in inland waters or are anadromous, moving inland to spawn. These species utilize bays, estuaries, marshes, rivers, and freshwater streams as habitats. Landlocked populations have formed as shads, alewives, and herrings moved into lakes or rivers and became trapped between dams.

Behavior

Clupeiformes are perhaps best recognized for the large schools they form. Schools may include hundreds or thousands of individuals ranging from the young to adults, but individuals in a school are usually of similar size. Schooling is

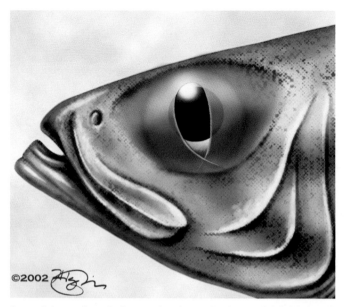

Some fish—mullets, jacks, mackerels, anchovies, and this herring, for example—have an adipose eyelid (highlighted here in red). Rather than being "fatty" as the name suggests, this immobile covering consists primarily of a matrix of extremely fine microfibrils of collagen. Its purpose is unknown. (Illustration by Jonathan Higgins)

a form of organization in which behavior is synchronized; large numbers of fish may swim parallel to each other in the same direction with fairly uniform spacing. These synchronized aggregations are believed to confer swimming efficiency and, most importantly, to enable fishes to avoid or deter predators. Clupeiformes also congregate in smaller, less-organized shoals, particularly during spawning seasons. In addition to schooling, many Clupeiformes undertake some type of migration. Some clupeoid fishes are anadromous, migrating from the ocean to streams and rivers for spawning. They also may migrate inshore or latitudinally on a seasonal basis. Many clupeoids migrate in the water column on a diel basis, staying at deep depths during the day and moving to shallow depths at night.

Northern anchovies (Engraulis mordax) swim in tightly packed schools. (Photo by Tom McHugh/Steinhart Aquarium/Photo Researchers, Inc. Reproduced by permission.)

A school of dwarf herrings (Jenkinsia sp.) swimming near Andros, Bahamas. (Photo by Jerry McCormick/Photo Researchers, Inc. Reproduced by permission.)

Feeding ecology and diet

Most Clupeiformes filter feed by straining water through their long and numerous gill rakers. They consume plankton, particularly small crustaceans and the larval stages of larger crustaceans and fishes. Some herrings visually locate and target food particles. Clupeoid fishes are important prey for larger fishes, seabirds, and marine mammals.

Reproductive biology

Clupeiformes produce large numbers of offspring, either through a single seasonal spawning event or by spawning in seasonal peaks throughout the year. Most Clupeiformes spawn in shoals by broadcasting large numbers of small, buoyant eggs in waters near the surface. The eggs and larvae drift passively in currents as they develop. Herrings, on the other hand, produce demersal eggs that sink to the bottom, where they often adhere to the substrate until they hatch. After hatching, larvae become pelagic.

Conservation status

Two Clupeiformes are listed as Endangered by the IUCN: the Alabama shad (Alosa alabamae) and the Laotian shad (Tenualosa thibaudeaui). The Alabama shad is found in the northern portion of the Gulf of Mexico, from the Mississippi delta eastward to the Choctawhatchee River in Florida. It also occurs in inland rivers from Iowa to Arkansas and eastward to West Virginia. The Laotian shad occurs in the Mekong River basin, including inland waters of Thailand, Laos, and Cambodia. Most Clupeiformes are not threatened by severe population disruptions, but populations do show natural variability due to fluctuations in reproductive success. This natural variability is exacerbated by fishing pressure and global climate patterns.

Significance to humans

Clupeiformes are some of the most economically important fishes in the world's oceans. They have been widely exploited throughout human history, primarily for food but also as a source of oil, fertilizer, and animal feed. Herring fishing was one of the earliest occupations of coastal peoples, as first described in England in a chronicle that dates back to A.D. 709. The first commercial fishing establishment opened in Heligoland, a small island in the North Sea off the coast of Germany, in 1425.

Clupeiformes continue to constitute a large portion of world's commercial fisheries. Although 186 species are exploited by pelagic fisheries worldwide, 50% of the total landings in 1997 were represented by only seven species. Among these seven, four are Clupeiformes: the anchoveta (*Engraulis ringens*), Atlantic herring (*Clupea harengus*), Japanese pilchard (*Sardinops melanostictus*), and South American pilchard (*Sardinops sagax*). Herrings and anchovies constitute approximately 25% of the total fisheries harvest worldwide.

In addition to being heavily utilized by humans, Clupeiformes are an important component of the broader marine ecosystem. They serve as food items for larger predatory fishes, sea birds, and marine mammals. Thus, clupeoids sustain other organisms of importance to humans through ecosystem interactions.

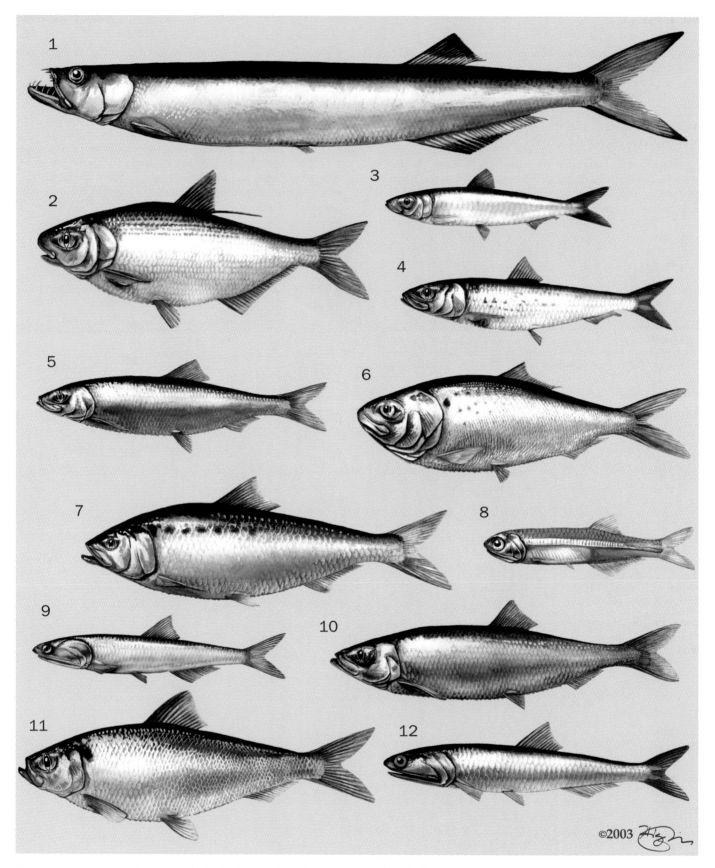

1. Dorab wolf herring (*Chirocentrus dorab*); 2. Gizzard shad (*Dorosoma cepedianum*); 3. European pilchard (*Sardina pilchardus*); 4. South American pilchard (*Sardinops sagax*); 5. Pacific herring (*Clupea pallasii*); 6. Atlantic menhaden (*Brevoortia tyrannus*); 7. American shad (*Alosa sapidissima*); 8. Bay anchovy (*Anchoa mitchilli*); 9. Northern anchovy (*Engraulis mordax*); 10. Atlantic herring (*Clupea harengus*); 11. Alewife (*Alosa pseudoharengus*); 12. Anchoveta (*Engraulis ringens*). (Illustration by Jonathan Higgins)

Grzimek's Animal Life Encyclopedia

Species accounts

Dorab wolf herring
Chirocentrus dorab

FAMILY
Chirocentridae

TAXONOMY
Chirocentrus dorab Forsskål, 1775, Jiddah, Saudi Arabia.

OTHER COMMON NAMES
English: Blackfin wolf herring, dorab, silver barfish, wolf herring; French: Chirocentre dorab, sabre; German: Wolfshering; Spanish: Arencón dorab, arenque lobo de la India, sabre; Arabic: Gairi, kerli, dorab, samak abu sayf.

PHYSICAL CHARACTERISTICS
Reaches lengths of approximately 3.3 ft (1 m). Elongate, highly compressed body. Silvery with a bright blue-gray back. Unlike species of the Clupeidae, the wolf herring has no scutes along the belly. Two distinctive fang-like canine teeth point forward in the upper jaw, and a series of canine teeth is present in the lower jaw.

DISTRIBUTION
Warm coastal waters of the Indo-Pacific, from the Red Sea and East Africa, to the Solomon Island, north to Japan, and south to Australia.

HABITAT
Nearshore portions of oceans and inshore brackish areas.

BEHAVIOR
Does not form large schools, but a small number of individuals may be found together. Known for its leaping powers.

FEEDING ECOLOGY AND DIET
Feeds on other small schooling fishes, such as herrings and anchovies. To a lesser extent, it also eats fish eggs, larvae, and crustaceans.

REPRODUCTIVE BIOLOGY
Researchers believe that Dorab wolf herrings broadcast eggs into the water column, from which pelagic larvae hatch.

CONSERVATION STATUS
Not listed by the IUCN.

SIGNIFICANCE TO HUMANS
Minor commercial fishery in the Indo-Pacific, with products marketed fresh, frozen, dried, and salted. ◆

Alewife
Alosa pseudoharengus

FAMILY
Clupeidae

TAXONOMY
Alosa pseudoharengus Wilson, 1811, Delaware River at Philadelphia, Pennsylvania, United States.

OTHER COMMON NAMES
English: Bigeye herring, branch herring, freshwater herring, golden shad, grayback, gray herring, green shad, kyack, mulhaden, sawbelly, spring herring, white herring; French: Alose gaspareau, gaspareau, gasparot, gasperot; German: Maifisch; Spanish: Alosa, pinchagua.

PHYSICAL CHARACTERISTICS
Sea-run alewives reach a maximum length of 15 in (38.1 cm); landlocked alewives are typically around 6 in (15.2 cm). Small bodies. Strongly compressed laterally. Silvery bodies with grayish green backs. A row of scutes runs along the ventral edge of the belly. A single black spot is present behind the head.

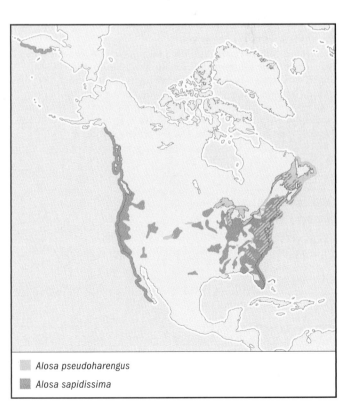

Alosa pseudoharengus
Alosa sapidissima

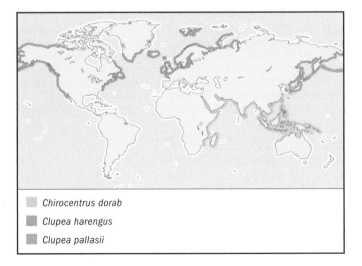

Chirocentrus dorab
Clupea harengus
Clupea pallasii

DISTRIBUTION

Atlantic coast of North America, from the St. Lawrence River and Nova Scotia, Canada, to North Carolina. Numerous translocations of this species have resulted in the establishment of landlocked populations in many inland water bodies, including all of the Great Lakes.

HABITAT

Spends a large portion of its life at sea and migrates into inland freshwater streams for spawning. Landlocked populations live entirely in inland lakes or river systems.

BEHAVIOR

Forms schools in open water. Sea-run alewives migrate from the sea to freshwater streams to spawn. Populations in lakes migrate on a diel basis, moving inshore at night and retreating to deeper offshore waters during the day.

FEEDING ECOLOGY AND DIET

Feeds on zooplankton, primarily copepods, cladocerans, mysids, and ostracods. Inshore adults eat insect larvae as a large part of their diet.

REPRODUCTIVE BIOLOGY

Moves from the sea to rivers for spawning, with the timing of migration dependent on water temperature. In lakes, alewives move onto shallow beaches and into ponds to spawn. Males and females mature at one and two years of age, respectively. Females usually move to the spawning site before males. Groups of 2–3 fish spawn at night over gravel or rocky substrates. Eggs are broadcast by females and fertilized in the water column by males. Immediately after spawning, the eggs sink and adhere to the substrate. Upon hatching, the young remain at the spawning grounds until the late larval stage and then move slowly into deeper water or the sea.

CONSERVATION STATUS

Not listed by the IUCN, but overfishing, pollution, and impassable dams contribute to stock declines.

SIGNIFICANCE TO HUMANS

Harvested commercially and used as fresh, dried, or salted meat, bait for crab or lobsters, and sometimes as animal feed. Also serves as a forage base for other commercial or recreational fish stocks. In lakes, alewives can be a nuisance to humans. They may become so abundant that they clog industrial water intake pipes, and they often die off in massive events related to fluctuations in water temperature and dissolved oxygen. ◆

American shad
Alosa sapidissima

FAMILY
Clupeidae

TAXONOMY
Alosa sapidissima Wilson, 1811, Delaware River at Philadelphia, Pennsylvania, United States.

OTHER COMMON NAMES
English: Atlantic shad, common shad, Connecticut river shad, herring jack, North River shad, Potomac shad, Susquehanna shad, white shad; French: Alose, alose canadienne, alose savoureuse; German: Amerikanische Finte, Amerikanischer Maifisch; Spanish: Sábalo americano.

PHYSICAL CHARACTERISTICS

Average size 15 in (38 cm) long. Silvery with blue or blue-green metallic shading on its back. Moderately compressed body with a distinct keel of ventral scutes along the belly. A large black spot is located behind the rear edge of the gill cover, followed by several smaller dark spots.

DISTRIBUTION

Occurs off the Atlantic coast of North America, from the St. Lawrence River and Nova Scotia to central Florida. Introduced to the Sacramento River in California by the U.S. Fish Commission in 1971. Since then, its range has expanded greatly, and it can now be found from Cook Inlet, Alaska, to Baja, Mexico.

HABITAT

Occurs in the ocean at depths to 820 ft (250 m). As an anadromous fish, it spends most of its life at sea but migrates into estuaries and freshwater streams to breed.

BEHAVIOR

Forms schools after reaching a juvenile size of approximately 8–12 in (20–30 cm). Anadromous species, migrating into freshwater areas for spawning.

FEEDING ECOLOGY AND DIET

Planktonic feeder that consumes primarily copepods and mysids (shrimp-like crustaceans), but occasionally eats small fishes.

REPRODUCTIVE BIOLOGY

Ascends freshwater rivers and streams in the spring to spawn. Males mature at four years of age; females first spawn between five and seven years. Spawning commences at water temperatures of 53.6°F (12°C) and peaks at 65°F (18.3°C). The males arrive at the spawning location before the females. During spawning, several males and a female swim close to the surface during the evening. Females release eggs in the open water, where they are fertilized by males. Eggs are 0.1–0.15 in (2.5–3.5 mm) in diameter after fertilization, and after 8–10 days, small larvae of 0.35–0.40 after spawning, but juveniles spend their first summer in the river and reach the sea by autumn. Most American shad spawn more than once, but some die after spawning at southern latitudes of the Atlantic coast.

CONSERVATION STATUS

Not listed by the IUCN, but the presence of dams in rivers and streams impedes spawning migrations and has contributed to the decline of some populations.

SIGNIFICANCE TO HUMANS

Commercially fished in rivers and estuaries during spawning migrations. The flesh is eaten fresh, salted, or smoked, and this species is prized for its tasty roe. American shad play a central role in shad planking parties, where the fish are slow-cooked on oak boards over campfires. Shad planking parties are central to political rallies and community gatherings, particularly in the mid-Atlantic region of the United States. ◆

Atlantic menhaden
Brevoortia tyrannus

FAMILY
Clupeidae

TAXONOMY
Brevoortia tyrannus Latrobe, 1802, Chesapeake Bay, United States.

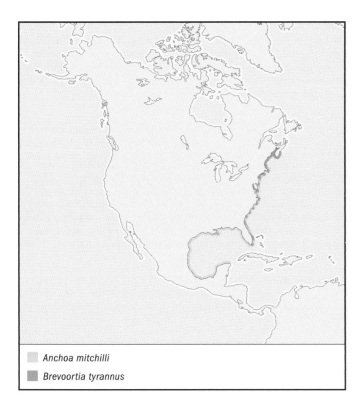

Anchoa mitchilli
Brevoortia tyrannus

OTHER COMMON NAMES
English: Bony fish, bugfish, bunker, fatback, menhaden, moss-bunker, pogy, whitefish; French: Alose tyran, menhaden tyran, menhaden; German: Menhaden; Spanish: Lacha tirana.

PHYSICAL CHARACTERISTICS
Adults typically 12–14 in (30.5–35.6 cm) in length. Deep bodied and laterally compressed. Silvery in color, with brassy sides and a dark blue-green back. A row of sharp scutes extends along the ventral edge of the belly. A large dark spot is located behind the gill cover, followed by several smaller spots.

DISTRIBUTION
Western Atlantic from Nova Scotia, Canada, to Indian River, Florida.

HABITAT
Primarily pelagic fish. Found in waters over the continental shelf. Moves inshore to bays, inlets, and estuaries in the summer.

BEHAVIOR
Forms large, compact schools of juveniles and adults. Stratifies by size along the Atlantic seaboard during annual north-south migrations. Fish of all ages congregate near Cape Hatteras, North Carolina, during the winter months. Most adults move northward after March, with the largest fish migrating as far as the Gulf of Maine; some adults move southward as far as to waters off of Florida. Fish of all ages and sizes then return to Cape Hatteras in late autumn. Atlantic menhaden also move in and out of inshore habitats with the tides, the season, and the weather.

FEEDING ECOLOGY AND DIET
Feeds by filtering phytoplankton and zooplankton, including diatoms, copepods, and euphausids.

REPRODUCTIVE BIOLOGY
Spawns in the open ocean throughout the year. Eggs are buoyant and hatch at sea. Over one to three months, larvae are

transported to estuaries by ocean currents, where they develop into juveniles. Most Atlantic menhaden first spawn during their third year of life. Females exhibit high fecundity levels, producing about 38,000–50,000 eggs per female.

CONSERVATION STATUS
Not listed by the IUCN.

SIGNIFICANCE TO HUMANS
Previously used by Native Americans and early European colonists as fertilizer. A fishery developed for menhaden during the late 1800s and early 1900s. Today used primarily to produce fish meal and oil, although some are marketed fresh, salted, canned, or smoked. Due to the large biomass represented by this species, menhaden constituted around 30% of commercial fisheries landings along the U.S. Atlantic coast in 2000. ◆

Atlantic herring
Clupea harengus

FAMILY
Clupeidae

TAXONOMY
Clupea harengus Linnaeus, 1758, European seas. This species was previously considered a subspecies, *Clupea harengus harengus*, but recent taxonomic classifications distinguish it as a separate species.

OTHER COMMON NAMES
English: Bank herring, brit, fall herring, hern, herning, herron, labrador herring, mesh herring, murman hering, Norwegian herring, sardine, sea Atlantic herring, sea stick, shore herring, split, spring herring, summer herring, yawling; French: Hareng atlantique, hareng de l'Atlantique; German: Allec, Hering, Silling; Spanish: Arenque, arenque del Atlántico.

PHYSICAL CHARACTERISTICS
Maximum size 17.7 in (45 cm), but most fish captured in fisheries are 11.8–13.8 in (30–35 cm). Elongate and slender. Back is dark blue-green (or bluish green), sides and belly are silvery, and snout is blackish blue. There are no distinct dark spots on the body or fins. The belly is rounded with scutes, but has no prominent keel. The lower jaw of the Atlantic herring is slightly longer than the upper jaw.

DISTRIBUTION
Eastern Atlantic Ocean, from the northern Bay of Biscay northward to Spitzbergen and Novaya Zemlya, and around Iceland and southern Greenland. Also found in the western Atlantic, from southwestern Greenland, around Labrador, and south to South Carolina.

HABITAT
Coastal pelagic species. Found at depths ranging from near the surface down to 656 ft (200 m).

BEHAVIOR
Schools in coastal waters. Exhibits complex feeding and spawning migrations, but the timing and extent of migrations varies by morphological race of the fish. Stays in deep water during the day but moves to the surface at night. Most migrate to coastal spawning grounds at the onset of spawning seasons. Also migrates north and south seasonally to feeding areas and for over wintering.

FEEDING ECOLOGY AND DIET

During their first year of life, Atlantic herring feed on small planktonic copepods. Thereafter, they consume mostly copepods, but also amphipods, mysids, small fishes, and ctenophores. Atlantic herring locate food particles visually, but they can switch to filter feeding if they encounter a high density of small food particles. This species is important as prey in the marine food chain, and it is consumed by larger fishes, squids, skates, whales, and seabirds.

REPRODUCTIVE BIOLOGY

Exhibits a wide range of spawning behaviors and strategies that vary by stock and race. At least one population spawns in any month of the year, but each race spawns at a different time and place. Off Greenland, these fishes spawn in up to 16 ft (5 m) of water; autumn-spawning herrings in the North Sea spawn at depths of 656 ft (200 m). The eggs are adhesive and are laid over rocks, stones, gravel, sand, algae, or vegetation on the seabed. Atlantic herring reach their sexual maturity at three to nine years of age, and females may produce 20,000–40,000 eggs.

CONSERVATION STATUS

Populations fluctuate widely. Some stocks are considered overfished, while others have shown recent increases in abundance. Not listed by the IUCN.

SIGNIFICANCE TO HUMANS

The commercial fishery for Atlantic herring ranks among the most important in the world in terms of biomass and value. Atlantic herring were a major item of trade between Scandinavian countries and western European countries as early as the twelfth century. The western Atlantic fishery developed in the 1800s and supplied fish that were sold as bait or sardines. In the United States and Canada, most Atlantic herring are sold as sardines or are converted to fish meal or oil. Smoked, salted, pickled, fresh, and frozen herring are more common in Europe. Herring form an important part of the cuisine of certain cultures; it is particularly central to the Jewish cuisine due to dietary and cooking restrictions observed by many Jews. ◆

Pacific herring
Clupea pallasii

FAMILY
Clupeidae

TAXONOMY
Clupea pallasii Valenciennes, 1847, Kamchatka, Russia. This species was previously considered a subspecies, *Clupea harengus pallasii*, but recent taxonomic classifications distinguish it as a separate species. One subspecies: *Clupea pallasii marisalbi* Berg, 1923.

OTHER COMMON NAMES
English: Herring, North Pacific herring, Oriental herring; French: Hareng du Pacifique, hareng pacifique; German: Pazifischer Hering; Spanish: Arenque del Pacifico.

PHYSICAL CHARACTERISTICS
Maximum size 18 in (46 cm), but most are less than 11.8 in (30 cm). Elongate and slender, with a rounded belly and scutes. Back is dark blue to olive, the sides and belly are silvery. There are no distinct dark spots on the body or fins. Distinguished from the Atlantic herring by having fewer vertebrae and fewer postpelvic scutes.

DISTRIBUTION

Arctic Sea from the White Sea eastward to the Ob Inlet. They range in the Western Pacific along the coast of Russia, south to Japan and Korea. In the eastern Pacific, they are found from the Kent Peninsula and Beaufort Sea, around Alaska, and south to Baja California, Mexico.

HABITAT

Coastal pelagic species that uses estuaries and bays for spawning.

BEHAVIOR

Begins schooling as a juvenile and is found in inshore waters during this life history stage. Adults form schools at sea, but they migrate to inshore waters, including bays and estuaries, to spawn. They migrate from the sea to inshore waters, entering bays and estuaries to spawn. They also migrate daily from deep waters during the day to shallower portions of the water column at night. Pacific herring do not undertake extensive latitudinal migrations along the coast.

FEEDING ECOLOGY AND DIET

Feeds on zooplankton, including euphausids, copepods, mysids, and amphipods. Serves as food for a wide variety of predatory fishes and seabirds.

REPRODUCTIVE BIOLOGY

Breeds from December to July, with the precise timing of spawning varying with latitude. Adults congregate near inshore spawning grounds several weeks before spawning. Herring spawn over a variety of substrates, including rocks, algae, vegetation, and flat surfaces. The female deposits rows of eggs along the substrate; the eggs are then fertilized by the males. Spawning occurs several days at a time, with events separated by a day to several weeks. In California, these herring spawn at age two to three, but members of the species that inhabit waters at higher latitudes reach maturity later.

CONSERVATION STATUS

Not listed by the IUCN, but populations are sensitive to shoreline development and fishing pressures.

SIGNIFICANCE TO HUMANS

Previously used by Native Americans as fresh or salted food and for bait. In the early 1900s, commercial fisheries developed along the west coast of the United States and Canada for salted and canned herring and for the reduction of the fish to meal and oil. These fishes are valued in Russia as food and are utilized fresh, dried, smoked, canned, and frozen. In the eastern Pacific, the roe and eggs are taken to supply Asian markets. Pacific herring also are used in Chinese medicine. ◆

Gizzard shad
Dorosoma cepedianum

FAMILY
Clupeidae

TAXONOMY
Dorosoma cepedianum Lesueur, 1818, Baltimore, Maryland and Philadelphia, Pennsylvania, United States.

OTHER COMMON NAMES
English: American gizzard shad, eastern gizzard shad, gizzard mud, gizzard nanny shad, hickory shad, mud shad, nanny shad, skipjack winter shad; French: Alose à gésier, alose a gésier américaine, aloser noyer; Spanish: Sábalo molleja.

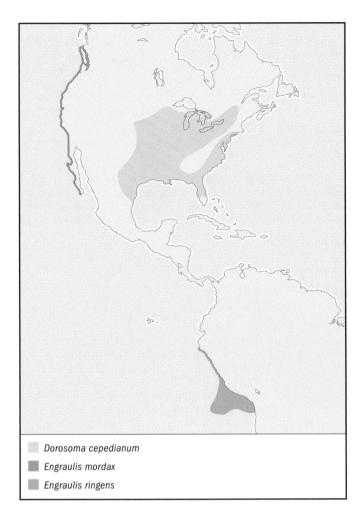

☐ *Dorosoma cepedianum*

■ *Engraulis mordax*

■ *Engraulis ringens*

PHYSICAL CHARACTERISTICS
Maximum size 22.4 in (57 cm). Silver to brassy body with a
bluish back and white belly. There is a large dark spot behind
the operculum and above the pectoral fin. Snout is rounded
and blunt. The last dorsal fin ray extends to a long filament.

DISTRIBUTION
Atlantic coast of the United States from New York, southward
into the Gulf of Mexico and to the basin of the Rio Pánuco in
Mexico. Found inland in the Great Lakes, St. Lawrence, Mississippi, Atlantic, and Gulf Slope drainages, with its east-west
range extending from Quebec to North Dakota in the north
and from Florida to New Mexico in the south.

HABITAT
Can tolerate salinities from fresh water to 33 or 34 parts per 1,000,
but is most common in freshwater lakes, reservoirs, swamps, and
slow-moving rivers. Adults are found in estuaries and protected
bays; the young are sometimes found far upstream in small streams.

BEHAVIOR
Forms schools. Individuals may be seen leaping out of the water and skipping along the surface on their sides, giving rise to
one common name for the species, the skipjack.

FEEDING ECOLOGY AND DIET
Young gizzard shad feed on zooplankton. As they mature, gizzard shad become herbivorous filter feeders, consuming phytoplankton and algae on the bottom of the water body.

REPRODUCTIVE BIOLOGY
Spawns from late winter through most of the summer; the exact spawning dates relate to latitude and water temperature.
Adults spawn near the surface, usually in groups of two males
and one female. The eggs, which are very small (0.03 in [0.75
mm] in diameter) sink and adhere to the bottom. They hatch
in three days to one week, depending on temperature.

CONSERVATION STATUS
Not listed by the IUCN.

SIGNIFICANCE TO HUMANS
Young gizzard shad are important forage fishes that sustain
other game and commercial fish species. In some areas, the
abundance of gizzard shad creates a nuisance for humans when
low temperatures, low oxygen, or disease trigger the death of a
large number of fish that often wash onto shores and decay.
Gizzard shad also create problems for fisheries; they entangle
the nets of commercial fishermen and out-compete many preferred recreational species. The gizzard shad is used for fertilizer and as a component of cattle and hog food. ◆

European pilchard
Sardina pilchardus

FAMILY
Clupeidae

TAXONOMY
Sardina pilchardus Walbaum, 1792, Cornwall, England.

OTHER COMMON NAMES
English: Pilchard, sardine, true sardine; French: Sardine, sar-
dine commune, sardine d'Europe; German: Pilchard, Sardine;
Spanish: Majuga, parrocha, sardina, sardiña, xouba.

PHYSICAL CHARACTERISTICS
Can attain lengths of 9.8 in (25 cm). Body elongate with a
rounded belly that has scutes but not a defined keel. Back dark
green or olive in color, with golden flanks and a silvery white
belly. There is a series of dark spots along the upper flanks.

DISTRIBUTION
Northeast Atlantic from Iceland and the North Sea south to
Senegal. Also found in the western Mediterranean and Adriatic
Seas.

HABITAT
Coastal, pelagic species that is found at water depths of 82–180
ft (25–55 m) during the day and 49–115 ft (15–35 m) at night.

BEHAVIOR
Forms schools and undertakes diel migrations in the water column, staying at deeper depths during the day and rising to
shallower waters at night.

FEEDING ECOLOGY AND DIET
Feeds mainly on planktonic crustaceans, but may also consume
larger planktonic organisms.

REPRODUCTIVE BIOLOGY
Breeds at 66–82 ft (20–25 m) depth up to 62 mi (100 km) offshore. Spawning time varies by geographic location, taking
place in April in the English Channel, from June to August in
the North Sea, from September to May along European coasts
in the Mediterranean, and from November to June along

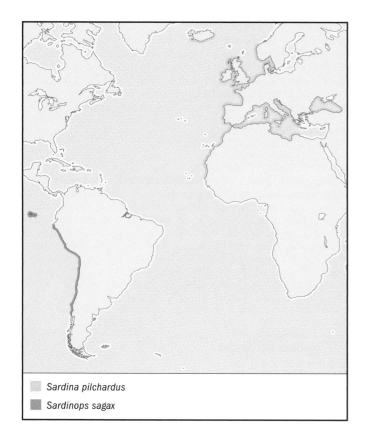

Sardina pilchardus
Sardinops sagax

DISTRIBUTION
Eastern Pacific along the coasts of Peru and Chile; also found near the Galápagos Islands.

HABITAT
Coastal, pelagic species that is found to depths of around 131 ft (40 m).

BEHAVIOR
Forms large schools and migrates up and down in the water column on a diel cycle.

FEEDING ECOLOGY AND DIET
Feeds mainly on planktonic crustaceans but may also consume phytolankton. Bethnic organisms, including ostracods and polychaete worms make up a minor portion of the diet of adults.

REPRODUCTIVE BIOLOGY
Spawns twice each year by broadcasting eggs. Eggs and larvae are pelagic and develop in the water column.

CONSERVATION STATUS
Not listed by the IUCN.

SIGNIFICANCE TO HUMANS
Forms the basis of a substantial fishery off Peru and Chile. The fishery reached a peak of 6.5 million tons (5.9 million tonnes) in 1985, corresponding to the crash of the Peruvian anchoveta fishery caused by El Niño. However, catches have declined since then, and the total harvest in 1999 was less than 450,000 tons (408,233 tonnes). ◆

African coasts in the Mediterranean. The eggs are buoyant and develop in the water column.

CONSERVATION STATUS
Not listed by the IUCN.

SIGNIFICANCE TO HUMANS
Constitutes an important fishery throughout its range. Catches have steadily risen since 1950, and in 1999, over 900,000 tons (816,466 tonnes) was harvested. The fish is marketed fresh, frozen or canned but may also be utilized in dried or smoked forms. ◆

South American pilchard
Sardinops sagax

FAMILY
Clupeidae

TAXONOMY
Sardinops sagax Jenyns, 1842, Lima, Peru.

OTHER COMMON NAMES
English: Chilean pilchard, Chilean sardine, Pacific American sardine, pilchard, sardina, sardine; French: Hareng, Pilchard sudaméricain, Sardinops du Chili; German: Chilenishe Sardine, Südeamerikanische Sardine; Spanish: Pilchard chileña, Sardina, Sardina española.

PHYSICAL CHARACTERISTICS
May grow to 12 in (30 cm), but sizes of 8 in (20 cm) are more common. Bluish on the back and silvery on the sides. Round-bodied and has a keel with scutes along the belly.

Bay anchovy
Anchoa mitchilli

FAMILY
Engraulidae

TAXONOMY
Anchoa mitchilli Valenciennes, 1848, New York, United States.

OTHER COMMON NAMES
English: Common anchovy; French: Anchois américain; Spanish: Anchoa de caleta.

PHYSICAL CHARACTERISTICS
Small species, typically 3–4 in (7.6–10.2 cm) total length. Nearly transparent and greenish in color, with a silvery band along the side of the body. Snout overhangs the mouth and low jawbone extends well beyond the eye.

DISTRIBUTION
Atlantic coast of North America from Casco Bay, Maine, to the Florida Keys, and westward around the Gulf of Mexico south to the Yucatán peninsula. ◆

HABITAT
Primarily an estuarine and inshore coastal species. Utilizes a wide variety of habitats including bays, sandy beaches, marshes, islands, and spoil banks. Typically found over muddy bottoms or in vegetation. Tolerates a wide range of salinities but is often found in brackish water.

BEHAVIOR
Swims in schools. Migrates seasonally from deep waters in winter to shallow shores and wetlands in summer.

FEEDING ECOLOGY AND DIET
Plankton feeder; primarily consumes mysids and copepods. Small fishes, gastropods, and isopods are occasionally taken.

REPRODUCTIVE BIOLOGY
Spawns from late winter to early fall when water temperatures are around 68°F (20°C). Spawning takes place during the evening hours in shallow waters near barrier islands, in bays, and in estuaries. Spawning usually occurs in large schools. Females broadcast eggs, which are fertilized in the water column by males. Eggs float near the water surface for approximately 24 hours after fertilization before they hatch. Bay anchovies mature to adults in two and one-half months.

CONSERVATION STATUS
Not listed by the IUCN.

SIGNIFICANCE TO HUMANS
Used as bait and to make anchovy paste. Important as forage fishes in food chains that sustain other commercial and recreational fishery species in estuaries and coastal areas. ◆

Northern anchovy
Engraulis mordax

FAMILY
Engraulidae

TAXONOMY
Engraulis mordax Girard, 1854, San Francisco, California, United States.

OTHER COMMON NAMES
English: Anchovy, Californian anchovy, North Pacific anchovy, pinhead; French: Anchois de California, anchois du nord, anchois du Pacifique, anchois du Pacifique nord; German: Amerikanische Nordpazifische, Amerikanische Sardelle; Spanish: Anchoa de California, anchoa del Pacifico, anchoveta, anchoveta de California, anchoveta norteña.

PHYSICAL CHARACTERISTICS
Grows to 4 in (10 cm). Body slender and elongate but round in cross section. Back is green and sides are silvery, including a stripe along the flank in young individuals. Large head and mouth with a pointed snout.

DISTRIBUTION
Northern Pacific from Vancouver Island, Canada, to Baja California, Mexico.

HABITAT
Pelagic marine species found to depths of 984 ft (300 m). Usually stays in coastal waters within 18.6 mi (30 km) of shore but may range as far as 298 mi (480 km) offshore. Enters estuaries, bays, and inlets during the spring and summer.

BEHAVIOR
Forms schools during most of the year, although these become smaller or break up in late spring, typically around the end of spawning in April or May. Moves to inshore waters during spring and summer and migrates offshore in the fall and winter. Diel migrations also occur, with the northern anchovy remaining at depths during the day and approaching the surface in low-density schools at night.

FEEDING ECOLOGY AND DIET
Obtains food both by filter feeding and particulate biting. Feeds on plankton, primarily euphausids, copepods, and decapod larvae. Northern anchovies are an important forage species for other fishes, birds, and marine mammals.

REPRODUCTIVE BIOLOGY
Spawns in inlets and offshore, with spawning activity occurring at night. Two major spawning areas exist in coastal waters off southern California and Baja California, Mexico. Eggs are broadcast and fertilized in the water column, then float and incubate for two to four days before hatching. Spawning takes place throughout the year, and individual females may spawn several times each year. However, as a whole, the species exhibits a clear peak in spawning during the winter and early spring.

CONSERVATION STATUS
Not listed by the IUCN.

SIGNIFICANCE TO HUMANS
Supports a commercial and bait fishery, which developed after the collapse of the Pacific sardine fishery in the 1940s. Approximately 25 million pounds were landed in 2000, with most used for fish meal, fertilizer, and animal feed. A small portion is consumed by humans in pickled or salted forms. ◆

Anchoveta
Engraulis ringens

FAMILY
Engraulidae

TAXONOMY
Engraulis ringens Jenyns, 1842, Iquique, Chile.

OTHER COMMON NAMES
English: Anchovy, Peruvian anchoveta; French: Anchois du pérou, anchois péruvien; German: Perusardelle, Südamerikanische Sardelle; Spanish: Anchoa, anchoa bocona, anchoveta, anchoveta peruana, atunera, chicora, manchuma, manchumilla, peladilla, sardina bocona.

PHYSICAL CHARACTERISTICS
Grows to 7 in (18 cm) and has a slender, elongated body that is round in cross section. Its large snout and mouth are similar to those of other anchovies. Silvery in color; juveniles have a stripe along the flank.

DISTRIBUTION
West coast of South America along the Peru Current, typically from around Aguja Point, Peru, south to Chiloë, Chile.

HABITAT
Pelagic species. Occurs in surface waters, usually within 49 mi (80 km) of the shore, but occasionally offshore to 99 mi (160 km).

BEHAVIOR
Forms huge schools. Descends to deeper waters during the day but rises to the surface at night.

FEEDING ECOLOGY AND DIET
Filter feeder. Depends entirely on the plankton of the Peru Current for food. Diet consists largely of diatoms, which make up as much as 98% of its consumption according to some studies. It also consumes copepods, euphausids, fish eggs, and dinoflagellates. Seabirds prey heavily on schools of anchoveta.

REPRODUCTIVE BIOLOGY
Matures at around one year of age. Breeds throughout the year, but two peaks of spawning occur in late winter and early autumn. Their buoyant ellipsoidal eggs are broadcast into the water column and fertilized.

CONSERVATION STATUS
Not listed by the IUCN, but populations vary greatly with climatic conditions. El Niño, the oceanographic condition that results in warmer water in the Pacific Ocean, slows and may stop the upwelling of nutrient rich waters in the Peru Current. This can have devastating effects on the anchoveta by dramatically reducing its planktonic food base.

SIGNIFICANCE TO HUMANS
Previously the world's largest fishery in terms of biomass in 1971, with over 13 million tons (11.7 million tonnes) harvested. Populations declined during the 1970s and 1980s due to overfishing and the occurrence of severe El Niño events in 1972–1973 and 1982–1983. Since that time, anchoveta populations have recovered and the species again constitutes a major fishery. In 2000 the anchoveta comprised over 12 million tons (11 million tonnes) of a total harvest in the southeastern Pacific of around 16.5 million tons (15 million tonnes) for all fishes. Peru and Chile rely upon this fishery as a major export and source of income. It is utilized primarily as fish meal and oil.

Resources

Books
Laws, Edward A. *El Niño and the Peruvian Anchovy Fishery.* Sausalito, CA: University Science Books, 1997.

Lecointre, G., and G. Nelson. "Clupeomorpha, Sister-Group of Ostariophysi." In *Interrelationships of Fishes,* edited by Melanie L. J. Stiassney, Lynne R. Parenti, and G. David Johnson. San Diego: Academic Press, 1996.

Periodicals
Petitgas, P., D. Reid, P. Carrera, M. Iglesias, S. Georgakarakos, B. Liorzou, and J. Masse. "On the Relation Between Schools, Clusters of Schools, and Abundance in Pelagic Fish Stocks." *ICES Journal of Marine Science* 58 (2001): 1,150–1,160.

Whitehead, P. J. P. "Clupeoid Fishes of the World (Suborder Clupeoidei). An Annotated and Illustrated Catalogue of the Herrings, Sardines, Pilchards, Sprats, Shads, Anchovies, and Wolf-Herrings. Part 1—Chirocentridae, Clupeidae, and Pristigasteridae." *FAO Fisheries Synopsis* 125, no. 7 (1985): 303.

Whitehead, P. J. P, G. J. Nelson, and T. Wongratana. "Clupeoid Fishes of the World (Suborder Clupeoidei). An Annotated and Illustrated Catalogue of the Herrings, Sardines, Pilchards, Sprats, Shads, Anchovies, and Wolf-Herrings." Part 2—Engraulidae." *FAO Fisheries Synopsis* 125, no. 7 (1988): 274.

Organizations
Menhaden Resource Council. 1901 N. Fort Myer Drive, Suite 700, Arlington, VA 22209 USA. Phone: (703) 796-1793. E-mail: resource@menhaden.org Web site: <http://www.menhaden.org>

Other
FAO Fisheries Department. (13 Nov. 2002) <http://www.fao.org/fi/default.asp>

FishBase. 8 Aug. 2002 (13 Nov. 2002). <http://www.fishbase.org/search.cfm>

Katherine E. Mills, MS

Gonorynchiformes

(Milkfish and relatives)

Class Actinopterygii

Order Gonorynchiformes

Number of families 3

Photo: School of milkfish (*Chanos chanos*) in the warm waters along the continental shelves and around islands in the Indo-Pacific. (Photo by Stuart Westmorland/Corbis. Reproduced by permission.)

Evolution and systematics

The placement of the order Gonorynchiformes within Actinopterygii (the bony fishes) has been problematic at best. This may have been due to the enormous amount of morphological variation among gonorynchiform subgroups. For example, individuals in the genus *Phractolaemus* are facultative air breathers and show many morphological specializations of the head, while members of genera *Cromeria* and *Grasseichthys* are paedomorphic, and show extreme cranial miniaturization. As a result, few researchers could believe that these genera could belong to the same taxonomic group. Pivotal studies by Rosen and Greenwood in 1970 and Fink and Fink in 1981, however, demonstrated that the Gonorynchiformes do form an evolutionary assemblage and belong to the superorder Ostariophysi, as the sister-group to those fishes with a functioning Weberian apparatus (e.g., carps, minnows, and catfishes). In the studies of Gayet in 1993 and Grande and Poyato-Ariza in 1999, the taxonomic composition and evolutionary relationships within this unusual group of fishes was investigated. These researchers corroborated previous studies in the placement of Gonorynchiformes within Ostariophysi, and Grande and Poyato-Ariza divided the group into three families: Chanidae, Gonorynchidae, and Kneriidae. Characteristics identifying gonorynchiform fishes include a unique arrangement of intermuscular bones, as well as a specialized articulation between the first neural arch of the vertebral column and the back of the skull.

The Gonorynchiformes is an ancient group, with a fossil record dating back to the early Cretaceous period (about 100 million years ago). Although controversial, the habitats, or paleoecologies, of the fossil forms were most likely subtropical. This is indicated by the plants and animals collected in these deposits, whose living relatives have subtropical habitats today. Because many fossil gonorynchiforms are endemic to specific geographic localities, an understanding of their distribution patterns can provide insight into the early history of Earth. For example, well-preserved fossil chanids have been collected from the Santana Formation of eastern Brazil and the bituminous shales of Equatorial Guinea, supporting a hypothesized physical connection between the continents of South America and Africa. This connection may have lasted until the early Tertiary period. Additional chanid fossils have been collected from marine and lacustrine deposits of Germany and Spain, while fossils belonging to the family Gonorynchidae have been found in early Cretaceous marine deposits of Israel and Lebanon and from freshwater deposits of North America. This interesting fossil-distribution pattern has lead researchers (such as Jerzmanska in 1977 and Gaudant in 1993) to hypothesize a possible Tethys Sea (an ancient sea that once separated northern Africa from Asia) origin for the Gonorynchiformes, with subsequent dispersal routes throughout the Pacific Ocean. This model, however, does not discount the possible Pangaean origin of the group as proposed by Patterson in 1975.

As of 1999, 20 nominal gonorynchiform genera (7 living and 13 fossil) and about 50 species had been described. These genera are grouped into three families. The first, the Chanidae, consists of one living representative, *Chanos chanos*, and five extinct forms. Chanids have similar body shapes and are identifiable by a distinctively shaped permaxilla, a notch in the anterior border of the dentary, and an anteroventral process of the hyomandibular, a bone that connects the jaws to the cranium. The second family, the Gonorynchidae, is represented by the marine Indo-Pacific form *Gonorynchus*, its extinct sister group *Notogoneus*, and four Cretaceous marine groups from the Middle East. These share multiple fusions of the caudal fin skeleton; there exists a fusion of hypurals 1 and 2, as well as the parhypural with preural centrum 1. In addition, all gonorynchids have a patch of conical teeth on the gill arches, indicating that they can crush and presumably eat crustaceans and organisms with hard shells and carapaces. The third family, the Kneriidae, consists of an interesting assemblage of morphologically diverse fishes: *Phractolaemus*, which breathes atmospheric air; *Kneria*, which sports an elaborate opercular structure on the side of its body that is used as an adhesive device; and *Cromeria* and *Grasseichthys*, two miniature paedomorphs thought at one time to be juveniles. All kneriids are endemic to Africa, they live in freshwater streams and rivers and have no known fossil record. They are grouped taxonomically by distinctive modifications of the back of the skull and anterior neural arches.

Physical characteristics

The Gonorynchiformes is a morphologically diverse assemblage of fishes, ranging in body shape from the silvery herringlike chanids, to the long and slender eel-like gonorynchids, to the tiny minnowlike kneriids. The fishes also vary in size. *Gonorynchus* is one of the larger genera, and can achieve a standard length of over 19.7 in (50 cm), while the miniature *Grasseichthys* achieves an adult body length of a mere 0.71 in (1.8 cm). With the exception of *Kneria*, no sexual dimorphism is evident. In this species, however, males sport a predominant an opercular apparatus, a suckerlike structure. This feature, although present in females, is rudimentary. Its apparent sexually dimorphic occurrence in males has led researchers to assume that its central role is in reproduction. It seems clear that this structure is an adhesive devise and that males can attach themselves to females. Males can also attach themselves to rocks or substrate if necessary.

Distribution

Gonorynchiforms exhibit a widespread geographic distribution, with representatives found on virtually all continents except Antarctica. The milkfish (*Chanos chanos*) and species within the genus *Gonorynchus* inhabit waters of the Indian and Pacific Oceans. Unlike *Chanos* and *Gonorynchus*, fossil representatives are known from separate localities. Fossil chanids, for example, are restricted to Brazil, western Africa, and Europe. Fossil gonorynchids have a more complex distribution, with several members restricted to marine Cretaceous deposits of Lebanon and Israel, whereas *Notogoneus* is found in freshwater deposits of Europe, North America, Mexico, Asia, and Australia. The

freshwater kneriids are restricted to specific river systems surrounding the Gulf of Guinea, and the central and southern parts of the Africa. Species of the genera *Kneria* and *Parakneria* are the most geographically widespread, with ranges overlapping throughout Zaire, Angola, Tanzania, Zambia, and the Congo Basin. The species *Kneria auriculata* reaches the southern tip of South Africa. *Phractolaemus* is found in the Niger and Congo River tributaries, whereas *Grasseichthys* is known from streams deep in the forests of Gabon and central Congo. *Cromeria*, the sister species to *Grasseichthys*, is found in tributaries and sandy river banks of the Nile and Niger Rivers.

Habitat

Gonorynchiform species inhabit both marine and freshwater systems. As adults, milkfishes live in marine open-water habitats of the Indian and Pacific Oceans. The milkfish's diadromous nature enables it to breed in inshore waters, where it produces pelagic eggs. When the larvae reach about 0.4 in (1 cm), they enter brackish pools and creeks that have limited contact with the ocean. As mature fishes they return to the sea.

Gonorynchus species often live in coastal sandy habitats. They are nocturnal and remain buried in the sand during the day, thus their common name of sandfishes. Morphologically these fishes are well adapted for living in very dark, open, deep water on the continental shelf. They have a modified lateral line system that extends posterior to the hypural plate, and large eyes that are covered by transparent skin. Sandfishes have been recorded down to a depth of 525 ft (160 m) off Tasmania, and at depths of 340–2,225 ft (104–678 m) on the Chatham Rise and Challenger Plateau off New Zealand. *Gonorynchus* species are thought to breed in deep water. The young are transparent and have a long pelagic postlarval stage. Not until the fish reach a standard length of about 3.5 in (9 cm) do they become benthic. This long pelagic stage in their life cycle allows for the wide dispersal of juveniles.

Little is known about the habitat and ecology of the African kneriids. *Phractolaemus* is thought to inhabit quiet, shaded waters and to be an epiphytic feeder. It also has a gas bladder that is divided into many alveoli, enabling it to breathe atmospheric air. Like *Gonorynchus*, *Cromeria* is found near sandy riverbanks and apparently spends much of its time buried in the sand. The habitat of *Cromeria* is quite different from that of *Grasseichthys*, in that it is found further north in more arid environments. *Grasseichthys* inhabits forested areas farther south.

Behavior

Chanos is a schooling species, both as a juvenile and as an adult. Collection data for *Gonorynchus*, however, suggests that this species is solitary. A solitary behavior is also inferred from collection data for *Phractolaemus*.

Feeding ecology and diet

Feeding ecology seems to be variable, in that only the gonorynchids have teeth and are known to eat crustaceans. *Chanos*, like *Phractolaemus*, has a well-developed epibranchial organ and consumes planktonic prey, most often plant material.

Reproductive biology

All gonorynchiform species are oviparous, i.e., fertilization and hatching of eggs occurs outside the body. A variety of egg types exists. *Chanos* and *Gonorynchus* produce pelagic eggs, whereas *Kneria* is thought to produce demersal eggs. Sexual dimorphism is clearly evident in *Kneria* and *Phractolaemus*. In all *Kneria* species, adult males develop a characteristic opercular apparatus on the side of the head. Male fishes have been observed swimming attached to females during courtship and mating. By doing this, the male is in close proximity to the female during egg production. In *Phractolaemus*, large thickened keratinized breeding tubercles form on the head and along the sides of adult males. Although the presence of breeding tubercles is characteristic of ostariophysans, tubercules are particularly well developed in *Phractolaemus*.

Conservation status

No gonorynchiform species is listed by the IUCN. However, the South African government has designated at least one *Kneria* species as endangered, and the specialized requirements and extremely limited ranges of other species of *Kneria* and *Parakneria* render them vulnerable to the degradation of their habitats by humans.

Significance to humans

The milkfish is commercially farmed in Southeast Asia. These fishes feature in an extensive aquiculture industry in the Philippines and in Indonesia, where the young are caught close to shore and then reared in coastal ponds. The milkfish is also the subject of a targeted fishery throughout its extensive range.

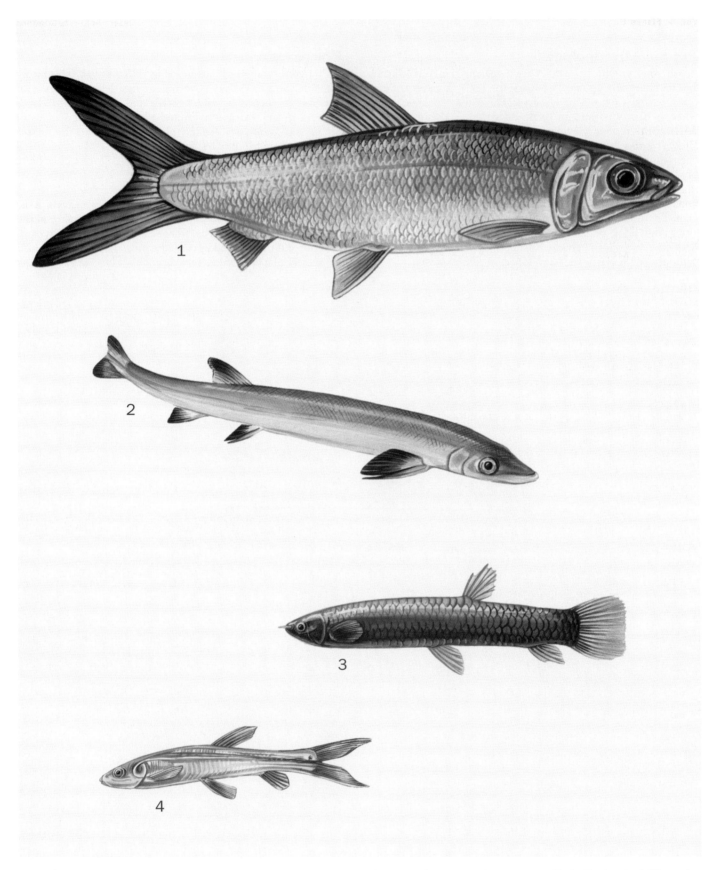

1. Milkfish (*Chanos chanos*); 2. Sandfish (*Gonorynchus gonorynchus*); 3. African mudfish (*Phractolaemus ansorgii*); 4. *Kneria wittei*. (Illustration by Patricia Ferrer)

Species accounts

Milkfish
Chanos chanos

FAMILY
Chanidae

TAXONOMY
Chanos chanos Forsskål, 1775, Red Sea.

OTHER COMMON NAMES
English: Bangos; French: Chanos; German: Milchfisch; Spanish: Chano, sabalote.

PHYSICAL CHARACTERISTICS
Standard length of over 70.9 in (180 cm). Adults are silvery herringlike fishes with a forked tail, large eyes, pointed snout with terminal mouth, cycloid scales, and an epibranchial organ. The mouth is small and terminal. The jaws are toothless. The dorsal fin has 13–17 rays; the anal fin has 6–8; the pectoral fins 15–17, and the pelvic fins 10–11. Four or five branchiostegal rays are present on each side.

DISTRIBUTION
Throughout the Indian and Pacific Oceans.

HABITAT
Diadromous; adults occur in marine open waters, larvae inhabit brackish inland ponds.

BEHAVIOR
Schooling fishes, both as juveniles and adults.

FEEDING ECOLOGY AND DIET
Larvae in coastal ponds consume diatoms and copepods. Adults have well-developed epibranchial organ used as an extension of the alimentary canal and may live on plant material.

REPRODUCTIVE BIOLOGY
Breeds in inshore waters and produces pelagic eggs. Larvae of about 0.4 in (1 cm) enter brackish waters and as young adults return to the sea.

CONSERVATION STATUS
Not threatened.

SIGNIFICANCE TO HUMANS
Commercially raised for food in the Philippines and Indonesia and fished extensively throughout its range. Local fishermen use cormorants with rings around the birds' necks to fish for milkfish. The rings prevent the birds from fully swallowing the fish.◆

Sandfish
Gonorynchus gonorynchus

FAMILY
Gonorynchidae

TAXONOMY
Gonorynchus gonorynchus Linnaeus, 1766, Cape of Good Hope, South Africa.

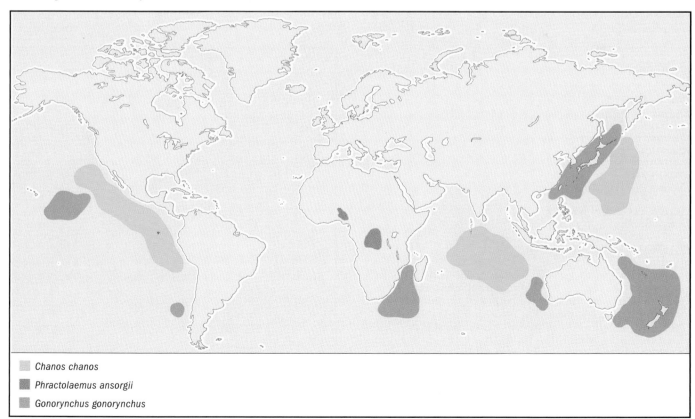

Chanos chanos
Phractolaemus ansorgii
Gonorynchus gonorynchus

OTHER COMMON NAMES
English: Beaked sandfish, mousefish, sand eel; French: Caduchon; Spanish: Caduchón; Afrikaans: Spitsbek-sandvis.

PHYSICAL CHARACTERISTICS
Standard length 23.6 in (60 cm). Long slender fishes; mouth inferior, dorsal fin positioned posteriorly (predorsal length at least 70% that of standard length), swim bladder absent, lateral line extends to tip of caudal fin rays, number of lateral line scales ranges from 200–220, presence of median sensory barbel on ventral side of snout, presence of a clover-shaped barbel within the mouth extending from its roof, presence of ctenoid scales that cover the entire body. The young are transparent. Uniformly brown in color, with black patches at the tips of fins.

DISTRIBUTION
Indian and Pacific Oceans.

HABITAT
Marine environments, coastal sandy habitats as well as benthic waters, can reach over 1,970 ft (600 m) in depth.

BEHAVIOR
Does not school. Young have long pelagic postlarval stage that may allow for a wide dispersal of juveniles. Nocturnal, and can be found buried in sand or mud during the day.

FEEDING ECOLOGY AND DIET
A benthic feeder of small decopod crustaceans.

REPRODUCTIVE BIOLOGY
Breeds in deep water. Young are pelagic, subadults of about 3.5 in (9 cm) become benthic.

CONSERVATION STATUS
Not threatened.

SIGNIFICANCE TO HUMANS
None known. ◆

No common name
Kneria wittei

FAMILY
Kneriidae

TAXONOMY
Kneria wittei Poll, 1944, Congo River basin.

OTHER COMMON NAMES
None known.

PHYSICAL CHARACTERISTICS
Standard length 3.4 in (8.6 cm). Small minnowlike fish. Males have large opercular apparatus, females have rudimentary opercular apparatus. There is a modification and expansion of the epicentral intermuscular bones in both sexes, cycloid scales, tail forked, and mouth terminal. The color pattern is darker on top, lighter on bottom, with darker stripe along side.

DISTRIBUTION
Rivers and tributaries throughout Zaire, Angola, Tanzania, Zambia (i.e., central and southern Africa).

HABITAT
Quiet pools, but most often fast-moving streams with waterfalls.

BEHAVIOR
Schooling fishes, at least during mating periods.

FEEDING ECOLOGY AND DIET
Feeds mostly on algae and plant material. Like most gonorynchiforms, has an epibranchial organ which enables it to filter particulate matter from the water column.

REPRODUCTIVE BIOLOGY
Males attach themselves to females via the opercular apparatus during courtship, ensuring that the largest number of demersal eggs laid by the female will be fertilized.

CONSERVATION STATUS
Not listed by the IUCN. However, like all freshwater African gonorynchiforms, habitat destruction may eventually result in an endangered status.

SIGNIFICANCE TO HUMANS
None known. ◆

African mudfish
Phractolaemus ansorgii

FAMILY
Kneriidae

TAXONOMY
Phractolaemus ansorgei Boulenger, 1901, Niger delta. Species of the genus *Phractolaemus* have traditionally been placed within their own family, Phractolaemidae, but according to current research, they are closely related to kneriids and have been placed in Kneriidae by Grande and Poyato-Ariza.

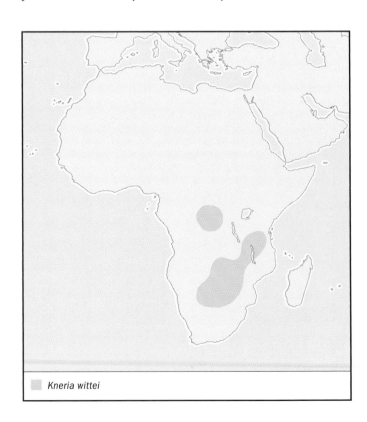

Kneria wittei

OTHER COMMON NAMES
English: Hingemouth, snake mudhead; German: Afrikanischer Schlammfisch.

PHYSICAL CHARACTERISTICS
Standard length 6 in (15 cm). Elongate, cylindrical body with large cycloid scales. Head is small, broad, and strongly ossified; eyes are small and laterally positioned. Infraorbital bones two, three, and four are greatly enlarged. The mouth is highly projectile and capable of being thrust forward; at rest the mouth folds over into a depression on the upper surface of the snout. The mouth has no teeth except for a conical tooth on each dentry near the symphysis. There is a single narial opening preceded by a barbel. The opercular openings are narrow due to a sealing of the opercular boarder to the body wall. The interopercle is spinelike, and the preopercles are greatly enlarged, overlapping along the ventral midline of the body. There are six pelvic fin rays, six dorsal rays, and six anal fin rays. Three slender branchiostegal rays are evident. Unlike other gonorynchiform fishes, the swim bladder is divided into alveoli, which enables the fish to breathe atmospheric air. Body is uniformly gray above, light brown on the sides, pale ventrally with darkly colored fins.

DISTRIBUTION
Tropical Africa ranging through the lower Niger drainage and central Zaire basin.

HABITAT
Quiet, low-oxygenated muddy waters.

BEHAVIOR
Little is known about the behavior of this species. Unlike other gonorynchiforms, it is able to breathe atmospheric air, an ability that has made this fish of interest to aquarists.

FEEDING ECOLOGY AND DIET
Feeds on small, mud-dwelling organisms. Based on the presence of a moderately developed epibranchial organ, some researchers believe this species is an epiphytic feeder.

REPRODUCTIVE BIOLOGY
Utilizes external fertilization. Exhibits clear sexual dimorphism, as males sport conspicuous whitish breeding tubercles on the head, along the lateral line, and on the caudal peduncle.

CONSERVATION STATUS
Not listed by the IUCN. However, continued habitat destruction will undoubtedly affect the population dynamics and future of this fish.

SIGNIFICANCE TO HUMANS
Not an economically important food fish, but has been imported into the United States as an aquarium fish.

Resources

Books
Aizawa, M. "Gonorynchidae." In *Fishes Collected by the Shinkai Maru Around New Zealand*, edited by K. Amaoka. Tokyo: Japan Marine Fishery Resource Research Center, 1990.

Grande, T. "Distribution Patterns and Historical Biogeography of Gonorynchiform Fishes (Teleostei: Ostariophysi)." In *Mesozoic Fishes: Systematics and the Fossil Record: Proceedings of the International Meeting, Buckow, 1997*, edited by G. Arratia and H. P. Schultze. Munich: Pfeil, 1999.

Roberts, C. D., and T. C. Grande. "The Sandfish *Gonorynchus forsteri* (Gonorynchidae), from Bathyal Depths off New Caledonia, with Notes on New Zealand Specimens." In *Proceedings of the 5th Indo-Pacific Fish Conference*, edited by B. Séret and J.-Y. Sire. Paris: Société Française d'Ichtyologie; Institut de Recherche pour le Développement, 1999.

Periodicals
Bertmar, G., B. G. Kapoor, and R. V. Miller. "Epibranchial Organs in Lower Teleostean Fishes: An Example of Structural Adaptation." *Trop. Atlan. Biol. Lab., Bureau of Comm. Fish.* 76 (1969): 149.

Cope, E. D. "On Two New Forms of Polydont and Gonorhynchid Fishes from the Eocene of the Rocky Mountains." *Memoirs of the National Academy of Sciences* 3 (1885): 161–165.

Fink, S. V., and W. L. Fink. "Interrelationships of Ostariophysan Fishes (Teleostei)." *Zoological Journal of the Linnean Society* 72, no.4 (1981): 297–353.

Forsskål, P. "Descriptones animalium, avium, amphibiorum, piscium, insectorum, vermium: quae in itinere orientali observait." *Postmortem auctoris edidit Carsten Niebuhr* 20, no. 34 (1777): 1–164.

Gaudant, J. "The Eocene Freshwater Fish Fauna of Europe: From Paleobiogeography to Paleoclimatology." *Kaupia* 3 (1993): 231–244.

Gayet, M. "Relations Phylogénétiques de Gonorhynchiformes (Ostariophysi)." *Belgian Journal of Zoology* 123, no. 2 (1993): 165–192.

Grande, T. "Revision of the Genus *Gonorynchus* Scopoli, 1777 (Teleostei: Ostariophysi)." *Copeia* 2 (1999): 453–469.

Grande, T., and F. J.Poyato-Ariza. "Phylogenetic Relationships of Fossil and Recent Gonorynchiform Fishes (Teleostei: Ostariophysi)." *Zoological Journal of the Linnean Society* 125 (1999): 197–238.

Grande, T., and B. Young. "Morphological Development of the Opercular Apparatus in *Kneria wittei* (Ostariophysi: Gonorychiformes) with Comments on Its Possible Function." *Acta Zoologica* 78, no. 2 (1997): 145–162.

Jerzmanska, A. "Süßwasserfische des alteren Tertiärs von Europe." In *Eozäne Wirbeltier des Geiselatles*, edited by H. W. Matthes and B. Thaler. (1977): 67–76.

Linnaeus, C. "Systema Naturae." *Laurentii Salvii* 12, no.1 (1766): 528.

Morioka, S., A. Ohno, H. Kohno, and Y. Taki. "Recruitment and Survival of Milkfish *Chanos chanos*." *Japanese Journal of Ichthyology* 40, no. 2 (1993): 247–260.

Patterson, C. "The Distribution of Mesozoic Fishes." *Mèmoires du Muséum national d'Histoire naturelle, Paris* 88, sèr. A (1975): 156–173.

Resources

Poll, M. "Descriptions de poissons nouveaux recueillis dans la region d'Albertville (Congo Belge) par le Dr. G. Pojer." *Bulletin du Musée Royal d'Histoire naturelle de Belgique* 20, no. 3 (1944): 1–12.

Rosen, D. E., and P. H. Greenwood. "Origin of the Weberian Apparatus and the Relationships of the Ostariophysan and Gonorynchiform Fishes." *American Museum Novitates* 2428 (1970): 1–25.

Seegers, L., "Revision of the Kneriidae of Tanzania with Description of Three New *Kneria* Species (Teleostei: Gonorynchiformes)." *Ichthyol. Explor. Freshwaters* 6, no.2 (1995): 97–128.

Terry Grande, PhD

Cypriniformes I
(Minnows and carps)

Class Actinopterygii

Order Cypriniformes

Number of families 1 of minnows and carps

Photo: Roach (*Rutilus rutilus*), a freshwater fish of Europe. (Photo by Tom McHugh/Photo Researchers, Inc. Reproduced by permission.)

Evolution and systematics

Cypriniforms are typical freshwater fishes in which the upper jaw is usually protractile; the mouth (jaws and palate) is always toothless; the adipose fin is absent; the head almost always scaleless; and barbels are either present or absent. These fishes have Weberian ossicles (four small bones and their ligaments connecting the swim bladder to the inner ear for sound transmission). The fifth ceratobranchial is enlarged as the pharyngeal bone, with teeth ankylosed (joined) to it. For cyprinids, pharyngeal teeth are in one to three rows, and there are never more than eight teeth in any one row; for non-cyprinid cypriniforms, pharyngeal teeth are usually greater in number but only in one row.

Generally, Cypriniformes is divided into two monophyletic groups: the family Cyprinidae and the non-cyprinid cypriniforms. The Cyprinidae includes different kinds of minnows and carps. The non-cyprinid cypriniforms are composed of the family Catostomidae (suckers), family Gyrinocheilidae (algae eaters), and many different loaches. The relationships among the non-cyprinid cypriniforms are still in debate. Recent molecular data suggest that suckers could be at the basal position of this group, followed by the algae eaters and then the different loaches. This chapter focuses on the family Cyprinidae.

The recognition and composition of the subfamilies in the Cyprinidae is still in question. Several proposals have been provided based on morphological characters, and the recent molecular data support a combination of them. Thus, 9 subfamilies, forming two phyletic lineages, are recognized. The first lineage consists of subfamilies Cyprininae, Barbinae (including Schizothoracinae), and Labeoninae. The second lineage consists of subfamilies Rasborinae (Danioninae), Leuciscinae, Tincinae, Acheilognathinae, Gobioninae (including Gobiobotinae), and Xenocyprinae (the east Asian group,

including Cultrinae, Hypophthalmichthyinae, etc.). All these different subfamilies could have evolved from *barbus*-like cyprinids, with parallel evolutions of certain characters, such as the loss of three-rowed pharyngeal teeth, loss of barbels, and so on.

The earliest cyprinid fossils are known from South China from the Eocene period, and could represent cyprinini and rasborini; the earliest European and North American ones are of Oligocene age. The cyprinids might have originated in Asia and dispersed to North America through the Bering land bridge and to Europe before the upheaval of the Tibetan Plateau. The earliest record of cyprinids in Africa was in the Miocene period. Cyprinids may have migrated from Southeast Asia to Africa through the Near East during the Miocene.

Physical characteristics

Normally, carps are fusiform or streamlined, with the body somewhat compressed. The dorsal fin is long (in *Cyprinus*) or short, and the last unbranched fin ray is soft, hard, or spine-like, with serrations on the posterior edge in Cyprininae and some barbinins. Pectoral fins and ventral fins are in the normal position. The anal fin has five soft, branched fin rays in Cyprininae, Barbinae, and Labeoninae species; six in Gobioninae species; and seven or more in other groups. The last unbranched anal fin ray is soft, hard, or spine-like. The caudal fin is forked in all species.

Breams and some cultrin species have a very deep body, which may protect them from predators' bites. Living on the river bottom and adapted to fast running water, most Gobionine species have a round and slender body and are called "stick fish" by fisherman. The head of *Luciobrama* is strongly elongated, forming a pipe shape.

Redside dace (*Clinostomus elongatus*) swimming in Ohio. (Photo by Gary Maeszaros/Photo Researchers, Inc. Reproduced by permission.)

A northern pikeminnow (*Ptychocheilus oregonensis*) resting on the rocks of the Middle Fork Boise River in Idaho, USA. (Photo by William H. Mullins/Photo Researchers, Inc. Reproduced by permission.)

Most carps and minnows are covered with scales. However, the leather carp, one variation of the common carp, has no scales. Some Schizothoracin fishes are half or completely naked. The sawbwa barb, *Sawbwa resplendens*, which is endemic to Lake Inle in Myanmar, is completely naked and a little transparent. The lateral line is complete in most species but incomplete in some small fishes, such as *Aphyocypris* and *Rhodeus*.

The most common coloration for carps is dark green on the back and whitish on the belly, which makes the fishes difficult to spot both from above the water and under the water. However, color variations are also very common. Some *gobio* species have different dots; *Danio* and *Zacco* species have beautiful stripes; and some barbin species and bitterlings are even more colorful.

The largest species are the tetraploid barbine *Catlocarpio siamensis* of Thailand, which is known to reach at least 8.2 ft (2.5 m) and probably 9.84 ft (3 m), and *Tor putitora* of the Brahmaputra River (eastern India), which reaches about 8.86 ft (2.7 m); in North America the Colorado pikeminnow, *Ptychocheilus lucius*, can reach 5.9 ft (1.8 m); other large Asian species (6.6 ft [2 m] or larger) include *Elopichthys bambusa* and *Barbus esocinus*. The smallest cyprinid is *Danionella translucida*, distributed in Burma, in which females are mature at about 0.4–0.43 in (10–11 mm) and the longest specimen known is 0.47 in (12 mm).

Distribution

Cyprinidae is the largest family of freshwater fishes, with about 210 genera and about 2,010 species. Of this figure, about 1,270 species are native to Eurasia (the greatest generic diversity and number of species is in China and Southeast Asia; China alone contains about 532 species in 132 genera); about 475 species in 23 genera are native to Africa; and about 270 species in 50 genera are native to North America. In North America there are only phoxinine species, while in Europe there are mainly leuciscine species in addition to one species of bitterling, the monotypic *Tinca*, some *Gobio* species,

and some *Barbus* species. In Africa there are only three subfamilies present: barbine, labeonine, and rasborine. In East Asia, especially China, all types are distributed. The specialized schizothoracine fishes are mainly found in and around the Tibetan Plateau.

Habitat

Carps can live in a large variety of habitats, from small streams and ponds to large rivers and lakes. Almost all carps live in freshwaters, although the European roach and bream (*Rutilus* and *Abramis*, respectively) have populations in the brackish part of the Baltic Sea; the Japanese *Tribolodon* spends part of its life at sea; and the Chinese carp *Cyprinus acutidorsalis* can live in the river mouth of Qingjiang River near Vietnam. *Phoxinus* and gudgeons like to stay in small streams. Big fishes like *Elopichthys bambusa* are mainly found in large rivers. *Danio* species must live in waters with temperatures higher than 64.4°F (18°C), while *Leuciscus* species only live in cold waters (39.2–71.6°F [4–22°C]). *Garra* and *Labeo* species like to adhere to the bottoms of streams and rivers with fast running water. Bitterlings prefer still or slow running waters such as ponds and lakes. *Culter* fishes often swim in the upper parts of waters or near the surface to catch insects, but the common carp and crucian carp mainly stay at the bottom sucking worms.

Behavior

Cyprinid fishes have good vision, including color vision, and use visual displays. The use of pheromones in cyprinid social communication is well established. For instance, they have an ability called fright reaction. When threatened by a predator, an individual may release alarm substances through specialized goblet or club cells in the skin. These secretions cause the other fishes nearby to disperse and hide. In this way, the rest of the group can avoid the predator. Cypriniform fishes also have excellent hearing. Shoaling minnows find food more quickly in groups and are less vulnerable to

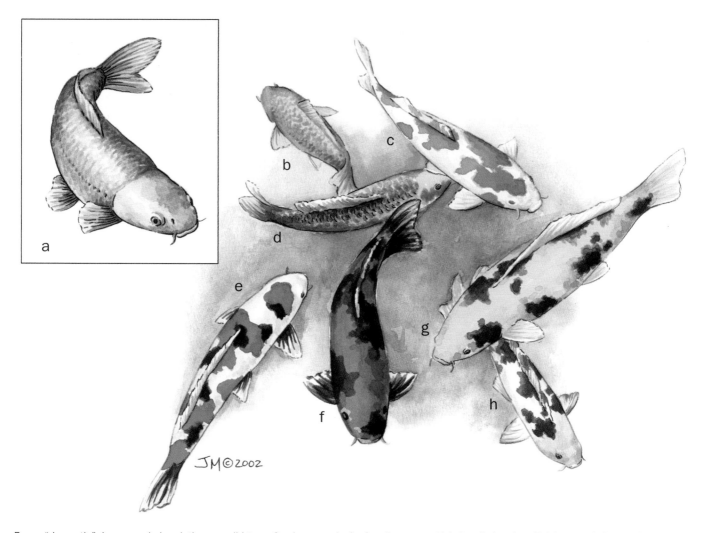

Fancy "domestic" Japanese koi varieties: a. wild type *Cyprinus carpio*; b. Orenji ogon; c. Kohaku; d. Asagi; e. Taisho sanshoku (sanke); f. Hi utsuri; g. Ki bekko; h. Shiro bekko. (Illustration by Jacqueline Mahannah)

predation. Though some species like to search for food solitarily, in the spring they may form schools for reproduction, and in fall or winter they may transfer to deeper water as water levels decline.

Some species exhibit territorial behavior. For example, in breeding season, the male of the bitterling species, *Rhodeus ocellatus*, may find a good mussel and protect the area around this mussel as his territory.

Living only in freshwater, cyprinids need not migrate between fresh and marine waters. But some species can swim for very long distances (up to 1,012 mi [1,629 km]). Some East Asian groups undergo river and lake migration. These fishes spawn in the middle or upper reaches of rivers when heavy flood occurs. Their eggs float with the running water and hatch. The fries also float with the running water in the first few days after hatching, before running into lakes that connect to the rivers. The young fish then may stay in the lakes to take advantage of the abundance of food. When they mature, they will migrate to the rivers to breed.

Feeding ecology and diet

Cyprinids comprise a wide variety of specialists and generalists feeding on all trophic levels. Most feed on secondary producers: zooplankton, crustaceans, larvae, pupae and adults of insects, oligochaetes, bryozoans, snails, and mussels. Some also consume primary (macrophytes and phytoplankton) or tertiary (fishes) producers. According to feeding behavior, cyprinids can be categorized into three modes: herbivores, pelagic feeders, and benthic feeders. Herbivores like grass carp eat not only aquatic plants but also the land grasses submerged by flood water. *Condrostoma* and *Xenocypris*, for example, use the horny edge on their lower jaws to scrape the algae on the bottom. Pelagic feeders mainly catch zooplankton and surface insects, but *Elopichthys bambusa* and the Colorado pikeminnow are very ferocious and feed on fishes. Some species, like silver carp and bighead carp, have evolved special gill organs for filtering plankton. *Pectenocypris balaena* of the Kapuas River in Borneo has up to at least 212 gill rakers for filtering phytoplankton. Benthic feeders suck in the sediment particles together with the organisms and separate the

A central stoneroller (*Campostoma anomalum*) male moves pebbles from the riverbed with his mouth to build a breeding nest. A group of females watches at a distance and will rush in to spawn on the nest at the same time. Another male builds his nest in the distance. (Illustration by Bruce Worden)

organisms in the pharyngeal slit. Sediment particles pass through the sieve, whereas food organisms are retained. Substratum particles too large to pass the basket are spit out.

None of the cyprinids are strictly monophagous, but many may feed on only one type of food organism, depending on

Carp bream (*Abramis brama*) live in Asia and Europe. (Photo by Y. Lanceau/Photo Researchers, Inc. Reproduced by permission.)

availability. The feeding of European cyprinids includes all diets and feeding modes. The cyprinids from Asia seem to have the greatest variety in feeding specialists with both small and large species, whereas the cyprinids in North America have the smallest variety. Cyprinids in Africa are comprise a relatively small variety of feeding types.

There is an interesting ontogenetic switch of the feeding mode in cyprinids. Almost all cyprinids start to feed on plankton shortly after hatching. As individuals increase in size, their prey choice changes, and they differentiate into the specialized feeding modes (herbivore, piscivore, and benthivore).

Reproductive biology

Carps and minnows mostly spawn in spring and summer, because the larvae will get food easily. One bitterling species breeds in autumn, which is an alternative strategy. The water temperature for reproduction may be as low as 44.6–48.2°F (7–9°C) for cold water species but must be above 64.4°F (18°C) for most East Asian groups. During breeding season, the males usually have beautiful color, for example an orange tail or anal fin to stimulate or attract females. Some species may have tubercles on the head or pectoral fin called

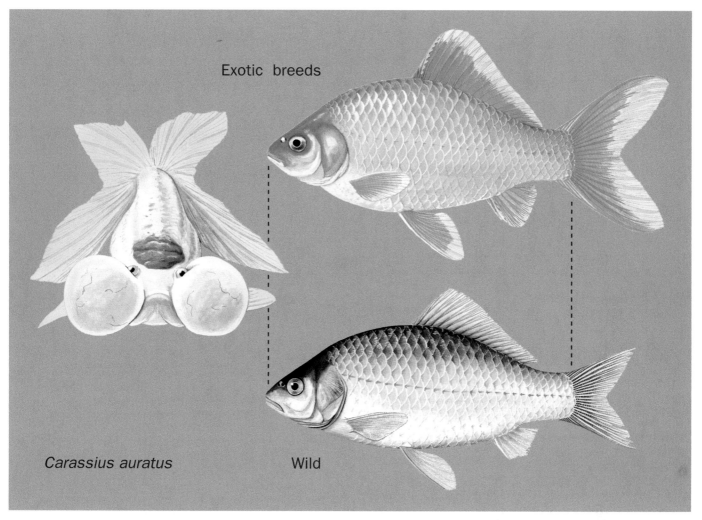

Exotic breeds

Carassius auratus   Wild

A colorful aquarium goldfish, *Carassius auratus*, selected for its color and fin size (top and left), compared to a much duller wild one (bottom) of the same species. (Illustration by Emily Damstra)

pearl organs, which can be used to provide tactile stimulation during courtship by pushing the female.

Usually, when their conditions are suitable, the males chase the females and press up against their abdomens. During this activity the fishes swim at very high speeds. The females then lay eggs, and the males release their sperm. Environmental conditions that impact fertilization include water temperature as well as the nature of the substances that the eggs adhere to, whether they are aquatic plants, stones, or other substances. Grass carp only spawn after heavy floods and when the water temperature is above 64.4°F (18°C); the flood surge is needed to carry eggs and larvae. Otherwise, the eggs may sink to the bottom and die.

Some cyprinids have adopted unusual breeding habits. A small minnow in southwestern China, *Gobiocypris rarus*, pushes the eggs to adhere to walls above water level through the beating of its tail. The female bitterling species lay their eggs into the gill chambers of mussels through a long tube (ovipositor).

In breeding season, the males of North American minnow known as the stoneroller dig spawning pits by driving their heads into the gravel. They transport gravel from the pits by nudging stones out with their snouts, or by transporting them with their mouths. The males compete aggressively for favored spawning areas. Male fathead minnows select nest sites under rocks or logs and they excavate the area to increase the available space, and then defend the nests aggressively from all other fatheads. After the female lays her sticky eggs on the underside of the nest object, the male fertilizes the eggs and guards the incubating eggs. He even fans them with his fins and massages them with his back pad to keep them clean and well oxygenated.

Spawning minnows of some species of *Luxilus*, *Cyprinella*, *Notemigonus*, and *Notropis* use the nests of other species of cyprinids or of species of the family Centrarchidae to deposit their eggs and leave the embryos to the protection of the host.

Many species spawn only one time in a single breeding season, but some species (e.g., common carps, the bitterling

The female fathead minnow (*Pimephales promelas*) lays her eggs on the underside of a lilypad. The male (shown here) cleans and aerates the eggs with the spongy tissue and tubercles it acquires on its head and neck during the breeding season. (Illustration by Emily Damstra)

fishes) have developed a strategy to spawn more than one time in a breeding season. Some species can even spawn for the whole year at an interval of 3 to 15 days.

The eggs of many species are adhesive, sticking to stones or aquatic plants. Eggs are semi-pelagic in some East Asian groups (grass carp, silver carp): the eggs sink to the bottom in still water but float with the current in running water.

The duration of development differs from species to species depending on water temperature. At 44.6°F (7°C), roach eggs take 30 days to hatch and dace eggs take 44 days. At 59°F (15°C), by contrast, only 14 days are needed for dace eggs to hatch. When temperature is 68–77°F (20–25°C), common carp eggs take 2.5–3 days to hatch. For the first few days after hatching, the larvae can not swim. Several days later, the air-bladder begins to fill with air, the yolk-sac is nearly gone,

The warty, furrowed snout of some cyprinids (including *Garra orientalis,* shown here) may provide a hydrodynamic advantage, allowing the fish to "sit" more securely in rapidly moving water. Arrow indicates direction of water flow. (Illustration by Emily Damstra)

and the larva begins to swim freely and catch food, usually plankton.

Conservation status

The IUCN Red List contains 252 species of cyprinids. Of these, 15 are categorized as Extinct; 1 as Extinct in the Wild; 39 as Critically Endangered; 31 as Endangered; 89 as Vulnerable; 6 as Lower Risk/Conservation Dependent; 23 as Lower Risk/Near Threatened; and 48 as Data Deficient. Of the 15 extinct species, 12 are from the Americas, 1 from East Asia, 1 from the Middle East, and 1 from Europe.

Major threats to cyprinids are habitat destruction, such as the construction of dams that cut off the migration routes; the eutrophication in lakes that destroys the aquatic plants necessary for cyprinid spawning; and the decrease of water area due to economic development. In addition, overfishing and competition for water resources with agricultural irrigation are also important factors threatening cyprinids. In recent years, the threat due to the introduction of exotic non-native species has become more serious.

Significance to humans

Cyprinids are an important food fish. According to an FAO (Food and Agriculture Organization of the United Nations) production report for the year 1996, there were 4 cyprinid species among the top 10 species or species groups: silver carp, grass carp, common carp, and bighead carp. Cyprinids are also important in sport fishing, especially the barbel in Europe and the common carp, crucian carp, and grass carp in Asia. One cyprinid, the zebrafish, has become one of the most important model fishes in genetics and medical research.

Many cyprinids are important aquarium pets. Good examples are goldfish, zebrafish, and other danios, small barbs, rasboras, and bitterlings. The Japanese colored carp, koi, is cultured in ponds as an ornamental fish. After many years of selection, the Japanese koi and Chinese goldfish have become very different from their wild types.

Grass carp have been introduced to many countries to control aquatic vegetation. Phytoplankton eaters such as the silver carp, have been used to control eutrophication in some countries. However, the introduction of carp species has also had negative effects, including the destruction of native fish fauna because of the competition for food and/or habitat changes, such as the decrease of aquatic vegetation.

1. Dagaa (*Rastrineobola argentea*); 2. Stoneroller (*Campostoma anomalum*); 3. Ningu (*Labeo victorianus*); 4. Harlequin (*Rasbora heteromorpha*); 5. Rudd (*Scardinius erythrophthalmus*); 6. Fathead minnow (*Pimephales promelas*); 7. Tiger barb (*Puntius tetrazona tetrazona*); 8. Colorado pikeminnow (*Ptychocheilus lucius*). (Illustration by Emily Damstra)

1. Upper mouth (*Culter alburnus*); 2. Smallscale yellowfin (*Xenocypris microlepis*); 3. Rosy bitterling (*Rhodeus ocellatus*); 4. Silver carp (*Hypophthalmichthys molitrix*); 5. Tench (*Tinca tinca*); 6. Grass carp (*Ctenopharyngodon idellus*). (Illustration by Emily Damstra)

1. Gudgeon (*Gobio gobio*); 2. Black stick (*Garra pingi*); 3. Common carp (*Cyprinus carpio*); 4. Eurasian minnow (*Phoxinus phoxinus*); 5. Common dace (*Leuciscus leuciscus*); 6. Crucian carp (*Carassius auratus*); 7. Zebrafish (*Danio rerio*); 8. *Schizothorax prenanti*; 9. Barbel (*Barbus barbus*). (Illustration by Emily Damstra)

Species accounts

Barbel
Barbus barbus

FAMILY
Cyprinidae

TAXONOMY
Cyprinus barbus Linnaeus, 1758, Europe.

OTHER COMMON NAMES
French: Barbet; German: Barbern.

PHYSICAL CHARACTERISTICS
Size large, usually up to 29.92 in (76 cm) in length. Body long. Snout pointed. Mouth inferior. Lips fleshy. Barbels 2 pairs. Pharyngeal teeth in three rows. Dorsal fin with 4 unbranched, 7–9 branched rays; anal fin with 3 unbranched, 5 branched rays. Lateral line complete, with 56–65 scales. Vertebrae 46–47. Brown-green above, green-yellow lower sides, white-yellow belly. Covered with dark-brown spots.

DISTRIBUTION
West and Central Europe excluding Italian, Greek and Iberian peninsulas.

HABITAT
Deep, fast-flowing upper reaches of rivers with stony or gravel bottoms (barbel zones). Common temperature is 59–71.6°F (15–22°C).

BEHAVIOR
Barbels normally occur in groups of several individuals close to the river bed, but they do not congregate in schools. They migrate in rivers with a home range of 1.24–12.43 mi (2–20 km).

FEEDING ECOLOGY AND DIET
Feeds chiefly on benthic invertebrates such as small crustaceans, insect larvae, mollusks, mayfly and midge larvae, and small fishes.

REPRODUCTIVE BIOLOGY
Males mature in the fourth year and females in the fifth year of life. After the fish have migrated upriver, spawning occurs from May to July when water temperature is 57.2–68°F (14–20°C) and the bottom is filled with sand and pebbles. Eggs are firmly attached to stones. Fecundity is 8,000–12,000 eggs per kilogram of body weight.

CONSERVATION STATUS
Not listed by IUCN.

SIGNIFICANCE TO HUMANS
An important food fish and sport fish, particularly in Europe. ◆

Stoneroller
Campostoma anomalum

FAMILY
Cyprinidae

TAXONOMY
Rutilus anomalous Rafinesque, 1820, Licking River, Kentucky, Ohio River drainage, United States.

OTHER COMMON NAMES
English: Central stoneroller, largescale stoneroller.

PHYSICAL CHARACTERISTICS
Size small to moderate, maximum 7.87 in (20 cm) in total length. Body stout and moderately compressed, with the nape

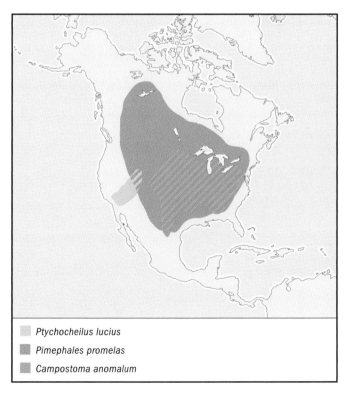

Ptychocheilus lucius
Pimephales promelas
Campostoma anomalum

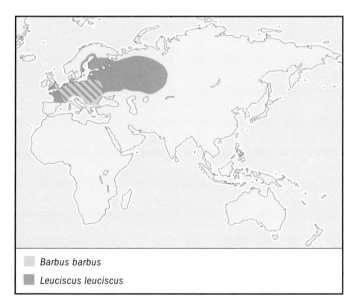

Barbus barbus
Leuciscus leuciscus

region becoming swollen and prominent in adults. Snout bluntly rounded and projecting beyond the nearly horizontal mouth. Lower jaw has spade-like extension. Scales deep, rather small, and crowed anteriorly; more or less mottled with dark background; scales in lateral line 53. Dorsal fin with 8 branched rays; anal fin with 7 branched rays. Color brownish, with a brassy luster above. Dusky vertical bar behind the opercle; dorsal and anal fins each with a dusky crossbar about half way up, the rest of the fin is olive in females and fiery red in males in spring. In the spring, the head and sometimes the entire body of males are covered with large rounded tubercles.

DISTRIBUTION
North America, widespread across most of eastern and central United States in Atlantic, Great Lakes, Mississippi River, and Hudson Bay basins from New York west to North Dakota and Wyoming and south to South Carolina and Texas; Thames River system in Canada; from Galveston Bay in Texas to Rio Grande in Mexico.

HABITAT
Moderate to high gradient streams with sandy to gravely substrate. Prefers riffle areas where riffles and pools alternate in rapid succession. However, can survive in almost any stream with a food supply.

BEHAVIOR
Shoaling species.

FEEDING ECOLOGY AND DIET
Primarily herbivorous, feeding diurnally on filamentous algae and diatoms but also taking detritus and aquatic insects from the periphyton assemblage on rock surfaces. Because of its long intestine (up to 8 times its body length), this species is incredibly efficient at digesting detritus and algae.

REPRODUCTIVE BIOLOGY
Matures in second or third summer of life. Adults spawn between March and late May, when water temperatures are from 55.4–80.6°F (13–27°C). Males dig spawning pits in shallow, swift riffles and occasionally in quiet pools by driving their heads into the gravel. They transport gravel from the pits by nudging stones out with their snouts (hence the name *stoneroller*) or by transporting them with their mouths. Males compete aggressively for favored spawning areas. Females remain in deeper water near the spawning pits and enter the pits individually or in groups to deposit eggs. The adhesive eggs become lodged in the gravel and are abandoned prior to hatching.

CONSERVATION STATUS
Not listed by IUCN.

SIGNIFICANCE TO HUMANS
Not sought by anglers. They do make good bait but are difficult to culture. ◆

Crucian carp
Carassius auratus

FAMILY
Cyprinidae

TAXONOMY
Cyprinus auratus Linnaeus, 1758, China, Japanese rivers.

OTHER COMMON NAMES
English: Goldfish, golden carp.

☐ *Carassius auratus*
■ *Schizothorax prenanti*

PHYSICAL CHARACTERISTICS
Size small to moderate, normally 5.12–7.48 in (13–19 cm) in standard length. Body deep and stout, moderately compressed. Snout pointed. Mouth terminate, oblique. Barbels absent. Pharyngeal teeth in one row. Gill rakers 37–43. Dorsal fin long, 4 spines, 15–19 rays. Anal fin short, 3 spines, 5 rays. Back of last dorsal and anal spines serrated. Lateral line complete, with 27–30 scales. Wild forms are usually olive-green in the back, gray-white on belly.

There are many aquarium varieties in different forms and colors. These can be divided into four types: (1) Grass type: primitive with slender body, pointed head, small eyes, and single or double tails; (2) Fancy type, with double tails and all fins very long; (3) Dragon or Eye type, with large eyes that protrude out; (4) Egg-shaped type, with the dorsal fin absent.

DISTRIBUTION
Native to Asia. The wild type has been introduced to Europe and North America. Aquarium varieties have been introduced all over the world.

HABITAT
Shallow, warm waters with dense vegetation such as lakes, reservoirs, and streams. Adults are generally found near the bottom, but they sometimes appear in schools at the surface.

BEHAVIOR
The social behavior of crucian carp is similar to that of the common carp, but under some conditions it can attain a greater population density.

FEEDING ECOLOGY AND DIET
Omnivorous, consuming a variety of larvae and aquatic insects, mollusks, crustaceans, aquatic worms, and aquatic vegetation.

REPRODUCTIVE BIOLOGY
The crucian carp matures after the body length reaches 3.54 in (9 cm) in the first or second year. Spawning occurs in spring when the water temperature reaches 60.08°F (15.6°C) and heavy rains occur. The eggs are released in batches, and are

usually attached to aquatic plants and other fixed objects. The male fertilizes the eggs immediately. The incubation may take 4 days at 62.6–66.2°F (17–19°C). After hatching, the larvae cling to plants or remain quietly on the bottom, but after 1–2 days they become free-swimming. Fecundity varies from 12,000 to 28,000 eggs per individual.

CONSERVATION STATUS
Not listed by the IUCN. Due to artificial hybridization and transplantation, local types have been seriously damaged.

SIGNIFICANCE TO HUMANS
An important food fish, although its production is much less than that of the common carp. Its greatest value is as an aquarium fish. ◆

Grass carp
Ctenopharyngodon idellus

FAMILY
Cyprinidae

TAXONOMY
Leuciscus idella Cuvier and Valenciennes, 1844, China.

OTHER COMMON NAMES
English: White amur; French: Amour blanc; German: Graskarpfen; Spanish: Carpa herbivora.

PHYSICAL CHARACTERISTICS
Size large, commonly 9.84–35.43 in (25–90 cm) in body length. Body long, cylindrical in the front, compressed in the hind. Belly round. Snout short and blunt. Mouth terminal, large and wide. Barbels absent. Pharyngeal teeth in 2 rows, larger ones compressed like a comb. Gill rakers 14–18. Dorsal fin short, with 3 unbranched, 7 branched rays. Anal fin with 3 unbranched, 8–9 branched rays. Scales moderate, lateral line complete, with 35–42 scales. Coloration brassy olive above, white on lower sides and belly. Edge of scales dark gray. Pectoral and ventral fins gray-yellow, other fins gray.

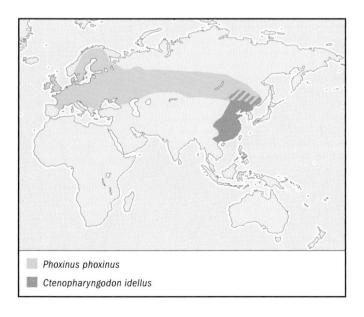

　　Phoxinus phoxinus
　　Ctenopharyngodon idellus

DISTRIBUTION
East Asia, from Amur River to Yellow River, Yangtze River, and Pearl River. Widely transported around the world.

HABITAT
Lakes, ponds, pools, and backwaters of large rivers. Prefers large, slow-moving or standing water bodies with vegetation.

BEHAVIOR
Usually stay in lower depths and are solitary in nature. They mature in lakes and migrate to rivers for reproduction.

FEEDING ECOLOGY AND DIET
Mainly feeds on aquatic plants and submerged land grasses; also takes detritus, insects, and other invertebrates.

REPRODUCTIVE BIOLOGY
Grass carps usually mature in the fourth year of life. Spawning occurs in late April and early May, when water temperatures reach 64.4°F (18°C), and with the onset of heavy floods. Mature fishes swim upstream. When the water level rises suddenly, males may chase females and push them. The females then lay eggs, and males release sperm. The eggs are semi-pelagic, floating with the currents. Incubation may take 35–40 hours when temperature is 66.92–70.16°F (19.4–21.2°C). Fecundity is 306,578–1,162,920 eggs, depending on the size of the adult female.

CONSERVATION STATUS
Not listed by IUCN.

SIGNIFICANCE TO HUMANS
One of the world's most important aquaculture species. Also used for weed control in rivers, fish ponds, and reservoirs. ◆

Upper mouth
Culter alburnus

FAMILY
Cyprinidae

TAXONOMY
Culter alburnus Basilewsky, 1855, northern China.

OTHER COMMON NAMES
English: Whitefish, lookup.

PHYSICAL CHARACTERISTICS
Size moderate to large, 5.9–25.6 in (15–65 cm) in body length. Body long, compressed. Dorsal straight, abdomen curved. Belly is keeled from ventral base to anus. Snout blunt. Mouth extremely superior, almost vertical. Barbels absent. Pharyngeal teeth sharp, in three rows. Gill rakers long, 24–28. Dorsal fin short, with 3 unbranched, 7 branched rays; last unbranched ray is spine-like. Anal fin long, with 3 unbranched, 21–24 branched rays. Scales small. Lateral line complete, with 80–92 scales. Air bladder has 3 chambers. Back dark gray, lower sides and belly silver-white. Fins gray.

DISTRIBUTION
East Asia from Amur River to the Pearl River, and into northern Vietnam.

HABITAT
Rivers and floodplain lakes with aquatic macrophytes and slow-running water.

Hypophthalmichthys molitrix

Culter alburnus

BEHAVIOR
Often lives in middle and upper parts of water bodies in small groups. Swims fast and likes to leap.

FEEDING ECOLOGY AND DIET
Carnivorous. Feeds on zooplankton, insects, and small fishes.

REPRODUCTIVE BIOLOGY
Males mature in the second, females in third year of life in the Yangtze River. Spawning occurs in mid-June in rivers or in shallow areas of lakes. Eggs are slightly attached to aquatic plants and may be detached and sink to the bottom due to wave movements. Fecundity varies from 51,490 to 532,350 eggs.

CONSERVATION STATUS
Not listed by IUCN.

SIGNIFICANCE TO HUMANS
An important food fish in the floodplain area of China. ◆

Common carp
Cyprinus carpio

FAMILY
Cyprinidae

TAXONOMY
Cyprinus carpio Linnaeus, 1758, Europe.

OTHER COMMON NAMES
English: Carp, German carp, European carp, mirror carp, leather carp, leatherback, German bass.

PHYSICAL CHARACTERISTICS
Size moderate to large, usually 11.8–15.74 in (30–40 cm) in body length. Body robust, compressed laterally. Snout long; mouth of moderate size reaching to below nostril. Two barbels on each side of upper jaw, smaller one from edge of snout,

larger one near corner of mouth. Pharyngeal teeth in three rows; larger teeth molarlike. Gill rakers 21–27. Dorsal fin long, 4 spines, 15–23 branched rays; anal fin short, 3 spines, 5 branched rays. Back of last dorsal and anal spines serrated. Lateral line complete, with 32–41 scales. The usual longevity of the carp is 9–15 years; maximum observed longevity is 47 years. Brassy olive above, lower sides golden yellow; belly yellow-white. Basal half of caudal and anal fins often reddish; stronger coloration in adults.

DISTRIBUTION
Native to Asia from the Amur River to North Vietnam. It was carried to Europe just before and after the beginning of the common era. Its introduction to the American continent took place during the first half of the nineteenth century. By now it has been transplanted all over the world.

HABITAT
Lives in a wide variety of habitats, including ponds, lakes, streams, and large rivers. It can tolerate a very low concentration of oxygen and high salinity. Normally, it prefers shallow, warm waters with aquatic plants over cold, small streams with fast-running water.

BEHAVIOR
Usually live in lower part or bottom of waters. In spring and autumn, they form schools. Though they need not migrate to rivers for reproduction, some fish can swim very long distances (up to 1,012 mi [1,629 km]).

FEEDING ECOLOGY AND DIET
Typically omnivorous and a benthic feeder. Food diet includes macrophytes, detritus and algae, molluscs, aquatic insects and their larvae, minute crustaceans, and small fishes.

REPRODUCTIVE BIOLOGY
Males mature usually by the second year of life in Asia, third or fourth year in Europe. Females require an additional year for maturation. Spawning may occur when water temperature reaches 64.4°F (18°C). Another prerequisite for spawning is

Cyprinus carpio

Garra pingi

the vegetation. Flood waters usually stimulate spawning. Spawning groups are composed of one female and one or more males. The males initiate the spawning act by repeatedly pushing their heads against the body of the female. On stimulation by the males, the female responds by raising her caudal peduncle and tail. Her tail lashes violently, and as she propels herself forward, she scatters the eggs over the vegetation. Simultaneously, the males come along the side of the female with their tail region proximate to the female genital opening and, by violent movements of their tail region, discharge their milts. The eggs are released gradually, in batches of 3–4, within a period of 3–4 days if the weather is good, or 2 to 3 weeks if spawning is interrupted by cold, cloudy, or windy spells. The eggs are attached to the vegetation and hatch after 2.5 to 3 days (water temperature 68–77°F [20–25°C]). For the first two days after hatching, the larvae stay on the grass quietly. On the third day, their air-bladders begin to get air. On the fourth day, the yolk-sac is nearly gone and the larvae begin to feed. The period of planktonic feeding is short; juveniles take invertebrate food from the bottom after reaching a length of 0.79 in (2 cm). Forty days later, with the body covered with scales and barbels appearing, the fish look very much like adults. Fecundity varies from 59,000 to 1,579,000 eggs per individual.

CONSERVATION STATUS
Listed as Data Deficient by the IUCN. Due to artificial hybridization and transplantation, the genetic resources of common carp have been seriously damaged. Many different local varieties (e.g., red carp, glass carp) have been contaminated genetically.

SIGNIFICANCE TO HUMANS
This is the earliest domesticated fish species. Ancient Chinese began to culture the common carp around 200 B.C. As this species is considered a symbol of happiness and good fortune in China, it is still common as wedding gift, particularly in rural areas. The Japanese colored carp, koi, had its origin in Japan between A.D. 794 and 1184. It is now one of the most common ornamental fishes in the world.

Some local varieties are important as a source of food, while others are important in the aquarium trade. The common carp is also an important game fish.

Zebrafish
Danio rerio

FAMILY
Cyprinidae

TAXONOMY
Cyprinus rerio Hamilton, 1822, Kosi River, Utter Pradesh, India.

OTHER COMMON NAMES
English: Zebra danio, striped danio; German: Zebrabärbling.

PHYSICAL CHARACTERISTICS
Small fish, rarely grows beyond 1.97–2.36 in (50–60 mm) in length. Body slender, slightly compressed. Two pairs of barbels: rostral barbels extend to anterior margin of orbit; maxillary barbels end at about middle of opercle. Dorsal fin with 3 unbranched, 7 branched rays. Anal fin striped, with 3 unbranched, 10–12 branched rays. Lateral line incomplete or absent. Vertebrae 31–32. Body silvery, sometimes tinted with

□ *Rhodeus ocellatus*
■ *Danio rerio*

gold, with five blue horizontal stripes on the sides. Stripes also present in the anal and caudal fins.

DISTRIBUTION
Native to tributaries of the Ganges River, along the Coromandel Coast of India, from Calcutta to Masulipatam, Bengal, Nepal, Pakistan, and Bangladesh. As aquarium fish and experimental animal, it has been introduced worldwide.

HABITAT
Slow-moving and still water bodies such as streams, canals, ditches, and ponds, particularly rice fields.

BEHAVIOR
Very active, usually swimming in schools.

FEEDING ECOLOGY AND DIET
Feeds mainly on worms and small crustaceans. It also feeds on insect larvae and can be used for mosquito control.

REPRODUCTIVE BIOLOGY
Matures in 4–6 months. Females are larger and have less vibrant coloration than males. Typically, spawning occurs in the early hours of the morning. Eggs are semiadhesive, relatively large, and released into open waters. Under experimental conditions, this species spawns at intervals of 1.9 to 2.7 days. However, spawning intervals in nature are typically much longer, varying from 5 days to several weeks. Between 21 and 60 eggs are released per spawning event. In captivity, the total number of eggs spawned is usually between 400 and 500. Eggs hatch approximately 20–48 hours after spawning. Larvae live from 48 to 72 hours on their yolk-sac provision.

CONSERVATION STATUS
Not listed by IUCN.

SIGNIFICANCE TO HUMANS
Because of their small size, zebrafish are of no value as a food fish. However, they are very popular aquarium fish and also very important as experimental subjects for genomic study. ◆

Black stick
Garra pingi

FAMILY
Cyprinidae

TAXONOMY
Garra pingi Tchang, 1929, Katin, Szechuan, China.

OTHER COMMON NAMES
German: Pings Saugbarbe.

PHYSICAL CHARACTERISTICS
Size small to moderate, normally 2.76–10.6 in (7–27 cm) in body length. Body slender, slightly cylindrical in the front, and compressed in the hind. Snout round, blunt. Mouth inferior, transverse. Lower lip specialized as disc. Barbels absent in adults. Pharyngeal teeth very small, in three rows. Gill rakers 36–44. Dorsal fin short, 2 unbranched, 9 branched rays; anal fin has 2 unbranched and 5 branched rays. Lateral line complete, with 48–52 scales. Dark black above, gray black on the lower sides; gray-white on belly. Bases of scales have black spots.

DISTRIBUTION
Upper reaches of the Yangtze, Pearl, and Mekong Rivers in Southeast Asia.

HABITAT
Lives in upper reaches of rivers with fast-running water and stony bottoms. Limited to warm waters.

BEHAVIOR
Normally adheres to the stony bottom with the sucking disc formed by lower lip. Very active.

FEEDING ECOLOGY AND DIET
Mainly scrapes algae and fragmental plants, but also feeds on insect larvae.

REPRODUCTIVE BIOLOGY
Little is known.

CONSERVATION STATUS
Not listed by IUCN.

SIGNIFICANCE TO HUMANS
Used for food, but the production is very low. ◆

Gudgeon
Gobio gobio

FAMILY
Cyprinidae

TAXONOMY
Cyprinus gobio Linnaeus, 1758, Britain and surrounding countries.

OTHER COMMON NAMES
French: Gofi, goujon; German: Grässling; Spanish: Gobio.

PHYSICAL CHARACTERISTICS
Small fish, length rarely exceeding 5.9 in (15 cm); maximum age 8 years. Body slender, round in front, compressed laterally toward the tail. Snout blunt. Mouth inferior. Barbels one pair. Pharyngeal teeth in 2 rows. Dorsal fin with 3 unbranched, 5–7

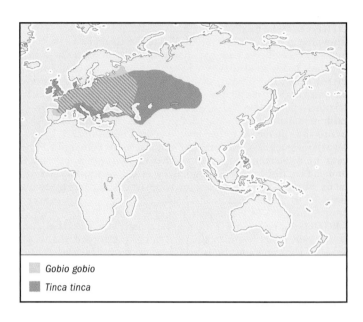

Gobio gobio
Tinca tinca

branched rays; anal fin with 3 unbranched, 6 branched rays. Lateral line complete, with 40–45 scales. Vertebrae 39–41. Brown-grey above, sides lighter, yellowish-white belly. On the sides there is a row of large, indistinct dark marks. Dorsal and tail fins are spotted.

DISTRIBUTION
Europe.

HABITAT
Fast-flowing rivers with sand or gravel bottoms; may also occur in still waters.

BEHAVIOR
Often appear in large numbers. Normally active during the day, but if they are disturbed, in particular by predators, they can defer their activity to periods when light intensity is weak. They are capable of emitting squeaking sounds.

FEEDING ECOLOGY AND DIET
Bottom feeder. Usually active during summer months. Its diet includes insect larvae, mollusks, and freshwater shrimps.

REPRODUCTIVE BIOLOGY
Matures in 2–3 years. The spawning time is from April to August, when temperatures are above 62.6°F (17°C). The fecundity is 2,500–6,500 eggs per individual.

CONSERVATION STATUS
Not listed by IUCN.

SIGNIFICANCE TO HUMANS
The gudgeon's chief value is as bait for larger fish. Its flesh is tasty, but its small size reduces its value as food. ◆

Silver carp
Hypophthalmichthys molitrix

FAMILY
Cyprinidae

TAXONOMY
Leuciscus molitrix Cuvier and Valencinnes, 1844, China.

OTHER COMMON NAMES
English: Chinese schemer; French: Carpe argentée; German: Silberkarpfen, Tolstolob; Spanish: Carpa plateada.

PHYSICAL CHARACTERISTICS
Size moderate to large, usually 11.8–15.74 in (30–40 cm) in body length. Body robust, compressed laterally. Keels from isthmus to anus on the belly. Head large. Snout blunt. Mouth terminal, wide and large. Barbels absent. Pharyngeal teeth in 1 row. Contains spiral gill-organ for feeding. Dorsal fin short, with 3 unbranched, 7 branched rays; anal fin long, with 3 unbranched 11–13 branched rays. Scales small. Lateral line complete, with 91–120 scales. Coloration is brassy olive above, and silver-white on lower sides and belly.

DISTRIBUTION
China and eastern Siberia. Introduced around the world for aquaculture and control of algal blooms.

HABITAT
Can live in standing or flowing waters such as ponds, lakes, and rivers. Prefers large waters with abundant plankton.

BEHAVIOR
Active species well known for its habit of leaping clear of the water when disturbed. It often swims just beneath the water surface. Undergoes lake-river migration.

FEEDING ECOLOGY AND DIET
Typical filter feeder that lives near the surface and feeds on phytoplankton and microzooplankton with its big mouth and the gill organ.

REPRODUCTIVE BIOLOGY
In Yangtze River, this fish matures in 3–4 years. Breeding season is from April to July. The fish migrate to middle or upper reaches of the river to breed. Spawning occurs near the surface of water. The eggs are semipelagic. Eggs and larvae float downstream to floodplain zones. The fecundity varies from 207,083 to 1,610,440 eggs.

CONSERVATION STATUS
Not listed by IUCN.

SIGNIFICANCE TO HUMANS
The silver carp is an important food fish and has been introduced to many countries. It is among the 4 species of cyprinids whose world production in aquaculture exceeds 1 million tons per year. However, in some places it is mainly used for cleaning reservoirs and other waters where eutrophication is a problem. ◆

Ningu
Labeo victorianus

FAMILY
Cyprinidae

TAXONOMY
Labeo victorianus Boulenger, 1901, Lake Victoria Nyanza, East Africa.

OTHER COMMON NAMES
None known.

PHYSICAL CHARACTERISTICS
Size small to moderate, maximum 15.75 in (40 cm) in standard length. Body long, slightly compressed. Snout profile smooth and rounded. Jaws with horny cutting ridge, with flap of skin in front of upper lip. Barbels hidden. Dorsal fin with 9–10 branched fin rays. Lateral line running along middle of the flank and the caudal peduncle; scales in lateral line 36–39. Olive dorsally, cream-colored ventrally. Dorsal, anal, and pelvic fins often orange-tipped.

DISTRIBUTION
East Africa: Lake Victoria (Nile drainage basin).

HABITAT
Shallow inshore waters and influent rivers.

BEHAVIOR
Spends most of its life span in lakes, but migrates to spawn in flooded grasslands beside both permanent and temporary streams.

FEEDING ECOLOGY AND DIET
Feeds on detritus and algae. Also feeds on rotifers growing on the bodies of other fishes.

REPRODUCTIVE BIOLOGY
Migrates to rivers to spawn in flood season.

CONSERVATION STATUS
Not listed by IUCN. This species has been adversely affected by overfishing and predation by the introduced Nile perch.

SIGNIFICANCE TO HUMANS
Important food fish. ◆

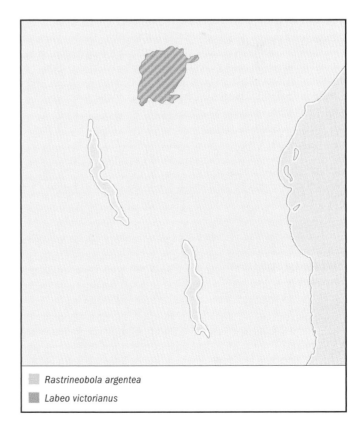

   ▨ *Rastrineobola argentea*
   ▨ *Labeo victorianus*

Common dace
Leuciscus leuciscus

FAMILY
Cyprinidae

TAXONOMY
Cyprinus leuciscus Linnaeus, 1758, Central Europe.

OTHER COMMON NAMES
English: Dace; French: Acourcie, assée, aubour, gandoise; German: Hasel.

PHYSICAL CHARACTERISTICS
Size small to moderate, length rarely exceeding 10.2 in (26 cm), maximum length 15.75 in (40 cm); maximum reported age is 16. Body long. Snout pointed. Mouth narrow, slightly inferior. Barbels absent. Pharyngeal teeth in 2 rows. Dorsal fin has 3 unbranched, 7–9 branched rays; anal fin has 3 unbranched, 8–9 branched rays. Lateral line complete, with 47–52 scales. Vertebrae 42–46. Dorsal part dark, sides silver, belly white.

DISTRIBUTION
Native to Europe and northern Asia. It has become widespread in Europe and gained access to Ireland as a bait fish.

HABITAT
Prefers upper reaches of rivers and especially clear, cool lakes with fast-flowing water and sand or gravel substrate.

BEHAVIOR
Usually swims near the surface in large numbers, with large home range and localized feeding migrations.

FEEDING ECOLOGY AND DIET
Feeds on insects, worms, snails, and only rarely plants. It may feed all year, intensely so at dawn and dusk.

REPRODUCTIVE BIOLOGY
Matures in 3–4 years. Reproduction occurs in March and April, when water temperature is 48.2–50°F (9–10°C), with single spawning. Eggs are pale yellow and attach to gravel and stones in shallow, flowing water. Fecundity for females is 6,500–9,500 eggs.

CONSERVATION STATUS
Not listed by IUCN.

SIGNIFICANCE TO HUMANS
Not good as food fish because they are too small and bony, but they make good bait for larger fish. ◆

Eurasian minnow
Phoxinus phoxinus

FAMILY
Cyprinidae

TAXONOMY
Cyprinus phoxinus 1758 Europe.

OTHER COMMON NAMES
English: Common minnow; French: Amarante, arlequin, vairon; German: Blutelritze.

PHYSICAL CHARACTERISTICS
Small fish, length rarely exceeding 3.94 in (10 cm); maximum age 6 years. Body slender, slightly compressed. Snout short and blunt. Mouth small, terminal. Barbels absent. Pharyngeal teeth in 2 rows. Gill rakers 7–8. Dorsal fin short, with 3 unbranched, 7 branched rays; anal fin with 3 unbranched, 7 branched rays. Scales small. Lateral line incomplete, with 32–41 scales, ending before anal origin. Coloration is brassy olive above, golden yellow on lower sides, and yellow-white on belly. Basal half of caudal and anal fins often reddish; stronger coloration in adults.

DISTRIBUTION
Eurasia, from British Isles and eastern Spain to eastern Siberia and Amur River.

HABITAT
Cold yet well-oxygenated waters (running or still) over gravel substrate, mainly in rivers and streams, occasionally in lakes and canals.

BEHAVIOR
Often found in large numbers, never solitary. They may migrate upstream for spawning in shallow gravel areas.

FEEDING ECOLOGY AND DIET
Feeds on algae, plant debris, mollusks, crustaceans, zooplankton, and insect larvae. It may forage all day, and feeding takes place throughout the year.

REPRODUCTIVE BIOLOGY
Matures in 2–3 years. The spawning times are mainly from June to July, with multiple spawning. Spawning takes place over gravel and weeds. Eggs are fixed on plants or stones. Fecundity is 200–1,000 eggs per female.

CONSERVATION STATUS
Not listed by IUCN.

SIGNIFICANCE TO HUMANS
No commercial value but an important laboratory fish. ◆

Fathead minnow
Pimephales promelas

FAMILY
Cyprinidae

TAXONOMY
Pimephales promelas Rafinesque, 1820, Lexington, Kentucky, United States.

OTHER COMMON NAMES
English: Black-head minnow, rosy-red.

PHYSICAL CHARACTERISTICS
Small fish, maximum 3.94 in (10 cm) in total length. Body moderately compressed. Snout short and blunt. Mouth subterminal. Scales deep, closely overlapping. Lateral line incomplete or nearly complete, with 43–44 scales. Both dorsal and anal fin with 7 branched rays. Olive to brown on the upper body and silvery white on the lower body with a dark midlateral stripe. Nuptial males tend to be larger than females with horny tubercles on the snout and a prominent pad of spongy wrinkled tissue on the nape. One variety with light orange color named *rosy red*.

DISTRIBUTION
Native to central North America, northeastern United States, and northeastern Mexico. It has been introduced widely throughout much of North America as a bait minnow.

HABITAT
Found in a wide variety of habitats in rivers, streams, lakes, and ponds, particularly in waters with abundant floating and submerged vegetation. It has a high tolerance for turbid waters, low oxygen, and high temperatures.

BEHAVIOR
Shoaling species. Males are territorial in breeding season.

FEEDING ECOLOGY AND DIET
Feeds in soft bottom mud, taking a variety of items from algae and plant fragments to insect larvae and microscopic crustaceans.

REPRODUCTIVE BIOLOGY
Matures by about six months of age. Spawning occurs from late May to early June when water temperatures exceed 60.8°F (16°C). In breeding season, a male develops dark coloration and breeding tubercles on his head, and a soft mucus-like pad on his nape. The male selects the nest site under an object such as a log, rock, or stick, and excavates the area around the object. He then defends it aggressively from all other fatheads except egg-laden females. The female enters the nest, turns upside down, lays her sticky eggs on the underside of the nest object, and leaves the nest. The male then fertilizes the eggs and guards the incubating eggs. He fans them with his fins and massages them with his back pad to keep them clean and well oxygenated. Other females may add eggs to the nest as the spawning season progresses. The male continues his care until all of the eggs hatch. Females produce clutches of eggs. Each clutch may contain 80–370 eggs. Most females probably spawn several clutches in a season. The embryos hatch in about 4–6 days.

CONSERVATION STATUS
Not listed by IUCN.

SIGNIFICANCE TO HUMANS
Often sold as bait and in aquarium stores to be used as "feeder fish." Several countries report adverse ecological impact after introduction, such as causing the spread of enteric redmouth disease, which has infected wild and cultured trouts and eels. The fathead minnow was used in the past as a form of mosquito control in some places and is still widely used as a bioassay subject. ◆

Colorado pikeminnow
Ptychocheilus lucius

FAMILY
Cyprinidae

TAXONOMY
Ptychocheilus lucius Girard, 1856, Colorado River, California, United States.

OTHER COMMON NAMES
English: Colorado squawfish.

PHYSICAL CHARACTERISTICS
Largest cyprinid in North America, maximum 71 in (180 cm) in total length. Body slender, elongate, with long, depressed head. Maxillary reaching past anterior margin of the eye. Fins moderate, both dorsal and anal fins with 9 branched rays. Scales very small; lateral line very strongly decurved; lateral line scales 83–87. Color plain, dark above. Young always have a black caudal spot and a faint pale lateral band below a darker one.

DISTRIBUTION
North America, in the Colorado River drainage in Wyoming, Colorado, Utah, New Mexico, Arizona, Nevada, and California, from the United States to Mexico. Now mostly restricted to Utah and Colorado and extirpated from the southern portion of the range by the construction of large dams on the Colorado and Gila Rivers.

HABITAT
Juveniles utilize backwater and side channel areas with little or no current and silt or sand substrates. Adults inhabit medium to large rivers, with larger ones found in deep pools with a strong current flowing over rocky or sandy substrates.

BEHAVIOR
Adults are largely solitary, except during spawning or during low flows that crowd them into reduced habitat.

FEEDING ECOLOGY AND DIET
Younger individuals feed primarily on insects and crustaceans, whereas older fish are piscivores, consuming only other fish.

REPRODUCTIVE BIOLOGY
Spawns from early July through mid-August, after water temperatures have exceeded 64.4°F (18°C) for about a month. Preferred spawning sites are apparently gravel and cobble-bottomed riffles, where the interstitial spaces are free of organic matter and sediment. Newly hatched larvae drift downstream to quiet backwaters, where they grow rapidly and then return to the main-channel habitats when they are about 3 in (7.6 cm) long.

CONSERVATION STATUS
Classified as Vulnerable by the IUCN. The near extinction of this species is due to a combination of factors, the most significant being those associated with water development projects that have altered stream morphology, flow patterns, temperatures, water chemistry, and silt loads of most major streams throughout the Colorado River basin. Also, several exotic species may prey upon and compete with Colorado pikeminnows, particularly juveniles. Finally, there have been several fish eradication projects that may have had a severe impact on local populations. A captive propagation is underway; reintroductions into the Salt and Verde drainages began in 1985.

SIGNIFICANCE TO HUMANS
This species used to be an important food fish, but now its population is too small to sustain such activities. ◆

Tiger barb
Puntius tetrazona

FAMILY
Cyprinidae

■ *Puntius tetrazona*

■ *Rasbora heteromorpha*

TAXONOMY
Capoeta tetrazona Bleeker, 1855, Lahat, Palembang Province, Sumatra, Indonesia.

OTHER COMMON NAMES
English: Green tiger barb, Sumatra barb, partbelt barb; German: Moosbarbe; Malay: Ikan baja.

PHYSICAL CHARACTERISTICS
Small fish, maximum 2.76 in (7 cm) in total length. Body stocky, deep, and compressed. Mouth obtuse. Barbels absent. Dorsal fin with 8–9 branched rays, anal fin with 5 branched rays. Brilliantly colored. Dorsal brown to olive in color, flanks with a delicate reddish brown luster. Scales splendidly edged with shining gold. Body covered with four black vertical bars, with the first one passing through the eye, the second just anterior to the insertion of the dorsal fin, the third posterior to the dorsal, and the fourth passing through the caudal peduncle. Dorsal and anal fins blood-red; remaining fins more or less reddish. Ventral fins are occasionally black. Besides the traditional form, albino, black, and green morphs are also seen in the aquarium industry.

DISTRIBUTION
Southeast Asia: Sumatra and Borneo. Introduced widely and has been reared in several countries in facilities for breeding aquarium fishes.

HABITAT
Shallow, warm rivers and streams.

BEHAVIOR
Lively shoaling fish. Aquarium specimens are notorious for their habit of picking at the fins of other fish.

FEEDING ECOLOGY AND DIET
Feeds on worms, small crustaceans, and plant matter.

REPRODUCTIVE BIOLOGY
Matures between 9–12 months of age. Males are smaller and more brightly colored than females. Generally from 300 to 1,000 eggs are spawned. The eggs are large and yellowish, and the fry hatch within 36 hours.

CONSERVATION STATUS
Not listed by IUCN.

SIGNIFICANCE TO HUMANS
Very popular as an aquarium fish. ◆

Harlequin
Rasbora heteromorpha

FAMILY
Cyprinidae

TAXONOMY
Rasbora heteromorpha Duncker, 1904, Kuala Lumpur, Selangor, Malaysia.

OTHER COMMON NAMES
English: Red rasbora; German: Keilfleckrasbora; Finnish: Kiilakylki; Russian: Rasbora krasnaya.

PHYSICAL CHARACTERISTICS
Small fish, maximum 1.97 in (5 cm) in total length. Lateral line incomplete. Anterior part silver in color. Starting beneath the dorsal fin running to the tail, individuals are marked with a blackish blue triangular-shaped patch. This patch is slightly rounded at the bottom and ends in an extended tip in males but is straight in females. Eyes have a bright red glow. Dorsal fin vivid red with a yellow tip; tip of caudal fin bright red; inner rays yellow.

DISTRIBUTION
Thailand to Sumatra, Indonesia.

HABITAT
Small waters in rainforests.

BEHAVIOR
Shoaling species.

FEEDING ECOLOGY AND DIET
Feeds on worms, crustaceans, and insects.

REPRODUCTIVE BIOLOGY
The usual site for spawning is the underside of a broad-leafed aquatic plant. Females deposit their eggs rather than merely scattering them in the water.

CONSERVATION STATUS
Not listed by IUCN.

SIGNIFICANCE TO HUMANS
Very popular as an aquarium fish. ◆

Dagaa
Rastrineobola argentea

FAMILY
Cyprinidae

TAXONOMY
Neobola argentea Pellegrin, 1904, Lake Victoria, Africa.

OTHER COMMON NAMES
English: Silver cyprinid, mukene, sardine; German: Viktoria-Sardine; Swahili: Omena.

PHYSICAL CHARACTERISTICS
Small fish, rarely reaching a length greater than 3.15 in (8 cm) in standard length. Body slender, compressed. Barbels absent. Cheeks covered by thin suborbital bones. Lateral line low on the body and running along lower part of caudal peduncle; scales in lateral line 42–56. Body silver-white with an overall nacreous sheen. Caudal fin yellow. White-blue stripe along the middle of the body side.

DISTRIBUTION
East Africa: Lakes Victoria and Kyoga and Nile River drainage basin.

HABITAT
Found inshore and offshore. Adults stay near the bottom during the day and near the surface at night.

BEHAVIOR
Shoaling species.

FEEDING ECOLOGY AND DIET
Feeds on zooplankton and surface insects.

REPRODUCTIVE BIOLOGY
Matures in the second year of life and spawns near the shore. Juvenile fish migrate away from the shore after spending their larval stage in shallow areas.

CONSERVATION STATUS
Not listed by IUCN.

SIGNIFICANCE TO HUMANS
One of the three most important commercial species in Lake Victoria. It is a cheap source of protein food for direct human consumption and also an important food item for larger fishes in the lake. ◆

Rosy bitterling
Rhodeus ocellatus

FAMILY
Cyprinidae

TAXONOMY
Pseudoperilampus ocellatus Kner, 1866, Shanghai, China.

OTHER COMMON NAMES
German: Hongkong Bitterling.

PHYSICAL CHARACTERISTICS
Small fish, body length usually shorter than 2.76 in (7 cm). Body very deep, oval-shaped, compressed laterally. Snout short and blunt. Mouth small, terminal. Barbels absent. Pharyngeal teeth compressed, in 1 row. Gill rakers 10–14. Dorsal fin moderate, with 3 unbranched, 10–12 branched rays; anal fin with 3

unbranched, 9–12 branched rays. Lateral line incomplete, with 2–6 scales. Olive green above, lower sides white. Males very colorful, with a red dot on the base of caudal fin; one horizontal blue stripe from below dorsal origin to caudal base; two transverse blue bands above shoulder area; and eyes that are reddish on their upper halves.

DISTRIBUTION
China and Vietnam. Has been introduced into Japan.

HABITAT
Shallow lakes, ponds, and streams. Prefers clear, slow-running water, especially when it includes aquatic plants.

BEHAVIOR
Normally found in small groups near the shore. In breeding season, males are territorial.

FEEDING ECOLOGY AND DIET
Pelagic feeder, taking zooplankton, algae, and fragmental plants.

REPRODUCTIVE BIOLOGY
Matures in the first year, spawns in the second year from April to May. During spawning, the male finds a mussel, then attracts a female to the site. They both swim around the mussel, and the female next lays eggs into the mussel with a tube (ovipositor). The male then releases sperm, which is carried by the current into the gill chamber of the mussel, where it fertilizes the eggs. The eggs hatch after one day, but the larvae stay in the gill chamber for about 20 days. When the air-bladders fill with air and the egg yolks are nearly gone, the larvae swim out of the gill chamber to live independently.

CONSERVATION STATUS
Not listed by IUCN.

SIGNIFICANCE TO HUMANS
Bitterlings are interesting aquarium fishes because of their beautiful color and special breeding mode. They can be used to control mosquitos. In some places, they are used as indicators of environmental conditions, since they are sensitive to pollution. ◆

Rudd
Scardinius erythrophthalmus

FAMILY
Cyprinidae

TAXONOMY
Cyprinus erythrophthalmus Linnaeus, 1758, northern Europe.

OTHER COMMON NAMES
French: About, gardon rouge; German: Rotfeder; Spanish: Gardí; Russian: Krasnoperka.

PHYSICAL CHARACTERISTICS
Size moderate to large, maximum 20 in (51 cm) in total length. Body stocky and compressed. Mouth narrow, directed forward with the lower lip protruding. Belly strongly keeled between ventral fins and anus. Dorsal fin with 8–9 branched rays. Anal fin with 9–12 branched rays. Color brown-olive dorsally, sometimes with a brassy sheen; flanks brassy colored; belly silver-

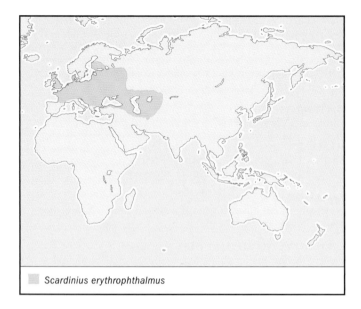

Scardinius erythrophthalmus

white. Ventral, anal, and lower part of caudal fins are a brilliant golden-red. Males develop spawning tubercles.

DISTRIBUTION
Widespread in Europe and middle Asia in the basins of the North, Baltic, Black, Caspian (from Emba, Ural, and Volga to the rivers of the southern coast) and Aral Seas. Introduced to several countries.

HABITAT
Dwells mainly in waters with a soft bottom and strong plant growth, especially where there are reedy margins including lakes, rivers, marshlands, canals, and ponds.

BEHAVIOR
Shoaling species.

FEEDING ECOLOGY AND DIET
Omnivorous. Feeds on invertebrates (including insect larvae and adults) and plants; particularly partial to insects from the water surface.

REPRODUCTIVE BIOLOGY
Matures after three to four years of age and spawns from mid-May to early June. The eggs adhere to water plants. The average rudd produces 108,000 to 211,000 eggs per kilogram of body weight. The young, which hatch after six to eight days, are no larger than a pinhead and spend the first part of their lives sheltering in shoals in the water margins. As the fry grow larger, they move out into deeper water and begin to feed on insects.

CONSERVATION STATUS
Although not listed by the IUCN, this species is threatened due to the introduction of other species.

SIGNIFICANCE TO HUMANS
Cultured as food in some countries; also used as bait for fishing. Although adults actively feed on macrophytes present in abundance in the environment, they are not an effective species for the biological clearing of weeds. Several countries report adverse ecological effects after introduction. ◆

No common name
Schizothorax prenanti

FAMILY
Cyprinidae

TAXONOMY
Oreinus prenanti Tchang, 1930, Omei Mountain, Szechuan, China.

OTHER COMMON NAMES
None known.

PHYSICAL CHARACTERISTICS
Size moderate, often 5.9–7.87 in (15–20 cm) in body length. Body long, slightly compressed. Snout somewhat pointed. Mouth inferior, transverse. Sharp horny edge on lower jaw. Lower lip papillae. Barbels 2 pairs, length similar to eye diameter. Pharyngeal teeth are sharp and in three rows. Gill rakers 14–23. Dorsal fin short, 3 unbranched, 8 branched rays; anal fin 3 unbranched, 5 branched rays. Back of last unbranched fin ray weakly serrated. Scales small. Lateral line complete, with 90–109 scales. Two rows of large scales along the anus. Brown or blue-brown above, sometimes with black spots; belly light yellow. Fins also light yellow.

DISTRIBUTION
Upper reaches of the Yangtze River in China. However, the group Schizothoracin, adapted to the plateau environment, is mainly distributed in and around the Tibetan Plateau.

HABITAT
Lives on the bottom of large rivers. Normally stays in succession areas between swift- and slow-running waters. Prefers low temperatures.

BEHAVIOR
Populations of this species mainly stay in large rivers, but some migrate to tributaries with swift waters to spawn eggs.

FEEDING ECOLOGY AND DIET
Normally uses the sharp horny edge on its lower jaw to scrape algae. Its diet also includes insects.

REPRODUCTIVE BIOLOGY
Males mature in the third, females in the fourth year of life. Spawning occurs from March to April. Eggs are slightly adhesive, staying in apertures between stones to hatch. Fecundity varies from 20,000 to 40,000 eggs per individual.

CONSERVATION STATUS
Not listed by IUCN.

SIGNIFICANCE TO HUMANS
Important food fish in local areas and very expensive. ◆

Tench
Tinca tinca

FAMILY
Cyprinidae

TAXONOMY
Cyprinus tinca Linnaeus, 1758, European lakes.

OTHER COMMON NAMES
French: Aiguillons, tanche; German: Alia.

PHYSICAL CHARACTERISTICS
Size moderate, normally 11.8 in (30 cm) in length, and rarely exceeding 23.6 (60 cm), and 3.97 lbs (1.8 kg) in weight; maximum age 14 years. Body robust, slightly compressed. Snout somewhat blunt. Mouth moderate, terminal, and oblique, with thick lips. Barbels 1 pair, very short. Pharyngeal teeth compressed, in 1 row. Gill rakers short, 12–14. Dorsal fin short, with 3 unbranched, 8 branched rays; anal fin with 3 unbranched, 7 branched rays. Scales small. Lateral line complete, with 100–105 scales. The caudal peduncle is characteristically deep and short. Skin thick and very slimy. Olive green above, dark green or almost black, with golden reflections, on ventral surface; the fins are always dark; eyes orange-red.

DISTRIBUTION
Eurasia, found throughout Europe to northwestern China.

HABITAT
Warm lakes and pools with weed and mud bottoms. It can tolerate low oxygen levels. In winter, this fish stays in the mud without feeding itself.

BEHAVIOR
Tench have limited home range. They are mostly solitary, occasionally occurring in small groups.

FEEDING ECOLOGY AND DIET
Omnivorous. Feeds on bottom invertebrates and aquatic insect larvae. Young tench also feed on algae. Foraging is often active at dawn and dusk.

REPRODUCTIVE BIOLOGY
Matures in 3–5 years. Breeds in shallow water among dense vegetation, laying numerous sticky green eggs in the period from May to August. After hatching, the larvae remain attached to plants for several days. The fecundity is 300,000–400,000 eggs per kilogram of body weight.

CONSERVATION STATUS
Not listed by IUCN.

SIGNIFICANCE TO HUMANS
Because its flesh is highly esteemed, the tench has considerable value, although it grows very slow. A golden color variety is a popular ornamental pond fish. ◆

Smallscale yellowfin
Xenocypris microlepis

FAMILY
Cyprinidae

TAXONOMY
Xenocypris microlepis Bleeker, 1871, Yangtze River, China.

Xenocypris microlepis

OTHER COMMON NAMES
English: Fine-scaled yellowfin; Russian: Melkocheshuinyi zheltoper.

PHYSICAL CHARACTERISTICS
Size moderate, normally 4.67–8.8 in (11.9–22.4 cm) in body length; maximum 27.56 in (70 cm) in total length. Body long and compressed. Snout short and blunt. Mouth inferior, with sharp horny edge on the lower jaw. Barbels absent. Pharyngeal teeth in three rows; teeth in the main row compressed. Gill rakers 36–48. Dorsal fin short, with 3 unbranched, 7 branched rays; anal fin with 3 unbranched, 10–14 branched rays. Scales small. Lateral line complete, with 72–84 scales. Back gray-black, lower sides and belly silver-white. Dorsal fin dark; pectoral and ventral fins gray-white; anal and caudal fins orange-yellow.

DISTRIBUTION
East Asia from Amur River to Yellow River, Yangtze River, and Pearl River.

HABITAT
Large rivers and lakes. Prefers clear waters with vegetation.

BEHAVIOR
Congregates in schools in winter in deep waters. Moves to shore to catch food in spring. Migrates to upper reaches of rivers to breed.

FEEDING ECOLOGY AND DIET
Normally uses the sharp horny edge on its lower jaw to scrape algae on the bottom. Its diet includes algae, fragmental plants, and insect larvae.

REPRODUCTIVE BIOLOGY
Matures in the second year of life. Spawning occurs from April to June. When floods occur and the water level rises rapidly, mature fish swim against the current and spawn in areas with fast-flowing water and stony bottoms. Eggs are attached to stones and hatch there. Fecundity varies from 42,000 to 292,000 eggs.

CONSERVATION STATUS
Not listed by IUCN.

SIGNIFICANCE TO HUMANS
Important food fish. Also used to keep fishery ponds clean.

Resources

Books

Becker, G. C. *Fishes of Wisconsin.* Madison: The University of Wisconsin Press, 1983.

Chen, Y., ed. *Fauna Sinica, Osteichthys: Cypriniformes, Part II.* Beijing: Science Press, 1998.

Department of Ichthyology, Hubei Institute of Hydrobiology. *Fishes of the Yangtze River.* Beijing: Science Press, 1976.

Nelson, J. S. *Fishes of the World,* 3rd ed. New York: John Wiley & Sons, 1994.

Mayden, R. L., ed. *Systematics, Historical Ecology and North American Freshwater Fishes.* Stanford: Stanford University Press, 1992.

Wheeler, A. *The Fishes of the British Isles and North-west Europe.* London: Macmillan, 1969.

Winfield, I., and J. Nelson, eds. *Cyprinid Fishes: Systematic Biology and Exploitation.* New York: Chapman & Hall, 1991.

Yue, P., ed. *Fauna Sinica, Osteichthys: Cypriniformes, Part III.* Beijing: Science Press, 2000.

Periodicals

Briolay, J., N. Galtier, R. M. Brito, and Y. Bouvet. "Molecular Phylogeny of Cyprinidae Inferred from Cytochrome bDNA Sequences." *Molecular Phylogenetics and Evolution* 9 (1998): 100–108.

Chen, X., P. Yue, and R. Lin. "Major Groups Within the Family Cyprinidae and Their Phylogenetic Relationships." *Acta Zootaxonomica Sinica* 9 (1984): 424–440.

Gilles, A., G. Lecointre, A. Miquelis, M. Loerstcher, R. Chappaz, and G. Brun. "Partial Combination Applied to Phylogeny of European Cyprinids Using the Mitochondrial Control Region." *Molecular Phylogenetics and Evolution* 19 (2001): 22–33.

Gilles, A., G. Lecointre, E. Faure, R. Chapaz, and G. Brun. "Mitochondrial Phylogeny of the European Cyprinids: Implications for Their Systematics, Reticulate Evolution and Colonization Time." *Molecular Phylogenetics and Evolution* 10 (1998): 132–143.

Gosline, W. A. "Unbranched Dorsal-fin Rays and Subfamily Classification of the Fish Family Cyprinidae." *Occasional Papers of the Museum of Zoology, University of Michigan* 684 (1978): 1–21.

Harris P. M., and R. L. Mayden. "Phylogenetic Relationships of Major Clades of Catostomidae (Telesotei: Cypriniformes) as Inferred from Mitochondrial SSU and LSU rDNA sequences." *Molecular Phylogenetics and Evolution* 20 (2001): 225–237.

Liu, H., C. S. Tzeng, and H. Y. Teng. "Sequence Variations in the mtDNA Control Region and Their Implications for the Phylogeny of the Cypriniformes." *Canadian Journal of Zoology* 80 (2002): 596–581.

Zardoya, R., and I. Doadrio. "Molecular Evidence on the Evolutionary and Biogeographical Patterns of European Cyprinids." *Journal of Molecular Evolution* 49 (1999): 227–237.

Huanzhang Liu, PhD

Cypriniformes II
(Loaches and relatives)

Class Actinopterygii

Order Cypriniformes

Number of families 4 of loaches and relatives

Photo: The tiger loach (*Botia dario*) is originally from India and Bangladesh. (Photo by Mark Smith/Photo Researchers, Inc. Reproduced by permission.)

Evolution and systematics

Fossils of this order date as far back as the Oligocene (about 38 million years ago), and have been found in North America, Europe, and Asia.

Cypriniformes are in the superorder Ostariophysi, together with the Gonorhynchiformes, Characiformes, Siluriformes, and Gymnotiformes. This superorder is characterized by having a Webberian apparatus, i.e., the first four or five vertebrae, called ossicles, are modified and connect the inner ear with the swim bladder. Because of this, they can hear very well, which is an advantage in murky fresh waters. This may explain why this group has been so successful inland, while being absent from the marine environment. Another distinguishing characteristic of fishes in this superorder is the production of an alarm substance, a chemical released by their skin when damaged, which helps to warn conspecifics of possible danger.

There have been a number of classifications for the order Cypriniformes. The most commonly employed is that of Nelson (1994), in which the order is divided into two superfamilies. The first superfamily, Cyprinoidea, includes one family and eight subfamilies (Cyprininae, Gobioninae, Rasborinae [=Danioninae], Acheilognathinae, Leuciscinae, Cultinae, Alburninae, and Psilorhynchinae), and has a total of 210 genera and about 2,010 species. The second superfamily, Cobitoidea, includes four families: 1) Gyrinocheilidae, or algae eaters; 2) Catostomidae, or suckers, which includes three subfamilies (Ictiobinae, Cycleptinae, and Catostominae), 3) Cobitidae, or loaches, which includes two subfamilies (Cobitinae and Botiinae); and 4) Balitoridae (=Homalopteridae), or river loaches, which includes two subfamilies (Nemacheilinae and Balitorinae). The superfamily Cobitoidea has a total of about 70 genera and 690 subspecies. This is considered the most primitive within this order. The Cobitoidea represent the scope of this chapter.

Physical characteristics

From a morphological viewpoint, the Cobitoidea is not a very well-defined group (it could be said that it was largely put in place to differentiate all noncyprinid families of the Cypriniformes). Therefore, it is difficult to generalize in terms of major morphological characters. For example, the algae eaters, fishes in the family Gyrinocheilidae, lack pharyngeal teeth, while the number of teeth in fishes in other families is extremely variable, reflecting a great deal of ecological adaptations to different types of food. Sometimes the lateral line is complete, in other families it is incomplete or even absent. The algae eaters are further characterized by having a ventral mouth that has been modified for feeding on algae on hard substrate and by having inhalant and exhalant gill openings. The suckers (Catostomidae) have one row of 16 or more pharyngeal teeth and have four sets of chromosomes, a condition called tetraploidy. The loaches (Cobitidae) have an elongated body with a subterminal mouth and three to six pairs of barbels; the river loaches (Balitoridae) have three or more pairs of barbels near the mouth.

Distribution

Suckers occur in China, northeastern Siberia, and North America. The remaining families in this group are native to Eurasia and Africa, with some species having been introduced in other parts of the world. The algae eaters are found in parts of Southeast Asia and Borneo; loaches are found in Europe, Asia, and in Morocco (North Africa), with one species introduced in North America; the river loaches are found throughout most of Eurasia.

Habitat

Like the rest of the Cypriniformes, the families included in this chapter contain all freshwater fishes. They are mostly benthic, feeding and reproducing at or near the bottoms of

A sucker feeds on the sea floor. (Illustration by Wendy Baker)

rivers and streams, especially those that are medium to small in size. Several species are troglomorphic, i.e., blind and depigmented, and living in caves.

Behavior

Fishes in this group can be effective swimmers because of their need to adapt to fast-moving currents. Some members of the family Catostomidae use their ventral mouths to ad-

Clown loaches (*Botia macracanthus*) originate from Borneo and Sumatra, where they live in moving waters. (Photo by M. H. Sharp/Photo Researchers, Inc. Reproduced by permission.)

here to the bottom and thus avoid being swept away by currents. These fishes usually have a reduced swim bladder that prevents them from rising to the surface of the water (their least preferred habitat).

Feeding ecology and diet

There is a great variety of feeding habits and items in this group. Some fishes eat algae, others, such as those in the *Chasmistes* spp. in the family Catostomidae, are midwater planktivores, and most feed on aquatic insect larvae, worms, crustaceans, and detritus from the bottom. A large proportion of species in several families, including the Cyprinidae, have a mouth that is essentially ventral, which allows them to feed from the bottom. Others, such as some representatives of the family Balitoridae, have barbels around the mouth that help them locate food. These barbels can be so sensitive that they can sense an imminent thunderstorm. Members of this superfamily are preyed upon by other fishes and some aquatic reptiles and mammals.

Reproductive biology

Some species, such as the longnose sucker (*Catostomus catostomus*), can engage in rather extensive upstream migrations prior to spawning. In many cases, two males compete for a single female, but not much is known about the genetic role played by each male. Some fishes, like the members of the family Catostomidae, are tetraploid, having four sets of chromosomes instead of the usual two, but only two sets are functional. The evolutionary meaning of this is unknown. Hybridization occurs among species in all the families, raising the question of the validity of the systematics.

Conservation status

The IUCN Red List includes 65 species from these four families: two (*Chasmistes muriei* and *Moxostoma lacerum*) are Extinct; two are Critically Endangered; six are Endangered; 27 are Vulnerable; seven are Lower Risk/Near Threatened; one is Lower Risk/Conservation Dependent; and 20 are Data Deficient. Many of these species have a very restricted range and live in small streams that are very sensitive to pollution.

Significance to humans

Many species, particularly those in the Gyrinocheilidae, are commonly sold in pet shops because they are believed to be useful in keeping aquaria clean of algae. Others, such as the hillstream loaches of the family Balitoridae, are important as ecological indicators because of their sensibility to minor environmental change. A few species can occasionally be seen in fish markets, but since most are small, they are rarely of any commercial value. Others, such as those living in caves, are of scientific value because of their interest from an evolutionary standpoint.

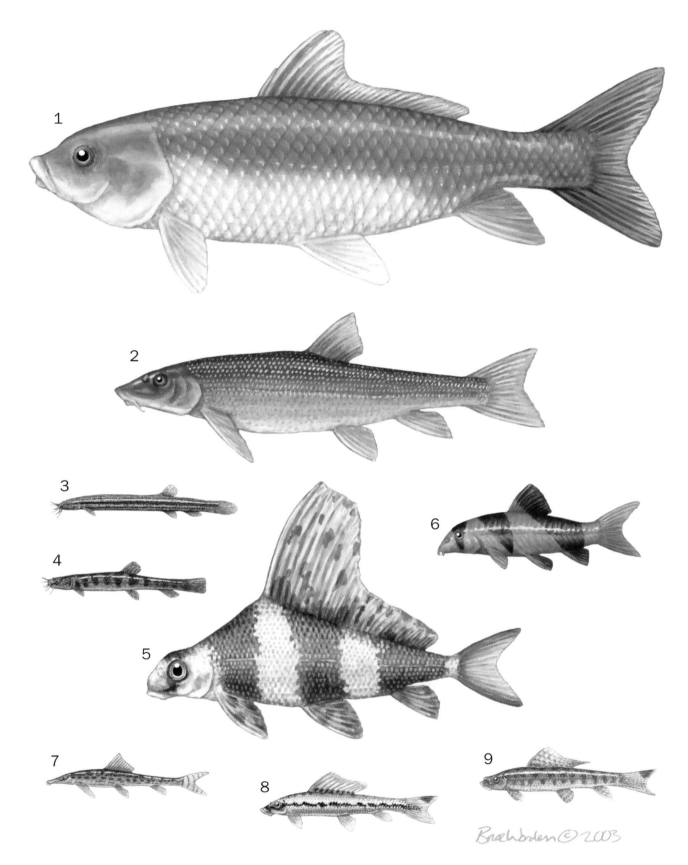

1. Bigmouth buffalo (*Ictiobus cyprinellus*); 2. Longnose sucker (*Catostomus catostomus*); 3. Weatherfish (*Misgurnus fossilis*); 4. Stone loach (*Barbatula barbatula*); 5. Chinese sucker (*Myxocyprinus asiaticus*); 6. Clown loach (*Botia macracanthus*); 7. Horseface loach (*Acanthopsis choirorhynchus*); 8. Chinese algae eater (*Gyrinocheilus aymonieri*); 9. Spotted algae eater (*Gyrinocheilus pennocki*). (Illustration by Bruce Worden)

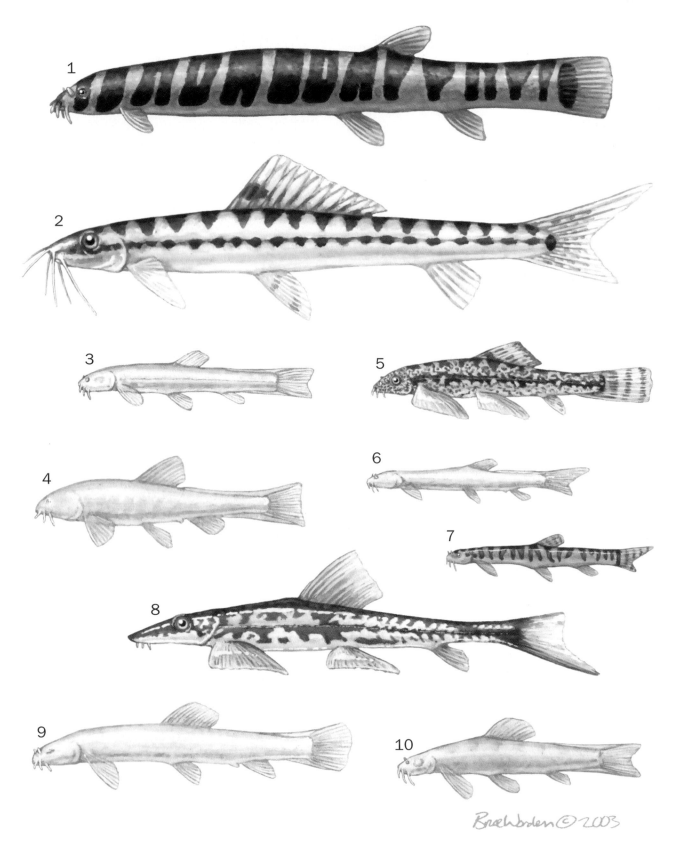

1. Coolie loach (*Pangio kuhlii*); 2. Arrow loach (*Nemacheilus masyae*); 3. Cavefish of Nam Lang (*Schistura oedipus*); 4. Kughitang blind loach (*Nemacheilus starostini*); 5. Chinese hillstream loach (*Pseudogastromyzon cheni*); 6. Blind loach (*Paracobitis smithi*); 7. Hillstream loach (*Nemacheilus evezardi*). 8. Anamia (*Annamia normani*); 9. Blind cave loach of Xiaao (*Protocobitis typhlops*); 10. Siju blind cavefish (*Schistura sijuensis*). (Illustration by Bruce Worden)

Species accounts

Anamia
Annamia normani

FAMILY
Balitoridae

TAXONOMY
Parhomaloptera normani Hora, 1931, Annam.

OTHER COMMON NAMES
Laotian: Pa tit hin.

PHYSICAL CHARACTERISTICS
Standard length 3.07 in (7.8 cm). Characterized by a flat belly, a small, horseshoe shaped mouth located far from the tip of the snout. Only one simple ray in the pelvic and pectoral fins.

DISTRIBUTION
Mekong River basin.

HABITAT
Steeped streams and rivers, on shallow and fast waters, prefers rocky bottoms.

BEHAVIOR
Swims slowly over the bottom.

FEEDING ECOLOGY AND DIET
Feeds on small benthic animals, primarily insect larvae.

REPRODUCTIVE BIOLOGY
No information available.

CONSERVATION STATUS
Not listed by the IUCN.

SIGNIFICANCE TO HUMANS
None known. ◆

Stone loach
Barbatula barbatula

FAMILY
Balitoridae

TAXONOMY
Cobitis barbatula Linnaeus, 1758, Europe. May be the same species as *Nemacheilus toni*, the Siberian stone loach.

OTHER COMMON NAMES
English: European stone loach, groundling; French: Loche franche; German: Schmerle; Spanish: Gobio de río.

PHYSICAL CHARACTERISTICS
Maximum total length 8.3 in (21 cm); maximum weight 0.4 lb (200 g). Fishes in this species have an elongated, slender body that is somewhat flattened in the first half but laterally compressed in the second half. The species is characterized by three pairs of mouth barbels and no erectile spine below the eye.

DISTRIBUTION
Very common from Ireland (where it was introduced) throughout Europe (except in the Italian and Iberian peninsula), and Asia to China. There are reports of this species from Siberia and Japan, but the taxonomy of the specimens collected there is not clear.

HABITAT
Shallow, fast-flowing creeks with gravel bottoms, as well as in shallow areas of clear lakes.

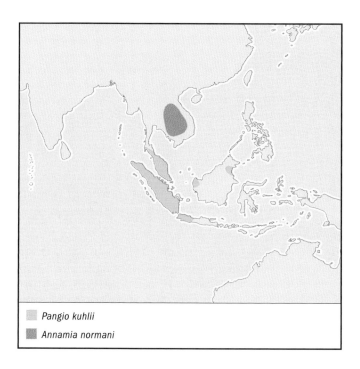

Pangio kuhlii
Annamia normani

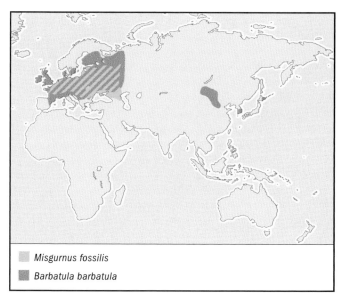

Misgurnus fossilis
Barbatula barbatula

BEHAVIOR
Usually swims near the bottom. Mostly active at dusk and night, shelters underneath rocks or burrows in sand.

FEEDING ECOLOGY AND DIET
Feeds on small crustaceans, insect larvae, and benthic invertebrates. They are fed upon by birds.

REPRODUCTIVE BIOLOGY
Breeds from April to June, spawning more than once. Deposits eggs among rocks and aquatic plants.

CONSERVATION STATUS
Not listed by the IUCN.

SIGNIFICANCE TO HUMANS
These fishes are sensitive to pollution and low oxygen levels, so their presence in a river can be taken as an indication of good water quality. ◆

Hillstream loach
Nemacheilus evezardi

FAMILY
Balitoridae

TAXONOMY
Nemacheilus evezardi Day, 1872, "A cave in India." Kottelat (1990) changed the generic status of this species to *Indoreonectes,* but without explanation. Singh and Yazdani (1993) gave it the name of *Oreonectes evezardi.*

OTHER COMMON NAMES
None known.

PHYSICAL CHARACTERISTICS
Maximum total length 2 in (3.8 cm). Characterized by small eyes and minute scales. They are mostly depigmented.

DISTRIBUTION
Kotumsar Cave (18°52'09"N, 81°56'05"E) of the Bastar District, Madhya Pradesh State, India.

HABITAT
The cave in which they occur is at 1,837 ft (560 m) above sea level and is subject to frequent flooding during the monsoon season.

BEHAVIOR
They have a low oxygen consumption, but despite living in a cave, they still exhibit circadian (day-to-day) and circannual (year-to-year) rhythmicity, which suggests that the cave population recently (in terms of evolution) invaded the cave environment.

FEEDING ECOLOGY AND DIET
Little is known.

REPRODUCTIVE BIOLOGY
Little is known.

CONSERVATION STATUS
Not listed by IUCN.

SIGNIFICANCE TO HUMANS
Of scientific value because of its adaptation to its cave environment. ◆

Arrow loach
Nemacheilus masyae

FAMILY
Balitoridae

TAXONOMY
Nemacheilus masyae Smith, 1933, Siam.

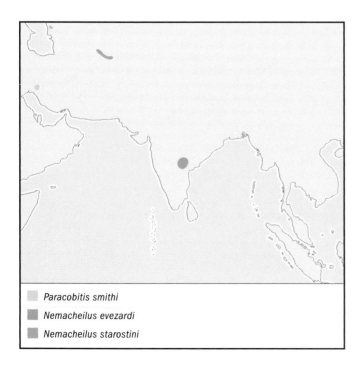

Paracobitis smithi
Nemacheilus evezardi
Nemacheilus starostini

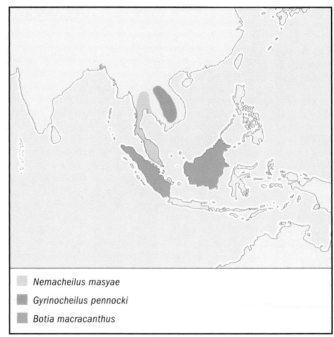

Nemacheilus masyae
Gyrinocheilus pennocki
Botia macracanthus

OTHER COMMON NAMES
None known.

PHYSICAL CHARACTERISTICS
Standard length 5.31 in (13.5 cm). What separates this species from others in the Nemacheilinae in Indochina is the color pattern, which consists of a light background with 14–18 dark blotches along the sides and 12–17 dark saddles along the back. They also have a black spot at the posterior extremity of the lateral line and a black spot on the anterior dorsal rays at about one-fourth of ray length. They have a slender body with the eyes situated more on the top of the head.

DISTRIBUTION
Western Malaysia, Peninsular Thailand, and the Meklong basin, as well as in parts of the Mae Nam Chao Phraya basin.

HABITAT
Mostly forest streams with clear or more turbid waters, in depths of 6.6 ft (2 m) or less in rivers and streams with a moderate current and muddy to sandy bottoms.

BEHAVIOR
Adults can be aggressive toward conspecifics.

FEEDING ECOLOGY AND DIET
Feeds on insect larvae and aquatic invertebrates.

REPRODUCTIVE BIOLOGY
No information is available.

CONSERVATION STATUS
Not listed by the IUCN.

SIGNIFICANCE TO HUMANS
Found in the aquarium trade. ◆

Kughitang blind loach
Nemacheilus starostini

FAMILY
Balitoridae

TAXONOMY
Noemacheilus (Troglobitis) starostini Parin, 1983, Turkmenistan. Weber (2000) gave it the name of *Paracobitis starostini*.

OTHER COMMON NAMES
English: Starostin's loach.

PHYSICAL CHARACTERISTICS
Has no externally visible eyes, no pigmentation, no scales, and no swim bladder, all convergent features among many hypogean species.

DISTRIBUTION
Several sinkholes (ca. 37°55' N, 66°23' E) of the Khrebet Kughitang (mountains), in the Chardzhou province of the Republic of Turkmenistan.

HABITAT
Found in a sinkhole about 62 ft (19 m) deep with a seasonally fluctuating water level. A large portion of the pool is exposed to light part of the time. The pool is connected to an underwater stream. Up to 40 fish can be observed at a time in the pool.

BEHAVIOR
No information is available.

FEEDING ECOLOGY AND DIET
Feeds on insect larvae and small benthic crustaceans.

REPRODUCTIVE BIOLOGY
Other than reproducing via external fertilization, nothing else is known.

CONSERVATION STATUS
Classified as Vulnerable by the IUCN because of the susceptibility of their narrow habitat to underground pollution.

SIGNIFICANCE TO HUMANS
Of scientific interest because of their adaptation to the hypogean environment. ◆

Blind loach
Paracobitis smithi

FAMILY
Balitoridae

TAXONOMY
Noemacheilus smithi Greenwood, 1976, Iran.

OTHER COMMON NAMES
None known.

PHYSICAL CHARACTERISTICS
Lacks externally visible eyes, pigmentation, and scales.

DISTRIBUTION
Found exclusively in a natural well in an oasis at Kaaje-Ru (33°05' N, 48°36' E), in Ab-i-Serum Valley near Tang-e-haft railway station, at the Zagros Mountains, in the Lorestan (Khorramabad) province of Iran.

HABITAT
Occurs in a well-like water resurgence that seems to be the result of a collapsed subterranean system. This is probably part of a larger, but complex, narrow, and inaccessible network of underground waters. This species is syntopic with *Iranocypris typhlops*, that is, it shares the same habitat within the same geographical range.

BEHAVIOR
No information is available.

FEEDING ECOLOGY AND DIET
Most likely feeds on small invertebrates and/or detritus.

REPRODUCTIVE BIOLOGY
Nothing is known about its reproductive habits except that fertilization is external.

CONSERVATION STATUS
Not listed by IUCN.

SIGNIFICANCE TO HUMANS
None known. ◆

Chinese hillstream loach
Pseudogastromyzon cheni

FAMILY
Balitoridae

TAXONOMY
Pseudogastromyzon cheni Liang, 1942, Fukien, China.

OTHER COMMON NAMES
German: Chinesischer Flossensauger.

PHYSICAL CHARACTERISTICS
Maximum total length at least 2 in (5 cm). They have fins as modified suckers, and their mouths are below their bodies, like saltwater rays.

DISTRIBUTION
Inland waters of China.

HABITAT
Benthic; prefers waters of 68–77°F (20–25°C).

BEHAVIOR
Not aggressive, and thus do not defend well against other fishes trying to take away territory.

FEEDING ECOLOGY AND DIET
Feeds on a variety of sources, including algae, plants, and invertebrates.

REPRODUCTIVE BIOLOGY
Breeds in pits built under rocks.

CONSERVATION STATUS
Not listed by the IUCN.

SIGNIFICANCE TO HUMANS
Common in the aquarium trade. ◆

Cavefish of Nam Lang
Schistura oedipus

FAMILY
Balitoridae

TAXONOMY
Nemacheilus oedipus Kottelat, 1988, Tham Nam Lang, Thailand.

OTHER COMMON NAMES
None known.

PHYSICAL CHARACTERISTICS
Standard length 2.2 in (5.4 cm). It is microphthalmic, almost totally depigmented, and has embedded scales. Also characterized by a forked caudal fin.

DISTRIBUTION
Thailand, in the Mae Hong Son Province in the Tham (cave) Nam Lang (19°31' N, 98°09' E).

HABITAT
Has been found in an outflow of a cave with a stream that is probably part of Nam Lang, part of a karstic endoreic basin, usually on muddy bottoms.

BEHAVIOR
Scotophilic (moves away from light) and may have some residual "biological clock" (day/night rhythmicity).

FEEDING ECOLOGY AND DIET
Feeds on insects and roundworms.

REPRODUCTIVE BIOLOGY
Scatters eggs on the bottom.

CONSERVATION STATUS
Listed as Vulnerable by the IUCN.

SIGNIFICANCE TO HUMANS
No commercial value; of scientific interest. ◆

Pseudogastromyzon cheni
Schistura sijuensis
Acanthopsis choirorhynchus

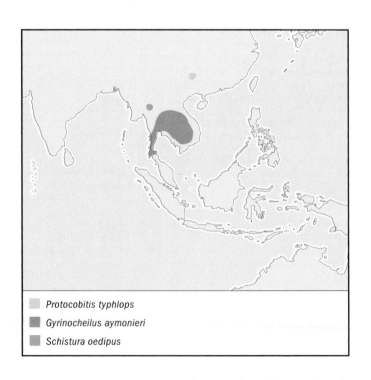

Protocobitis typhlops
Gyrinocheilus aymonieri
Schistura oedipus

Siju blind cavefish

Schistura sijuensis

FAMILY
Balitoridae

TAXONOMY
Noemacheilus sijuensis Menon, 1987, Siju Cave, India.
Noemacheilus Mesonoemachilus sijuensis appears on the plate caption of the original description. Pillai and Yazdani (1977) refer to this fish as *Nemacheilus multifasciatus*. However, Talwar and Jhingran (1991) consider these two separate species, and when referring to *N. multifasciatus* make no mention of it being found in a cave.

OTHER COMMON NAMES
None known.

PHYSICAL CHARACTERISTICS
Standard length 2 in (5.1 cm). The first cave individuals belonging to this species may have been reported by Hora (1924), who described three specimens as "*Nemacheilus*" sp. from the "Siju Cave, Assam, India." One specimen caught within 115 ft (35 m) of the cave mouth showed coloration comparable to the epigean (not cave-restricted) forms. Two others netted about 1,800 ft (550 m) from the entrance of the cave were paler in color, had reduced scales and reduced eyes.

DISTRIBUTION
Siju Cave (ca. 25°25' N, 90°30' E), Garo Hills, Meghalaya, India.

HABITAT
Found only in caves.

BEHAVIOR
No information is available.

FEEDING ECOLOGY AND DIET
Nothing known.

REPRODUCTIVE BIOLOGY
No information is available.

CONSERVATION STATUS
Listed as Vulnerable by the IUCN.

SIGNIFICANCE TO HUMANS
Of scientific interest because of their adaptations to the cave environment. ◆

Longnose sucker

Catostomus catostomus

FAMILY
Catostomidae

TAXONOMY
Cyprinus catostomus Forster, 1773, tributaries of the Hudson's Bay.

OTHER COMMON NAMES
English: Black sucker, red-sided sucker; French: Meunier rouge; German: Maulbiden-Saugdöbel.

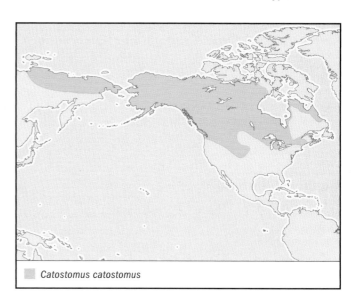

Catostomus catostomus

PHYSICAL CHARACTERISTICS
Maximum total length 21.2 in (64 cm); maximum weight 7.3 lb (3.3 kg). This species is characterized by a ventrally placed sucking mouth with thick papillose lips. The fishes also have an inconspicuous but complete lateral line, short gill rakers, and slightly rounded caudal tips. The color varies with age and reproductive state: young fishes are usually dark gray with small black spots; adults may be reddish brown, dark brassy green or black above, paler on the lower sides, with the ventral parts white. Breeding males are usually dark above with a brilliant reddish stripe along each side, and show prominent tubercles on the rays of the anal and caudal fins and also on the head. Breeding females are greenish gold to copper, with a less brilliant red stripe.

DISTRIBUTION
North America throughout most of Canada and Alaska; and in the drainages of the Delaware River (New York State, United States), Columbia River, upper Monongahela River (Maryland and West Virginia, United States), and Missouri River (south to Nebraska and Colorado, United States), and in the Great Lakes basin and the Arctic basin of Siberia (Russia).

HABITAT
Clear, cold, deep waters of lakes and tributary streams; occasionally brackish water in the Arctic.

BEHAVIOR
Moves from lakes into inlet streams or from slow, deep pools into shallow, gravel-bottomed portions of streams to spawn.

FEEDING ECOLOGY AND DIET
Feeds on benthic invertebrates. Young are preyed upon by other fishes and fish-eating birds; adults in spawning streams are eaten by otters, bears, ospreys, and eagles.

REPRODUCTIVE BIOLOGY
Spawning takes place during daytime over coarse gravel bottoms. Males, which do not engage in nest-building behavior, stay close to the bottom of fast-moving streams. Females keep themselves to the banks of still water from which they swim to the areas where the males are. Females are soon escorted by up to five (but usually two) males to the center of the stream. Males then engage in a ritual that lasts up to six seconds, either clasping the females with their pelvic fins or vibrating with

their anal fins against her. Egg deposition occurs at this time. After this, males and females return to their place of origin.

CONSERVATION STATUS
Not listed by the IUCN.

SIGNIFICANCE TO HUMANS
Utilized as food for humans or dogs. ◆

Bigmouth buffalo
Ictiobus cyprinellus

FAMILY
Catostomidae

TAXONOMY
Sclerognathus cyprinella Valenciennes, 1844, near New Orleans, Lake Pontchartrain, Louisiana, United States.

OTHER COMMON NAMES
English: Bigmouth buffalo, bullhead buffalo, common buffalofish; Romanian: Buffalo cu gura mare.

PHYSICAL CHARACTERISTICS
Maximum total length 48.4 in (123 cm); maximum weight 70.3 lb (31.9 kg). Characterized by a large, robust, and somewhat deep body. They are dorsally lighter in color on the rest of the body. Other characteristics include an anteriorly curved ventral line, a bluntly rounded snout, and a terminate mouth with thin lips, of which the lower one is striated. They have a total of approximately 165 short pharyngeal teeth. The dorsal fin is sickle shaped, with between 24 and 32 rays; the anal fin rays have between 8 and 10 rays. The lateral line is complete, with between 34 and 39 scales.

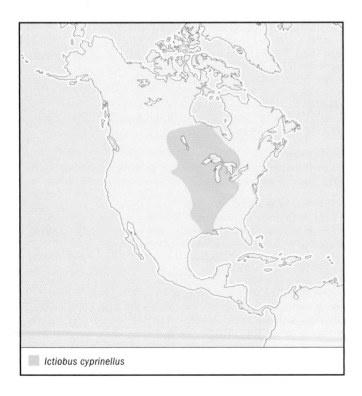

Ictiobus cyprinellus

DISTRIBUTION
North America: in the Hudson Bay (Nelson River drainage), lower Great Lakes, and Mississippi River basins from Ontario to Saskatchewan in Canada, and from Montana south to Louisiana in the United States. Successfully introduced to California, Arizona, Virginia, and North Carolina, in the United States; and in Uzbekistan, the Jordan River, Romania, and Cuba.

HABITAT
Occurs in a great variety of habitats, including still waters such as ponds, pools, impoundments, and floodplain lakes with shallow waters, as well as large rivers, main channels, and backwaters of small to large rivers.

BEHAVIOR
Usually forms schools.

FEEDING ECOLOGY AND DIET
Benthic feeders that feed on copepods, caldocerans, bottom plants, aquatic insects, mollusks, small fishes, and fish eggs. They are preyed upon by northern pikes, black bullheads, and walleyes.

REPRODUCTIVE BIOLOGY
Usually mature at age three. Spawning often occurs between April and May (at least in North America), when water temperatures rise suddenly and remain between 60.1 and 65°F (15.6 and 18.3°C). Adults migrate into the shallow bays and inlets of lakes or into sloughs and flooded marshes of large rivers. At times they swim over the top of one another as they move through small, shallow inlets into the sloughs. They can spawn at any time of the day over low, sparse vegetation, rocks, or even mud, but the water must be clear. Spawning takes place in groups of one female surrounded by two to four males. During spawning, the female swims, making huge ripples and hitting the surface with her tail. Then she sinks to the bottom and releases her eggs. The males then push in all around in order to move into a position to fertilize the eggs. Then they make a tremendous rush, causing the water to foam. The whole spawning process is very noisy and splashy. Depending on her size, a single female can lay 100,000–750,000 eggs. The eggs adhere to any object they contact. Hatching occurs after 9 or 10 days and the larvae remain in shallow water to feed.

CONSERVATION STATUS
Not listed by the IUCN, but not as abundant as in the early twentieth century. This decline was due to overexploitation for commercial purposes.

SIGNIFICANCE TO HUMANS
Important food fishes; this species has been artificially propagated. Fishes in this species may compete with native minnows and suckers, as well as with juvenile sport fishes, for food and space. The meat is nutritious; the taste of the flesh is considered to be inferior to that of the flesh of the catfish but superior to that of carp. ◆

Chinese sucker
Myxocyprinus asiaticus

FAMILY
Catostomidae

TAXONOMY
Carpiodes asiaticus Bleeker, 1865, China.

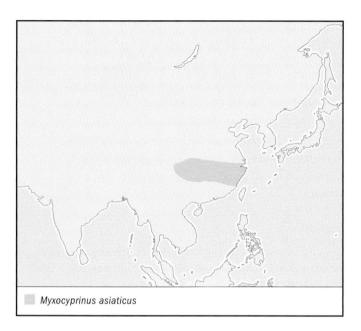

Myxocyprinus asiaticus

OTHER COMMON NAMES
English: High fin banded shark, freshwater batfish; German:
Wimpelkarpfen; Finnish: Kiinanimukarppi.

PHYSICAL CHARACTERISTICS
At least 23.6 in (60 cm) in length. The most important characteristic of this species is its unusual deep, flat body, that elon-
gates somewhat as the fish grows.

DISTRIBUTION
Upper and middle tributaries of the Yangtze River basin
in China.

HABITAT
Unlike most other Catostomidae, they occur in large (and
cool) rivers of temperate areas.

BEHAVIOR
Forms schools.

FEEDING ECOLOGY AND DIET
Feeds on algae and plants (at least under captive conditions).

REPRODUCTIVE BIOLOGY
Spawns in small groups in quiet waters.

CONSERVATION STATUS
Not listed by the IUCN, but considered endangered in China
due to habitat destruction.

SIGNIFICANCE TO HUMANS
Sold in the aquarium trade. Prior to construction of Gezhouba
Dam on the Yangtze River, the fish was a popular food fish in
Sichuan Province. After construction of the dam, catches declined dramatically and attempts to reestablish reproducing
populations have not been successful. Small breeding populations remain below the dam. ◆

Horseface loach
Acanthopsis choirorhynchus

FAMILY
Cobitidae

TAXONOMY
Cobitis choirorhynchus Bleeker, 1854, Sumatra.

OTHER COMMON NAMES
English: Banana fish, longnose loach; German: Lanstirnschmerle;
Laotian: Pa it; Malay: Jeler; Spanish: Locha de hocico largo;
Thai: Pla sai, pla chon sai; Vietnamese: Cá chia voi.

PHYSICAL CHARACTERISTICS
Maximum total length 12 in (30 cm). Mimetic fishes with a
great ability to change coloration to appear similar to almost
all natural backgrounds.

DISTRIBUTION
India, Myanmar, Thailand, Malaya, Indonesia (Sumatra and
Java), Borneo, and Vietnam (basinwide mainstream of the
lower Mekong River).

HABITAT
Demersal, freshwater fishes found mostly in swift, clear
streams, as well as in large rivers and flooded fields. Burrows in
sandy or gravelly bottoms.

BEHAVIOR
Nocturnal; likes to bury during the day. Not aggressive.

FEEDING ECOLOGY AND DIET
Omnivorous.

REPRODUCTIVE BIOLOGY
Oviparous; lays eggs underneath small rocks.

CONSERVATION STATUS
Not listed by the IUCN.

SIGNIFICANCE TO HUMANS
Common in the aquarium trade. ◆

Clown loach
Botia macracanthus

FAMILY
Cobitidae

TAXONOMY
Cobitis macracanthus Bleeker, 1852, Sumatra. Sometimes known
as *Botia macracantha*.

OTHER COMMON NAMES
English: Tiger botia; German: Prachtsschmerle; Malay: Ikan
macan.

PHYSICAL CHARACTERISTICS
Maximum total length 12 in (30 cm). They have a rather
deep body, which is orange with three black stripes. A fourth
stripe is sometimes present. The caudal fin is forked. Males
have a larger tail that hooks inward rather than pointing
straight out from the body. Females are smaller and more
slender.

DISTRIBUTION
Sumatra and Borneo in Indonesia. It has been introduced in Thailand and the Philippines, but the success of such introductions is unknown.

HABITAT
Demersal freshwater fishes that prefer streams.

BEHAVIOR
Nonaggressive; spends most of the time at the bottom swimming or stationary among vegetation.

FEEDING ECOLOGY AND DIET
Feeds on benthic algae and weeds, benthic crustaceans, and worms.

REPRODUCTIVE BIOLOGY
Breeds only at the beginning of the rainy season and does so in fast-flowing rivers. Rarely breeds under captive conditions, which is why so little is known about their reproductive behavior.

CONSERVATION STATUS
Not listed by the IUCN.

SIGNIFICANCE TO HUMANS
Popular aquarium fishes; sometimes consumed as food. ◆

Weatherfish
Misgurnus fossilis

FAMILY
Cobitidae

TAXONOMY
Cobitis fossilis Linnaeus, 1758, Europe.

OTHER COMMON NAMES
French: Loche d'étang; German: Europäischer Schlammpeitzger; Czech: Cik; Finnish: Mutakala; Romanian: Chiscar.

PHYSICAL CHARACTERISTICS
Maximum total length 12 in (30 cm). Very elongated bodies. The total number of spines per fins is dorsal: three; dorsal (soft rays): five to six; pectoral: one; pectoral (soft rays): eight; anal: three; anal (soft rays): five; and caudal: 14. They have 16–17 gill rakers. Total number of vertebrae 49–50.

DISTRIBUTION
Central and eastern Europe, from France to Russia; successfully introduced in Spain and Croatia.

HABITAT
Mainly the lower reaches of slow-flowing rivers, in still pools, over sandy bottoms.

BEHAVIOR
Although it is a facultative air-breather, spends most of the day buried in the sand and is considered nocturnal. Not aggressive.

FEEDING ECOLOGY AND DIET
Feeds on small invertebrates, mostly insect larvae and mollusks.

REPRODUCTIVE BIOLOGY
Reaches sexual maturity when two years old. Spawns between April and June. The developing eggs and embryos of this species interact with each other in such a manner that the embryos of more advanced stages are in most cases suppressing the development of younger ones.

CONSERVATION STATUS
Classified as Lower Risk/Near Threatened by the IUCN. Also listed in the Appendix III of the Bern Convention.

SIGNIFICANCE TO HUMANS
Valuable as an ecological indicator because of its sensitivity to pollutants accumulated in sediments. ◆

Coolie loach
Pangio kuhlii

FAMILY
Cobitidae

TAXONOMY
Cobitis kuhlii Valenciennes, 1846, Malay Peninsula. It may represent a species complex (Kottelat et al., 1993); commonly used synonyms include *Acanthophthalmus kuhli*, *A. semicinctus*, and *Pangio semicincta*.

OTHER COMMON NAMES
English: Leopard loach, slimy loach; German: Halbgebändertes Dornauge; Spanish: Culebrita.

PHYSICAL CHARACTERISTICS
Maximum total length 4.7 in (12.0 cm). This species is characterized by an elongated and scaleless body. The basic color is dark brick red, with six to ten lighter, irregular stripes running down the body. They also have a dark and large quadrangular blotch occupying the proximal half of the caudal fin. All fins are small and transparent.

DISTRIBUTION
Indonesia in West Java, Sumatra, Kalimantan Timur and Kalimantan Barat in Borneo, and the Malay Peninsula at least as far north as Phangnga. Introduced in the Philippines with unknown success.

HABITAT
Sandy bottoms of clear, fresh waters.

BEHAVIOR
Solitary and nocturnal, they like to burrow in fine sand.

FEEDING ECOLOGY AND DIET
Omnivorous bottom feeders.

REPRODUCTIVE BIOLOGY
Scatters adhesive, floating green eggs that attach to floating vegetation. Growth rate constant; reaches maturity between 2.44 and 3.22 in (6.2–8.2 cm) in length.

CONSERVATION STATUS
Not listed by the IUCN.

SIGNIFICANCE TO HUMANS
They can be found in the aquarium trade. ◆

Blind cave loach of Xiaao
Protocobitis typhlops

FAMILY
Cobitidae

TAXONOMY
Protocobitis typhlops Yang Chen, and Lan, 1994, cave in Guangxi, Duan County, China.

OTHER COMMON NAMES
None known.

PHYSICAL CHARACTERISTICS
Adult length up to 1.77 (5.4 cm). These fishes are depigmented, and lack visible eyes and a lateral line. Scales are rudimentary and only along the midline of the sides of the body. Also lack a bony swim-bladder capsule, which may be a primitive characteristic.

DISTRIBUTION
Asia in a cave located in the province of Guangxi, Duan county, near the town of Xiaao (24°15'N, 107°05' E), China.

HABITAT
The cave where they live is 689 ft (210 m) above sea level.

BEHAVIOR
Swims slowly near the bottom.

FEEDING ECOLOGY AND DIET
Nothing is known.

REPRODUCTIVE BIOLOGY
Nothing is known.

CONSERVATION STATUS
Classified as Vulnerable by the IUCN.

SIGNIFICANCE TO HUMANS
Of scientific value because of its adaptations to the cave environment. ◆

Chinese algae eater
Gyrinocheilus aymonieri

FAMILY
Gyrinocheilidae

TAXONOMY
Psilorhynchus aymonieri Tirant, 1883, mountains of Samrong-Tong, Cambodia.

OTHER COMMON NAMES
English: Sucker loach; German: Siamesische Saugschmerle; Spanish: Pez ventosa, chupa-algas; Khmer: Chun chuok dai; Thai: Pla e-dood.

PHYSICAL CHARACTERISTICS
Standard length 11.02 in (28 cm). This species, like other members of the family, is characterized by a ventral mouth modified for sucking. They also have nine branched dorsal rays, 36 to 43 lateral line scales, no dark spots on the pelvic and anal fins, a small dark spot always present behind the spiracle, and sometimes tiny tubercles on the side of the head and large tubercles on the snout. It has a long black stripe laterally and a pair of bright blue spots, one underneath the head and one on top of the anal fin.

DISTRIBUTION
Asia, specifically the Mekong, Chao Phraya, Meklong, and Xe Bangfai basins, and also the northern Malay Peninsula.

HABITAT
Solid surfaces in flowing waters of medium to large-sized rivers and flooded fields.

BEHAVIOR
Under fast-flowing current conditions, holds on to fixed objects with its suckerlike mouth. Has a small spiracle that operates as an inhalant opening for absorbing oxygen from the water. Because of the small size of the gill cavity, lives in highly oxygenated waters with a breathing rate of up to 240 times per minute. Peaceful as juveniles, but become territorial and aggressive as adults.

FEEDING ECOLOGY AND DIET
Almost exclusively herbivorous; feeds mostly on algae but occasionally consumes phytoplankton, zooplankton, and insect larvae.

REPRODUCTIVE BIOLOGY
No information is available.

CONSERVATION STATUS
Not listed by the IUCN.

SIGNIFICANCE TO HUMANS
Popular among aquarists because of their reputed ability to clean algae in home aquaria. Large individuals are sold in fish markets of Southeast Asia. Small ones are used to make *prahoc*, a fermented fish paste consumed mostly in Cambodia. ◆

Spotted algae eater
Gyrinocheilus pennocki

FAMILY
Gyrinocheilidae

TAXONOMY
Gyrinocheilops pennocki Fowler, 1937, Siam.

OTHER COMMON NAMES
Khmer: Trey smok; Laotian: Ko, pa wa.

PHYSICAL CHARACTERISTICS
Standard length 11 in (28.0 cm). They are characterized by having large tubercles on sides of head (in larger specimens), 10 branched dorsal rays, with all fins strongly spotted, and a dark spot just behind the spiracle. Brownish coloration, darker dorsally. They have approximately 12 diffused, vertical, dark stripes laterally.

DISTRIBUTION
Mekong basin.

HABITAT
Solid surfaces of the bottom of fast-moving waters.

BEHAVIOR
No information is available.

FEEDING ECOLOGY AND DIET
Feeds mostly on algae, but occasionally on aquatic plants, zooplankton, and small invertebrates.

REPRODUCTIVE BIOLOGY
No information is available.

CONSERVATION STATUS
Not threatened.

SIGNIFICANCE TO HUMANS
Large individuals are occasionally sold in fish markets. ◆

Resources

Books

Berra, T. M. *Freshwater Fish Distribution.* San Diego: Academic Press, 2001.

Kottelat, M. *Indochinese Nemacheilines: A Revision of Nemacheiline Loaches (Pisces: Cypriniformes) of Thailand, Burma, Laos, Cambodia, and Southern Viet Nam.* Munich: Pfeil, 1990.

Kottelat, M., A. J. Whitten, S. N. Kartikasari, and S. Wirjoatmodjo. *Freshwater Fishes of Western Indonesia and Sulawesi.* Republic of Indonesia: Periplus Editions, Ltd., 1993.

Lee, D. S., C. R. Gilbert, C. H. Hocutt, R. E. Jenkins, D. E. McAllister, and J. R. Stauffer, Jr. *Atlas of North American Freshwater Fishes.* Raleigh: North Carolina State Museum of Natural History, 1980.

Menon, A. G. K. *The Fauna of India and the Adjacent Countries.* Madras: Amra Press, 1987.

Nelson, J. S. *Fishes of the World.* New York: John Wiley & Sons, Inc., 1994.

Page, L. M., and B. M. Burr. *A Field Guide to Freshwater Fishes of North America North of Mexico.* Boston: Houghton Mifflin Company, 1991.

Siebert, D. J. *Interrelationships Among Families of the Order Cypriniformes (Teleostei).* Unpublished Ph.D. Dissertation, 1987.

Talwar, P. K., and A. G. Jhingran. *Inland Fishes of India and Adjacent Countries.* New Delhi: Oxford & I.B.H. Publishing Co., 1991.

Tomelleri, J., and M. Eberle. *Fishes.* Lawrence: University of Kansas Press, 1990.

Weber, A. "Fish and Amphibia." In *Subterranean Ecosystems,* edited by H. Wilkens, D. C. Culver, and W. F. Humphrey. Amsterdam: Elsevier, 2000.

Wu, Y. "On the Present Status of Cyprinid Fish Studies in China." *Proceedings of the Fifth Congress of European Ichthyology,* edited by S. O. Kullander and B. Fernholm. Stockholm, 1985.

Periodicals

Brabrand, A., B. Faafeng, and J. Nilssen. "Relative Importance of Phosphorus Supply to Phytoplankton Production: Fish Excretion Versus External Loading." *Canadian Journal of Fisheries and Aquatic Sciences* 47 (1990): 364–372.

Greenwood, P. H. "A New and Eyeless Cobitid Fish (Pisces: Cypriniformes) from the Zagros Mountains, Iran." *Journal of Zoology* 180 (1976): 129–137.

Hora, S. L. "Fish of the Siju Cave, Garo Hills, Assam." *Records of the Indian Museum* 26 (1924): 27–31.

Jun-Xing, Yang, Chen Yin-Rui, and Lan Jia-Hu. "*Protocobitis typhlops,* a New Genus and Species of Cave Loach from China (Cypriniformes: Cobitidae)." *Ichthyological Explorations in Freshwaters* 5 (1994): 91–96.

Kottelat, M., and K. K. P. Lim. "A Review of the Eel-Loaches of the Genus *Pangio* (Teleostei: Cobitidae) from the Malay Peninsula, with Descriptions of Six New Species." *Raffles Bulletin of Zoology* 41 (1993): 208–210.

Liu, Jia-Shou, Jing-Shen Chen, and Zhi-Tang Yu. "The Effect of Common Fungicides on Chinas Sucker *Myxocyprinus asiaticus* Eggs." *Journal of the World Aquaculture Society* 26 (1995): 84–87.

Pillai, R. S., and G. M. Yazdani. "Ichthyo-Fauna of Garo Hills, Meghalaya (India)." *Records of the Zoological Survey of India* 71 (1977): 1–22.

Romero, A., and K. M. Paulson. "It's a Wonderful Hypogean Life: A Guide to the Troglomorphic Fishes of the World." *Environmental Biology of Fishes* 62 (2001): 13–41.

Singh, D. F., and G. M. Yazdani. "Studies on the Ichthyofauna of Nasik District, Maharashtra, India." *Records of the Zoological Survey of India* 90 (1993): 195–201.

Other

Romero, Aldemaro. "Guide to Hypogean Fishes" [cited January 21, 2003]. <http://www.macalester.edu/envirost/ARLab/HypogeanFishes/synbranchidae.htm>

Aldemaro Romero, PhD

Characiformes
(Characins)

Class Actinopterygii

Order Characiformes

Number of families 11

Photo: Red-bellied piranhas (*Pygocentrus nattereri*) in the Amazon. Red-bellied piranhas always face the same direction when they hunt together. (Photo by Davis M. Schleser/Nature's Images/ Photo Researchers, Inc. Reproduced by permission.)

Evolution and systematics

Characiformes are members of the superorder Ostariophysi, which contains nearly three-quarters of all freshwater fishes in the world. Characiformes possess several specialized adaptations, common to most Ostariophysi, which enable them to thrive in freshwater environments. One specialized anatomical feature is the presence of the Weberian apparatus, a linkage of bones called ossicles, derived from the vertebrae immediately following the skull, that connect the inner ear and the swim bladder; this structure improves the hearing ability of the fishes. Another evolved characteristic is the production of *Schreckstoff*, an alarm pheromone released into the water by an injured fish that triggers an escape response in other members of the species. Certain Characiformes, such as blind cavefish, do not produce these pheromones, as such a substance would not enhance a blind fish's ability to escape predators. Other species do not respond to the presence of the pheromones; for example, many characins preyed on by piranhas produce *Schreckstoff*, but piranhas do not flee when feeding on these species. Finally, the presence of a moveable upper jaw confers a feeding advantage to Characiformes. In some species, the jaw is protrusible, enabling the fishes to use suction pressure to capture prey.

Lineages of Characiformes date back more than 100 million years. Morphological and genetic evidence suggests that much of the species diversity of this group likely developed prior to the time when Africa and South America split into separate continents. The classification of the Characiformes into families has proven complex and controversial. The taxonomic relationships have been revised considerably in the last 30 years, and the classifications continue to change as new evidence becomes available. Although some taxonomists consider there to be up to 16 distinct families, most recognize 11 families: Alestiidae, Anostomidae, Characidae, Citharinidae, Ctenoluciidae, Curimatidae, Erythrinidae, Gasteropelecidae, Hemiodontidae, Hepsetidae, and Lebiasinidae. These families encompass species that are diverse in life form and behavior, with examples including the blind cavefish and predatory piranhas, as well as popular aquarium species such as tetras.

African pike (*Hepsetus odoe*) foam nest in the Okavango Delta in Botswana. (Photo by Animals Animals ©C-Par Farnetti, OSF. Reproduced by permission.)

Piranha spawning behavior. (Illustration by Patricia Ferrer)

A silver tetra (*Poptella orbicularis*) living in the Peruvian Amazon. (Photo by David M. Schleser/Nature's Images, Inc./Photo Researchers, Inc. Reproduced by permission.)

Physical characteristics

Most species of Characiformes are small fishes, although size and shape vary widely throughout the order. The smallest member of the order is the Bolivian pygmy blue characin (*Xenurobrycon polyancistrus*), which attains a maximum size of around 0.5 in (1.4 cm); the largest species, the giant tigerfish (*Hydrocynus goliath*) grows to 4.3 ft (1.33 m).

Most Characiformes are brightly colored and often silvery, but some are also brilliant shades of red or blue. Characiformes vary widely in body form. Some species have long and slender bodies, while others, such as hatchetfishes, are deep-bodied and laterally compressed. The body is covered in scales, and the lateral line is often decurved or may be incomplete in some species. Most species have an adipose fin, a short fin between the dorsal and caudal fins. Pelvic fins generally have five to twelve rays, and the anal fin may be short to moderately long, with 45 rays or fewer. Characiformes lack sensory barbels, but they typically have large eyes to heighten visual acuity.

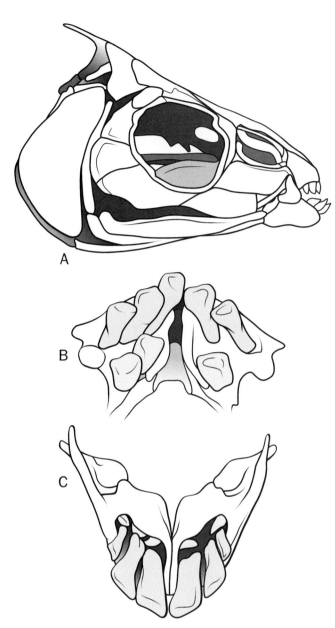

stagnant ponds, rushing streams, and even underground caves. Certain species also possess morphological adaptations, including the ability to breath air or swollen lips that facilitate gas exchange in the upper portion of the water column, that enable them to move into shallow waters of floodplains and flooded forests.

Behavior

The diversity of species represented by the Characiformes necessitates a diversity of behaviors. Many species travel in schools under certain conditions. Some species, such as members of the Ctenoluciidae, form schools as juveniles but become solitary as adults. Other species, including members of the Curimatidae, piranhas, and tetras, travel in large aggregations during all life stages. However, certain species, such as the blind cavefish, rarely gather in groups, and do not form organized aggregations. Although many of the smaller species of Characiformes live their whole lives in a limited geographic area, some species undertake extensive seasonal or spawning migrations.

Courtship behaviors and migrations are associated with spawning in some members of the Characiformes. These unique courtship behaviors often involve fin displays to attract mates as well as elaborate swimming patterns to lure the mate to the spawning site. In addition to complex courtship rituals, some species possess morphological adaptations to attract mates. Males of certain species may utilize sensory cues, including pheromones and visual lures, to gain the attention of females.

Feeding ecology and diet

As reflected by their diversity of body shapes, Characiformes exhibit remarkable feeding specializations and exploit all available trophic modes. Many predatory species, particularly those of the Characidae, have well-developed teeth that enable them to feed on other fish. Some species, such as tigerfishes, solitarily stalk their prey: others, such as piranhas, engage in voracious group predation. Other carnivorous species

Skull showing replacement teeth (in blue)—an important feature of this group of fish. (Illustration by Patricia Ferrer)

Distribution

Characiformes are found in fresh waters of Texas and Mexico in North America and are widely distributed in Central America, South America, and Africa. Over 1,300 species, 252 genera, and 10 families inhabit South America. At least 176 species in 23 genera and 4 families occur throughout Africa.

Habitat

As a dominant group of freshwater fishes, Characiformes inhabit all types of fresh waters, including weedy river edges,

African pike (*Hepsetus odoe*) adults guard their foam nest in the Okavengo Delta in Botswana (Photo by Animals Animals ©Partridge OSF. Reproduced by permission.)

Tetra mating: 1. Tetra pair (*Copella arnoldi*) swim toward the surface, seeking suitable site to lay eggs (male left, female right); 2. The pair attaches briefly to the underside of a leaf, where the female lays eggs and the male fertilizes them. They repeat until all eggs are laid; 3. The female leaves, but the male cares for eggs until they hatch, using his long caudal fin to splash them and keep them moist. (Illustration by Gillian Harris)

pick or suck invertebrates from the substrate. In contrast, some Characiformes are strictly herbivorous, feeding on plants, fruits, or seeds. Other species filter plankton or are detritivores that feed on mud, algae, and ooze. Among the most remarkable feeding adaptations observed in freshwater fishes, certain Characiformes survive by eating the scales or pieces of fins from other fishes.

Many Characiformes are small-bodied and, thus, fall prey to larger fish species. In fact, predatory piranhas may indiscriminately attack a variety of food items, including smaller characins. Humans are the major predator of some characins. Small species are harvested as bait for fisheries and as trade items for the aquarium industry. A variety of the larger species, particularly tigerfishes, pacus, and piranhas, sustain important local fisheries in South America and Africa.

Reproductive biology

Most Characiformes broadcast their eggs and devote little parental care to their young. The eggs are often scattered among aquatic plants, and the vegetative cover confers some protection to young by providing shelter from predators. In a few species, including members of the Characidae, the male inseminates the female, and she may retain the sperm cells in her ovary for a period of days to months. However, fertilization does not take place until the eggs and sperm are shed into the water column at the same time.

Certain Characiformes exhibit an assortment of more specialized breeding behaviors. Members of the Erythrinidae construct nests for their eggs, and the African pike characins of the family Hepsetidae deposit eggs in a bubble-like nest of floating foam. To avoid egg predation, female splash tetras (*Copella arnoldi*) jump out of the water and lay their eggs on the underside of overhanging vegetation or rocks. The males follow, fertilize the eggs, and remain in the area to splash water onto the eggs until they hatch. Members of another species, *Brycon petrosus*, crawl onto the banks of rivers to lay their eggs.

Conservation status

The IUCN Red List includes 7 Characiformes. Six are listed as Data Deficient, and one, the naked characin (*Gymnocharacinus bergii*), is listed as Endangered. Although international conservation concerns have not been recognized for other

The "flying" hatchetfish: 1. River hatchetfish (*Gasteropelecus sternicla*), front view showing its thin body and long pectoral fins; 2. Hatchetfish swims just beneath the surface; 3. Launches itself out of the water, leaping into the air and staying aloft for some distance by rapidly beating its pectoral fins. (Illustration by Gillian Harris)

species, many Characiformes are harvested for the aquarium trade, an industry that is monitored and regulated only in certain countries or localities.

Significance to humans

Many Characiformes, including tetras, hatchetfishes, and pencilfishes, are popular aquarium fishes. Others are im-portant as a food resource for humans. Although some species are traded commercially, many are relied upon to meet the subsistence food needs of communities living along tropical rivers. Examples include members of the Citharinidae (dorados) and Serrasalminae (tambaqui and pacus). In addition, some Characiformes, such as tigerfishes, are popular recreational targets that attract anglers from around the world.

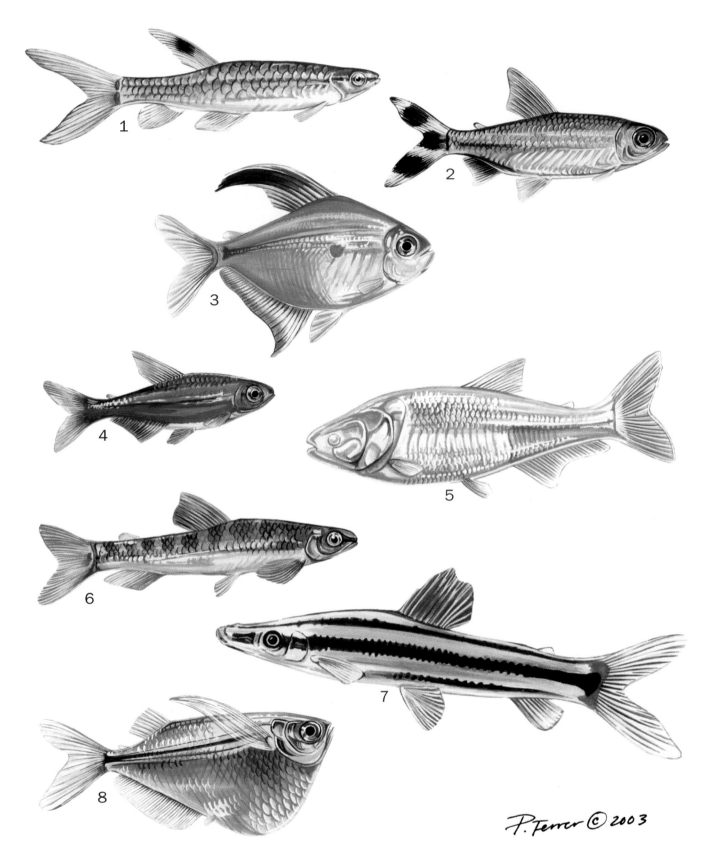

1. Splash tetra (*Copella arnoldi*); 2. False rummynose tetra (*Petitella georgiae*); 3. Bleeding-heart tetra (*Hyphessobrycon erythrostigma*); 4. Cardinal tetra (*Paracheirodon axelrodi*); 5. Blind cavefish (*Astyanax mexicanus jordani*); 6. Striped African darter (*Nannocharax fasciatus*); 7. Striped headstander (*Anostomus anostomus*); 8. River hatchetfish (*Gasteropelecus sternicla*). (Illustration by Patricia Ferrer)

1. Red-bellied piranha (*Pygocentrus nattereri*); 2. Pirapitinga (*Piaractus brachypomus*); 3. Trahira (*Hoplias malabaricus*); 4. Silver dollar (*Myleus pacu*); 5. Flagtail prochilodus (*Semaprochilodus taeniurus*); 6. Golden pike characin (*Boulengerella lucius*); 7. Giant tigerfish (*Hydrocynus goliath*). (Illustration by Patricia Ferrer)

Species accounts

Giant tigerfish
Hydrocynus goliath

FAMILY
Alestiidae

TAXONOMY
Hydrocynus goliath Boulenger, 1898, Kinshasa and Umangi, Upper Congo.

OTHER COMMON NAMES
German: Wolfsalmler.

PHYSICAL CHARACTERISTICS
The largest characin, reaching a maximum reported fork length of 4.3 ft (1.33 m) and weight of 110 lb (50 kg). The fish has an elongate, fully scaled body, with a high dorsal fin and deeply forked caudal fin. The body appears silvery, but the back is a darker gray. The fins are often orange or red, and the fish may become brightly colored during the breeding season. The sharp canine teeth are placed alternately on the upper and lower jaws.

DISTRIBUTION
Africa in the Congo River basin, Lualaba River, Lake Upemba, and Lake Tanganyika.

HABITAT
Large rivers and nearshore areas of lakes.

BEHAVIOR
Tigerfish aggregate with similar-sized individuals of the same species, sometimes in schools. Smaller fish occur in large aggregations; large fishes gather with few members in the group. This species migrates in rivers to find appropriate spawning sites.

FEEDXING ECOLOGY AND DIET
Adults are voracious predators that consume a wide variety of smaller fish. Larvae eat zooplankton, but they quickly move to larger prey as they grow. Juveniles and subadults are vulnerable to crocodiles, otters, and a great variety of predatory fishes and fish-eating birds. Humans are the only known predators of aduts.

REPRODUCTIVE BIOLOGY
Spawning takes place during the summer along the shores of lakes and flooded banks of large rivers. Females disperse hundreds of thousands of eggs into flooded vegetation, where they hatch and develop with no parental care.

CONSERVATION STATUS
Not listed by the IUCN.

SIGNIFICANCE TO HUMANS
This species is known internationally as a sport fish and attracts anglers from around the world. It also is sought in local commercial and subsistence fisheries in several African countries, including Tanzania and Zambia. ◆

Striped headstander
Anostomus anostomus

FAMILY
Anostomidae

TAXONOMY
Anostomus anostomus Linnaeus, 1758, South America.

OTHER COMMON NAMES
English: Striped anostomus; German: Prachtkopfsteher; Spanish: Anostomus rayados, lisa; Portuguese: Anostomo.

PHYSICAL CHARACTERISTICS
Maximum total length 6.3 in (16 cm). Has elongated body with head that tapers to pointed nose, and small upward-facing mouth. Three dark longitudinal stripes start at front of body and extend to tail area; stripes widen as the fish ages. Area between the stripes is gold or mustard color. All fins have red bases; a red spot is apparent on the dorsal fin.

DISTRIBUTION
South America in inland streams and rivers of the Amazon, Orinoco, and Essequibo River basins.

HABITAT
Mainly weeds, rocks, and woody debris of tropical rivers.

BEHAVIOR
Often nips at fins of other fishes under aquarium conditions. Swim in slow-moving schools, generally with their heads down, although may tilt the head backward at times while feeding.

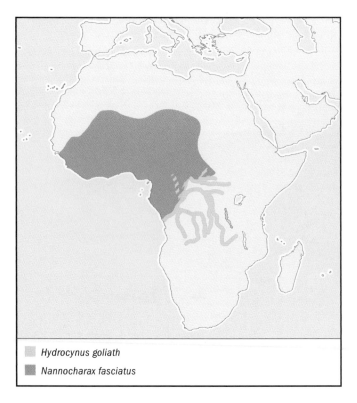

Hydrocynus goliath

Nannocharax fasciatus

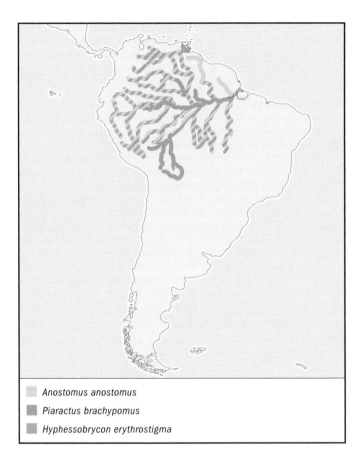

Anostomus anostomus

Piaractus brachypomus

Hyphessobrycon erythrostigma

FEEDING ECOLOGY AND DIET
Feeds on worms, crustaceans, insects, and plant material.

REPRODUCTIVE BIOLOGY
Scatters eggs and fertilizes them in the water column. Few additional details are known.

CONSERVATION STATUS
Not listed by the IUCN.

SIGNIFICANCE TO HUMANS
Popular aquarium fish. ◆

Blind cavefish
Astyanax mexicanus jordani

FAMILY
Characidae

TAXONOMY
Astyanax mexicanus jordani Hubbs and Innes, 1936, San Luis Potosí, Mexico.

OTHER COMMON NAMES
English: Blind cave characin, blind cave tetra; German: Blinder Höhlensalmler; Spanish: Sardina ciega; Portuguese: Peixe-caverna, Peixe-cego.

PHYSICAL CHARACTERISTICS
Maximum length 3.5 in (8.9 cm). The species evolved in the darkness of caves. Perhaps because of the absence of light in the environment, the body lacks pigments; instead, it is peach-colored with a silvery sheen. Has a complete lateral line along the body. Fins are colorless and transparent. Another characteristic attributed to the dark environment are vestigial eyes; the eye depressions are covered by skin and other tissue.

DISTRIBUTION
Mexico.

HABITAT
Subterranean lakes, streams, and pools of underground caves.

BEHAVIOR
Despite a lack of vision, this species exhibits exceptional navigation abilities. The lateral line is highly sensitive to vibrations, which allows the fish to find food and avoid obstacles as they swim. A solitary species, there is little interaction between individuals.

FEEDING ECOLOGY AND DIET
General, omnivorous feeders that primarily consume benthic invertebrates and algae. Use their excellent sense of smell to locate food items.

REPRODUCTIVE BIOLOGY
Spawns by broadcasting eggs. The male and female swim alongside each other near the surface of the water, with their ventral surfaces pressed together; the female scatters eggs, which are then fertilized by the male. The eggs fall to the bottom and hatch in two to three days. The larvae have normal eyes when they hatch, but these become enclosed in tissue after a few weeks.

CONSERVATION STATUS
Not listed by the IUCN.

SIGNIFICANCE TO HUMANS
Commonly kept in aquaria, but not as popular as most other characins. Studies of this species have provided useful insights into the evolution of eyes and sight, which can also help improve scientific understanding of human vision. ◆

Bleeding-heart tetra
Hyphessobrycon erythrostigma

FAMILY
Characidae

TAXONOMY
Hyphessobrycon erythrostigma Fowler, 1943, Peru-Brazil border.

OTHER COMMON NAMES
English: Tetra perez; German: Fahnen-Kirschflecksalmler, Perez Salmler; Spanish: Mojarita, punto rojo.

PHYSICAL CHARACTERISTICS
Standard length 2.4 in (6.1 cm). Deep-bodied fishes that are orange to brown in color on the back and belly. The scales and fins are reddish pink, with a red spot behind the gill cover along the deeply colored lateral line. Dark spots are prominent on the dorsal fin, and the long rays of the dorsal fin are highly arched in males. The anal fin is elongate, stretching from mid-body to near the caudal fin. This fin is white in males, a feature that distinguishes them from females, in which the fin is transparent gray.

DISTRIBUTION
Upper Amazon basin of South America.

HABITAT
Occur in the bottom portion of the water column in freshwater tropical habitats, preferring temperatures of approximately 73–82°F (23–28°C). Individuals often live in vegetation and woody debris.

BEHAVIOR
Schooling fishes; little is known of their activities, social behavior, or reproductive behavior in nature.

FEEDING ECOLOGY AND DIET
Omnivore; primarily consumes insects and crustaceans. Because of its small size, succumbs to predation by a variety of larger fishes.

REPRODUCTIVE BIOLOGY
Females deposit 20–30 eggs in vegetation during each mating act. The eggs incubate for approximately 48 hours in aquaria; fry emerge in three days.

CONSERVATION STATUS
Not listed by the IUCN.

SIGNIFICANCE TO HUMANS
Popular aquarium fish that are valued for their bright colors. ◆

Myleus pacu
Astyanax mexicanus jordani
Paracheirodon axelrodi

Silver dollar
Myleus pacu

FAMILY
Characidae

TAXONOMY
Myleus pacu Jardine and Schomburgk, 1841, Guyana.

OTHER COMMON NAMES
English: Pacu, pacupeba; German: Brauner Mühlsteinsalmler; Portuguese: Pacu dente; Creole: Koumarou-nwé, pakou.

PHYSICAL CHARACTERISTICS
Maximum length 8 in (20 cm). This species is deep-bodied with a small head and a broad mottled reddish brown body. A row of scutes occurs along the edge of the belly, and the body scales are small. Two rows of teeth are found on the upper jaw; the front row is incisor-like, the back row contains molars.

DISTRIBUTION
South America in Guyana, French Guiana, Brazil, Suriname, and Bolivia.

HABITAT
Inhabits streams and main stems of rivers.

BEHAVIOR
Gregarious and rarely aggressive. During the rainy season, large groups migrate to small creeks to reproduce.

FEEDING ECOLOGY AND DIET
Although their powerful dentition can cause serious bites, silver dollars are herbivores that feed primarily on aquatic plants (often of the family Podostemaceae). Occasionally, they also consume seeds and fruits, which they crush with their large molars. Predators include larger carnivorous fishes, caimans,

otters, river dolphins, and a wide variety of fish-eating birds. Some fish are also taken by humans.

REPRODUCTIVE BIOLOGY
Reproduce by external fertilization. Females and males swim through the water together; females broadcast eggs into the water column, where they are fertilized by males. No parental care.

CONSERVATION STATUS
Not listed by the IUCN.

SIGNIFICANCE TO HUMANS
Common aquarium fishes throughout the world. Also provide an important food source for many communities living near streams inhabited by pacus. ◆

Cardinal tetra
Paracheirodon axelrodi

FAMILY
Characidae

TAXONOMY
Paracheirodon axelrodi Schultz, 1956, Rio Negro, Brazil.

OTHER COMMON NAMES
English: Red neon, neon tetra, scarlet characin; German: Kardinaltetra, Roter Neon.

PHYSICAL CHARACTERISTICS
Maximum standard length 1 in (2.5 cm). One of the most colorful of the characoids. Although the dorsal portion is gray,

the sides are marked with an iridescent turquoise-blue stripe and the lower flanks are bright red. The dorsal fins exhibit a peculiar blue-green iridescence.

DISTRIBUTION
Upper Rio Orinoco and upper Rio Negro of South America.

HABITAT
This species inhabits a wide variety of habitats. It is often found in shaded or vegetated areas of small, slow-moving, clear, and blackwater creeks. During the rainy season, the fish move into headwaters and flooded forests of rivers. At times of low water, it lives along the margins of lakes or becomes concentrated in creeks and backwaters.

BEHAVIOR
Often aggregates in shoals of 12–30 individuals in middle portions of the water column. Individuals move onto the floodplain during the rainy season. They are considered "annual fishes," spawning during the spring floods and often perishing due to starvation when foraging habitats retract during the subsequent low-water season.

FEEDING ECOLOGY AND DIET
Feeds mainly on worms and small crustaceans. Due to its small size, indiscriminately consumed by many larger piscivores, including piranhas.

REPRODUCTIVE BIOLOGY
Spawns on floodplains of rivers as water levels rise during the rainy season. Females bear 300–500 large eggs; these are broadcast into the water column and fertilized by males. In captivity, eggs hatch in 24–30 hours; fry become free-swimming after three to four days.

CONSERVATION STATUS
Not listed by the IUCN. However, this species accounts for 80% of the total catch of ornamental species in some areas, such as the state of Amazonas in Brazil, and this high level of extraction to support the aquarium trade has resulted in local regulations that restrict harvests during the spawning season.

SIGNIFICANCE TO HUMANS
One of the most popular aquarium fishes throughout the world, its commercial harvest for the aquarium trade also supports the economic and social structure of many human communities along the rivers it inhabits in South America. ◆

Petitella georgiae

Pygocentrus nattereri

Semaprochilodus taeniurus

False rummynose tetra

Petitella georgiae

FAMILY
Characidae

TAXONOMY
Petitella georgiae Gery and Boutiere, 1964, Rio Huallago, Peru.

OTHER COMMON NAMES
English: False rednose tetra, false rummy-nosed tetra; German: Rotmaulsalmler; Finnish: Punapää tetra; Polish: Czerwonogłówka.

PHYSICAL CHARACTERISTICS
Maximum size 1.5 in (3.9 cm). These fish appear iridescent and shiny in water. The body is silver to olive-brown, with a gold stripe extending from the head to the base of the tail. The

head and the iris of the eye are bright red. The tail is deeply forked and striped, with three black and four white bands. All fins other than the caudal fin are transparent.

DISTRIBUTION
South America in the upper Rio Amazonas basin in Peru, as well as in the Purus, Negro, and Madeira basins in Brazil.

HABITAT
Swift-moving headwater streams, often preferring the cover of vegetation if available.

BEHAVIOR
A schooling species; other behavioral features have not been described.

FEEDING ECOLOGY AND DIET
Feeds on benthic algae, plants, and crustaceans in the plankton and benthos. Its small size makes it susceptible to predation by many larger species of fish as well as water snakes and many different fish-eating birds.

REPRODUCTIVE BIOLOGY
Reproduces by scattering eggs; females broadcast eggs into the water column, and these eggs are fertilized by the males. Mating activity typically occurs near vegetation in the evening.

CONSERVATION STATUS
Not listed by the IUCN.

SIGNIFICANCE TO HUMANS
An important aquarium species. ◆

Pirapitinga
Piaractus brachypomus

FAMILY
Characidae

TAXONOMY
Piaractus brachypomus Cuvier, 1818, Brazil.

OTHER COMMON NAMES
English: Cachama, freshwater pompano, pacu, red pacu, red-bellied pacu; German: Gamitana-Scheibensalmler, Riesenpacu; Spanish: Cachama, cachama blanca, morocoto, paco; Portuguese: Caranha, pirapitinga.

PHYSICAL CHARACTERISTICS
Total length 34.6 in (88 cm). Deep-bodied; appears silvery gray on back and sides, but areas under the throat, along the belly, and at the edges of the pelvic and anal fins are red. Unlike in other Characiformes, the adipose fin contains rays in adults.

DISTRIBUTION
Amazon and Orinoco River basins in South America. Feral individuals have been caught in 16 states throughout the United States, likely as a result of releases from personal aquariums or fish farms.

HABITAT
Pelagic; occurs in streams and tributaries to main river channels, also in ponds and oxbow lakes.

BEHAVIOR
Young live in flooded savanna vegetation until about two years old, then move into flooded forests with adults. Feeds during rainy seasons. During low water and just before the start of the second rainy season of the year, adults move upstream in schools to spawn.

FEEDING ECOLOGY AND DIET
Primarily herbivorous, although it possesses powerful teeth that could inflict severe bites. Feeds mostly on plants and detritus, but may also consume nuts, fruits, seeds, and sometimes insects. Juveniles are vulnerable to larger predatory fishes, notably giant pimelodid catfishes, caimans, otters, and a variety of fish-eating birds. Adults have few predators because of their large size but can be preyed upon by jaguars.

REPRODUCTIVE BIOLOGY
Spawns as waters begin to rise during the second rainy season. Do not feed during spawning but live off fat stored during the wet season. Eggs are scattered and fertilized in the water column. Males and females mature around seven years of age.

CONSERVATION STATUS
Not listed by the IUCN.

SIGNIFICANCE TO HUMANS
One of the most important commercial species in the Amazon basin. Popular food fishes, they are taken in wild fisheries in their native range and raised in fish farms in North and South America. Also popular in aquaria. ◆

Red-bellied piranha
Pygocentrus nattereri

FAMILY
Characidae

TAXONOMY
Pygocentrus nattereri Kner, 1858, Brazil.

OTHER COMMON NAMES
English: Red piranha, redbelly piranha; German: Diamant-piranha, Schulterfleck-Piranha; Natterers Sägesalmler; Spanish: Caribe boca de locha, palometa, paña; Portuguese: Piranha caju; piranha-quexicuda.

PHYSICAL CHARACTERISTICS
Average length of adult 6–8 in (15.2–20.3 cm), but can grow up to 12 in (30.5 cm). Male and female red piranhas are alike externally, with body height about one-half body length. The stocky bodies have reddish bellies, though overall coloration varies depending on location and age. Sides often pale brown to slightly olive; back bluish gray to brownish; throat belly areas bright red. Forked caudal fin usually gray. Pectoral, pelvic, and anal fins typically bright red. Powerful jaws with triangular, interlocking sharp teeth.

DISTRIBUTION
Widely distributed in South America and in basins of the Amazon, Paraguay-Paraná, and Essequibo Rivers, as well as coastal rivers of northeast Brazil.

HABITAT
Creeks and interconnected ponds; prefers areas with dense vegetation.

BEHAVIOR
Diel activity varies by age; adults forage mainly at dusk and dawn, medium-sized individuals are most active at dawn, late afternoon, and night. Smaller fishes feed during the day. Exhibits a "lurking, then dashing" sequence of behaviors during the day. Hierarchical structure often exists in small schools.

FEEDING ECOLOGY AND DIET
Feeds communally, with groups of 20–30 individuals waiting in vegetation to ambush prey. Prey are attacked in a feeding frenzy, further induced by the presence of blood in the water. Highly predaceous carnivores, but also scavenge for food and consume insects, snails, worms, plants, and fins of other fishes. Can feed continuously and maintain a voracious bite by replacing teeth on alternate sides of the jaw. Preyed upon by other fishes (including large pimelodid catfishes), crocodilians, fish-eating birds, and large mammals (including jaguars).

REPRODUCTIVE BIOLOGY
Spawns through external fertilization. After elaborate courtship display involving swimming in circles, the female deposits layers of eggs on plants in the water, and the male fertilizes them. The male guards and fans the egg masses until they hatch in 9 to 10 days. Annual reproductive success varies, but is dependent upon the degree to which the savanna is flooded.

CONSERVATION STATUS
Not listed by the IUCN.

SIGNIFICANCE TO HUMANS
This species is kept as an aquarium fish. This activity is illegal in some states of the United States to prevent irresponsible hobbyists from releasing the species into the wild, where it may multiply and prey upon indigenous fish species. These fish can also inflict serious bites, although they are not as aggressive as once believed, and they are unlikely to attack humans unless the human is bleeding or in water near congregations of other prey species. The species is commonly caught and eaten by river dwellers throughout its extensive range. Large numbers are also caught for use as trot-line bait for large catfishes. ◆

Striped African darter
Nannocharax fasciatus

FAMILY
Citharinidae

TAXONOMY
Nannocharax fasciatus Guenther, 1867, Gabon.

OTHER COMMON NAMES
English: African darter tetra; German: Afrikanischer Boden-salmler.

PHYSICAL CHARACTERISTICS
Maximum length 2.6 in (6.6 cm). These fish have a short, blunted snout and a thin, elongate body. The body is golden brown on the back, with a dark lateral line midbody, below which is light brown to white. The pigmentation is marked by seven to eleven dark brown transverse bands across the back along the length of the body. The fins are also marked with spots forming transverse bands; a large dark spot is on the anal and adipose fins. The tip of the dorsal fin is red in adults. An elongate pectoral fin extends to the origin of the pelvic fins.

DISTRIBUTION
Throughout equatorial West Africa.

HABITAT
Clear, swift-moving waters in forested regions; often found over sandy substrates or in similar open areas.

BEHAVIOR
These fishes live near the water bottom, often in contact with the substrate. They swim with their heads up at an angle of 45° and use long pectoral fins for support while searching for food along the bottom. They rest for longer periods on their pectoral, pelvic, and caudal fins.

FEEDING ECOLOGY AND DIET
Consumes small invertebrates, including insect larvae, worms, ostracods, and zooplankton near the substrate. Because of their small size, likely consumed by a variety of larger predators.

REPRODUCTIVE BIOLOGY
Reproduces externally; females scatter eggs in the water column, where they are fertilized by males. Eggs are released in partial batches during several spawning events. There is no parental care.

CONSERVATION STATUS
Not listed by the IUCN.

SIGNIFICANCE TO HUMANS
Sometimes kept as aquarium fishes and traded commercially as part of the ornamental fish industry. ◆

Golden pike characin
Boulengerella lucius

FAMILY
Ctenoluciidae

TAXONOMY
Boulengerella lucius Cuvier, 1816, Brazil.

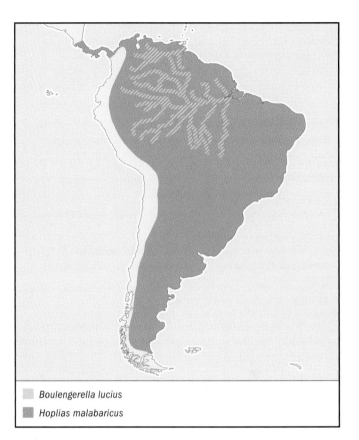

■ *Boulengerella lucius*
▨ *Hoplias malabaricus*

OTHER COMMON NAMES
English: Cuvier's pike characin; German: Cuviers Hechtsalm-ler; Portuguese: Pirapacu.

PHYSICAL CHARACTERISTICS
Standard length 16.5 in (42 cm). Elongated, slender fishes resembling North American pikes. Dorsal fin is located toward the back of the body; the anal fin is short. Fishes are metallic green on the back, lighter on sides, and silvery on belly. A small dark spot on caudal fin helps distinguish this species from others in the genus. The jaw terminates in a long fleshy filament and contains numerous teeth.

DISTRIBUTION
Inland waters of the Amazon, Rio Negro, and Rio Orinoco River basins in South America.

HABITAT
Pelagic; inhabits open waters of large rivers.

BEHAVIOR
Migrates upriver to spawn during the rainy season.

FEEDING ECOLOGY AND DIET
Predatory; eats many species of smaller fishes available in open water areas. Predators have not been specified.

REPRODUCTIVE BIOLOGY
Breeds by external fertilization of eggs. Females release eggs into the water column, where they are fertilized by males. Eggs are demersal, meaning they sink to the bottom.

CONSERVATION STATUS
Not listed by the IUCN.

SIGNIFICANCE TO HUMANS
Exploited by humans as a subsistence and commercial food fish; also exported as part of the aquarium trade. ◆

Flagtail prochilodus
Semaprochilodus taeniurus

FAMILY
Curimatidae

TAXONOMY
Semaprochilodus taeniurus Valenciennes, 1817, Amazon River.

OTHER COMMON NAMES
English: Silver prochilodus; German: Nachtsalmler, Schwanzstreifensalmler; Finnish: Juovahuulitetra; Portuguese: Jaraqui, jaraqui escama fina.

PHYSICAL CHARACTERISTICS
Standard length 9.4 in (24 cm). Like other members of this family, has many small teeth on the jaws and a mouth with enlarged fleshy lips that can be turned outward to form a rasping or suction disk. The lips are often lined with fine papillae, thus the name "flannel mouths." Deep-bodied; dark gray on the back fading to silvery hues on the flanks. Dorsal, anal, and adipose fins are dark gray; pectoral and pelvic fins are nearly clear. Edges of the dorsal and anal fins appear yellow. The caudal fin is most striking, yellow with prominent black horizontal stripes.

DISTRIBUTION
Central portion of the Amazon River basin in Brazil.

HABITAT
Main channels of streams and rivers, as well as floodplain lakes.

BEHAVIOR
Undertakes annual migrations for spawning, feeding, and dispersal. At the beginning of the flood season, schools of mature fish migrate downstream from tributaries to the Amazon River to spawn. Migration occurs during the day; spawning takes place at night. After spawning, small groups move into flooded forests of their home tributaries to feed. As soon as the water levels begin to fall, adults undertake a complex dispersal migration. They move back into the main channel of the Amazon and migrate upstream, before dispersing into tributaries other than those in which they lived the year before.

FEEDING ECOLOGY AND DIET
Primarily consumes plants, algae, and other detritus. It can use its lips as suction cups to eat detritus attached to trees and other submerged vegetation in the flooded forest. Large quantities of mud are often reported in the stomach. Although the fishes are moderate in size, they are preyed upon by larger predatory fishes, such as catfishes.

REPRODUCTIVE BIOLOGY
Spawning may occur in shallow water along the course of large rivers, sometimes below barriers such as waterfalls. Males emit loud, grunting noises to attract females to the spawning areas. Females scatter a single batch of small eggs into well-oxygenated waters. The eggs and fry are carried passively downstream and eventually onto the floodplain, where they begin feeding. After two years, these offspring spawn for the first time.

CONSERVATION STATUS
Not listed by the IUCN.

SIGNIFICANCE TO HUMANS
Popular worldwide as an aquarium species. As one of the most common species in the Amazon basin, it is an important food fish for local villages and towns. It is also traded commercially to areas outside the Amazon basin. ◆

Trahira
Hoplias malabaricus

FAMILY
Erythrinidae

TAXONOMY
Hoplias malabaricus Bloch, 1794, South America.

OTHER COMMON NAMES
English: Haimara, tararira, tararura, wolffish, wolf characin; French: Patagaye; German: Kleiner Trahira, Tigersalmler; Spanish: Perro de aqua, tararira; Portuguese: Dorme-dorme, lobó, traira, trairitinga; Palikur: Iigl.

PHYSICAL CHARACTERISTICS
Standard length 19 in (48.5 cm). Elongated, cylindrical body with short anal fin and large scales. Adipose fin is absent. Bodies of young have reddish brown back and yellowish belly, red bands across the head, and green band along the sides. Adults are mottled dull green and brown.

DISTRIBUTION
Inland waters of most river basins from Costa Rica to Argentina.

HABITAT
Lives in diverse habitats, ranging from clear streams to slow, turbid waters. Tolerates stagnant conditions in irrigation ditches, drainage channels, and floodplain ponds. Utilizes atmospheric oxygen, enabling it to inhabit low-oxygen habitats, including small brooks or ponds.

BEHAVIOR
Sedentary species, often remains in a relatively confined geographic area for life. It rests during the day and becomes active at night. These fish may form schools, typically in low numbers. They are ambush predators, waiting quietly under cover until suitable prey items approach.

FEEDING ECOLOGY AND DIET
Adults are voracious predators that feed on a wide variety of fish and shrimp. Juveniles eat crustacean and insect larvae, shrimp, and other small invertebrates. Trahiras are preyed upon by larger fishes, including piranhas and catfishes, as well as crocodilians, fish-eating birds, and otters.

REPRODUCTIVE BIOLOGY
Spawns over a protracted reproductive period, with multiple spawnings and size classes distributed uniformly throughout the year. Eggs are adhesive and placed in nests made in nearshore vegetation of shallow, slow watercourses. Male provides a high level of parental care.

CONSERVATION STATUS
Not listed by the IUCN.

SIGNIFICANCE TO HUMANS
Supports important commercial fisheries in many countries of
South America. Although displayed in some aquaria, not very
popular as an aquarium species. ◆

River hatchetfish
Gasteropelecus sternicla

FAMILY
Gasteropelecidae

TAXONOMY
Gasteropelecus sternicla Linnaeus, 1758, Suriname.

OTHER COMMON NAMES
English: Common hatchetfish, silver hatchetfish; French: Poisson hachette argenté; German: Silberbeilbauchfisch; Spanish:
Pechito, pechito plateado de raya negra; Portuguese: Borboleta,
sapopema, voador.

PHYSICAL CHARACTERISTICS
Standard length 1.5 in (3.8 cm), but adults in aquaria may
reach 2.4 in (6.5 cm). Females are larger than males. Deep,
highly compressed body; belly profile creates a semicircular
arc containing sternum and strong chest muscles. Pectoral fins
are long and located high on the body near the head. Yellow
to silver in color, with dark stripe running along the length
of the body; fins are transparent.

DISTRIBUTION
South America in the upper Amazon basin, the Guyanas, and
Venezuela.

HABITAT
Pelagic; lives near surface of slow waters in creeks and
swamps. Often inhabits vegetated areas.

BEHAVIOR
Typically gregarious; lives in groups near surface of water.
May be aggressive or calm. To avoid predators and capture
insect prey, swims very fast to raise its body out of the water,
then flies above the surface using its long pectoral fins.
Unlike other "flying fish" that rely on gliding, it rapidly beats
its pectoral fins to remain airborne for distances up to 10
feet (3 m).

FEEDING ECOLOGY AND DIET
Feeds on worms, crustaceans, and insects from the surface of
the water, but also captures aerial insects. Predators are not
specified in current literature, although likely eaten by a variety
of larger fishes.

REPRODUCTIVE BIOLOGY
Spawns after a lengthy courtship. The female scatters eggs
in the water or onto floating plants, where they are fertilized by males. The eggs then fall to the bottom or onto
vegetation.

CONSERVATION STATUS
Not listed by the IUCN.

SIGNIFICANCE TO HUMANS
Common aquarium species. ◆

▢ *Copella arnoldi*

▨ *Gasteropelecus sternicla*

Splash tetra
Copella arnoldi

FAMILY
Lebiasinidae

TAXONOMY
Copella arnoldi Regan, 1912, Amazon River.

OTHER COMMON NAMES
English: Copeina, jumping characin, splashing tetra, spotted
characin, spraying characin; French/Creole: Ti-yaya; German:
Spritzsalmler; Finnish: Roiskuttajatetra; Polish: Smuklen
pryskacz; Portuguese: Piratanta.

PHYSICAL CHARACTERISTICS
Small species, reaches a length of only 3 in (8 cm). The slender
body is yellowish brown on the back, lighter on the sides, and
almost white on the belly. The reddish, long, fanlike fins are
particularly pronounced on males. There is a yellow spot at
the base of the dorsal fin. Large scales with dark edges cover
the body.

DISTRIBUTION
South America in the lower Amazon River basin and coastal
portions of the Guyanas.

HABITAT
Swampy, slow-moving waters that often contain little oxygen.

BEHAVIOR
Remains near the surface of the water, likely a behavioral adaptation to life in stagnant, low-oxygen waters.

FEEDING ECOLOGY AND DIET
Feeds on a wide variety of insects and plant matter. Because of its small size and surface-living habits, it is likely preyed on by numerous larger fishes and fish-eating birds.

REPRODUCTIVE BIOLOGY
Male and female leap out of the water simultaneously and deposit a few eggs in a gelatinous mass on the underside of overhanging vegetation. They repeat this process until 60 or more eggs have been deposited. The male then uses its tail to splash water onto the eggs every 20–30 minutes until they hatch in two to three days. After hatching, the fry fall into the water.

CONSERVATION STATUS
Not listed by the IUCN.

SIGNIFICANCE TO HUMANS
Popular as an aquarium fish throughout the world. ◆

Resources

Books

Berra, T. M. *Freshwater Fish Distribution.* San Diego: Academic Press, 2001.

Gery, J. *Characoids of the World.* Neptune City, NJ: Tropical Fish Hobbyist Publications, Inc., 1977.

Periodicals

Araujo-Lima, C. A. R. M., and E. C. Oliveira. "Transport of Larval Fish in the Amazon." *Journal of Fish Biology* 53, Supplement A (1998): 297–306.

Loubens, G. and J. Panfili. "Biologie de *Piaractus brachypomus* (Teleostei: Serrasalmidae) dans le bassin du Mamoré (Amazonie bolivienne)." *Ichthyological Explorations of Freshwaters* I 12, no. 1 (2001): 51–64.

Orti, Guillermo, and Axel Meyer. "The Radiation of Characiform Fishes and the Limits of Resolution of Mitochondrial Ribosomal DNA Sequences." *Systematic Biology* 46, no. 1 (March 1997): 75–100.

Ribeiro, M. C. L. B., and M. Petrere, Jr. "Fisheries Ecology and Management of the Jaraqui (*Semaprochilodus taeniurus, S. insignis*) in Central Amazonia." *Regulated Rivers Research and Management* 5, no. 3 (1990): 195–215.

Roberts, T. R. "Osteology and Relationships of the Prochilodontidae, a South American Family of Characoid Fishes." *Bulletin of the Museum of Comparative Zoology* 145, no. 4 (1973): 213–235.

Ruffino, M. L., and V. J. Isaac. "Life Cycle and Biological Parameters of Several Brazilian Amazon Fish Species." *NAGA, The ICLARM Quarterly* 18, no. 4 (October 1995): 41–45.

Schrieber, R. "The African Darter Tetra, *Nannocharax fasciatus.*" *Tropical Fish Hobbyist* 41, no. 4 (1992): 132–135.

Other

FishBase [cited January 22, 2003]. <http://www.fishbase.org/search.cfm>

Ortí, Guillermo, and Richard P. Vari. "Characiformes." [cited January 22, 2003]. <http://www.museum.unl.edu/research/systematics/Orti/>

Katherine E. Mills, MS
Elizabeth Mills, MS

Siluriformes

(Catfishes)

Class Actinopterygii

Order Siluriformes

Number of families 34

Photo: Flathead catfish (*Pylodictis olivaris*) genus fossils have been found dating back 15 million years. (Photo by Gary Retherford/Photo Researchers, Inc. Reproduced by permission.)

Evolution and systematics

The late nineteenth century work of Carl H. Eigenmann and Rosa Smith Eigenmann on South American catfishes established the foundation upon which later researchers, such as William A. Gosline, George S. Myers, and Mario C. C. de Pinna, have built to clarify the classification of catfishes. However, the classification is still uncertain, and there is no full agreement among the authors about the relationships of the families. Most knowledge about the phylogeny of siluriforms comes from the research on Neotropical groups carried out in the last three decades of the twentieth century.

During the International Symposium on Phylogeny and Classification of Neotropical Fishes (Porto Alegre, Brazil, July 1997), de Pinna presented a talk about the interrelationships of Neotropical catfishes. He proposed a preliminary cladistic hypothesis that accepted 12 monophyletic groups of siluriforms. In at least three cases, the relationships between New World and Old World clades are well established: the South American diplomystids, the sister group to all other catfishes; the African mochokids, the sister group to the South American doradoid superfamily, which includes the doradids and the auchenipterids; and the South American aspredinids, the sister group to the Asian erethistids. Nevertheless, de Pinna's cladogram implies the existence of four more transcontinental phylogenetic relationships. The discovery of close interrelationships between systematic groups otherwise isolated in a continental block is relevant, due to the fact that most groups of ostariophysean fishes, such as siluriforms, primarily inhabit freshwaters.

Catfish fossils include *Andinichthys, Incaichthys,* and *Hoffstetterichthys* (family Andinichthyidae) from the Upper Cretaceous and early late Paleocene of Bolivia; *Corydoras revelatus* (family Callichthyidae) from the late Paleocene of Argentina;

Hypsidoris farsonensis, the only known species of the family Hypsidoridae, from the early middle Eocene of Wyoming, United States; late Eocene or early Oligocene (about 37 million years ago) material from Antarctica; *Hoplosternum* sp. (family Callichthyidae) from the middle Miocene of Colombia; and some material of the arioid (from the Arioida group) material from coastal marine Eocene deposits from Camden (Arkansas, United States).

It has been shown, based on cladistic grounds, that the gymnotiforms, or electric knifefishes, are the group most closely related to the siluriforms. Some authors have chosen to include both orders under the Siluriformes, a lead not followed here.

Physical characteristics

Catfishes are well known and recognized worldwide by a few characteristics. They normally show one to four pairs of barbels on the head around the mouth: one nasal, one maxillary, and two on the lower jaw or mandible. The nasal and/or lower jaw barbels may be absent, but in any case, the maxillary barbels are the longest. As these threadlike structures have plenty of taste buds, they have chemosensory and tactile functions, allowing the small-eyed catfishes to gather food efficiently.

Catfishes usually have spiny rays at the front of the dorsal and pectoral fins. These thornlike structures are large, unsegmented, and strong elements, technically similar to the other lepidotrichia, the rays that support the fins. Most families of siluriform have two dorsal fin spines; the first is tiny, but very important, because it locks the second, larger spine in an erect position. A catfish with erected, pointed spines is nicely protected from most potential predators. The spines in some families also deliver a toxin produced in an associated venom

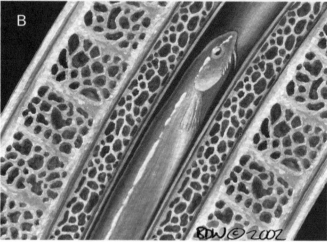

The parasitic catfish of the Amazon and Oranoco Rivers, known as the candiru (*Vandellia cirrhosa* and related species) usually feeds on blood from other fishes' gills (A). It detects the outward flow of nitrogen-rich water from the gills, but may confuse that with the outward flow of urine from a person urinating underwater. In this case, it has been known to swim into the urethra (B), where it feeds on blood, causing excrutiating pain. But the tight space and backward-pointing spines on the sides of its head make it impossible for the fish to turn around or back out, and it eventually dies inside the person. If not removed, it causes restriction or complete blockage of urination, massive infection, and often death. Native South Americans who live in its range typically wear protective garments when bathing or swimming in rivers. (Illustration by Bruce Worden)

gland. Although formidable weapons, spines are lacking in some catfish groups.

Most catfish species have a "naked" body because their leatherlike skin is completely scaleless. However, the so-called armored catfishes are covered at least partially by deeply embedded bony plates or by tubercles. Other external characteristics of catfishes are an adipose fin, usually present, sometimes rayed or preceded by a spine, and a small and toothless maxilla.

There are several internal characteristics worth mentioning, such as reductions, fusions, or absences in the set of skull

bones typical of lower teleosts, as well as of intermuscular bones. The Weberian apparatus of catfishes involves five vertebrae, and their caudal skeleton is variable, ranging from six independent plates to the complete fusion of all its elements.

Distribution

Catfishes are mainly inhabitants of freshwaters, reaching their highest diversity in South America. There are more than 1,650 catfish species in South America distributed in eight monophyletic and endemic groups, including the richest family, the suckermouth armored catfishes (Loricariidae), with around 800 species. Loricariids and five other families form the armored catfishes, the largest monophyletic group of the order, with at least 1,150 species and a range between Costa Rica and Argentina. Included in this clade is the most speciose catfish genus, the plated catfishes *Corydoras* (Callichthyidae) with about 142 species. Other families of armored catfishes are the monotypic Nematogenyidae, the pencil or parasitic catfishes (Trichomycteridae), the spiny dwarf catfishes (Scoloplacidae), and the climbing, or South American hillstream, catfishes (Astroblepidae). The long-whiskered or antenna catfishes (Pimelodidae) used to be the second largest family, with about 300 species. However, since it is not a natural group, it was officially split in 2001 into three monophyletic families: the pimelodids (in the narrow sense), the heptapterids (Heptapteridae), and the pseudopimelodids (Pseudopimelodidae). In any case, Heptapteridae, the most species-rich group of the "old" long-whiskered catfishes with about 150 species, is widely distributed from Southern Mexico to Argentina. The remaining monophyletic groups of South American catfishes are the velvet or Patagonian catfishes (Diplomystidae), the whalelike catfishes (Cetopsidae), the doradoids (including the thorny or talking catfishes, Doradidae, and the driftwood catfishes, Auchenipteridae), and the banjo catfishes (Aspredinidae). The bullhead or North American freshwater catfishes (Ictaluridae), with 44 species, is officially the only family of the order endemic to North America, ranging as far south as Guatemala. The so-called Chiapas catfish was presented to the scientific community at the 2000 meeting of the American Society of Ichthyologists and Herpetologists; this unusual catfish seems to deserve its own family. So far, it is only known from southern Mexico.

Old World catfishes are also highly diverse; more than 1,000 species are known from Africa and Eurasia. The bagrid catfishes (Bagridae), the airbreathing or labyrinth catfishes (Clariidae), the glass catfishes (Schilbeidae), and the eeltail or tandan catfishes (Plotosidae) are the four families occurring in both Africa and Asia. Bagridae is the richest family of these widely distributed catfish clades, with about 130 species. The other three families include about 170 species. There are six catfish families endemic to Africa. Of them, the squeakers or upside-down catfishes (Mochokidae) is the richest, since it contains about180 species; two-thirds of which are included in the genus *Synodontis*. The loach or African hillstream catfishes (Amphiliidae) has about 60 species. Other endemic African families are the claroteids (Claroteidae), the auchenoglanidids (Auchenoglanididae), the electric catfishes (Malapteruridae), and the austroglanidids (Austroglanidae); together these four groups contain more than 100 species.

Asia is next to South America in the number of endemic catfish families. However, these tend to be poorly speciated groups, usually with fewer than 30 species each. The sucker or Asian hillstream catfishes (Sisoridae) is the richest family, with a few more than 110 species. Other endemic Asian families are the stream or shortfin catfishes (Akysidae) with 27 species, the shark catfishes (Pangasiidae) with 26 species, the torrent or Asian loach catfishes (Amblycipitidae) with 25 species, the erethistids (Erethistidae) with 13 species, and the armorhead or Chinese catfishes (Claroglanididae) and the squarehead or frogmouth catfishes (Chacidae), both with 3 species. The sheatfishes (Siluridae) is the only family of catfishes extending its range to Europe; it includes about 100 species, only 2 of which occur in Europe.

Finally, two families have widely invaded the marine realm; interestingly, both groups include the only catfishes known from Australia. The eeltail catfishes, a rather small family of 27 species, is known from the Indian and the western Pacific Oceans and extends to Japan and Fiji. The sea catfishes (Ariidae) includes about 200 species in all the tropical continental shelves.

Habitat

Catfishes occupy practically all freshwater environments, where they are often the dominant group of fishes. Catfishes are mostly benthic and freshwater inhabitants, and they can be ubiquitous in the rivers of tropical continents such as South America and Asia, invading habitats such as riffles and waterfalls, as well as marginal, almost stagnant ponds. Some groups occupy the water column, swimming above the bottom. In those hovering catfishes some bizarre adaptations are noteworthy. Some sheatfishes and glass catfishes, for example, hover tail down, and both groups are schooling, transparent, with long barbels (one or four pairs), a long anal fin, a minute or absent adipose fin, a forked tail, and a dorsal fin completely lacking or represented by a sole reduced ray. Other catfishes, such as some banjo catfishes and most eeltail and sea catfishes, are estuarine and may venture even to deep-bottom continental-shelf environments or to offshore islands.

An African polka-dot catfish (*Synodontis angelicus*) from the Zaire River in the Congo. (Photo by Mark Smith/Photo Researchers, Inc. Reproduced by permission.)

Behavior

Catfishes are generally bottom-dwelling, nocturnal, and solitary inhabitants of freshwater environments. However, an important minority are relatively specialized. For example, some sheatfishes, glass catfishes, and long-whiskered catfishes live far away from the bottom, are diurnally active, and may form schools. Eeltail catfishes and some thorny catfishes are well known for forming relatively large schools close to the bottom. Many South American species of armored catfishes are benthic and diurnal.

Feeding ecology and diet

Catfishes exhibit a wide array of feeding strategies. Most species live intimately linked to the bottom and feed mainly on invertebrates. Fish eaters are also abundant. For example, the squarehead, angler, and frogmouth catfishes (Chacidae) belong to a small group of Asian species that may use their maxillary barbels to attract smaller fishes to their enormous mouths. On the other hand, the Neotropical suckermouth armored catfishes (Loricariidae) are basically plant eaters, taking aquatic and terrestrial materials, such as algae and fallen leaves and trees. Catfishes, mainly those included in the pencil catfish family (Trichomycteridae), also engage in parasitic activities. Parasitic catfishes may be lepidophagous, targeting mucus, scales, and associated tissues (stegophyline trichomycterids), or hematophagous, swallowing blood (vandelliine trichomycterids). Some lepidophagous parasitic catfishes eat large pieces of flesh and enter the body cavities of their host. These species attack and devour commercially important catfishes trapped in nets or hooks and are therefore considered pests.

Catfishes are under the predatory pressure of almost all carnivores sharing their habitat. Even large predatory species are eaten as eggs, larvae, and juveniles. Their main fish predators are gars, bony tongues, trahiras, Nile perch, cichlids, and of course, other catfishes. Non-fish predators include crocodiles and caimans, and freshwater and coastal dolphins.

A bulldog pleco (*Chaetostoma sp.*) from the Amazon River in Brazil. (Photo by Mark Smith/Photo Researchers, Inc. Reproduced by permission.)

Schouteden's African squeaker catfish (*Synodontis schoutedeni*) from the Congo, West Africa. (Photo by Mark Smith/Photo Researchers, Inc. Reproduced by permission.)

Reproductive biology

The usual process of catfish reproduction involves adult specimens with reduced or absent external sexual dimorphism and some courtship activity before the demersal spawning, followed by moderate engagement in parental care of the rather small eggs. Most guarding of eggs and fry is carried out by males. Nevertheless, females of some relatively large banjo catfish species, such as *Aspredinichthys tibicen* and *Aspredo aspredo*, manage to adhere the eggs after fertilization to spongy tentacles or cuplike depressions present seasonally on their bellies. It has been said that the mother nurtures the litter. Another highly derived system, brood parasitism, has been developed by a squeaker species, *Synodontis multipunctatus*, which could be referred to as an aquatic "cuckoo." This Lake Tanganyika mochokid shares its habitat with a mouth brooding species of Cichlidae. The catfish synchronizes its spawning activity to that of the cichlid, allowing the perciform female to ingest its eggs for mouth brooding. The process does not end when the squeaker eggs hatch. Since the squeaker eggs hatch before the cichlid eggs, the catfish fry feed upon the host's fry inside her mouth!

In many cases parental care by the male is highly significant. Perhaps the best example of this is the mouth-breeding sea catfish. Sea catfishes are relatively dimorphic: females develop pads or claspers on the pelvic fins and males shed the teeth on the roof of their mouth. The process by which the female transfers the fertilized eggs to the male's mouth is not clear. In any case, he fasts for as long as two months until the fry are released.

Conservation status

The IUCN Red List includes 66 siluriform species. One, the bagre graso (*Rhizosomichthys totae*), a pencil catfish from the Tota Lake, 9,845 ft (3,000 m) high in the Colombian Andes, is classified as Extinct. In addition, 8 species are listed as Critically Endangered; 7 as Endangered; 22 as Vulnerable; 4 as Lower Risk/Near Threatened; and 24 as Data Deficient.

Significance to humans

Catfishes are considered one of the more important orders of vertebrates, not only because more than one-tenth of living fishes are siluriforms, but because their relationship with humans covers many aspects of life and culture. Several hundred species are now used as food, and other species will be looked upon as a source of protein in the future. Some species are greatly appreciated as game and commercial fishes and reach high market prices, but some are simply the difference between starvation and survival for millions of humans. In addition, most catfishes can be used as aquarium fishes, a business that is not only worth millions of dollars, but gives urban populations the opportunity to get in touch with a significant part of nature.

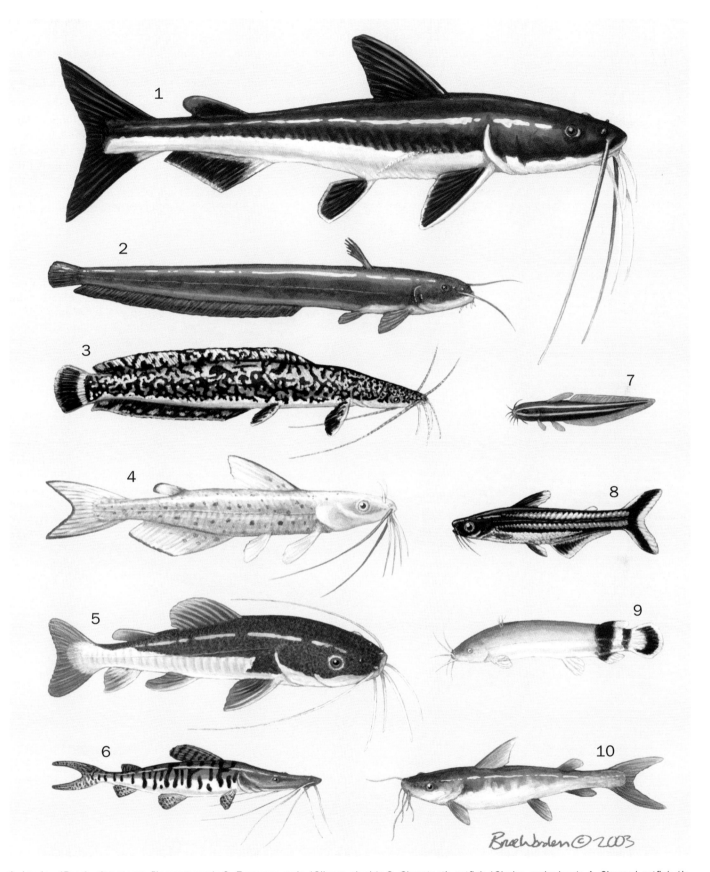

1. Lau-lau (*Brachyplatystoma filamentosum*); 2. European wels (*Silurus glanis*); 3. Sharptooth catfish (*Clarias gariepinus*); 4. Channel catfish (*Ictalurus punctatus*); 5. Redtail catfish (*Phractocephalus hemioliopterus*); 6. Tiger shovelnose catfish (*Pseudoplatystoma fasciatum*); 7. Coral catfish (*Plotosus lineatus*); 8. Iridescent shark-catfish (*Pangasius hypophthalmus*); 9. Electric catfish (*Malapterurus electricus*); 10. New Granada sea catfish (*Ariopsis bonillai*). (Illustration by Bruce Worden)

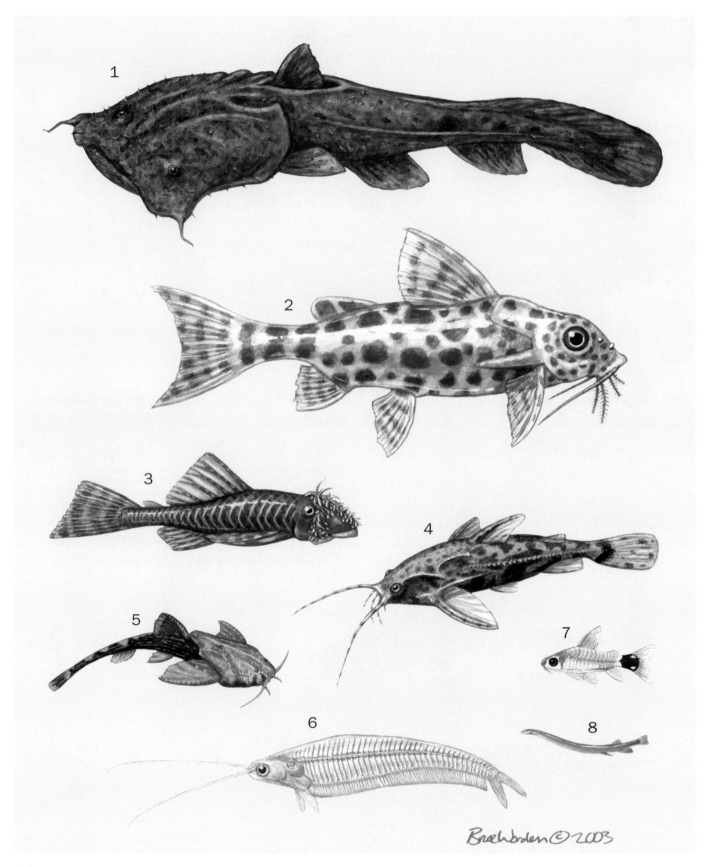

1. Squarehead catfish (*Chaca chaca*); 2. Blotched upsidedown catfish (*Synodontis nigriventris*); 3. Branched bristlenose catfish (*Ancistrus triradiatus*); 4. Blue-eye catfish (*Amblydoras hancockii*); 5. Guitarrita (*Dysichthys coracoideus*); 6. Glass catfish (*Kryptopterus bicirrhis*); 7. Dwarf corydoras (*Corydoras hastatus*); 8. Candiru (*Vandellia cirrhosa*). (Illustration by Bruce Worden)

Species accounts

New Granada sea catfish
Ariopsis bonillai

FAMILY
Ariidae

TAXONOMY
Galeichthys bonillai Miles, 1945, Magdalena River, Honda, Colombia.

OTHER COMMON NAMES
French: Mâchoiron requin; Spanish: Chivo cabezón, chivo cazón.

PHYSICAL CHARACTERISTICS
Length 31.5 in (80 cm). Body naked, elongate, and robust; dorsal fin with one spine and seven soft rays; adipose fin present; pectoral fins with one spine and 10 soft rays; pelvic fins with six soft rays, inner rays strongly modified to form a hook in mature females; caudal fin deeply forked; head covered by a very rugose bony shield, well visible beneath the skin and extended anteriorly to opposite eyes, the posterior portion of the shield extends backward to meet the predorsal plate, which is large and crescent-shaped; three pairs of barbels; posterior pair of nostrils partly covered by a flap of skin; eye large, with free orbital rims. Dark gray to bluish gray dorsally, bluish white ventrally.

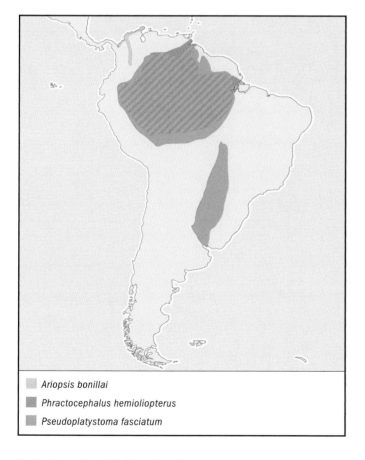

■ *Ariopsis bonillai*
■ *Phractocephalus hemioliopterus*
■ *Pseudoplatystoma fasciatum*

DISTRIBUTION
Southern Caribbean between Colombia and the Gulf of Venezuela.

HABITAT
Mainly inhabits coastal, brackish water, mangrove-lined lagoons, but has been collected in riverine environments as well as in shallow turbid marine waters.

BEHAVIOR
Benthic over muddy bottoms; solitary or forms schools.

FEEDING ECOLOGY AND DIET
Omnivorous; feeds mainly on benthic invertebrates, including crustaceans and polychaetes, but also consumes insects, small fishes, algae, and detritus.

REPRODUCTIVE BIOLOGY
As with all studied members of this family, engages in oral brooding; the male incubates the relatively large eggs in his mouth until hatching. Spawning occurs almost year round, with a peak in the Ciénaga Grande de Santa Marta between April and July. Fecundity is rather low, each female produces 24–39 eggs of about 0.5 in (1.3 cm). Sexual maturity is reached at sizes around 17 in (43 cm), but sexes can be told apart using external morphology at 7.5 in (19 cm). Males carry eggs and young for about two months.

CONSERVATION STATUS
Classified as Endangered by the IUCN. This status is due to its endemicity to the southern Caribbean, combined with enormous pressure from artisan fisheries and significant habitat alteration. As a consequence of overfishing, the medium size of capture (12 in/30.5 cm) is well below the size of sexual maturity.

SIGNIFICANCE TO HUMANS
Fished for with hook and line, cast nets, and beach seines. Widely used as food by fishing communities and low-income populations of cities and towns in the Colombian Caribbean. By 2002 attempts to develop a specific aquaculture procedure were being made. Young may be used also as an aquarium species. ◆

Guitarrita
Dysichthys coracoideus

FAMILY
Aspredinidae

TAXONOMY
Dysichthys coracoideus Cope, 1874, Nauta, Peru. Nomenclatural situation is confused due to lack of clarity of the type species of genus *Bunocephalus*. Sometimes divided in two subspecies: *D. coracoideus coracoideus* for the Amazonas, and *D. coracoideus amaurus* for the Guyanas; the last subspecies may be treated as a separate species (*D. amaurus*) under the common name camouflaged catfish.

Vandellia cirrhosa

Dysichthys coracoideus

OTHER COMMON NAMES
English: Guitarrero, little guitar; French: Poisson banjo; Spanish: Catalina, guitarrita; Portuguese: Rabeca; Guyanas indigenous languages: Grongron, kronkron.

PHYSICAL CHARACTERISTICS
Length 4.3 in (11.0 cm). Body naked but covered by papillae or bumps; dorsal fin spineless, with five rays; adipose fin absent; anal fin short, with seven rays; pectoral fins with a strong serrated spine; caudal peduncle long and slender; caudal fin rounded; head and anterior part of body depressed; head covered by thickened bones; mouth small and anterior; three pairs of relatively short barbels; eyes small and on top of head; gill opening a short slit; anterior and posterior borders of pectoral fin spines with strong and recurved teeth.

DISTRIBUTION
South America in the Amazon River basin and Guyanas.

HABITAT
Fresh waters in ponds and small forest streams rich in plant remains.

BEHAVIOR
Bottom dwelling; swims by wavelike motion of body and tail; ejection of water through the gill slits results in a jerky forward movement.

FEEDING ECOLOGY AND DIET
Omnivorous, even feeds on organic matter from the bottom.

REPRODUCTIVE BIOLOGY
Spawns in groups; female does not carry her eggs, but lays 4,000–5,000 eggs each time on sandy bottom nests made by the male, which guards the eggs and fry.

CONSERVATION STATUS
Not listed by the IUCN.

SIGNIFICANCE TO HUMANS
Too small to be of interest as a food fish. However, it is an important aquarium fish which has been spawned in captivity. ◆

Dwarf corydoras
Corydoras hastatus

FAMILY
Callichthyidae

TAXONOMY
Corydoras hastatus Eigenmann and Eigenmann, 1888, Villa Bella (=Parintins), Amazonas, Brazil.

OTHER COMMON NAMES
English: Pygmy corydoras; French: Corydoras nain; Spanish: Coridoras enano, corredora.

PHYSICAL CHARACTERISTICS
Length 1.4 in (3.5 cm). Body compressed, with two rows of overlapping bony plates on each side, the nuchal scutes not meeting dorsally; dorsal fin with a strong spine and seven soft rays; spine at anterior border of adipose fin; pectoral fins with a strong spine and eight rays; caudal fin forked; head compressed; mouth small and ventral; two pairs of well-developed barbels.

DISTRIBUTION
South America in Amazon and Paraguay River basins.

Corydoras hastatus

Ancistrus triradiatus

HABITAT
Fresh waters in ponds.

BEHAVIOR
Forms small schools in midwater among aquatic vegetation. It rests on any kind of plant leaves in normal position or with the ventral surface up against the bottom of a leaf, high above the bottom. Normally swims by rapidly moving the pectoral fins, which, combined with a rapid breathing rate, gives it a rather nervous appearance.

FEEDING ECOLOGY AND DIET
Feeds among plants, as well as from the bottom, on small invertebrates and on detritus.

REPRODUCTIVE BIOLOGY
Sexually dimorphic; males are more elongate and their dorsal fins are more pointed. Near the time of spawning, both sexes clean their surroundings, mainly plant leaves. Courtship consists first of dashing about back and forth and up and down. Then the so-called "T" position is assumed close to the bottom; the female positions herself at the side of the male, close to his vent, pushing him with her head while constantly moving her barbels. After some "trembling" by both sexes, the female lays an egg. After fertilization the egg is carried in the female's ventral pouch and deposited over one of the leaves cleaned before. The female continues to spawn single eggs for one to two hours over three or four consecutive days. At this rate, a total of 30–60 eggs are spawned by each female. Since group spawning is normal, 300 eggs or more are frequently deposited. Fry 0.20 in (0.5 cm) long hatch usually after three to four days, and mature in about 200 days. The parents can spawn again after two weeks.

CONSERVATION STATUS
Not listed by the IUCN.

SIGNIFICANCE TO HUMANS
Commercially important for the aquarium trade. ◆

Squarehead catfish
Chaca chaca

FAMILY
Chacidae

TAXONOMY
Platystacus chaca Hamilton, 1822, northeastern Bengal.

OTHER COMMON NAMES
English: Angler catfish, chaca, chega, Indian frog-mouth catfish; Spanish: Bagre cabezicuadrado.

PHYSICAL CHARACTERISTICS
Length 9.1 in (23 cm). Body compressed posteriorly; with many fine granulations; dorsal fin with a short spine and four soft rays; adipose fin a low ridge confluent with caudal fin; pectoral fin with one serrated spine and five soft rays; pelvic fins large, with six rays; a row of cirri above the lateral line; head broad and depressed, almost square, with a deep longitudinal groove located dorsally; mouth terminal, very wide, provided with appendages resembling barbels; three pairs of small barbels; eyes very small.

DISTRIBUTION
Asia in India, Bangladesh, and Nepal.

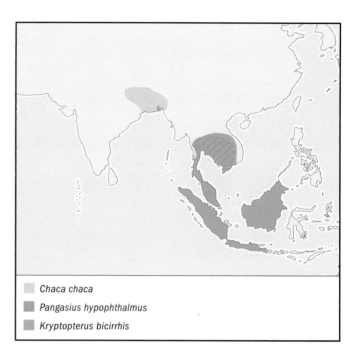

Chaca chaca
Pangasius hypophthalmus
Kryptopterus bicirrhis

HABITAT
Freshwater; bottom dweller in rivers, canals, ponds, and flood plains.

BEHAVIOR
Prefers soft bottoms, where it hides lying still on the river bed. Depends completely on this concealment for protection, remaining motionless even after being touched.

FEEDING ECOLOGY AND DIET
Sometimes moves its tiny maxillary barbels in a jerky motion to lure small fishes, including gouramies and cyprinids, near its large mouth; may take fishes as large as one-half its length.

REPRODUCTIVE BIOLOGY
Nothing known.

CONSERVATION STATUS
Not listed by the IUCN.

SIGNIFICANCE TO HUMANS
Commonly fished but not eaten probably due to its odd appearance. Sometimes used as an aquarium fish. The dorsal spine can inflict painful wounds if the fish is stepped on. ◆

Sharptooth catfish
Clarias gariepinus

FAMILY
Clariidae

TAXONOMY
Silurus (Heterobranchus) gariepinus Burchell, 1822, Smidtsdrift, Vaal River, Cape Province, South Africa.

OTHER COMMON NAMES
English: Mubondo, North African catfish, Zambezi barbel; Spanish: Bagre dientón.

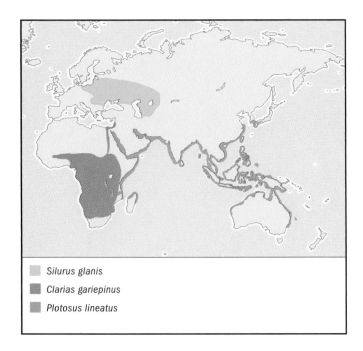

Silurus glanis
Clarias gariepinus
Plotosus lineatus

PHYSICAL CHARACTERISTICS
Length 5.6 ft (1.7 m); weight 132 lb (60 kg). Body naked and
elongate; dorsal fin base very long, with 61–80 rays, not preceded
by a spine, not continuous with caudal fin; adipose fin absent;
pectoral fin spine serrated only on its external border; anal fin
long, separated from caudal fin; caudal fin rounded; head depressed, covered with rugose bony plates; mouth terminal and
transverse; four pairs of barbels; eyes superior, relatively small,
and with a free orbital rim; gill openings wide; air-breathing
labyrinthic organ arising from second, third, and fourth gill
arches. Coloration countershaded, dark gray dorsally, milky
white ventrally, ventral surface of head in adults with a black
longitudinal band in each side; band absent in juveniles.

DISTRIBUTION
Africa in Niger and Nile River basins; also the Limpopo, Or-
ange-Vaal, Okavango, and Cunene Rivers, South Africa, and in
the Middle East, including Israel, Jordan, Lebanon, and Syria.

HABITAT
Benthopelagic in fresh waters and upper estuaries 13–262 ft
(4–80 m) deep, but more common in shallow environments.

BEHAVIOR
Widely resistant to challenging environmental conditions; the
accessory labyrinthic organ allows clariid catfishes to breathe
air under dry conditions. Favors shallow marginal areas, but
may appear in the open; nocturnal. Capable of walking
through dry land using tough pectoral fin spines. May also discharge electricity during intraspecific agonistic behavior.

FEEDING ECOLOGY AND DIET
Omnivorous bottom feeder, feeds on insects, fishes, crustaceans, mollusks, plankton, fruits, plants, small birds, and carrion. Occasionally feeds at the surface.

REPRODUCTIVE BIOLOGY
Migrates upstream or to lakeshores in large numbers to spawn
immediately after the first heavy showers of the rainy season.
Sexually dimorphic, both males and females have elongated
sexual organs, but male organ has cone-shaped tip. Spawning is

nocturnal in shallow waters; eggs, which have an adhesive disk
and are less than 0.10 in (0.25 cm) long, are not protected.
Eggs hatch in one or two days. Juveniles stay in shallow, protected waters for about six months, migrating downstream before their nursery area dries up.

CONSERVATION STATUS
Not listed by the IUCN.

SIGNIFICANCE TO HUMANS
Although of relatively minor importance for fisheries, this
is a valuable aquaculture species, due to its hardiness, rapid
growth, ease of feeding and handling, and flesh quality. Also
considered a game fish. Widely introduced to almost all Africa,
and to some European, Asian, and South American countries;
negative ecological impact has been reported. Trade restricted
in Germany. ◆

Blue-eye catfish
Amblydoras hancockii

FAMILY
Doradidae

TAXONOMY
Doras hancockii Valenciennes, 1840, Demerara, Guyana.

OTHER COMMON NAMES
English: False talking catfish, Hancock's amblydoras, purring
catfish; Spanish: Bagre ojiazul; Portuguese: Quiri-quiri.

PHYSICAL CHARACTERISTICS
Length 5.9 in (15 cm). Body somewhat elongated, with a row
of 24–28 lateral bony plates, each provided with a posteriorly
directed hook; anterior granulations of body confluent into a
large buckler; dorsal fin spine grooved, but lacking teeth, six
dorsal soft rays; interdorsal plate absent; adipose fin present;
pectoral fins' spines serrated on both edges, capable, as the
dorsal spine, of being erected and locked; caudal fin truncated;
head large and depressed; mouth terminal; three pairs of unbranched barbels; eye located in the middle of head. Dark
brown to violet dorsally, white ventrally, a white strip along
the body; eyes bright blue.

DISTRIBUTION
South America in Guyana and eastern Brazil to Peru and Bolivia.

HABITAT
Freshwater in coastal swamps, flood plains, ponds, and creeks.

BEHAVIOR
Lives in schools, sometimes by hundreds, hiding during the
day in places with the bottom covered by plant material,
mainly fallen trees and leaves. Resists desiccation by crawling
to a suitable body of water. Can produce sounds by different
combinations of pectoral spines, swim bladder, and the sophis-
ticated internal elastic spring apparatus.

FEEDING ECOLOGY AND DIET
Nocturnal plant and detritus eater in the wild, but reported to
eat animal material, such as worms, in captivity.

REPRODUCTIVE BIOLOGY
Spawns in the rainy season. The male builds a nest of bubbles,
leaves, and other plant material at the surface, where the female lays the eggs in a flattened cluster. The male guards the
nest until egg hatching.

CONSERVATION STATUS
Not listed by the IUCN.

SIGNIFICANCE TO HUMANS
Not important to fisheries due to its small size. Commercially important as an aquarium fish. ◆

Channel catfish
Ictalurus punctatus

FAMILY
Ictaluridae

TAXONOMY
Silurus punctatus Rafinesque, 1818, Ohio River, United States.

OTHER COMMON NAMES
English: Blue channel catfish, eel catfish, fiddler, forked-tail cat, government cat, Mississippi cat, river cat, sand cat, sharpie, silver cat, speckled cat, spot, white cat, willow cat; Spanish: Bagre de canal.

PHYSICAL CHARACTERISTICS
Length 4.3 ft (1.32 m); weight 58 lb (26.3 kg). Body naked and elongate; dorsal fin with a spine and six or seven soft rays; adipose fin small, remote from caudal fin, which is deeply forked; pectoral fins with a spine with large posterior serrae; pelvic fins with eight soft rays; mouth subterminal to inferior; four pairs of barbels; eye large and oval. Young mottled, brownish above, whitish below; adults mainly deep slate brown.

DISTRIBUTION
North America, native to central drainages of the United States and Mexico into southern Canada east of the Rocky Mountains, and possibly parts of the Atlantic coast. Original distribution somewhat unclear due to extensive introductions to the Atlantic coast and elsewhere. Fossils 10–20 million years old have been reported.

HABITAT
Freshwater in clear, well-oxygenated streams, and medium to large rivers with swift currents over relatively hard bottoms, including sand, gravel, and rocks. Also in quiet waters of lakes, reservoirs, and ponds. May enter brackish waters. Taken as deep as 49 ft (15 m).

BEHAVIOR
Benthic. Capable of relatively long migrations. Travels upstream in the spring and downstream in the fall. Young school close to the bottom at light hours during their first year, scattering at night.

FEEDING ECOLOGY AND DIET
A nocturnal carnivore, feeds mainly on small fishes, crustaceans, insects, and mollusks. Reported to also eat filamentous green algae. Well known for taking almost any bait, including rotten items.

REPRODUCTIVE BIOLOGY
Spawning occurs during the day in nests guarded by the male, beginning early spring in the south but later in the northern parts of its range, usually when the water warms to 60–75°F (16–24°C). Eggs hatch in 6–10 days. Maturity is reached in 2–5 years at about 1 ft (0.3 m) long. Reported to live 16 years.

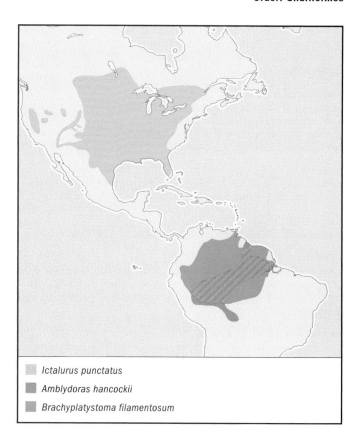

■ *Ictalurus punctatus*
■ *Amblydoras hancockii*
■ *Brachyplatystoma filamentosum*

CONSERVATION STATUS
Not threatened.

SIGNIFICANCE TO HUMANS
Highly important fishery, aquaculture, aquarium, and game species. It is a top sport fish, and is actively farmed. Used in "pay as you fish" ponds. Mississippi produces over 90% of the channel catfish farm-raised in the United States. In 1996, Mississippi ponds produced 334 million lb (about 152 million kg) valued at $265 million. Trade is restricted in Germany, where it is considered a potential pest. ◆

Branched bristlenose catfish
Ancistrus triradiatus

FAMILY
Loricariidae

TAXONOMY
Ancistrus triradiatus Eigenmann, 1918, Quebrada Gramalote, Barrigona, Villavicencio, Departamento del Meta, Colombia.

OTHER COMMON NAMES
Spanish: Cucha barbuda.

PHYSICAL CHARACTERISTICS
Length 4.7 in (12 cm). Body broad and depressed, covered with bony plates, 23–26 in the lower lateral series; belly naked; dorsal fin with one spine and eight rays; adipose fin present, not attached to adipose fin by a membrane; caudal fin truncate; anal fin with one spine and three to four rays; pectoral fin spine short, not reaching the origin of pelvic fin; anterior

portion of upper snout surface naked and with well-developed branching barbels in males, barbels small and simple in females; mouth ventral; teeth small, numerous, and bifid; opercular bones armed with about a dozen spines.

DISTRIBUTION
South America in middle and lower Orinoco River tributaries; Venezuelan and Colombian Caribbean coastal drainages, including the Magdalena River; Lake Maracaibo Basin, including the Catatumbo River.

HABITAT
Fresh waters. Benthic, in fast-running mountain streams with coarse pebble bottoms; lives as high up as 3,300 ft (1,000 m).

BEHAVIOR
Relatively shy fishes, usually live in caves or below rocks.

FEEDING ECOLOGY AND DIET
Omnivorous, feeds on filamentous algae and microorganisms growing on rocks, logs, and leaves. Also rasps rotten wood and ingests the cellulose. The lignin is reportedly important for their digestive processes.

REPRODUCTIVE BIOLOGY
The male selects an area such as under a log or a rocky cave and after some hours of courtship the female lays 50–150 yellowish to orange eggs, about 0.10 in (2.5 mm) long, which are guarded by the male and hatch in one week or less.

CONSERVATION STATUS
Not listed by the IUCN.

SIGNIFICANCE TO HUMANS
Unimportant for fisheries. Used as an aquarium fish, it has been spawned in captivity. ◆

Electric catfish
Malapterurus electricus

FAMILY
Malapteruridae

TAXONOMY
Silurus electricus Gmelin, 1789, Rosetta, Nile River.

OTHER COMMON NAMES
French: Silure électrique; Spanish: Bagre eléctrico.

PHYSICAL CHARACTERISTICS
Length 4 ft (1.22 m); weight 59.5 lb (27 kg). Body naked and sausage-shaped; a thick subdermal electrogenic organ derived from pectoral musculature extends the entire length and circumference of the fish; dorsal fin absent; adipose and anal fins far back; pectoral fin spine absent; swim bladder with an elongate posterior chamber; caudal fin rounded; mouth terminal; three pairs of long and fleshy barbels; eyes small and without a free border. Very large spots on the body and vertical bars on posterior part of the body and caudal fin.

DISTRIBUTION
Africa in Nile, Chad, and Niger basins.

HABITAT
Benthic in freshwater.

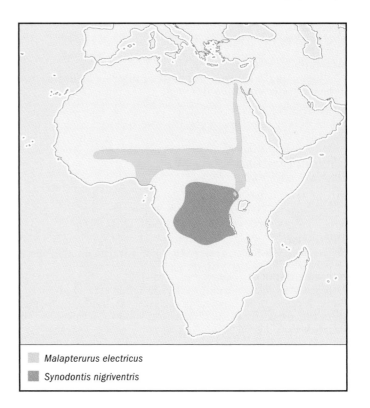

☐ *Malapterurus electricus*
■ *Synodontis nigriventris*

BEHAVIOR
Rests around rocks and roots in quiet waters during the day, becoming active at night. Intermittently produces electric discharges of basically two types: high frequency, used for self-defense and prey capture, and low frequency, associated with prey detection. Discharges can be 100–400 V, depending on catfish size. Generally, the initial shock, short and sharp, is followed by less intense secondary discharges.

FEEDING ECOLOGY AND DIET
A poor swimmer, feeds mainly on aquatic invertebrates and small fishes stunned by its electric discharges. Food items reported from stomachs are shrimps and fishes, such as cichlids, clupeids, schilbeids, bagrids, cyprinids, and characids.

REPRODUCTIVE BIOLOGY
Forms pairs during the rainy season; spawning occurs in excavated holes. Reported to live 10 years.

CONSERVATION STATUS
Not listed by the IUCN.

SIGNIFICANCE TO HUMANS
Used in subsistence fisheries and considered a game fish; also important for show aquariums. However, it is better known for its strong electric discharges, which are supposed to be capable of knocking down a person if the catfish is stepped on. Humans have known about these electric capabilities for over 5,000 years. ◆

Blotched upsidedown catfish

Synodontis nigriventris

FAMILY
Mochokidae

TAXONOMY
Synodontis nigriventris David, 1936, Buta, Congo Democratic Republic (=Zaire).

OTHER COMMON NAMES
Spanish: Bagre manchado al revés.

PHYSICAL CHARACTERISTICS
Length 3.8 in (9.6 cm). Body naked and relatively elongated; dorsal and pectoral fin spines strong and with a locking mechanism; adipose fin very large, not rayed; caudal fin forked; head bony, with a nape shield; mouth relatively small, lips not suckerlike; three pairs of barbels, mandibular barbels branched, eye with a free border. Color reverse countershaded, dorsal surface light, ventral surface dark.

DISTRIBUTION
Africa in Central Congo basin.

HABITAT
Benthopelagic in fresh water.

BEHAVIOR
Schools by the thousands. Large fishes swim upside down and may be able to escape predation by hiding with the ventral surface up under a log. Young swim in a conventional way. The pectoral fin spine produces a squeaking sound when it moves in its socket.

FEEDING ECOLOGY AND DIET
Nocturnal. Uses its inverted swimming position to feed on filamentous algae that grow on cave ceilings and on the undersides of leaves and other aquatic vegetation, and also to take terrestrial insects and other material from the water surface. Also feeds on bottom material, such as plant and fish remains, crustaceans, and insect larvae.

REPRODUCTIVE BIOLOGY
Females lay about 100 eggs under a rock or on the roof of a cave; eggs hatch in about seven days.

CONSERVATION STATUS
Not listed by the IUCN.

SIGNIFICANCE TO HUMANS
Unimportant for fisheries because of its small size. A widely used aquarium fish, has been spawned in captivity. ◆

Iridescent shark-catfish

Pangasius hypophthalmus

FAMILY
Pangasiidae

TAXONOMY
Helicophagus hypophthalmus Sauvage, 1878, Laos.

OTHER COMMON NAMES
English: Candy-striped catfish, sutchi catfish, sharkfin catfish, silver catfish, swai; Spanish: Bagre sutchi.

PHYSICAL CHARACTERISTICS
Length 4.3 ft (1.3 m); weight 34 lb (15.5 kg). Body naked, compressed; predorsal area relatively keeled; dorsal fin far forward, with a sharp serrated spine and six soft rays; adipose fin small; anal fin long, with about 30–40 rays; pectoral fins with a sharp spine; caudal fin forked; head covered by soft skin; mouth small, subterminal; two pairs of barbels (maxillary and mandibular) that shorten with age; eyes above and below the corner of the mouth with a free orbital rim. Young with two black stripes, one along the lateral line and another below it; adults mainly gray.

DISTRIBUTION
Southeast Asia (Indochina) in Mekong and Chao Phraya basins. Introduced into streams for aquaculture.

HABITAT
Freshwater in large rivers.

BEHAVIOR
Benthic. Undergoes long upstream migrations, both reproductive and trophic, in late fall and early winter, migrates downstream in late spring and summer. Reported to be always on the move when in confinement; capable of sound production.

FEEDING ECOLOGY AND DIET
Omnivorous, feeds on fishes, crustaceans, mollusks, and plant litter. Young tend to be carnivorous, eating more plants while growing.

REPRODUCTIVE BIOLOGY
Since eggs appear from March to August, it is believed that the downstream migration is not only trophic, but also reproductive.

CONSERVATION STATUS
Not listed by the IUCN.

SIGNIFICANCE TO HUMANS
Important commercially for fisheries, aquaculture, and the aquarium trade. ◆

Lau-lau

Brachyplatystoma filamentosum

FAMILY
Pimelodidae

TAXONOMY
Pimelodus filamentosus Lichtenstein, 1819, Brazil.

OTHER COMMON NAMES
French: Torch; Spanish: Lechero, plumita, valentón, zungaro salton; Portuguese: Filhote, piraíba, piratinga; Guyanas indigenous languages: Axakwan, lektaima, pilau.

PHYSICAL CHARACTERISTICS
Length 11.8 ft (3.6 m); weight 440 lb (200 kg). Body naked; dorsal and pectoral fins with one spine; adipose fin long based; caudal fin deeply forked, upper and lower fin rays very long; head broad; snout very depressed; mouth slightly ventral; maxillary barbels very long, one to two times the length of the body in juveniles, two-thirds the length of the body in adults; eyes small, dorsolaterally located, with a free orbital rim. Color dark gray dorsally, whitish ventrally.

DISTRIBUTION
South America in Amazon and Orinoco basins, also main rivers of the Guyanas and northeastern Brazil.

HABITAT
Channels of large rivers, including muddy, black-water and clear-water tributaries, and upper reaches of estuaries. Young specimens occur in flood plains and ponds as well as on the main river channels and even in brackish water.

BEHAVIOR
Soft-bottom dweller, but a good swimmer capable of long migrations.

FEEDING ECOLOGY AND DIET
Piscivorous, mainly consuming characids, catfishes, and knife-fishes it swallows whole. Monkeys have also been reported from the stomachs of large specimens.

REPRODUCTIVE BIOLOGY
Medium sizes at maturity in the lower Caquetá (=Japurá) river are 59 in (1.5 m) standard length for females, 51 in (1.3 m) for males. Estimations of fecundity are variable, figures range from 60,000 eggs for a 143 lb (65 kg) specimen to one million eggs for a 106 lb (48 kg) female. Spawning in Venezuela coincides with the rainy season; young specimens are found in July and August.

CONSERVATION STATUS
Not listed by the IUCN. However, it is considered endangered in Colombia due to overfishing, and most specimens fished at the turn of the twentieth century in that South American country were immature. Minimum legal capture size in Colombia is 40 in (1 m).

SIGNIFICANCE TO HUMANS
Highly important commercial fish captured using nets and longlines, flesh is considered of excellent quality. Also a valuable game fish. Older literature reports that lau-lau occasionally prey on humans. ◆

Redtail catfish
Phractocephalus hemioliopterus

FAMILY
Pimelodidae

TAXONOMY
Silurus hemioliopterus Bloch and Schneider, 1801, Maranham River, Brazil.

OTHER COMMON NAMES
Spanish: Bagre colirrojo, guacamayo, pez torre; Portuguese: Bigorilo, guacamayo, peixe-arara, pirarara.

PHYSICAL CHARACTERISTICS
Length 4.4 ft (1.34 m); weight at least to 176.2 lb (80 kg). Body stout and naked; dorsal fin with one spine and seven rays; adipose fin short and high; caudal fin broad and almost square; head as broad as high, flattened dorsally; maxillary barbels extending at least to dorsal fin level; eyes small, with free margins; nostrils widely separated; pectoral fins spine thick with sharp, recurved teeth posteriorly. Coloration countershaded, vertical fins with bright orange tips, white ventrally.

DISTRIBUTION
South America in Amazon and Orinoco River basins. Reported from Venezuelan fossils six million years old.

HABITAT
Freshwater.

BEHAVIOR
Bottom dwelling. In the upper Caquetá River (Colombia) its upstream migration occurs between March and June.

FEEDING ECOLOGY AND DIET
Omnivorous, feeds on animals (fishes and crabs) and plant material (fruits).

REPRODUCTIVE BIOLOGY
Nothing known.

CONSERVATION STATUS
Not listed by the IUCN.

SIGNIFICANCE TO HUMANS
Important for subsistence fisheries, also a game fish. Commercially important as an aquarium fish; introduced, but not established, in Florida. ◆

Tiger shovelnose catfish
Pseudoplatystoma fasciatum

FAMILY
Pimelodidae

TAXONOMY
Silurus fasciatus Linnaeus, 1766, Brazil, Suriname. The population restricted to the Magdalena-Cauca basin (Colombia), usually included under this name, seems to deserve status as a species.

OTHER COMMON NAMES
English: Tiger catfish, bared sorubim, tumare; French (Creole): Poson-tig, torch-tig; Spanish: Bagre pintadillo, bagre rayado, bagre tigre, zungaro doncella; Portuguese: Pintado, surubim, surubim-lenha.

PHYSICAL CHARACTERISTICS
Length 3.4 ft (1.05 m); weight 154.2 lb (70 kg). Body naked; dorsal and pectoral fin spines well developed and covered by thick skin; well-developed adipose fin; caudal fin forked; head large and depressed; upper jaw overhanging the lower jaw; maxillary barbels relatively short, not extending to anal fin; eyes relatively small and dorsally located; dorsal skull fontanel relatively short and shallow. Coloration dark gray dorsally, whitish ventrally, crossed by dark bands or loops.

DISTRIBUTION
South America in Amazon, Corintijns, Essequibo, Orinoco, and Paraná River basins.

HABITAT
Fresh waters in lakes, flooded plains, and forests, main river channels, and floating meadows. Absent from estuaries. Occurs at maximum depth of 16 ft (4.9 m).

BEHAVIOR
Bottom dweller. Strongly migratory. In the Amazon basin, undergoes two annual migrations: a feeding one in the dry

season and another at the beginning of the rainy season. In Venezuela, follows the migratory pattern of the bocachicos (Prochilodontidae) and other nomadic fishes.

FEEDING ECOLOGY AND DIET
Basically a nocturnal fish eater, consumes mainly characiform fishes, as well as other catfishes and cichlids, which it swallows whole. In the Apure River (Venezuela), 34 species of migratory fishes have been found in stomach contents. Spiders, crabs, and seeds are also taken.

REPRODUCTIVE BIOLOGY
Reproductive season in the Orinoco basin runs between March and June. Median sizes of gonad maturity in eastern Colombia estimated at 32.7 in (83 cm) standard length for females, and 23.6–41.3 in (60–105 cm) for males; relative fecundity was estimated as well as about 30,000 eggs per pound (about 60,000 per kg). Sizes at first sexual maturation reported as 22 in (56 cm) for females and 17.7 in (45 cm) for males.

CONSERVATION STATUS
Not listed by the IUCN. However, it is considered threatened in Colombia. The population from the Magdalena-Cauca basin is considered critically endangered. That country has tried to regulate the exploitation of the barred sorubim, as well as that of the northern Colombian population, establishing minimum sizes of capture and fishing seasons.

SIGNIFICANCE TO HUMANS
Enormously important for commercial fisheries in most countries where it occurs because of its excellent tasting yellowish flesh. Cultured in Colombia and Venezuela. Also a game fish and traded for show aquariums. ◆

Coral catfish
Plotosus lineatus

FAMILY
Plotosidae

TAXONOMY
Silurus lineatus Thunberg, 1787, Indian Ocean.

OTHER COMMON NAMES
English: Bumblebee catfish, striped eel-catfish; Spanish: Bagre-anguila rayado; Afrikaans: Streep-baberpaling.

PHYSICAL CHARACTERISTICS
Length 12.6 in (32 cm). Body naked, eel-like; dorsal fin short, with a sharp, highly venomous, spine armed with relatively small teeth on both borders, and four rays; no adipose fin; pectoral fins spines serrated and highly venomous; tail pointed, caudodorsal fin rays extending forward ahead of midbody, lower caudal rays joining the long anal fin, forming a continuous fin with 139–200 rays; lateral line fully developed; dendritic organs present behind vent; mouth transverse; four pairs of relatively short barbels; eye relatively small with free orbital margins. Young black with two or three white or yellow stripes, adults mainly gray brown.

DISTRIBUTION
Indian and western Pacific Oceans, from the Red Sea and East Africa, including Madagascar, to Samoa, southern Japan and Korea, Australia, Lord Howe, Palau and Yap Islands.

HABITAT
Marine. The only catfish observed in coral reefs, also in tide pools, brackish environments, and freshwater; bottom dweller to about 200 ft (61 m) deep.

BEHAVIOR
Solitary or forms tight shoals of hundreds in shallow water and tide pools, sometimes covering the bottom; schools may break into smaller pods that move coordinately. Schooling confuses and discourages predators.

FEEDING ECOLOGY AND DIET
Eats mainly invertebrates, such as crustaceans, mollusks, and worms, and fishes taken from the soft bottoms.

REPRODUCTIVE BIOLOGY
Oviparous; the eggs are spherical, a little larger than 0.10 in (0.25 cm), nonadhesive, demersal; larvae are planktonic. In Japan reproduction occurs in early summer, the male builds and guards a nest with debris and rocks where the female deposits her eggs; the eggs hatch in 10 days or less. Females reach maturity in about one year, at a length of 5.5 in (14 cm). Maximum reported age is seven years.

CONSERVATION STATUS
Not listed by the IUCN.

SIGNIFICANCE TO HUMANS
The venomous fin spines are very dangerous (but rarely fatal); therefore all users of coastal waters, including bathers, swimmers, divers, and fishermen should be very cautious. Pectoral fin spines receive venom from the axillary and pectoral spine glands. Nevertheless, these fishes are used as food, mainly in Africa, and as aquarium fishes; in Japan they are kept and spawned in public and private aquaria. ◆

Glass catfish
Kryptopterus bicirrhis

FAMILY
Siluridae

TAXONOMY
Silurus bicirrhis Valenciennes, 1840, Java, Indonesia.

OTHER COMMON NAMES
English: Phantom glass catfish; Spanish: Bagre vítreo.

PHYSICAL CHARACTERISTICS
Length 5.9 in (15 cm). Body naked, strongly compressed; dorsal profile arched with a nuchal concavity; dorsal fin rudimentary, represented by only one short ray; adipose fin absent; anal fin base very elongate, with 55–68 rays; pectoral fins longer than head; pelvic fins small; caudal fin forked; mouth very small, oblique; one pair of barbels on lower jaw, maxillary barbels elongate, reaching to the anal fin; eye large, subcutaneous. Body translucent, internal features such as blood vessels and backbone visible, gut covered by a silvery peritoneum; depending on light the body may reflect iridescent colors, ranging from yellow to violet, or even fluorescent blue in twilight.

DISTRIBUTION
Asia in Mekong, Chao Phraya and Xe Bangfai basins, Malay Peninsula, Java, Sumatra, and Borneo.

HABITAT
Freshwater. Benthopelagic in large rivers and flooded plains, favors turbid shady waters.

BEHAVIOR
Free-swimming, shoaling species. In captivity, the group stays in a dark corner during light hours, only disturbed by individuals changing places.

FEEDING ECOLOGY AND DIET
A diurnal carnivore, feeds on pelagic and planktonic invertebrates, such as hemipterans, other insects, worms, and crustaceans, and small fishes. In captivity only takes floating items, it never disturbs the bottom looking for food.

REPRODUCTIVE BIOLOGY
Has not been spawned in captivity. No information from the wild.

CONSERVATION STATUS
Not listed by the IUCN.

SIGNIFICANCE TO HUMANS
An important aquarium species. Also used to make the fish sauce *prahoc*. ◆

REPRODUCTIVE BIOLOGY
Spawning takes place in shallow lakes and on flooded areas during warm seasons, and even in saltwater in the Aral Sea. No nest building activity, but males do protect the adhesive eggs, 0.10 in (0.25 cm) each, which stick to aquatic vegetation or to large pieces of detritus, sometimes in numbers over 100,000. Hatching occurs in about 20 days. Reported to live 30 years.

CONSERVATION STATUS
Not listed by the IUCN; however, it appears in Appendix III of the Bern Convention (protected fauna). Considered a potential pest that produces a negative ecological impact when introduced, but it does not always become well established after transplantation.

SIGNIFICANCE TO HUMANS
Important commercial and sport fish throughout its range. Marketed fresh, canned, and frozen. An industry based on this species in eastern Europe and in countries of the former Soviet Union produces leather from the skin, glue from the gas bladder and bones, and cheap caviar from the eggs. ◆

European wels
Silurus glanis

FAMILY
Siluridae

TAXONOMY
Silurus glanis Linnaeus, 1758, Orient, European lakes.

OTHER COMMON NAMES
English: Catfish, Danube catfish, European catfish, sheatfish; German: Wels; French: Silure glane; Spanish: Bagre europeo.

PHYSICAL CHARACTERISTICS
Length 16.5 ft (5 m); weight 726 lb (330 kg). Body naked, robust, anteriorly compressed; dorsal fin with 1 spine and 4–5 rays; adipose fin absent; anal fin base very elongate, with 90–95 rays covered by integument for most of their length; pectoral fin spine stout; pectoral and pelvic fins short; caudal fin rounded; head broad and depressed; mouth terminal; maxillary barbels heavy, flattened.

DISTRIBUTION
Eurasia, including Baltic, Black, and Aral Seas estuaries.

HABITAT
Fresh and brackish waters, including quiet, shallow places such as marshes, lagoons, backwaters, and large lakes, and as deep as 98 ft (30 m) in large rivers and dams.

BEHAVIOR
Benthic, occurring mainly over mud and sandy bottoms, where it hides in holes and under logs or tree roots during day hours; active during the night. Considered nonmigratory; however, in the Baltic Sea it reportedly undergoes short spawning migrations upriver, returning to spend winter months in deeper waters close to the river mouth.

FEEDING ECOLOGY AND DIET
Nocturnal and highly carnivorous, feeds on almost any aquatic animal. Juveniles consume bottom-living invertebrates; adults ingest large crustaceans, fishes, amphibians, ducks and other birds, as well as water voles and other small mammals.

Candiru
Vandellia cirrhosa

FAMILY
Trichomycteridae

TAXONOMY
Vandellia cirrhosa Valenciennes, 1846, probably South America.

OTHER COMMON NAMES
Spanish: Candirú.

PHYSICAL CHARACTERISTICS
Length 1 in (2.5 cm). Body naked and very elongate; dorsal fin short, without a pungent spine, its origin posterior to pelvic fins; adipose fin absent; anal fin origin behind dorsal fin base, short; mouth narrow and suckerlike, inferiorly located; lower jaw toothless; mental barbels absent; eyes without free margins; opercular bones with spines. With a yellowish tinge or almost transparent.

DISTRIBUTION
South America in the Amazon River basin.

HABITAT
Freshwaters. Benthic, burrows in sandy bottoms.

BEHAVIOR
Parasitic, enters the gill cavity of larger fishes, frequently long-whiskered catfishes, to suck blood.

FEEDING ECOLOGY AND DIET
An obligate parasite that in captivity refuses any kind of food. However, it attacks a living fish by entering the gill chamber when inhaled water is expelled. Once in place, the candiru lodges itself using its opercular spines, bites off tips of host's gill filaments, and gorges with flowing blood. The candiru's body can distend considerably, and after few minutes of feeding it drops off to the bottom, where it burrows or just remains quiet.

REPRODUCTIVE BIOLOGY
Nothing known.

CONSERVATION STATUS
Not listed by the IUCN.

SIGNIFICANCE TO HUMANS
Known to enter the urogenital openings of bathers, usually if they happen to urinate under water. The candiru swims up the flow of urine, possibly mistaking it for the water flow from a gill cavity. After penetrating as far as possible, the fish locks its opercular spines into position. This has obviously serious consequences for both the fish and the person, since the candiru can only be removed by surgery. Humans in its range protect themselves by wearing tight clothing when swimming (and by refraining from urinating underwater). ◆

Resources

Books

Barthem, Ronaldo, and Michael Goulding. *The Catfish Connection*. New York: Columbia University Press, 1997.

Berra, Tim M. *Freshwater Fish Distribution*. San Diego: Academic Press, 2001.

Burgess, Warren E. *An Atlas of Freshwater and Marine Catfishes*. Neptune City, NJ: T.F.H, 1989.

Ferraris, Carl J., Jr. "Catfishes and Knifefishes." In *Encyclopedia of Fishes*, edited by John R. Paxton and William N. Eschmeyer. San Diego: Academic Press, 1995.

Galvis, Germán, José Iván Mojica, and Mauricio Camargo. *Peces del Catatumbo*. Bogotá: Asociación Cravo Norte, 1997.

Helfman, Gene S., Bruce B. Collette, and Douglas E. Facey. *The Diversity of Fishes*. Malden, MS: Blackwell Science, 1997.

Mejía, Luz Stella, and Arturo Acero P., eds. *Libro Rojo de Peces Marinos de Colombia*. Bogotá: Invemar, Instituto de Ciencias Naturales, Ministerio del Medio Ambiente, 2002.

Mojica, José Iván, Claudia Castellanos, José Saulo Usma, and Ricardo Álvarez, eds. *Libro Rojo de Peces Dulceacuícolas de Colombia*. Bogotá: Instituto de Ciencias Naturales, Ministerio del Medio Ambiente, 2002.

Nelson, Joseph S. *Fishes of the World*. New York: John Wiley & Sons, 1994.

De Pinna, Mario C. C. "Phylogenetic Relationships of Neotropical Siluriformes (Teleostei: Ostariophysi): Historical Overview and Synthesis of Hypotheses." In *Phylogeny and Classification of Neotropical Fishes*, edited by L. Malabarba, R. Reis, R. Vari, Z. Lucena, and C. Lucena. Porto Alegre, Brazil: Edipucrs, 1998.

Ross, Stephen T. *Inland Fishes of Mississippi*. Jackson: University Press of Mississippi, 2001.

Other

"All Catfish Species Inventory" [cited February 7, 2003]. <http://clade.acnatsci.org/allcatfish>

"Fishbase" [cited February 7, 2003]. <http://www.fishbase.org>

"Planet Catfish" [cited February 7, 2003]. <http://www.planetcatfish.com>

"ScotCat" [cited February 7, 2003]. <http://www.scotcat.com>

Arturo Acero, MSc

Gymnotiformes

(South American knifefishes and electric eels)

Class Actinopterygii

Order Gymnotiformes

Number of families 5

Photo: A group of blackspot green knifefish (*Eigenmannia* sp.). (Photo by Wally and Burkard Kahl. Reproduced by permission.)

Evolution and systematics

South American freshwaters are dominated by fishes in the superorder Ostariophysi: the siluriforms, characiforms, and gymnotiforms together comprise more than 75% of all the known freshwater species in South America. Recent data based on cladistic methodology strongly indicate that Gymnotiformes and Siluriformes are sister groups within the Ostariophysi. Of the three Neotropical ostariophysan orders, the gymnotiforms have the smallest number of species. It is also the least investigated group as far as systematics and ecology are concerned. This is largely because many species are difficult to distinguish, hard to catch, especially in their preferred deep riverine benthic habitats, and finally yet importantly are of little commercial interest. The first known gymnotiform fish, now known as *Gymnotus carapo*, was described in 1648. In 1758 Linnaeus described four species, today known as *Electrophorus electricus*, *Gymnotus carapo*, *Rhamphichthys rostratus*, and *Apteronotus albifrons*. Our main understanding of the systematics of these fishes is based on work performed during the past 20 years. Science now recognizes 105 described and at least 32 undescribed species, grouped within 29 genera and 5 families. The family Gymnotidae includes 17 described species, the family Rhamphichthyidae 13, and the family Hypopomidae 15 species. The fishes of these families have retained the plesiomorphic pulse-type electric organ discharge. Twenty-four species are grouped in the family Sternopygidae and 43 in the family Apteronotidae; these knifefishes produce the derived tone-type electric organ discharge.

Sternopygus astrabes most closely resembles the ancestral gymnotiform phenotype. Gayet and Meunier in 1991 described the first fossil gymnotiform fish, *Ellisella kirschbaumi*, from the Upper Miocene (about 10 million years ago) from Bolivia. However, the morphological characteristics of *E.*

kirschbaumi—the absence of dorsal and pelvic fins and the replacement of the caudal fin by a long bony cartilaginous rod (as seen in sternopygids) in the regenerated caudal part—indicate that knifefishes are much older than 10 million years.

Physical characteristics

The Gymnotiformes constitute a group of rather specialized fishes. In most species, the body is compressed laterally, very elongated and slender, with long caudal appendages and a long anal fin tht resembles a knife (thus the name knifefishes). The fishes lack pelvic and dorsal fins; the dorsal filament of the Apteronotidae is sometimes interpreted as a rudiment of an adipose fin, but it might also be modified muscle. The caudal fin is replaced by the caudal appendage except in the apteronotids, which possess a reduced caudal fin that is interpreted by some authors as being derived from the caudal appendage and not representing an intermediate state in the loss of a caudal fin. All gymnotiform fishes possess an extremely well-developed ability to regenerate the caudal, or hind, parts of their bodies. In the apteronotids, however, this feature is less pronounced.

The caudal appendage represents an extension at the end of the anal fin, including the axial skeleton (a bony rod in most cases), the spinal cord, the remaining soft tissues, and the electric organ. A derived feature of gymnotiform fishes is their head morphology, which apparently reflects an adaptation for specialized feeding. For example, some knifefishes possess large mouths (*Gymnotus*, some *Apteronotus*) and feed on large prey; some possess terminal mouths (*Rhabdolichops*, *Eigenmannia*) and feed on insect larvae and on plankton; some posses external teeth (*Oedemognathus*) and feed on scales; some posses a long curved mouth (*Sternarchorhynchus*) and search

Electric eels (*Electrophorus electricus*) live the Orinoco River and in rivers throughout the Amazon Basin. (Photo by Richard T. Nowitz/Baltimore MD National Aquarium/Photo Researchers, Inc. Reproduced by permission.)

for insect larvae in holes and crevices. The same variations in head morphology, related to feeding, are found in the African mormyrid fishes.

The most outstanding feature of the knifefishes is the possession of electroreceptors and electric organs in the skin. In some *Rhabdolichops* species, the large transparent electric cells in the tail region can be easily seen by external inspection. *Electrophorus electricus* is characterized by strong electric discharges up to 700 volts (V) at more than 1 ampere, or amp (A); in addition, this species can emit weakly electrical discharges for orientation. All the remaining knifefishes produce weakly electrical discharges around 1 V. The electric organs are derived from muscle, except in apteronotids, where modified spinal axons constitute the electric organ. Some species, such as *Steatogenys elegans* and *Hypopygus lepturus*, possess accessory electric organs in the head region. Larvae of some species (families Sternopygidae, Apteronotidae) differentiate larval electric organs early during their development.

Distribution

Gymnotiform fishes occur in all Neotropical river systems, from the Rio Salado in the pampas of Argentina to the Rio San Nicolas of Chiapas in Mexico. Species diversity is highest in the Amazon river system (89 species), followed by the Orinoco river system (61 species), rivers found in the Guyanas (35 species), and the Parana-Paraguay river system (26 species). Fourteen species occur in northwestern South America, twelve in southeastern Brazil and in Uruguay, nine in northeastern Brazil, seven on the Pacific slope of South America, seven in Central America, and just one species in the Rio Sali-Dulce of northwestern Argentina and on the island of Trinidad.

Largely because of their poorly understood systematics, the distribution of gymnotiforms at the species level is not

well known. The most widely distributed species seem to be *Gymnotus carapo*, *Brachyhypopomus pinnicaudatus*, *Apteronotus albifrons*, *Eigenmannia lineata*, and *Sternopygus macrurus*. Not surprisingly, these are the species most commonly encountered in the tropical fish trade.

Habitat

Most rivers in tropical South America can be classified according to water conditions: white water (e.g., the Amazon), black water (e.g., the Rio Negro) or clear water (e.g., the Rio Tapachos). All three types are characterized by low pH values (from 7 down to 3.5) and low levels of salts and nutrients. The freshwater systems of Central America have higher levels of mineral content due to their different geology. Gymnotiform fishes occur in all three major water types. In the extremely acidic and nutrient-poor black waters of the Rio Negro, 36 gymnotiform species have been found. Knifefishes occur in small streams, large rivers, lakes, and various types of backwaters. The enormous abundance of gymnotiforms that occur in deep, main-river channels was only recently discovered (1980s): in the Orinoco River system, for example, 86% of the fish species are gymnotiforms. It is still not well understood why this predominance occurs in this type of habitat.

Some knifefishes show adaptations to habitats with low oxygen. The electric eel (*Electrophorus electricus*; modified buccal cavity) and the banded knifefish (*Gymnotus carapo*; swim bladder) can take air from the water surface, as can several species of *Brachyhypopomus*. The electric eel is an obligatory air breather and drowns if denied access to atmospheric air. However, as long as its skin is kept moist, it can survive for several hours out of water. A large embryonic fin fold of the gymnotiform free embryos serves as a respiratory organ during early development.

Most gymnotiform fishes tolerate temperatures between 68 and 92°F (20 and 35°C); a few species, however, have been found in colder waters. *Apteronotus* sp. and the glass knifefish (*Eigenmannia lineata*) occurred in considerable quantities in the Huallaga River (about 64.4°F or 18°C in the tropical mountain rainforest near Tingo Maria, Peru). Under experimental conditions, *Sternarchorhynchus* sp. did not survive below 77°F (25°C), but tolerated temperatures up to 98.6°F (37°C). For *Sternopygus* sp., the range was 66.2–86°F (19–30°C).

In the field, gymnotiforms can tolerate pH values of 7 (Amazon) to below 4 (black water). In captivity, the glass knifefish tolerates a pH range between 3.5 and 8.4!

Behavior

Knifefishes are nocturnal, hiding during daytime between plants, in floating meadows, in crevices and holes, and under various kinds of shelter. During the day some species, such as *Rhamphichthys rostratus* and *Steatogenys elegans*, lie flat and motionless on the bottom, imitating marbled leaves. Species of the genus *Gymnorhamphichthys* burrow in the sand during daytime.

Gymnotiforms move by undulating the elongate anal fin that extends along most of the ventral part of the body. In association with this form of locomotion, the anal fin rays articulate directly with proximal pterygophores, providing the fin rays unrestricted mobility. The caudal, elongate portion of the body is maintained in a rigid posture by numerous intermuscular bones. This facilitates the use of the integument as a sensory sheet due to the presence of numerous electroreceptors. This mechanism is used for object location. The presence of an individual nearby with an electric organ discharge frequency of similar magnitude disturbs the object location performance of the fish: a shift in frequency occurs, which is termed the jamming avoidance response.

Feeding ecology and diet

The food of knifefishes is of indigenous origin, mostly insect larvae, annelid worms, and crustaceans. The banded knifefish predominantly feeds on shrimp and smaller fishes, as do some larger knifefishes. Electric eels are also piscivorous and include their own weakly electric relatives in their diet. Regardless of seasonal changes, *Rhamphichthys marmoratus* does not change its diet, which consists mostly of chironomids (Diptera) and polymitarcids (Ephemeroptera).

Ontogenetic change in the diet is documented for the banded knifefish: up to 7.9 in (20 cm) the principal food items are aquatic insects, predominantly chironomid larvae; beyond 7.9 in (20 cm) the fish prefer large insect larvae (Odonates), shrimp, and fish, preferably small characids (genera *Ctenobrycon* and *Curimata*). This change in diet coincides with the time when the banded knifefish females reach first maturity (9.8 in/25 cm, 2 years old).

Species of the genus *Rhabdolichops* possess well-developed gill rakers and are known to be effective plankton feeders. The knifefish *Oedemognathus exodon* feeds on scales of fishes. The knifefish *Magosternarchus duccis*, from the murky channel of the Amazon, chiefly eats the tails of other knifefishes (these do regenerate afterward).

The importance of gymnotiforms as part of the food chain is demonstrated in the case of top predators of the pimelodid family, the catfishes; fifteen commercially important pimelodid species of the Rio Apure in Venezuela include gymnotiforms in their diet. Most pimelodids feed on only one or two species; however, some pimelodids, such as *Brachyplatystoma rousseauxii*, feed on a maximum of five species.

Reproductive biology

Reproduction in most species occurs during the high-water season. Experimental studies have shown that gonad maturation is triggered mainly by decreasing water conductivity and increasing water level. Information on reproductive biology has been gathered from a limited number of species (8-9), but has consistently revealed diverse reproductive strategies. The electric eel breeds during the dry season in small ponds. The male builds a foam nest at the water surface between the adventitious roots of the plant *Montrichardia arborescens*. The eggs are deposited in the

A glass knifefish (*Eigenmannia lineata*) from the Amazon River in Brazil. (Photo by Mark Smith/Photo Researchers, Inc. Reproduced by permission.)

foam nest, and at the beginning of exogenous feeding, the larvae feed on eggs of subsequent ovipositions. Juveniles are guarded up to a length of 3.9 in (10 cm). The males of the banded knifefish are mouth breeders and guard the fry afterward. The longtail knifefish (*Sternopygus macrurus*) is a substrate spawner and guards the eggs until hatching. Other knifefishes hide the eggs in plants, in between roots, or in crevices. All species are fractional spawners, with spawning intervals of a few days up to several weeks. Egg diameter varies between 0.08 and 0.1 in (0.2 cm and 0.3 cm). Some species live in pairs during reproduction; others form complicated social hierarchies that are based on size, motor components of aggressive behavior, and seasonal conditions. Best studied in this respect is the glass knifefish (*Eigenmannia lineata*).

During courtship behavior, the electrical discharges serve various purposes: males and females often produce different electrical discharge frequencies, and frequency modulations of the male discharge can trigger oviposition. Large amplitudes characterize dominant males, differences in the form of the discharge occur between males and females.

During the breeding season the males of the Rosen knifefish (*Sternarchorhynchus roseni*) develop external teeth; these knifefishes are very territorial and rather aggressive. The males of gymnotiforms often grow larger than the females. Males of several species of the genus *Apteronotus* develop longer heads than the females.

Conservation status

No gymnotiform species are listed by the IUCN.

Significance to humans

The strongly electric *Electrophorus electricus* was for decades a preferred animal for the study of basic bioelectric phenomena. The physiology, anatomy, and behavior of weakly electric knifefishes have been studied intensively, particularly in the context of orientation (object location) and electrocommunication. Some larger knifefishes, such as *Rhamphichthys rostratus* and the longtail knifefish are locally of some economical importance in fisheries.

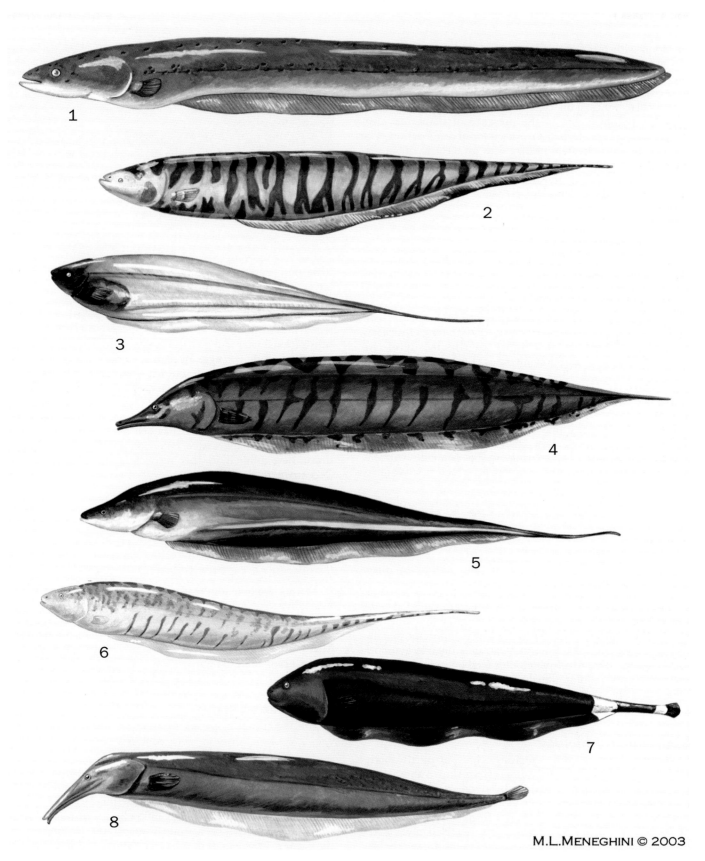

1. Electric eel (*Electrophorus electricus*); 2. Banded knifefish (*Gymnotus carapo*); 3. Glass knifefish (*Eigenmannia lineata*); 4. Bandfish (*Rhamphichthys rostratus*); 5. Longtail knifefish (*Sternopygus macrurus*); 6. *Brachyhypopomus pinnicaudatus*; 7. Black ghost (*Apteronotus albifrons*); 8. *Sternarchorhynchus curvirostris*. (Illustration by Michelle Meneghini)

Species accounts

Black ghost
Apteronotus albifrons

FAMILY
Apteronotidae

TAXONOMY
Apteronotus albifrons Linnaeus, 1766, South America.

OTHER COMMON NAMES
English: Apteronotid eel; French: Poisson-couteau; German: Amerikanischer Weißstirnmesserfisch.

PHYSICAL CHARACTERISTICS
Size 11.8 to 19.7 in (30 to 50 cm). They are completely black, except for a yellow frontal longitudinal stripe and two whitish transversal bands at the end of the anal fin and just before the black caudal fin. The males have longer heads than the females. The species is weakly electric.

DISTRIBUTION
Amazon, Orinoco, Parana-Paraguay river systems; rivers in French Guiana.

HABITAT
Occurs near undercut banks and in wooden debris along river and stream margins, as well as in densely vegetated habitats.

■ *Brachyhypopomus pinnicaudatus*
■ *Apteronotus albifrons*

BEHAVIOR
Gregarious and nocturnal. During swimming movements they are not always in an upright position, often turning their body parallel to the ground.

FEEDING ECOLOGY AND DIET
Various kinds of aquatic insect larvae, as well as ants and termites. The larger fishes also eat shrimps and fishes.

REPRODUCTIVE BIOLOGY
Breeding groups of black ghosts are composed of several males and females. As oviposition does not occur in regular intervals, single eggs of 0.1 in (0.3 cm) are deposited in holes and crevices, rather irregularly. The embryos hatch on day three, feeding starts on day 10.

CONSERVATION STATUS
Not listed by the IUCN

SIGNIFICANCE TO HUMANS
Black ghosts are the most common knifefishes in the tropical-fish trade. They are regarded by indigenous people with superstition, as they are reputed to be inhabited by a ghost or an evil spirit. ◆

No common name
Sternarchorhynchus curvirostris

FAMILY
Apteronotidae

TAXONOMY
Sternarchorhynchus curvirostris Boulenger, 1887, Canelos, Ecuador.

OTHER COMMON NAMES
None known.

PHYSICAL CHARACTERISTICS
Size from 15.7 to 19.7 in (40 to 50 cm). Their bodies are extremely compressed, with a pronounced curved snout; they lack dorsal and pelvic fins. The skin is uniformly brown. The electric organ is weak as it is derived from nerve cells (neurogenic) as opposed to muscle tissue (myogenic).

DISTRIBUTION
Amazon and Orinoco river systems; rivers in the Guyanas.

HABITAT
Channel bottom of large rivers.

BEHAVIOR
Nocturnal, feeding (and probably spawning) at night. Territorial and rather aggressive.

FEEDING ECOLOGY AND DIET
Insect larvae and possibly smaller crustaceans.

REPRODUCTIVE BIOLOGY
Probably spawn during the rainy season. Other details of their reproductive biology are not known.

Sternopygus macrurus
Sternarchorhynchus curvirostris

Electrophorus electricus
Gymnotus carapo

CONSERVATION STATUS
Not listed by IUCN. They are common throughout their range of distribution but at low densities.

SIGNIFICANCE TO HUMANS
None known. ◆

Electric eel
Electrophorus electricus

FAMILY
Gymnotidae

TAXONOMY
Electrophorus electricus Linnaeus, 1766, South America.

OTHER COMMON NAMES
French: Anguille électrique; anguille trembleuse; German: Zitteraal; Spanish: Anguilla, anguilla electrica; Portuguese: Poraquê.

PHYSICAL CHARACTERISTICS
Electric eels are the largest of all knifefishes, up to 8 ft (2.4 m) long. They lack dorsal, caudal, and pelvic fins, and do not have scales. The color is a uniform dull olive to almost black, yellowish to orange underneath the head and throat. These fishes produce strong electric discharges up to 700 volts.

DISTRIBUTION
Rivers in the Guyanas; the Amazon and Orinoco river systems.

HABITAT
Occurs in creeks and ponds, and along the banks of lakes.

BEHAVIOR
Nocturnal. They hide during the day under shelter or in holes. They are sometimes gregarious.

FEEDING ECOLOGY AND DIET
These fishes are mostly piscivorous (they stun prey with electric shocks), but they also eat amphibians.

REPRODUCTIVE BIOLOGY
These fishes breed during the dry season in small ponds. The male builds a foam nest. The larvae first eat eggs of subsequent spawnings, then change diet to insect larvae. Piscivorous feeding starts at around 3.9 in (10 cm); the males guard juveniles up to this size.

CONSERVATION STATUS
Not listed by the IUCN. They are common throughout their range of distribution but at low densities.

SIGNIFICANCE TO HUMANS
Electric eels have been used for basic studies on bioelectric phenomena. ◆

Banded knifefish
Gymnotus carapo

FAMILY
Gymnotidae

TAXONOMY
Gymnotus carapo Linnaeus, 1758, South America.

OTHER COMMON NAMES
English: Eel knifefish; gymnotid eel; cutlass fish; French: Coutelas; German: Gebänderter Messerfisch.

PHYSICAL CHARACTERISTICS
Banded knifefishes range in size from 19.7 to 23.6 in (50 to 60 cm). They have an eel-like body shape, with a blunt head, and extend to a fine point at the tip of the tail. The anal fin originates close behind the anus and continues to the tip of the tail. These fishes are variable in color, with many dark bands, and are sometimes mottled. The bands become increasingly narrow toward the tail. Banded knifefishes are air-breathing (with swim bladder), and are weakly electric.

DISTRIBUTION
Tropical South America from Venezuela to Uruguay, including Trinidad.

HABITAT
Creeks and ponds and along the banks of lakes and rivers.

BEHAVIOR
Nocturnal, hiding during daytime under various types of shelter. They are territorial, defending areas of several square yards (meters).

FEEDING ECOLOGY AND DIET
Banded knifefishes up to 7.9 in (20 cm) long eat insect larvae; afterwards odonats (dragonflies), shrimps, and fish.

REPRODUCTIVE BIOLOGY
Banded knifefishes spawn during the rainy season. Sticky eggs are deposited in between the stalks of plants. Males build nests from plant material and guard the free embryos in their mouths until the start of exogenous feeding. The first reproduction in captivity was in 1999.

CONSERVATION STATUS
Not listed by the IUCN. They are widespread and common but at low densities.

SIGNIFICANCE TO HUMANS
These weakly electric fishes are studied regularly for scientific purposes. They are extensively fished for and eaten in their native range. ◆

No common name
Brachyhypopomus pinnicaudatus

FAMILY
Hypopomidae

TAXONOMY
Brachyhypopomus pinnicaudatus Hopkins 1991 coastal swamp in French Guiana, 3.5 km northwest of Kourou.

OTHER COMMON NAMES
French: Poisson-couteau; Creole/French: Poson-sab, bloblo; Portuguese: Tuvira, itui.

PHYSICAL CHARACTERISTICS
These fishes range in size from 4.7 to 5.9 in (12 to 15 cm). Their bodies are compressed laterally; they lack dorsal, tail, and pelvic fins. Their bodies have a light-brown or reddish background with 22–25 narrow, dark brown bands running from the base of the anal fin dorsally to the lateral line. The

dorsal surface is dark brown and reticulated, with distinct contrasting borders. The tail filament is compressed in males, pointed in females. These fishes are weakly electric.

DISTRIBUTION
Rivers in French Guiana and southeastern Brazil; Amazon and Parana-Paraguay river systems.

HABITAT
Occurs along the edges of slow-moving rivers and in standing water, often in dense vegetation that becomes stagnant and partially deoxygenated.

BEHAVIOR
Nocturnal, resting during the day in dense vegetation.

FEEDING ECOLOGY AND DIET
Aquatic insects and copepods (small crustaceans).

REPRODUCTIVE BIOLOGY
Reproduction in these fishes occurs mainly during the high-water season; some populations seem to show aperiodical reproduction. Sticky eggs are deposited in crevices and holes, and in between plants. The embryos hatch on day three, exogenous feeding starts on day eight to nine; the juveniles grow quickly.

CONSERVATION STATUS
Not listed by the IUCN.

SIGNIFICANCE TO HUMANS
None known. ◆

Bandfish
Rhamphichthys rostratus

FAMILY
Rhamphichtyidae

TAXONOMY
Rhamphichthys rostratus Linnaeus, 1766, South America.

OTHER COMMON NAMES
Spanish: Bombilla (Argentina), cuchillo (Venezuela); French Guiana: Wabri.

PHYSICAL CHARACTERISTICS
Bandfishes range in size from 3.3 to 4.9 ft (1 to1.5 m). This most striking knifefish has a long, trunklike snout, ending in a small mouth. Otherwise, the body form is as in most knifefishes. Bandfishes lack dorsal, tail, and pelvic fins. They are medium brown in color, with blotches of dark brown and black on the back, and brown mottling on the sides. They are weakly electric.

DISTRIBUTION
Amazon, Parana-Paraguay, and Orinoco river systems; rivers of Guyana and northeastern Brazil.

HABITAT
Small and more open streams, and occasionally in large rivers.

BEHAVIOR
Nocturnal. During the day they hide motionless in plants or on the bottom.

FEEDING ECOLOGY AND DIET
Bandfishes are bottom feeders; they eat oligochaets (annelid worms) and insect larvae.

Rhamphichthys rostratus
Eigenmannia lineata

REPRODUCTIVE BIOLOGY
Bandfishes spawn during the rainy season. Data on reproduction in captivity known from one *Rhamphichtys* species, probably *R. rostratus*. About 1,000 sticky eggs are deposited in crevices; the embryos have attachment organs. Spawning intervals between three and four weeks. Juveniles show quick growth.

CONSERVATION STATUS
Not listed by the IUCN. They are widespread and common but at low densities.

SIGNIFICANCE TO HUMANS
Bandfishes are highly regarded as a food fish in areas where they are common. ◆

Glass knifefish
Eigenmannia lineata

FAMILY
Sternopygidae

TAXONOMY
Eigenmannia lineata Müller and Troschel, 1849, Lake Amucu, French Guiana.

OTHER COMMON NAMES
English: Green knifefish; German: Grüner Messerfisch; Spanish: Chucho, ratón, mayupa; Portuguese: Tuvira.

PHYSICAL CHARACTERISTICS
Female glass knifefishes are 7.9 (20 cm) long; males are 13.8 in (35 cm). Their bodies are slender and compressed, possessing

only pectoral fins and the very long anal fin. Most of the fish is very transparent, except in the head region. There are three black stripes running lengthwise along the body. Glass knifefishes are weakly electric.

DISTRIBUTION
Amazon, Orinoco, and Parana-Paraguay river systems.

HABITAT
Occurs near undercut banks and in wooden debris along river and stream margins; in the open-water areas of small creeks, lagoons, and marshes; and particularly in densely vegetated habitats of river margins.

BEHAVIOR
Gregarious and nocturnal. During daytime can be sometimes found in large numbers in hiding places.

FEEDING ECOLOGY AND DIET
Aquatic insect larvae and crustaceans.

REPRODUCTIVE BIOLOGY
Glass knifefishes spawn during rainy season; breeding groups of several fish establish complicated social hierarchies. The dominant male spawns with the ripe female at night. Sticky eggs are deposited on floating plants; they hatch on day three. Exogenous feeding begins on day eight.

CONSERVATION STATUS
Not threatened.

SIGNIFICANCE TO HUMANS
Glass knifefishes are the most intensively studied species of knifefish. They are often designated as *Eigenmannia virescens*. The systematics of the genus *Eigenmannia* is very difficult, which often leads to misidentification. The fish sold in the fish trade is in general called *E. virescens* (this name therefore often appears in scientific papers); however, the morphological features of this fish make it clear that it is not *E. virescens* but rather *E. lineata*. ◆

Longtail knifefish
Sternopygus macrurus

FAMILY
Sternopygidae

TAXONOMY
Sternopygus macrurus Bloch and Schneider, 1801, Brazil.

OTHER COMMON NAMES
English: Ghost knifefish; Spanish: Bio del rio.

PHYSICAL CHARACTERISTICS
Longtail knifefishes range in size from 23.6 to 31.5 in (60 to 80 cm), and are among the largest knifefishes. They have a pale yellow skin, a large black humeral spot; and a pale-yellow or white longitudinal stripe along the base of the anal fin, pterygiophores, and lateral midline posteriorly. They are weakly electric.

DISTRIBUTION
Amazon, Orinoco, and Parana-Paraguay river systems; rivers of northeastern and southeastern Brazil.

HABITAT
Widely distributed in streams, trenches, ditches, and the back waters of rivers in open savanna and in plantations.

BEHAVIOR
Mainly active at dusk and dawn, when the light level is low; they forage at night. They are territorial, slow-moving fishes.

FEEDING ECOLOGY AND DIET
When young, longtail knifefishes feed on small crustaceans and insect larvae. During growth, their diet changes to larger organisms such as shrimps and fishes. Adult insects are an important food item at all stages.

REPRODUCTIVE BIOLOGY
Longtail knifefishes mature when they are 11.8–15.7 in (30–40 cm) long. Several hundred 0.1 in (0.3 cm) eggs are deposited on a substrate. The male guards the eggs, which hatch after three days. The larval stage begins at 10 days.

CONSERVATION STATUS
Not listed by the IUCN. They are common throughout their range of distribution but at low densities.

SIGNIFICANCE TO HUMANS
Longtail knifefishes are fished for and eaten in their native range. The flesh is firm and considered to have a good flavor. ◆

Resources

Books
Albert, J. *Species Diversity and Phylogenetic Systematics of American Knifefishes (Gymnotiformes, Teleostei)*. Ann Arbor: Museum of Zoology (University of Michigan), 2001.

Bullock, T., and W. Heiligenberg. *Electroreception*. New York: John Wiley & Sons, 1986.

Gayet, M., and F. J. Meunier. *Première découverte de Gymnotiformes fossiles (Pisces, Ostariophysi) dans le Miocène supérieur de Bolivie*. C. R. Acad. Sci. Paris, t. 313, ser. 2,II, 1991.

Kirschbaum, F., and L. Wieczorek. *Entdeckung einer neuen Fortpflanzungsstrategie bei südamerikanischen Messerfischen (Teleostei: Gymnotiformes: Gymnotidae): Maulbrüten bei Gymnotus carapo*. Vol. 2. In *Verhalten der Aquarienfische*, edited by H. Greven and R. Riehl. Bornheim, Germany: Birgit Schmettkamp Verl., 2002.

Mago-Leccia, F. *Electric Fishes of the Continental Waters of America*. Vol. 29, *Biblioteca de la Academia de Ciencias Fisicas Matematicas y Naturales*. Caracas, Venezuela: FUDECI, 1994.

Moller, P., ed. *Electrical Fishes: History and Behavior. Fish and Fisheries Series 117*. London: Chapman & Hall, 1995.

Periodicals
da Silva Assunção, M. I., and H. O. Schwassmann. "Reproduction and Larval Development of *Electrophorus electricus* on Marajó Island (Pará, Brazil)." *Ichthyological Exploration of Freshwaters* 5 (1995): 1–10.

Kirschbaum, F. "Reproduction of the Weakly Electric Fish *Eigenmannia virescens* (Rhamphichthyidae, Teleostei) in Captivity." *Behavioral Ecology and Sociobiology* 4 (1979): 331–355.

———. "Electric Fishes." *Aqua Geographia* 1 (1992): 59–70.

Kirschbaum, F., and F. J. Meunier. "Gymnotiform Fishes As Model Systems for Regeneration Experiments." *Arch. Anat. Mic. Morph. Exper.* 75, no. 4 (1986): 307.

Frank Kirschbaum, PhD

Esociformes

(Pikes and mudminnows)

Class Actinopterygii

Order Esociformes

Number of families 1 or 2

Photo: A central mudminnow (*Umbra limi*) burrows into the mud to escape predators. (Photo by Gregory K. Scott/Photo Researchers, Inc. Reproduced by permission.)

Evolution and systematics

Three key questions to consider regarding the evolution of esociforms are: When did they originate? What other group of fishes is their closest relative? And, what are the evolutionary relationships among esociform species?

As of 2002, the earliest evidence of esociform fishes in the fossil record was in the form of fragmentary fossils found in deposits dating to the Upper Cretaceous from Alberta, Canada. The evidence consists of numerous fossilized fragments of bones that form the mouth skeleton, which displays a distinctive form of tooth attachment that is characteristic of living esociform species. Almost complete fossils of esociform fishes are found in Paleocene deposits from Alberta and Eocene deposits from Wyoming, United States. These fossils are extremely similar to living species of *Esox*. The fossil evidence suggests that esociform fishes belong to an ancient fish lineage that is more than 70 million years old.

Unfortunately, scientists have a poor understanding of the relationship of esociforms to other fish groups of similar age. The many primitive anatomical features that characterize esociforms hinder the development of a satisfactory theory on the classification of esociforms, despite a long history of phylogenetic studies on this area of fish evolution. Systematic studies based on anatomical comparisons have yielded tentative conclusions based on scant evidence. Studies based on comparisons of genetic sequences have produced the most strongly supported hypothesis of esociform relationships. These studies show unequivocal support for a close evolutionary relationship between the fishes in the order Salmoniformes (i.e., salmons, trouts, chars) and the esociforms. However, a consensus hypothesis is not available.

The order Esociformes includes four genera and 10 to 12 species, depending on the validity of two species described in 1982 from Siberia. The four genera of Esociformes are:

- *Dallia* (Alaska blackfish or dogfish)
- *Esox* (Northern pike, muskellunge, and pickerels)
- *Novumbra* (Olympic mudminnow)
- *Umbra* (European, central, and eastern mudminnows)

Until the late 1990s, esociform genera were divided into two families: Esocidae, for the genus *Esox*; and Umbridae, for the remaining three genera, which are collectively termed the mudminnows. This classification conflicts with molecular data first reported in 2000, which showed that these two families do not represent natural evolutionary groups. The molecular evidence suggests that the genera *Esox* and *Novumbra* are the closest evolutionary relatives, and that *Umbra* is the most primitive of the four esociform genera. In light of this evidence, the traditional two-family division of esociforms should be disregarded. A one-family scheme is followed in this chapter but, as of 2002, there was no general agreement on the classification of esociform genera.

Physical characteristics

One striking characteristic of all living esociforms is the posterior placement of the dorsal fin, which is located nearly opposite the anal fin in all species. Esociforms have elongate bodies with round cross sections. *Esox* species are further characterized by a flattened and elongated snout, which somewhat resembles a duck's bill.

Most species are small or medium sized; but *Esox* includes very significant exceptions. For example, individuals of *E. masquinongy* may reach over 5 ft (1.6 m) in length and weigh more than 66 lb (30 kg). At the other end of the scale, individuals of the four species in *Novumbra* and *Umbra* rarely reach 4 in (10 cm) and weigh less than 1 oz (28.3 g). In light of their marked contrast in size, it is interesting to note that *Esox* and *Novumbra* are thought to be very close relatives.

The coloration of esociforms is varied, but markings or mottled patterns on brown or olive green backgrounds are a common feature. A mottled coloration pattern in the densely vegetated habitats that esociforms favor likely hinders their detection by prey and/or predators. Males of some species show intensified coloration during the breeding season, with attractive iridescent hues on the body and fins.

Distribution

Living species of esociforms are found in all major Northern Hemisphere landmasses, with the exception of Greenland. North America is home to eight species of esociforms, three are found in northern Asia, and two inhabit European waters. The fossil record shows that the diversity of species of esociforms in Europe and Asia has been higher in the past. However, as of 2002, there was no evidence to indicate major changes in the global distribution of the order over its evolutionary history.

There have been changes in the distribution of some esociform species due to human intervention. The ranges of species in *Novumbra* and *Umbra* have undergone attrition due to habitat destruction. In contrast, the natural distributions of some *Esox* species have been expanded by human intervention because of their value as sport fishes.

Habitat

The understanding of the biology of esociforms is much more extensive for members of the genus *Esox* because these species are important in recreational fishing. However, the few studies conducted on the biology of the other esociform species point to broad similarities among all members of the order. Adults of the larger species of *Esox* move freely between shore and open water habitat. All esociforms show a similar preference for still or slow-moving water where dense vegetation allows them a place to hide. Other than dense vegetation, mudminnows seek areas with thick and loose muddy substrate, into which they quickly dive when startled. In addition, mudminnows have modified areas of the gut and swim bladder to extract oxygen from ingested atmospheric air. This allows them to withstand the widely fluctuating oxygen levels that may be associated with the heavy vegetation and rich organic matter substrate characteristic of their habitat.

Behavior

Outside the breeding season, esociforms are solitary and sedentary. They are most often found hovering among vegetation using elegant and economical pectoral and medial-fin movements to remain in place, with occasional pelvic-fin motions to correct body orientation. From this stance, the pikes wait for their prey, which they capture with a fast strike. Mudminnows may perch on the vegetation or rest on the substrate, eliminating the need for any fin movement. When dissolved oxygen levels are low, mudminnows occasionally swim up to the water surface and gulp air.

Feeding ecology and diet

Esociformes are carnivorous, predatory fishes. All *Esox* species prey most commonly on fishes, including smaller individuals of their own species. The larger species may also prey on frogs, waterfowl, and small mammals; their diets seem to be limited only by the size of the potential prey item and the opportunity for its capture. The mudminnows and the Alaska blackfish are also carnivorous, but due to their small size their diet consists almost entirely of aquatic invertebrates and, very rarely, juvenile fish.

Adult northern pike and muskellunge have only a few opportunistic natural predators, including bears, otters, and large birds of prey. Juveniles and fry of species of *Esox* are commonly preyed upon by larger members of the genus and other predatory fishes that share their habitat (e.g., centrarchid basses). The blackfish and the Olympic mudminnow have few predators because they occupy areas with few other species and are often found among thick vegetation that prevents terrestrial predators from targeting them. Pikes, pickerels, and other piscivorous fishes prey on mudminnows of the genus *Umbra*.

Reproductive biology

Most esociforms spawn early in the spring when water temperatures begin to rise. Some populations must undertake a migration to reach the spawning grounds. There is no evidence of nest building, but there are reports of territoriality around the spawning site in species of *Novumbra* and *Umbra*. Spawning most often involves one female and a few to several males. In some species, the males court the females through swimming displays or aggression. Egg deposition is usually preceded by exaggerated swimming motions and side-to-side contact. Eggs may stick to vegetation or drop to the ground. Fry do not receive parental care.

Conservation status

Habitat destruction is the main threat to populations of esociforms. Mudminnows are particularly vulnerable to this threat because they occur in areas with large human populations, and their habitat preference is not compatible with traditional land-development practices. The European mudminnow (*Umbra krameri*) is classified as Vulnerable by the IUCN, while the Olympic mudminnow (*Novumbra hubbsi*) is classified as Lower Risk/Near Threatened. Due to the interest of recreational anglers on all species of *Esox*, management policies and fishing regulations are in place to ensure stable populations.

Significance to humans

The greatest interest of humans in these fishes in modern times is limited to those species, all of which are members of the genus *Esox*, targeted by recreational anglers. There has been sporadic interest by aquarium enthusiasts in various mud-minnows. Historically, some human populations indigenous to Siberia and Alaska have included blackfish in their diet.

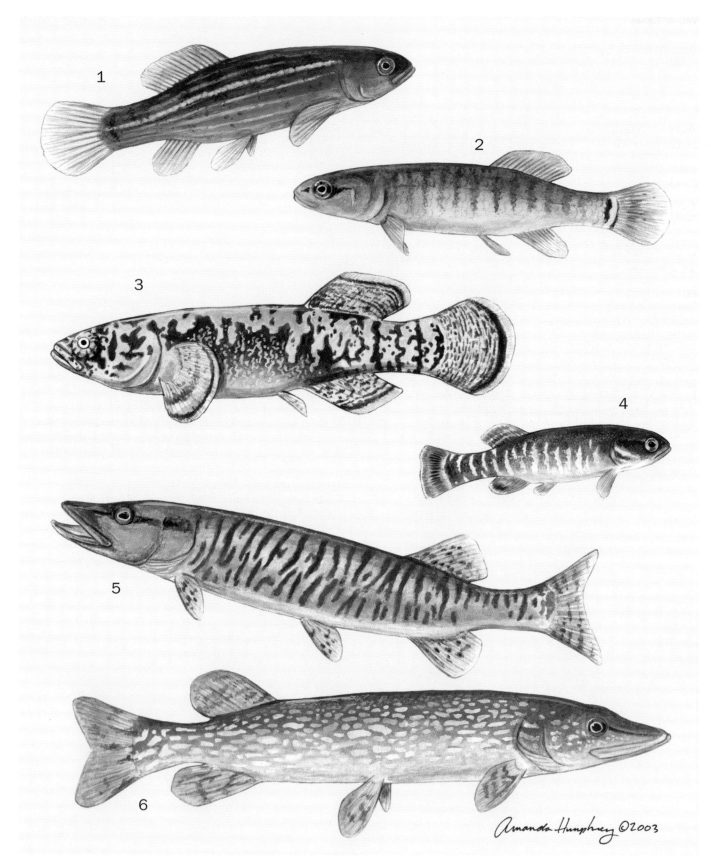

1. European mudminnow (*Umbra krameri*); 2. Central mudminnow (*Umbra limi*); 3. Alaska blackfish (*Dallia pectoralis*); 4. Olympic mudminnow (*Novumbra hubbsi*); 5. Muskellunge (*Esox masquinongy*); 6. Northern pike (*Esox lucius*). (Illustration by Amanda Humphrey)

Species accounts

Alaska blackfish
Dallia pectoralis

FAMILY
Esocidae

TAXONOMY
Dallia pectoralis Bean, 1880, Saint Michaels, Alaska, United States.

OTHER COMMON NAMES
English: Dogfish; French: Dallia; German: Fächerfisch; Russian: Chernaya ryba.

PHYSICAL CHARACTERISTICS
Length 1.9–6.3 in (4.8–16 cm); specimens as large as 13 in (33 cm) have been recorded; males reach larger sizes due to their longer lifespan. Stout head, well-developed fanlike pectoral fins, and much-reduced pelvic fins. Usually with a brown mottled pattern on a light cream background and light margins on the fins.

DISTRIBUTION
Lowlands of northern and western Alaska and the Chukot Peninsula of eastern Siberia. Introduced and established in Anchorage and St. Paul Island. Populations of *Dallia* that inhabit Arctic drainages of eastern Siberia may represent two different species of this genus.

HABITAT
In the spring and summer, occurs in shallow ditches and lakeshores with little or no water movement and thick vegetation. In winter, may move to deeper areas to avoid freezing.

Umbra limi
Novumbra hubbsi
Dallia pectoralis

BEHAVIOR
Mostly unknown. Their ability to tolerate extremely low temperatures is well known but not thoroughly understood; behavioral and/or physiological adaptations are probably involved. A gas-exchange organ associated with the esophagus allows them to obtain oxygen from gulped air, thus allowing them to occupy waters with low concentration of dissolved oxygen.

FEEDING ECOLOGY AND DIET
Preys predominantly on crustaceans (caldocerans, copepods, and ostracods) and aquatic insect larvae. Other aquatic invertebrates and, rarely, fish complement the diet.

REPRODUCTIVE BIOLOGY
The timing of spawning varies, depending on seasonal conditions at a given location. A population near Fairbanks, Alaska, spawns during May and June. Some populations undertake a migration to the spawning grounds in spring when thawing makes these areas available. Mature females carry 40–300 sticky, transparent eggs. Little is known about spawning behavior.

CONSERVATION STATUS
Not threatened. Widespread and abundant in their native range. Their remote distribution protects their habitat from destruction by human action. As of 2002 in the United States, it was illegal to transport live blackfish out of their native range without a permit.

SIGNIFICANCE TO HUMANS
Some indigenous communities have relied heavily on blackfish for food for themselves and their dogs. The nutritive value of blackfish was greatest in the winter when other sources of protein were scarce and these communities faced starvation. In modern times human consumption of blackfish has decreased, probably as a result of year-round availability of other foods. ◆

Northern pike
Esox lucius

FAMILY
Esocidae

TAXONOMY
Esox lucius Linnaeus, 1758, Europe.

OTHER COMMON NAMES
English: Jack, luce; French: Brochet; German: Hecht; Russian: Shchuka; Spanish: Lucio.

PHYSICAL CHARACTERISTICS
Standard length 35.4 in (90 cm); standard weight 24.2 lb (11 kg); flattened, elongate snout, scales on cheek and operculum. Long, cylindrical body with irregular yellow spots on a silvery green background that is lighter towards the belly. Stoutness of body varies depending on growing conditions, with better conditions resulting in deeper-bodied fish.

DISTRIBUTION
Northern North America, Europe, and Asia. Introduced by humans to Africa, Ireland, and Spain.

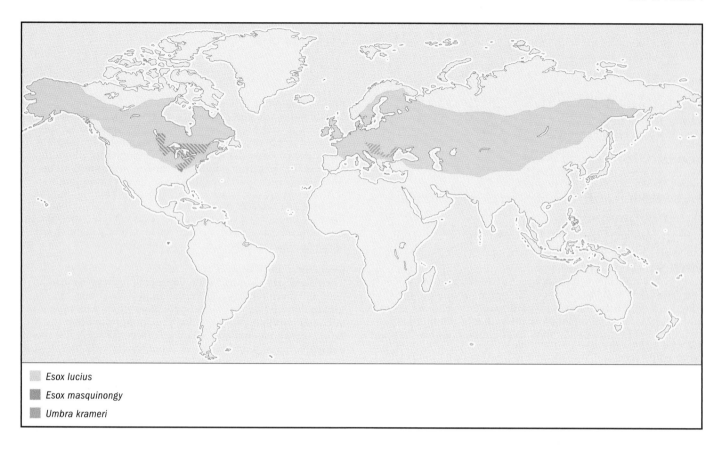

Esox lucius
Esox masquinongy
Umbra krameri

HABITAT

Prefers cool lakes, ponds, and slow-moving rivers with abundant vegetation. Young and smaller adults occupy shallow shore areas and move to greater depths as they grow.

BEHAVIOR

Solitary outside the breeding season. Hovers almost motionless among vegetation in wait for unsuspecting prey. May remain in same location for long periods of time if food is plentiful, otherwise will range over large distances in habitat with low prey density.

FEEDING ECOLOGY AND DIET

Voracious sit-and-wait predator; feeds on any animal large enough to be worth the effort and small enough to be subdued. Fish are the largest component of the diet. Fry feed on invertebrates until they are able to take fish. Fry and juveniles fall prey to larger individuals of the species, as well as to frogs and large aquatic insects. Adults are vulnerable to osprey and other large fish-eating birds, as well as otters, bears, and commercial and recreational fishing.

REPRODUCTIVE BIOLOGY

Spawning occurs in early spring. Males and females move close to shore or from streams to marshy areas. One female and a few smaller males swim into shallow vegetated areas. Females release a small number of eggs, which are immediately fertilized by attendant males. Egg release is repeated a varying number of times. Females may carry as many as 250,000 eggs. Both males and females may spawn with different mates during a spawning season. Adults do not guard spawning sites or provide care to the young.

CONSERVATION STATUS

Not threatened. Many populations are managed to maintain strong recreational fishing. Habitat destruction is a threat in some parts of their distribution, but worldwide they are not at risk.

SIGNIFICANCE TO HUMANS

In modern times, they are highly valued sport fishes and the focus of a recreational fishery of substantial economic importance. In the past, they were also valued as food fishes, and a sizable commercial fishery existed. Some praise the flavor of their flesh, but demand is not widespread and the commercial fishery is small. ◆

Muskellunge

Esox masquinongy

FAMILY

Esocidae

TAXONOMY

Esox masquinongy Mitchill, 1824, Lake Erie, North America.

OTHER COMMON NAMES

English: Lunge, muskie, mascalonge.

PHYSICAL CHARACTERISTICS

Standard weight 33 lb (15 kg); scales only on upper half of cheek and operculum. Prominent sensory pores on lower jaw. Dark spots or mottled markings on a silver or light green body, but color variation exists between populations in different parts of their distribution and markings may be absent.

DISTRIBUTION

Northeastern North America; centered around Great Lakes region.

HABITAT

As with *E. lucius*, prefers slow-moving or still waters with abundant vegetation, but with slightly warmer water temperatures.

BEHAVIOR

Mostly sedentary among vegetation. Hovers among aquatic vegetation, striking at prey with a fast, powerful lunge. For unknown reason, sometimes floats just beneath the water surface with the back exposed to the air.

FEEDING ECOLOGY AND DIET

Voracious predators; predominantly feed on fishes, but are known to eat crayfishes, frogs, waterfowl, and small mammals. Fry begin eating small invertebrates until they are able to capture larger prey. Fry are often eaten by *E. lucius* fry, which hatch some weeks earlier. Subadults are vulnerable to ospreys and other large fish-eating birds, as well as otters, bears, and commercial and recreational fishing. Large adults have no predators other than humans.

REPRODUCTIVE BIOLOGY

Spawns in spring when water temperature reaches 49–59°F (9.4–15°C). Similar courtship and spawning behavior as *E. lucius*, but avoids areas where that species spawns by releasing eggs in deeper water. In nature, these two species form sterile hybrids that are known as tiger muskies.

CONSERVATION STATUS

Not threatened. However, this species is rare in some areas and these populations are protected. Recreational fishing is strictly managed.

SIGNIFICANCE TO HUMANS

Numerous businesses profit from activities related to recreational angling in Canada and the United States. Winter fishing for these and other fishes through the ice using spears is a valued part of the cultural tradition of some Native American communities, and is still practiced in part of Wisconsin. ◆

Olympic mudminnow

Novumbra hubbsi

FAMILY

Esocidae

TAXONOMY

Novumbra hubbsi Schultz, 1929, near Satsop, Washington, United States.

OTHER COMMON NAMES

None known.

PHYSICAL CHARACTERISTICS

Length 1–3.4 in (2.5–8.5 cm); light to dark brown with light green vertical stripes along the entire length of the fish. Dark vertical bar across eye. Males develop a striking breeding coloration consisting of green to blue iridescent vertical stripes on an almost black background.

DISTRIBUTION

Restricted to lowlands of south and west drainages of the Olympic Peninsula in Washington State, United States, and a few drainages of the Puget Sound. Not known whether the Puget Sound populations are native or introduced.

HABITAT

Ponds, ditches, or creeks with still or slow-moving water, abundant submerged vegetation, and, usually, a thick layer of loose mud substrate

BEHAVIOR

Can survive in water with low oxygen content and under a wide range of temperatures, but shows little tolerance for salinity.

FEEDING ECOLOGY AND DIET

Feeds on aquatic invertebrates; primarily ostracods, isopods, worms and aquatic insects. Tends to coexist with only a few other fish species; primarily the reticulate sculpin (*Cottus perplexus*) and the three-spined stickleback (*Gasterosteus aculeatus*). Reticulate sculpins may prey on juvenile mudminnows. Although vulnerable to fish-eating birds, adults appear to have few predators.

REPRODUCTIVE BIOLOGY

Peak spawning occurs in April and May, but can take place from November to June with a lull in the winter. Males defend territories and perform a courtship display for passing females. Mating pairs wiggle in place side by side and spawn repeatedly, releasing a few eggs at a time. Eggs stick to vegetation or the substrate. Fry remain in place for about one week before dispersing. Parental care has not been documented.

CONSERVATION STATUS

Listed as Lower Risk/Near Threatened by the IUCN, as habitat destruction is a significant threat due to a very restricted distribution. The state of Washington gives the species special management consideration.

SIGNIFICANCE TO HUMANS

None known. ◆

European mudminnow

Umbra krameri

FAMILY

Esocidae

TAXONOMY

Umbra krameri Walbaum, 1792, Danube River, Europe.

OTHER COMMON NAMES

German: Hundsfisch; Hungarian: Lápi póc; Russian: Evdoshka; Ukrainian: Boboshka, lezheboka.

PHYSICAL CHARACTERISTICS

Length 2–3.5 in (5–9 cm); elongated body is slightly flattened dorsoventrally. Dorsal fin has almost rectangular profile. Drab brown coloration with a light horizontal stripe through the middle of the body.

DISTRIBUTION

Lowlands of the Danube River drainage from Vienna to the Black Sea, as well as the Prut and Dniester Rivers.

HABITAT

Oxbow lakes, ditches, and ponds with dense vegetation and debris.

BEHAVIOR

Few field studies have examined behavior.

FEEDING ECOLOGY AND DIET
Feeds mainly on aquatic crustaceans and insect larvae; rarely fish fry and vegetation. Main predators are piscivorous waterfowl, predatory fishes (including young of *E. lucius*), and large insects.

REPRODUCTIVE BIOLOGY
Spawning takes place from February to April. Several males pursue ripe females, which become progressively more receptive to their attention. Females that are ready to spawn find a suitable patch of vegetation and swim into it, followed by several attentive males. Males position themselves next to the females and eggs and sperm are released. The eggs may stick to the vegetation or sink to the bottom. After an interval of varying length, the process is repeated until the female has exhausted her egg supply. The female subsequently guards her spawning territory, but does not seem to take special care of the eggs or fry.

CONSERVATION STATUS
Habitat destruction has led to extirpation from a large part of the known historical distribution. Flood control of rivers and wetland development destroy its required habitat. Pollution affects all drainages where it exists. As a result, classified as Vulnerable by the IUCN.

SIGNIFICANCE TO HUMANS
Of no interest to commercial or recreational fisheries. In the past, used to feed livestock and as crop fertilizer. Because they represent a unique evolutionary lineage, countries where viable populations exist use this species to stimulate wildlife awareness and education. ◆

Central mudminnow
Umbra limi

FAMILY
Esocidae

TAXONOMY
Hydrargira limi Kirtland, 1840, Yellow Creek, Trumbull County, Ohio, United States.

OTHER COMMON NAMES
English: Dogfish, mudfish, mudpuppy; German: Amerikanischer Hundsfisch.

PHYSICAL CHARACTERISTICS
Length 1.6–3.9 in (4–10 cm); short cylindrical body. Dark brown to green, with lighter, irregular vertical stripes along the length of the body.

DISTRIBUTION
Centered around the Great Lakes region from northern Tennessee, United States, to southern Manitoba and Quebec, Canada.

HABITAT
Still or slow-moving water in ponds and ditches with a thick layer of mud on the bottom and abundant aquatic vegetation.

BEHAVIOR
Little information available. Sedentary in vegetation or resting on substrate, moving only to forage or to gulp air from surface. Capable of obtaining oxygen from air through the swim bladder. Evades predators by diving into the muddy substrate.

FEEDING ECOLOGY AND DIET
Carnivorous; feeds on aquatic invertebrates, particularly crustaceans and insect larvae. Important predators are juveniles of *Esox* species. Also vulnerable to water snakes and fish-eating birds.

REPRODUCTIVE BIOLOGY
Spawning takes place in the early spring on flooded banks. Mature females carry 200–2,200 eggs. Eggs are sticky and deposited in dense vegetation, where they hatch after about six days. Fry are not guarded or cared for by parents.

CONSERVATION STATUS
Not listed by IUCN. Rare in the western and southern limits of their distribution, probably as a result of habitat alteration. Widespread and abundant where their habitat is available. Habitat destruction by flood control and draining of wetlands is an ongoing threat.

SIGNIFICANCE TO HUMANS
Of no interest as a food fish. Because they are hardy, considered good for bait in some regions, but drab coloration reduces their value to the angler. Of some interest to aquarists. ◆

Resources

Books
Craig, John F., ed. *Pike Biology and Exploitation*. New York: Chapman & Hall, 1996.

Morrow, James E. "Mudminnows and Blackfish." In *The Freshwater Fishes of Alaska*. Anchorage: Alaska Northwest Publishing Co., 1980.

Scott, William B., and E. D. Crossman. "Mudminnows: Umbridae." In *Freshwater Fishes of Canada*. Oakville, Ontario: Galt House Publishing, 1998.

———. "Pikes: Esocidae." In *Freshwater Fishes of Canada*. Oakville, Ontario: Galt House Publishing, 1998.

Periodicals
Grande, L. "The First *Esox* (Esocidae: Teleostei) from the Eocene Green River Formation, and a Brief Review of Esocid Fishes." *Journal of Vertebrate Paleontology* 19 (1999): 271–292.

López, J., T. Pietsch, and P. Bentzen. "Phylogenetic Relationships of Esocoid Fishes (Teleostei) Based on Partial Cytochrome b and 16S Mitochondrial DNA Sequences." *Copeia* (2000): 420–431.

Martin-Bergmann, K. A., and J. H. Gee. "The Central Mudminnow, *Umbra limi* (Kirtland), a Habitat Specialist and Resource Generalist." *Canadian Journal of Zoology* 63 (1985): 1,753–1,764.

Mikschi, E., J. Wanzenböck, compilers. "Proceedings of the First International Workshop on *Umbra krameri* Walbaum, 1792." *Annalen des Naturhistorischen Museums in Wien Serie B Botanik und Zoologie* 97B (1995): 437–508.

Organizations

Muskies Canada Sport Fishing and Research, Inc.. P.O. Box
814, Station C, Kitchener, Ontario N2G 4C5 Canada. Web
site: <http://www.trentu.ca/muskie/mc.html>

Other

"Alaska Blackfish." Alaska Department of Fish and Game,
Wildlife Notebook Series [cited February 6, 2003]. <http://
www.state.ak.us/local/akpages/FISH.GAME/notebook/fish/
blackfsh.htm>

"Blackfish: A Cultural Mini-Unit." Alaska Native Knowledge
Network [cited February 6, 2003]. <http://www.ankn.uaf
.edu/UNITS/blackfish.html>

"The International Muskie Homepage." [cited February 6,
2003]. <http://www.trentu.ca/muskie/muskie.html>

"Understanding Northern Pike and Muskie." The Content
Well [cited February 6, 2003]. <http://www.thecontentwell
.com/Fish_Game/Northern_Pike/Pike_index.html>

Juan Andrés López, MS

Osmeriformes

(Smelts, galaxiids, and relatives)

Class Actinopterygii

Order Osmeriformes

Number of families 9

Illustration: 1. Black-stripe minnow (*Galaxiella nigrostriata*); 2. Barreleye (*Macropinna microstoma*); and 3. Australian smelt (*Retropinna semoni*). (Illustration by Jonathan Higgins)

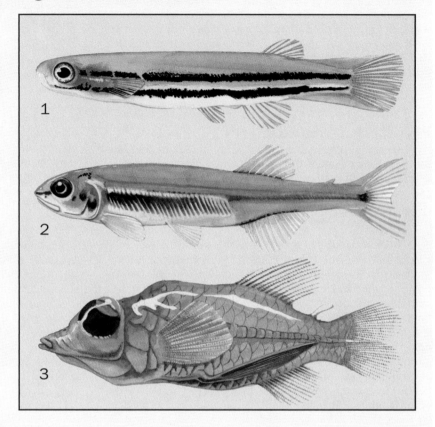

Evolution and systematics

Fossil Osmeriformes include *Speirsaenigma lindoei*, the oldest fossil osmerid from the Paleocene formations of Alberta, Canada, and a presumed close relative of the ayu (*Plecoglossus altivelis*; suborder Osmeroidei), *Enoplophthalmus* (Osmeroidei) from Europe, and possibly the family Pattersonellidae within the Argentinoidei.

The systematics of the "lower" euteleosts, particularly the Protacanthopterygii (the pikes, salmons, and smelts), has been in a constant state of flux since about the middle of the 1970s. Many phylogenies developed since that time considered that the Protacanthopterygii included three main lineages, one that included the pikes, pickerels, and mudminnows (Esociformes), another including the salmons, trouts, chars, graylings, and whitefishes (Salmoniformes), and a third comprising the freshwater and diadromous smelts, the galaxiids, the southern graylings, the marine argentines, deep-sea smelts, barreleyes, and slickheads (Osmeriformes). The ancestor of the remaining teleosts, the "higher teleosts," or the Neoteleosti, is thought to be either the common ancestor of the Protacanthopterygii and the Neoteleosti, or to have risen within the Protacanthopterygii (Johnson and Patterson, 1996). However, many researchers did not necessarily formally recognize any of the above groups at the order level. Furthermore, the precise relationships of the various members of these

groups has been a source of constant debate among ichthyologists researching the "lower" euteleosts (Johnson and Patterson [1996], as well as Rosen [1974], Fink [1984], Nelson [1994], and Begle [1991 and 1992]), who base their research primarily on anatomical and morphological data, and the most extensive study utilizing molecular data of Waters et al. (2000).

In light of the lack of consensus among these studies, it seems prudent, for the time being at least, to retain the traditional groupings/orders of the Esociformes, Salmoniformes, and Osmeriformes, while modifying them slightly to take into account the thorough studies of Johnson and Patterson (1996) and Waters et al. (2000). Such a scheme has been used in this chapter. However, it is likely that, with the collection of additional molecular data from a suite of taxa that represents all the major groupings, a further revision to the classification of the "lower" euteleostean fishes will occur.

If the classification given above is accepted, the Osmeriformes contains two suborders, the Osmeroidei (the smelts, galaxiids, etc.), which includes four families and approximately 70 species, and the Argentinoidei (the argentines, deep-sea smelts, etc.), which includes five families and approximately 170 species. Just as it is likely that changes to the classification of this group will occur with the collection of more data, so it is also likely that the number of species will increase.

The nine families of Osmeriformes are: smelts (Osmeridae); salamanderfishes (Lepidogalaxiidae); New Zealand smelts or southern graylings (Retropinnidae); galaxiids or southern minnows (Galaxiidae); Argentines or herring smelts (Argentinidae); barreleyes or spookfishes (Opisthoproctidae); deep-sea smelts (Microstomatidae); tubeshoulders (Platytroctidae); and deep-sea slickheads or slickheads (Alepocephalidae).

Physical characteristics

Morphological characteristics that are unique to, and present in, all members of the Osmeriformes have not yet been determined, and this is reflected in the placement of the group and its constituent members at different taxonomic levels by many researchers, including Fink (1984), Eschmeyer (1990), Johnson and Patterson (1996), and Helfman et al. (1997).

It is often stated that members of this group are generally silvery and elongate fishes. However, this probably reflects the fact that most authors are more familiar with the Northern Hemisphere species of the family Osmeridae (smelts) than with other groups. For example, even though members of the Southern Hemisphere families Galaxiidae and Lepidogalaxiidae are generally elongate, only a few have a silvery belly and none could be considered silvery overall, the majority having a brownish or olive base color overlaid with stripes, spots, or mottling of various colors. In addition, although many Argentinidae possess silvery sides, many species found in the other families of the suborder Argentinoidei, and in particular those that inhabit the deep oceans, are black or brown in color. In regard to shape, most members of the order are elongate, but the Opisthoproctidae (barreleyes) are often deep bodied and laterally compressed.

Members of the Osmeriformes are usually very small to small fishes, ranging in size from 1.2 to 27.5 in (3–70 cm); 0.0016 oz to 3 lb (0.22 g–1.3 kg). The maximum length recorded for the eastern Australian galaxiid (*Galaxiella pusilla*) is only 1.5 in (3.9 cm), and the mean length of males and females at maturity is only 0.9 and 1.1 in (2.4 and 2.8 cm), respectively. However, some representatives of the argentines (Argentinidae) and slickheads (Alepocephalidae) may reach considerably larger sizes than the diminutive eastern Australian galaxiid, including the slickhead (*Alepocephalus rostratus*) of the western Mediterranean and eastern Atlantic, which attains a length of at least 27.5 in (70 cm).

There are several physical characteristics of the major groups of Osmeriformes that help to distinguish them. Helfman noted that Osmeriformes have a single dorsal fin made up of soft rays, have abdominal pelvic fins, often have an adipose fin, and generally have a toothless maxilla. The smelts (superfamily Osmeroidea) are typical osmeriforms, in that they are elongate fishes with bright silvery scales, and have a dorsal fin situated toward the center of body, as well an adipose fin and a forked caudal fin. In most species, fingerlike filaments rising from the pyloric region of the stomach (pyloric ceca) are present, these structures aid digestion. Although most osmeroids are covered with bright silvery scales,

others, such as the ayu (subfamily Plecoglossinae) are olive dorsolaterally and white ventrally. An adhesive membrane that acts as an anchoring mechanism surrounds the eggs of all members of the Osmeroida, and several species smell of cucumber instead of fish.

In contrast to the predominantly silvery Osmeroidea, fishes in the superfamily Galaxioidea (galaxiids and relatives) often have a brown-to-olive base color overlaid with stripes, bands, or spots of various colors. Although members of the Retropinnidae and some species of the Galaxiidae, i.e., the single species *Lovettia sealii* (Tasmanian whitebait), and both species of *Aplochiton*, have an adipose fin and a deeply forked caudal fin, the other galaxiids and the salamanderfish (*Lepidogalaxias salamandroides*) have no adipose fin and a truncate or rounded caudal. Pyloric ceca are present in both species of *Aplochiton* and some other members of the Galaxiidae. The majority of the Galaxiidae are scaleless.

Like the Osmeroidea, most members of the superfamily Argentinoidea (argentines, deep-sea smelts, and relatives) are generally silvery, and have a dorsal fin situated toward the middle of the body and a forked caudal fin. Although some species within the Microstomatidae and the Bathylagidae (deep-sea smelts) possess an adipose fin, others do not. Most members of this group have small mouths with no teeth on the upper jaw. As is the case with many fishes that are nocturnal or inhabit deep waters, many members of the superfamilies Argentinoidea and Alepocephaloidea (slickheads and relatives) have very large eyes, an adaptation that maximizes the collection of the little light available in poorly lit waters.

Unlike the other marine, and often deep-sea, members of the Osmeriformes, such as the Argentinoidea, the Alepocephaloidea are generally black/brown in color, have a dorsal fin situated toward the back of the body, do not possess an adipose fin, and generally have large mouths that, except for the three species of *Leptochilichthyes*, have teeth on the upper jaw.

Sexual dimorphism has been recorded in some members of the osmeriforms. For example, female salamanderfish attain lengths of over 2.7 in (7 cm), but the males are much smaller and have never been recorded at lengths greater than 2 in (5 cm). Furthermore, the males develop a series of enlarged scales, or a scale sheath, around their anal fin, which is absent in females. Males of many osmerids develop tubercles during the breeding season. In addition to exhibiting sexual dimorphism with regard to differences in size, some osmeriforms develop nuptial coloration. For example, during the breeding season, some galaxiids become particularly brightly colored. The black-stripe minnow (*Galaxiella nigrostriata*) develops a bright orange to vermillion lateral stripe, which is often highlighted by the development of dark brown to black stripes on either side.

Distribution

The first suborder of the Osmeriformes, the Osmeroidei (smelts, galaxiids, etc.), is restricted to latitudes north and south of the tropics. Within this suborder, those taxa that constitute the superfamily Osmeroidea are, with the notable exception of some members of the tribe Salangini (icefishes and noodle-fishes), restricted to the cooler regions of the Northern Hemisphere. The Osmeroidea have a wide distribution, occurring in the Arctic, Atlantic and Pacific Oceans, as well as the rivers that drain into these oceans. However, they are absent from the Indian Ocean and its drainages, presumably a reflection of the fact that the vast majority of species comprising this superfamily occur in cool temperate waters. They are also absent from the Mediterranean and its drainages. The tribe Osmerini (smelts) has a circumpolar distribution, with representatives in the Arctic, Atlantic, and particularly the Pacific Oceans, but the ranges of the Plecoglossinae (the ayu, or sweetfish) and the Salangini are far more restricted; both these groups are found only in the coastal waters and drainages of the eastern Pacific. However, while the Plecoglossinae and the Salangini do not occur in the Arctic or Atlantic Oceans or their drainages, their extension into the lower latitudes is far greater than any other member of the Osmeroidei. For example, the Plecoglossinae, represented solely by the ayu, has a range that extends from northern China to Korea and Taiwan in the south; the Salangini, represented by 11 species, has a range that that has a similar northern extent of its range but extends into Vietnam.

In contrast to the Northern Hemisphere Osmeroidea, the three families in the superfamily Galaxioidea are found only in the Southern Hemisphere. The most widespread family is the Galaxiidae, representatives of which are found in Australia (where 22 of the 35 species occur), Lord Howe Island, New Caledonia, Tasmania, Chile, Argentina, and the tip of South Africa. Two of the remaining three families, the Retropinnidae and Prototroctidae, are found in Australia and New Zealand. Two species of the Retropinnidae, the Australian smelt (*Retropinna semoni*) and the Tasmanian smelt (*Retropinna tasmanica*), together have an extensive range that encompasses much of subtropical and temperate eastern Australia. The cucumberfish (*Retropinna retropinna*) and Stokell's smelt (*Stokellia anisodon*) are found in New Zealand. The Prototroctidae are represented in Australia by a single species, the Australian grayling, found in the coastal streams of New South Wales, Victoria, and Tasmania. The New Zealand grayling (*Prototroctes oxyrhynchus*) was last recorded in the 1920s. The third family of Southern Hemisphere Osmeroidei, Lepidogalaxiidae, consists of the single species, the salamanderfish, which is confined to the acid/peat flat region of the southwestern tip of Western Australia.

In contrast to the suborder Osmeroidei, whose members occur in fresh or coastal marine waters and essentially have a holarctic distribution, the obligate marine members of the suborder Argentinoidei can be found in all ocean waters from the Arctic in the north, through the Atlantic, Indian, and Pacific Oceans (including tropical and equatorial regions), to the southern ocean around Antarctica in the south. Species of the family Opisthoproctidae, subfamily Microstomatinae, are found in tropical to temperate regions of the Atlantic, Indian, and Pacific Oceans. Fishes in the subfamily Bathylaginae are represented in the subarctic to Antarctic, Atlantic, Indian, and Pacific Oceans; members of the family Argentinidae are found throughout Atlantic, Indian, and Pacific waters; and members of the families Platytroctidae and Alepocephalidae are found in all oceans.

Habitat

Osmeriforms are found in most aquatic habitats, in fresh waters in torrential mountain streams, slow-flowing rivers, ponds, lakes, and even pools that become completely dry for several months of the year, to highly variable salinity of estuaries, out into shallow inshore marine habitats, and on to the meso- and bathypelagic regions of the open ocean (deeper than 9,840 ft/3,000 m). Thus, with the exception of the freshwater systems of the lower latitudes of the Americas, Africa, the Indian subcontinent, and Australia, members of this order may be found in marine and freshwater environments off or in all of the continents. Furthermore, in the more temperate regions of the Southern Hemisphere, they are likely to be encountered in any aquatic habitat that fish biologists sample.

For example, in the fresh waters of Australia, species such as the Australian grayling and the climbing galaxias (*Galaxias brevipinnis*) tend to be restricted to the upper reaches of river systems, where they are found in shady areas over a rocky substrate in clear waters of moderate- to fast-flowing tributary streams. Other species, such as the barred galaxias (*Galaxias fuscus*), although also restricted to the headwaters of river systems, appear to prefer the slower-flowing pools below riffles and rapids. Further downstream, species such as the Australian smelt and the trout minnow (*Galaxias truttaceus*) tend to be found in the still to slow-flowing backwaters of streams and rivers, and to lakes and swamps, where they are found amid vegetation, logs, rocks, and other structure around the margins. Other species of Southern Hemisphere osmeriforms, such as swan galaxias (*Galaxias fontanus*), Clarence galaxias (*Galaxias johnstoni*), and Pedder galaxias (*Galaxias pedderensis*) are most commonly found in lakes and pools; Tasmanian mudfish (*Galaxias cleaveri*) and swamp galaxias (*Galaxias parvus*) occur in muddy marshes and swamps; and black-stripe minnow and salamanderfish inhabit small tannin-stained and acidic pools that often become completely dry in summer. Other Southern Hemisphere forms, such as the western minnow (*Galaxias occidentalis*), are found in all the above habitats, as well as in many of the brackish, salt-affected, streams, rivers, and lakes that are unfortunately becoming more common in southwestern Australia. Some populations of trout minnow and common jollytail (*Galaxias maculatus*) spend their adult lives in rivers and lakes, but move down into the brackish areas of estuaries to spawn. The larvae then move out to sea, where they spend the first year of life before moving back into rivers as juveniles.

In the Northern Hemisphere, most osmeriforms in fresh waters are usually anadromous forms on their annual spawning run, however, some species, such as the rainbow smelt (*Osmerus mordax*) of North America and ayu of Asia, have both anadromous or amphidromous and freshwater populations. Freshwater populations of both these species tend to spend their adult life in lakes.

As with the freshwater osmeriforms, the marine forms also occur in a wide range of marine habitats. The majority of the anadromous, amphidromous, and strictly marine Osmeroidea, such as the capelin (*Mallotus villosus*), are found in the shallow coastal waters out to the continental shelf, while members of the superfamily Argentinoidei are found in these waters and also out over the continental shelf and into the open ocean. Although representatives of the Argentinoidei occur in all major habitats of the open ocean, only one species of the family Alepocephalidae, *Microphotolepis schmidti*, is known to inhabit the epipelagic zone (the top 656 ft/200 m of the water column). However, the Argentinoidei is well represented in the deeper zones of the open oceans. For example, in the Western Central Pacific Ocean (based on the FAO designation), *Nansenia ardesiaca*, *Nansenia pelagica*, and *Xenophthalmichthys* (subfamily Microstomatinae); *Bathylagus nigrigenys* and *Dolicholagus longirostris* (longsnout blackmelt, subfamily Bathylaginae); and *Dolichopteryx longipes*, *Opisthoproctus grimaldii*, *Opisthoproctus soleatus*, and *Winteria telescopa* (family Opisthoproctidae), are meso- to bathypelagic, meaning that they live in open waters above the bottom at depths of 656–13,120 ft (200–4,000 m). In the same region, *Argentina* spp. and *Glossanodon struhsakeri* (family Argentinidae) are benthopelagic, living on and just above the bottom, on the outer continental shelf and slopes to depths of about 4,590 ft (1,400 m). Approximately 59 species of the family Alepocephalidae and approximately 21 species of the family Platytroctidae are found above the continental shelf and ocean ridges, and are considered to be engybenthic, as they inhabit a layer of water that extends from just above the bottom to several feet above it. Although most of the species of the families Alepocephalidae and Platytroctidae are known from depths of 2,625–8,200 ft (800–2,500 m), a few species of the former are inhabitants of the abyssal depths, having been caught from more than 16,400 ft (5,000 m).

Not only are the osmeriforms found in a diverse array of habitats, they also live in some of the most inhospitable environments in which fishes are found. For example, fishes that live in the deep oceans, such as members of the suborder Argentinoidei, live in one of the most hostile environments on Earth, where little, if any, light visible to the human eye penetrates below 656–2,625 ft (200–800 m). Thus, deep-sea fishes, depending on the amount of light entering their domain, must either have eyes that maximize light collection or rely on other senses in such a dark environment. Furthermore, such a lack of light in the ocean depths has two major effects; first, due to the lack of the warming light, temperatures in the bathypelagic zone are 35.6–41°F (2–5°C); second, as no light enters this deep zone, photosynthesis cannot occur and phytoplankton cannot survive, thus productivity is dependent on the "rain" of dead organisms and feces from the euphotic zone (the zone of light penetration) and food is consequently scarce. Third, low productivity is one thing, but the massive volume of the ocean deep means that the little food available is spread very thinly. If these factors were not enough to create a harsh environment, the added difficulty of living in deep water, where pressure increases by one atmosphere for every increase in depth of 33 ft (10 m), cannot be underestimated. Not only are chemical reactions affected by increases in pressure, but the gas solubility in the blood and tissues and the difficulty of secreting gas into the swim bladder are also increased.

In addition to the severe environmental conditions of deep-sea environments, other osmeriforms, such as salamanderfish

and black-stripe minnow, live in small, tannin-stained and acidic pools that often become completely dry in summer, habitats which are equally extreme.

Behavior

Published works describing the behavior of osmeriform species are scarce. The few that do exist generally describe the spawning migrations of the freshwater, anadromous, and coastal forms, or the migrations of juveniles of the amphidromous forms back into the rivers. For example, with the onset of winter rains, many galaxiids move upstream into tributaries and feeder streams where they spawn amongst flooded vegetation (western minnow), or on rocks (marbled galaxias, *Galaxias olidus*). During these upstream migrations, some species, such as climbing galaxias and western minnows, can "climb" obstacles as high as 33 ft (10 m) by jumping and wriggling in an eel-like fashion, using their pelvic and pectoral fins to lever themselves forward and upward. This is quite a feat, considering that individuals of these species are rarely larger than 6 in (15 cm). Species that have amphidromous populations, such as the common jollytail, undertake migrations. The adults move downstream to spawn in estuaries, the larvae that result then move out to sea before returning to the rivers in large schools of so-called whitebait. The spawning migrations of the amphidromous, anadromous, and coastal forms are described in the species descriptions of the ayu, eulachon, and the capelin, respectively.

Virtually nothing is known about the behavior of the deep-sea osmeriforms. However, two behavior patterns are likely in at least some species. First, most mesopelagic fishes undergo vertical feeding migrations, in which they move to the epipelagic zone at night to feed on the abundant zooplankton found in this region. Such migrations have been recorded for at least one species within the Argentinoidea, the longsnout blacksmelt (*Dolicholagus longirostris*), so it is likely that other species also undertake such daily migrations. Second, many species of deep-sea fishes have bioluminescent light-emitting organs called photophores, which are thought to be used for attracting mates (by flashing in particular pulse patterns), for attracting prey items, and for "hiding" the fishes from predators. The latter mechanism is thought to work in those species that live in regions into which some light just penetrates, whereby rows of photophores along the ventral surface reduce the chance that the fish is silhouetted when viewed from below.

Feeding ecology and diet

Although some representatives from most groups, including the capelin and the slender argentine (*Microstoma microstoma*), prey predominantly on small benthic and pelagic arthropods, others, such as the surf smelt (*Hypomesus pretiosus*) and European smelt (*Osmerus eperlanus*), have much wider diets, ingesting arthropods, fishes, mollusks, and worms. The Agassiz's slickhead (*Alepocephalus agassizii*) feeds on crustaceans, worms, echinoderms, and comb jellies. Other species of osmeriforms are, or become, specialist feeders. The mirrorbelly (*Opisthoproctus grimaldii*) is reported to feed on the stinging cells and tentacles of jellyfishes, and others, such as the ayu and black-stripe minnow, both exhibit marked shifts in diet at the end of their larval/juvenile stages.

Osmeriforms are preyed on by a wide variety of predators. For example, sharks, as well as a variety of bony fishes, seals, birds, and dolphins, have been reported to prey on the eulachon (*Thaleichthys pacificus*); bony fishes and squids prey on the argentines.

Reproductive biology

Fishes in the suborder Osmeroidei, such as the surf smelt and capelin, show a diverse array of breeding strategies, undergoing extensive migrations from coastal seas to surf beaches and estuaries, where they often form massive spawning aggregations, males pressing against females until they release their eggs, the males then releasing their sperm. The eggs are then fertilized and wave action buries them below the sand. Other members of the Osmeroidei, such as the eulachon and ayu, move up into rivers to spawn. Freshwater populations of Northern Hemisphere osmeroids tend to undergo similar spawning migrations, and may spawn in habitats similar to those of the migratory forms, such as the ayu. They may also spawn in different habitats. The freshwater forms of the European smelt tend to spawn in lakes, the anadromous forms spawn in rivers.

Within the Southern Hemisphere osmeroids, the Tasmanian smelt is anadromous, spending most of its life in the sea before returning to estuaries and rivers to spawn. In other species, such as the common jollytail, most populations are amphidromous. The adults of this species live in fresh water, but during autumn move downstream to spawn in estuaries. Spawning takes place on spring tides amid inundated vegetation, the eggs developing out of water and hatching two weeks later on the next spring tide. The larvae are then washed out to sea, where they feed and develop into juveniles before returning to the rivers to complete their life cycle. In purely freshwater forms such as the western minnow, migrations into tributaries or into the shallows of lakes are usual.

Within the Argentinoidei, stout blacksmelt (*Pseudobathylagus milleri*), and the longsnout blacksmelt produce planktonic eggs and larvae, and it is likely that other members of this suborder do likewise. Little other data are available on the reproductive behavior in the deep-sea osmeriforms.

The age at first maturity, whether the adults survive to spawn in more than one year, and the spawning pattern,

spawning period, and number and size of eggs vary greatly within the order. For example, anadromous forms of the European smelt do not spawn until their third or fourth year, when during a short breeding season of one to three months they produce 8,000 to 50,000 small (0.02–0.03 in/0.06–0.09 cm) demersal eggs. Adults usually die after spawning. Other species, such as the mudminnow (*Galaxiella munda*), which becomes mature at the end of its first year of life, has a spawning period that extends for three to four months, during which they produce several small batches (37–92) of smallish (0.06–0.05 in/0.09–0.14 cm) demersal eggs. Most mudminnow die within one or two months of spawning. Unlike the above two species, the western minnow spawns at the end of its first year and can live to spawn a further four times.

The salamanderfish is unique in the Osmeriformes in that, rather than broadcasting its eggs and sperm, it undergoes internal fertilization.

There are no reports of parental care in the Osmeriformes.

Conservation status

Eight species of Osmeriformes are listed by the IUCN. One species is classified as Extinct, one species as Endangered, one species as Critically Endangered, three species as Vulnerable, and two species as Data deficient. In addition, many Australian forms are considered by the Australian Society for Fish Biology to be data deficient to endangered, largely because of land clearing and the subsequent degradation of freshwater habitats. Many deep-sea forms are small and rare, and thus are likely to be overlooked by conservationists.

Significance to humans

Due to their generally small size and/or inaccessibility (in deep-sea forms), most Osmeriformes are of little importance to humans, either commercially or recreationally. However, in the Northern Hemisphere, eulachon, capelin, European smelt, and ayu form the basis of important local commercial and recreational fisheries during their spawning migrations, and in the Southern Hemisphere, the juveniles of the anadromous species, such as the Tasmanian whitebait and common jollytail, form the basis of important commercial whitebait fisheries in New Zealand and Chile. In Australia, massive falls in whitebait capture during the 1970s led to the closure of the fishery. However, it has since been reopened as a tightly regulated recreational fishery.

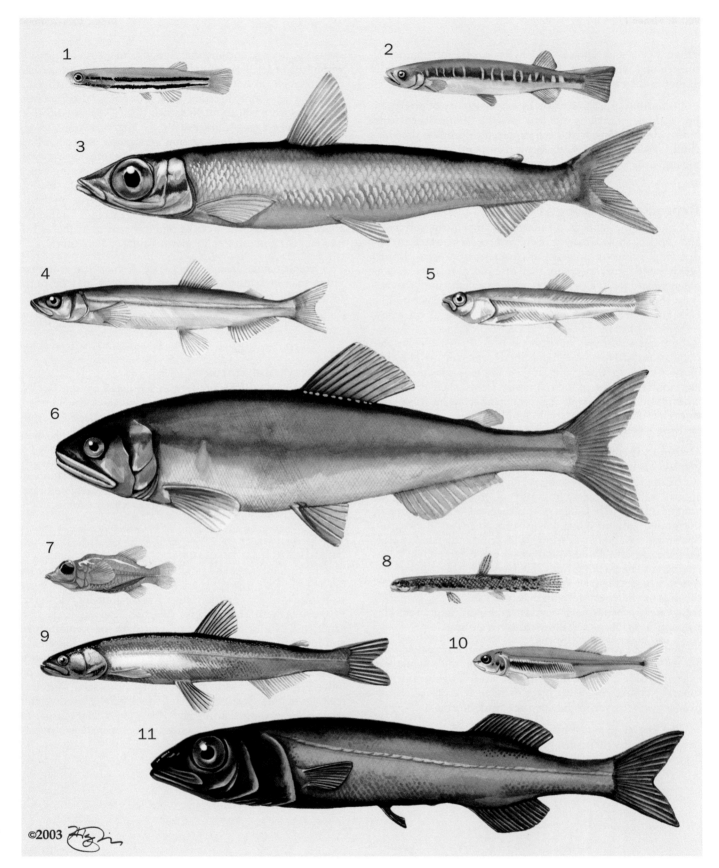

1. Black-stripe minnow (*Galaxiella nigrostriata*); 2. Western minnow (*Galaxias occidentalis*); 3. Greater argentine (*Argentina silus*); 4. Capelin (*Mallotus villosus*); 5. California smoothtongue (*Leuroglossus stilbius*); 6. Ayu (*Plecoglossus altivelis*); 7. Barreleye (*Macropinna microstoma*); 8. Salamanderfish (*Lepidogalaxias salamandroides*); 9. Eulachon (*Thaleichthys pacificus*); 10. Australian smelt (*Retropinna semoni*); 11. California slickhead (*Alepocephalus tenebrosus*). (Illustration by Jonathan Higgins)

Species accounts

California slickhead
Alepocephalus tenebrosus

FAMILY
Alepocephalidae

TAXONOMY
Alepocephalus tenebrosus Gilbert, 1892, Santa Barbara Channel, California, United States, 359–822 fathoms (2,154–4,932 ft/ 655–1,500 m).

OTHER COMMON NAMES
None known.

PHYSICAL CHARACTERISTICS
Total length about 23.6 in (60 cm), moderately sized, fusiform, laterally compressed; dorsal fin posteriorly placed; adipose fin absent. Skull translucent, body brown to black. Swim bladder absent.

DISTRIBUTION
North Pacific from Bering Sea to California; southeastern Pacific, Chile.

HABITAT
Marine. Deep waters, oceanic at depths 150–18,045 ft (46–5,500 m).

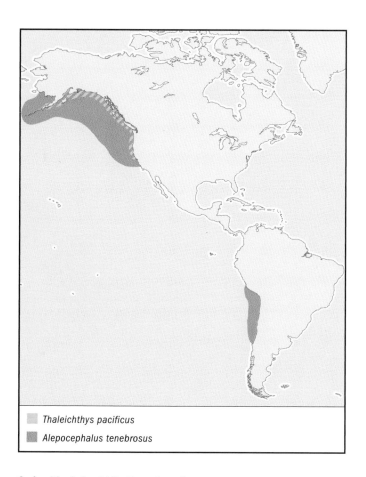

☐ *Thaleichthys pacificus*
■ *Alepocephalus tenebrosus*

BEHAVIOR
Little is known about the biology and behavior. However, other slickheads are gregarious, forming large schools close to the bottom.

FEEDING ECOLOGY AND DIET
Larvae and juveniles ingest benthic and planktonic crustaceans. Others feed on squids, arrow worms, comb jellies, and other deep-sea fishes such as anglerfishes.

REPRODUCTIVE BIOLOGY
Eggs and larvae are pelagic.

CONSERVATION STATUS
Not listed by the IUCN.

SIGNIFICANCE TO HUMANS
Not commercially sought after due to soft texture. ◆

Greater argentine
Argentina silus

FAMILY
Argentinidae

TAXONOMY
Salmo silus Ascanius, 1775, Bergen, Norway.

OTHER COMMON NAMES
English: Atlantic argentine; French: Grande argentine, saumon doré; German: Goldlachs; Spanish: Pez plata, sula.

PHYSICAL CHARACTERISTICS
Standard length about 27.5 in (70 cm), moderately sized, elongate; dorsal fin medially to anteriorly placed, anterior to origin of pelvic fins. Adipose fin present. Brown to olive dorsally, silver laterally and ventrally. Mouth small, and lacks teeth on jaws.

DISTRIBUTION
Western Atlantic, from Davis Strait in the north, to Georges Bank in the south. Eastern Atlantic from Denmark Strait to west coast of Scotland and deeper parts of the North Sea.

HABITAT
Marine. Deep waters, oceanic at depths of 460–4,725 ft (140–1,440 m).

BEHAVIOR
Gregarious, forms large schools close to the bottom.

FEEDING ECOLOGY AND DIET
Feeds on benthic and planktonic crustaceans, squids, arrow worms, and comb jellies. Preyed upon by bony fishes, such as swordfishes and hakes, and seals.

REPRODUCTIVE BIOLOGY
Attains sexual maturity at the end of the fourth year. Spawns in late winter off Canada, and in April and May in the North Sea. Eggs and larvae are pelagic. Juveniles settle on the bottom of waters up to about 410 ft (125 m) deep before moving out into deeper water.

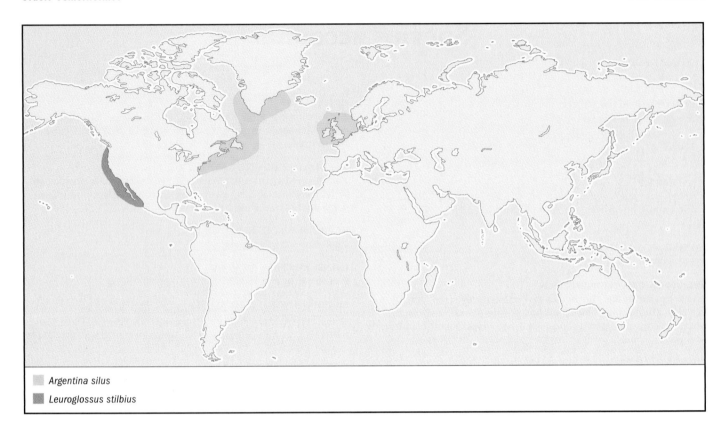

Argentina silus

Leuroglossus stilbius

CONSERVATION STATUS
Not listed by the IUCN. May be particularly susceptible to overfishing because they have a fairly long population doubling time of between 4.5 and 14 years.

SIGNIFICANCE TO HUMANS
A commercial fishery was started by Russia in 1963, joined by Japan in 1968. Estimated sustainable yield is 5,500–24,250 tons (5,000 to 20,000 metric tons). ◆

Western minnow
Galaxias occidentalis

FAMILY
Galaxiidae

TAXONOMY
Galaxias occidentalis Ogilby, 1900, streams south of Perth, Western Australia.

OTHER COMMON NAMES
English: Western galaxias.

PHYSICAL CHARACTERISTICS
Length about 7.9 in (20 cm), small, elongate, scaleless; dorsal fin posteriorly placed. Olive green dorsally, fading to cream/white ventrally. Dark lateral bars often present.

DISTRIBUTION
Restricted to the southern corner of Western Australia, from approximately 155 mi (250 km) north of Perth to 50 mi (80 km) east of Albany.

HABITAT
Fresh waters in streams, rivers, lakes, and ponds/pools; also in brackish habitats. Schools amid riparian vegetation and open waters at all depths.

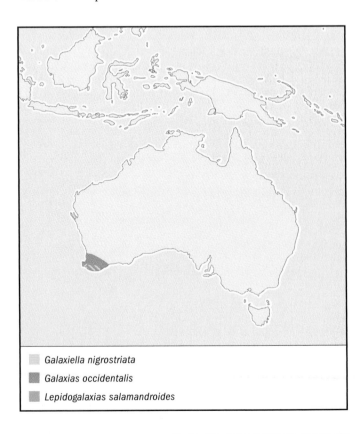

Galaxiella nigrostriata

Galaxias occidentalis

Lepidogalaxias salamandroides

BEHAVIOR
Little is known.

FEEDING ECOLOGY AND DIET
During the day most food eaten is from the water surface (e.g., flying insects and spiders), although insect larvae and small crustaceans from the substrate and within the water column are also taken. At night switches diet to freshwater shrimps.

REPRODUCTIVE BIOLOGY
Attains sexual maturity at the end of the first year at about 2.9 in (7.5 cm). Breeding occurs during wet winter months between June and September, peaks in August (winter). Migrates into creeks or the inundated shores of lakes to spawn. Mean number of mature eggs is 905; diameter of mature eggs is 0.05 in (0.13 cm). Can live up to five years and will spawn at the end of each of these years.

CONSERVATION STATUS
Not listed by the IUCN.

SIGNIFICANCE TO HUMANS
Fishers use them as bait for introduced trouts and European perches. ◆

Black-stripe minnow
Galaxiella nigrostriata

FAMILY
Galaxiidae

TAXONOMY
Galaxias pusillus nigrostriatus Shipway, 1953, a small drain running into Marbellup Creek on a property owned by Mr. Byland, Elleker, near Albany, Western Australia.

OTHER COMMON NAMES
German: Schwartzstreifen-Zwerggalaxie.

PHYSICAL CHARACTERISTICS
Total length about 2 in (5 cm), tiny, elongate, scaleless; dorsal fin posteriorly placed. Between June and September, adults develop nuptial coloration of two longitudinal black stripes separated by a brilliant yellow (in light-colored waters) to red stripe (in dark-colored waters). In dark-colored waters, the dorsal-most black stripes extend upward and meet. Larvae show the same nuptial coloration. Outside breeding season, color is uniform brown to black, with a faint dull-yellow to red lateral stripe.

DISTRIBUTION
Restricted to a small area of peat flats on the southern coast of Western Australia, from Augusta in the west (about 220 mi/ 350 km south of Perth) to Albany in the east. Two disjunct populations about 125 mi and 250 mi (200 and 400 km) to the north were found in the 1990s, suggesting that the loss of habitat caused by massive urban and rural development during the previous hundred years had a significant impact.

HABITAT
Fresh waters in streams, ponds, and shallow pools. Most commonly found in highly tannin-stained and acidic ephemeral pools, in which salamanderfish also occur. Small schools found amid riparian vegetation and open waters and at all depths.

BEHAVIOR
Estivates when pools become dry in summer. Large numbers are found in even the smallest pools (11.8 in/30 cm diameter) before they completely dry up. When pools were filled artifi-

cially, black-stripe minnow and salamanderfish emerged from the substrate within two hours. Unlike salamanderfish, does not apparently have any specific anatomical, physiological, or behavioral adaptations to aid estivation, and must presumably survive in crustacean burrows (and similar) that contain water through the dry season.

FEEDING ECOLOGY AND DIET
Active carnivores at all stages. The smallest larvae, 0.19–0.43 in (0.5–1.1 cm), feed predominantly on rotifers; larger larvae, 0.5–0.9 in (1.2–2.3 cm) consume fewer rotifers, but far more small crustaceans and insect larvae. Juveniles and adults continue to take crustaceans and insect larvae, but a large part of their diet consists of terrestrial insects and spiders taken from the surface. Larvae and small juveniles assume a head-down, tail-up stance (between 30 and 45°), forming their body into an "S" shape as prey approaches, striking at high speed when the prey is within about half a body length.

REPRODUCTIVE BIOLOGY
Sexually mature at the end of the first year, when males are about 1.3 in (3.3 cm) and females 1.4 in (3.7 cm). Individuals spawn several times between June and September, peak activity in late June/early July. Mean number of mature eggs is 62; diameter is 0.03 in (0.07 cm). Most adults die within two months of spawning.

CONSERVATION STATUS
Not listed by the IUCN. Regarded as vulnerable by the Australian Society for Fish Biology.

SIGNIFICANCE TO HUMANS
The Western Australian Department of Fisheries and commercial breeders of ornamental fishes are considering its suitability as an aquarium fish. ◆

Salamanderfish
Lepidogalaxias salamandroides

FAMILY
Lepidogalaxiidae

TAXONOMY
Lepidogalaxias salamandroides Mees, 1961. Tiny creek about 6 mi (9.6 km) east-northeast of Shannon River, Western Australia.

OTHER COMMON NAMES
English: Shannon mudminnow; German: Salamanderfisch.

PHYSICAL CHARACTERISTICS
Length about 2.9 in (7.5 cm), small, elongate; dorsal fin posteriorly placed. Scales absent on head, back, and belly. Pelvic fins well developed, even at hatching. Males have enlarged anal fin sheathed with scales. Brown to gray overall, dark saddlelike markings dorsally, and blotches laterally, the latter forming a pair of stripes during breeding.

DISTRIBUTION
Restricted to a small area of peat flats on the southern coast of Western Australia. Used to extend from Margaret River in the west (about 186 mi/300 km) south of Perth) to Albany in the east; in 2003 its range had contracted to central 93–125 mi/ 150–200 km) of this area.

HABITAT
Fresh waters in streams, ponds, and shallow pools. Most common in ephemeral, tannin-stained, highly acidic (pH 3.0) pools that can exceed 95°F (35°C) in summer. Solitary and benthic,

rests on pelvic fins amid riparian vegetation and in open waters. Larvae and juveniles most common in very shallow water (less than 3.9 in/10 cm), tend to move into deeper waters (usually less than 39.4 in/1 m) as they grow.

BEHAVIOR
Exhibits several behaviors rare among fishes. Like lungfishes, estivates when pools in which they live become dry, burrowing into the substrate, forming an "S" shape, and producing a mucous cocoon, emerging when pools refill with winter rain. Unlike black-striped minnow, buries itself before the pool becomes almost dry.

FEEDING ECOLOGY AND DIET
Active carnivore at all stages, feeds on small crustaceans and insect larvae. Individuals rest on the substrate with their head and body raised up, supported by the large pelvic fins present at hatching. Unlike any other species of fishes, salamanderfish have very large gaps between the vertebrae immediately behind the head. These gaps allow them to move their head from side to side when scanning for prey, which is crucial, as muscles that allow movement of the eyes are absent or greatly reduced. When prey are within about half a body length, darts forward, engulfing and swallowing prey whole.

REPRODUCTIVE BIOLOGY
About 27% attain sexual maturity and spawn at the end of the first year, when males are 1.5 in (3.9 cm) and females are about 1.7 in (4.3 cm). All two-year-old fishes mature and spawn. Each individual spawns several times between late May and late August, peak activity in late July/early August. Mean number of mature eggs is approximately 82; diameter is 0.07 in (0.18 cm). Many fish survive to spawn in two seasons, but only a few survive to spawn at the end of the third and fourth years. Spawning behavior is unique. Males develop an enlarged anal fin surrounded by a sheath of scales. On finding a receptive female, a male will approach and nudge her sides, if she does not swim away or react aggressively, he will approach from the side, and using his scale sheath grasp the female around her anal fin and cloaca. Sperm is introduced into the female via his enlarged anal fin. In the laboratory, males may remain attached to a female for several hours. When physically removed, a male always reattaches from the same side; apparently, just like humans, salamanderfish are either right or left "handed."

CONSERVATION STATUS
Not listed by the IUCN. Regarded as vulnerable by the Australian Society for Fish Biology.

SIGNIFICANCE TO HUMANS
None, although the Western Australian Department of Fisheries and commercial breeders of ornamental fishes are considering its suitability as an aquarium fish. ◆

California smoothtongue
Leuroglossus stilbius

FAMILY
Microstomatidae

TAXONOMY
Leuroglossus stilbius Gilbert, 1890, Pacific off northwestern Mexico, Albatross sta. 2997, 221 fathoms (1,326 ft/400 m).

OTHER COMMON NAMES
English: Esperlan, smoothtongue, southern smoothtongue; Spanish: Esperlan, esperlan de lengua suave.

PHYSICAL CHARACTERISTICS
Length about 7.9 in (20 cm), small to moderately sized, elongate; dorsal fin medially placed; adipose fin present; eyes large. Body light to dark, may be silvery.

DISTRIBUTION
Eastern Pacific from Oregon, United States, in the north, to Gulf of California, Mexico, in the south.

HABITAT
Marine. Deep waters, oceanic at depths between surface and 2,265 ft (690 m).

BEHAVIOR
Gregarious, forms large schools.

FEEDING ECOLOGY AND DIET
Benthic and planktonic crustaceans, worms, and tunicates have been recorded in the guts of juveniles. Reported to be preyed upon by tunas off the coast of the United States. Likely to have specialized eyes as those of some close relatives. Eyes are positioned on the top of the head to increase binocular vision, and in each eye a gap in front of the lens increases illumination of the retina, which itself has specialized photoreceptors. These specializations ensure great visual sensitivity in the forward plane and spatial perception, attributes particularly relevant when searching for prey in dim environments.

REPRODUCTIVE BIOLOGY
Spawning occurs in the California Current between December and May. Eggs and larvae are pelagic.

CONSERVATION STATUS
Not listed by the IUCN.

SIGNIFICANCE TO HUMANS
None known. ◆

Barreleye
Macropinna microstoma

FAMILY
Opisthoproctidae

TAXONOMY
Macropinna microstoma Chapman, 1939, northeastern Pacific, 53°50"00"N, 134°20"00", 2,950–2,300 ft (900–700 m) wire out.

OTHER COMMON NAMES
English: Pacific barreleye; Japanese: Demenigisu.

PHYSICAL CHARACTERISTICS
Length about 1.8 in (4.5 cm). Small, deep-bodied, laterally compressed; dorsal fin posteriorly placed; adipose fin present. Eyes telescopic and large. Head clear, body black dorsally and ventrally, may be silver laterally.

DISTRIBUTION
North Pacific from Bering Sea in the north to Japan and Mexico in the south.

HABITAT
Marine. Deep waters, oceanic at depths about 330–2,950 ft (100–900 m).

BEHAVIOR
Nothing known.

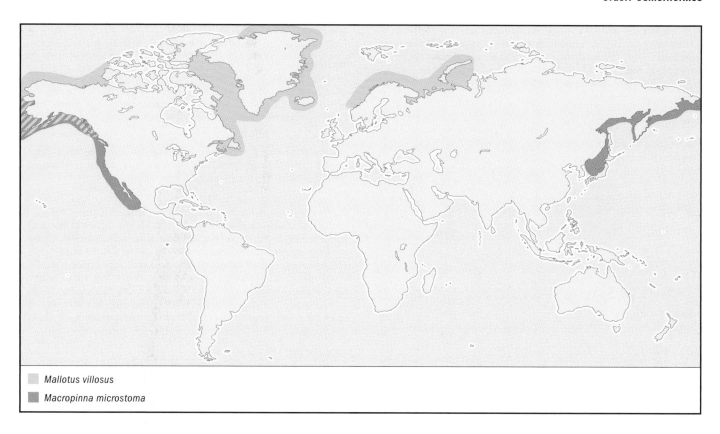

Mallotus villosus

Macropinna microstoma

FEEDING ECOLOGY AND DIET
Not much is known. Other barreleye feed on stinging cells of
jellyfishes. Each stalked eye in the closely related brownsnout
spookfish (*Dolichopteryx longipes*) has two retinas, one directly
below the lens, the other to the side of the lens. The first retina
contains a golden to silvery pigment that ensures any light is
reflected backward and forward, passing through photoreceptors several times before its energy is expended, and thus maxi-
mizing the efficiency of the eye. The second retina, although
not as efficient, permits the barreleye to distinguish the movement of brightly lit objects, and thus see prey from below. Due
to the excellent binocular vision provided by these tubular eyes,
the barreleye can determine position precisely, and can also see
predators that may be approaching from its side. Such optic
specializations thus aid in both feeding and predator avoidance.

REPRODUCTIVE BIOLOGY
Larvae, and probably eggs, are pelagic.

CONSERVATION STATUS
Not listed by the IUCN.

SIGNIFICANCE TO HUMANS
None known. ◆

Capelin
Mallotus villosus

FAMILY
Osmeridae

TAXONOMY
Mallotus villosus Muller, 1776, Iceland.

OTHER COMMON NAMES
English: Whitefish, lodde; French: Capelan atlantique; German: Lodde; Spanish: Capelan.

PHYSICAL CHARACTERISTICS
Total length about 9.8 in (25 cm). Small, moderately deep
bodied; dorsal fin medially placed; adipose fin with long
base. Body covered in small scales. Olive dorsally, silver laterally and ventrally. Enlarged pectoral and anal fins; two
ridges of enlarged scales along each side of body in breeding
males.

DISTRIBUTION
Circumpolar in the Arctic, North Atlantic, and North Pacific
Oceans.

HABITAT
Marine. Oceanic moving into coastal seas to spawn. Found
down to about 985 ft (300 m).

BEHAVIOR
Little is known.

FEEDING ECOLOGY AND DIET
All stages feed on zooplankton and small benthic organisms
such as euphausiid shrimps, copepods, amphipods, and worms.
Predators such as rays, a range of bony fishes, seabirds, seals,
whales, and dolphins take an estimated 3.3 million tons (3 million metric tons) in the northwest Atlantic.

REPRODUCTIVE BIOLOGY
Forms massive spawning aggregations over shallow banks.
Spawns over a two to three month period between March and
October, depending on location. Two males flank and hold a
female with enlarged fins and scale ridges until she releases her

eggs, they then release the sperm. Females release up to 12,000 adhesive eggs. Many individuals survive to spawn in more than one year, although most spawning animals are in year three or four. Many fish spawning along shorelines, particularly males, become stranded on beaches and die.

CONSERVATION STATUS
Not listed by the IUCN.

SIGNIFICANCE TO HUMANS
Very important in commercial, recreational, and traditional fisheries throughout the range. In 1998 the commercial fishery in the northwest Atlantic was estimated at 121,000 tons (110,000 metric tons). Marketed fresh, frozen, salted, dried, smoked, as fish meal (males), and roe (females). ◆

Ayu
Plecoglossus altivelis

FAMILY
Osmeridae

TAXONOMY
Salmo (Plecoglossus) altivelis Temminck and Schlegel, 1846, Japan.

OTHER COMMON NAMES
English: Ko-ayu, sweetfish; French: Ayu; Japanese: Koayu.

PHYSICAL CHARACTERISTICS
Length about 11.8 in (30 cm). Small, moderately deep bodied; dorsal fin medially placed; adipose fin present. Body covered in small scales. Olive dorsolaterally, white ventrally.

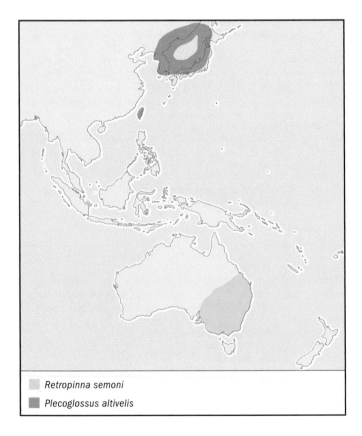

▨ *Retropinna semoni*
▮ *Plecoglossus altivelis*

DISTRIBUTION
Coastal seas and rivers in Japan, China, Korea, and Taiwan.

HABITAT
Marine, brackish, and fresh waters. Amphidromous, demersal in coastal seas, estuaries, rivers, streams, and lakes to a depth of about 33 ft (10 m). Landlocked forms also exist. In rivers and lakes appears to prefer clear waters.

BEHAVIOR
Except for feeding and reproduction little is known regarding specific behavior patterns. In rivers, form territories, which they guard by attacking and nipping other ayu.

FEEDING ECOLOGY AND DIET
Larvae and juveniles feed primarily on small benthic crustaceans such as copepods and amphipods. Juveniles of the amphidromous form move into fresh waters in winter and spring, moving up into the higher reaches of rivers and streams. Adults use specially modified jaws and teeth to scrape algae from rocks. Adults have also been reported to eat small pebbles, although it seems likely that these were inadvertently ingested while grazing on algae. The few adults that survive spawning and return to the sea have been reported to feed on zoobenthos.

REPRODUCTIVE BIOLOGY
Sexual maturity usually attained at end of first year at about 7.9–11.8 in (20–30 cm). Spawns in freshwater during autumn, when adults move downstream to the spawning grounds. At night, excavates small pits in sand or gravel banks into which about 10,000 adhesive eggs (0.04 in/0.1 cm) are released. Eggs hatch 14–20 days later. Larger individuals spawn once, after which most die; smaller individuals have about a 50% chance of surviving to spawn again two weeks later.

CONSERVATION STATUS
Not listed by the IUCN.

SIGNIFICANCE TO HUMANS
Ayu form important commercial, aquaculture, recreational, and traditional fisheries. In Japan the commercial fishery and aquaculture accounted for 16,500 tons (15,000 metric tons) and 8,820 tons (8,000 metric tons), respectively, in 1979. Sport fishing in Japan involves baited hooks, flies, as well as a traditional method. In the traditional method, an ayu that has a small free-swinging treble hook fastened to it is attached to the line. Anglers swing the attached fish to ayu guarding territory, and when the resident fish nips the back of the "invader" it is caught on the free-swinging hook. In Japan the ayu forms the basis of the cormorant fishery, whereby cormorants are trained to dive for fish. This fishery is believed to date back at least 2,500 years, but is now for tourists. ◆

Eulachon
Thaleichthys pacificus

FAMILY
Osmeridae

TAXONOMY
Salmo (Mallotus?) pacificus Richardson, 1836, Columbia River, no higher than Katpootl, northwestern United States.

OTHER COMMON NAMES
English: Candlefish, oilfish, fathom fish, hooligan; French: Eulachon.

PHYSICAL CHARACTERISTICS
Total length 11.8 in (30 cm), small, slender bodied; dorsal fin medially placed; adipose fin present. Body covered in small scales. Brown to blue dorsally, silver laterally, white ventrally. Raised ridge along middle of body; large tubercles on fins of breeding males are absent or small in females.

DISTRIBUTION
Northwest Pacific between 61° and 36°N. From west of Saint Matthew Island and Kuskokwim (Bering Sea) in the north to Monterey Bay in northern California in the south.

HABITAT
Marine, brackish, and fresh waters. Anadromous, found in coastal seas, estuaries, rivers, and streams to a depth of 2,050 ft (625 m). Landlocked forms exist.

BEHAVIOR
Little is known.

FEEDING ECOLOGY AND DIET
All stages feed on zooplankton and small benthic crustaceans such as mysid and euphausiid shrimps, as well as copepods and amphipods. Adults do not feed in freshwater. Reported to be preyed upon by dogfishes, salmonids, cods, flatfishes, sturgeons, seagulls, seals, and porpoises.

REPRODUCTIVE BIOLOGY
Sexual maturity is attained at end of third year. Spawning migrations from the sea to rivers occur when river temperatures rise above 40°F (4.4°C), spawns in spring after the ice melt. Sticky eggs with a short stalk are broadcast over sand or gravel. Larvae hatch after about 30 days and are swept downstream and out to sea. Most adults die following spawning, a few survive for five years. Adults may return to natal streams to spawn.

CONSERVATION STATUS
Not listed by the IUCN.

SIGNIFICANCE TO HUMANS
Traditionally an important fishery for Native Americans for food and oil. Fish are so oily they were dried and used as candles. Used as food for minks and other animals farmed for fur. ◆

Australian smelt
Retropinna semoni

FAMILY
Retropinnidae

TAXONOMY
Prototroctes semoni Weber, 1895, Burnett River, Queensland, Australia.

OTHER COMMON NAMES
German: Australischer Stint; Polish: Rakietniczka semona.

PHYSICAL CHARACTERISTICS
Total length 3.9 in (10 cm), small, elongate; dorsal fin posteriorly placed; adipose fin present. Scales absent on head. Olive green dorsally, golden to orange or purple laterally, silvery ventrally. Fins enlarged in breeding males. Has only a left gonad. Often smells of cucumber when fresh.

DISTRIBUTION
Widespread in coastal drainages of southeastern Australia, from the Fitzroy River in southern Queensland to eastern South Australia; also in the Cooper Creek drainage of Lake Eyre.

HABITAT
Slow-flowing streams and rivers, lakes, and ponds/pools; also in brackish habitats. Schools among riparian vegetation and open waters at all depths.

BEHAVIOR
Gregarious, forms large schools from midwater to the surface in large open waters.

FEEDING ECOLOGY AND DIET
Feeds on insects, microcrustaceans, and algae. An important component of the diets of other fishes.

REPRODUCTIVE BIOLOGY
Sexual maturity usually attained at end of first year at about 2–3.9 in (5–10 cm). Some fish from inland and northern drainages may reach maturity in 9–11 months at less than 1.6 in (4 cm). Spawns in fresh waters during spring. Spawning individuals develop nuptial tubercles on scales and fin rays. Releases 100–1,000 adhesive eggs (0.03 in/0.08 cm) over the streambed and/or aquatic vegetation. Larvae of about 0.18 in (0.45 cm) hatch after 10 days. Coastal populations may be amphidromous.

CONSERVATION STATUS
Not listed by the IUCN.

SIGNIFICANCE TO HUMANS
Introduced to Tasmania as forage for introduced trouts.

Resources

Books
Allen, G. R. *Freshwater Fishes of Australia.* Neptune City, NJ: T. F. H. Publications, 1989.

Allen, G. R., S. H. Midgley, and M. Allen. *Guide to the Freshwater Fishes of Australia.* Perth: Western Australian Museum, 2002.

Berra, T. M. *An Atlas of Distribution of the Freshwater Fish Families of the World.* Lincoln: University of Nebraska Press, 1981.

Coad, B. W. *Guide to the Marine Sport Fishes of Atlantic Canada and New England.* Toronto: University of Toronto Press, 1992.

Eschmeyer, W. N. *Catalog of the Genera of Recent Fishes.* San Francisco: California Academy of Sciences, 1990.

Fink, W. L. "Basal Euteleosts: Relationships." In *Ontogeny and Systematics of Fishes,* edited by H. G. Moser, W. J. Richards, D. M. Cohen, M. P. Fahay, A. W. Kendall, Jr., and S. L. Richardson, Special Publication Number 1. Lawrence, KS:

Resources

American Society of Ichthyologists and Herpetologists, Allen Press, 1984.

Glover, J. C. M. "Argentinidae, Bathylagidae, Opisthoproctidae, Alepocephalidae, and Platytroctidae." In *The Fishes of Australia's South Coast*, edited by M. F. Gomon, J. C. M. Glover, and R. H. Kuiter. Adelaide, Australia: State Print, 1994.

Helfman, G. S., B. B. Collette, and D. E. Facey. *The Diversity of Fishes*. Malden, MA: Blackwell Science, 1997.

Johnson, G. D., and C. Patterson. "Relationships of Lower Euteleostean Fishes." In *Interrelationships of Fishes*, edited by M. L. J. Stiassny, L. R. Parenti, and G. D. Johnson. San Diego: Academic Press, 1996.

Koehn, J. D., and W. G. O'Connor. *Biological Information for Management of Native Freshwater Fish in Victoria.* Melbourne: Victorian Government Printing Office, 1990.

Marshall, N. B. *Developments in Deep-Sea Biology*. Poole, Australia: Blandford Press, 1979.

McDowall, R. M. *Diadromy in Fishes: Migrations Between Freshwater and Marine Environments*. London: Croom Helm, 1988.

———. *Freshwater Fishes of South-Eastern Australia*. Sydney: Reed Books, 1996.

Morgan, D. L., H. S. Gill, and I. C. Potter. *Distribution, Identification and Biology of Freshwater Fishes in South-Western Australia.* Perth, Australia: Western Australian Museum, 1998.

Nelson, J. S. *Fishes of the World*, 3rd edition. New York: John Wiley & Sons, 1994.

Paxton, J. R., and D. M. Cohen. "Argentinidae." In *Living Marine Resources of the Western Central Pacific.* Vol. 3, *Batoid Fishes, Chimaeras and Bony Fishes, Part 1 (Elopidae to Linophrynidae)*, edited by K. E. Carpenter and V. H. Niem. Rome: FAO, 1999.

———. "Alepocephalidae." In *Living Marine Resources of the Western Central Pacific.* Vol. 3, *Batoid Fishes, Chimaeras and Bony Fishes, Part 1 (Elopidae to Linophrynidae)*, edited by K. E. Carpenter and V. H. Niem. Rome: FAO, 1999.

———. "Bathylagidae" In *Living Marine Resources of the Western Central Pacific.* Vol. 3, *Batoid Fishes, Chimaeras and Bony Fishes, Part 1 (Elopidae to Linophrynidae)*, edited by K. E. Carpenter and V. H. Niem. Rome: FAO, 1999.

———. "Opisthoproctidae." In *Living Marine Resources of the Western Central Pacific.* Vol. 3, *Batoid Fishes, Chimaeras and Bony Fishes, Part 1 (Elopidae to Linophrynidae)*, edited by K. E. Carpenter and V. H. Niem. Rome: FAO, 1999.

Sazonov, Y. I. "Platytroctidae." In *Living Marine Resources of the Western Central Pacific.* Vol. 3, *Batoid Fishes, Chimaeras and Bony Fishes, Part 1 (Elopidae to Linophrynidae)*, edited by K. E. Carpenter and V. H. Niem. Rome: FAO, 1999.

Sazonov, Y. I., and D. F. Markle. "Alepocephalidae." In *Living Marine Resources of the Western Central Pacific.* Vol. 3, *Batoid Fishes, Chimaeras and Bony Fishes, Part 1 (Elopidae to Linophrynidae)*, edited by K. E. Carpenter and V. H. Niem. Rome: FAO, 1999.

Wheeler, A. *The World Encyclopedia of Fishes*. London: Macdonald, 1985.

Periodicals

Allen, G. R., and T. M. Berra. "Life History Aspects of the West Australian Salamanderfish, *Lepidogalaxias salamandroides* Mees." *Records of the Western Australian Museum* 14 (1989): 253–267.

Begle, D. P. "Monophyly and Relationships of Argentinoid Fishes." *Copeia* (1992): 350–366.

———. "Relationships of the Osmeroid Fishes and the Use of Reductive Characters in Phylogenetic Analysis." *Systematic Zoology* 40 (1991): 33–53.

Berra, T. M., and G. R. Allen. "Burrowing, Emergence, Behaviour, and Functional Morphology of the Australian Salamanderfish, *Lepidogalaxias salamandroides*." *Fisheries* 14 (1989): 2–10.

———. "Population Structure and Development of *Lepidogalaxias salamandroides* (Pisces: Salmoniformes) from Western Australia." *Copeia* (1991): 845–850.

———. "Inability of Salamanderfish, *Lepidogalaxias salamandroides*, to Tolerate Hypoxic Water." *Records of the Western Australian Museum* 17 (1995): 117.

Gill, H. S., and D. L. Morgan. "Larval Development in the Salamanderfish, *Lepidogalaxias salamandroides*." *Copeia* (1999): 219–224.

———. "Ontogenetic Changes in the Diet of *Galaxiella nigrostriata* (Shipway, 1953) (Galaxiidae) and *Lepidogalaxias salamandroides* Mees, 1961 (Lepidogalaxias)." *Ecology of Freshwater Fish* (in press).

Gill, H. S., and F. J. Neira. "Larval Descriptions of Three Galaxiid Fishes Endemic to South-Western Australia: *Galaxias occidentalis, Galaxiella munda* and *Galaxiella nigrostriata* (Salmoniformes: Galaxiidae)." *Australian Journal of Marine and Freshwater Research* 45 (1994): 1,307–1,317.

Iguchi, K., and Y. Tsukamoto. "Semelparous or Iteroparous: Resource Allocation Tactics in the Ayu, an Osmeroid Fish." *Journal of Fish Biology* 58 (2001): 520–528.

McDowall, R. M., and R. S. Frankenberg. "The Galaxiid Fishes of Australia (Pisces: Galaxiidae)." *Records of the Australian Museum* 33 (1981): 443–605.

McDowall, R. M., and B. J. Pusey. "*Lepidogalaxias salamandroides* Mees: A Redescription, with Natural History Notes." *Records of the Western Australian Museum* 11 (1983): 11–23.

Morgan, D. L., and H. S. Gill. "Fish Associations Within the Different Inland Habitats of Lower South-Western Australia." *Records of the Western Australian Museum* 20 (2000): 31–37.

Morgan, D. L., H. S. Gill, and I. C. Potter. "Age Composition, Growth and Reproductive Biology of the Salamanderfish *Lepidogalaxias salamandroides*: A Re-examination." *Environmental Biology of Fishes* 57 (2000): 191–204.

Morgan, D. L., D. C. Thorburn, and H. S. Gill. "Salinization of South-western Western Australian Rivers and the Implications for the Inland Fish Fauna: The Blackwood River, a Case Study." *Pacific Conservation Biology* (in press).

Pen, L. J. and I. C. Potter. "Biology of the Western Minnow, *Galaxias occidentalis* Ogilby (Teleostei: Galaxiidae), in a South-Western Australian River. 1. Reproductive Biology." *Hydrobiologia* 211 (1991): 77–88.

Resources

————. "Biology of the Western Minnow, *Galaxias occidentalis* Ogilby (Teleostei: Galaxiidae), in a South-Western Australian River. 2. Size and Age Composition, Growth and Diet." *Hydrobiologia* 211 (1991): 89–100.

Pen, L. J., H. S. Gill, I. C. Potter, and P. Humphries. "Growth, Age Composition, Reproductive Biology and Diet of the Black-Stripe Minnow *Galaxiella nigrostriata* (Shipway), Including Comparisons with the Other Two *Galaxiella* Species." *Journal of Fish Biology* 43 (1993): 847–863.

Pen, L. J., I. C. Potter, and R. W. Hilliard. "Biology of *Galaxiella munda* McDowall (Teleostei: Galaxiidae), Including a Comparison of the Reproductive Strategies of This and Three Other Local Species." *Journal of Fish Biology* 39 (1991): 717–731.

Pusey, B. J. "The Shannon Mud Minnow." *Fishes of Sahul: Journal of Australia and New Guinea Fishes Association* 1 (1983): 9–11.

————. "Aestivation in the Teleost Fish *Lepidogalaxias salamandroides* Mees." *Comparative Biochemistry and Physiology* 92A (1989): 137–138.

————. "Seasonality, Aestivation and the Life History of the Salamanderfish *Lepidogalaxias salamandroides* (Pisces: Lepidogalaxiidae)." *Environmental Biology of Fishes* 29 (1990): 15–26.

Rosen, D. E. "Phylogeny and Zoogeography of Salmoniform Fishes and Relationships of *Lepidogalaxias salamandroides*." *Bulletin of the American Museum of Natural History* 153 (1974): 265–326.

Waters, J. M., J. A. Lopez, and G. P. Wallis. "Molecular Phylogenetics and Biogeography of Galaxiid Fishes (Osteichthyes: Galaxiidae): Dispersal, Vicariance, and the Position of *Lepidogalaxias salamandroides*." *Systematic Zoology* 49 (2000): 777–795.

Howard Stamper Gill, PhD

Salmoniformes
(Salmons)

Class Actinopterygii

Order Salmoniformes

Number of families 1

Photo: Sockeye salmon (*Oncorhynchus nerka*) in spawning area of Alaska, USA. (Photo by Animals Animals ©Victora McCormick. Reproduced by permission.)

Evolution and systematics

Salmoniform fossils are known from the Upper Cretaceous (about 100 million years ago) until the Pleistocene (about 11,000 years ago), mostly from North America and Europe. This order possesses several primitive anatomical features representative of an early stage in the evolution of modern bony fishes. There have been a number of systematic arrangements for this group, ranging from a single order with a single family (Salmonidae), to two orders with several families (see, for example, Johnson and Patterson 1996). For the purpose of this chapter, we will follow Nelson (1994), who considers the Salmonidae the only family for this order, with 3 subfamilies, 11 genera, and 66 species.

Physical characteristics

The largest of the salmoniform fishes are the Chinook salmon (*Oncorhynchus tshawytscha*) and the Danube and Siberian huchens (*Hucho hucho*) both about 59 in (1.5 m) in length, with the former weighing up to 136 lb (62 kg). As expected from such strong swimmers, the salmoniform body and fins are streamlined and symmetrical, being covered with small and smooth cycloid scales. All fins have soft rays. They have a small, fleshy adipose fin (which is a primitive character) located between the dorsal fin and the powerful caudal fin. The dorsal fin is located midway along the body; the paired pectoral fins are ventral and located directly posterior to the head. They also have a pair of pelvic fins directly beneath the dorsal fin, and a single anal fin located beneath the adipose fin. The swim bladder is connected to the gut.

Distribution

Salmoniformes were originally found only in cool and cold waters of the Northern Hemisphere, where they are one of the most dominant freshwater fishes. In the North American continent, they are found from tributaries of the Arctic Ocean to tributaries of the Gulf of California in northwestern Mexico. They are also found from Eurasia to all Asia north to the Himalayas from the Arctic Circle south to Bangladesh, Manchuria, and the Korean Peninsula and the Japanese Islands and Taiwan. In Africa they are only found in the northwestern margin (Atlas mountains of Morocco and Algeria). The Arctic char (*Salvelinus alpinus*) is the world's most northerly occurring freshwater fish. Many species of this order have been successfully introduced in many other parts of the world.

Habitat

Many salmoniform species (such as trouts and salmons) have an anadromous life cycle, in that they spawn in fresh

Brook trout (*Salvelinus fontinalis*) in their vibrant autumn spawning colors. Note the tail of the male, lower left, showing a bite mark received during one of the many battles between males for the right to court a female. (Photograph. AP/Wide World Photos. Reproduced by permission.)

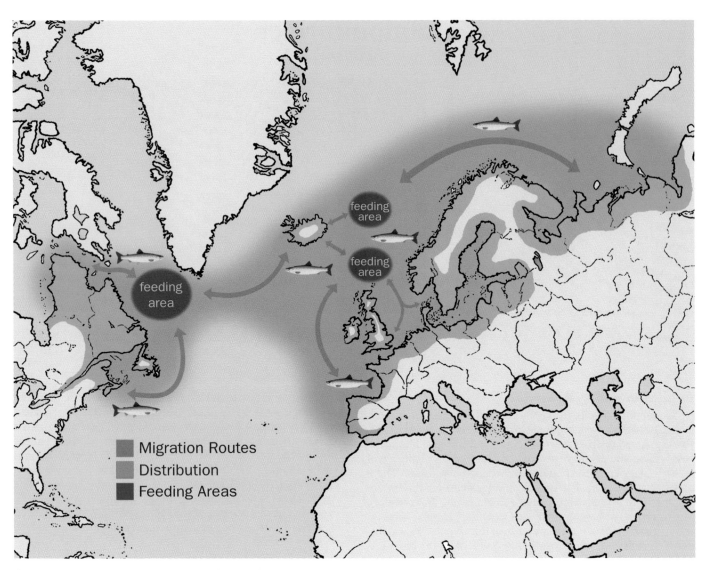

Salmon migration routes. (Illustration by John Megahan)

Grizzly bear fishing for salmon at a waterfall in Katmai National Park in Alaska. (Photo by Galen Rowell/Corbis. Reproduced by permission.)

waters, but migrate to sea for feeding and maturation. The pink salmon (*Oncorhynchus gorbuscha*) is the least anadromous species in the Pacific, since it has reduced the freshwater stage to the spawning migration and incubation of the eggs. Some individuals even spawn in the intertidal zone with no real freshwater phase. However, others are entirely freshwater, with completely landlocked local populations.

Behavior

Some species of Salmoniformes are fiercely territorial; others form schools shortly after hatching before they initiate their seaward migration. the most remarkable behavioral characteristic in these fishes is their strong swimming ability (some can leap over obstacles such as waterfalls as high as 10 ft/3 m) and their migratory capability. Almost all, if not all, Salmoniformes can return to the stream of their birth after migrating thousands of miles (or kilometers) in the ocean for one or more years, a behavior known as homing. They use their sense of smell to orient themselves, but some experi-

Sockeye salmon (*Oncorhynchus nerka*) spawning in North Fork Payette River, Idaho. (Photo by Willam H. Mullins/Photo Researchers, Inc. Reproduced by permission.)

Reproductive biology

All salmoniforms lay eggs that are externally fertilized. The egg size is related to the amount of nutrients in the water, with the largest eggs (0.16–0.31 in/4–8 mm), found in nutrient-poor waters. This allows individuals born under those conditions to have enough nutrients to survive, and is followed by direct development. This means that the young look very much like small adults. The size of the egg is also conversely proportional to the number of the eggs. In the case of smaller eggs, the young are less developed after hatching.

Conservation status

The IUCN Red List includes 36 salmoniform species. Of these, four are categorized as Extinct (*Coregonus alpenae, C. johannae, C. nigripinnis,* and *Salvelinus agassizi*); four are Critically Endangered; five are Endangered; 10 are Vulnerable; and 13 are Data Deficient.

Because all salmoniforms depend upon migrations into clear, highly oxygenated waters, they are very sensitive to water pollution as well as interruptions in the watercourse by means of dams. This problem is particularly acute for the Pacific salmon, whose numbers and genetic diversity have both declined dramatically.

Significance to humans

Many species of trouts, salmons, chars, whitefishes, and graylings are among the best-known and most intensively studied species of fishes. They have tremendous economic importance in many areas because of their value in both sport and commercial fisheries. This is the reason they have been introduced all over the world. Information about the life cycle and reproduction, some of it based on DNA analysis, has been used to settle disputes between nations regarding the origin of fishes caught in the open ocean, as well as for their management.

ments suggest that vision and magnetic clues may also play a role in this behavior. This is not surprising, since other migratory species of animals, such as some birds, can use more than one source of information to achieve their migratory paths. Because of their migrations between fresh and salt waters, salmoniforms have developed a number of physiological adaptations to cope with changes in salinity. This is achieved via osmotic regulation by excreting excess salts through cells in the gills and by having well-developed kidneys, which, in fresh waters, excrete the excess water that diffuses into their blood via the gills.

Feeding ecology and diet

There is variation in feeding habits in this order. Some species feed upon plankton and benthic invertebrates, while others are top predators of other fish species.

Close-up photo of newly hatched brown trout (*Salmo trutta*) on a river bed. (Photo by Science Pictures Ltd/Science Photo Library/Photo Researchers, Inc. Reproduced by permission.)

1. Brook trout (*Salvelinus fontinalis*); 2. Brown trout (*Salmo trutta*); 3. Lake trout (*Salvelinus namaycush*); 4. Charr (*Salvelinus alpinus*); 5. Atlantic salmon (*Salmo salar*); 6. Arctic grayling (*Thymallus arcticus arcticus*); 7. Lake whitefish (*Coregonus clupeaformis*). (Illustration by John Megahan)

1. Cherry salmon (*Oncorhynchus masou*); 2. Rainbow trout (*Oncorhynchus mykiss*); 3. Golden trout (*Oncorhynchus aguabonita*); 4. Chinook salmon (*Oncorhynchus tshawytscha*); 5. Pink salmon (*Oncorhynchus gorbuscha*); 6. Sockeye salmon (*Oncorhynchus nerka*); 7. Coho salmon (*Oncorhynchus kisutch*). (Illustration by John Megahan)

Species accounts

Lake whitefish
Coregonus clupeaformis

FAMILY
Salmonidae

TAXONOMY
Salmo clupeaformis Mitchill, 1818, Falls of St. Mary's River, Chippewa County, Michigan, United States. May be conspecific with *Coregonus lavaretus*.

OTHER COMMON NAMES
English: Common whitefish, eastern whitefish, lake whitefish; French: Grand corégone; German: Felchen; Spanish: Corégono; Inuktitut: Kavisilik.

PHYSICAL CHARACTERISTICS
Length 40 in (100 cm). A well-developed adipose fin, usually larger among males, characterizes this species. Coloration is dark brown to midnight blue above fading to silver on sides and white beneath.

DISTRIBUTION
North America throughout Alaska and most of Canada, south into New England, the Great Lakes basin, and central Minnesota. Successfully introduced in Chile and Argentina.

HABITAT
Primarily a lake dweller; can also be found in large rivers and enters brackish water.

BEHAVIOR
When not migrating, tends to be sedentary in small lakes. Migration in large lakes consists of movement from deep to shallow water in the spring, movement back to deep water in the summer as the shoal water warms, migration to shallow-water spawning areas in the fall and early winter, and post-spawning movement back to deeper water.

FEEDING ECOLOGY AND DIET
Adults feed on aquatic insect larvae, amphipods, mollusks, as well as smaller fishes and fish eggs, including their own. They are vulnerable to larger fishes, otters, bears, and fish-eating birds.

REPRODUCTIVE BIOLOGY
Spawns annually at night from October through December. One female and one or more males rise to the surface where the eggs are released and fertilized. Spawning fish are very active, sometimes leaping out of the water. Eggs are demersal.

CONSERVATION STATUS
Not listed by the IUCN.

SIGNIFICANCE TO HUMANS
Extensively hatchery-reared in the Great Lakes and other areas because of their value as meat and roe. ◆

Golden trout
Oncorhynchus aguabonita

FAMILY
Salmonidae

TAXONOMY
Salmo mykiss aguabonita Jordan, 1892, Kern River California.

OTHER COMMON NAMES
French: Troite dorée; Spanish: Aguabonita.

PHYSICAL CHARACTERISTICS
Length 28 in (71 cm); maximum weight 11 lb (5 kg). One of the smallest species of trouts, characterized by a golden color with orange to red stripes along the side. In both the dorsal fin

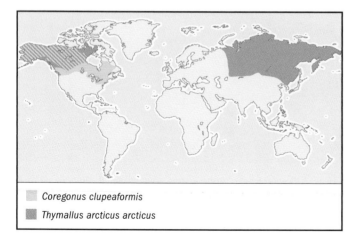

Coregonus clupeaformis	
Thymallus arcticus arcticus	

Oncorhynchus aguabonita	
Oncorhynchus mykiss	

and the area right before the tail, coloration is speckled with dark spots. This species is also known for its small scales.

DISTRIBUTION
High altitude, freshwater bodies in the western area of the United States, particularly in Wyoming, Idaho, Washington State, and most abundantly in California, where it was first discovered. Introduced in Canada.

HABITAT
Freshwater lakes and rivers at altitudes of 9,000–12,000 ft (2,740–3,700 m), generally in hard-to-access mountain areas. Because the waters they inhabit are very clear, with little vegetation, and of great beauty, the Spanish name is *aguabonita* (pretty water).

BEHAVIOR
Unlike other species of salmoniforms, largely a social species that travels in small schools. Active all summer. Because of their lack of aggressiveness and small size, they cannot compete with larger, more aggressive salmonids. This accounts for their restriction to headwaters of streams, where physical barriers such as waterfalls prevent other fish species from entering.

FEEDING ECOLOGY AND DIET
Diet includes surface water-dwelling insects, such as caddisflies and midges, as well as small crustaceans and terrestrial insects floating on the surface. Feeds mostly between May and September due to the scarcity of insects during winter. Practices a form of filter feeding, in which it opens the gills and inhales the prey whole, after which the food remains in the mouth. Vulnerable to larger fishes, otters, bears, and fish-eating birds.

REPRODUCTIVE BIOLOGY
The development of this small species is rather brief when compared to that in other salmoniform species. The egg is almost completely ripe by the time temperatures arrest development at the beginning of winter. The fishes are then ready to spawn the following spring. The spawning routine begins at the time the snow starts to melt, sometime from March through July. As with other trout species, the female prepares the nest and lays her eggs in specific areas. After some behavioral displays, the males then come to fertilize them. Egg development, hatching, and early growth stages are virtually the same as in other spring spawners.

CONSERVATION STATUS
Translocated rainbow trouts aggressively out compete golden trouts and also hybridize with them. The *whitei* population has been classified as threatened since 1978 by the U.S. federal government. The California Department of Fish and Games Committee on Threatened Trout has been working to conserve and enhance the survival of this species. Although many attempts to stock this species in other appropriate waters in the western United States have taken place, most have not been successful.

SIGNIFICANCE TO HUMANS
Highly prized by sport fishers and much sought after as food. ◆

Pink salmon
Oncorhynchus gorbuscha

FAMILY
Salmonidae

TAXONOMY
Salmo gorbuscha Walbaum, 1792, rivers of Kamchatka, Russia.

OTHER COMMON NAMES
English: Humpback salmon; French: Saumon rose; German: Buckelkopflachs; Spanish: Salmón rosado.

PHYSICAL CHARACTERISTICS
Length 30 in (76 cm); weight 15 lb (6.8 kg). The smallest of the true salmon species. Like most salmoniforms, has a streamlined, fusiform body, somewhat laterally compressed. The mouth is terminal, and among breeding males is greatly deformed by being very oblique, with the lower jaw enlarged and turned up at the tip, preventing the mouth from closing. Another characteristic is the presence of large black spots on the back and on both lobes of the caudal fin. The general coloration varies. Individuals at sea are steel blue to blue-green on the back, silver on the sides, and white on the belly, with large oval spots present on the back, the adipose fin, and on both lobes of the caudal fin. Breeding males are dark on the back, and red with brownish green blotches on the sides. Breeding females are similar to males, although less distinctly colored.

DISTRIBUTION
Arctic and Eastern Pacific, from eastern Korea and Hokkaido, Japan, to the Bering and Okhotsk Seas, to Alaska and the Aleutian Islands, to southern California, United States. Successfully introduced in Canada, Ireland, Norway, Greenland, Poland, Finland, and the United Kingdom.

HABITAT
Spends 18 months at sea before returning either to its native river or some other river to spawn. Unique because the homing behavior is not as strong as that of other salmoniform species. After emerging from the gravel, fry move downstream, remaining inshore for a few months before going out to sea.

BEHAVIOR
Reaches sexual maturity at two years of age. Both male and female die up to a few weeks after spawning.

FEEDING ECOLOGY AND DIET
The diet varies with age. Fry feed on nymphal and larval insects while in fresh water, but once at sea may not feed at all until they become juveniles, when they eat copepods and other zooplankton. As they continue to grow the food items shift toward larger crustaceans and fishes. They are preyed upon by other salmonids as fry, and by larger fishes (including sharks), fish-eating birds, and mammals as they grow.

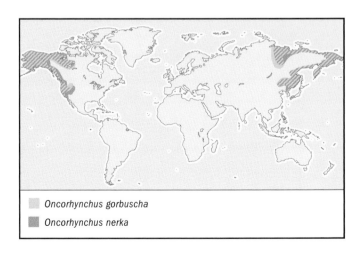

■ *Oncorhynchus gorbuscha*
■ *Oncorhynchus nerka*

REPRODUCTIVE BIOLOGY

Upstream migration takes place from June to late September, triggered by high water. The female builds the redd or spawning trench by lying on one side and using her tail to displace silt and light gravel. The accompanying male spends most of the time defending its territory. When the nest is complete, the female drops into it, followed immediately by the male. As for other salmoniforms, both male and female open their mouths, vibrate, and release eggs (1,200–1,800) and sperm. Then the eggs are covered by the female digging a new redd at the upstream edge of the previous one.

CONSERVATION STATUS

Not listed by the IUCN.

SIGNIFICANCE TO HUMANS

The flesh is highly prized; eggs are highly valued as caviar, particularly in Japan. ◆

Coho salmon

Oncorhynchus kisutch

FAMILY

Salmonidae

TAXONOMY

Salmo kisatch Walbaum, 1792, rivers and lakes of Kamchatka, Russia.

OTHER COMMON NAMES

English: Blueback, silver salmon; French: Saumon argenté; German: Chumlachs; Spanish: Salmón plateado.

PHYSICAL CHARACTERISTICS

Length 42.5 in (108 cm) in males, 25.8 in (65.5 cm) in females; weight 35.5 lb (15.2 kg). Characterized by the presence of small black spots on the back and on the upper lobe of the caudal fin, as well as by the lack of dark pigment along the gum line of the lower jaw. Coloration varies according to environmental and reproductive conditions, as well as with sex. Females are generally more brightly colored than males. At sea all are dark metallic blue or greenish on the back and upper sides, with a brilliant silver color on middle and lower sides, and white below. In fresh water they display small black spots on the back and upper sides, and on upper lobe of the caudal

fin. When ready for breeding they become dark to bright green on head and back and bright red on the sides, and often dark on the belly. The lateral line is nearly straight.

DISTRIBUTION

North Pacific in Asia from the Anadyr River in Russia in the north to Hokkaido, Japan, in the south. In North America from Point Hope in Alaska to Chamalu Bay, Baja California, Mexico. Successfully introduced in Chile, France, and the Laurentian Great Lakes in North America.

HABITAT

Oceans or lakes, returns to streams for spawning.

BEHAVIOR

Reach sexual maturity between two and four years of age.

FEEDING ECOLOGY AND DIET

Food items vary with age. Young in freshwater streams feed mainly on insects. Smolts feed on planktonic crustaceans upon reaching the sea. As they mature, they venture further into the ocean and feed on larger organisms. They are preyed upon by lampreys and various other species of fishes, birds, marine mammals, and bears.

REPRODUCTIVE BIOLOGY

Adults migrate from the sea or large lakes to the mouths of rivers, where they aggregate in large numbers forming schools. As rains increase the rivers' flow, they start swimming upstream. The reproductive behavior is very similar to that of the Chinook salmon. Females are in charge of finding the appropriate spot and digging a pit, they brush off any other female trying to do the same thing at the same spot. Attending males court females even while they are still digging the pit, As soon the pit is complete, the female drops into it, immediately followed by the male. After that, a ritualistic behavior takes place, which includes staying side by side and opening their mouths. This is followed by quivering and the release of eggs and sperm, with other males moving in and releasing sperm into the nest. The female then moves to the upstream edge of the nest and starts digging a new pit, covering the eggs. The entire process is repeated several times for several days, until the female deposits all her eggs and then dies. Meanwhile the male may pursue other females.

CONSERVATION STATUS

Not listed by the IUCN. The Alaskan fishery has been certified by the Marine Stewardship Council as well managed and sustainable.

SIGNIFICANCE TO HUMANS

Highly esteemed for its meat and supports important recreational fisheries in many parts of the world. ◆

Cherry salmon

Oncorhynchus masou

FAMILY

Salmonidae

TAXONOMY

Salmo masou Brevoort, 1856, Japan. There are two subspecies: one, *Oncorhynchus masou masou*, spawns in the sea; the other, restricted to freshwater, is yamame, or *Oncorhynchus masou ishikawae*.

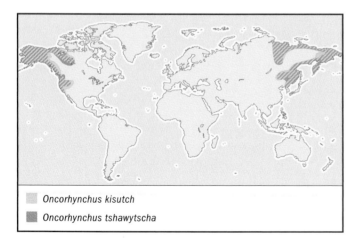

☐ *Oncorhynchus kisutch*
■ *Oncorhynchus tshawytscha*

Salvelinus alpinus

Oncorhynchus masou

OTHER COMMON NAMES
English: Salmon trout; French: Saumon masou; German: Masu-Lachs; Japanese: Masu.

PHYSICAL CHARACTERISTICS
Length 28 in (71 cm); weight 22 lb (10 kg). Troutlike shape. Coloration variable depending upon age and habitat or sub-species. Many adults have a general yellowish coloration and two series of gray oval spots: one fully laterally consisting of about 10 spots and one more dorsally consisting of five to six smaller spots. There are additional smaller spots over most of the rest of the body.

DISTRIBUTION
Northwest Pacific in Okhotsk Sea and Sea of Japan; northern Japan and eastern Korea Peninsula. Unlike other Pacific species of salmon, does not enter North American waters, and is said to be the most southerly salmonid. Introduced in Chile.

HABITAT
The river form generally inhabits headwaters.

BEHAVIOR
Like many other salmonids, defends a feeding territory.

FEEDING ECOLOGY AND DIET
Feeds mainly on insects and also on small crustaceans and fishes. The sea-run form goes downstream forming schools. After a short stay in the brackish zone it enters the sea, where it feeds on small fishes and pelagic crustaceans.

REPRODUCTIVE BIOLOGY
Not much is known about the reproduction of this species in nature, except that it is anadromous and spawns between Au-

gust and October. Japanese hatcheries have promoted a number of programs designed to accelerate growth rate and fecundity by adding hormones to the diet.

CONSERVATION STATUS
Not listed by the IUCN. This species is monitored by The North Pacific Anadromous Fish Commission (NPAFC) established under the Convention for the Conservation of Anadromous Stocks in the North Pacific Ocean, signed in 1992.

SIGNIFICANCE TO HUMANS
The common name derives from the Japanese nickname *sakurasmasu* (cherry salmon), a reference to the fact that it spawns at cherry blossom time. Marketed fresh and frozen; eaten broiled and baked. ◆

Rainbow trout
Oncorhynchus mykiss

FAMILY
Salmonidae

TAXONOMY
Salmo mykiss Walbaum, 1792, mouth of Columbia River at Fort Vancouver, Washington State, United States.

OTHER COMMON NAMES
English: Rainbow trout (North American landlocked populations), steelhead (sea populations); French: Truite arc-en-ciel; German: Regenbogenforelle; Spanish: Trucha arco iris.

PHYSICAL CHARACTERISTICS

Length 47.2 in (120 cm); weight 56 lb (25.4 kg). Body elongate and somewhat compressed, especially in larger individuals. Brightly colored, varies in color (especially males) depending upon habitat, size, and sexual condition. Stream residents and spawners are darker with more intense colors, lake residents tend to be lighter, brighter, and more silvery.

DISTRIBUTION

Eastern Pacific from Alaska to Baja California, Mexico. This is one of the most widely introduced fishes in the world in at least 50 countries, which makes its present distribution virtually global. In tropical countries where it has been introduced it is found only above 4,000 ft (about 1,200 m) of altitude above sea level. Introduction has had a negative ecological impact in many parts of the world.

HABITAT

Fresh waters where the water temperature is not higher than 53.6°F (12°C) in summer. Although they can be found in cold lakes, they require moderate-to-fast flowing, well-oxygenated waters for breeding. Yet, their survivorship is better in lakes than in streams.

BEHAVIOR

Adults aggressively defend feeding territories. All stocks of rainbow trout are opportunistic regarding migration, since they are able to migrate to, or at least to adapt to sea water, according to environmental factors. This may be another case of extreme behavioral plasticity.

FEEDING ECOLOGY AND DIET

Benthic feeders. Adults consume mostly aquatic and terrestrial insects, mollusks, crustaceans, fish eggs, minnows, and other small fishes (including other trouts). Young feed mostly on zooplankton. Ocean-going populations are vulnerable to larger fishes, pinnipeds, and toothed whales. Freshwater populations are preyed upon by larger fishes, otters, bears, and fish-eating birds.

REPRODUCTIVE BIOLOGY

As for other salmonids, growth rate varies according to environmental conditions. Usually reach sexual maturity between two and three years of age, with some extreme cases becoming sexually mature at five years. In this species, the female finds a spot and digs a pit. However, while she digs she is accompanied by an attendant male, which courts her and also drives away other males. Once the pit is completed, the female drops into it, immediately followed by the male. When the pair is side by side, they open their mouths, quiver, and release egg and sperm. A total of 700–4,000 eggs are produced per spawning event, which are then fertilized by the subordinate male. The female then quickly moves to the upstream edge of the nest and starts digging a new pit, covering the eggs. This process goes on for several days until the female has deposited all her eggs. The young move downstream at night, shortly after they emerge.

CONSERVATION STATUS

Not listed by the IUCN. However, in May 2002 the National Marine Fisheries Service (under the Endangered Species Act) issued a ruling redefining the geographic range of the listed anadromous population of this species to include all steelheads and their progeny occurring in coastal river basins from the Santa Maria River (inclusive) to the United States/Mexico Border. Within the redefined geographic range, only anadromous, naturally spawned populations and their progeny, which reside below naturally occurring and man-made impassable barriers, such as impassable waterfalls and dams, are listed as endangered.

SIGNIFICANCE TO HUMANS

Perhaps the most often bred fish species in the world because of its adaptability and value as the subject of commercial and sports fisheries. Anglers find it very interesting because of its spectacular leaps and hard fighting when hooked. ◆

Sockeye salmon

Oncorhynchus nerka

FAMILY

Salmonidae

TAXONOMY

Salmo nerka Walbaum, 1792, rivers and seas of Kamchatka, Russia.

OTHER COMMON NAMES

English: Blueback salmon, land-locked sockeye, red salmon; French: Saumon rouge; German: Blaurücken; Spanish: Salmon rojo.

PHYSICAL CHARACTERISTICS

Length 33 in (84 cm) in males, 28 in (71 cm) in females; weight 17 lb (7.7 kg). Like most salmoniforms, has a streamlined, fusiform body that is laterally compressed, but unlike other species, the body depth is moderate, slightly deeper among breeding males. Head is conical, very pointed, with small eyes. Lateral line is straight. Has long, fine, serrated, and closely spaced gill rakers on the first arch. Coloration varies with sex and reproductive stage. Pre-spawning individuals are dark steel blue to greenish blue on the head and back, silvery on the sides, and white to silvery on the belly. Spawning individuals (particularly males) have a bright to olive green head with black on the snout and upper jaw; the adipose and anal fins turn red and the paired fins and tail generally become grayish to green or dark. Females are less brightly colored than males.

DISTRIBUTION

North Pacific, from northern Japan to Bering Sea and to Los Angeles, California, United States, with landlocked populations in Alaska, Yukon Territory, and British Columbia in Canada, and Washington and Oregon, United States. Successfully introduced in New Zealand.

HABITAT

There are two forms of this species: the sockeye, an anadromous form (a marine form that migrates to fresh water for spawning), and the kokanee, a landlocked form (with a much smaller maximum size).

BEHAVIOR

Age structure and morphology differ among populations. Sexual selection and reproductive capacity (fecundity and egg size) generally favor large (old), deep-bodied fish, thus the sizes and shapes of salmon vary among spawning habitats. Stream width is positively correlated with age at maturity and negatively correlated with the level of predation by bears. Therefore, sexual maturity can be reached as early as at one year of age or as late as five. Adults show a great inclination for homing, which takes place during summer and fall, as late as December.

FEEDING ECOLOGY AND DIET

Young in lakes feed largely on crustaceans and insect larvae; adults in lakes become pelagic and feed on plankton in the upper 60 ft (20 m) or so of the water column. As they grow, their diet starts to include other fishes. The landlocked form feeds mainly on plankton, insects, and benthic organisms. Ocean-going populations are preyed upon by larger fishes, including sharks, as well as by pinnipeds and killer whales. Freshwater populations are vulnerable to larger fishes, otters, bears, and fish-eating birds.

REPRODUCTIVE BIOLOGY
Like other salmoniforms, the female selects a site to dig a nest, where she is attended by a dominant male along with a few subordinate males. Both females and males protect their site by aggressively shoving off individuals of the same sex. After courtship by the dominant male, the female enters the nest, followed immediately by the dominant male, which comes close beside her. After the ritual of mouth gaping, the pair vibrates to release eggs and sperm, and one or more subordinate males comes to the other side of the female and joins in the spawning. The female then moves to the upstream edge of the nest and digs again, covering the old nest, while at the same time creating a new one just upstream. After three to five days of depositing eggs, the female is exhausted and, along with the male, dies. The fry emerging from the gravel is very photophobic, and becomes mostly nocturnal.

CONSERVATION STATUS
Not listed by the IUCN.

SIGNIFICANCE TO HUMANS
One of the most commercially important Pacific salmons. The kokanee form is much sought after as a sport fish and is valued as food. ◆

Chinook salmon
Oncorhynchus tshawytscha

FAMILY
Salmonidae

TAXONOMY
Salmo tshawytscha Walbaum, 1792, rivers of Kamchatka, Russia.

OTHER COMMON NAMES
English: King salmon, Pacific salmon; French: Saumon chinook; German: Königslachs; Spanish: Salmón chinook.

PHYSICAL CHARACTERISTICS
Length 59 in (150 cm); weight 136.36 lb (61.4 kg). The fusiform body is streamlined and laterally compressed among large adults. Important distinguishing characteristics are small black spots on the back and the upper and lower lobes of the caudal fin, as well as on the black gums of the lower jaw. The gill rakers are widely spaced and rough. Coloration varies with environmental conditions. At sea, adults are dark greenish to blue black on top of head and back and silvery to white on the lower sides and belly; they also present numerous small and dark spots along back and upper sides and on both lobes of the caudal fin. In fresh water, they acquire an olive brown to red or purplish coloration, which is particularly noticeable among males.

DISTRIBUTION
Arctic and Pacific regions, in the drainages from Point Hope, Alaska to California, United States, as well as Japan, the Bering Sea, and the Okhotsk Sea. Successfully introduced in Australia, New Zealand, and Chile.

HABITAT
Similar to that of other salmoniforms; however, in lakes they may inhabit depths down to 1,230 ft (375 m).

BEHAVIOR
Like other salmoniforms, adults show a strong homing behavior. Their migratory behavior varies greatly. Usually, after three months in fresh water, most fry migrate to the sea, although some may stay in fresh waters for as long as three years. Some individuals remain close inshore throughout their lives, others make extensive migrations.

FEEDING ECOLOGY AND DIET
Food in streams consists mainly of terrestrial insects and small crustaceans; at sea major food items include fishes, crustaceans, and other invertebrates. The young are preyed upon by fishes and birds, such as mergansers and kingfishers; adults are preyed upon by sharks, large mammals (including killer whales), and birds.

REPRODUCTIVE BIOLOGY
Anadromous adults can migrate nearly 3,100 mi (5,000 km) from the ocean upstream to spawn. In December adults start to migrate from the sea, so that by early spring the first individuals arrive near the river mouths. The female is in charge of selecting the spawning spot where she will dig her nest, and aggressively drives away other females competing for the same spot. Once she has established her territory, a dominant male and several smaller males join her, at the same time driving away other males that compete for her. Male courtship behavior includes resting beside the female, quivering, swimming around over her, touching her dorsal fin with his body and fins, and occasionally nudging her side gently with the snout. Once the nest is complete, the female drops into it, and is immediately joined by the dominant male. Both open their mouths, vibrate, and eggs and sperm are released. At this point smaller males may swim into the nest and release their own sperm. (It is unclear to what extent these smaller males play a role in fertilizing the eggs.) The female then quickly covers the eggs by moving to the upstream edge of the nest and digging small pebbles for a new nest. This process, which may last several days, is repeated several times, until the female releases all her eggs. The female guards the nest for as long as she can. The male leaves the female and may mate with another female. Spent adults usually die a few days after spawning.

CONSERVATION STATUS
Not listed by the IUCN. The Alaskan fishery of this species has been certified by the Marine Stewardship Council as well managed and sustainable.

SIGNIFICANCE TO HUMANS
Highly regarded commercial and game fishes, whose red meat commands a high price. The viscera are rich in vitamin A and are used as food for hatchery fish. ◆

Atlantic salmon
Salmo salar

FAMILY
Salmonidae

TAXONOMY
Salmo salar Linnaeus, 1758, "Seas of Europe."

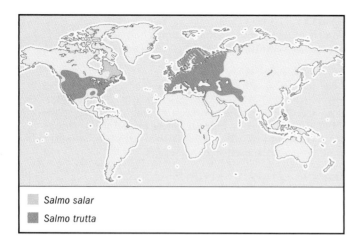

Salmo salar

Salmo trutta

OTHER COMMON NAMES
English: Black salmon, sea salmon; French: Saumon Atlantique; German: Atlantischer Salmon; Spanish: Salmón.

PHYSICAL CHARACTERISTICS
Length in males 59.1 in (150 cm), in females 47.2 in (120 cm); weight 103 lb (46.8 kg). Individuals display different morphology and coloration depending upon the phase of the breeding season or the habitat they are in. Body usually has black spots, caudal fin usually is unspotted, and adipose fin is not black bordered. Other times they are blue-green overlaid with a silvery guanine coating in salt water, losing silvery coat to become greenish or reddish brown mottled with red or orange in freshwater.

DISTRIBUTION
Western Atlantic in the coastal drainages from northern Quebec, Canada, to Connecticut, United States; Eastern Atlantic in drainages from Arctic Circle to Portugal. Some landlocked stocks can be found in North America, Norway, Finland, Sweden, and Russia. Successfully introduced in Chile, Argentina, Finland, Australia, and New Zealand.

HABITAT
Juveniles inhabit freshwater; adults inhabit marine waters except to spawn. Usually found in rocky runs and pools of rivers, large and small, as well as lakes.

BEHAVIOR
Young remain in freshwater for one to six years, after which they migrate to the ocean. They remain there for one to four years, before returning to fresh water to spawn, then they return to the sea. Active mostly during the day.

FEEDING ECOLOGY AND DIET
Juveniles feed on mollusks, crustaceans, insects, and fishes; adults at sea feed on squids, shrimps, and fishes. Ocean-going populations are vulnerable to larger fishes, pinnipeds, and toothed whales. Freshwater populations are preyed upon by larger fishes, otters, bears, and fish-eating birds.

REPRODUCTIVE BIOLOGY
Growth rate depends on food availability and quality, as well as on water temperature and photoperiods. They reach sexual maturity between three and seven years of age. Adults reaching sexual maturity return to their home rivers, usually to the same areas where they were hatched and spent their initial freshwater life. Once there, the female selects a spawning site with ap-

preciable current based on depth (usually 1.6–9.6 ft (0.5–3 m) and gravel size. Then she excavates a hole by turning on her side and flexing her body up and down creating a current and never touching the stones. After the female releases 8,000–26,000 eggs, the males visit the area, fertilize them, and cover the eggs. Spawning takes between two and three days. Early maturing or sneaker males return to their home stream every year, older males do so after several years in the ocean. The older males are not only larger, but also more colorful. Aggregations around a female are composed of both sneaker (smaller, younger) and older males. Once the female releases her eggs, all males release their sperm, with the greater number of eggs being fertilized by the first male that enters the nest. Young salmons fathered by precocious males grow faster than those fathered by anadromous males. Juvenile salmons (known as parr) spend most of their freshwater life in shallow riffles, mostly at the southern end of their range, until they reach 4–5 in (12–15 cm) in length, when they transform themselves into smolt and are ready for migration in spring the first year after hatching.

CONSERVATION STATUS
Not listed by the IUCN. Listed in Appendix III of The (Bern) Convention on the Conservation of European Wildlife and Natural Habitats as long as the species is in fresh waters.

SIGNIFICANCE TO HUMANS
Commercially valued for meat and consumed in many forms. ◆

Brown trout
Salmo trutta

FAMILY
Salmonidae

TAXONOMY
Salmo trutta Linnaeus, 1758, "Europe."

OTHER COMMON NAMES
English: German brown trout, herling, sea trout; French: Truite brune de mer; German: Lassföhren; Spanish: Trucha común.

PHYSICAL CHARACTERISTICS
Length 55.1 in (140 cm); weight 110.4 lb (50 kg). Coloration varies only according to the breeding season. They are dorsally black, usually orange on sides, surrounded by pale halos. The adipose fin has always a red margin. Troutlike body. The upper jaw reaches below the center of the eye in juveniles and well beyond the eye in larger individuals.

DISTRIBUTION
Originally from Eurasia. Now introduced all over the world, including Europe, Latin America, Australia, and New Zealand. Widely transplanted because it thrives in warmer waters than most other species of trouts. As with other species of salmonids, the introductions have had a negative impact on the local fauna.

HABITAT
Prefers cold, well-oxygenated upland waters. Favorite habitat is large streams in mountain areas with submerged rocks, undercut banks, and overhanging vegetation. Preferences in terms of high temperature tend to be looser than that of the rainbow trout.

BEHAVIOR
Mainly diurnal. Very territorial, aggressively defends feeding areas from conspecifics (members of the same species) and other trout species.

FEEDING ECOLOGY AND DIET
Diverse diet including small aquatic and terrestrial insects, mollusks, crustaceans, and small fishes. When eating very small prey, utilizes gill rakers on the surface of the gill arches. The plasticity in feeding preferences is accompanied by morphological plasticity in the feeding apparatus. For example, the mouth is fairly large and has nonspecialized teeth on the jaws and on several bones within the mouth, which serve for eating any almost creature. Very voracious, adults prey on items up to one-fourth their own length. The well-defined muscular stomach opens by a valve into the intestine. The intestine has a series of fingerlike appendages (pyloric ceca) that open off the intestine, immediately posterior to the stomach. These appendages secrete enzymes to facilitate food digestion. This character is typical of many very predacious species, and the more predacious they are, the higher the number of pyloric ceca. Ocean-going populations are vulnerable to larger fishes, pinnipeds, and toothed whales. Freshwater populations are preyed upon by larger fishes, otters, bears, and fish-eating birds.

REPRODUCTIVE BIOLOGY
Life history is similar to that of the Atlantic salmon, reproducing in rivers, producing about 10,000 eggs, and taking between three and four years for maturation.

CONSERVATION STATUS
Not listed by the IUCN.

SIGNIFICANCE TO HUMANS
Much sought after as a food item, particularly in Europe. Ocean-going individuals are called sea trout, and are larger than freshwater forms. They provide good sport, as do those that enter large lakes. ◆

Charr
Salvelinus alpinus

FAMILY
Salmonidae

TAXONOMY
Salmo alpinus Linnaeus, 1758, "Europe."

OTHER COMMON NAMES
English: Alpine char, Arctic charr; French: Omble arctique; German: Schwarzreuter; Spanish: Trucha alpina; Inuktitut: Akalukpik; Italian: Salmerino alpino; Portuguese: Truta-das-fontes.

PHYSICAL CHARACTERISTICS
Length 42 in (107 cm); weight 33 lb (15 kg). Distinguished from other species by light rather than black spots and by a boat-shaped bone (vomer) that is toothed only in front, on the roof of the mouth. Another distinguishing characteristic is the presence of 23–32 gill rakers and pink to red spots on the sides and back, the largest of which are usually larger than the pupil of the eye. Unlike other salmoniforms, the lateral line is not straight, but curves slightly downward from the head. Coloration varies with location, time of year, and degree of sexual development. Generally the back is dark, mostly brown, but also somewhat green, with lighter sides and a rather pale belly.

Spawning adults, especially males, acquire very bright orange to red coloration on the ventral side and on the pectoral, pelvic, and anal fins.

DISTRIBUTION
North America and Europe in the Arctic and adjacent oceans, also rivers and lakes they enter to breed. This distribution includes the North Atlantic south to southern Norway, including Iceland and southern Greenland. There are isolated populations in northern United Kingdom, Scandinavia, Finland, and the Alps. Other populations are restricted to freshwater lakes (colonized in glacial times) in Quebec, Canada, and in Maine and New Hampshire, United States. Successfully introduced in Yugoslavia and France.

HABITAT
Like other salmoniforms this is an anadromous species found in fresh, brackish; and marine waters ranging in depth from 100 to 230 ft (30–70 m).

BEHAVIOR
Little is known besides feeding and reproductive behavior.

FEEDING ECOLOGY AND DIET
Freshwater populations feed on planktonic crustaceans, amphipods, mollusks, insects, and fishes (particularly finfish). Large individuals are piscivorous; dwarf and small fishes feed on a wide range of invertebrates. The proportion of plankton in the diet of dwarf and small individuals correlates positively with the number and length of gill rakers. Ocean-going populations are preyed upon by larger fishes and pinnipeds. Freshwater populations are vulnerable to larger fishes, otters, bears, and fish-eating birds.

REPRODUCTIVE BIOLOGY
Males reach sexual maturity between 4 and 5 years of age, females do so between 5 and 10 years old. The males establish a territory in order to attract females. When the female enters the territory, she looks for a suitable spot for a redd (nest) and starts digging. As she does, the male begins its courting behavior by circling around her and then gliding along her side and quivering. Once the redd is complete, the pair release egg and sperm. The female then covers the eggs by digging at the edge of the pit, thus starting the next redd. In the subarctic Lake Fjellfrosvatn, northern Norway, there are two morphs of Arctic charr that are reproductively isolated because they spawn five months apart. The smaller morph (less than or equal to 5.5 in [14 cm]) is confined to the deeper zones of the lake, the larger morph is mainly littoral.

CONSERVATION STATUS
Not listed by the IUCN.

SIGNIFICANCE TO HUMANS
Highly prized as both a food and sport fish. ◆

Brook trout
Salvelinus fontinalis

FAMILY
Salmonidae

TAXONOMY
Salmo fontinalis Mitchill, 1814, vicinity of New York City, New York, United States.

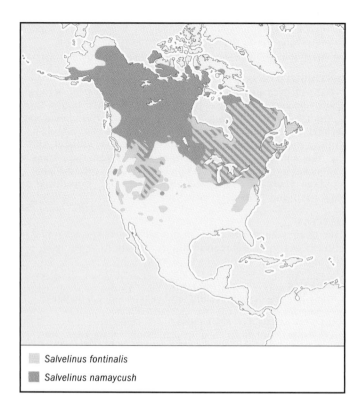

Salvelinus fontinalis

Salvelinus namaycush

OTHER COMMON NAMES
English: Brook charr, speckled trout, squaretail; French: Omble de fontaine; German: Bachsaibling; Spanish: Salvelino; Inuktitut: Iqaluk tasirsiutik.

PHYSICAL CHARACTERISTICS
Length 33.9 in (86.0 cm); maximum weight 20.7 lb (9.39 kg). Characterized by combination of dark green marbling on the back and dorsal fin, and by red spots with blue halos on the sides. Coloration can vary among populations and in reproductive states, with lower sides and fins red in spawning fishes. While migrating, dark green above with silvery sides, and white bellies and pink spots.

DISTRIBUTION
North America in most of eastern Canada from Newfoundland to western side of Hudson Bay; south in Atlantic, Great Lakes, and Mississippi River basins to Minnesota and northern Georgia, United States. In general, south of the Hudson River, distribution is correlated to altitude. For example, populations in North Carolina and Georgia are restricted to headwaters of streams in the Piedmont region of the Appalachians. Introduced in temperate regions all over the world.

HABITAT
Clear, cool, well-oxygenated creeks, small to medium rivers, and lakes.

BEHAVIOR
Migrates upstream in early spring, summer and late fall, migrates downstream in late spring and fall. As stream temperatures rise in the spring, may run to the sea (never more than a few miles [kilometers] from river mouths) and stay there for up to three months.

FEEDING ECOLOGY AND DIET
Opportunistic feeders, eating worms, leeches, crustaceans, insects, mollusks, fishes, amphibians, and even small mammals

and plant matter. Littoral individuals exhibit lower physiological performance than do pelagic individuals, when restricted to feeding in the pelagic zones. Ocean-going populations are preyed upon by larger fishes and pinnipeds. Freshwater populations are vulnerable to larger fishes, otters, bears, and fish-eating birds.

REPRODUCTIVE BIOLOGY
Reaches sexual maturity between one and three years, with variable growth rates depending upon temperature conditions. Spawning takes place from October through December. Hatching takes about 100 days or longer if waters are below 41°F (5°C). The male courts females by attempting to drive them toward a suitable spawning gravel site that he will defend aggressively. If a female is receptive, she will choose a spot and dig a redd. Even while the female is digging, the male continues its courtship by darting alongside the female, swimming over and under her, and rubbing her with his fins. When the redd is complete, the pair enter the nest and deposit eggs and milt (sperm). The female then covers the eggs with small pebbles. Once the eggs are completely covered, she moves to the upstream end of the redd and begins digging a new redd. Early maturing, dwarf "jack" or sneaker males return to their home stream every year; "hooknose" males do so after several years in the ocean. The latter are not only larger, but also more colorful. Aggregations surrounding a female are composed of both "jacks" and "hooknose" males. Once the female releases her eggs, all males release their sperm, with the greater number of eggs being fertilized by the first male that enters the nest. Individuals of this species can reach 15 years of age in captivity.

CONSERVATION STATUS
Not listed by the IUCN.

SIGNIFICANCE TO HUMANS
Commercially farmed because of their value as food, also used extensively as an experimental animal. Anglers regard them highly because of their fighting qualities when hooked. ◆

Lake trout
Salvelinus namaycush

FAMILY
Salmonidae

TAXONOMY
Salmo namaycush Walbaum, 1792, Hudson Bay.

OTHER COMMON NAMES
English: Great Lake trout, lake charr, Mackinaw trout, salmon trout; French: Omble d'Amérique; German: Amerikanische Seeforelle; Spanish: Trucha lacustre; Inuktitut: Isuuq.

PHYSICAL CHARACTERISTICS
Average length 59 in (150 cm); average weight 72 lb (32.7 kg). Body typically troutlike, elongate, somewhat rounded, with a stout head that is dorsally broad. The deeply forked tail distinguishes them from other species. The lateral line is slightly curved at the front. Coloration characterized by white or yellowish spots on a dark green to grayish background, but also have pale spots on dorsal, adipose, and caudal fins, and usually on base of anal fin; sometimes orange-red on paired fins. During spawning males develop a dark lateral stripe and become paler on the back.

DISTRIBUTION
North America, from northern Canada and Alaska, south to New England in United States, and Great Lakes basin in Canada and the United States. Successfully introduced in many other areas, including South America, Europe, and New Zealand.

HABITAT
Shallow and deep waters of northern lakes and streams, rarely brackish waters.

BEHAVIOR
Sexually mature fishes return to the rocky creek where they were spawned in the same manner that river-living salmonids return home to their natal stream.

FEEDING ECOLOGY AND DIET
Extremely voracious. Most populations feed on freshwater sponges, crustaceans, insects, fishes, and small mammals; others feed on plankton throughout their lives. Planktivorous trouts show characteristics typical of plastic (variable) populations. For example, they grow more slowly, mature earlier and at smaller size, die sooner, and attain smaller maximum size than do their fish-eating counterparts. Vulnerable to larger fishes, otters, bears, and fish-eating birds.

REPRODUCTIVE BIOLOGY
More or less dispersed away from the spawning beds during the day, returning in late afternoon and spawning mostly at night, particularly between dusk and 9 or 10 P.M. Males establish their territory by rocks on the substrate, females arrive a few days later when males court them. From one to seven males will approach one to three females in the same area and press themselves against the sides of one or more females. Then the eggs fall into the crevices and the spawners disperse. This behavior is repeated until the female releases all her eggs. The eggs are heavy and sink to the bottom.

CONSERVATION STATUS
Not listed by the IUCN, yet they are highly susceptible to insecticides. In the 1930s sea lampreys invading the Great Lakes reduced this species almost to extinction. Great Lakes populations are largely sustained by extensive stocking of hatchery-reared fry.

SIGNIFICANCE TO HUMANS
Fished by both commercial and sport fishers. ◆

Arctic grayling
Thymallus arcticus arcticus

FAMILY
Salmonidae

TAXONOMY
Salmo arcticus Pallas, 1776, Ob River, Siberia, Russia.

OTHER COMMON NAMES
English: American grayling, Black's grayling; French: Ombre arctique; Russian: Kharius sibirskiy; Inuktitut: Sulukpauga.

PHYSICAL CHARACTERISTICS
Length 30 in (76.0 cm); weight 8.45 lb (3,830 g). They have a dark, enlarged dorsal fin (especially among males) and a small mouth with fine teeth on both jaws. The pelvic fins are rather long, reaching the anal fin in adult males, but not in females. The lower lobe of caudal fin is usually longer than the upper. Bodies are brightly colored, dark purple or blue-black to blue-gray in dorsal areas; sides are gray to dark blue with pinkish iridescence, gray to white in ventral region. Scattered dark spots on sides, particularly the young, with a dark longitudinal stripe along lower sides between pectoral and pelvic fins.

DISTRIBUTION
North America in the Arctic drainages from Hudson Bay, Canada to Alaska, and in Arctic and Pacific drainages to central Alberta and British Columbia in Canada. Relict populations are also found in the upper Missouri River drainage in Montana, United States. Past distribution included the Great Lakes basin in Michigan, United States, and in Siberia, Russia.

HABITAT
Clear, open waters of cold, medium-to-large rivers and lakes. Spawns in rocky creeks.

BEHAVIOR
Forms schools of moderate numbers.

FEEDING ECOLOGY AND DIET
Young feed on zooplankton but later prefer immature insects. Adults feed mainly on surface insects as well as other fishes, fish eggs, lemmings, and planktonic crustaceans. Vulnerable to larger fishes, otters, bears, and fish-eating birds.

REPRODUCTIVE BIOLOGY
Sexual maturity comes between two and six years of age. Spawning takes place between April and June. Once spawning adults move into tributaries, males establish territories. When a female enters a territory, males court her with displays of the dorsal fin. Then he positions himself beside the female and curves his extended dorsal fin over her. The pair releases eggs and milt while vibrating, with that vibration stirring up the substrate to produce a slight depression. After spawning, adults establish summer territories in pools farther upstream from the spawning site, returning downstream in mid-September.

CONSERVATION STATUS
Not listed by the IUCN.

SIGNIFICANCE TO HUMANS
Utilized as food and as an object of recreational fishery.

Resources

Books

Berra, T. M. *Freshwater Fish Distribution.* San Diego: Academic Press, 2001.

Elliott, J. M. *Quantitative Ecology and the Brown Trout.* Oxford: Oxford University Press, 1994.

Johnson, G. D., and C. Patterson. "Relationships of Lower Teleostean Fishes." In *Interrelationships of Fishes,* edited by M. L. J. Stiassny, L. R. Parenti, and G. D. Johnson. San Diego: Academic Press, 1996.

Lee, D. S., C. R. Gilbert, C. H. Hocutt, R. E. Jenkins, D. E. McAllister, and J. R. Stauffer, Jr. *Atlas of North American Freshwater Fishes.* Raleigh, NC: North Carolina State Museum of Natural History, 1980.

Lichatowich, J. *Salmon Without Rivers: A History of the Pacific Salmon Crisis.* Washington, DC: Island Press, 1999.

Page, L. M., and B. M. Burr. *A Field Guide to Freshwater Fishes of North America North of Mexico.* Boston: Houghton Mifflin Company, 1991.

Taylor, J. E. *Making Salmon: An Environmental History of the Northwest Fisheries Crisis.* Seattle: University of Washington Press, 1999.

Periodicals

Alekseyev, S. S., V. P. Samusenok, A. N. Matveev, and M. Y. Pichugin. "Diversification, Sympatric Speciation, and Trophic Polymorphism of Arctic Charr, *Salvelinus alpinus* Complex, in Transbaikalia." *Environmental Biology of Fishes* 64 (2002):97–114.

Garant, D., P. M. Fontaine, S. P. Good, J. J. Dodson, and L. Bernatchez. "The Influence of Male Parental Identity on Growth and Survival of Offspring in Atlantic Salmon (*Salmo salar*)." *Evolutionary Ecology Research* 4 (2002): 537–549.

Klemetsen A., J. M. Elliott, R. Knudsen, and P. Sorensen. "Evidence for Genetic Differences in the Offspring of Two Sympatric Morphs of Arctic Charr." *Journal of Fish Biology* 60 (2002): 933–950.

Levin, P. S., and M. H. Schiewe. "Preserving Salmon Biodiversity." *American Scientist* 89 (2002): 220–227.

Proulx, R., and P. Magnan. "Physiological Performance of Two Forms of Lacustrine Brook Charr, *Salvelinus fontinalis,* in the Open-Water Habitat." *Environmental Biology of Fishes* 64 (2002): 127–136.

Quinn, T. P., L. Wetzel, S. Bishop, K. Overberg, and D. E. Rogers. "Influence of Breeding Habitat on Bear Predation and Age at Maturity and Sexual Dimorphism of Sockeye Salmon Populations." *Canadian Journal of Zoology* 79 (2001): 1,782–1,793.

Winfield, I. J., C. W. Bean, and D. P. Hewitt. "The Relationship Between Spatial Distribution and Diet of Arctic Charr, *Salvelinus alpinus,* in Loch Ness, UK." *Environmental Biology of Fishes* 64 (2002): 63–73.

Organizations

Salmon and Trout Association (UK). Fishmongers' Hall, London Bridge, London, EC4R 9EL UK. Phone: (020) 7283 5838. Fax: (020) 7626 5137. E-mail: salmon .trout@virgin.net Web site: http://www.salmon-trout.org

United States Trout Farmers Association. 111 West Washington St., Suite One, Charles Town, WV 25414-1529 USA. Phone: (304) 728 2189. Fax: (304) 728 2196. E-mail: ustfa@intrepid.net Web site: http://www.ustfa.org

Other

The Northwest Salmon Crisis: A Documentary History [cited January 29, 2003]. <http://www.netLibrary.com/urlapi .asp?action=summary>

"Pictures available for *Oncorhynchus masou*" [cited February 7, 2003]. <http://www.fishbase.org/Photos/ThumbnailsSummary .cfm?ID=242>

"*Oncorhynchus masou masou*" [cited February 7, 2003]. <http://www.city.chitose.hokkaido.jp/tourist/salmon/event/ miniatrure/masou.html>

Stomiiformes
(Dragonfishes and relatives)

Class Actinopterygii

Order Stomiiformes

Number of families 4

Photo: The sabertooth viperfish (*Chauliodus sloani*) lives in the deep sea. (Photo by Dr. Paul A. Zahl/Photo Researchers, Inc. Reproduced by permission.)

Evolution and systematics

The fossil record of the stomiiform fishes is scant, as is the case with most groups of thin-boned deep-sea pelagic fishes. The oldest known stomiiform-like fossils (*Paravinciguerria*) date back to the Upper Cretaceous, some 70 million years ago, though critical evaluation has shed some doubt as to whether it is related. Fossil evidence suggests that the hatchetfish family Sternoptychidae arose during the early Tertiary and reached its present evolutionary grade by the mid-Miocene, 16 million years ago. Several other fossils from the Miocene have been placed in modern stomiiform genera. In general, the fossils that have been described and verified as stomiiform-like differed little from recent forms. As such, they have not proven highly useful in determining the origin of stomiiform fishes. In a 1984 treatise on the origin of pelagic fishes, Nikolay Parin suggested that the Stomiiformes, an order of entirely pelagic, open-ocean fishes, arose from fishes living near the bottom of the deep sea. In *Interrelationships of Higher Euteleostean Fishes*, published in 1973, Donn Rosen stated that the ancestral form of the Stomiiformes is thought to be allied with the salmon-like fishes.

Among living fishes, the Stomiiformes are thought to be most closely related to the "jellynose fishes" of the order Ateleopodiformes. (Ateleopodiformes is a closely related order that has not been resolved completely [Olney et al. 1993], and is not discussed here.) These two orders comprise the Stenopterygii, which is the basal taxon of the Neoteleostei ("new bony fishes"). With respect to the entire spectrum of fishes, the stomiiforms can be thought of as more advanced than eels, herrings, minnows, catfishes, and trout, but more primitive than lanternfishes, cod, bass, and flounders.

Currently, four families are recognized: Gonostomatidae (bristlemouths), Phosichthyidae (lightfishes), Sternoptychidae (marine hatchetfishes), and Stomiidae (dragonfishes and rel-

atives). The classification of families within this order was radically revised in 1985. At that time William Fink combined the 230 or so species of "barbeled stomiiforms," previously recognized in six families, into one large family, the Stomiidae. At the close of the twentieth century, approximately 321 species of stomiiform fishes were described, grouped in two suborders, four families, nine subfamilies, and 51 genera. The majority of the species occur within the Stomiidae, and almost half of these species belong to a single dragonfish genus, *Eustomias*.

Physical characteristics

The stomiiform fishes are small to moderate in size, ranging from 0.6 in (15 mm) to 20 in (508 mm). The high species diversity of the Stomiiformes is matched only by its morphological diversity. Some species are extremely elongate and slender, while others are deep-bodied and laterally compressed. In one group there are radical differences in the form of males and females (called sexual dimorphism; see *Idiacanthus fasciola* species account). The order Stomiiformes includes some of the most morphologically specialized bony fishes. Much of this specialization relates to their deep-sea milieu:

- Both food and potential mates are scarce in the vast expanse of the deep sea.

- The interior of the sea is dimly lit or completely dark.

- The structureless pelagic environment provides no refuge from predation.

One of the trademark physical features of many stomiiform fishes is their fearsome dentition; their large mouths are filled with enormous fang-like teeth. This allows them to efficiently capture relatively large prey that are infrequently

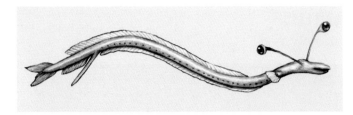

Larval dragonfish. (Illustration by Joseph E. Trumpey)

encountered. On average, fishes possessing these huge fangs take prey about a third their own size. This would be equivalent to an adult human eating more than 100 hamburgers in a single sitting—and such an equivalent would be an average meal for these fishes! In some cases, the weight of their prey has been known to exceed that of the predator. These fishes have a suite of other adaptations for this type of diet: long, sac-like stomachs; reduced ossification of the anterior vertebrae, allowing the mouth to expand dramatically; and a lack of gill rakers in adults.

Another feature possessed by all but one stomiiform species is the ability to create light (known as bioluminescence) with specialized organs called photophores. Bioluminescence, while rare on land, is quite common in the middle depths of the deep ocean's interior. Bioluminescence results from an oxidative reaction in which an organic molecule, called a luciferin, is raised to a chemically excited state in the presence of the enzyme luciferase. The excited luciferin then decays to the stable state by releasing light (most oxidations release heat). This energy can also be transferred to a fluorescent molecule, which releases light of its own color. Some fishes (e.g., deep-sea anglerfishes) rely on bacterial symbionts for light production. The stomiiforms, however, rely on self-generated, self-regulated luciferin/luciferase reactions occurring within their photophores for light production. The bodies of most stomiiform species bear two serial rows of photophores along each flank. The majority of stomiiform species also bear barbels on their chin. At the end of these barbels are often elaborate, bulb-like bioluminescent organs, thought to serve as "fishing lures." These chin-barbels range in size from less than the head length to as much as 10 times the length of the fish. Many species, particularly those of the genus *Eustomias*, are differentiated solely on the form of the barbel.

In general, stomiiform fishes fall into three major body plans. Plan A, exhibited by the dragonfishes (Stomiidae), is characterized by an elongate body, usually black or dark brown, with dorsal and anal fins placed far back on the body, the previously mentioned chin-barbel, and large teeth that are hinged to accommodate the passage of large prey items. Plan B, exhibited by the bristlemouths (Gonostomatidae), the lightfishes (Phosichthyidae), and maurolicines (Sternoptychidae: Maurolicinae), is characterized by a moderately elongate, minnow-like body, with dorsal and anal fins near mid-body, and large mouths studded with bristle-like teeth. Plan C, exhibited by most marine hatchetfishes (Sternoptychidae: Sternoptychinae), is characterized by a short, deep, highly compressed body with an abdominal keel that gives the body a hatchet-like shape, silvery flanks with a dark back, and a

mouth nearly vertical in position. Some species of the latter body plan have tubular eyes directed upwards.

Distribution

Stomiiform fishes are found over great depths throughout the world's oceans, including the Antarctic seas (but absent in the Arctic Ocean). Highest abundance and diversity is found in tropical seas. Distribution near landmasses is dictated by water depth. Stomiiform fishes are rarely found in waters shallower than 1,650 ft (500 m).

Habitat

Most species are mesopelagic (residing between 660 and 3,300 ft [200 and 1,000 m] deep), while some others are bathypelagic (residing below 3,300 ft [1,000 m] deep). Stomiiform fishes have been recorded to a depth of more than 14,000 ft (4,270 m). Most species undertake a daily vertical migration, swimming from a daytime depth of 1,600 to 3,300 ft (490 to 1,000 m) to near the surface at night, and then back down again before sunrise. This upward migration is thought to be mainly one for feeding—food in the form of plankton, fish, and other invertebrates is much more plentiful near the surface. Daytime depths, on the other hand, provide refuge from visual predators as well as an energy savings in the colder deep water. This vertical migration, undertaken by most mesopelagic fishes, shrimp, and squid, is the largest animal migration on Earth, and it happens every day in the deep sea.

Behavior

Very little is known of the behavior of stomiiform fishes, mainly because the vast majority of species have never been seen alive. Most specimens have been observed only after being caught in trawls, and the combination of physical trauma,

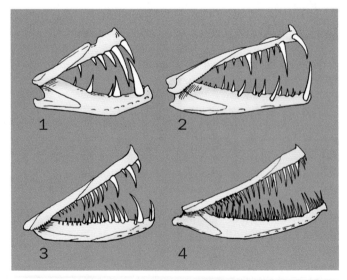

Jaws and teeth of four dragonfish: 1. *Leptostomias gladiator*; 2. *Grammatostomias dentatus*; 3. *Echiostoma barbatum*; 4. *Tactostoma macropus*. (Illustration by Joseph E. Trumpey)

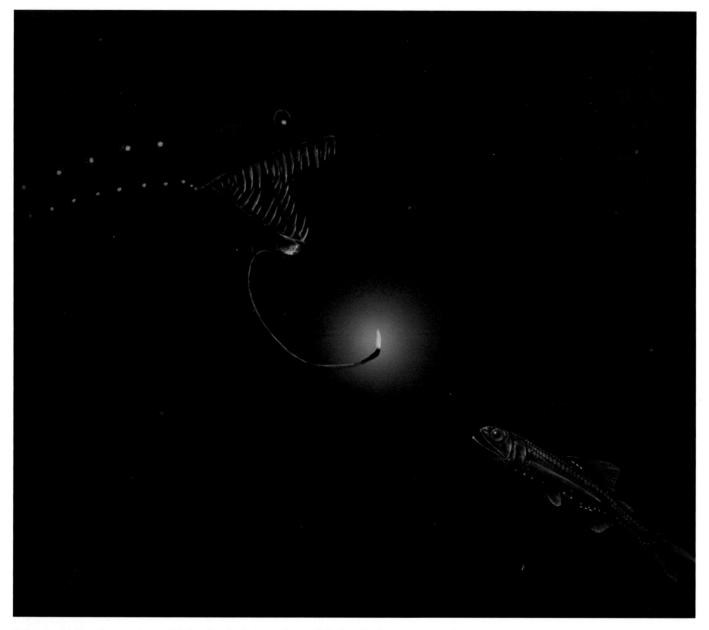

A female *Melanostomias* dragonfish hunting by luring a lanternfish (*Lepidophanes guentheri*) with her extended luminescent chin barbel. (Illustration by Joseph E. Trumpey)

pressure trauma, and temperature shock is usually lethal. The best-known behavior is that of diel vertical migration, and that is indirectly inferred from differences in abundance as a function of depth and time of day. As of 2001, no stomiiform fishes have been kept alive in aquaria for any length of time.

Feeding ecology and diet

For their size, the stomiiforms are voracious predators. Their feeding ecology falls into two main categories. First is predation on other fishes, shrimps, and squid. Within this category, predation on fishes, particularly lanternfishes (see Myctophidae, this volume), overwhelmingly predominates. Most of the larger stomiiform species feed in this manner (e.g.,

dragonfishes, loosejaws, viperfishes, and snaggletooths of the family Stomiidae). The second category is planktivory (feeding on microscopic organisms). Within this pattern, predation on copepods (small crustaceans less than 1/8 of an inch [3 mm]) predominates. Most of the smaller stomiiform species feed in this manner (e.g., bristlemouths, lightfishes, and hatchetfishes). As a general rule, planktivorous fishes are always more abundant than higher-level predators. It should be noted that even the higher-level predators are planktivorous during their early life stages.

The manner of feeding of planktivorous stomiiforms is similar to that of all planktivorous fishes—these fishes search small volumes of water, snapping up prey within an appropriate size range as they are encountered. The predacious dragonfishes, however, use a novel means of capturing prey. Though it has not been directly observed, it is thought that the barbeled forms use their luminescent lures to attract their prey, thus conserving their own energy in a food-limited environment. The posterior fin position of these fishes resembles that of "lie-in-wait" predators, such as pike (Esocidae). That, combined with the muscular control of the barbel, and the otherwise weak body musculature of these fishes, suggests such a strategy. This strategy is most likely the mechanism that has allowed the stomiids to become the top predators of the mesopelagic zone of the world ocean.

Little is known of the predators of stomiiform fishes. Some of the smaller, planktivorous species, such as *Vinciguerria* and *Maurolicus*, are important prey of both pelagic species (e.g., tunas) and outer continental shelf demersal fishes (e.g., Acadian redfish). It is likely that the vast numbers of stomiiform larvae (e.g., *Cyclothone*, *Vinciguerria*) are an important food source in the early life stages of many oceanic fishes. Larger stomiiform species (e.g., dragonfishes) have been found in the diets of benthopelagic fishes (e.g., rattails, roughies) as well as epipelagic fishes (e.g., swordfish) and mammals (e.g., Fraser's dolphin).

Reproductive biology

As with behavior, very little is known of the reproduction of stomiiform fishes. Because of the gear used to capture specimens, few sexually mature individuals of most species have been available for study. It has been theorized that larger, sexually mature individuals are more able to avoid midwater trawls than smaller individuals. That which is known suggests that stomiiforms exhibit a wide range of reproductive modes, mirroring their taxonomic and morphological diversity. It is thought that most stomiiforms spawn at their deeper daytime depths. Higher-latitude species appear to have more discrete spawning seasons than do tropical species, which spawn year-round. Some species spawn several times, while others spawn once and die. Some species have separate sexes, while others

mature into males, produce sperm to fertilize eggs, and then later develop into females, producing eggs that are fertilized by younger males. In species with separate sexes, males often have greatly developed olfactory (smell) organs to help in locating females.

Conservation status

Degradation of the marine environment is a global phenomenon, but as of 2000 no stomiiform fishes are known to be endangered. No species are on the IUCN Red List.

Significance to humans

Due to their bizarre and fearsome appearance, stomiiform fishes have been depicted in myth, literature, and art. Often the forms are exaggerated for effect, but the similarity to extant species suggests that they are often the inspiration.

No direct fishery exists for stomiiform species, though they are an important component of marine food webs. The main value of the stomiiforms is that they are key trophic mediators in the overall economy of the sea. A few isolated ecosystems notwithstanding, life in the ocean begins near the surface, where microscopic plants turn sunlight and carbon dioxide into organic carbon. The consumption of this "primary production" by small planktic animals (zooplankton) serves as the primary means of transforming plant life into animal life. As stated previously, many stomiiform species consume zooplankton directly, while others are the key predators of fishes that consume zooplankton. Thus, through their own consumption, stomiiform fishes are a key link in the transfer of microscopic organic matter to higher trophic levels. For example, the lightfish *Vinciguerria lucetia* is one of the most abundant pelagic fishes in the North Pacific and in turn is an important prey item of commercially fished tunas. To emphasize their importance, it has been claimed that the stomiiform fishes are the most abundant vertebrates on Earth.

1. Viperfish (*Chauliodus sloani*); 2. Silver hatchetfish (*Argyropelecus aculeatus*); 3. Brauer's bristlemouth (*Cyclothone braueri*); 4. Black dragonfish (*Idiacanthus fasciola*); 5. Scaleless dragonfish (*Eustomias schmidti*); 6. Rat-trap fish (*Malacosteus niger*). (Illustration by Joseph E. Trumpey)

Species accounts

Brauer's bristlemouth
Cyclothone braueri

FAMILY
Gonostomatidae

TAXONOMY
Cyclothone braueri Jesperson and Tåning, 1926, Mediterranean Sea.

OTHER COMMON NAMES
English: Bristlemouth

PHYSICAL CHARACTERISTICS
Females to 1.5 in (38.1 mm), males to 1 in (25.4 mm); elongate body, with dorsal and anal fins near midbody; scaleless; nearly transparent, save a few scattered black spots on head; very small eyes; relative to its size it has one of the largest gapes (mouth opening) of any fish; teeth small, bristle-like; males have well-developed olfactory organs to locate females.

DISTRIBUTION
Species occurs throughout Atlantic, but confined to a band at 15–50°S in Pacific and Indian Oceans.

HABITAT
Oceanic and mesopelagic. Depth distribution is age-specific: larvae occur from 30 to 160 ft (9 to 50 m). Adults occur from 825 to 3,000 ft (250 to 900 m) and juveniles in between. This is one of the shallowest living species in an otherwise deep-living genus.

BEHAVIOR
This is one of the few stomiiforms that does not vertically migrate.

FEEDING ECOLOGY AND DIET
Small crustaceans, mainly copepods, are the primary prey.

REPRODUCTIVE BIOLOGY
Thought to spawn once and die; spawning occurs from spring to fall.

CONSERVATION STATUS
Not threatened.

SIGNIFICANCE TO HUMANS
There is no commercial significance. It is claimed to be the most abundant vertebrate on Earth. ◆

Silver hatchetfish
Argyropelecus aculeatus

FAMILY
Sternoptychidae

TAXONOMY
Argyropelecus aculeatus Valenciennes, in Cuvier and Valenciennes, 1830, Azores.

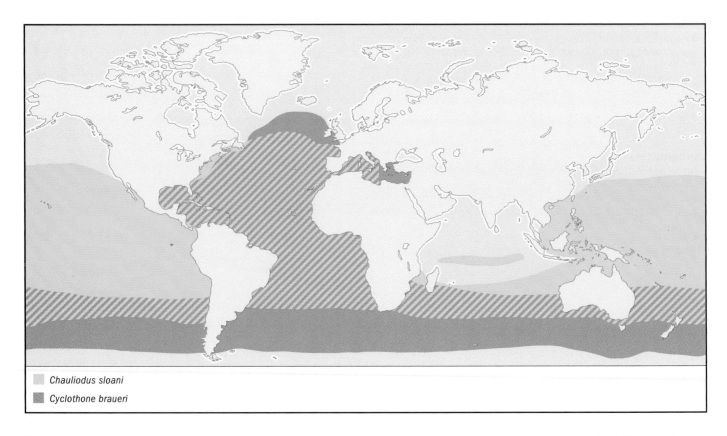

Chauliodus sloani
Cyclothone braueri

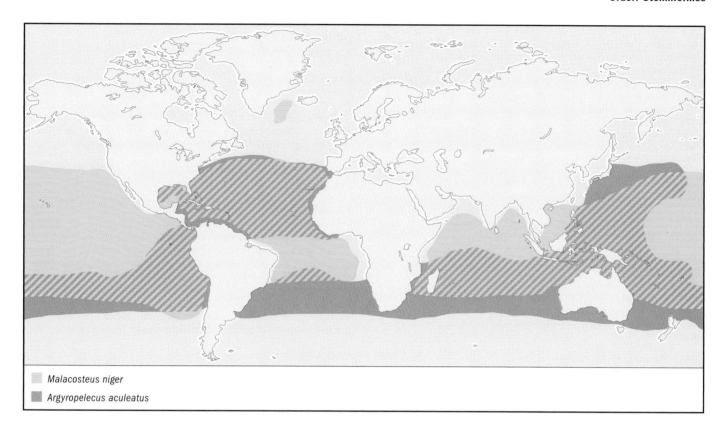

☐ *Malacosteus niger*
■ *Argyropelecus aculeatus*

OTHER COMMON NAMES
One of many species called "marine hatchetfish."

PHYSICAL CHARACTERISTICS
Females to 3.3 in (83.8 mm), males to 2.4 in (61 mm); body very deep, laterally compressed, hatchet-like shape derived from abdominal keel-like structure; abdomen bearing well-developed, ventrally directed photophores (these photophores actually account for 21% of the animal's surface area and 15% of its volume); silvery flanks with black back; eyes are tubular and directed upwards; mouth nearly vertical; teeth large, re-curved, with two very enlarged canine teeth in lower jaws; scales present, but easily lost.

DISTRIBUTION
This species exhibits an antitropical distribution in the Atlantic, Pacific, and Indian Oceans (i.e., it occurs on each side of the equator, but not directly along the equator). It is much more abundant in the western Pacific than in the east.

HABITAT
Oceanic, mesopelagic; most abundant between 1,000 and 2,000 ft (300 and 600 m) during daylight, 330 and 1,000 ft (100 and 300 m) at night.

BEHAVIOR
This species is a strong vertical migrator. It is known to form discrete, well-defined layers within its vertical range, but it does not form schools.

FEEDING ECOLOGY AND DIET
Feeds mainly at dusk; young fish eat small crustaceans, mainly ostracods, while older fish take larger crustaceans such as eu-phausiids and decapod shrimp.

REPRODUCTIVE BIOLOGY
Sexes are separate; thought to spawn once and die; spawning takes place year-round, peaking in summer.

CONSERVATION STATUS
Not threatened.

SIGNIFICANCE TO HUMANS
No commercial significance; conspicuous component of the "deep scattering layer" registered by the depth sounders of ocean-going ships. ◆

Viperfish
Chauliodus sloani

FAMILY
Stomiidae

TAXONOMY
Chauliodus sloani Bloch and Schneider, 1801, Gibraltar.

OTHER COMMON NAMES
English: Sloane's viperfish.

PHYSICAL CHARACTERISTICS
Length can be up to 13.8 in (350.5 mm). The body is very elongate and compressed with a large head. The body is en-closed in a gelatinous sheath. The dorsal fin is well forward on the body, the second ray of which is elongated and thought to serve as a fishing lure. The anal fin is well back, near the caudal fin. There are adipose dorsal and anal fins. The teeth are rigid and large to enormous (so large, in fact, so as not to fit within

the confines of the mouth, giving this fish a unique perspective—it sees the world through its teeth). There is iridescent yellowish to blue green on the flanks, with a dark back. There are more than 1,500 photophores on the body of this fish. There are five rows of large scales on each side of the body. Juveniles have a short barbel, which degenerates in adults.

DISTRIBUTION
Worldwide in tropical to temperate seas, including the western Mediterranean; absent in the northern Indian Ocean.

HABITAT
Oceanic, meso- to bathypelagic; most abundant between 1,650 and 9,200 ft (500 and 2,800 m) during daylight, with some of the population migrating up to between 70 and 660 ft (20 and 200 m) at night. The rest (non-hungry component) stay at daylight depths.

BEHAVIOR
Nothing known.

FEEDING ECOLOGY AND DIET
Juveniles eat small crustaceans, mainly euphausiids, while adults eat fishes, mainly lanternfishes, and occasionally decapod shrimp.

REPRODUCTIVE BIOLOGY
Little is known; spawning takes place year-round with a peak in late winter/early spring.

CONSERVATION STATUS
Not threatened.

SIGNIFICANCE TO HUMANS
No commercial significance. Fearsome appearance has inspired depiction in myth, literature, and art. ◆

Scaleless dragonfish
Eustomias schmidti

FAMILY
Stomiidae

TAXONOMY
Eustomias schmidti Regan and Trewavas, 1930, North Atlantic. One of more than 100 species called "scaleless dragonfish."

OTHER COMMON NAMES
None known.

PHYSICAL CHARACTERISTICS
It is recorded to 8 in (203.2 mm). It is very elongate and slender with a long, protrusible snout, as well as a scaleless body uniformly black to dark brown. There are dorsal and anal fins placed far back on the body, with an anal fin that is twice as long as the dorsal. There are fang-like, depressible teeth. The pectoral fins are encased in a black membrane. The belly bears a deep but short groove. The chin-barbel is present and is about half as long as the head, with a main stem and three branches. The main stem bears a bulbous light organ constricted near the tip.

DISTRIBUTION
Worldwide in tropical to subtropical seas; not known from the northern Indian Ocean.

HABITAT
It is oceanic, meso- to bathypelagic. It is most abundant between 2,000 and 3,300 ft (600 and 1,000 m) during daylight, with some of the population migrating up to between 330 and 500 ft (100 and 150 m) at night. The rest (non-hungry component?) stay at daylight depths.

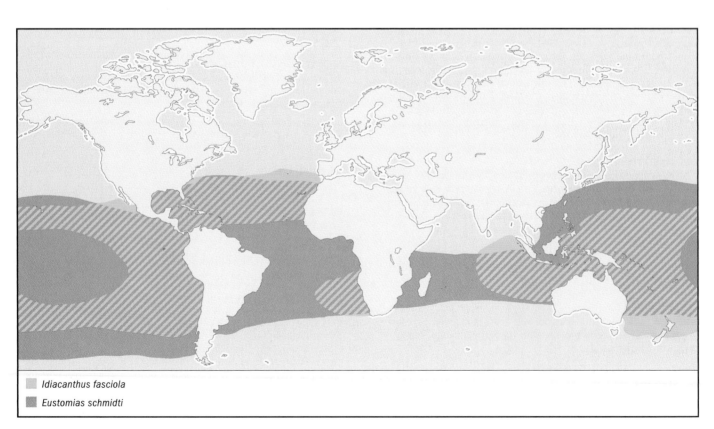

Idiacanthus fasciola
Eustomias schmidti

BEHAVIOR
Nothing known.

FEEDING ECOLOGY AND DIET
Preys on fishes, primarily lanternfishes.

REPRODUCTIVE BIOLOGY
Nothing known.

CONSERVATION STATUS
Not threatened.

SIGNIFICANCE TO HUMANS
No commercial significance, but of particular interest to deep-sea taxonomists. The true number of valid species in this genus may exceed 100. ◆

Black dragonfish
Idiacanthus fasciola

FAMILY
Stomiidae

TAXONOMY
Idiacanthus fasciola Peters, 1877, Pacific Ocean north of New Guinea. One of several species called "black dragonfish."

OTHER COMMON NAMES
None known.

PHYSICAL CHARACTERISTICS
This is among the most extreme cases of sexual dimorphism known. Females are up to 15 in (381 mm); they are extremely elongate and slender, with a long dorsal fin originating before mid-body and an anal fin about half the length of the dorsal. Pectoral fins are present in the larvae and are absent in adults. Pelvic fins are present; there are fanglike, depressible teeth. A chin-barbel is present with an elaborate leaf-like structure at the tip. Males up are to 2 in (50.8 mm). The body resembles postlarval females. Pectoral fins are present in the larvae and absent in adults. The eyes are larger than those of females. Males lack teeth, internal organs, and pelvic fins. Both sexes are uniformly black to dark brown. Females have three longitudinal rows of white tissue on the flanks. There is extraordinary larval development—the eyes of the larvae are borne on long stalks that retreat as the fish grows until the eyes are nested in sockets.

DISTRIBUTION
There is worldwide but disjunct distribution. It is known from the tropical/subtropical Atlantic and Pacific, the eastern South Atlantic, and the eastern Indian Ocean/Indo-Pacific. It is apparently absent or uncommon in the central Pacific.

HABITAT
It is oceanic, meso- to bathypelagic. Females occur between 1,800 and 6,600 ft (550 and 2,000 m) during daylight, while migrating up to as shallow as 165 ft (50 m) at night. Males apparently stay between 3,300 and 6,600 ft (1,000 and 2,000 m).

BEHAVIOR
Females vertically migrate, while males stay deep.

FEEDING ECOLOGY AND DIET
Females prey on fishes, primarily lanternfishes. Males have no teeth or digestive organs—apparently they do not feed after larval transformation, and may live only a few weeks.

REPRODUCTIVE BIOLOGY
Little is known; it may spawn in the summer, probably at great depths. Males have enlarged light organs behind the eyes, apparently to attract females.

CONSERVATION STATUS
Not threatened.

SIGNIFICANCE TO HUMANS
None known. ◆

Rat-trap fish
Malacosteus niger

FAMILY
Stomiidae

TAXONOMY
Malacosteus niger Ayres, 1848, southeast of Nova Scotia, 42°N–50°W.

OTHER COMMON NAMES
English: Loosejaw.

PHYSICAL CHARACTERISTICS
It is recorded to 9.5 in (241.3 mm). It is somewhat elongate but deeper bodied than dragonfishes, with dorsal and anal fins well back near the caudal fin. The body is uniformly black. It has an enormous mouth, much longer than the skull. There are four pairs of large fangs in the lower jaw. Like other members of this subfamily, this species has no floor in its mouth (i.e., no ethmoid membrane), hence the common name "loosejaw." The body photophores are very small. The eyes are very large, with a large, deep-red photophore under each eye. There is no chin-barbel.

DISTRIBUTION
Worldwide in tropical to temperate seas; has been taken as far north as Iceland in the North Atlantic, but generally occurs south of 40°N.

HABITAT
Oceanic, mesopelagic; occurs between 1,650 and 3,000 ft (500 and 900 m) day and night. This species is the only known non-vertically migrating stomiid.

BEHAVIOR
This is one of the very few animals that can create red light via the large eye photophores, and even more amazing is that it has the ability to see red light (most deep-sea fishes can see only blue/green wavelengths). This confers the ability to search for prey without making itself visible to potential predators. It has been suggested that this fish acquires its red-sensitive visual pigment from the chlorophyll (the green photosynthetic pigment in plants) packaged in the guts of its copepod prey.

FEEDING ECOLOGY AND DIET
Despite its fearsome appearance, this species mainly eats small crustaceans (copepods), while occasionally taking fish and decapod shrimp.

REPRODUCTIVE BIOLOGY
Nothing known.

CONSERVATION STATUS
Not threatened.

SIGNIFICANCE TO HUMANS
There is no commercial significance, but the protein responsible for its red light production has been investigated for possible medical uses. This protein, if it could be synthesized (this species is rare throughout its range) and attached to an antibody, would provide a means of locating and treating cancerous tumors within a human body without the need for invasive surgery. ◆

Resources

Books

Gartner, J., R. Crabtree, and K. Sulak. "Feeding at Depth." In *Deep-Sea Fishes*, edited by David Randall and Anthony Farrell. Boston: Academic Press, 1997.

Hoyt, E. *Creatures of the Deep: In Search of the Sea's "Monsters" and the World They Live In.* Buffalo, NY: Firefly Books, Ltd., 2001.

Periodicals

Fink, W.L. "Phylogenetic Interrelationships of the Stomiid Fishes (Teleostei: Stomiiformes)." *Miscellaneous Publications Museum of Zoology, The University of Michigan* no. 171 (1985): 1–127.

Olney, J.E., G.D. Johnson, and C.C. Baldwin. "Phylogeny of Lampridiform Fishes." *Bulletin of Marine Science* 52 (1993): 137–169.

Sutton, T.T. and T.L. Hopkins. "Trophic Ecology of the Stomiid (Pisces: Stomiidae) Fish Assemblage of the Eastern Gulf of Mexico: Strategies, Selectivity, and Impact of a Top Meso-pelagic Predator Group." *Marine Biology* 127 (1996): 179–192.

Tracey T. Sutton, PhD

Aulopiformes

(Lizardfishes and relatives)

Class Actinopterygii

Order Aulopiformes

Number of families 12

Photo: A slender lizardfish (*Saurida gracilis*) inside the reef in a Tahiti lagoon. (Photo by Fred McConnaughey/Photo Researchers, Inc. Reproduced by permission.)

Evolution and systematics

The Aulopiformes have been the subject of numerous systematic revisions over the past 30 years. The most recent one, by Carole C. Baldwin and G. David Johnson, published in 1996, recognized 4 suborders, 12 families, and 43 genera. The suborders include the Synodontoidei (three families, including the lizardfishes and Bombay ducks, the latter group formerly placed in the Harpadontidae and now synonomized), the Chlorophthalmoidei (three families plus two species, *Bathysauropsis gracilis* and *B. malayanus*, with no family assignment), the Alepisauroidei (four families), and the Giganturoidei (two families plus a single species, *Bathysauroides gigas*, with no family assignment). The order is largely monophyletic. (Some familial relationships remain to be resolved.) It traces its ancestry to the Upper Cretaceaous period (135 million years ago). The Aulopiformes share many features with midwater and deep-sea-dwelling Myctophiformes (lanternfishes and relatives), and most species may have evolved over time from life in the pelagic realm to a bottom-dwelling existence.

Physical characteristics

The Aulopiformes are characterized by both primitive and derived or advanced characters. Examples of primitive characters include abdominal pelvic fins with 8–12 rays, the absence of fin spines, the presence of an adipose fin (a structure consisting of fatty tissue located between the dorsal fin and the caudal fin), and scales that are cycloid or rounded in shape. Derived characters include a ductless swim bladder and a maxillary bone that has been excluded from the gape of the jaw. Other characters, present in some groups, include highly modified eyes, hermaphroditism, and well-developed metamorphic stages between larval and juvenile phases. Specialized gill arches further distinguish them from the Myctophiformes. Within the order, body sizes and shapes are diverse and a function of habitat and depth distribution.

Lizardfishes (25 species within the Synodontidae) are cryptically colored, cigar-shaped bottom-dwelling fishes with large, reptilian heads, large mouths, numerous teeth, and an eye placed nearly halfway along the length of the mouth. Bombay ducks (15 species within the Synodontidae) are distinguished by a short and rather rounded snout, long pelvic and pectoral fins, and a medially placed tail lobe formed from lateral line scales. Synodontids are cryptically colored, usually in mottled shades of brown, gray, or red. The Aulopidae (about 10 species) resemble lizardfishes but possess a large dorsal fin; in males, this fin has a first dorsal fin ray that extends well beyond the others. Aulopids also have two small jawbones, the supramaxillae, located along the upper edge of the jawbone. They also are cryptically colored, primarily in shades of green, brown, gray, and red. The monotypic Pseudotrichonotidae are elongate and cylindrical in shape, with a pointed snout, a single elongate dorsal fin, and elongate pelvic fins. Body coloration is exquisite, with a series of red to orangish red patches laterally, and resembling bars when viewed from above. Yellow spots are present just above the lateral line and bordering the red patches, with rows of

Engleman's lizardfish (*Synodus englemani*), eating a fish, near Indonesia, in the Pacific Ocean. (Photo by Animals Animals ©W. Gregory Brown. Reproduced by permission.)

A lizardfish (*Synodus intermedius*) on the sandy bottom, near Bonaire Island in the Netherlands Antilles. (Photo by Mary Beth Angelo/Photo Researchers, Inc. Reproduced by permission.)

faint electric blue spots above and below the lateral line. The fins are pale white to clear with traces of faint iridescent blue.

The Bathysauridae (two species) also resemble lizardfishes but have curved, barbed teeth and a flattened head. Members of the Chlorophthalmidae (12 species), commonly referred to as greeneyes or cucumberfishes (allegedly because they smell like cucumber) have iridescent green or yellow eyes of various sizes. Body color varies from pale green or light brown to a mottled pattern of green, silver, white, brown, and black. The Ipnopidae (about 26 species) have pencil-shaped bodies, flat heads with tiny or poorly developed eyes, and long, thickened fin rays. Body color is variably black, bronze, brown, or dark gray, or mottled in a combination of two or more of these colors. These fishes include the tripod fishes, whose extended pelvic fin rays form two legs of a tripod (with the caudal fin forming the third leg) that allows them to perch above the abyssal bottom. Among the midwater aulopiformes are the large lancetfishes (two species within Alepisauridae), which are distinguished by a long, sail-like dorsal fin that stretches from the head nearly to the adipose fin and may grow to nearly 7 ft (more than 200 cm) in length. These fishes are silvery, pale bronze, or brownish black in color, darker dorsally, and iridescent. Almost as large is the daggertooth (one species within the Anotopteridae), another midwater predator that reaches nearly 5 ft (150 cm) in length but lacks the long sail-like dorsal fin. The daggertooth lacks scales. Body color ranges from silvery to dusty with black on the tips of the pectoral fins, jaws, and also onto the caudal fin.

Also in midwater depths are the telescope fishes (Giganturidae) and pearleyes (Scopelarchidae), which have specialized tubular eyes whose form and function resemble that of

a telescope (useful for detecting light in dim surroundings) and allow for forward- or upward-directed vision. Telescope fishes are generally silvery in color, while pearleyes are pale brown. Midwater and deep-water aulopiform fishes that may or may not have tubular eyes and are distinguished by fanglike teeth and a lack of scales include the sabertooth fishes (seven species within Evermannellidae) and the hammerjaw (Omosudidae, one species). The sabertooth fishes range from pale brown to black in color. Hammerjaws are translucent silvery brown, dark brown, or olive-black with hints of silver, white, and pinkish brown. Similar in shape are the waryfishes (Notosudidae), which have both scales and, uniquely, teeth on the upper jawbones or maxillae. Waryfishes are generally pale brown in color, and some species have black fins. The barracudinas (50–60 species within the Paralepididae) are slender and elongate in shape and superficially resemble barracudas; they may grow as large as 39 in (100 cm). These fishes also possess fanglike teeth; a long, pointed snout; and a dorsal fin placed midway along the body. They may or may not have scales. Body color ranges from pale brown to tan or yellowish olive.

Distribution

Members of the Aulopiformes may be found throughout the Atlantic, Pacific, and Indian Oceans, with the exception of most polar waters. They also occur in the Mediterranean, but they appear to be absent from the Black Sea. The Synodontidae typically dwell in warm, shallow waters, although at least one species is found as deep as 1,300 ft (400 m). The Bathysauridae live at depths in excess of 5,250 ft (1,600 m), whereas the Aulopidae is distributed widely, except for the eastern Pacific, in waters near shore to a depth of 3,300 ft (1,000 m). The Pseudotrichonotidae is limited to the shallow waters of Japan and the southeastern Pacific, while the Chlorophthalmidae also may be found in shallow, warm, temperate waters worldwide. Both the Ipnopidae and Giganturidae are distributed widely in deep or abyssal waters. Midwater and bathypelagic species of the families Alepisauridae, Scopelarchidae, Evenmannellidae, Omosudidae, Notosudidae, and Paralepididae all tend to be distributed widely at depths in excess of 1,625 ft (500 m), and some are found down to 3,300 ft (1,000 m). The Anotopteridae range in shallow to deep temperate waters and are rarely found in the tropics.

Habitat

With respect to patterns of habitat use, aulopiform fishes are divided neatly into two camps: bottom-dwelling and water-column or pelagic-dwelling species. Bottom-dwelling families include the Synodontidae (Bombay ducks may be secondarily pelagic), Aulopidae, Pseudotrichonotidae, Chlorophthalmidae, Bathysauridae, and the Ipnopidae. These fishes typically perch or rest upon substrates that include rubble, sand, coral, rock, algae, and even abyssal silt and mud. Some fishes, such as the synodontids, aulopids, and pseudotrichonotids, may bury themselves in sand or rubble. Pelagic families include the Alepisauridae, Anotopteridae, Scopelarchidae, Evenmannellidae, Omosudidae, Notosudidae, and

Paralepididae. These fishes swim actively in the water column even at depths where light is virtually absent.

Behavior

The behavior of most aulopiform fishes is unknown. Many species live at depths that make direct observation of their behavioral patterns extremely difficult if not impossible. Generalizations about some types of behavior can be made, however. For example, bottom-dwelling species are often cryptic and have minimal movement. These characteristics allow them to ambush prey and to avoid predation. Swimming movements of these fishes often are accomplished in short bursts, followed by burying movements in the substrate. Only when they are ambushing prey or escaping predators are their swimming movements rapid. The tripodfishes, which dwell on the abyssal bottom, may "walk" along the bottom with the support of extended pelvic and caudal fins.

Fishes that swim pelagically in the midwater and deep-water realms may do so rapidly in pursuit of prey. Both lancetfishes and daggertooths are powerfully built for swimming and hunting prey. Alternately, some smaller species might hover or rest in the water column and reserve swimming for hunting prey or for vertical migrations at night. The barracudinas swim vertically, with their tails down and their heads up.

Courtship behavior has been observed for some shallow-water species. Observations of lizardfishes have revealed a repertoire of ritualized patterns, some of which also may be used in territorial displays. Among midwater and deep-dwelling species, simultaneous hermaphroditism is common, and self-fertilization is possible, thus precluding the need for courtship behavior.

Feeding ecology and diet

Aulopiform fishes are predators. Bottom-dwelling species, such as the lizardfishes and Bombay ducks, typically feed upon smaller fishes and shrimps that they ambush and capture with their large, tooth-filled mouths. These fishes may bury themselves under the substrate and ambush passing prey just above the bottom and well into the water column. Alternately, they may ambush from perched positions on corals, rocks, or other objects, or they may launch an attack while hovering above the bottom. Regardless, the ambush is executed rapidly and is remarkably successful. Midwater and deep-water pelagic species probably capture prey by hunting or, under dim light conditions, by ambushing passing fishes or larger invertebrates, such as cephalopods or crustaceans. Owing to the colder water temperatures at greater depths, the digestive processes of deep-dwelling species appear to be slow. Thus, these fishes probably feed much less frequently than their shallow-water relatives. Aulopiformes are likely preyed upon by other predatory fishes and also by toothed whales.

Reproductive biology

Members of the Aulopiformes have two mating systems, gonochorism and simultaneous hermaphroditism. Members

A lizardfish (*Synodus* sp.) captures a fangblenny (*Plagiotremus* sp.) that was trying to avoid detection by mimicking a cleaner wrasse. (Photo by David Hall/Photo Researchers, Inc. Reproduced by permission.)

of the suborder Synodontoidei are gonochoristic, that is, they have separate male and female sexes, and no sex change occurs. Although few detailed studies of reproduction in this suborder exist, its shallow water-dwelling members probably spawn pelagically in the water column and release gametes, which, as fertilized eggs that become larvae, drift in the water column until the larvae settle on suitable habitat and recruit to a local population as a functional member. Courtship and spawning may be seasonal and likely is affected by water temperature. Thus, at higher latitudes courtship is limited to the warmer months of the year. In the tropics, courtship and spawning may proceed all year long.

The lizardfishes mate in pairs before or just after sunset; if the local population size is relatively large and males outnumber females, an alternative tactic may be used. For example, *Synodus dermatogenys* employs lek-like courtship behavior during reproduction. In this system, males form mobile territories in the presence of one or more females. Then they attempt to court a female while defending courtship space around her. Females control spawning. The female may choose to spawn with a single male, or she may allow the participation of additional males who contribute a proportion of their gametes to the fertilization of the gelatinous egg mass that is produced by the female. Furthermore, and unlike most pelagic spawning species that reproduce around sunset, the female may engage in more than one spawning event during a given evening.

The female also may choose to abort a spawning ascent if she determines that too few males are participating in the ascent. By allowing multi-male spawning, the female may ensure greater fertilization success for her eggs and promote genetic diversity as well. The remaining suborders appear to be capable of being both male and female at the same time, and they can reproduce by self-fertilization. This trait may have allowed for the radiation of species throughout the deep-water realm. Upon maturity, members of the Notosudidae leave the continental slope and spawn pelagically in the open ocean.

Conservation status

No species are listed by the IUCN.

Significance to humans

Although the flesh of shallow-water aulopiform fishes may be generally bony, some species are taken as food in both subsistence and commercial fisheries. Bombay ducks, for example, are caught in the estuaries of India and sold dried in the markets of Mumbai (Bombay), from which their common name is derived. A commercial fishery for other synodontids exists on the trawling grounds of Australia, Hong Kong, and parts of Southeast Asia (where they also are taken with monofilament nets and hook and line). Lizardfishes are sold fresh or salted or in processed form, such as fish balls, fish crackers, or fishmeal. The sergeant baker, *Aulopus purpurissatus* (Aulopidae), is taken by recreational fishers in Australia as well. Similarly, lancetfishes are caught accidentally on lures trolled by anglers in South African waters and elsewhere.

1. Longnose lancetfish (*Alepisaurus ferox*); 2. Greeneye (*Chlorophthalmus acutifrons*); 3. Painted lizardfish (*Trachinocephalus myops*); 4. Slender lizardfish (*Saurida gracilis*); 5. Tripodfish (*Bathypterois quadrifilis*); 6. Clearfin lizardfish (*Synodus dermatogenys*). (Illustration by Patricia Ferrer)

Species accounts

Longnose lancetfish
Alepisaurus ferox

FAMILY
Alepisauridae

TAXONOMY
Alepisaurus ferox Lowe, 1833, off Madeira.

OTHER COMMON NAMES
None known.

PHYSICAL CHARACTERISTICS
Body elongate and compressed, with a very high dorsal fin and a large mouth. The dorsal fin has 30–45 soft rays and the anal fin 13–18 soft rays. The mouth has two erect fangs. The body color is a pale iridescent silver, white, or cream, but darker along the back and upper flanks; at times coloring can contain hints of light blue, green, or red. The fins are brown or black and the lateral adipose keel darkly colored. The adipose fin is positioned posteriorly over the anal insertion. Lacks both a swim bladder and light organs. Reaches over 84.6 in (215 cm) in length.

DISTRIBUTION
In the Atlantic from the Gulf of Maine south to the northern Gulf of Mexico and South America in the west and from Iceland to southern Portugal and the Mediterranean in the east; reappears from Nambia to South Africa. In the Pacific, range extends over the Kuril Islands, Okhotsk Sea, and Japan and in Australia in the west and from Alaska south to Chile in the east. Apparently not recorded from the Indian Ocean.

HABITAT
Mesopelagic and epipelagic waters from more than 5,900 ft (1,800 m) to the surface, depending upon the time of day. This species occasionally ventures inshore.

BEHAVIOR
Not much is known about the behavior of this species. A nocturnal species that migrates vertically toward the surface at night and descends into deep water during the day.

FEEDING ECOLOGY AND DIET
Preys upon cephalopods, crustaceans, and fishes; will also take tunicates from the bottom. Preyed upon by sharks, albacore, and yellowfin tuna, opahs, and, close to shore, fur seals.

REPRODUCTIVE BIOLOGY
Not well known. Likely produces pelagic eggs and larvae.

CONSERVATION STATUS
Not listed by the IUCN.

SIGNIFICANCE TO HUMANS
Taken incidentally but not esteemed as food. ◆

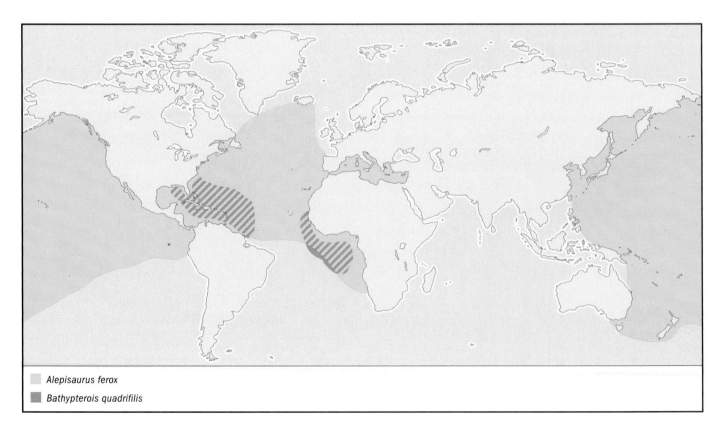

Alepisaurus ferox
Bathypterois quadrifilis

Greeneye
Chlorophthalmus acutifrons

FAMILY
Chlorophthalmidae

TAXONOMY
Chlorophthalmus acutifrons Hiyama, 1940, Kumano-nada, Japan from 1,640 ft (500 m).

OTHER COMMON NAMES
None known.

PHYSICAL CHARACTERISTICS
Body elongate; the contour of the head concave dorsally and elevated to the dorsal fin. The dorsal fin has 10–12 soft rays and the anal fin 9–11 soft rays. Eyes are greenish. There is a light organ surrounding the anus. The body color is a pale, iridescent green, with a somewhat pale belly. The caudal fin is forked and, like the other fins, pale in color. Reaches 11.8 in (30 cm) in length.

DISTRIBUTION
Western Pacific, from southern Japan to the Philippines.

HABITAT
Dwells on the bottom of moderately deep waters.

BEHAVIOR
Not well known. Probably somewhat similar to that of shallow water dwelling lizardfishes.

FEEDING ECOLOGY AND DIET
Carnivorous, feeding upon fishes.

REPRODUCTIVE BIOLOGY
Not well known. Likely produces pelagic eggs and larvae.

CONSERVATION STATUS
Not listed by the IUCN.

SIGNIFICANCE TO HUMANS
Of minor commercial importance as a food fish. ◆

Tripodfish
Bathypterois quadrifilis

FAMILY
Ipnopidae

TAXONOMY
Bathypterois quadrifilis Günther, 1878, off Brazil.

OTHER COMMON NAMES
None known.

PHYSICAL CHARACTERISTICS
Body elongate and compressed, with the tip of the upper jaw extending posteriorly beyond the orbit of the eye. Pectoral, pelvic, and caudal fin rays are all quite elongate (although the latter less so) and firm; when erect, they function like a tripod, thus allowing this species to balance itself above the bottom. Color somewhat bronze to pale, with grayish hues on the head and ventral surface and along the lower back. Fins somewhat pale. The dorsal fin has 14–15 soft rays and the anal fin 9–10 soft rays. Two of the pelvic rays reach the origin of the anal fin. The species reaches 7 in (18 cm) in length.

DISTRIBUTION
Deep water of the Atlantic Ocean, in the Gulf of Guinea off Africa and also from the western Atlantic in the region

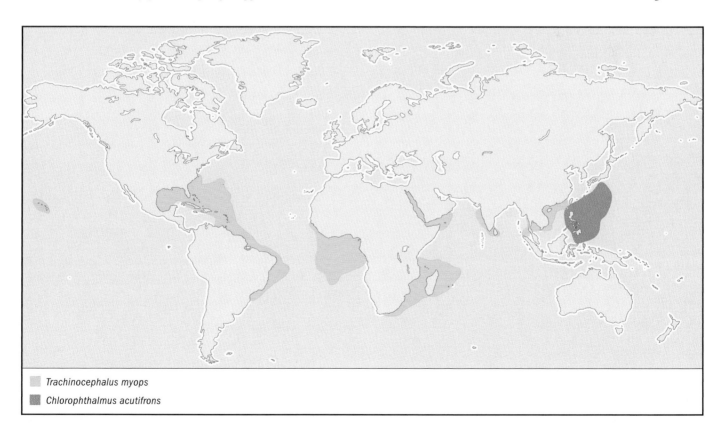

☐ *Trachinocephalus myops*
■ *Chlorophthalmus acutifrons*

between 35° north and 10° south latitude (includes both the Gulf of Mexico and the Caribbean).

HABITAT
This is a bathydemersal species that dwells between 1,310 and 4,650 ft (400–1,410 m) on sand, rubble, silt, and other bottom sediments.

BEHAVIOR
This species is capable of "walking" on the bottom by using its modified fins, which act like a tripod.

FEEDING ECOLOGY AND DIET
Carnivorous, feeding upon smaller fishes and crustaceans that it ambushes on the bottom.

REPRODUCTIVE BIOLOGY
A synchronous hermaphrodite. Eggs are likely pelagic, as are the larvae.

CONSERVATION STATUS
Not listed by the IUCN.

SIGNIFICANCE TO HUMANS
No commercial importance but an obvious curiosity to science. ◆

Slender lizardfish
Saurida gracilis

FAMILY
Synodontidae

TAXONOMY
Saurida gracilis Quoy and Gaimard, 1824, Hawaiian Islands and Mauritius.

OTHER COMMON NAMES
English: Graceful lizardfish, slender grinner; German: Graziler Eidechsenfisch; Japanese: Madara-eso.

PHYSICAL CHARACTERISTICS
Grows to 12.4 in (31.5 cm). Has 11–12 dorsal rays, 9–10 anal rays, and 12–14 pectoral fin rays (usually 13). There are 50–52 lateral line pores. Head distinguished by several rows of villiform teeth.

DISTRIBUTION
Tropical waters from the Red Sea east to the Hawaiian Islands, the Marquesas, Rapa Island, and Ducie Island; also from the Ryukyu Islands (Japan) south to southeastern Australia and Lord Howe Island.

HABITAT
This lizardfish is found in depths of 3–405 ft (about 1–135 m) but most commonly in shallow sand and rubble flats of lagoon and protected seaward reefs.

BEHAVIOR
Usually solitary but may be seen in pairs. Perches on ledges and rocks or buries itself in the sand or rubble. Reportedly active at night.

FEEDING ECOLOGY AND DIET
An ambush predator that feeds upon smaller fishes near the bottom or at depths to 15 ft (5 m) into the water column.

REPRODUCTIVE BIOLOGY
The reproductive behavior and ecology of this species are largely unknown; it likely engages in paired courtship and pelagic spawning just before or after sunset. Eggs and larvae are pelagic.

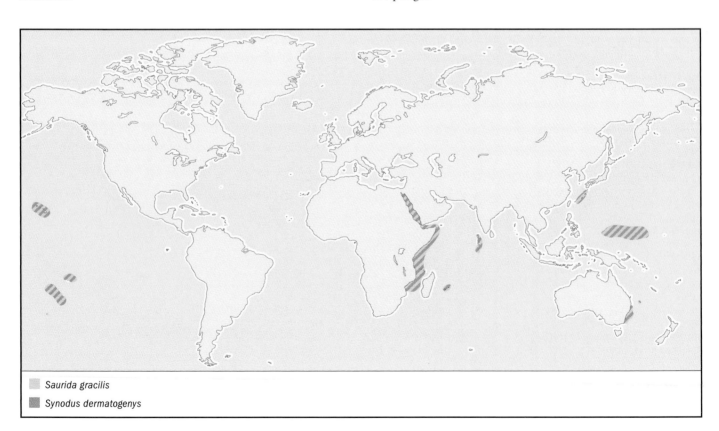

◼ *Saurida gracilis*
◼ *Synodus dermatogenys*

CONSERVATION STATUS
Not listed by the IUCN.

SIGNIFICANCE TO HUMANS
Taken only incidentally in subsistence, artesanal, or commercial fisheries. Not generally valued as a food fish but may be processed into fish meal, fish cakes, or other processed forms. ◆

Clearfin lizardfish
Synodus dermatogenys

FAMILY
Synodontidae

TAXONOMY
Synodus dermatogenys Fowler, 1912, Hawaiian Islands.

OTHER COMMON NAMES
English: Sand lizardfish, two-spot lizardfish; German: Sand-Eidechsenfisch.

PHYSICAL CHARACTERISTICS
Grows to 7.1 in (18 cm). Has 10–13 dorsal rays, 8–10 anal rays, and 11–13 pectoral fin rays. There are 56–61 lateral line pores and 5.5 scale rows above the lateral line. Yellowish streaks between each pelvic fin ray and at least eight diamond-shaped marks along each flank.

DISTRIBUTION
Tropical waters from the Red Sea east to the Hawaiian Islands and southeast to the Marquesas, the Tuamotu Archipelago, and Rapa Island. Also from the Ryukyu Islands (Japan) south to southeastern Australia and Lord Howe Island.

HABITAT
This lizardfish is found in depths of 3–210 ft (about 1–70 m) on sand and rubble of lagoon and seaward reefs.

BEHAVIOR
Usually solitary but may be seen in mating groups. Buries in sand to ambush prey and escape predators.

FEEDING ECOLOGY AND DIET
An ambush predator that feeds upon smaller fishes.

REPRODUCTIVE BIOLOGY
If population densities are relatively high and males outnumber females, this species engages in lek-like courtship and group or paired spawning; otherwise, it engages in paired courtship and spawning. Males in leks defend mobile territories against rival males while courting females. Females may spawn more than once per day and appear to regulate the number of males participating in a spawning event and, hence, the relative proportion of eggs fertilized by each participating male. Spawning is pelagic, just before or after sunset, and the spawning ascent results in the release of a gelatinous cloud of eggs. The eggs, as in all lizardfishes, typically are large and spherical, are marked by a pattern of hexagonal meshing on the surface, and have only a few oil droplets or none at all. The eggs hatch after two to three days. Spawning may be daily (low tropics) or seasonal

(higher latitudes), depending on water temperature. Larvae are long, slender, and pelagic and undergo a well-developed metamorphosis into juveniles.

CONSERVATION STATUS
Not listed by the IUCN.

SIGNIFICANCE TO HUMANS
Taken only incidentally in subsistence or artesanal fisheries. Not generally valued as a food fish, except in Southeast Asia. ◆

Painted lizardfish
Trachinocephalus myops

FAMILY
Synodontidae

TAXONOMY
Trachinocephalus myops Forster, 1801, Saint Helena.

OTHER COMMON NAMES
German: Gemalter Eidechsenfisch; Japanese: Oki-eso.

PHYSICAL CHARACTERISTICS
Grows to nearly 14 in (38 cm). Has 11–14 dorsal rays; 13–18 anal rays, and 11–13 pectoral fin rays. There are 51–61 lateral line scales and 3.5 scales above the lateral line. Pectoral fin reaches line from origin of dorsal fin to the origin of the pelvic fin. Has a single series of slender needlelike teeth along entire jaws.

DISTRIBUTION
Tropical waters of the Atlantic, Pacific, and Indian Oceans.

HABITAT
Usually on sand in estuaries and subtidal shores to a depth of more than 600 ft (200 m).

BEHAVIOR
Solitary, paired, or in loose aggregations (the latter possible mating groups). Buries itself in the sand with only the eyes emerging.

FEEDING ECOLOGY AND DIET
Ambushes smaller fishes near the bottom or in the water column.

REPRODUCTIVE BIOLOGY
The reproductive behavior and ecology of this species are largely unknown. Pairs have been observed courting in the water column, a departure from the benthic courtship seen among members of the pelagic-spawning genus *Synodus*. Eggs and larvae are probably pelagic.

CONSERVATION STATUS
Not listed by the IUCN.

SIGNIFICANCE TO HUMANS
Taken incidentally in subsistence, artesanal, or commercial fisheries. Not generally valued as a food fish, except in Southeast Asia. ◆

Resources

Books

Allen, G. R., and R. Swainston. *The Marine Fishes of North-Western Australia: A Field Guide for Anglers and Divers.* Perth, Australia: Western Australian Museum, 1988.

Eichler, Dieter, and Robert F. Myers. *Korallenfische Zentraler Indopazifik.* Hamburg: Jahr Verlag, 1997.

Helfman, Gene S., Bruce B. Collette, and Douglas E. Facey. *The Diversity of Fishes.* Malden, MA: Blackwell Science, 1997.

Kuiter, Rudie H. *Coastal Fishes of South-Eastern Australia.* Honolulu: University of Hawaii Press, 1993.

Leis, J. M., and B. M. Carson-Ewart, eds. *The Larvae of Indo-Pacific Coastal Fishes: An Identification Guide to Marine Fish Larvae.* Leiden, Netherlands: Brill, 2000.

Masuda, H., K. Amaoka, C. Araga, T. Uyeno, and T. Yoshino, eds. *The Fishes of the Japanese Archipelago.* Tokyo: Tokai University Press, 1984.

Myers, R. F. *Micronesian Reef Fishes: A Field Guide for Divers and Aquarists.* 3rd edition. Barrigada, Guam: Coral Graphics, 1999.

Nelson, Joseph S. *Fishes of the World.* 3rd edition. New York: John Wiley & Sons, 1994.

Paxton, John R., and William N. Eschmeyer, eds. *Encyclopedia of Fishes.* San Diego: Academic Press, 1995.

Sadovy, Y., and A. S. Cornish. *Reef Fishes of Hong Kong.* Hong Kong: Hong Kong University Press, 2000.

Smith, M. M., and P. C. Heemstra, eds. *Smiths' Sea Fishes.* Berlin: Springer-Verlag, 1986.

Thresher, R. E. *Reproduction in Reef Fishes.* Neptune City, NJ: T. F. H. Publications, 1984.

Periodicals

Donaldson, T. J. "Lek-like Courtship by Males and Multiple Spawnings by Females of *Synodus dermatogenys* (Synodontidae)." *Japanese Journal of Ichthyology* 36 (1990): 439–458.

Sweatman, H. P. A. "A Field Study of the Predatory Behavior and Feeding Rate of a Piscivorous Coral Reef Fish, the Lizardfish *Synodus englemani.*" *Copeia* 1984 (1984): 187–194.

Organizations

IUCN/SSC Coral Reef Fishes Specialist Group. International Marinelife Alliance-University of Guam Marine Laboratory, UOG Station, Mangilao, Guam 96921 USA. Phone: (671) 735-2187. Fax: (671) 734-6767. E-mail: donaldsn@uog9.uog.edu Web site: <http://www.iucn.org/themes/ssc/sgs/sgs.htm>

Terry J. Donaldson, PhD

Myctophiformes

(Lanternfishes)

Class Actinopterygii
Order Myctophiformes
Number of families 2

Photo: A brooch lanternfish (*Benthosema fibulatum*) seen in the daytime in Bali. (Photo by Animals Animals ©R. Kuiter, OSF. Reproduced by permission.)

Evolution and systematics

The earliest fossil records for the family Myctophidae are reported from the Miocene epoch of the Upper Tertiary period approximately 23 million years ago.

The current order has undergone a series of taxonomic revisions over the years. Until about 1940, the neoscopelids were included as a subfamily within the Myctophidae. Both were included as members of the order Myctophiformes (two suborders containing some 15 families), first erected in 1911. The Myctophiformes were relegated to a suborder within the Salmoniformes in the sweeping systematic revision of living teleost fishes in the mid-1960s (the original *Grzimek's* account of myctophids places them in that order), but this received little support. By the early 1970s, the Myctophiformes were reestablished as a separate order, with the neoscopelid and myctophid families being recognized as a distinct suborder, the Myctophoidei. An extensive re-view of the myctophoid families in the mid-1990s resulted in the establishment of a separate superorder, the Scopelomorpha, containing the order Myctophiformes. As constructed, the Myctophiformes were reduced to the families Neoscopelidae and the Myctophidae, with the other genera of the former Myctophiformes placed in a sister group, the Acanthomorpha.

As the order is currently organized, the Neoscopelidae ("blackchins") represent the more generalized morphology, and only three genera and six species are currently recognized. The Myctophidae ("lanternfishes") are believed to be derived from the neoscopelids, and are a far more diverse family of some 230 to 250 species in 32 currently recognized genera. In the most recent review of the family, two subfamilies are recognized, the Myctophinae (13 genera) and the Lampanyctinae (19 genera).

Physical characteristics

Neoscopelids and myctophids are generally small fishes, with the largest species in both families not exceeding 11.8 in (300 mm) standard length (SL, measured from tip of the snout to base of tail). Most myctophid species are much smaller, usually less than 3.1 in (80 mm) SL at maximum size. The most common members of both families resemble anchovies, with very large terminal to subterminal mouths and large eyes. In fact, in some Russian papers, myctophids are referred to as "luminescent anchovies." The jaws bear numerous tiny teeth, and in most species, the gills bear enlarged bladelike gill rakers along the first gill arch, the number of which is used as a taxonomic character for differentiating species.

Probably the most distinctive characteristic of most species of both families are the numerous light-producing organs (known as photophores) covering the body. The photophores are concentrated along the lower sides and ventral (belly) region, as well as on the head. The photophores are arranged in patterns that are distinctive to each species. In addition to photophores, many species of myctophids bear luminescent scales and glands in particular regions. The production of light in the glands and photophores is the result of a chemical oxidation reaction that can be triggered and regulated by the animal's nervous system. The light produced is in the blue-green range and in some of the glands can produce a burst of light like a camera flashbulb going off, enough to illuminate the interior of a darkened submersible!

Both families exhibit two general body morphologies among different genera and species. One is a robust, firmly muscled body, the other, a watery, flabby body with weakly developed skeletons and muscles. These body forms are believed to relate to vertical distribution and habits.

The rays of all fins are soft. An adipose fin is present in all species. Pelvic fins are forward of the mid-body, usually under the beginning of the single dorsal fin. Pectoral fins are quite variable in length among different genera, but in all cases are attached along the lower one-third of the body. In neoscopelids and some species of myctophids, the pectorals are long and sweeping, in some species extending almost the length of the body, while a number of soft-bodied species in the genus *Lampanyctus* have such small, delicate pectorals that they are almost unnoticeable.

The bodies are fully covered with deciduous scales, which are cycloid (smooth) in most species, although members of a few genera (most notably, *Myctophum*) bear ctenoid (ctenii are small toothlike structures) scales. These scales blow off the body in a glittering cloud when the fish is struck by objects such as a net, submersible, or predator. Specialized scales cover the photophores. Some scales in myctophid species are covered with a luminescent gland.

Coloration observed in living specimens ranges from a brilliant metallic bronze in the neoscopelid genus *Neoscopelus* to a countershaded dark blue-black on the dorsal surface and mirror-like silvery sides and ventral surfaces in most myctophid species. However, some deep-dwelling myctophid species, including members of the genus *Lampanyctus* and *Taaningichthys*, are dark brown or black. Photophores on freshly captured specimens appear silver, reddish, or in species of the genus *Diaphus*, deep sapphire blue.

Distribution

The neoscopelid species are mainly found in deep waters in tropical and subtropical regions. Two of the three genera have representatives found in all three oceans, but the monotypic genus *Solivomer* is thus far only known from deep waters off the Philippines.

In contrast, myctophids are found in all oceanic waters of the world, with endemic species occurring in all but the Arctic Ocean. In his comprehensive treatise *Developments in Deep-Sea Biology*, Marshall (1980) stated, "...lantern-fishes are so successful that a sizeable net towed at mesopelagic levels almost everywhere in the ocean is almost bound to catch them. If, as seems very likely, they have not colonized the Arctic Sea, then it is improbable that other groups of midwater fishes have gained a footing."

Species diversity of myctophids is highest at tropical-subtropical latitudes, while abundances of individual species appear greatest at temperate and higher latitudes.

Habitat

Both neoscopelids and myctophids are considered oceanic, that is, they normally inhabit waters oceanward of the continental shelf edge. As adults, species of both families are primarily mesopelagic inhabitants, usually found between 660 and 3,330 ft (200 and 1,000 m). Distribution of some species (*Scopelengys tristis* in the Neoscopelidae, and all *Taaningichthys* species in the Myctophidae) may extend into the upper bathypelagic,

perhaps to about 3,960 ft (1,200 m) or so. Several neoscopelid and a number of myctophid species are thought to be (or to become at some point in their lives) benthopelagic, spending part of their life cycle near, but not in contact with, the bottom.

Larval myctophids are primarily epipelagic. The few studies that have been done on their distribution shows them to mainly occur between about 165 to 825 ft (50 to 250 m) depth.

Behavior

One of the most notable behaviors observed among many mesopelagic animals, especially myctophid fishes, is that of diel vertical migration. With the exception of the deep mesopelagic genus *Taaningichthys*, adolescents and adults of all myctophid species move from mesopelagic depths into very shallow epipelagic waters at night. In fact, a number of species can be dip-netted from the very surface waters. They do this to feed and to lay their eggs in the food-rich shallower surface waters, usually above 330 ft (100 m) depth. Ascent and descent have been estimated to take about 2 hours each for a species population to shift from daytime to nighttime depths and *vice versa*.

Let us put this remarkable achievement into a human perspective. A 2-in (50-mm) myctophid migrating from 2,300 ft (700 m) to 330 ft (100 m) and back travels about 12,000 body lengths in four hours. This would be equivalent to a 6-ft (180-cm) human having to travel 72,000 ft or 13.6 mi (2.2 km) on foot in 4 hours *every night*.

Within any migratory population, it is believed that most individuals undertake this movement over each diel cycle. However, in most species, larvae are nonmigratory. Newly metamorphosed juveniles also have been shown to either remain shallow and nonmigratory, or to descend into deep water, after which they remain nonmigratory for a period of time. Other restrictions on migration also have been demonstrated. Researchers have shown that in some species, old individuals may restrict or cease migrating, and gravid females often reduce the extent of migration. In addition, it has been shown that physical conditions such as solar eclipses, the degree of moon fullness, clarity of the night sky and the presence of strong current boundaries all affect the extent of vertical migration.

Observed swimming behavior among myctophids takes two forms, depending on the body type. Robust-bodied species swim in short bursts, propelled by a rapid closing of the tail fin rays and flick of the tail. In general, these are the strongest migrators, with the widest difference in day and night vertical distribution. The flabby-bodied forms tend to move with a slow eel-like wriggling of the entire body. These species tend to be found in the deeper mesopelagic depths and migrate shorter distances.

Neoscopelids are not common animals, and little is known about their behavior or habits. An individual of the neoscopelid *Neoscopelus microchir* was observed from a submersible in a head down position with its large pectorals held straight out from the sides. It evaded the submersible by swimming downward in short bursts similar to the robust myctophid behavior and remaining in a head down posture.

Feeding ecology and diet

As with other aspects of their biology, specific feeding habits of neoscopelids are unknown.

Historically, myctophids have been characterized as opportunistic feeders with a generalized diet of small crustacean zooplankton. While it is true that almost all myctophid species can be characterized as crustacean zooplanktivores, a few detailed studies have shown that myctophids actually exhibit a high degree of size and species selectivity. Prey are consumed whole.

Post-metamorphic and adult myctophids are mainly nocturnal feeders. Indeed, a longstanding hypothesis regarding their diel migration is to enter the food-rich surface waters at night to feed. In contrast, a recent study conducted on 14 species of myctophid larvae showed that all but one species fed during the day, with one species showing no temporal response to feeding. Most larvae fed on very small crustacean zooplankton, but several species also included gelatinous zooplankton such as salps as an important or primary prey resource.

While the presence of neoscopelids as prey items of other species has rarely been reported, myctophids are important diet components for a number of different predators. Myctophids form consistent to significant prey items for a number of mesopelagic predatory fishes including many of the stomiiforms (dragonfishes, snaggletooths, and viperfishes). Squids have been observed feeding on myctophids on numerous occasions, and myctophid otoliths have been recovered from squid stomachs as well. A number of marine mammal species have been found with significant numbers of myctophid remains in their stomachs. Reported stomach contents of the beaked whale genus *Mesoplodon* from the western North Atlantic include large numbers of the myctophid *Ceratoscopelus maderensis* as well as squid remains.

Myctophids have also been reported as regular diet components for a number of oceanic seabirds, especially species that feed at night and presumably capture the myctophids as they themselves feed in shallow waters at night. One group of predatory animals in which lanternfishes are rarely reported are epipelagic fishes such as tunas (family Scombridae) and their relatives. This is because scombrids are daytime visual predators, and except for some deep diving species, such as large bluefin tuna, they are not feeding when myctophids are present in their habitat. A second hypothesis regarding the diel vertical migrations of myctophids and other mesopelagic migrators is that the migration is driven by a downward descent in daytime to avoid epipelagic predation, with a return to food-rich epipelagic waters at night for feeding and reproduction.

Reproductive biology

Neoscopelids are apparently dioecious, but beyond this, no information is available on their reproductive habits.

The myctophids are also dioecious, and a number of genera also exhibit sexual dimorphism. The type of dimorphism varies among these genera and even among populations.

In a number of genera (e.g. *Benthosema*, *Hygophum*, *Myctophum*), males possess a supracaudal luminescent gland (just in front of tail on the upper surface of body), while females possess an infracaudal gland (in front of tail on lower surface). In the diverse genus *Diaphus*, most species have enlarged luminescent glands on their heads (called supraorbital glands or "headlights"), which are typically much larger in males. Males and females of the monotypic genus *Notolychnus valdiviae* can be distinguished by the noticeably larger eyes of the males. Another pattern of dimorphism is found in some species such as *Ceratoscopelus warmingii* and *Electrona antarctica*, in which males are distinctly smaller than females at maximum size.

Although there are exceptions to the rule, lower latitude species tend to spawn year-round, whereas higher latitude species typically spawn during a restricted period once per year. Lower latitude species tend to live 1 year or less, whereas higher latitude species may live 3–4 years.

Conservation status

There are no known conservation measures specific to these families. No species from either family is listed by the IUCN.

Significance to humans

From a commercial standpoint, the significance of these families is minimal and mostly indirect. Several fisheries have been or are being conducted for myctophids, in particular by the former Soviet Union, Republic of South Africa, and most recently, in the Arabian Sea by Iran. Myctophid catches are mainly processed to form fish meal, which is used as poultry feed and also as a crop fertilizer. In a fashion similar to sardines and anchovies, one may purchase tins of myctophids packed in oil.

In an ecosystem sense, however, the impact of myctophids may be especially enormous. Studies show that myctophids occupy a central role in open ocean energy transfer, especially between surface and deep ocean waters. They are important prey (in some cases, the most important) for a number of marine mammals. Damage to, or loss of, myctophid populations could have serious repercussions for many other organisms in open ocean food webs.

Species accounts

Skinnycheek lanternfish
Benthosema pterotum

FAMILY
Myctophidae

TAXONOMY
Scopelus pterotus Alcock, 1890, Bay of Bengal.

OTHER COMMON NAMES
Japanese: Iwa-hadaka.

PHYSICAL CHARACTERISTICS
Deep, slightly pointed head with very large eyes. Pectoral fins are moderately long, reaching to about the middle of the dorsal fin. This is one of the

Benthosema pterotum

smaller species of lanternfishes, reaching a maximum of 2 in (50 mm) SL, although relatively few specimens of greater than 1.6 in (40 mm) have been reported. Coloration is highly reflective silver on sides and ventral surfaces, with a blue-black dorsal surface.

DISTRIBUTION
Found primarily in the northern Indo-Pacific region, although larval specimens have been reported in the south Atlantic off South Africa. There are also some reported specimens from the eastern Pacific near Costa Rica and El Salvador as well. It is especially abundant in the northern Indian Ocean and Arabian Sea, and is the only myctophid species captured in the Gulfs of Aden and Oman.

HABITAT
Oceanic, mesopelagic. It is often reported to be associated with continental shelf edges and around islands. Some studies have suggested that it is benthopelagic. Reported daytime depth ranges are between 426 and 1,640 ft (130 and 500 m), with night depth ranges from the surface to about 984 ft (300 m).

BEHAVIOR
This species is a vertical migrator. It also is known to form extremely dense aggregations. Catch rates using large commercial pelagic fishing trawls have captured as much as 88 tons

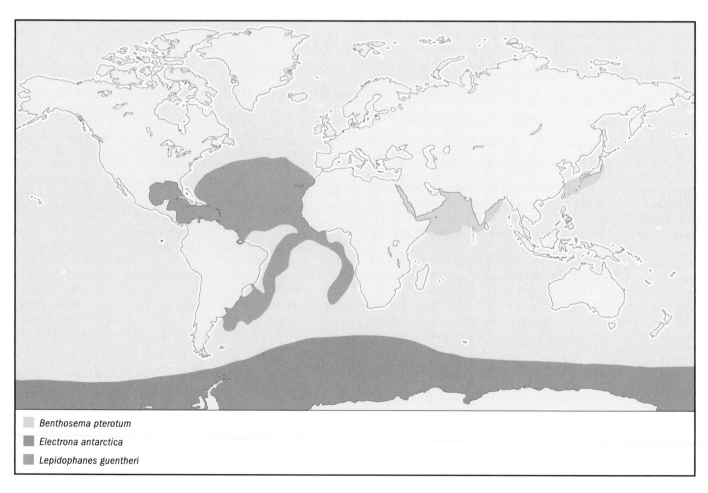

Benthosema pterotum
Electrona antarctica
Lepidophanes guentheri

(80 metric tons) per hour of fish in the northern Arabian Sea, where this species is especially abundant.

FEEDING ECOLOGY AND DIET
Crustacean zooplankton, especially copepods. There is a shift towards larger crustacean prey as the fish grow larger. Feeding occurs at night.

REPRODUCTIVE BIOLOGY
Sexes are separate and dimorphic. Like other members of this genus, sexually mature males have a series of luminescent scales in a supracaudal position, whereas in females, the luminescent scales are infracaudal. Sexual maturity in females occurs at around 1 in (25.4 mm) SL. Release of eggs and sperm is external, and reproduction occurs in the late afternoon and evening as the animals are migrating upwards; the buoyant eggs then float into the food-rich zone while developing. *Benthosema pterotum* is a serial batch spawner, releasing from 200 to 3,000 eggs per batch, depending on body size. Spawning appears to continue from the onset of sexual maturity until death, with an estimated life span of slightly less than one year.

CONSERVATION STATUS
Not threatened.

SIGNIFICANCE TO HUMANS
Benthosema pterotum is a target species for commercial capture by Iran. Plans were to use a processing vessel of 165 tons (150 metric tons)/day processing capacity to convert *B. pterotum* captured in the Gulf of Oman (a region of especially high density for this species) into fish meal for fertilizer. It is unknown whether these operations are still continuing. ◆

No common name
Electrona antarctica

FAMILY
Myctophidae

TAXONOMY
Scopelus antarcticus Günther, 1878, Antarctic Ocean, Stations 156 and 157.

OTHER COMMON NAMES
None known.

PHYSICAL CHARACTERISTICS
The species is characterized by a relatively large, deep head, with large eyes, and a long series of photophores above the anal fin.

Electrona antarctica

Pectoral fins are moderately long, reaching to about the middle of the dorsal fin. It is a large species, attaining a maximum reported size of more than 4 in (101.6 mm) SL. Sides and ventral surfaces are highly reflective and silver in color, with a deep blue-black along the dorsal surface.

DISTRIBUTION
Considered to be endemic to Antarctic waters, although some small individuals, thought to be waifs, have been collected

north of the Antarctic convergence. Considered one of the few truly polar species of myctophids.

HABITAT
Oceanic, mesopelagic. During daylight hours, this species resides below 1,640 to over 2,952 ft (500 to over 900 m) depth, while at night it can be found between 0 and 990 ft (0 and 300 m). During periods of 24-hour sunlight, "nighttime" depths of capture usually are more than 660 ft (200 m).

BEHAVIOR
The species is a strong vertical migrator, and is known to enter the very surface waters during night hours, where it is often eaten by seabirds.

FEEDING ECOLOGY AND DIET
This species is a crustacean zooplanktivore. It feeds on copepods, ostracods, and euphausiids. Larger individuals are known to have a greater incidence of euphausiids in their diet, especially the abundant krill, *Euphausia superba*.

REPRODUCTIVE BIOLOGY
Sexes are separate, and males are distinctly smaller than females at a maximum size of 3.3 in (83.8 mm), vs. 4 in (101.6 mm) SL. Although age has been estimated to be as much as 11 years, the most reliable report has suggested a maximum age of 3.5 years for this species. It is believed to attain sexual maturity in its second year. Based on the staging of oocytes in collected females, it is thought that this species has a restricted spawning period during the southern spring.

CONSERVATION STATUS
Not threatened.

SIGNIFICANCE TO HUMANS
Of no known commercial use, but is an important link in southern ocean ecosystem dynamics. ◆

No common name
Lepidophanes guentheri

FAMILY
Myctophidae

TAXONOMY
Lampanyctus güntheri Goode and Bean, 1896, based on single specimen "obtained by the Gloucester fleet."

OTHER COMMON NAMES
None known.

PHYSICAL CHARACTERISTICS
Males and females both exhibit elongated, slender bodies with relatively pointed heads. Maximum length is about 2.75 in (69.9 mm) SL. There is no ap-

Lepidophanes guentheri

parent sexual dimorphism. In addition to body photophores, numerous luminescent scales are found along the bases of the dorsal, anal, and adipose fins, plus along the ventral region of the tail. A distinctive feature of this species is the length of the pectoral fins, which are very long, extending to the posterior end of the anal fin. Freshly preserved specimens are deep blue-black along the dorsal surface and iridescent silver along the sides and ventral region.

DISTRIBUTION
Confined to the Atlantic Ocean. It is one of the most abundant myctophid species in the Gulf of Mexico, and according to the comprehensive zoogeographic analysis of Atlantic myctophids by Backus et al. (1977), it is "... the ranking myctophid in the Atlantic Tropical Region."

HABITAT
Oceanic, mesopelagic. *Lepidophanes guentheri* has been reported between 1,394 and 2,460 ft (425 and 750 m) during the day and between 131 to 410 ft (40 and 125 m) at night in the tropics, but at higher latitudes (off Bermuda and in the Gulf of Mexico), its daytime and nighttime depths are usually deeper (2,296–3,300 ft, or 700–1,000 m during the day; 165–574 ft, or 50–175 m, at night). A detailed study on the Gulf of Mexico population reported nonmigration by juveniles.

BEHAVIOR
Throughout its range in tropical environments, this species is a dominant component of the vertically migrating mesopelagic fauna. No other specifics of behavior are available.

FEEDING ECOLOGY AND DIET
This species is a crustacean zooplanktivore. Primary prey are copepods and at larger sizes, euphausiids. In a detailed study conducted in the eastern Gulf of Mexico, this species was one of three myctophid species that were found to remove more than 10% of the zooplankton biomass in epipelagic waters at night.

REPRODUCTIVE BIOLOGY
Males and females are separate sexes, or dioecious. This species is a serial batch spawner, releasing from 600 to more than 2,000 eggs per batch, depending on fish size. Sexual maturity in females is attained between 1.6 and 1.8 in (40.6 and 45.7 mm) SL. Eggs are released at night when the adults are at their shallow nighttime depths. Once sexual maturity is attained, spawning is repeated continuously every few days until death. Maximum age for this species has been shown to be about 14 months in the Gulf of Mexico.

CONSERVATION STATUS
Not threatened.

SIGNIFICANCE TO HUMANS
None known. ◆

Resources

Books
Backus, Richard H., et al. "Atlantic Mesopelagic Zoogeography." In *Fishes of the Western North Atlantic*. Vol. 7, edited by Robert H. Gibbs, Jr. New Haven, CT: Sears Foundation for Marine Research, 1977.

Gartner, John V., Jr., Roy E. Crabtree, and Kenneth J. Sulak. "Feeding At Depth." In *Fish Physiology. Deep-Sea Fishes*. Vol. 16, edited by David Randall and Anthony Farrell. San Diego: Academic Press, 1997.

Marshall, Norman B. *Developments in Deep-Sea Biology*. London: Blandford Press, 1980.

Nafpaktitis, Basil G. "Family Neoscopelidae." In *Fishes of the Western North Atlantic*. Vol. 7, edited by Robert H. Gibbs, Jr. New Haven, CT: Sears Foundation for Marine Research, 1977.

Nafpaktitis, Basil G., et al. "Family Myctophidae." In *Fishes of the Western North Atlantic*. Vol. 7, edited by Robert H. Gibbs, Jr. New Haven, CT: Sears Foundation for Marine Research, 1977.

Stiassny, Melanie L.J. "Basal Ctenosquamate Relationships and the Interrelationships of the Myctophiform (Scopelomorph) Fishes." In *Interrelationships of Fishes*, edited by Melanie Stiassny, Lynne Parenti, and G. Johnson. San Diego: Academic Press, 1996.

Periodicals
Dalpadado, Patricia. "Reproductive Biology of the Lanternfish *Benthosema pterotum* from the Indian Ocean." *Marine Biology* 98 (1988): 307–316.

Dalpadado, Patricia, and Jakob Gjösaeter. "Feeding Ecology of the Lanternfish *Benthosema pterotum* from the Indian Ocean." *Marine Biology* 99 (1988): 555–567.

Gartner, John V., Jr. "The Life Histories of Three Species of Lanternfishes (Pisces: Myctophidae) from the Eastern Gulf of Mexico. II. Age and Growth Patterns." *Marine Biology* 111, no. 1 (1991): 21–28.

———. "Patterns of Reproduction in the Dominant Lanternfish Species (Pisces: Myctophidae) of the Eastern Gulf of Mexico, with a Review of Reproduction Among Tropical-Subtropical Myctophidae." *Bulletin of Marine Science* 52, no. 2 (1993): 721–750.

———. "Review of the Fisheries Biology of the Mesopelagic Fishes of the Northern Arabian Sea and Gulf of Oman." *Food and Agriculture Organization of the United Nations Technical Report* (1996): 43–60.

Gjösaeter, Jakob, and Kouichi Kawaguchi. "A Review of the World Resources of Mesopelagic Fish." *Food and Agriculture Organization of the United Nations Technical Report* 93 (1980): 1–151.

Greely, Teresa M, John V. Gartner, Jr., and Joseph J. Torres. "Age and Growth of *Electrona antarctica* (Pisces: Myctophidae), the Dominant Mesopelagic Fish of the Southern Ocean." *Marine Biology* 133 (1999): 145–158.

Hopkins, Thomas L., and John V. Gartner, Jr. "Resource Partitioning and Predation Impact of a Low Latitude Myctophid Community." *Marine Biology* 114, no. 2 (1999): 185–198.

Other
"A Global Information System on Fishes." *FishBase*. June 14, 2002 [cited Oct. 27, 2002]. <http://www.fishbase.org/home.htm>

"A Collaborative Internet Project Containing Information About Phylogeny and Biodiversity." *Tree of Life Web Project*. June 14, 2002 [cited Oct. 27, 2002]. <http://tolweb.org/tree>

John V. Gartner Jr., PhD

Lampridiformes
(Opah and relatives)

Class Actinopterygii

Order Lampridiformes

Number of families 7

Photo: The oarfish (*Regalecus glesne*) is a deep sea dweller and rarely seen alive. (Photo by Robert E. Pelham. Bruce Coleman, Inc. Reproduced by permission.)

Evolution and systematics

The order Lampridiformes contains some of the most uniquely evolved, morphologically diverse, and colorful forms of living fishes. The imaginative common names of these species (oarfish, ribbonfish, tube-eye, inkfish, and unicornfish), their extreme rarity, and their fanciful reputations combine to promote the assemblage as subjects of public curiosity. In turn, scientists have particular interest in the group, because its members include species that are basal sister taxa in phylogenetic analysis of higher perchlike fishes and because many lampridiform species have evolved highly specialized traits. The icon of the order, *Regalecus glesne*, is a fabled species whose tremendous length (up to 55 ft, or 17 m—the longest of all bony fishes), bright crimson fins, long dorsal rays, silvery form, and tendency to appear suddenly at the ocean surface after wind storms brought fearful cries of "sea monster!" from ancient mariners. Recently, living oarfish have been sighted and filmed by divers in the clear waters of the Caribbean, and these accounts have stirred considerable popular interest. The order includes the deep-bodied *Lampris guttatus*, a brightly colored and heavy species (up to 600 lb, or 275 kg) that brings good prices in commercial markets, as well as other, lesser known and much smaller species that are never seen by the public and only infrequently collected on scientific expeditions.

In all, the order Lampridiformes (formerly known as the Allotriognatha) comprises 21 extant species classified in 12 genera and seven families. Skeletal evidence of another six or so extinct species has been found in fossil beds around the world. The current lampridiform classification is as follows: family Veliferidae, with two monotypic genera (containing one species), *Velifer* and *Metavelifer*; family Lamprididae, with one genus, *Lampris* (two species); family Stylephoridae, with one monotypic genus, *Stylephorus*; family Radiicephalidae, with one monotypic genus, *Radiicephalus*; family Lophotidae, with two monotypic genera, *Lophotus* and *Eumecichthys*; family Trachipteridae, with three genera, *Trachipterus* (four species), *Zu* (two species), and *Desmodema* (two species); and family Regalecidae, with two monotypic genera, *Regalecus* and *Agrostichthys*.

The oarfish and its relatives form a natural assemblage of fishes that evolved from a common *Velifer*-like ancestor sometime in the late Cretaceous or early Eocene. Independent investigations of lampridiform evolution, one using only morphological characters and another using both morphological and molecular characters in a total-evidence analysis, agree on the following shorthand pattern of phylogenetic relationships of lampridiform families: Veliferidae (Lamprididae (Stylephoridae ((Lophotidae + Radiicephailidae) (Regalecidae + Trachipteridae)))). In this shorthand scheme, sequentially, the taxon on the left and outside the parentheses is the sistergroup to the taxa within the parentheses. Thus, Veliferidae is the sistergroup to Lamprididae plus all other lampridiforms. Likewise, Lamprididae is the sistergroup to Stlylephoridae and all other lampridiforms. Finally, Stylephoridae is the sistergroup to two clades, each containing two families. To date, there have been no phylogenetic studies of species-level relationships, and, in some families, species-level and genera-level revisions are required.

Until recently, the limits of Lampridiformes and the systematic position of the order within the larger context of other bony fishes have been difficult, unresolved issues. Some scientists have provisionally included other families of deep-sea fishes (e.g., Ateleopodidae, Mirapinnidae, and Megalomycteridae) in Lampridiformes, but recent phylogenetic analysis confirms that these additional families are unrelated. Lampridiform fishes lack certain specializations of the axial skeleton (principally, patterns of small bones and ligaments

The oarfish (*Regalecus glesne*) is a long serpent-like fish that was thought to be a sea serpent 100 years ago. It becomes disoriented in rough seas and may writhe around on the surface. (Illustration by Wendy Baker)

associated with anterior vertebrae) that characterize more advanced fishes. These higher taxa, collectively classified as the Acanthomorpha, include the beardfish (*Polymixia*) and numerous major categories, including the codlike fishes (Paracanthopterygii) and perchlike fishes (Perciformes). Otherwise, Lampridiformes possess many other acanthomorph traits. Thus, the order Lampridiformes is considered the primitive sister taxon to all higher fishes, a significant group making up approximately 60% of all known teleost species. This systematic position makes the order, especially its basal members, veliferid lampridiforms, important to scientists hoping to unravel the intricate evolutionary histories of the diverse assemblage of acanthomorph fishes.

Abrupt transitions from shallow, nearshore habitats to the open ocean and from deep-bodied (termed bathysomous forms) to elongate (taeniosomous) forms are two major events hypothesized to have shaped the evolution of Lampridiformes. The basal and most generalized lampridiform family is the Veliferidae, moderate-sized, shallow-dwelling fishes that reside in coastal seas. All other lampridiforms are open-ocean, epipelagic or mesopelagic fishes. Since veliferids are the basal sister group to the Lamprididae (plus all other lampridiforms), it appears that an early important event in the

evolution of the lampridiform lineage was the invasion and subsequent adaptation of veliferid-like progeny to pelagic life in the deep ocean. *Lampris*, a ponderous and deep-bodied species, is the sister group to the tube-eyes (Stylephoridae) and all other lampridiforms. The second significant evolutionary transition occurs at this level in the phyletic sequence, because tube-eyes and their relatives are all slender, elongate, ribbon-shaped species. As the most highly evolved taxa of the order, *Stylephorus* is the sister taxon to two distinct clades in the lampridiform lineage, the inkfishes (represented by Lophotidae and Radiicephalidae) and the oarfishes and ribbonfishes (Regalecidae and Trachipteridae).

Numerous fossil forms have been attributed to lampridiform ancestry, some of which may be incorrectly classified. The oldest known lampridiform fish is *Nardovelifer altipinnis*, a recently described fossil found in Cretaceous deposits of Nardo near Lecce, in southern Italy. Eocene deposits near Verona, Italy, have yielded two species, *Veronavelifer* and *Bajaichthys*. Additional fossil taxa include several very large (up to 10 ft, or 3 m, in length) *Lampris*-like species discovered in Miocene deposits in California (*Lampris zatima*) and Oligocene deposits in New Zealand (an new, unnamed taxon) as well as two Oligocene lophotids, *Protomecicthys* and *Protolophotus*.

Physical characteristics

Unlike all other teleostean fishes, lampridiforms lack a ligamentous connection between the bones of the upper jaws (maxillae) and the cheekbones. In addition, the nasal cartilage that partially supports the maxillae is large and inserted in a groove on the skull in the frontal region. These two features allow the upper jaws of lampridiforms to be carried far forward during feeding, a modification that results in extreme jaw protrusion and expansion. In one species, *Stylephorus chordatus*, jaw protrusion allows the mouth cavity to expand suddenly to 40 times that of the closed mouth and permits *Stylephorus* to capture small planktonic prey items with considerable suction. Other unique physical characteristics of the order include crimson red fins, brightly colored bodies, and anterior placement of the thin bones (pterygiophores) that support the first dorsal fin. The elongate forms possess 60–150 vertebrae, whereas veliferids and lamprids have 33–46 vertebrae.

Distribution

Examples of most lampridiform families are rarely seen alive, and distribution records usually result from dead or dying animals that become stranded on beaches or are observed injured or disoriented at the sea surface. Some small species have been taken in scientific expeditions using midwater trawls. One species in the family Veliferidae inhabits shallow, nearshore waters of the Indian and Pacific Oceans. Another veliferid has been taken in bottom trawls on Pacific continental slopes and sea mounts. The remaining lampridiform families contain mesopelagic and epipelagic species that are distributed widely in temperate, subtropical, and tropical regions of all oceans except polar seas. Some species in the family Trachipteridae have somewhat restricted ranges. Radiicephalids have been collected only in the Atlantic Ocean and are extremely rare. One species of the family Lamprididae, the rare southern opah, occurs around the world in the subantarctic zone below 45° south latitude.

Habitat

Lampridiforms are strictly marine fishes. Veliferids occur at depths of 130–360 ft (40–110 m) in nearshore regions. *Lampris* occasionally is taken near the surface in hook-and-line fisheries off the California coast but also is known from deeper waters. Other lampridiform fishes inhabit the deep ocean, from surface waters to depths of hundreds of feet. Very little is known about the specific habitats of these species or their associations in pelagic communities.

Behavior

Observations of living specimens are extremely limited. Submersible sightings and diver observations suggest that elongate forms, such as *Regalecus* and trachipterids, normally orient in a vertical, head-up position and move vertically in the water column by undulating their long fin bases. In a short video of an oarfish sighted near the Bahamas, the specimen hovered almost motionless near the surface and then descended vertically and rapidly out of sight when disturbed by

Thirteen feet (4 m) in length with pelvic and anterior dorsal-fin rays extended, an oarfish (*Regalecus glesne*) rises from the deep ocean to sunlit surface waters in the Bahamas. The elongate rays bear fleshy, pigmented swellings but this specimen lacks the bright, crimson red fins that are typical of lampridiform fishes. (Photograph by Brian J. Skerry; National Geographic Society. Reproduced by permission.)

divers. Veliferids and lamprids swim horizontally in a typical teleost posture. Lamprids are powerful swimmers, using large pectoral fins for forward propulsion. Lampridiform fishes probably do not exhibit schooling behavior—all sightings and strandings have been of individuals. Feeding and spawning have never been observed. One regalecid, *Agrostichthys parkeri*, has been reported to deliver a mild electric shock when handled. *Lophotus* and *Radiicephalus* have specialized organs that empty into the cloaca and expel large quantities of a dark, squidlike ink when disturbed.

Feeding ecology and diet

Lampridiform fishes consume small planktonic crustaceans, especially euphausids; small to moderate-sized squids; and small to medium-sized fishes. These data come from anecdotal observations made of the stomach contents of single specimens. There have been no detailed studies of the feeding ecology of any lampridiform species. Very little is

An opah (*Lampris guttatus*) hunts squid. (Illustration by Bruce Worden)

known about predators of lampridiform fishes. Some species have been found in the stomachs of tunas.

Reproductive biology

Spawning has not been observed, but lampridiform fishes probably are broadcast spawners, because eggs are planktonic and have been identified for many lampridiform species in scientific collections. Eggs are large, 0.08–0.24 in (2–6 mm) in diameter, and brightly colored, usually in hues of red, pink, or amber. Incubation takes place in surface waters for approximately three weeks. At hatching, larvae have fully developed mouths and digestive tracts and begin to feed immediately on minute plankton. Larval lampridiforms are distinguished further by long, ornamented dorsal and pelvic rays. Some species, especially trachipterids, undergo abrupt metamorphosis from the larval to the juvenile form.

Conservation status

No lampridiform species are listed by the IUCN, but all are very rare. Care should be taken to ensure conservation of veliferid lampridiforms, because of their unique role in phylogenetic studies as basal taxa.

Significance to humans

Lampridiform fishes, especially the large, elongate species, attract considerable public attention when they are stranded or beached, owing to their rarity and unusual morphological features. *Lampris guttatus* has excellent flesh and is prized as bycatch of certain offshore fisheries. Most other lampridiform fishes are considered inedible by those who have recorded attempts to eat them.

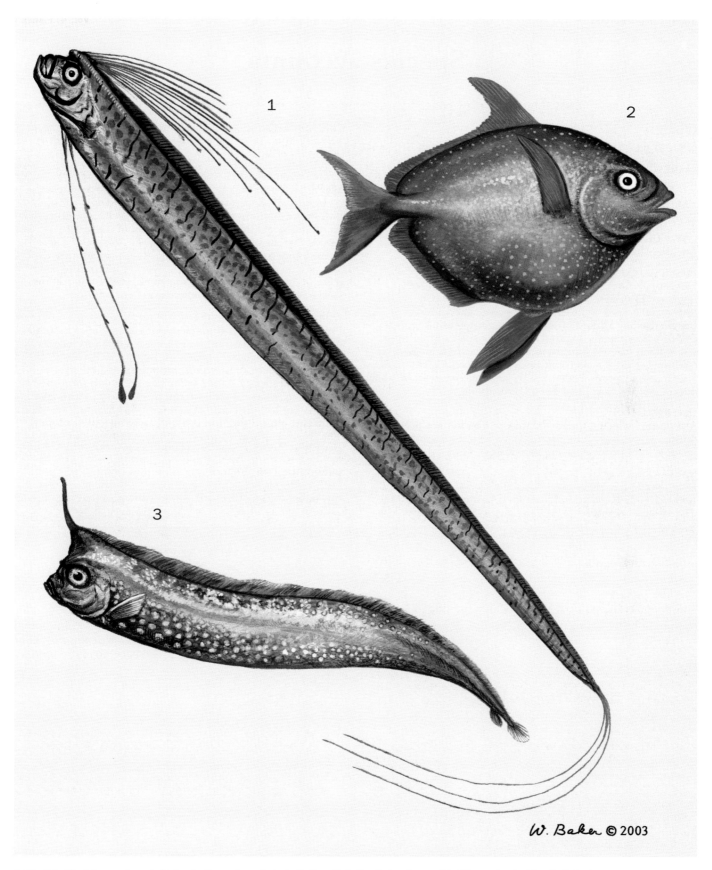

1. Oarfish (*Regalecus glesne*); 2. Opah (*Lampris guttatus*); 3. Crestfish (*Lophotus lacepede*). (Illustration by Wendy Baker)

Species accounts

Opah
Lampris guttatus

FAMILY
Lamprididae

TAXONOMY
Lampris guttatus Brunnich, 1788, North Sea.

OTHER COMMON NAMES
English: Moonfish, Jerusalem haddock.

PHYSICAL CHARACTERISTICS
Large, oval, colorful species with silvery blue, iridescent body; bright red fins and snout; large eyes; and white spots covering the body and the bases of the dorsal, anal, pelvic, and caudal fins. Length typically 3.3 ft (1 m) and total weight approximately 600 lb (275 kg). Pectoral fins large and horizontally oriented. The dorsal and anal fins have elongate anterior portions, and the caudal fin is moderately forked. Superficially resembles the ocean sunfish, *Mola mola*, a pelagic species that is similarly large and somewhat oval but not colorful and lacks pelvic fins and a caudal fin.

DISTRIBUTION
All oceans, including the Mediterranean Sea, but not polar oceans. Most distribution records come from catches of pelagic longline gear set for tuna and swordfish.

HABITAT
Lower epipelagic zone, generally in surface waters to perhaps 1,640 ft (500 m). Analysis of data from Japanese longline fish-

eries shows that catches of opah increase with depth to about 920 ft (280 m) in the central and eastern Pacific Ocean.

BEHAVIOR
There are no published observations of living *Lampris*. Swimming behavior has been inferred from the anatomical and morphological features of the pectoral fin. The massive shoulder girdle; the presence of extensive pectoral musculature, including a large red muscle mass; and the horizontal placement of the pectoral fin base suggest that the opah swims by pectoral-fin "flapping." This unusual swimming mode apparently allows the opah to swim rapidly, since it is an effective predator of active prey. Records of occurrence of opah are sporadic and widespread, suggesting that the species is a solitary wanderer in the world's oceans.

FEEDING ECOLOGY AND DIET
Predatory species that consumes squids and octopus as well as small to medium-sized fishes. Fish species eaten include hake, rockfish (Scorpaenidae), ophidiiforms, small dolphins (*Coryphaena* species), berycoids, argentines, and other mesopelagic species. There is one unusual report of opah feeding on small clams in shallow waters off Florida. Opah take live and cut bait on hooks as well as trolled fishing lures off the California coast. Opah can consume large quantities of prey. The stomach of one 126-lb (57-kg) specimen was found to contain 63 fishes, 8 cephalopods, and 7 crabs.

REPRODUCTIVE BIOLOGY
Little is known of the spawning behavior, spawning seasons, or spawning habitats of *Lampris*. Advanced-stage oocytes removed

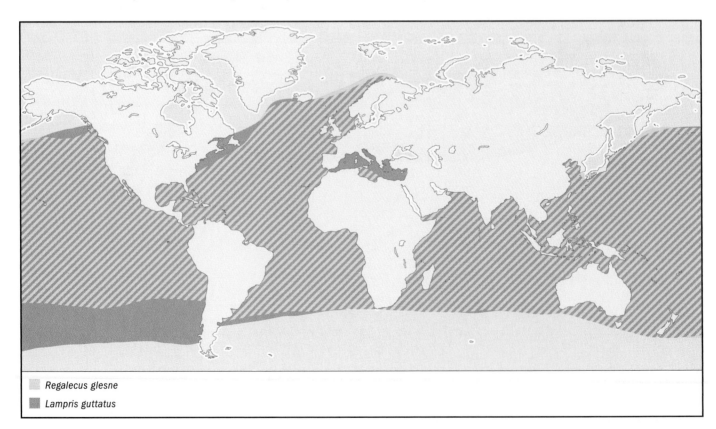

Regalecus glesne
Lampris guttatus

from the ovaries of ripe fish have a thick shell (chorion) that is amber to pink in color. Fertilized eggs are probably free floating but have not been identified in plankton collections. Only a few larval examples have been collected, all in the Atlantic Ocean. Larvae are slender and somewhat elongate initially but rapidly become deep-bodied. Pigment is confined to the head region and above the gut in the smallest specimens, but larger larvae have extensive scattered melanophores (pigmented cells) covering the entire body except the fins. The anterior rays of the dorsal and pelvic fins are elongate. Rates of mortality and growth, feeding habits, and predators of larvae are largely unknown.

CONSERVATION STATUS
Not listed by the IUCN.

SIGNIFICANCE TO HUMANS
The species is a marketable bycatch of longline fisheries and is pursued by recreational anglers. The all-tackle game fish record is 162.9 lb (73.9 kg), and this specimen was captured off California in 1998. Its flesh is considered good, and it is thought to be excellent as a smoked product. ◆

Crestfish
Lophotus lacepede

FAMILY
Lophotidae

TAXONOMY
Lophotus lacepede Giorna, 1809, type locality unknown.

The taxonomy of *Lophotus* is confused, and the genus is in need of revision. There is probably only one widely distributed species, but a number of nominal forms exist (*L. cepedianus*, *L. capellei*, and *L. cristatus*), all of which may be synonyms of *L. lacepede*.

OTHER COMMON NAMES
English: Highbrow crestfish, inkfish, crested oarfish.

PHYSICAL CHARACTERISTICS
The supraoccipital (a cranial bone situated above the eye) of *L. lacepede* bears a strong and enlarged anterior process that projects anteriorly over the forehead. This skeletal feature also is strongly developed in the crestfish's closest relative, *Eumecichthys fiski*, the unicornfish. The anterior process lies directly beneath the anterior dorsal fin rays and supports a large fleshy protuberance (the crest or horn) that gives the forehead of these fishes their distinct appearance. In *L. lacepede* the crest is thick and deep, and the forehead is either blunt or slightly projecting. In the unicornfish the fleshy protuberance is slender and greatly elongate, forming a long horn on the forehead. The body of *L. lacepede* is relatively long, attaining a total length of about 4.9–6.6 ft (1.5–2.0 m) and ribbon-like. Fresh specimens have been described as blackish blue, light brown, or silvery, sometimes with scattered white spots. Fins red. Dorsal fin extends the full length of the body. Caudal fin small and rounded. Pectoral fin low on body. Pelvic fins minute and located posterior to the pectoral fins. Anal fin small and located posteriorly. Possesses an elongate organ that lies above the lower intestine and produces thick, black, squidlike ink that is discharged through the cloaca. The unicornfish apparently does not possess this trait, although the ink gland is found in another lampridiform fish that is related closely to the lophotids, *Radiicephalus elongatus*.

DISTRIBUTION
Distributed widely in warm waters of all oceans. Specimens are rare, and most either have come from research cruises or have

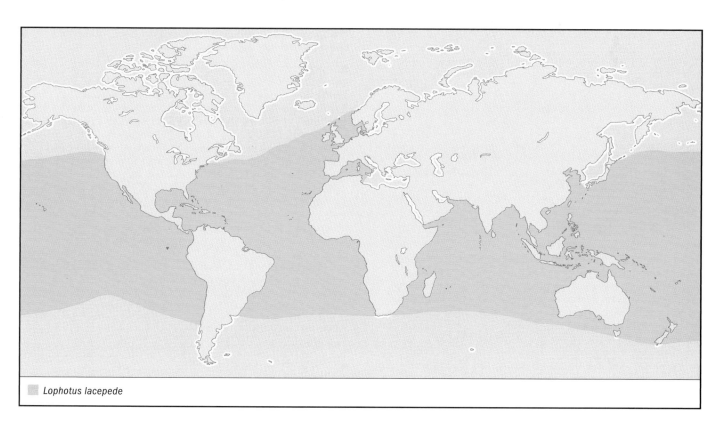

Lophotus lacepede

been captured close to shore as stranded individuals in weakened condition.

HABITAT

Very little is known of the habitats of these fishes, owing to their rarity. Crestfish are apparently members of the epipelagic community and have been collected in surface waters to depths of only a few hundred feet (or meters). Some specimens have been taken in very shallow water, but they were probably dying.

BEHAVIOR

Almost nothing is known of the natural behavior of this species. Crestfish probably discharge ink when frightened. A large adult specimen captured and gaffed during a research cruise expelled a copious amount of ink. Crestfish may develop this capability at a young age, since late-stage larvae possess the ink gland.

FEEDING ECOLOGY AND DIET

Apparently consume small fishes, squid, and pelagic octopus.

REPRODUCTIVE BIOLOGY

Little is known of the spawning behavior, spawning seasons, or spawning habitats of *Lophotus* or its relatives. Eggs are pelagic and have been described from plankton collections in the Mediterranean Sea as relatively large, about 0.01 in (2.5 mm) in diameter, with many tiny spines covering the chorion. Early larvae have elongate dorsal and pelvic rays that bear large pigmented swellings. These rays often are damaged in capture and are absent in preserved specimens. Late larvae are sparsely pigmented and possess the internal ink gland, with the dark contents visible through the body wall. Vital rates, feeding habits, and predators of larvae are largely unknown.

CONSERVATION STATUS

Not listed by the IUCN.

SIGNIFICANCE TO HUMANS

None known. ◆

Oarfish

Regalecus glesne

FAMILY

Regalecidae

TAXONOMY

Regalecus glesne Ascanius, 1772. Most ichthyologists recognize a single species with worldwide distribution, but there are several nominal species that may be valid.

OTHER COMMON NAMES

English: King of the herrings.

PHYSICAL CHARACTERISTICS

Spectacular animals with long, slender, usually silver bodies; brilliant red fins; a large, cockscomb-like plume of 6–10 anterior dorsal-fin rays; and long pelvic fins. When not damaged, the pelvic ray bears a large red swelling at its tip, from which its name is derived. Often attaining a length of 26–33 ft (8–10 m), oarfish are the longest of all bony fishes. (There are unconfirmed reports of some specimens reaching 56 ft, or 17 m.) The long body gradually tapers to a point. The caudal fin (when present and not damaged) is tiny, and its rays bear small, laterally projecting spinules. No anal fin. About 400 dorsal fin rays and 150 vertebrae. Some confusion exists concerning the distinction between regalecids and their close relatives,

the trachipterids or ribbonfishes. Fishes in both families are elongate, have crimson red fins, lack an anal fin, and possess laterally projecting, small spines on the individual rays of the caudal and pelvic fins. The oarfish attains a greater length, has only two pelvic rays (one very elongate and one small and splintlike) and a very elongate anterior dorsal fin, and is relatively slender. The trachipterids are shorter, have more than two pelvic rays (though pelvic rays are absent in one genus), lack the extremely elongate dorsal rays, are relatively deeper-bodied, and have small spines on the lateral-line scales.

DISTRIBUTION

All oceans, including the Mediterranean Sea, but not polar seas. Most distribution records result from strandings of weak or dying fish in shallow water, although some fish have been captured by trawl.

HABITAT

Epipelagic species, normally inhabiting surface waters to depths of about 656 ft (200 m). The species is capable of occupying greater depths, however. Little is known of its specific habitat requirements or associations.

BEHAVIOR

Recent sightings of live fish by divers confirm that swimming is not accomplished by rowing the pelvic fins, as one might suppose from the name oarfish. Instead, oarfish move in the water column by undulations of the dorsal fin. The natural body position is vertical, with the head up and dorsal fin rays and pelvic rays extended outward. The pelvic fin may have a sensory capability, allowing the oarfish to "taste" its surrounding habitat. Its close relative, *Agrostichthys parkeri*, has been reported to be electrogenic, delivering a mild shock when handled. It is unknown whether the oarfish shares this capability.

FEEDING ECOLOGY AND DIET

Apparently consumes planktonic crustaceans and small fishes, but little is known of its feeding ecology. The stomach of a 9.8-ft (3-m) oarfish stranded in shallow waters off California contained a large volume of krill (euphausids), numbering about 10,000 individuals.

REPRODUCTIVE BIOLOGY

The spawning of oarfish has not been observed, and little is known of its spawning habits or seasons. The ovaries of a dying female, about 9.8 ft (3 m) in length, that was cast ashore on the west coast of Florida weighed 3.1 lb (1.4 kg) and contained approximately 140,000 unspawned eggs, some of which were ovulated. The female was observed in March, along with other dead male fish, presumably following a spawning occurrence. Oarfish eggs are pelagic and have been identified in plankton collections in the Mediterranean Sea and the waters off New Zealand. Eggs are large, 0.08–0.16 in (2–4 mm) in diameter, with pink to red chorions, and they contain numerous scattered oil droplets. Researchers report that eggs can take up to three weeks to incubate and that the emerging larva is highly developed at hatching, with a functional mouth and gut, a pigmented eye, and long dorsal and pelvic fin rays. Young oarfish resemble adults in most details and have been captured on research cruises in the Atlantic and Pacific Oceans. Vital rates, feeding habits, and predators of larvae are largely unknown.

CONSERVATION STATUS

Not listed by the IUCN.

SIGNIFICANCE TO HUMANS

Sightings of the oarfish usually stir considerable public attention, but the species has no commercial value, and its flesh is reported to be unpalatable. ◆

Resources

Books

Eschmeyer, W. N., E. S. Herald, and H. Hammann. *A Field Guide to the Pacific Coast Fishes.* Boston: Houghton Mifflin, 1983.

Fitch, J. E., and R. J. Lavenberg. *Deep-Water Teleostean Fishes of California.* Berkeley: University of California Press, 1968.

Heemstra, P. C. "Order Lampridiformes." In *Smiths' Sea Fishes,* edited by M. M. Smith and P. C. Heemstra. Johannesburg, South Africa: MacMillan South Africa, 1986.

Periodicals

Olney, J. E., G. D. Johnson, and C. C. Baldwin. "Phylogeny of Lampridiform Fishes." *Bulletin of Marine Science* 52 (1993): 137–169.

Patterson, C. "An Overview of the Early Fossil Record of Acanthomorphs." *Bulletin of Marine Science* 52 (1993): 29–59.

Saloman, C. H., M. A. Moe, and J. L. Taylor. "Observations on a Female Oarfish (*Regalecus glesne*)." *Florida Scientists* 36 (1973): 187–189.

Skerry, B. "Eye-to-Eye with the Sea Serpent: First Photos of the Mysterious Oarfish." *Sport Diver* 5, no. 4 (August 1997): 40–43.

Sorbini, C., and L. Sorbini. "The Cretaceous Fishes of Nardo. 100. *Nardovelifer altipinnis*, gen. nov. et spec. nov. (Teleostei, Lampridiformes, Veliferidae)." *Museo Civico di Storia Naturale di Verona, Studi e Ricerche sui Giamenti Terziari do Bolca* 8 (1999): 11–27.

Wiley, E. O., G. D. Johnson, and W. W. Dimmick. "The Phylogenetic Relationships of Lampridiform Fishes (Teleostei: Acanthomorpha), Based on a Total-Evidence Analysis of Morphological and Molecular Data." *Molecular Phylogenetics and Evolution* 10 (1998): 417–425.

John E. Olney, PhD

For further reading

Able, Kenneth W., and Michael P. Fahay. *The First Year in the Life of Estuarine Fishes in the Middle Atlantic Bight.* New Brunswick, NJ: Rutgers University Press, 1998.

Alexander, R. M. *Functional Design in Fishes.* London: Hutchinson, 1967.

Allan, J. David. *Stream Ecology: Structure and Function of Running Waters.* New York: Chapman & Hall, 1995.

Allen. G. R. *Freshwater Fishes of Australia.* Neptune City, NJ: T. H. F. Publications, 1989.

———. *Marine Fishes of Tropical Australia and South-east Asia.* Perth: Western Australian Museum, 1997.

Allen, G. R., and D. R. Robertson. *Fishes of the Tropical Eastern Pacific.* Honolulu: University of Hawaii Press, 1994.

Allen, G. R., and R. Swainston. *The Marine Fishes of North-Western Australia: A Field Guide for Anglers and Divers.* Perth: Western Australian Museum, 1988.

Allen, G. R., S. H. Midgley, and M. Allen. *Field Guide to the Freshwater Fishes of Australia.* Perth: Western Australian Museum, 2002.

Andriashev, A. P. *Fishes of the Northern Seas of the USSR.* Moscow: Academy of Sciences, USSR, 1954.

Banarescu, P. *Zoogeography of Fresh Waters.* 3 vols. Wiesbaden, Germany: AULA-Verlag, 1990–1995.

Beard, J. A. *James Beard's New Fish Cookery.* New York: Galahad Books, 1976.

Bemis, William E., Warren W. Burggren, and Norman E. Kemp. *The Biology and Evolution of Lungfishes.* New York: A. R. Liss, 1987.

Benton, M. J., ed. *The Fossil Record 2.* London: Chapman and Hall, 1993.

Berra, Tim M. *An Atlas of Distribution of the Freshwater Fish Families of the World.* Lincoln: University of Nebraska Press, 1981.

———. *Freshwater Fish Distribution.* San Diego: Academic Press, 2001.

Bigelow, H. B., ed. *Fishes of the Western North Atlantic.* New Haven: Sears Foundation for Marine Research, Yale University, 1963.

Bohlke, James E., and Charles C. G. Chaplin. *Fishes of the Bahamas and Adjacent Tropical Waters.* 2nd edition. Austin: University of Texas Press, 1993.

Bond, Carl E. *Biology of Fishes.* 2nd edition. New York: Harcourt Brace College Publishers, 1996.

Bone, Q., N. B. Marshall, and J. H. S. Blaxter. *Biology of Fishes.* 2nd edition. Glasgow: Blackie Academic and Professional, 1995.

Boschung, H. T., Jr., J. D. Williams, D. W. Gotshall, D. K. Caldwell, and M. C. Caldwell. *The Audubon Society Field Guide to North American Fishes, Whales, and Dolphins.* New York: Alfred A. Knopf, 1983.

Breder, C. M., Jr., and D. E. Rosen. *Modes of Reproduction in Fishes.* Garden City, NY: Natural History Press, 1966.

Briggs, John C. *Marine Zoogeography.* New York: McGraw-Hill, 1974.

———. *Global Biogeography.* Amsterdam: Elsevier, 1995.

Brooks, D. R., and D. A. McLennan. *Phylogeny, Ecology, and Behavior.* Chicago: University of Chicago Press, 1991.

Bullock, T., and W. Heiligenberg. *Electroreception.* New York: John Wiley & Sons, 1986.

Butler, Ann B., and William Hodos. *Comparative Vertebrate Neuroanatomy: Evolution and Adaptation.* New York: John Wiley and Sons, 1996.

Carwardine, Mark, and Ken Watterson. *The Shark Watcher's Handbook: A Guide to Sharks and Where to See Them.* Princeton: Princeton University Press: 2002.

Coad, B. W. *Guide to the Marine Sport Fishes of Atlantic Canada and New England.* Toronto: University of Toronto Press, 1992.

Collette, Bruce B., and Grace Klein-MacPhee, editors. *Bigelow and Schroeder's Fishes of the Gulf of Maine.* 3rd edition. Washington, D.C.: Smithsonian Institution Press, 2002.

Compagno, L. J. V. *Sharks of the World: An Annotated and Illustrated Catalogue of Shark Species Known to Date.* FAO Species Catalogue, vol. 4, part 1. Rome: Food and Agriculture Organization of the United Nations, 1984.

Cushing, Colbert E., and J. David Allen. *Streams: Their Ecology and Life.* San Diego: Academic Press, 2001.

Darlington, Philip J., Jr. *Zoogeography: The Geographical Distribution of Animals.* New York: John Wiley and Sons, 1957.

Deloach, Ned. *Reef Fish Behavior: Florida, Caribbean, Bahamas.* Jacksonville: New World Publications, Inc., 1999.

Demski, Leo S. and John P. Wourms, eds. *Reproduction and Development of Sharks, Skates, Rays and Ratfishes.* Boston, MA: Kluwer Academic Publishers, 1993.

Diana, J. S.*Biology and Ecology of Fishes.* Carmel, IN: Cooper Publishing, 1995.

di Prisco, G., B. Maresca, and B. Tota, eds. *Biology of Antarctic Fish.* Berlin: Springer-Verlag, 1991.

di Prisco, G., E. Pisano, and A. Clarke, ed. *Fishes of Antarctica: A Biological Overview.* Milan: Springer-Verlag, 1998.

Dobson, Mike, and Chris Frid. *Ecology of Aquatic Systems.* Essex, U.K.: Addison Wesley Longman, 1998.

Eastman, J. T. *Antarctic Fish Biology: Evolution in a Unique Environment.* San Diego: Academic Press, 1993.

Echelle, A. A., and I. Kornfield, eds. *Evolution of Fish Species Flocks.* Orono: University of Maine at Orono Press, 1984.

Eddy, S., and J. C. Underhill. *Northern Fishes,* 3rd edition. Minneapolis: University of Minnesota Press, 1974.

Eschmeyer, W. N. *Catalog of the Genera of Recent Fishes.* San Francisco: California Academy of Sciences, 1990.

———, ed. *Catalog of Fishes.* 3 vols. San Francisco: California Academy of Sciences, 1998.

Eschmeyer, W. N., E. S. Herald, and H. Hammann. *A Field Guide to Pacific Coast Fishes of North America.* Boston: Houghton Mifflin Company, 1983.

Etnier, David A., and Wayne C. Starnes. *The Fishes of Tennessee.* Knoxville: University of Tennessee Press, 1993.

Evans, D. H., ed.*The Physiology of Fishes,* 2nd edition. Boca Raton, FL: CRC Press, 1998.

Everhart, W. H., and W. D. Youngs.*Principles of Fishery Science,* 2nd edition. Ithaca, NY: Cornell University Press, 1981.

Fitch, J. E., and R. J. Lavenberg. *Deep-Water Teleostean Fishes of California.* Berkeley: University of California Press, 1968.

———. *Tidepool and Nearshore Fishes of California.* Berkeley: University of California Press, 1975.

Frickhinger, Karl Albert. *Fossil Atlas: Fishes.* Blacksburg, VA: Tetra Press, 1996.

Fryer, G., and T. D. Iles. *The Cichlid Fishes of the Great Lakes of Africa.* Edinburgh: Oliver and Boyd, 1972.

Fuller, P. L., L. G. Nico, and J. D. Williams. *Nonindigenous Fishes Introduced into Inland Waters of the United States.* Bethesda, MD: American Fisheries Society, Special Publication 27, 1999.

Gery, J. *Characoids of the World.* Neptune City, NJ: Tropical Fish Hobbyist Publications, Inc., 1977.

Giller, Paul S., and Bjorn Malmqvist. *The Biology of Streams and Rivers.* Oxford: Oxford University Press, 1998.

Gloerfelt-Tarp, T., and P. J. Kailola. *Trawled Fishes of Southern Indonesia and Northwestern Australia.* Jakarta: Directorate General of Fisheries (Indonesia), German Agency for Technical Cooperation, Australian Development Assistance Bureau, 1984.

Gomon, M. F., J. C. M. Glover, and R. H. Kuiter, eds. *The Fishes of Australia's South Coast.* Adelaide: State Print, 1994.

Gon, O, and P. C. Heemstra, editors. *Fishes of the Southern Ocean.* Grahamstown, South Africa: J. L. B. Smith Institute of Ichthyology, 1990.

Goodson, G. *Fishes of the Pacific Coast.* Stanford: Stanford University Press, 1988.

Gordon, B. L.*The Secret Lives of Fishes.* New York: Grosset and Dunlap, 1977.

Graham, Jeffrey B. *Air-breathing Fishes: Evolution, Diversity, and Adaptation.* San Diego, CA: Academic Press, 1997.

Groombridge, B., and M. Jenkins. *Freshwater Biodiversity: A Preliminary Global Assessment.* Cambridge, U.K.: World Conservation Monitoring Centre—World Conservation Press, 1998.

Halstead, Bruce W. *Poisonous and Venomous Marine Animals of the World.* Washington, DC: Government Printing Office, 1965–1970.

Hamlett, William C., ed. *Sharks, Skates, and Rays: The Biology of Elasmobranch Fishes.* Baltimore, MD: Johns Hopkins University Press, 1999.

Hart, J. L. *Pacific Fishes of Canada.* Ottawa: Fisheries Research Board of Canada, Bulletin 180, 1973.

Helfman, Gene S., Bruce B. Collette, and Douglas E. Facey. *The Diversity of Fishes.* Malden, MA: Blackwell Science, 1997.

Hennemann, Ralf M.*Elasmobranch Guide of the World: Sharks and Rays.* Frankfurt: Ikan, 2001.

Hoar, W. S., and D. J. Randall, eds. *Fish Physiology.* Vols. 1–20. New York: Academic Press, 1969–1993.

Hoese, H. D., and R. H. Moore. *Fishes of the Gulf of Mexico, Texas, Louisiana, and Adjacent Waters.* 2nd edition. College Station, TX: Texas A & M University Press, 1998.

Horn, Michael H., Karen L. M. Martin, and Michael A. Chotkowski, eds. *Intertidal Fishes: Life in Two Worlds*, San Diego: Academic Press, 1999.

Hoyt, E. *Creatures of the Deep: In Search of the Sea's "Monsters" and the World They Live In.* Buffalo: Firefly Books, Ltd., 2001.

Janvier, Philippe. *Early Vertebrates.* New York: Oxford University Press, 1996.

Kock, Karl-Hermann. *Antarctic Fish and Fisheries.* Cambridge: Cambridge University Press, 1992.

Kottelat, M., A. J. Whitten, S. N. Kartikasari, and S. Wirjoatmodjo. *Freshwater Fishes of Western Indonesia and Sulawesi.* Jakarta: Periplus Editions, 1993.

Kramer, D. E., W. H. Barss, B. C. Paust, and B. E. Bracken. *Guide to Northeast Pacific Flatfishes: Families Bothidae, Cynoglossidae, and Pleuronectidae.* Marine Advisory Bulletin no. 47. Fairbanks: University of Alaska Fairbanks, Alaska Sea Grant College Program, 1995.

Kuiter, Rudie H. *Guide to Sea Fishes of Australia.* London: New Holland, 1996.

Kuiter, Rudie H. *Coastal Fishes of South-Eastern Australia.* Honolulu: University of Hawaii Press, 1993.

———. *Seahorses, Pipefishes and Their Relatives: A Comprehensive Guide to Syngnathiformes.* Chorleywood, U.K.: TMC Publishing, 2000.

La Rivers, Ira. *Fish and Fisheries of Nevada.* Reno: University of Nevada Press, 1994.

Lagler, Karl F. *Ichthyology.* 2nd edition. New York: John Wiley & Sons, 1977.

Lampert, Winfried, and Ulrich Sommer. *Limnoecology: The Ecology of Lakes and Streams.* New York: Oxford University Press, 1997.

Last, P. R., and J. D. Stevens. *Sharks and Rays of Australia.* Melbourne, Australia: CSIRO, 1994.

Lee, D. S., C. R. Gilbert, C. H. Hocutt, R. E. Jenkins, D. E. McAllister, and J. R. Stauffer, Jr. *Atlas of North American Freshwater Fishes.* Raleigh: North Carolina State Museum of Natural History, 1980.

Leis, Jeffrey M., and Brooke M. Carson-Ewart, eds. *The Larvae of Indo-Pacific Coastal Fishes: An Identification Guide to Marine Fish Larvae.* Boston: Brill, 2000.

Lieske, Ewald, and Robert Myers. *Coral Reef Fishes: Caribbean, Indian Ocean and Pacific Ocean: Including the Red Sea.* London: Harper Collins, 1994.

Loiselle, Paul V. *The Cichlid Aquarium.* Melle, Germany: Tetra-Press, 1985.

Long, John A. *The Rise of Fishes: 500 Million Years of Evolution.* Baltimore, MD: Johns Hopkins University Press, 1995.

Lourie, S. A., A. C. J. Vincent, and H. J. Hall. *Seahorses: An Identification Guide to the World's Species and Their Conservation.* London: Project Seahorse, 1999.

Love, Milton S. *Probably More Than You Want to Know About the Fishes of the Pacific Coast.* 2nd edition. Perm: Izd-vo Permskogo Universiteta, 2001.

Love, Milton S., Mary Yoklavich, and Lyman Thorsteinson. *The Rockfishes of the Northeast Pacific.* Berkeley: University of California Press, 2002.

Lythgoe, J., and G. Lythgoe. *Fishes of the Sea.* Garden City, NJ: Anchor Press/Doubleday, 1975.

Mago-Leccia, F. *Electric Fishes of the Continental Waters of America.* Vol. 29, Biblioteca de la Academica de Ciencias Fisicas Matematicas y Naturales. Caracas, Venezuela: FUDECI, 1994.

Maissey, J. G. *Santana Fossils: an Illustrated Atlas.* Neptune City, NJ: T. F. H. Publishers, 1991.

Marshall, N. B. *Aspects of Deep Sea Biology.* London: Hutchinson, 1954.

Masuda, H., K. Amaoka, C. Araga, T. Uyeno, and T. Yoshino, eds. *The Fishes of the Japanese Archipelago.* Tokyo: Tokai University Press, 1984.

Matthews, William J. *Patterns in Freshwater Fish Ecology.* New York: Chapman & Hall, 1998.

Mayden, R. L., ed. *Systematics, Historical Ecology and North American Freshwater Fishes.* Stanford: Stanford University Press, 1992.

McDowall, R. M. *Freshwater Fishes of South-Eastern Australia.* Sydney: Reed Books, 1996.

———. *Diadromy in Fishes: Migrations Between Freshwater and Marine Environments.* London: Croom Helm, 1988.

———. *Freshwater Fishes of New Zealand.* Auckland: Reed, 2001.

McEachran, John D., and Janice D. Fechhelm. *Fishes of the Gulf of Mexico.* Vol. 1, Myxiniformes to Gasterosteiformes. Austin, TX: University of Texas Press, 1998.

Mecklenburg, Catherine W., T. Anthony Mecklenburg, and Lyman K. Thorsteinson. *Fishes of Alaska.* Bethesda: American Fisheries Society, 2002.

Menon, A. G. K. *The Fauna of India and the Adjacent Countries.* Madras: Amra Press, 1987.

Merrick, J. R., and G. E. Schmida. *Australian Freshwater Fishes: Biology and Management.* North Ryde, N.S.W., Australia: J. R. Merrick, 1984.

Michael, Scott W. *Reef Fishes: A Guide to Their Identification, Behavior and Captive Care.* Shelburne, VT: Microcosm Ltd., 1998.

Miller, Richard Gordon. *History and Atlas of the Fishes of the Antarctic Ocean.* Carson City, NV: Foresta Institute for Ocean and Mountain Studies, 1993.

FOR FURTHER READING

Minckley W. L., and James E. Deacon, eds. *Battle Against Extinction, Native Fish Management in the American West.* Tucson: University of Arizona Press, 1991.

Moller, P., ed. *Electrical Fishes: History and Behavior. Fish and Fisheries Series 117.* London: Chapman & Hall, 1995.

Morrow, James E. *The Freshwater Fishes of Alaska.* Anchorage: Alaska Northwest Publishing, 1980.

Moyle, Peter B., and Joseph J. Cech, Jr. *Fishes: An Introduction to Ichthyology.* 4th edition. Upper Saddle River, NJ: Prentice Hall, 2000.

Murdy, E. O., R. Birdsong, and J. A. Musick. *Fishes of the Chesapeake Bay.* Washington, DC: Smithsonian Institution, 1997.

Myers, R. F. *Micronesian Reef Fishes: A Field Guide for Divers and Aquarists.* 3rd edition. Barrigada, Guam: Coral Graphics, 1999.

Neira, F. J., A. G. Miskiewicz, and T. Trnski, eds. *The Larvae of Temperate Australian Fishes: A Laboratory Guide for Larval Fish Identification.* Perth: University of Western Australia Press, 1998.

Nelson, J. S. *Fishes of the World.* 3rd edition. New York: John Wiley and Sons, 1994.

Page, L. M., and B. M. Burr. *A Field Guide to Freshwater Fishes of North America North of Mexico.* Boston: Houghton Mifflin Company, 1991.

Paulin, C., A. Stewart, C. Roberts, and P. McMillan. *New Zealand Fish: A Complete Guide.* National Museum of New Zealand Miscellaneous Series No. 19. Wellington: New Zealand, 1989.

Paxton, John R., and William N. Eschmeyer, eds. *Encyclopedia of Fishes.* 2nd edition. San Diego: Academic Press, 1998.

Payne, A. I. *The Ecology of Tropical Lakes and Rivers.* New York: John Wiley & Sons, 1986.

Perrine, D. *Sharks and Rays of the World.* Stillwater, MN: Voyager Press, 1999.

Pethiyagoda, R. *Freshwater Fishes of Sri Lanka.* Sri Lanka: The Wildlife Heritage Trust of Sri Lanka, 1991.

Pietsch, T. W., and D. B. Grobecker. *Frogfishes of the World: Systematics, Zoogeography, and Behavioral Ecology.* Stanford: Stanford University Press, 1987.

Pitcher, T. J., ed. *The Behaviour of Teleost Fishes.* London: Chapman and Hall, 1993.

Potts, D. T., and J. S. Ramsey. *A Preliminary Guide to Demersal Fishes of the Gulf of Mexico Continental Slope (100 to 600 fathoms).* Mobile, AL: Alabama Sea Grant Extension, 1987.

Potts, G. W., and R. J. Wootton, eds. *Fish Reproduction: Strategies and Tactics.* London: Academic Press, 1984.

Raasch, Maynard S. *Delaware's Freshwater and Brackish-Water Fishes: A Popular Account.* Neptune City, NJ: T.F.H. Publications, 1996.

Randall, John E. *Surgeonfishes of the World.* Honolulu: Bishop Museum Press/Mutual Publishing, 2001.

Randall, John E., Gerald R. Allen, and Roger C. Steene. *Fishes of the Great Barrier Reef and Coral Sea.* Honolulu: Crawford House Publishing/University of Hawaii Press, 1997.

Riehl, R., and H. A. Baensch. *Aquarium Atlas.* Melle, West-Germany: Baensch, 1986.

Robins, C. Richard, and G. Carleton Ray. *A Field Guide to Atlantic Coast Fishes of North America.* Boston: Houghton Mifflin, 1986.

Romero, Aldemaro, ed. *The Biology of Hypogean Fishes.* Dordrecht: Kluwer Academic Publishers, 2001.

Ross, Stephen T. *Inland Fishes of Mississippi.* Jackson: University Press of Mississippi, 2001.

Sadovy, Y., and A. S. Cornish. *Reef Fishes of Hong Kong.* Hong Kong: Hong Kong University Press, 2000.

Sale, Peter F., ed. *Coral Reefs Fishes: Dynamics and Diversity in a Complex Ecosystem.* San Diego: Academic Press, 2001.

Schultze, Hans-Peter, and Linda Trueb, eds. *Origins of the Higher Groups of Tetrapods: Controversy and Consensus.* Ithaca, NY: Comstock Publishing Associates, 1991.

Scott, W. B., and E. J. Crossman. *Freshwater Fishes of Canada.* Ottawa: Fisheries Resource Board of Canada, 1973.

Skelton, P. *Freshwater Fishes of Southern Africa,* 2nd edition. Cape Town: Struik, 2001.

Smith, C. Lavett. *The Inland Fishes of New York State.* Albany: New York State Department of Environmental Conservation, 1985.

———. *National Audubon Society Field Guide to Tropical Marine Fishes of the Caribbean, Gulf of Mexico, Florida, the Bahamas, and Bermuda.* New York: Knopf, 1997.

Smith, M. M., and P. C. Heemstra, eds. *Smiths' Sea Fishes.* Berlin: Springer-Verlag, 1986.

Snyderman, Marty, and Clay Wiseman. *Guide to Marine Life: Caribbean, Bahamas, Florida.* New York: Aqua Quest Publications, 1996.

Springer, Victor G., and Joy P. Gold. *Sharks in Question: The Smithsonian Answer Book.* Washington, DC: Smithsonian Institution Press, 1989.

Stiassny, Melanie L. J., Lynne R. Parenti, and G. David Johnson. *Interrelationships of Fishes.* San Diego, CA: Academic Press, 1996.

Talwar, P. K., and A. G. Jhingran. *Inland Fishes of India and Adjacent Countries.* New Delhi: Oxford & I.B.H. Publishing Co., 1991.

Grzimek's Animal Life Encyclopedia

Thomson, Donald A., Lloyd T. Findley, and Alex N. Kerstitch. *Reef Fishes of the Sea of Cortez.* 2nd edition. Tucson: University of Arizona Press, 1987.

Thresher, Ronald E. *Reef Fishes: Behavior and Ecology on the Reef and in the Aquarium.* Saint Petersburg, FL: Palmetto Publishing Company, 1980.

———. *Reproduction in Reef Fishes.* Neptune City, NJ: T.F.H. Publications, Inc., 1984.

Tomelleri, J., and M. Eberle. *Fishes* Lawrence: University of Kansas Press, 1990.

Wheeler, A. *The Fishes of the British Isles and North-west Europe.* London: Macmillan, 1969.

———. *The World Encyclopedia of Fishes.* London: Macdonald, 1985.

Whitworth, W. R. *Freshwater Fishes of Connecticut.* State Geological and Natural History Survey of Connecticut Bulletin 114. Hartford: Connecticut Department of Environmental Protection, 1996.

Winfield, I., and J. Nelson, eds. *Cyprinid Fishes: Systematic Biology and Exploitation.* New York: Chapman & Hall, 1991.

Wischnath, Lothar. *Atlas of Livebearers of the World.* Neptune City, NJ: T.F.H. Publications, 1993.

Wootton, R. J. *Ecology of Teleost Fishes.* London: Chapman and Hall, 1990.

Organizations

American Elasmobranch Society
114 Hofstra University
Hempstead, NY 11549-1140
USA
http://www.flmnh.ufl.edu/fish/Organizations/aes/aes.htm

American Fisheries Society
5410 Grosvenor Lane
Bethesda, MD 20814
USA
Phone: 301–897–8616
Fax: 301–897–8096
main@fisheries.org
http://www.fisheries.org

American Livebearer Association
5 Zerbe Street,
Cressona, Pennsylvania 17929-1513
United States
Phone: 570-385-0573
Fax: 570-385-2781
tjbrady@uplink.net
http://livebearers.org/

American Killifish Association
280 Cold Springs Drive
Manchester, Pennsylvania 17345-1243
United States
garrybartell@sprintmailcom
http://www.aka.org

American Society of Ichthyology and Herpetology
http//www.asih.org/

American Society of Ichthyologists and Herpetologists
donnelly@fiu.edu
Phone: (305) 919-5651
http://199.245.200.110/

American Sportsfishing Association
225 Reinekers Lane, Suite 420
Alexandria, VA 22314
USA
Phone: 703–519–9691
Fax: 703–519–1872
info@asafishing.org
http://www.asafishing.org

American Zoo and Aquarium Association
8403 Colesville Road, Suite 710

Silver Spring, MD 20910
http://www.aza.org

Australian Society for Fish Biology Inc.
123 Brown Street (PO Box 137)
Heidelberg, Victoria 3084
Australia
Phone: 61 3 9450 8669
Fax: 61 3 9450 8730
john.koehn@nre.vic.gov.au
http://www.asfb.org.au

Commission for the Conservation of Antarctic Marine Living
Resources (CCAMLR)
137 Harrington Street
Hobart, Tasmania 7000
Australia
Phone: 61 3 6231 0366
Fax: 61 6234 9965
webmaster@ccamlr.org
http://www.ccamlr.org

Desert Fishes Council
315 East Medlock Drive,
Phoenix, Arizona 85012
United States
602-274-5544.
stefferud@cox.net
http://www.desertfishes.org/

Food and Agriculture Organization of the United Nations
(FAO) Fisheries
Viale delle Terme di Caracalla
Rome 00100
Italy
Phone: +39 06 5705 1
Fax: +39 06 5705 3152
FAO-HQ@fao.org
http://www.fao.org/fi/default.asp

Grouper and Wrasse Specialist Group, Species Survival
Commission, IUCN
Department of Ecology and Biodiversity,
The University of Hong Kong
Hong Kong
China
Phone: 852 2859 8977
Fax: 852 2517 6082
yjsadovy@hkusua.hku.hk
http://www.hku.hk/ecology/GroupersWrasses/iucnsg/
index.html

Inter-American Tropical Tuna Commisssion
8604 La Jolla Shores Drive
La Jolla, CA 92037-1508
United States
Phone: 619-546-7133
Fax: 619-546-7159
www.iattc.org

International Commission for the Conservation of
Atlantic Tunas
Calle Corazón de Maria, 8, 6th floor
Madrid E-28002
Spain

IUCN: The World Conservation Union
Rue Mauverney 28
1196
Gland
Switzerland
Phone: 41-22-999-0000
mail@hq.iucn.org
http://www.iucn.org

IUCN/SSC Coral Reef Fishes Specialist Group
c/o IMA-Integrative Biological Research Program,
University of Guam Marine Laboratory, UOG Station
Mangilao, Guam 96913
USA
Phone: 1-671-735-2187
Fax: 1-671-734-6767
donaldsn@uog9.uog.edu
http://www.iucn.org

Menhaden Resource Council
1901 N. Fort Myer Drive, Suite 700
Arlington, VA 22209
USA
Phone: 703-796-1793
resource@menhaden.org
http://www.menhaden.org

Muskies Canada Sport Fishing and Research, Inc.
P.O. Box 814, Station C
Kitchener

Ontario N2G 4C5
Canada
http://www.trentu.ca/muskie/mc.html.

North American Native Fishes Association
1107 Argonne Dr.
Baltimore, MD 21218
USA
nanfa@att.net
http://www.nanfa.org

Salmon and Trout Association (UK)
Fishmongers' Hall, London Bridge
London EC4R 9EL
UK
Phone: 0207 283 5838
Fax: 0207 626 5137
http://www.salmon-trout.org

South African Coelacanth Conservation and
Genome Resource
Somerset Street
Private Bag 1015
Grahamstown 6140
South Africa
Phone: +27 (0)46 636 1002
Fax: +27 (0)46 622 2403

TRAFFIC International
219c Huntingdon Road
Cambridge CB3 0DL
United Kingdom
Phone: 44 1223 277427
Fax: 44 1223 277237
traffic@trafficint.org
http://www.traffic.org

United States Trout Farmers Association
111 West Washington St., Suite One
Charles Town, WV 25414-1529
USA
Phone: (304) 728 2189
Fax: (304) 728 2196
http://www.ustfa.org

Contributors to the first edition

The following individuals contributed chapters to the original edition of Grzimek's Animal Life Encyclopedia, *which was edited by Dr. Bernhard Grzimek, Professor, Justus Liebig University of Giessen, Germany; Director, Frankfurt Zoological Garden, Germany; and Trustee, Tanzanian National Parks, Tanzania.*

Dr. Michael Abs
Curator, Ruhr University
Bochum, Germany

Dr. Salim Ali
Bombay Natural History Society
Bombay, India

Dr. Rudolph Altevogt
Professor, Zoological Institute,
University of Münster
Münster, Germany

Dr. Renate Angermann
Curator, Institute of Zoology,
Humboldt University
Berlin, Germany

Edward A. Armstrong
Cambridge University
Cambridge, England

Dr. Peter Ax
Professor, Second Zoological Institute
and Museum, University of Göttingen
Göttingen, Germany

Dr. Franz Bachmaier
Zoological Collection of the State
of Bavaria
Munich, Germany

Dr. Pedru Banarescu
Academy of the Roumanian Socialist
Republic, Trajan Savulescu Institute
of Biology
Bucharest, Romania

Dr. A. G. Bannikow
Professor,
Institute of Veterinary Medicine
Moscow, Russia

Dr. Hilde Baumgärtner
Zoological Collection of the State

of Bavaria
Munich, Germany

C. W. Benson
Department of Zoology,
Cambridge University
Cambridge, England

Dr. Andrew Berger
Chairman, Department of Zoology,
University of Hawaii
Honolulu, Hawaii, U.S.A.

Dr. J. Berlioz
National Museum of Natural History
Paris, France

Dr. Rudolf Berndt
Director,
Institute for Population Ecology,
Hiligoland Ornithological Station
Braunschweig, Germany

Dieter Blume
Instructor of Biology,
Freiherr-vom-Stein School
Gladenbach, Germany

Dr. Maximilian Boecker
Zoological Research Institute and
A. Koenig Museum
Bonn, Germany

Dr. Carl-Heinz Brandes
Curator and Director, The Aquarium,
Overseas Museum
Bremen, Germany

Dr. Donald G. Broadley
Curator, Umtali Museum
Mutare, Zimbabwe

Dr. Heinz Brüll
Director; Game, Forest, and Fields

Research Station
Hartenholm, Germany

Dr. Herbert Bruns
Director, Institute of Zoology and the
Protection of Life
Schlangenbad, Germany

Hans Bub
Heligoland Ornithological Station
Wilhelmshaven, Germany

A. H. Chrisholm
Sydney, Australia

Herbert Thomas Condon
Curator of Birds,
South Australian Museum
Adelaide, Australia

Dr. Eberhard Curio
Director,
Laboratory of Ethology,
Ruhr University
Bochum, Germany

Dr. Serge Daan
Laboratory of Animal Physiology,
University of Amsterdam
Amsterdam, The Netherlands

Dr. Heinrich Dathe
Professor and Director, Animal Park
and Zoological Research Station,
German Academy of Sciences
Berlin, Germany

Dr. Wolfgang Dierl
Zoological Collection of the State
of Bavaria
Munich, Germany

Dr. Fritz Dieterlen
Zoological Research Institute,

A. Koenig Museum
Bonn, Germany

Dr. Rolf Dircksen
Professor, Pedagogical Institute
Bielefeld, Germany

Josef Donner
Instructor of Biology
Katzelsdorf, Austria

Dr. Jean Dorst
Professor, National Museum of
Natural History
Paris, France

Dr. Gerti Dücker
Professor and Chief Curator,
Zoological Institute,
University of Münster
Münster, Germany

Dr. Michael Dzwillo
Zoological Institute and Museum,
University of Hamburg
Hamburg, Germany

Dr. Irenäus Eibl-Eibesfeldt
Professor and Director, Institute of
Human Ethology, Max Planck
Institute for Behavioral Physiology
Percha/Starnberg, Germany

Dr. Martin Eisentraut
Professor and Director,
Zoological Research Institute and
A. Koenig Museum
Bonn, Germany

Dr. Eberhard Ernst
Swiss Tropical Institute
Basel, Switzerland

R. D. Etchecopar
Director, National Museum of
Natural History
Paris, France

Dr. R. A. Falla
Director, Dominion Museum
Wellington, New Zealand

Dr. Hubert Fechter
Curator, Lower Animals, Zoological
Collection of the State of Bavaria
Munich, Germany

Dr. Walter Fiedler
Docent, University of Vienna, and
Director, Schönbrunn Zoo
Vienna, Austria

Wolfgang Fischer
Inspector of Animals, Animal Park
Berlin, Germany

Dr. C. A. Fleming
Geological Survey Department of
Scientific and Industrial Research
Lower Hutt, New Zealand

Dr. Hans Frädrich
Zoological Garden
Berlin, Germany

Dr. Hans-Albrecht Freye
Professor and Director, Biological
Institute of the Medical School
Halle a.d.S., Germany

Günther E. Freytag
Former Director, Reptile and
Amphibian Collection,
Museum of Cultural History
in Magdeburg
Berlin, Germany

Dr. Herbert Friedmann
Director, Los Angeles County
Museum of Natural History
Los Angeles, California, U.S.A.

Dr. H. Friedrich
Professor, Overseas Museum
Bremen, Germany

Dr. Jan Frijlink
Zoological Laboratory,
University of Amsterdam
Amsterdam, The Netherlands

Dr. H .C. Karl Von Frisch
Professor Emeritus and former
Director, Zoological Institute,
University of Munich
Munich, Germany

Dr. H. J. Frith
C.S.I.R.O. Research Institute
Canberra, Australia

Dr. Ion E. Fuhn
Academy of the Roumanian Socialist
Republic, Trajan Savulescu Institute
of Biology
Bucharest, Romania

Dr. Carl Gans
Professor, Department of Biology,
State University of New York
at Buffalo
Buffalo, New York, U.S.A.

Dr. Rudolf Geigy
Professor and Director,
Swiss Tropical Institute
Basel, Switzerland

Dr. Jacques Gery
St. Genies, France

Dr. Wolfgang Gewalt
Director, Animal Park
Duisburg, Germany

Dr. H. C. Viktor Goerttler
Professor Emeritus, University of Jena
Jena, Germany

Dr. Friedrich Goethe
Director, Institute of Ornithology,
Heligoland Ornithological Station
Wilhelmshaven, Germany

Dr. Ulrich F. Gruber
Herpetological Section,
Zoological Research Institute and
A. Koenig Museum
Bonn, Germany

Dr. H. R. Haefelfinger
Museum of Natural History
Basel, Switzerland

Dr. Theodor Haltenorth
Director, Mammalology, Zoological
Collection of the State of Bavaria
Munich, Germany

Barbara Harrisson
Sarawak Museum, Kuching, Borneo
Ithaca, New York, U.S.A.

Dr. Francois Haverschmidt
President, High Court (retired)
Paramaribo, Suriname

Dr. Heinz Heck
Director, Catskill Game Farm
Catskill, New York, U.S.A.

Dr. Lutz Heck
Professor (retired), and Director,
Zoological Garden, Berlin
Wiesbaden, Germany

Dr. Dr. H.C.Heini Hediger
Director, Zoological Garder
Zurich, Switzerland

Dr. Dietrich Heinemann
Director, Zoological Garden, Münster
Dörnigheim, Germany

Dr. Helmut Hemmer
Institute for Physiological Zoology,
University of Mainz
Mainz, Germany

Dr. W. G. Heptner
Professor, Zoological Museum,
University of Moscow
Moscow, Russia

Dr. Konrad Herter
Professor Emeritus and Director
(retired), Zoological Institute, Free
University of Berlin
Berlin, Germany

Dr. Hans Rudolf Heusser
Zoological Museum,
University of Zurich
Zurich, Switzerland

Dr. Emil Otto Höhn
Associate Professor of Physiology,
University of Alberta
Edmonton, Canada

Dr. W. Hohorst
Professor and Director,
Parasitological Institute,
Farbwerke Hoechst A.G.
Frankfurt-Höchst, Germany

Dr. Folkhart Hückinghaus
Director,
Senckenbergische Anatomy,
University of Frankfurt a.M.
Frankfurt a.M., Germany

Francois Hüe
National Museum of Natural History
Paris, France

Dr. K. Immelmann
Professor, Zoological Institute,
Technical University of Braunschweig
Braunschweig, Germany

Dr. Junichiro Itani
Kyoto University
Kyoto, Japan

Dr. Richard F. Johnston
Professor of Zoology,
University of Kansas
Lawrence, Kansas, U.S.A.

Otto Jost
Oberstudienrat,
Freiherr-vom-Stein Gymnasium
Fulda, Germany

Dr. Paul Kähsbauer
Curator, Fishes, Museum of
Natural History
Vienna, Austria

Dr. Ludwig Karbe
Zoological State Institute
and Museum
Hamburg, Germany

Dr. N. N. Kartaschew
Docent, Department of Biology,
Lomonossow State University
Moscow, Russia

Dr. Werner Kästle
Oberstudienrat, Gisela Gymnasium
Munich, Germany

Dr. Reinhard Kaufmann
Field Station of the Tropical Institute,
Justus Liebig University,
Giessen, Germany
Santa Marta, Colombia

Dr. Masao Kawai
Primate Research Institute,
Kyoto University
Kyoto, Japan

Dr. Ernst F. Kilian
Professor,
Giessen University and Catedratico
Universidad Austral,
Valdivia-Chile
Giessen, Germany

Dr. Ragnar Kinzelbach
Institute for General Zoology,
University of Mainz
Mainz, Germany

Dr. Heinrich Kirchner
Landwirtschaftsrat (retired)
Bad Oldesloe, Germany

Dr. Rosl Kirchshofer
Zoological Garden, University of
Frankfort a.M.
Frankfurt a.M., Germany

Dr. Wolfgang Klausewitz
Curator, Senckenberg Nature
Museum and Research Institute
Frankfurt a.M., Germany

Dr. Konrad Klemmer
Curator, Senckenberg Nature
Museum and Research Institute
Frankfurt a.M., Germany

Dr. Erich Klinghammer
Laboratory of Ethology,
Purdue University
Lafayette, Indiana, U.S.A.

Dr. Heinz-Georg Klös
Professor and Director,
Zoological Garden
Berlin, Germany

Ursula Klös
Zoological Garden
Berlin, Germany

Dr. Otto Koehler
Professor Emeritus,
Zoological Institute,
University of Freiburg
Freiburg i. BR., Germany

Dr. Kurt Kolar
Institute of Ethology, Austrian
Academy of Sciences
Vienna, Austria

Dr. Claus König
State Ornithological Station
of Baden-Württemberg
Ludwigsburg, Germany

Dr. Adriaan Kortlandt
Zoological Laboratory,
University of Amsterdam
Amsterdam, The Netherlands

Dr. Helmut Kraft
Professor and Scientific Councillor,
Medical Animal Clinic, University
of Munich
Munich, Germany

Dr. Helmut Kramer
Zoological Research Institute and
A. Koenig Museum
Bonn, Germany

Dr. Franz Krapp
Zoological Institute,
University of Freiburg
Freiburg, Switzerland

Dr. Otto Kraus
Professor,
University of Hamburg, and Director,
Zoological Institute
and Museum
Hamburg, Germany

Dr. Hans Krieg
Professor and First Director (retired),

Scientific Collections of the State
of Bavaria
Munich, Germany

Dr. Heinrich Kühl
Federal Research Institute for
Fisheries, Cuxhaven Laboratory
Cuxhaven, Germany

Dr. Oskar Kuhn
Professor, formerly University
Halle/Saale
Munich, Germany

Dr. Hans Kumerloeve
First Director (retired), State
Scientific Museum, Vienna
Munich, Germany

Dr. Nagamichi Kuroda
Yamashina Ornithological Institute,
Shibuya-Ku
Tokyo, Japan

Dr. Fred Kurt
Zoological Museum of
Zurich University,
Smithsonian Elephant Survey
Colombo, Ceylon

Dr. Werner Ladiges
Professor and Chief Curator,
Zoological Institute and Museum,
University of Hamburg
Hamburg, Germany

Leslie Laidlaw
Department of Animal Sciences,
Purdue University
Lafayette, Indiana, U.S.A.

Dr. Ernst M. Lang
Director, Zoological Garden
Basel, Switzerland

Dr. Alfredo Langguth
Department of Zoology,
Faculty of Humanities and Sciences,
University of the Republic
Montevideo, Uruguay

Leo Lehtonen
Science Writer
Helsinki, Finland

Bernd Leisler
Second Zoological Institute,
University of Vienna
Vienna, Austria

Dr. Kurt Lillelund
Professor and Director, Institute for
Hydrobiology and Fishery Sciences,
University of Hamburg
Hamburg, Germany

R. Liversidge
Alexander MacGregor
Memorial Museum
Kimberley, South Africa

Dr. Konrad Lorenz
Professor and Director, Max Planck
Institute for Behavioral Physiology
Seewiesen/Obb., Germany

Dr. Martin Lühmann
Federal Research Institute for the
Breeding of Small Animals
Celle, Germany

Dr. Johannes Lüttschwager
Oberstudienrat (retired)
Heidelberg, Germany

Dr. Wolfgang Makatsch
Bautzen, Germany

Dr. Hubert Markl
Professor and Director, Zoological
Institute, Technical University
of Darmstadt
Darmstadt, Germany

Basil J. Marlow, B.SC. (Hons)
Curator, Australian Museum
Sydney, Australia

Dr. Theodor Mebs
Instructor of Biology
Weissenhaus/Ostsee, Germany

Dr. Gerlof Fokko Mees
Curator of Birds, Rijks Museum of
Natural History
Leiden, The Netherlands

Hermann Meinken
Director, Fish Identification Institute,
V.D.A.
Bremen, Germany

Dr. Wilhelm Meise
Chief Curator, Zoological Institute
and Museum, University of Hamburg
Hamburg, Germany

Dr. Joachim Messtorff
Field Station of the Federal Fisheries
Research Institute
Bremerhaven, Germany

Dr. Marian Mlynarski
Professor, Polish Academy of
Sciences, Institute for Systematic and
Experimental Zoology
Cracow, Poland

Dr. Walburga Moeller
Nature Museum
Hamburg, Germany

Dr. H.C.Erna Mohr
Curator (retired), Zoological State
Institute and Museum
Hamburg, Germany

Dr. Karl-Heinz Moll
Waren/Müritz, Germany

Dr. Detlev Müller-Using
Professor, Institute for Game
Management, University of Göttingen
Hannoversch-Münden, Germany

Werner Münster
Instructor of Biology
Ebersbach, Germany

Dr. Joachim Münzing
Altona Museum
Hamburg, Germany

Dr. Wilbert Neugebauer
Wilhelma Zoo
Stuttgart-Bad Cannstatt, Germany

Dr. Ian Newton
Senior Scientific Officer,
The Nature Conservancy
Edinburgh, Scotland

Dr. Jürgen Nicolai
Max Planck Institute for
Behavioral Physiology
Seewiesen/Obb., Germany

Dr. Günther Niethammer
Professor, Zoological Research
Institute and A. Koenig Museum
Bonn, Germany

Dr. Bernhard Nievergelt
Zoological Museum,
University of Zurich
Zurich, Switzerland

Dr. C. C. Olrog
Institut Miguel Lillo San Miguel
de Tucuman
Tucuman, Argentina

Alwin Pedersen
Mammal Research and
Arctic Explorer
Holte, Denmark

Dr. Dieter Stefan Peters
Nature Museum and Senckenberg
Research Institute
Frankfurt a.M., Germany

Dr. Nicolaus Peters
Scientific Councillor and Docent,
Institute of Hydrobiology and
Fisheries, University of Hamburg
Hamburg, Germany

Dr. Hans-Günter Petzold
Assistant Director,
Zoological Garden
Berlin, Germany

Dr. Rudolf Piechocki
Docent, Zoological Institute,
University of Halle
Halle a.d.S., Germany

Dr. Ivo Poglayen-Neuwall
Director, Zoological Garden
Louisville, Kentucky, U.S.A.

Dr. Egon Popp
Zoological Collection of the State
of Bavaria
Munich, Germany

Dr. H. C. Adolf Portmann
Professor Emeritus, Zoological
Institute, University of Basel
Basel, Switzerland

Hans Psenner
Professor and Director, Alpine Zoo
Innsbruck, Austria

Dr. Heinz-Siburd Raethel
Oberveterinärrat
Berlin, Germany

Dr. Urs H. Rahm
Professor, Museum of Natural History
Basel, Switzerland

Dr. Werner Rathmayer
Biology Institute,
University of Konstanz
Konstanz, Germany

Walter Reinhard
Biologist
Baden-Baden, Germany

Dr. H. H. Reinsch
Federal Fisheries Research Institute
Bremerhaven, Germany

Dr. Bernhard Rensch
Professor Emeritus, Zoological
Institute, University of Münster
Münster, Germany

Dr. Vernon Reynolds
Docent, Department of Sociology,
University of Bristol
Bristol, England

Dr. Rupert Riedl
Professor, Department of Zoology,
University of North Carolina
Chapel Hill, North Carolina, U.S.A.

Dr. Peter Rietschel
Professor (retired),
Zoological Institute,
University of Frankfurt a.M.
Frankfurt a.M., Germany

Dr. Siegfried Rietschel
Docent, University of Frankfurt;
Curator, Nature Museum and
Research Institute Senckenberg
Frankfurt a.M., Germany

Herbert Ringleben
Institute of Ornithology,
Heligoland Ornithological Station
Wilhelmshaven, Germany

Dr. K. Rohde
Institute for General Zoology,
Ruhr University
Bochum, Germany

Dr. Peter Röben
Academic Councillor, Zoological
Institute, Heidelberg University
Heidelberg, Germany

Dr. Anton E. M. De Roo
Royal Museum of Central Africa
Tervuren, South Africa

Dr. Hubert Saint Girons
Research Director,
Center for National
Scientific Research
Brunoy (Essonne), France

Dr. Luitfried Von Salvini-Plawen
First Zoological Institute,
University of Vienna
Vienna, Austria

Dr. Kurt Sanft
Oberstudienrat,
Diesterweg-Gymnasium
Berlin, Germany

Dr. E. G. Franz Sauer
Professor, Zoological Research
Institute and A. Koenig Museum,
University of Bonn
Bonn, Germany

Dr. Eleonore M. Sauer
Zoological Research Institute and
A. Koenig Museum,
University of Bonn
Bonn, Germany

Dr. Ernst Schäfer
Curator,
State Museum of Lower Saxony
Hannover, Germany

Dr. Friedrich Schaller
Professor and Chairman,
First Zoological Institute,
University of Vienna
Vienna, Austria

Dr. George B. Schaller
Serengeti Research Institute,
Michael Grzimek Laboratory
Seronera, Tanzania

Dr. Georg Scheer
Chief Curator and Director,
Zoological Institute,
State Museum of Hesse
Darmstadt, Germany

Dr. Christoph Scherpner
Zoological Garden
Frankfurt a.M., Germany

Dr. Herbert Schifter
Bird Collection,
Museum of Natural History
Vienna, Austria

Dr. Marco Schnitter
Zoological Museum,
Zurich University
Zurich, Switzerland

Dr. Kurt Schubert
Federal Fisheries Research Institute
Hamburg, Germany

Eugen Schuhmacher
Director, Animals Films, I.U.C.N.
Munich, Germany

Dr. Thomas Schultze-Westrum
Zoological Institute,
University of Munich
Munich, Germany

Dr. Ernst Schüt
Professor and Director (retired),
State Museum of Natural History
Stuttgart, Germany

Dr. Lester L. Short, Jr.
Associate Curator, American Museum
of Natural History
New York, New York, U.S.A.

Dr. Helmut Sick
National Museum
Rio de Janeiro, Brazil

Dr. Alexander F. Skutch
Professor of Ornithology,
University of Costa Rica
San Isidro del General, Costa Rica

Dr. Everhard J. Slijper
Professor, Zoological Laboratory,
University of Amsterdam
Amsterdam, The Netherlands

Bertram E. Smythies
Curator (retired),
Division of Forestry Management,
Sarawak-Malaysia
Estepona, Spain

Dr. Kenneth E. Stager
Chief Curator, Los Angeles County
Museum of Natural History
Los Angeles, California, U.S.A.

Dr. H. C. Georg H. W. Stein
Professor, Curator of Mammals,
Institute of Zoology and
Zoological Museum,
Humboldt University
Berlin, Germany

Dr. Joachim Steinbacher
Curator, Nature Museum and
Senckenberg Research Institute
Frankfurt a.M., Germany

Dr. Bernard Stonehouse
Canterbury University
Christchurch, New Zealand

Dr. Richard Zur Strassen
Curator, Nature Museum and
Senckenberg Research Institute
Frandfurt a.M., Germany

Dr. Adelheid Studer-Thiersch
Zoological Garden
Basel, Switzerland

Dr. Ernst Sutter
Museum of Natural History
Basel, Switzerland

Dr. Fritz Terofal
Director, Fish Collection, Zoological
Collection of the State of Bavaria
Munich, Germany

Dr. G. F. Van Tets
Wildlife Research
Canberra, Australia

Ellen Thaler-Kottek
Institute of Zoology,
University of Innsbruck
Innsbruck, Austria

Dr. Erich Thenius
Professor and Director,
Institute of Paleontolgy,
University of Vienna
Vienna, Austria

Dr. Niko Tinbergen
Professor of Animal Behavior,
Department of Zoology,
Oxford University
Oxford, England

Alexander Tsurikov
Lecturer, University of Munich
Munich, Germany

Dr. Wolfgang Villwock
Zoological Institute and Museum,
University of Hamburg
Hamburg, Germany

Zdenek Vogel
Director,
Suchdol Herpetological Station
Prague, Czechoslovakia

Dieter Vogt
Schorndorf, Germany

Dr. Jiri Volf
Zoological Garden
Prague, Czechoslovakia

Otto Wadewitz
Leipzig, Germany

Dr. Helmut O. Wagner
Director (retired),

Overseas Museum, Bremen
Mexico City, Mexico

Dr. Fritz Walther
Professor, Texas A & M University
College Station, Texas, U.S.A.

John Warham
Zoology Department,
Canterbury University
Christchurch, New Zealand

Dr. Sherwood L. Washburn
University of California at Berkeley
Berkeley, California, U.S.A.

Eberhard Wawra
First Zoological Institute,
University of Vienna
Vienna, Austria

Dr. Ingrid Weigel
Zoological Collection of the State
of Bavaria
Munich, Germany

Dr. B. Weischer
Institute of Nematode Research,
Federal Biological Institute
Münster/Westfalen, Germany

Herbert Wendt
Author, Natural History
Baden-Baden, Germany

Dr. Heinz Wermuth
Chief Curator,
State Nature Museum, Stuttgart
Ludwigsburg, Germany

Dr. Wolfgang Von Westernhagen
Preetz/Holstein, Germany

Dr. Alexander Wetmore
United States National Museum,
Smithsonian Institution
Washington, D.C., U.S.A.

Dr. Dietrich E. Wilcke
Röttgen, Germany

Dr. Helmut Wilkens
Professor and Director,
Institute of Anatomy,
School of Veterinary Medicine
Hannover, Germany

Dr. Michael L. Wolfe
Utah, U.S.A.

CONTRIBUTORS TO THE FIRST EDITION

Hans Edmund Wolters
Zoological Research Institute and
A. Koenig Museum
Bonn, Germany

Dr. Arnfrid Wünschmann
Research Associate, Zoological Garden
Berlin, Germany

Dr. Walter Wüst
Instructor, Wilhelms Gymnasium
Munich, Germany

Dr. Heinz Wundt
Zoological Collection of the State
of Bavaria
Munich, Germany

Dr. Claus-Dieter Zander
Zoological Institute and Museum,
University of Hamburg
Hamburg, Germany

Dr. Dr.Fritz Zumpt
Director,
Entomology and Parasitology,
South African Institute for
Medical Research
Johannesburg, South Africa

Dr. Richard L. Zusi
Curator of Birds,
United States National Museum,
Smithsonian Institution
Washington, D.C., U.S.A.

Glossary

Adipose fin—A small, fleshy fin without rays.

Afferent—Conducting impulses toward nerve centers or blood toward the gills. Compare efferent.

Agonistic behavior—Aggressive and submissive interaction between individuals of the same species.

Albinistic—Displaying the characteristics of an albino; an organism that has deficient pigmentation and white, colorless, or translucent skin and hair.

Amphidromous—Regular migration between fresh and seawater at different stages in their development.

Ampulla—A sac- or pouch-like anatomical swelling.

Anal fin—Fin located on the undersurface of the body, behind the anus.

Andropodium—Modified anal fin exhibited by some males.

Anoxic—Extreme deficiency of oxygen.

Anthropogenic—Caused by the activities of human beings.

Antitropical—Found in both Northern and Southern Hemispheres, but not in equatorial regions.

Aplacental—Without a placenta.

Axial skeleton—Skeleton of the main body and head.

Axillary process—Modified scale present at the upper or anterior base of the pectoral or ventral fins exhibited by some fishes.

Barbel—Fleshy, tactile projection resembling tentacles located near the mouth, chin, or snout.

Basiocapital—Bone located at the back of the head or skull; the occiput.

Basioccipital—Base of the head or skull.

Bathypelagic—Living and/or feeding in open waters at depths between 3,280 and 13,125 ft (1,000 and 4,000 m).

Benthic—Relating to, living on, or occurring at the bottom of a body of water.

Benthopelagic—Relating to, living on, or occurring on the bottom or midwaters of a body of water, feeding on benthic and free swimming organisms.

Benthos—The bottom of a body of water.

Bilobed—The division of matter into two lobes.

Branchial—Relating to the gills.

Branchiostegal membrane—The gill membrane; supported by the branchiostegal rays (bones).

Buccal cavity—Mouth cavity forward of the gills.

Bycatch—Species that are not targeted as catch, but are caught along with a target species during fishing.

Carapace—A hardened shell, such as turtles or crabs have.

Catadromous—Living in freshwater, but migrating to saltwater for spawning.

Caudal fin—Fin located at the end of the body, also known as the tail fin.

Caudal keels—Ridges on either side of the caudal peduncle that often function in stabilization during fast swimming.

Caudal peduncle—A narrow part of the body located at the base of the caudal fin.

Cavernicolous—Cave dwelling.

Cecum—Cavity or pouch extending off the intestine that receives undigested food.

Cephalic—Relating or belonging to the head.

Clade—A group of biological taxa, such as species, that includes all descendants of one common ancestor.

Cladist—One who classifies organisms based on their evolutionary history.

Cleithrum—The major bone of the pectoral girdle.

Cloaca—Chamber into which the intestinal, urinary, and reproductive ducts discharge.

Confluent fins—Fins that are joined or run together, having no true separation.

Crepuscular—Active in the twilight or evening.

Crypsis—Patterns and/or coloration that make an organism more difficult for predators to detect; protective patterns or coloration.

Ctenoid scales—Scales having minute spines on exposed surface.

Cupula—A cup-shaped structure.

Cycloid scales—Scales having smooth edges, absent of spines.

Demersal—Living near, laying on, or sinking towards the bottom of the ocean.

Dentary—Lower jawbone of vertebrates.

Dentine—Material similar to but harder than bone and is the principal mass of teeth.

Dermal denticles—Teeth-like scales, also known as placoid scales, on the skin of various elasmobranchs that acts as a protective barrier and also enables faster swimming.

Dextral—Occurring on or relating to the right side of the body.

Diadromous—Regular migration between freshwater and seawater.

Dichromatism—Partial color blindness; the ability to recognize only two colors.

Diel—Involving a 24 hour period of time; occurring on a daily basis.

Diploid—Two sets of chromosomes existing in a cell or organism.

Dorsal fin—Spined or rayed fin on the dorsal surface of body.

Dorsal—Relating or belonging to the back or top surface of the body.

Dorsolateral—Belonging to, or orientated between the dorsal and lateral surfaces.

Dorsoventral—Belonging to, or orientated between the dorsal and ventral surface.

Ectodermal—Formed from the outer germ layer of an embryo.

Ectothermic—Cold-blooded animal.

Efferent—Conducting impulses away from nerve centers or blood away from the gills. Compare afferent.

Elasmobrand—Relating to the group of fishes that includes the sharks, rays, and skates.

Electric organ—Organ capable of delivering an electric shock or used to emit electrical discharges to stun prey, repel predators, or detect objects.

Endemic—The restriction of a species to a particular geographic area or continent; native.

Engybenthic—Organisms living or occurring at the bottom of a body of water.

Epibenthic—Living on the bottom of the ocean.

Epigean—Organisms that are not cave-dwellers and do not live underground.

Epipelagic—Living or feeding in the uppermost layer of water; from the surface to midwater depths of 656.17 ft (200 m).

Euryhaline—Ability to live in waters of varying salinity.

Exogenous—Introduced from, or produced outside the organism or system.

Facultative parasite—An organism that can exist off of its host.

Falcate—Having a hooked or curved shape.

Filiform—Having the shape or form of a filament.

Fin base—Portion of a fin that attaches to the body.

Fin spine—Bony structure that supports the fin in more derived fishes.

Finlet—A small, isolated fin, usually without rays, that ususaly occurs dorsally or ventrally on the caudal peduncle.

Flange—A rib or rim that aids one object in attaching to another.

Flexion—The act of bending, extending or flexing; a physical structure having a bent shape.

Fry—Newly hatched juveniles, or very small adult fishes.

Fusiform—The tapering of each end.

Ganglia—Mass of nerve tissue containing nerve cells external to the brain or spinal cord.

Ganoid—Relating to, or having scales that are made of bone and an outer layer that resembles enamel.

Ganione—Substance that resembles enamel and makes up the outer layer of certain fishes' scales.

Gas bladder—Sac in the body cavity below the vertebral column; helps maintain buoyancy, may aid in respiration, and may help produce or receive sound. Also called swim bladder.

Gill—Organ for obtaining oxygen from water.

Gill cover—Flap made of bone or cartilage that covers and protects the gills. Also called operculum.

Gill rakers—Projections from the gill arch that help in retaining food particles.

Globose—Having the shape or form of a globule or ball.

Gregarious—Living in a group or colony.

Haploid—One set of chromosomes existing in a cell or organism.

Hermaphroditism—The presence of both male and female sexual organs in one individual. When both organs occur at the same time, the individual is bisexual or a simultaneous hermaphrodita; if they occur at different times, the individual is a sequential hermaphrodite.

Heterocercal—Upper lobe of the tail is larger than the lower lobe, and the vertebral column extends into the upper lobe.

Heterozygous—Having two alleles at corresponding loci on homologous chromosomes that are different.

Holarctic—Relating to, or being from the northern parts of the world.

Homocercal—A caudal fin in which all of the principal rays attach to the last vertebra.

Homologous—Structures or properties of organisms shared through common ancestry.

Hyoid—Belonging or pertaining to the tongue.

Hypogean—Lives underground.

Hypoxia—Deficiency of oxygen reaching the tissues of the body.

Hypural plate—Modified last vertebra, to which caudal fin rays attach.

Incisiform—Teeth that are flat with sharp edges.

Integument—A layer or membrane that encloses or envelopes an organism or one of its parts.

Intertidal—Shallow areas along the shore that are alternately exposed and covered by the tides.

Isosmotic—Having the same osmotic pressure on two sides of a membrane.

Isthmus—Narrow, triangular area on the underside of the body, between the gill openings.

Iteroparous—Successive production of offspring, annually or seasonal batches.

Lacustrine—Relating to, inhabiting, formed or growing in lake water.

Lamella—Thin plate or membrane; often refers to smallest divisions of gill.

Larvaceans—Small transparent animals found in marine plankton; belong to subphylum Urochordata.

Lateral line—A series of ampulla forming a sensory organ to detect movements in water. Scales are often modified with pores opening to a sensory canal on the side of a fish.

Lecithotrophic—Embryos feeding on the yolk stored in the yolk sac.

Littoral—Related to, inhabiting, or situated near a shore.

Lunate—Having the shape of a crescent.

Maxilla—Upper jaw bone.

Median fin—Fins located on the median plane.

Meiobenthic—Benthic organisms with dimensions less than 0.02 in (0.5 mm) but greater than or equal to 0.004 in (0.1 mm).

Melanistic—An organism that exhibits a high amount of melanin (black coloration) in the skin.

Melanophore—A cell containing melanin.

Mesentary—A membrane that attaches organ to the abdominal wall.

Mesopelagic—Relating to, inhabiting or feeding at midwater at depths between 656.17 ft (200 m) and 3,280.84 ft (1,000 m).

Microphthalmic—Having eyes noticeably reduced in size.

Micropredator—A predator smaller than its prey that comes into contact with its host only when needing to feed.

Midwater—The middle stratum of a body of water.

Milt—The combination of spermatozoa and seminal fluid in fishes.

Monogamy—Mating system in which a single pair joins together for spawning and may remain together for one or more seasons.

Monophyletic—Developed from or related to a single common ancestral form or stock.

Monotypic—A group containing a single representative.

Myoglobin—Protein pigment in muscles that contains iron.

Naked—A fish that has no scales.

Nares—Nostrils.

Nasohypophysical—Nostril opening.

Nektonic—Organisms that swim strongly enough to move against currents.

Neural spine—The uppermost spine of a vertebra.

Notched fin—A fin that has patterned indentation.

Obligate air breathers—An organism that must receive a certain amount of their oxygen directly from air.

Ocelli—An eye-like marking.

Ontogenetic—Changes that incur from growth or age.

Oophagy—The process of embryos feeding on eggs produced by the ovary while still inside the uterus.

Osmoregulation—The regulation of water in the body.

Otoliths—Calcareous deposit in the ear capsules of bony fishes that show daily, seasonal or annual checks, rings or layers that can be used to determine ages.

Oviduct—Duct that serves as the passage of eggs from the ovary.

Oviparous—Production of eggs that develop and hatch outside of the mother.

Ovoviviparous—Fertilized eggs are retained in the mother's body during development.

Paedomorphic—Phylogenetic retention of larval or juvenile characters in the adult stage.

Paired Fins—Fins that occur in pairs, on each side of the body.

Paleoecology—The study of ecological characteristics in ancient environments and their relationships to ancient plants and animals.

Palp (Palpus)—Segmented and tactile process on the mouth.

PCBs (polychlorinated biphenyls)—Substances used as coolants and lubricants; their manufacture was banned in the United States in 1977.

Pectoral—Relating or belonging to the forward pair of appendicular appendages.

Pectoral fins—Fins attached to the shoulder or pectoral girdle, just behind the head.

Peduncule—Narrow part by which a larger part or the whole body is attached.

Pelagic—Relating to, living or occurring in open ocean water.

Pelvic fins—Pair of fins attached to the pelvis or pelvic girdle.

Photophore—A luminous spot or light-producing organ.

Piscivorous—Diet consists solely of other fishes.

Planktivorous—Diet consists solely of passively floating or weakly swimming animal and plant life.

Polyandry—Females mate with more than one male in a season.

Polygamy—Mating system in which individuals mate with more than one partner in a season.

Polygyny—Males mate with more than one female in a season.

Polyphyletic group—An assemblage consisting of different ancestral taxa, i.e., a group based upon convergence rather than on common ancestry.

Precaudal pit—Cavity just anterior to the caudal fin.

Prehensile—Adapted specifically to enable seizing, grasping, or wrapping around.

Promiscuity—Males and females spawn together with little or no mate choice.

Protandrous—Sequential hermaphroditism in which the fish functions first as a male and then a female.

Protrusible mouth—Mouth which can project forward and out to help catch prey.

Protygynous—Sequential hermaphroditism in which the fish functions first as a female and then a male.

Ray—Segmented bony rod or element that supports a fin membrane.

Riffle stretches—Areas of rough water caused by submerged rocks or a sandbar.

Rostral—Located toward the mouth or nasal region.

Seamount—Submarine mountain rising above the deep-sea floor.

Sexual dichromism—Exhibiting both male and female forms and aspects.

Sinistral—Occurring on or relating to the left side of the body.

Sinusoidal—Relating to, or shaped like the sine curve or wave.

Sister group—Closest relative to a taxa or group. The two groups share a common ancestor.

Spinules—Minute or miniature spine.

Standard length—Standard scientific measure of a fish's length; found by measuring from the most anterior part of the snout, lip or chin to the end of the last vertebra.

Subtidal—Zone just below the low-water mark of the tide that is never exposed, even at low tide.

Swim bladder—See Gas Bladder.

Syntopic—Sharing the same habitat within the same geographical range.

Tetraploid—Four sets (two homologous pairs) of chromosomes existing in a cell or organism.

Translocated—Transferred or dislocated specimens.

Tubercle—Nodule, growth, or knob present in an organ or on the skin.

Ventral—Relating to, or located in the abdomen or belly.

Vermicular—Relating to, caused by, or resembling worms.

Vestigial—Body part that was functional in ancestral sources but has become reduced or nonfunctional descendants.

Vitellogenesis—Deposition of yolk within the growing egg.

Viviparous—Producing live young.

GLOSSARY

Fishes family list

Myxini [Class]
  Myxiniformes [Order]
    Myxinidae [Family]

Cephalaspidomorphi [Class]
  Petromyzoniformes [Order]
    Petromyzonidae [Family]
    Geotriidae
    Mordaciidae

Chondrichthyes [Class]
  Chimaerformes [Order]
    Callorhinchidae [Family]
    Chimaeridae
    Rhinochimaeridae

  Heterodontiformes [Order]
    Heterodontidae [Family]

  Orectolobiformes [Order]
    Parascyllidae [Family]
    Brachaeluridae
    Orectolobidae
    Hemiscyllidae
    Ginglymostomatidae
    Stegostomatidae
    Rhincodontidae

  Carcharhiniformes [Order]
    Scyliorhinidae [Family]
    Proscylliidae
    Pseudotriakidae
    Leptochariidae
    Triakidae
    Hemigaleidae
    Carcharhinidae
    Sphyrnidae

  Lamniformes [Order]
    Ondontaspididae [Family]
    Mitsukurinidae
    Pseudocarchariidae
    Megachasmidae
    Alopiidae
    Cetorhinidae
    Lamnidae

  Hexanchiformes [Order]
    Chlamydoselachidae [Family]
    Hexanchidae

  Squaliformes [Order]
    Echinorhinidae [Family]
    Dalatiidae
    Centrophoridae
    Squalidae
    Etmopteridae
    Somniosidae
    Oxynotidae

  Squatiniformes [Order]
    Squatinidae [Family]

  Pristiophoriformes [Order]
    Pristiophoridae [Family]

  Rajiformes [Order]
    Pristidae [Family]
    Torpedinidae
    Narcinidae
    Rhinidae
    Rhinobatidae
    Rajidae
    Plesiobatidae
    Hexatrygonidae
    Dasyatidae
    Urolophidae
    Gymnuridae
    Myliobatidae
    Potamotrygonidae
    Mobulidae
    Urotrygonidae
    Platyrhinidae
    Zanobatidae
    Narkidae
    Hypridae
    Rhinopteridae

Sarcopterygii [Class]
  Coelacanthiformes [Order]
    Latimeriidae [Family]

  Ceratodontiformes [Order]
    Ceratodontidae [Family]

  Lepidosireniformes [Order]
    Lepidosirenidae [Family]
    Protopteridae

Actinopterygii [Class]
  Polypteriformes [Order]
    Polypteridae [Family]

  Acipenseriformes [Order]
    Acipenseridae [Family]
    Polyodontidae

  Lepisosteiformes [Order]
    Lepisosteidae [Family]

  Amiiformes [Order]
    Amiidae [Family]

  Osteoglossiformes [Order]
    Osteoglossidae [Family]
    Pantodontidae
    Hiodontidae
    Notopteridae
    Mormyridae
    Gymnarchidae

  Elopiformes [Order]
    Elopidae [Family]
    Megalopidae

  Albuliformes [Order]
    Albulidae [Family]
    Halosauridae
    Notacanthidae

  Anguilliformes [Order]
    Anguillidae [Family]
    Heterenchelyidae
    Moringuidae
    Chlopsidae
    Myrocongridae
    Muraenidae
    Synaphobranchidae
    Ophichthidae
    Colocongridae
    Derichthydae
    Muraenesocidae

Nemichthydae
Congridae
Nettastomatidae
Serrivomeridae

Saccopharyngiformes [Order]
Cyematidae [Family]
Saccopharyngidae
Eurypharyngidae
Monognathidae

Clupeiformes [Order]
Denticipitidae [Family]
Engraulidae
Pristigasteridae
Chirocentridae
Clupeidae

Gonorynchiformes [Order]
Chanidae [Family]
Gonorynchidae
Kneriidae

Cypriniformes [Order]
Cyprinidae [Family]
Gyrinocheilidae
Catostomidae
Cobitidae
Balitoridae

Characiformes [Order]
Citharinidae [Family]
Hemiodontidae
Curimatidae
Anostomidae
Erythrinidae
Lebiasinidae
Ctenoluciidae
Hepsetidae
Gasteropelecidae
Characidae
Alestiidae

Siluriformes [Order]
Diplomystidae [Family]
Icaluridae
Bagridae
Claroteidae
Australoglanidae
Siluridae
Schilbeidae
Pangasiidae
Amphiliidae
Sisoridae
Amblycipitidae
Akysidae
Chacidae
Clariidae
Malapteruridae
Ariidae
Plotosidae
Mochokidae
Doradidae

Auchenipteridae
Pimelodidae
Cetopsidae
Aspredinidae
Trichomycteridae
Callichthyidae
Scoloplacidae
Loricariidae
Astroblepidae
Nematogenyidae
Heptapteridae
Pseudopimelodidae
Auchenoglanidae
Erthistidae
Claroglanididae

Gymnotiformes [Order]
Sternopygidae [Family]
Rhamphichthyidae
Hypopomidae
Apteronotidae
Gymnotidae

Esociformes [Order]
Esocidae [Family]

Osmeriformes [Order]
Argentinidae [Family]
Microstomatidae
Opisthoproctidae
Alepocephalidae
Platytroctidae
Osmeridae
Retropinnidae
Lepidogalaxiidae
Galaxiidae

Salmoniformes [Order]
Salmonidae [Family]

Stomiiformes [Order]
Gonostomatidae [Family]
Photichthyidae
Sternoptychidae
Stomiidae

Aulopiformes [Order]
Giganturidae [Family]
Aulopodidae
Synodontidae
Chlorophthalidae
Ipnopidae
Scopelarchidae
Notosuridae
Pseudotrichonotidae
Paralepididae
Anotopteridae
Evermannellidae
Omosudidae
Alepisauridae

Myctophiformes [Order]
Neoscopelidae [Family]
Myctophidae

Lampridiformes [Order]
Veliferidae [Family]
Lamprididae
Stylephoridae
Lophotidae
Radiicephalidae
Trachipteridae
Regalecidae

Polymixiiformes [Order]
Polymixiidae [Family]

Percopsiformes [Order]
Percopsidae [Family]
Aphredoderidae
Amblyopsidae

Ophidiiformes [Order]
Carapidae [Family]
Ophidiidae
Bythitidae
Aphyonidae
Parabrotulidae

Gadiformes [Order]
Macrouridae [Family]
Steindachneriidae
Moridae
Melanonidae
Bregmacerotidae
Muraenolepididae
Phycidae
Merlucciidae
Gadidae
Lotidae
Macruronidae
Raniciptidae

Batrachoidiformes [Order]
Batrachoididae [Family]

Lophiiformes [Order]
Lophiidae [Family]
Antennariidae
Lophichthyidae
Tetrabrachiidae
Caulophrynidae
Chaunacidae
Ogcocephalidae
Brachionichthyidae
Diceratiidae
Thaumatichthyidae
Centrophrynidae
Gigantactinidae
Neoceratiidae
Melanocetidae
Ceratiidae
Himantolophidae
Oneirodidae
Linophynidae

Mugiliformes [Order]
Mugilidae [Family]

Atheriniformes [Order]
Bedotiidae [Family]
Melanotaeniidae
Pseudomugilidae
Atherinidae
Notocheiridae
Telmatherinidae
Dentatherinidae
Phallostethidae

Beloniformes [Order]
Adrianichthyidae [Family]
Belonidae
Scomberesocidae
Exocoetidae
Hemiramphidae

Cyprinodontiformes [Order]
Aplocheilidae [Family]
Profundilidae
Fundulidae
Valenciidae
Anablepidae
Poeciliidae
Goodeidae
Cyprinodontidae
Rivulidae

Stephanoberyciformes [Order]
Melamphaidae [Family]
Gibberichthyidae
Stephanoberycidae
Hispidoberycidae
Rondeletiidae
Barbourisiidae
Cetomimidae
Mirapinnidae
Megalomycteridae

Beryciformes [Order]
Anoplogasteridae (also spelled
Anoplogastridae) [Family]
Diretmidae
Anomalopidae
Monocentridae
Trachichthyidae
Berycidae
Holocentridae

Zeiformes [Order]
Parazenidae [Family]
Macrurocyttidae
Zeidae
Oreosomatidae
Grammicolepidae
Caproidae
Zeniontidae

Gasterosteiformes [Order]
Hypoptychidae [Family]
Aulorhynchidae
Gasterosteidae

Pegasidae
Solenostomidae
Syngnathidae
Indostomidae
Aulostomidae
Fistulariidae
Macroramphosidae
Centriscidae

Synbranchiformes [Order]
Sychbranchidae [Family]
Chaudhuriidae
Mastacembelidae

Scorpeniformes [Order]
Dactylopteroidei [Suborder]
Dactylopteridae [Family]

Scorpaenoidei [Suborder]
Scorpaenidae [Family]
Caracanthidae
Aploactinidae
Pataecidae
Gnathanacanthidae
Congiopodidae
Peristediidae
Apistidae
Neosebastidae
Sebastidae
Setarchidae
Synanceiidae
Tetrarogidae
Triglidae

Platycephaloidei [Suborder]
Bembridae [Family]
Platycephalidae
Hoplichthyidae

Anoplopomatoidei [Suborder]
Anoplopomatidae [Family]

Hexgrammoidei [Suborder]
Hexagrammidae [Family]

Normanichthyiodei [Suborder]
Normanichthyidae [Family]

Cottoidei [Suborder]
Rhamphocottidae [Family]
Ereuniidae
Cottidae
Comephoridae
Abyssocottidae
Hemitripteridae
Agonidae
Psycholutidae
Bathylutichthyidae
Cyclopteridae
Liparidae
Cottocompheroidae

Perciformes [Order]
Percoidei [Suborder]
Centropomidae [Family]

Chandidae
Moronidae
Percichthyidae
Acropomatidae
Serranidae
Ostracoberycidae
Callanthiidae
Pseudochromidae
Grammatidae
Plesiopidae
Notograptidae
Opistognathidae
Dinopercidae
Banjosidae
Centrarchidae
Percidae
Priacanthidae
Apogonidae
Epigonidae
Sillaginidae
Malacanthidae
Lactariidae
Dinolestidae
Pomatomidae
Nematistiidae
Echeneidae
Rachycentridae
Coryphaenidae
Carangidae
Menidae
Leiognathidae
Bramidae
Caristiidae
Emmelichthyidae
Lutjanidae
Lobotidae
Gerreidae
Haemulidae
Inermiidae
Sparidae
Centracanthidae
Lethrinidae
Nemipteridae
Polynemidae
Sciaenidae
Mullidae
Pempheridae
Glaucosomatidae
Leptobramidae
Bathyclupeidae
Monodactylidae
Toxotidae
Coracinidae
Drepanidae
Chaetodontidae
Pomacanthidae
Enoplosidae
Pentacerotidae
Nandidae
Kyphosidae

Arripidae
Teraponidae
Kuhliidae
Oplegnathidae
Cirrhitidae
Chironemidae
Aplodactylidae
Cheilodactylidae
Latridae
Cepolidae
Elassomatidae
Paracorpididae
Dischistiidae
Percilidae
Nannopercidae
Gadopsidae

Labroidei [Suborder]
Cichlidae [Family]
Embiotocidae
Pomacentridae
Labridae
Odacidae
Scaridae

Zoarcoidei [Suborder]
Bathymasteridae [Family]
Zoarcidae
Stichaeidae
Cryptacanthodidae
Pholidae
Anarhichadidae
Ptilichthyidae
Zapororidae
Scytalinidae

Notothenioidei [Suborder]
Bovichthyidae [Family]
Nototheniidae
Harpagiferidae
Bathydraconidae
Channichthyidae
Pseudaphritidae
Eleginopidae
Artedidraconidae

Trachinoidei [Suborder]
Chiasmodontidae [Family]
Champsodontidae
Pholidichthyidae

Trichodontidae
Pinguipedidae
Cheimarrhichthyidae
Trichonotidae
Creediidae
Percophidae
Leptoscopidae
Ammodytidae
Trachinidae
Uranoscopidae

Blenniodei [Suborder]
Tripterygiidae [Family]
Dactyloscopidae
Labrisomidae
Clinidae
Chaenopsidae
Blenniidae

Icosteoidei [Suborder]
Icosteidae [Family]

Gobiesocoidei [Suborder]
Gobiesocidae [Family]

Callionymoidei [Suborder]
Callionymidae [Family]
Draconettidae

Gobioidei [Suborder]
Rhyacichthyidae [Family]
Odontobutidae
Eleotridae
Gobiidae
Kraemeriidae
Xenisthmidae
Microdesmidae
Ptereleotridae
Schindleriidae

Kurtoidei [Suborder]
Kurtidae [Family]

Acanthuroidei [Suborder]
Ephippidae [Family]
Scatophagidae
Siganidae
Luvaridae
Zanclidae
Acanthuridae

Scombrolabracoidei [Suborder]
Scombrolabracidae [Family]

Scombroidei [Suborder]
Sphyraenidae [Family]
Gempylidae
Trichiuridae
Scombridae
Xiphiidae
Istiophoridae

Stromateoidei [Suborder]
Amarsipidae [Family]
Centrolophidae
Nomeidae
Ariommatidae
Tetragonuridae
Stromateidae

Anabantoidei [Suborder]
Luciocephalidae [Family]
Anabantidae
Helostomatidae
Belontiidae
Osphronemidae

Channoidei [Suborder]
Channidae [Family]

Pleuronectiformes [Order]
Psettodidae [Family]
Citharidae
Bothidae
Achiropsettidae
Scophthalmidae
Paralichthyidae
Pleuronectidae
Samaridae
Achiridae
Soleidae
Cynoglossidae
Paralichthodidae
Poecilopsettidae
Rhombosleidae

Tetraodontiformes [Order]
Triacanthidae [Family]
Balistidae
Monacanthidae
Ostraciidae
Triodontidae
Tetraodontidae
Diodontidae
Molidae

FISHES FAMILY LIST

• • • • •

A brief geologic history of animal life

A note about geologic time scales: A cursory look will reveal that the timing of various geological periods differs among textbooks. Is one right and the others wrong? Not necessarily. Scientists use different methods to estimate geological time—methods with a precision sometimes measured in tens of millions of years. There is, however, a general agreement on the magnitude and relative timing associated with modern time scales. The closer in geological time one comes to the present, the more accurate science can be—and sometimes the more disagreement there seems to be. The following account was compiled using the more widely accepted boundaries from a diverse selection of reputable scientific resources.

Geologic time scale

Era	Period	Epoch	Dates	Life forms
Proterozoic			2,500-544 mya*	First single-celled organisms, simple plants, and invertebrates (such as algae, amoebas, and jellyfish)
Paleozoic	Cambrian		544-490 mya	First crustaceans, mollusks, sponges, nautiloids, and annelids (worms)
	Ordovician		490-438 mya	Trilobites dominant. Also first fungi, jawless vertebrates, starfish, sea scorpions, and urchins
	Silurian		438-408 mya	First terrestrial plants, sharks, and bony fish
	Devonian		408-360 mya	First insects, arachnids (scorpions), and tetrapods
	Carboniferous	Mississippian	360-325 mya	Amphibians abundant. Also first spiders, land snails
		Pennsylvanian	325-286 mya	First reptiles and synapsids
	Permian		286-248 mya	Reptiles abundant. Extinction of trilobytes
Mesozoic	Triassic		248-205 mya	Diversification of reptiles: turtles, crocodiles, therapsids (mammal-like reptiles), first dinosaurs
	Jurassic		205-145 mya	Insects abundant, dinosaurs dominant in later stage. First mammals, lizards, frogs, and birds
	Cretaceous		145-65 mya	First snakes and modern fish. Extinction of dinosaurs, rise and fall of toothed birds
Cenozoic	Tertiary	Paleocene	65-55.5 mya	Diversification of mammals
		Eocene	55.5-33.7 mya	First horses, whales, and monkeys
		Oligocene	33.7-23.8 mya	Diversification of birds. First anthropoids (higher primates)
		Miocene	23.8-5.6 mya	First hominids
		Pliocene	5.6-1.8 mya	First australopithecines
	Quaternary	Pleistocene	1.8 mya-8,000 ya	Mammoths, mastodons, and Neanderthals
		Holocene	8,000 ya-present	First modern humans

*Millions of years ago (mya)

Index

Bold page numbers indicate the primary discussion of a topic; page numbers in italics indicate illustrations.

1982 Convention on the Law of the Sea
  common dolphinfishes, 5:215
  pompano dolphinfishes, 5:214

A

Aba-aba, 4:57, 4:231, 4:232–233, 4:*235*, 4:*237*
Ablabys taenianotus. See Cockatoo waspfishes
Able, K. W.
  on red hakes, 5:28
  on white hakes, 5:28, 5:39
Abouts. *See* Rudds
Abramis spp., 4:298
Abramis brama. See Carp breams
Abyssobrotula galatheae, 5:15
Abyssocotidae, 4:57
Acanthemblemaria spp. *See* Barnacle blennies
Acanthemblemaria maria. See Secretary blennies
Acanthochromis polyacanthus. See Marine damselfishes
Acanthocybium solandri. See Wahoos
Acanthodii, 4:10
Acanthomorpha, 4:448
Acanthophthalmus kuhli. See Coolie loaches
Acanthophthalmus semicinctus. See Coolie loaches
Acanthopsis choirorhynchus. See Horseface loaches
Acanthopterygii, 4:12
Acanthuridae, 4:67–68, 5:391
Acanthuroidei, 5:**391–404**, 5:*394*, 5:*395*
  behavior, 5:392–393
  conservation status, 5:393
  distribution, 5:392
  evolution, 5:391
  feeding ecology, 5:393
  habitats, 5:392
  humans and, 5:393
  physical characteristics, 5:391
  reproduction, 5:393
  species of, 5:*396*–403
  taxonomy, 5:391
Acanthurus spp., 5:396
Acanthurus chirurgus. See Doctorfishes
Acanthurus coeruleus. See Blue tangs
Acanthurus lineatus. See Lined surgeonfishes
Acanthurus leucosternon. See Powder blue tangs
Acheilognathinae, 4:297
Achiridae, 5:450, 5:452, 5:453
Achiropsettidae, 5:450
Acipenser spp., 4:213

Acipenser brevirostrum. See Shortnose sturgeons
Acipenser fulvescens. See Lake sturgeons
Acipenser gueldenstaedtii, 4:*213*
Acipenser oxyrhinchus. See Atlantic sturgeons
Acipenser sturio. See Atlantic sturgeons
Acipenser transmontanus. See White sturgeons
Acipenseridae, 4:11, 4:58
Acipenseriformes, 4:18, 4:**213–220**, 4:*216*
Acronurus, 5:396
Acrotus willoughbyi. See Ragfishes
Actinopterygii, 4:10, 4:11–12
Acyrtops spp., 5:355
Acyrtus spp., 5:355
Adamson's grunters, 5:222
Adder pikes. *See* Lesser weevers
Adelphophagy, 4:133
Adrianichthyidae. *See* Ricefishes
Adrianichthyoidei, 5:79
Adrianichthys kruyti. See Duckbilled buntingis
Aeoliscus strigatus. See Common shrimpfishes
Aetobatus narinari. See Spotted eagle rays
AFGP (antifreeze glycopeptide), 5:321–322
African arawanas, 4:232, 4:*233*–234
African elephantfishes, 4:232, 4:234
African Great Lakes, cichlids, 5:275–276
African darter tetras. *See* Striped African darters
African hillstream catfishes, 4:352
African knifefishes, 4:232, 4:233
African lampeyes, 5:89
African lungfishes, 4:3, 4:57, 4:201–203, 4:*204*, 4:205–206
African mudfishes, 4:292, 4:294–295
African pikes, 4:*335*, 4:*337*, 4:338
African polka-dot catfishes, 4:*353*
African snakeheads, 5:*440*, 5:*446*
African snooks. *See* Nile perches
Afromastacembelinae, 5:151
Agassiz, Louis, 5:89
Agassiz's dwarf cichlids, 5:*282*, 5:*284*
Agassiz's slickheads, 4:392
Aggressive mimicry, 4:68, 5:342
Agnathans, 4:27, 4:77, 4:83
Agonidae. *See* Poachers
Agonistic behavior, 4:62–63
Agonostomus spp., 5:59
Agonostomus monticola. See Mountain mullets
Agrostichthys spp., 4:447
Agrostichthys parkeri, 4:449, 4:454
Aipichthyidae, 5:1
Airbreathing catfishes, 4:352
Akysidae, 4:353
Alabama cavefishes, 5:6, 5:7, 5:*9*, 5:10

Alabama shads, 4:278
Alabes spp. *See* Singleslits
Alabes dorsalis. See Common shore-eels
Alabes parvulus. See Pygmy shore-eels
Alaska blackcods. *See* Sablefishes
Alaska blackfishes, 4:379–380, 4:*382*, 4:*383*
Alaska pollocks, 5:27, 5:30, 5:*31*, 5:*35*–36
Albacore, 5:*408*, 5:*410*, 5:414–415
Albacores, 5:407
Albula spp., 4:249
Albula vulpes. See Bonefishes
Albulidae. *See* Bonefishes
Albuliformes, 4:11, 4:**249–253**, 4:*251*
Aldrichetta spp., 5:59
Aldrovandia spp., 4:249
Alepisauridae, 4:432
Alepisauroidei, 4:431
Alepisaurus ferox. See Longnose lancetfishes
Alepocephalidae. *See* Slickheads
Alepocephaloidea, 4:390
Alepocephalus agassizii. See Agassiz's slickheads
Alepocephalus rostratus. See Slickheads
Alepocephalus tenebrosus. See California slickheads
Alestiidae, 4:335
Alewives, 4:*280*, 4:*281*–282
Alfoncinos, 5:113, 5:115
Algae eaters, 4:57, 4:321, 4:*322*, 4:*326*, 4:*328*, 4:*333*
Allen, G. H., 5:353
Allen, G. R., 5:219
Allenbatrachus spp., 5:42
Alligator gars, 4:221, 4:*222*, 4:*223*, 4:*224*
Allodontichthys spp., 5:92
Allotriognatha. *See* Lampridiformes
Alopias vulpinus. See Thresher sharks
Alosa alabamae. See Alabama shads
Alosa pseudoharengus. See Alewives
Alosa sapidissima. See American shads
Alpine charrs. *See* Charrs
Alticus spp., 5:342
Aluterus monoceros, 5:469
Aluterus scriptus. See Scrawled filefishes
Amarsipidae. *See Amarsipus carlsbergi*
Amarsipus carlsbergi, 5:421
Amazon leaffishes, 5:*242*, 5:246, 5:251
Ambassidae. *See* Glassfishes
Amberjacks, 4:68
Ambloplites rupestris. See Rock basses
Amblycipitidae, 4:353
Amblydoras hancockii. See Blue-eye catfishes
Amblyeleotris wheeleri. See Gorgeous prawn-gobies
Amblyopinae, 5:373, 5:375

INDEX

INDEX

INDEX

INDEX

INDEX

INDEX

Mulloidichthys martinicus. See Yellow goatfishes
Mulloidichthys vanicolensis. See Yellowfin goatfishes
Mullus surmuletus. See Red goatfishes
Munehara, H., 5:182
Muraena anguilla. See European eels
Muraena rostrata. See American eels
Muraenesocidae, 4:255
Muraenidae, 4:44, 4:47, 4:68, 4:255
Muraenolepididae, 5:25–29, 5:27
Muraenolepis spp. See Muraenolepididae
Muraenophis sathete. See Slender giant morays
Murman herrings. See Atlantic herrings
Murray, A. M., 5:5
Murray breams. See Golden perches
Murray cods, 4:68, 5:223, 5:226, 5:231
Murray lampreys. See Short-headed lampreys
Muskellunges, 4:379, 4:380, 4:382, 4:384–385
Muskies. See Muskellunges
Mustelus antarcticus, 4:116
Mustelus henlei. See Brown smoothhound sharks
Muttonfishes. See Ocean pouts
Myctophidae. See Lanternfishes
Myctophiformes. See Lanternfishes
Myctophinae, 4:441
Myctophum spp., 4:442, 4:443
Myers, George S., 4:351, 5:421
Myleus pacu. See Silver dollars
Myliobatidae. See Eagle rays
Myliobatis californica. See Bat rays
Myliobatoidei, 4:11
Myllokunmingia, 4:9
Myrichthys maculosus. See Tiger snake eels
Myrichthys magnificus, 4:267
Myripristis jacobus. See Blackbar soldierfishes
Myripristis murdjan. See Red soldierfishes
Myripristus berndti. See Blotcheye soldierfishes
Myrocongridae, 4:255
Myxine spp., 4:77
Myxine glutinosa. See Atlantic hagfishes
Myxinidae, 4:4, 4:77
Myxiniformes. See Hagfishes
Myxinikela siroka, 4:77
Myxocyprinus asiaticus. See Chinese suckers
Myxus spp., 5:59

N

Nakabo, T., 5:365
Nakamura, I., 5:417
Naked characins, 4:338
Naked dragonfishes, 5:324, 5:325–326
Nandidae. See Leaffishes
Nandus nandus. See Gangetic leaffishes
Nannatherina spp. See Nannatherina balstoni
Nannatherina balstoni, 5:222
Nannocharax fasciatus. See Striped African darters
Nannoperca spp., 5:220
Nannoperca australis. See Southern pygmy perches
Nannoperca obscura. See Yarra pygmy perches
Nannoperca oxleyana. See Oxleyan pygmy perches
Nannoperca variegata. See Variegated pygmy perches
Nannopercidae. See Pygmy perches

Nanny shads. See Gizzard shads
Nansenia pelagica, 4:392
Napoleon wrasses. See Humphead wrasses
Narcine bancrofti. See Lesser electric rays
Narcinidae. See Electric rays
Nardovelifer altipinnis, 4:448
Naso lituratus. See Orangespine unicornfishes
Nassau groupers, 5:265, 5:269, 5:270–271
Natal moonies. See Monos
National Marine Fisheries Service (U.S.). See United States National Marine Fisheries Service
Natterer, Johann, 4:201
Naucrates ductor. See Pilotfishes
Nautichthys oculofasciatus. See Sailfin sculpins
Nazarkin, M. V., 5:331
Needlefishes, 4:12, 4:49, 4:70, 5:79–81, 5:82, 5:84–85
Needlenose gars. See Longnose gars
Neetroplus nematopus. See Poor man's tropheuses
Negaprion brevirostris. See Lemon sharks
Nelson, Joseph S.
  on Atheriniformes, 5:67
  on Cypriniformes, 4:321
  on fathead sculpins, 5:179
  on Osmeriformes, 4:389
  on Salmoniformes, 4:405
  on Scorpaeniformes, 5:157
  on Scorpaenoidei, 5:163
Nemacheilinae, 4:321
Nemacheilus evezardi. See Hillstream loaches
Nemacheilus masyae. See Arrow loaches
Nemacheilus multifasciatus. See Siju blind cavefishes
Nemacheilus oedipus. See Cavefishes of Nam Lang
Nemacheilus starostini. See Kughitang blind loaches
Nemacheilus toni. See Siberian stone loaches
Nemadactylus spp., 5:239, 5:240
Nemateleotris magnifica. See Fire gobies
Nematistiidae. See Roosterfishes
Nematistius pectoralis. See Roosterfishes
Nematogenyidae, 4:352
Nemichthydae, 4:255
Nemichthys scolopaceus. See Slender snipe eels
Nemipteridae, 5:255, 5:258–263
Neobola argentea. See Dagaa
Neobythitinae, 5:16, 5:18
Neoceratodus spp., 4:197–199, 4:201, 4:202, 4:203
Neoceratodus forsteri. See Australian lungfishes
Neocirrhites armatus, 4:42, 5:239, 5:240
Neocyema erythrostoma, 4:271
Neocynchiropus spp., 5:365
Neogobius melanostomus. See Round gobies
Neolamprologus callipterus, 5:279–280
Neon gobies, 5:374, 5:375, 5:376, 5:378, 5:384, 5:385
Neon tetras. See Cardinal tetras
Neoniphon spp., 5:113
Neopterygii, 4:11
Neoscopelidae. See Blackchins
Neoscopelus microchir, 4:442
Neosebastidae. See Gurnard perches
Neoselachii, 4:10, 4:11
Neostethus bicornis, 5:72, 5:75, 5:76
Nervous system, 4:20–23
  See also Physical characteristics

Nettastoma brevirostre. See Pignosed arrowtooth eels
Nettastomatidae, 4:255
Neurotoxins, 4:15, 4:179
New Granada sea catfishes, 4:355, 4:357
New Zealand graylings, 4:391
New Zealand smelts, 4:390
Ngege, 5:280, 5:283, 5:286, 5:287
Nightfishes, 5:221
Nigripinnis. See Blackfin pearl killifishes
Nile perches, 4:75, 4:313, 5:220, 5:224, 5:226–227, 5:280–281
Nineline gobies, 5:376
Ningu, 4:303, 4:313
Niphonini. See Japanese aras
Noemacheilus mesonoemachilus sijuensis. See Siju blind cavefishes
Noemacheilus sijuensis. See Siju blind cavefishes
Noemacheilus smithi. See Blind loaches
Nolf, D., 5:341
Nomeidae. See Driftfishes
Nomeus gronovii. See Man-of-war fishes
Nomorhamphus spp., 5:81
Noodlefishes, 4:391
Normanichthyidae. See Normanichthys crockery
Normanichthys crockery, 5:179
Noronha wrasses, 5:296
North African catfishes. See Sharptooth catfishes
North American basses, 5:195–206, 5:200, 5:201
North American freshwater catfishes, 4:352
North American mudsucker gobies, 4:6
North American paddlefishes. See American paddlefishes
North Pacific Anadromous Fish Commission, 4:413
North Pacific anchovies. See Northern anchovies
North Pacific herrings. See Pacific herrings
North River shads. See American shads
Northern anchovies, 4:4, 4:278, 4:280, 4:285, 4:287
Northern bluefin tunas. See Atlantic bluefin tunas
Northern cavefishes, 5:6, 5:7, 5:8–9
Northern clingfishes, 5:358, 5:360–361
Northern halibuts. See Pacific halibuts
Northern largemouth basses. See Largemouth basses
Northern longears. See Longear sunfishes
Northern mummichogs, 5:95, 5:97, 5:99
Northern pikeminnows, 4:7, 4:298
Northern pikes, 4:14, 4:379, 4:382, 4:383–384, 5:198
Northern red snappers, 5:264, 5:268, 5:269, 5:270
Northern rock basses. See Rock basses
Northern ronquils, 5:311, 5:312
Northern seadevils. See Krøyers deep sea anglerfishes
Northern sea robins, 5:164
Northern snakeheads, 5:438, 5:441, 5:442
Northern stargazers, 5:336, 5:337, 5:339–340
Norwegian herrings. See Atlantic herrings
Notacanthidae. See Marine spiny eels
Notacanthus spp., 4:249, 4:250
Notemigonus spp., 4:301
Notesthes robusta. See Bullrouts
Nothobranchius spp., 5:94

INDEX

Sphyrna lewini. See Scalloped hammerhead
sharks
Sphyrna mokarran. See Great hammerhead
sharks
Sphyrna tiburo. See Bonnethead sharks
Sphyrnidae, 4:27, 4:113
Spikefishes, 5:467–471, 5:*473*, 5:*482*–483
Spine-cheek anemonefishes, 5:*294*
Spined pygmy sharks, 4:151
Spinefoots. *See* Rabbitfishes
Spinicapitichthys spp., 5:365
Spiny butterfly rays, 4:*181*, 4:*183*
Spiny devilfishes. *See* Bearded ghouls
Spiny dogfishes. *See* Piked dogfishes
Spiny dwarf catfishes, 4:352
Spiny eels, 5:**151–156,** 5:*152*
Spiny flounders, 5:449–450, 5:452
Spiny plunderfishes, 5:321, 5:322
Spinyfins, 5:113, 5:115
Splash tetras, 4:338, 4:*340*, 4:*349*–350
Splashing tetras. *See* Splash tetras
Splendid coral toadfishes, 5:*41*, 5:*42*, 5:*43*,
5:*44*, 5:45–46
Splendid garden eels, 4:*259*, 4:*263*, 4:264–265
Splendid hawfishes, 5:238
Splendid perches, 5:255, 5:256, 5:258–263
Splitfin flashlightfishes, 5:*113*, 5:*117*, 5:*118*
Splitfins, 4:32, 4:57, 5:89, 5:92, 5:93
Splits. *See* Atlantic herrings
Spookfishes. *See* Barreleyes
Spoonbill cats. *See* American paddlefishes
Spot-fin porcupinefishes, 5:*472*, 5:*475*, 5:477
Spots. *See* Channel catfishes
Spotted algae eaters, 4:322, 4:*326*, 4:333
Spotted characins. *See* Splash tetras
Spotted coral crouchers, 4:42, 5:*168*, 5:170,
5:*171*
Spotted eagle rays, 4:*181*, 4:*184*–185
Spotted flounders. *See* Windowpane flounders
Spotted gars, 4:*221*, 4:222, 4:*223*, 4:225–226,
4:227
Spotted handfishes, 5:51
Spotted puffers. *See* Guinea fowl puffers
Spotted ragfishes. *See* Ragfishes
Spotted ratfishes, 4:*91*, 4:*93*, 4:*94*
Spotted sanddivers, 5:*332*
Spotted scats, 5:393, 5:*395*, 5:401–*402*
Spotted scorpionfishes, 5:*164*
Spotted sharpnose puffers. *See* Spotted tobies
Spotted snake eels. *See* Tiger snake eels
Spotted tinselfishes. *See* Tinselfishes
Spotted tobies, 5:*472*, 5:481, 5:*482*
Sprats, 4:277
Spraying characins. *See* Splash tetras
Spring cavefishes, 5:7, 5:*8*, 5:*9*–10
Spring herrings. *See* Alewives; Atlantic
herrings
Springer, V. G., 4:54, 4:191, 5:341
Spur-and-groove zones, 4:48
Spurdogs. *See* Piked dogfishes
Squalea, 4:11, 4:143, 4:161, 4:167
Squalidae. *See* Dogfish sharks
Squaliformes. *See* Dogfish sharks
Squaliolus spp., 4:151
Squalus spp., 4:15, 4:151, 4:152
Squalus acanthias. See Piked dogfishes
Squalus africanus. See Pajama catsharks
Squalus carcharias. See White sharks
Squalus cepedianus. See Broadnose sevengill
sharks

Squalus cuvier. See Tiger sharks
Squalus glauca. See Blue sharks
Squalus griseus. See Bluntnose sixgill sharks
Squalus longimanus. See Oceanic whitetip
sharks
Squalus maximus. See Basking sharks
Squalus microcephalus. See Greenland sharks
Squalus nasus. See Porbeagles
Squalus portus jacksoni. See Port Jackson sharks
Squalus vulpinus. See Thresher sharks
Squarehead catfishes, 4:*350*, 4:353, 4:*356*
Squaretails. *See* Brook trouts
Squatina spp., 4:161
Squatina aculeata, 4:162
Squatina africana, 4:162
Squatina argentina, 4:162
Squatina australis, 4:161–162
Squatina californica. See Pacific angelsharks
Squatina dumeril, 4:162
Squatina formosa, 4:162
Squatina guggenheim, 4:162
Squatina japonica, 4:162
Squatina nebulosa, 4:162
Squatina occulta, 4:162
Squatina oculata, 4:162
Squatina squatina, 4:162, 4:*163*, 4:*164*–*165*
Squatina tergocellata, 4:161, 4:162
Squatina tergocellatoides, 4:161, 4:162
Squatinidae, 4:161
Squatiniformes. *See* Angelsharks
Squeakers, 4:352, 4:354
Squirrel hakes. *See* Red hakes
Squirrelfishes, 4:*31*, 4:48, 4:67, 5:**113–120,**
5:*117*
St. Helena dragonets, 5:366–367
Staghorn sculpins, 5:183
Stargazers, 4:19, 4:48, 4:69, 5:331–335
Starksia spp., 5:343
Starksia lepicoelia. See Blackcheek blennies
Starostin's loaches. *See* Kughitang blind
loaches
Starry moray eels, 4:*3*
Steatogenys elegans, 4:370
Steelheads. *See* Rainbow trouts
Stegastes planifrons. See Threespot damselfishes
Stegastes sanctaehelenae, 5:298
Stegastes sanctipauli, 5:298
Stegostoma fasciatum. See Zebra sharks
Stegostomatidae. *See* Zebra sharks
Steindachner, F.
on *Cynolebias bellottii,* 5:90
on ragfishes, 5:351
on sailfin sandfishes, 5:339
Steindachneria spp. *See* Luminous hakes
Steindachneria argentea. See Luminous hakes
Steindachneriidae. *See* Luminous hakes
Stensiö, Erik A., 4:209
Stephanoberycidae. *See* Pricklefishes
Stephanoberyciformes, 5:**105–111,** 5:*107*
Stephanoberycoidea, 5:105, 5:106
Stephanoberyx monae. See Pricklefishes
Sternarchorhynchus spp., 4:369–370
Sternarchorhynchus curvirostris, 4:*372*,
4:*373*–*374*
Sternarchorhynchus roseni. See Rosen
knifefishes
Sternoptychidae. *See* Marine hatchetfishes
Sternoptychidae, 4:369, 4:370
Sternopygus spp., 4:370
Sternopygus astrabes, 4:369

Sternopygus macrurus. See Longtail knifefishes
Stethacanthid sharks, 4:10
Stiassny, Melanie L. J., 5:83, 5:94
Stichaeidae. *See* Pricklebacks
Stichaeinae, 5:309
Sticklebacks, 4:35, 5:**131–136,** 5:*137*, 5:*140*,
5:**141–143**
Stingrays, 4:57, 4:173–179, 4:*181*, 4:182–186,
4:188
distribution, 4:*52*, 4:176
evolution, 4:11, 4:173
feeding ecology, 4:68, 4:177–178
freshwater, 4:57, 4:*173*, 4:174, 4:176–177,
4:*180*, 4:*184*–186
physical characteristics, 4:174–176, 4:190
reproduction, 4:46, 4:178
Stings, by reef stonefishes, 5:177
Stizostedion canadense. See Saugers
Stizostedion vitreum. See Walleyes
Stokellia anisodon. See Stokell's smelts
Stokell's smelts, 4:391
Stomatorhinus spp., 4:233
Stomiidae, 4:421, 4:423
Stomiiformes, 4:**421–430,** 4:*425*
Stone loaches, 4:*322*, 4:*325*–326
Stonefishes, 4:15, 4:48, 4:49, 5:163, 5:164,
5:165
See also Reef stonefishes
Stonerollers, 4:*300*, 4:301, 4:*303*, 4:*307*–308
Stoplight parrotfishes, 5:*300*, 5:*306*–307
Stout beardfishes, 5:*3*
Stout blacksmelts, 4:393
Stratification, water, 4:38–39
Strawberry basses. *See* Black crappies
Stream catfishes, 4:353
Stream habitats, 4:39–40, 4:62
See also Habitats
Striped African darters, 4:*340*, 4:*342*, 4:347
Striped anostomus. *See* Striped headstanders
Striped basses, 4:49, 4:68
See also Striped sea basses
Striped boarfishes, 5:*245*, 5:251–252
Striped bristletooths, 5:*394*, 5:*398*–399
Striped catsharks. *See* Pajama catsharks
Striped danios. *See* Zebrafishes
Striped eel-catfishes. *See* Coral catfishes
Striped foureyed fishes. *See* Largescale
foureyes
Striped hardyheads. *See* Eendracht land
silversides
Striped headstanders, 4:*340*, 4:*342*–343
Striped mullets. *See* Flathead mullets
Striped parrotfishes, 5:*300*, 5:*306*
Striped poison-fang blennies, 5:*344*,
5:*345*–346
Striped sea basses, 5:196, 5:198, 5:*201*, 5:*203*,
5:206
Striped sea robins, 5:*167*, 5:177–178
Striped silversides. *See* Eendracht land
silversides
Striped snakeheads, 5:438, 5:*441*, 5:*443*,
5:445–446
Stripedfin ronquils, 5:311, 5:312
Striper basses. *See* Striped sea basses
Stromateidae. *See* Butterfishes (Stromateidae)
Stromateoidei, 5:**421–426,** 5:*423*
Strongylura exilis. See Californian needlefishes
Strophidon sathete. See Slender giant morays
Sturgeons, 4:*4*, 4:11, 4:18, 4:35, 4:58, 4:75,
4:**213–219,** 4:*216*